Python Programming for Biology

Bioinformatics and Beyond

Do you have a biological question that could be readily answered by computational techniques, but little experience in programming? Do you want to learn more about the core techniques used in computational biology and bioinformatics? Written in an accessible style, this guide provides a foundation for both newcomers to computer programming and those who want to learn more about computational biology. The chapters guide the reader through: a complete beginners' course to programming in Python, with an introduction to computing jargon; descriptions of core bioinformatics methods with working Python examples; scientific computing techniques, including image analysis, statistics and machine learning. This book also functions as a language reference written in straightforward English, covering the most common Python language elements and a glossary of computing and biological terms. This title will teach undergraduates, postgraduates and professionals working in the life sciences how to program with Python, a powerful, flexible and easy-to-use language.

Tim J. Stevens, a biochemist by training, is a Senior Investigator Scientist at the MRC Laboratory of Molecular Biology in Cambridge. He researches three-dimensional genome architecture and provides computational biology oversight, development and training within the Cell Biology Division.

Wayne Boucher, a mathematician and theoretical physicist by training, is a Senior Post-Doctoral Associate and computing technician for the Department of Biochemistry at the University of Cambridge. He teaches undergraduate mathematics and postgraduate programming courses. Wayne is currently developing software for the analysis of biological molecules by nuclear magnetic resonance spectroscopy.

Python Programming for Biology

Bioinformatics and Beyond

TIM J. STEVENS
MRC Laboratory of Molecular Biology

WAYNE BOUCHER
University of Cambridge

CAMBRIDGE
UNIVERSITY PRESS

University Printing House, Cambridge CB2 8BS, United Kingdom

One Liberty Plaza, 20th Floor, New York, NY 10006, USA

477 Williamstown Road, Port Melbourne, VIC 3207, Australia

314-321, 3rd Floor, Plot 3, Splendor Forum, Jasola District Centre, New Delhi - 110025, India

79 Anson Road, #06-04/06, Singapore 079906

Cambridge University Press is part of the University of Cambridge.

It furthers the University's mission by disseminating knowledge in the pursuit of
education, learning and research at the highest international levels of excellence.

www.cambridge.org
Information on this title: www.cambridge.org/9780521720090

© Tim J. Stevens and Wayne Boucher, 2015

First published 2015

A catalogue record for this publication is available from the British Library

Library of Congress Cataloging in Publication data
Stevens, Tim J., 1976–
Python programming for biology, bioinformatics, and beyond / Tim J. Stevens,
University of Cambridge, Wayne Boucher, University of Cambridge.
    pages   cm
Includes index.
ISBN 978-0-521-89583-5 (Hardback) – ISBN 978-0-521-72009-0 (Paperback)
1. Biology–Data processing.   2. Python (Computer program language)
I. Boucher, Wayne.   II. Title.
QH324.2.S727 2014
570.285–dc23   2014021017

ISBN 978-0-521-89583-5 Hardback
ISBN 978-0-521-72009-0 Paperback

Additional resources for this publication at www.cambridge.org/pythonforbiology

Contents

The colour plates are to be found between pages 342 and 343

Preface

Many years ago we started programming in Python because we were working on a large computational biology project. In those days choosing Python was not nearly as common as it is today. Nonetheless things worked out well, and as our expertise grew it seemed only natural that we should run some elementary Python courses for the School of Biology at the University of Cambridge, where we were employed. The basis for those courses is what turned into the initial idea for this book. While there were many books about getting started with Python and some that were tailored to bioinformatics, we felt that there was still some room for what we wanted to put across. We began with the idea that we could write some chapters in relatively straightforward English that were aimed at biologists, who might be complete novices at programming, and have other sections that are useful to a more experienced programmer. Also, given that we didn't consider ourselves to be typical bioinformaticians, we were thinking more broadly than just sequence-based informatics, though naturally such things would be included. We felt that although we couldn't anticipate all the requirements of a biological programmer there were nonetheless a number of key concepts and techniques which we could try to explain. The end result is hopefully a toolkit of ideas and examples which can be applied by biologists in a variety of situations.

Tim J. Stevens and
Wayne Boucher,
Cambridge, January 2014

Acknowledgements

We extend our sincere thanks to a group of intrepid volunteers who have been invaluable in the proof-reading and testing of this book: Olga Tkachenko, Magnus Lundborg, Neil Rzechorzek, Rasmus Fogh, Simon Fraser and Tom Drury.

Special thanks also go to David Judge, who has run the bioinformatics teaching facility at Cambridge for many years and who made it very easy to give the Python courses that eventually led to this book.

We acknowledge the support of the Medical Research Council and the Biotechnology and Biological Sciences Research Council, the UK funding bodies who have funded the scientific projects that we have been involved with over the years. This has allowed us to use and develop our Python programming skills while remaining gainfully employed.

1 Prologue

Contents

Python programming for biology

One of the main aims of this book is to empower the average researcher in the life sciences, who may have a pertinent scientific question that can be readily answered by computational techniques, but who doesn't have much, if any, experience with programming. For many in this position, the task of writing a program in a computer language is a bottleneck, if not an impassable barrier. Often, the task is daunting and seems to require a significant investment of time. The task is also subject to the barriers presented by a vocabulary filled with jargon and a seemingly steep learning curve for those people who were not trained in computing or have no inclination to become computer specialists. With this in mind for the novice programmer, one ought to start with the language that is the easiest to get to grips with, and at the time of writing we believe that that language is Python. This is not to say that we have made a compromise by choosing a language that is easy to learn but which is not powerful or fully featured. Python is certainly a very rich and capable way of programming, even for very large projects; otherwise we authors wouldn't be using it for our own scientific work.

A second main aim of this book is to use Python as a means to illustrate some of what is going on within biological computing. We hope our explanations will show you the scientific context of why something is done with computers, even if you are a newcomer to biology or medical sciences. Even where a popular biological program is not written in Python, or if you are a programmer who has good reason for using another language, we can still use Python as a way of illustrating the major principles of programming for biology. We feel that many of the most useful biological programs are based on combinations of simple principles that almost anyone can understand. By trying to separate the core concepts from the obfuscation and special cases, we aim to provide an overview of techniques and strategies that you can use as a resource in your own research. Virtually all of the examples in this book are working code that can be run and are based on real problems or programs within biological computing. The examples can then be adapted, altered and combined to enable you to program whatever you need.

We wish to make clear that this book intends to show you what sort of things can be done and how to begin. It does not intend to offer a deep and detailed analysis of specific biological and computational problems. This is not a typical scientific book, given that we don't always go for the most detailed or up-to-date examples. Given the choice, we aim to give a broad-based understanding to newcomers and avoid what some may consider pedantry.

No doubt some people will think our approach somewhat too simplistic, but if you know enough to know the difference then we don't recommend looking to this book for those kinds of answers. Likewise, there is only room for so many examples and we cannot cover all of the scientific methods (including Python software libraries) that we would want to. Hopefully though, we give the reader enough pointers to make a good start.

Choosing Python

It is perhaps important to include a short justification to say why we have written this book for the Python programming language; after all, we can choose from several alternative languages. Certainly Python is the language that we the authors write in on a daily basis, but this familiarity was actually born out of a conscious decision to use Python for a large biological programming project after having tried and considered a number of popular alternatives. Aside from Python, the languages that we have commonly come across in today's biological community include: C, C++, FORTRAN, Java, Matlab, Perl, R and Ruby. Specific comparison with some of these languages will be made at various points in the book, but there are some characteristics of Python that we enjoy, which we feel would not be available to the same level or in the same combination in any other language.

We like the clear and consistent layout that directs the programmer away from obfuscated program code and towards an elegantly readable solution; this becomes especially important when trying to work out what someone else's program does, or even what your own material does several years later. We like the way that Python has object orientation at its heart, so you can use this powerful way to organise your data while still having the easy look and feel of Python. This also means that by learning the language basics you automatically become familiar with the very useful object-oriented approach. We like that Python generally requires fewer lines of program code than other languages to do the equivalent job, and that it often seems so much less tedious to write.

It is important to make it clear that we would not currently use Python for every programming task in the life sciences. Python is not a perfect language. As it stands currently for some specialised tasks, particularly those that require fast mathematical calculations which are not supported by the numeric Python modules, we actively promote working with a Python extension such as Cython, or some faster alternative language. However, we heartily recommend that Python be used to administer the bookkeeping while the faster alternative provides extra modules that act as a fast calculation engine. To this end, in Chapter 27 we will show you how you can seamlessly mesh the Python language with Cython and also with the compiled language C, to give all the benefits of Python and very fast calculations.

Python's history and versions

The Python[1] programming language was the creation of Guido van Rossum. It is because of his innovation and continuing support that Python is popular and continues to grow. The Python programming community has afforded Guido the honour of the title 'benevolent

---

[1] The name itself derives from Monty Python, which is why you'll find the occasional honorary reference to 'spam', 'dead parrot' etc. when arbitrary examples are given.

dictator for life'. What this means is that despite the fact that many aspects of Python are developed by a large community, Guido has the ultimate say in what goes into Python. Although not bound in any legality, everyone abides by Guido's decisions, even if at times some people are surprised by what he decides. We believe that this situation has largely benefited Python by ensuring that the philosophy remains unsullied. Seemingly often, a committee decision has the tendency to try to appease all views and can become tediously slow with indecision; too timid to make any bold, yet improving moves. The Python programming community has a large role in criticising Python and guiding its future development, but when a decision needs to be made, it is one that everyone accepts. Certainly there could be a big disagreement in the future, but so far the benevolent dictator's decisions have always taken the community with him.

There have been several, and in our opinion improving, versions of the Python programming language. All versions before Python 3 share a very high degree of backward-compatibility, so that code written for version 1.5 will still (mostly) work with say version 2.7 with few problems. Python 3 is not as compatible with older versions, but this seems a reasonable price to be paid to keep things moving forward and eradicate some of the undesired legacy that earlier versions have built up. Rest assured though, version 3 remains similar enough in look and feel to the older Pythons, even if it is not exactly the same, and the examples in this book work with both Python 2 and Python 3 except where specifically noted. Also, included with the release of Python 3 is a conversion program '2to3' which will attempt to automatically change the relevant parts of a version 2 program so that it works with version 3. This will not be able to deal with every situation, but it will handle the vast majority and save considerable effort.

For this book we will assume that you are using Python version 2.6 or 2.7 or 3. Some bits, however, that use some newer features will not work with versions prior to 2.6 without alteration. We feel that it is better to use the best available version, rather than write in a deliberately archaic manner, which would detract from clarity.

Bioinformatics

The field of bioinformatics has emerged as we have discovered, through experimentation, large amounts of DNA and protein sequence information. In its most conservative sense bioinformatics is the discipline of extracting scientific information by the study of these biological sequences, which, because of the large amount of data, must be analysed by computer. Initially this encompassed what most biological computing was about, but we contend that this was simply where biomolecular computing began and that it has far to go. The informatics of biological systems these days includes the study of molecular structures, including their dynamics and interactions, enzymatic activity, medical and pharmacological statistics, metabolic profiles, system-wide modelling and the organisation of experimental procedures, to name only a subset. It is within this wider context that this book is placed.

At present the programming language that is historically most famous for being used with bioinformatics is probably Perl, which is notable for its ability to manipulate sequences, particularly when stored as letters within formatted text. It also has a library of modules available to perform many common bioinformatics tasks, collectively named BioPerl. In this arena Python can do everything that Perl can. There is a Python equivalent of BioPerl, unsurprisingly named BioPython, and at this time the uptake of Python within the bioinformatics community is growing, which is not surprising, given our belief that it is an easier but more powerful language to work with. It is important to note that although some of the BioPython modules will certainly be discussed in the course of this book (and we would generally advise using tested, existing code wherever possible to make your programs easier to write and understand) the explanations and examples will be more to do with understanding what is going on underneath. We aim to avoid this book simply becoming a brochure for existing programs where you don't have to know the inner workings.

Computer platforms and installations

Python is available for every commonly used computer operating system including versions of Microsoft Windows, Mac OS X, Linux and UNIX. With Windows you will generally have to download and install Python, as it is not included as standard. On most new Mac OS X, Linux and UNIX systems Python is included as standard (indeed some parts of Linux operating systems are themselves written with Python), although you should check to see which version of Python you have: typing 'python' at a command line reveals the version. For a list of website locations where you can download Python for various platforms see the reference section at the end of this book or the Cambridge University Press site: http://www.cambridge.org/pythonforbiology.

Precisely because Python is available for and can be run on many different computer platforms, any programs you write will generally be able to be run on all computer systems. However, there are a few important caveats you should be aware of. Although Python as a language is interpreted in the same way on every computer system, when it comes to interacting with the operating system (Windows, Mac OS X, Linux . . .), things can work differently on different computers. This is a problem that all cross-platform computing languages face. You will probably come across this in your Python programs when dealing with files and the directories that contain them. Although each operating system will have its own nuances, once you are aware of the differences it is a relatively simple job to ensure that your programs work just as well under any common operating system, and we will cover details of this as required in the subsequent chapters.

2 A beginners' guide

Programming principles

The Python language can be viewed as a formalised system of understanding instructions (represented by letters, numbers and other funny characters) and acting upon those directions. Quite naturally, you have to put something in to get something out, and what you are going to be passing to Python is a series of commands. Python is itself a computer program, which is designed to interpret commands that are written in the Python language, and then act by executing what these instructions direct. A programmer will sometimes refer to such commands collectively as 'code'.

Interpreting commands

So, to our first practical point; to get the Python interpreter to do something we will give it some commands in the form of a specially created piece of text. It is possible to give Python a series of commands one at a time, as we slowly type something into our computer. However, while giving Python instructions line by line is useful if you want to test out something small, like the examples in this chapter, for the most part this method of issuing commands is impractical. What we usually do instead is create all of the lines of text representing all the instructions, written as commands in the Python language, and store the whole lot in a file. We can then activate the Python interpreter program so that it reads all of the text from the file and acts on all of the commands issued within. A series of commands that we store together in such a way, and which do a specific job, can be considered as a computer program.[1] If you would like to try any of the examples given in the book the next chapter will tell you how to actually get started. The initial intention, however, is mostly to give you a flavour of Python and introduce a few key principles.

```
mass = 5.9736
volume = 1.08321
density = mass/volume
print(density)
```

An example of a very simple, four-line Python program that performs a calculation and displays the result.

Reusable functionality

When writing programs in the Python language, which the Python interpreter can then use, we are not restricted to reading commands from only one file. It is a very common practice to have a program distributed over a number of different files. This helps to organise writing of the program, as you can put different specialised parts of your instructions into different files that you can develop separately, without having to wade through large amounts of text. Also, and perhaps most importantly, having Python commands in multiple files enables different programs to share a set of commands. With shared files, the distinction between which commands belong to one program and which belong to another is mostly meaningless. As such, we typically refer to such a shared file as a *module*.

In Python you will use modules on a regular basis. And, as you might have already guessed, the idea is to have modules containing a series of commands which perform a function that would be useful for several programs, perhaps in quite different situations. For example, you could write a module which contains the commands to do a statistical analysis on some numeric data. This would be useful to any program that needs to run that kind of analysis, as hopefully we have written the statistics module in such a way that the precise amount and source of the numeric data that we send to the module is irrelevant. Whenever we use a module we are avoiding having to write new Python commands, and are hopefully using something that has been tried and tested and is known to work.

```
from Alignments import sequenceAlign
sequence1 = 'GATTACAGC'
sequence2 = 'GTATTAAT'
print(sequenceAlign(sequence1, sequence2))
```

A Python example where general functionality, to align two sequences of letters, is imported from a module called Alignments, *which was defined elsewhere.*

When working with Python there is already a long list of pre-made modules that you can use. For example, there are modules to perform common mathematical operations, to interact with

---

[1] Not 'programme', even in the UK.

the operating system and to search for patterns of symbols within text. These are all generally very useful, and as such they are included as standard whenever you have Python installed. You will still have to load, or *import*, these modules into a program to use them, but in essence you can think of these modules as a convenient way of extending the vocabulary of the Python language when you need to. By the same token, you don't have to load any modules that are not going to be useful, which might slow things down or use unnecessary computer memory.

Types of data

Before going on to give a more detailed tutorial we will first describe a little about the construction and makeup of commands written in the Python language. Writing the command code for a program involves thinking about items of data. There can be many different kinds of data, from different origins, that we would wish to manipulate with a computer. Typically we will represent the smallest units of this information as numbers or text. We can organise such numbers and text into structured arrangements, for example, to create a list of data, and we can then manipulate this entire larger container, with all of its underlying elements, as a single unit. For example, given a list containing numbers you could extract the first number from the list, or maybe get the list in reverse order.

```
numbers = [6, 0, 2, 2, 1, 4, 1, 5]
numbers.reverse()

print(numbers)
```
Defining a list of numbers as a single entity and then reversing its order, before printing the result to the screen.

In Python, as in many languages, there are some standard types of data-containing structures that form the basis of most programs, and which are very easy to create and fill with information. But you are not limited to these standard data structures; you can create your own data organisation. For example, you could create a data structure called a `Person`, which can store the name, sex, height and age of real people. In a program, just as you could get the first element of data stored in a list, so too could you extract the number that represents the age of a `Person` data structure. Going further, you could create many `Person` data structures and organise them further by placing them into lists. A data structure can appear inside the organisation of many other data structures, so a single `Person` could appear in several different lists (for example, organised by age, sex or whatever) or a `Person` could contain references to other `Person` data structures to indicate the relationships between parents and children.

Python objects

This is where we can introduce the concept of an *object*. The `Person` data structure described above would commonly be referred to as a `Person` object. Indeed, all of the organised data structures in Python, including the simple inbuilt ones, are referred to as Python objects. So numbers, text and lists are all kinds of objects. Not every programming language formalises things in this way, but it will start to feel natural once you are used to Python, and means that the form of the programming language is the same whatever type of object is being manipulated.

```
x = 3
y = 7
print(x + y)
print(x.__add__(y))
```

An example which shows the underlying object-oriented nature of numbers in Python: the last two lines do the same thing. Although we would normally write additions in a conventional way with a plus symbol, we are actually invoking the __add__ operation which all Python numbers possess.

An important concept when dealing with objects is inheritance. That is to say that we can make a new type of data structure by basing it on an existing one. Indeed, every object in Python, except the simplest data structure of them all (the base object), inherits its organisation from another object. Accordingly, you could take a Person object and use its specification to create a Scientist object. This would immediately give the Scientist object the same data organisation of a Person object, with its age, sex and height data, but we can go on to modify the Scientist object to also store different information, like a list of publications or current work institution. This can also be done for the built-in objects, so you could have your own version of a Python list that is only allowed to contain odd numbers, if you really, really wanted.

So far we have discussed the manipulation of data by a Python program in fairly loose terms, so it is about time to more properly introduce you to a few of the concepts that you will commonly use in Python programs. The examples that we give use operations and types of data that are built into the language as standard, i.e. that the Python interpreter will know how to handle without you having to add any special information.

Variables

As will already be apparent from the above Python snippets, when you refer to some data in your program you will often be assigning it with a name, like 'x' or 'sequence1' or 'dnaList'. Such names are commonly referred to as *variables*. They provide you with an identifiable label that you can use to track an individual item of data amongst many others within your program. The jargon term 'variable' is quite apt because you often want to keep the same name label but vary the value of the data it refers to. For example, you can write a program that calculates x+2 and x-2; where x can be set to any numeric value and both operations are performed on that same named item, whatever it may be. This concept is similar to algebra, where you can describe formulae, like $y = x^2 + 3$, without specifying what x or y actually are, and then use the formula on different values of x in order to compute y.

Note that in Python if you set the variable with the name 'x' to take a numeric value you can still set it to be some other type of data later on in the program, so initially it may be a number, but later be some text. Bearing this in mind, you must be careful that you only perform operations on the 'x' data that are valid for that type of data. Staying with the idea of data items having a particular *data type*, we next go through the basic types available in Python.

Basic data types

Numbers

There are two common types of numeric data in Python. These are *integers*, the whole numbers, and *floating point* numbers, numbers with decimal points.

Integers

Integers are whole numbers and can be positive, negative or zero in value. You would typically use integers to count things that only come as a whole, like the size of a list or number of people. You can naturally perform mathematical operations with integers, also in combination with other types of number object, but in Python 2 if you perform some mathematical operations with only integers the result is an integer too. While this makes sense for addition and multiplication, division will give you the perhaps surprising result of a whole number, rounding the answer (towards negative infinity to be precise). The advantage of integer operations is that they are quick and always precise; non-integer representation can give rise to small errors which can sometimes have serious consequences.

In Python 2 there are actually two types of integers, normal integers and long integers, although you usually don't have to pay much attention to this fact. The long integer variety is used when the number is so big[2] that it must be stored in a different way, as it takes up more memory slots to store the digits. Accordingly, you might see the 18-digit number 123,456,789,123,456,789 represented in Python (before version 3) as `123456789123456789L`, i.e. with an extra 'L' at the end giving a hint that it is the long variety. But otherwise you can simply treat it as a number and do all the usual operations with it. In Python 3 this distinction disappears and every integer is a long integer.

Floating point numbers

Floating point numbers, often simply referred to as *floats*, are numbers expressed in the decimal system, i.e. `2.1`, `999.998`, `0.000004` or whatever. The value `2.0` would also be interpreted as a floating point number, but the value `2`, without the decimal point, will not; it will be interpreted as an integer. Floating point numbers can also carry a suffix that states which power of ten they operate at. So, for example, you can express four point six million as 4.6×10^6, which in Python would be written as `4.6e6` (or as `46e5` or as `0.46e7`) and similarly one hundredth would be `1.0e-2`. A potential pitfall with floating point numbers is that they are of limited precision. Of course you would not expect to be able to express some fractions like ⅓ exactly, but there can otherwise be some surprises when you do certain calculations. For example, `0.1` plus `0.2` may sometimes give you something like `0.30000000000000004`, because of the way that the innards of computers work. The difference between this number and the desired value of `0.3` is what would be referred to as a *floating point error*. Often there is sufficient accuracy that a very small error doesn't matter, but sometimes it does matter and you should be aware of this issue. Common situations where the floating point errors could matter include: when you are repeatedly updating a value and the error grows, when you are interested in the small difference that results when subtracting two larger numbers and when

[2] Typically the long integers start at 2^{31} or 2^{63} depending on whether the system is 32 bit or 64 bit.

two values ought to be equal but they aren't exactly, e.g. after division you test for `1.0` but don't get the expected exact value.

Text strings

Strings are stretches of alphanumeric characters like `"abc"` or `'Hello world'`, in other words they represent text. In Python strings are indicated inside of single or double quotation marks, so that their text data can be distinguished from other data types and from the commands of the program. Thus if in Python we issue the command `print("lumberjack")` we know that `"lumberjack"` is the string data and everything else is Python command. Similarly, quotation marks will also distinguish between real numbers and text that happens to be readable as a number. For example, `1.71` is a floating point number but `"1.71"` is a piece of text containing four characters. You cannot do mathematics with the text string `"1.71"`, although it is possible to convert it to a number object with the value `1.71`.

String objects might contain elements that cannot be represented by the printable characters found on a keyboard, but which are nonetheless part of a piece of text. A good example of this is the way that you can split text over several lines. When you type into your computer you may use the *Return* key to do this. In a Python string you would use the special sequence `"\n"` to do this:[3] Python uses a combination of characters to provide the special meaning. For example, `"Dead Parrot"` naturally goes on one line, but `"Dead\nParrot"` goes on two, as if you had pressed *Return* between the two words.

Another concept that deserves some explanation is the empty string, written simply as `""`, with no visible characters between quotes. You can think of this in the same way as an empty list; as a data structure that is capable of containing a sequence, but which happens to contain nothing. The empty string is useful in situations where you must have a string object present but don't want to display any characters.

Text strings are made up of individual characters in a specific order, and in some ways you can think of them as being like lists. Thus, for example, you can query what the first character of a string is, or determine how long it is. In Python, however, you cannot modify strings once they are defined; if you want to make a change you have to recreate them in their entirety. This might seem stifling at first glance, but it rarely is in practice. The benefit of this system is that you can use strings to access items in a Python dictionary (which is a handy way to store data that we discuss below); if strings were internally alterable this would not be possible in Python. Python can readily perform operations to replace an existing string with a modified version. For example, if you wanted to convert some data that is initially stored as `"Dead Parrot"` into the text `"Ex-Parrot"` you could redefine the data as the string `"Ex-"` joined onto the last six characters of the original text. If at any point it really is painful to redefine a string entirely, a common trick is to convert the text into a list of separate characters (see list data type below) that you can manipulate internally, before converting the list of characters back into text.

Special objects

Booleans

The two Boolean objects are `True` and `False`, and they mean much what you might expect. Many objects can be examined to test whether they are logically false, like an

---

[3] To actually write the two characters `"\n"` without it being interpreted as a new line you would use `"\\n"`.

empty list or zero, or logically true, like `1.0` or a filled list. However, the `True` and `False` objects (note the capital letters) are special. They are the objects that you get back when you do a truth test. So, if you write a command to determine whether some number is equal to another number you will get a `True` object if they are equal or a `False` object if they are not equal. This differs from some languages where you might get `0` or `1` rather than dedicated Boolean objects. Also, you can set things to be `True` or `False` within your programs where you know that some data should only take one of two values.

```
hungry = True
chocolate = False
happiness = (not hungry) or (hungry and chocolate)
```
Performing Boolean logic operations with Python's True and False.

None

In Python there is a special object called `None`. When you use this object it means that something is totally undefined. This is in contrast to empty lists or empty text strings, which still exist as container objects of their respective kind. Accordingly the `None` object can be used, for example, to state that you don't have a list at all, rather than that you have an empty one. This distinction may seem tenuous, but it can be critical. For example, if I have records of people where I can store the names of their children, an empty list would indicate that a person has no children to be named, but a `None` object would indicate that I have been unable to determine whether the person had any children or not; not that they definitely had none.

Data collections

Python has several inbuilt collection data types, which are used to contain other items. The basic types of data containers in Python are *lists*, *tuples*, *sets* and *dictionaries*.

Lists

A list in Python is a data structure that can contain a sequence of other objects, of potentially different types,[4] in a specific order. Lists can have objects added to them and removed from them and they can be empty. Also, lists can refer to the same object more than once, at different positions in the sequence. For example, you could store the number of days in each month of a year as a list, as illustrated below. Often in Python programs you will be accessing the elements contained in a list by referring to a specific position (an index) within that list and by going through all the elements in a list in their given order. In Python we use square brackets to specify the beginning and end of a list:

```
days = [31, 28.243, 31, 30, 31, 30, 31, 31, 30, 31, 30, 31]
matrix = [[-1, 0, 0], [0, 1, 0], [0, 0, 3]]
```
Defining lists in Python; a simple list of numbers (both integers and floating point) and a second list which contains three sub-lists, each representing a row in a matrix.

---

[4] In Python 3 having different types in a list is discouraged.

Tuples

A tuple is a data structure that is very much like a list, but which you cannot change once it is created. In Python a tuple may be defined using round brackets ' () '. Although you cannot change its items, a tuple is used to contain a sequence of elements in a specific order, and the different positions of this sequence can be interrogated. The contents of a tuple are defined in their entirety when the tuple object is made. Having a kind of list that you cannot change may seem like a pointless data structure, but tuples are a surprisingly useful type of object. If you know that a sequence should definitely not have any elements modified, added or deleted, then you can use a tuple to ensure that it is not possible to deviate from this plan: for example, if you want to specify a vector with exactly three spatial coordinates, e.g. (x, y, z), using a tuple ensures that you can't have an invalid vector with too few or too many values. Similarly, tuples are used where you have elements that you know always go together; accordingly you could use tuples to specify a text font like ('helvetica', 10) or ('roman', 12), where you must have two elements to represent the name and the size of the font, and if you were to redefine the font you would have to specify both. Tuples, unlike lists, can be used as keys to refer to data in dictionary data structures (see the Dictionaries section below).

Sets

Sets, like lists and tuples, are data containers that encompass a collection of other objects. However, unlike lists and tuples, the elements are in no particular order and the elements cannot be repeated in a set. A notable use for sets is when you have some data that you know, or suspect, contains repeat objects. By placing such data within a set any duplication will be removed. For example, you might have a list containing the colours of different items; if you put these colours into a set object you can find out how many different colours were used. Also sets can be useful because you can easily perform set operations, for example, to find the items that two collections have in common; this would be trickier using lists or tuples.

```
females = set(['marge', 'maude', 'lisa', 'maggie', 'edna'])
simpsons = set(['homer', 'marge', 'bart', 'lisa', 'maggie'])

print(females & simpsons)
```

Defining Python sets and performing set operations; here finding an intersection (common elements).

There is actually another variety of set, called a *frozen set*. These are the same as regular sets with the exception, as the name suggests, that they cannot be altered once created (just like tuples). A useful consequence of this is that they can be used as keys to extract data from dictionary data structures (see below).

Dictionaries

A dictionary is a Python data structure which associates pairs of data objects to create a look-up table. The first object in the pair is called the *key*, and is unique inside a given dictionary, and the second object is its *value*. Unlike lists, where you refer to items by their position in a sequence, with dictionaries the data is not stored in any particular order and you access the values contained within by using the key. For example, you could have a dictionary which records the heights of various mountains, where you find the correct height for the correct mountain by using the name of the mountain (a string, the key) to look up the height (the value). In this instance if you were using a list you would have to know in which order the heights were stored, but with a dictionary you do not. You can have empty dictionaries and add and remove data from dictionaries, by adding and removing the pairs of key and value. The value that a particular key finds can be altered at any time, and although a key can only refer to one value, the value could be a data structure object, like a list or set, that contains other items. In Python we use curly brackets to specify the beginning and end of a dictionary.

```
ageDict = {'homer':36, 'marge':34, 'lisa':8, 'bart':10}
print(ageDict['lisa'])
```

Defining a dictionary in Python, which in this example allows an age value to be accessed using a name as the key.

An important point to be aware of is that while any type of data object can be a value in a dictionary, only certain kinds of object (those that cannot be internally modified to assume a different identity[5]) can be used as keys. Put verbosely: text strings,[6] tuples, the `True` and `False` objects, the `None` object, user-defined objects and frozen sets can be used as keys to access dictionary data, but lists, normal sets and other dictionaries cannot. The reason behind this rule is that the values in a dictionary are efficiently accessed based upon knowledge of the route from their keys, so Python must be able to consistently and uniquely identify each key and thus get the correct location of each value.

Converting between types

If you have some useful data in one type of data structure but need it to be in another, it can be a very quick operation to do the transfer. Indeed, many of the common transformations between the standard types of data are built into the language. Thus, for example, with single commands you can convert a floating point number to an integer, which ignores everything after the decimal point, or convert an unalterable tuple to a changeable list containing the same elements, or make a number from a text string that contains digits.

```
text = ' +0.783 '
floatNumber = float(text)
wholeNumber = int(floatNumber)

dnaTuple = ('A', 'C', 'G', 'T')
dnaList = list(dnaTuple)
```

Some data type conversions in Python.

Program flow

To form the logical flow of a computer program the component data objects and the commands that activate various operations are organised into a particular order. Just as this book is written to be read left to right, top to bottom, commands in Python are interpreted left to right and top to bottom. And it is also in this order that the operations are enacted.

---

[5] The jargon is *hashable*, and this point is discussed further in Chapter 3.
[6] It may seem surprising and limiting that text strings in Python are not internally modifiable, but in practice this causes few problems, given the right syntax.

Operations

So far we have described the common types of data that you will be dealing with in your programs. To make a working program, however, you must be able to do more than organise data; you have to work with it by performing operations that depend upon the content of the data. A simple example of an operation, and one which we have already used above, would be the addition of numbers. Operations are specific to the type of object that they work upon, so you can do mathematics with numbers, but not strings. Similarly, you can join two strings together to form a longer text. For the two operations of adding numbers and linking strings you can use the same '+' symbol in the Python code to perform the two operations, but the result is always appropriate for the type of object.

```
x = '22'
y = '78'
print(x + y)
```

Operations in Python are appropriate to the type of data involved, so for the above example the result is the text '2278' not the number 100.

If an object is not internally modifiable, like the number 4 or "abc", then when you perform an operation you get a different object as the result. So joining two text strings makes a new string. If an object is internally modifiable then an operation is allowed to (although it doesn't have to) alter the data within the object's structure without making any new structures. A clear example of this would be reversing the order of a list; the operation doesn't make any new data-containing object, it just moves the contents around internally.

Operations that are indicated within a Python program with symbols like + or * are really just shorthand ways of activating a procedure that is built into the fabric of the object being used. Thus when you do 2+5, the number object 2 has an internal procedure, activated by the + symbol, that deals with addition of the other number. Such procedures that are built into objects are called *methods* in the jargon, and these are a special kind of what we later describe as *functions*. Because there are only a limited set of symbols that can be sensibly used to indicate such inbuilt procedures, often you have to activate the procedure ('call the method' in jargon) directly with a dot notation. For example, to reverse the order of a list named myData, because there is no symbolic way of reversing the list, you would issue the command myData.reverse(). With this notation, notice that the method has a name[7] and that it is clearly associated with its list object using a dot '.'. We use the brackets at the end of the command to actually activate the procedure. If we simply issued the command myData.reverse then Python would interpret this as referring to the method (the procedure) without actually running it. The brackets at the end of an object's method may also contain some data that the operation is going to work with. For example, to put the number 6 onto the end of a list you can use the append operation that is built into list objects, and the extra number goes in brackets: myData. append(6).

Where there is actually a neat symbolic way of representing an operation there will also be an equivalent, albeit less elegant, version with the dot notation. As was illustrated in

---

[7] Which hopefully describes its purpose.

the example given earlier, x+y can be written as x.__add__(y). Here, you can see that the operation which the + symbol activates is internally called '__add__'. The plethora of underscore '_' symbols indicates that this method is inbuilt and normally hidden.

Control statements

You have more control in a program than just activating all of the written commands once in written order. You can use control commands to divert the flow of the program's execution under certain conditions, to add loops to repeat the execution of certain statements and to jump to a completely different part of the program, run some commands and then jump back again. It is very common to use all of these techniques, even in simple programs. As a simple example you might wish to look at all of the elements of a list in turn by using a repeating loop, performing the same operation on each of the values.

```
total = 0
numbers = [1,2,3,4,5,6]
for x in numbers:
   total = total + x * x

print(total)
```

An example of a loop in Python, specified using the for *keyword, where the indented command is repeated several times with different values. Here the total, which starts as zero, is redefined several times by adding a number squared (x * x) to its previous value, so that at the end the total is $1^2+2^2+3^2+4^2+5^2+6^2$.*

The ability to jump from executing the program flow in one place to another, execute some commands and then jump back again would be described in Python jargon as a *function*. In some older programming languages you can make the order of your program's execution jump about by using GOTO commands; which simply says that commands from now on are executed from a specified line of code in the program. In contrast, a function in Python is a section of code that is bundled together with a name. When the Python interpreter reads the commands that go together to make a function it does not activate those commands immediately. Only when the name of the function is used appropriately in the main flow of the program's execution are the commands from the function run. At the end of the function's execution the program flow goes back to the point where the function was activated, often sending back some data from the function.

```
def convertToFahrenheit(celsius):
   fahrenheit = celsius * 1.8 + 32.0
   return fahrenheit

print(convertToFahrenheit(37.0))
```

A simple example of a Python function: how the function works is defined in an abstract way after the def *keyword, but the operation is only actually performed on real numbers at the last line.*

Just as we write Python modules to store groups of commands as separate files that can be useful in many separate situations, so too functions are written because they perform a role that is useful in many different parts of a program. It is generally far better to write a function to do a particular job once, and then activate or *call* that function wherever that job needs to be performed, rather than writing several bits of code that do the same thing. One note of caution with using Python functions is that they can be proportionally slow to run compared to the regular flow of a Python program; so if speed is an issue things can often be helped by removing unnecessary calls to functions. Also, functions are generally only useful if you use them in more than one place in a program. If a procedure is only ever going to be run in one part of a program you would usually put the required commands directly into the program and not bother with a named function.

Although Python functions can exist on their own, they can also be linked to particular kinds of data structures. A function that is linked to an object becomes a method of that object (a procedure that belongs to the object), and can be executed in the same way as any other method with the dot notation, as discussed earlier.

```python
class UnitConverter:
  def metresToFeet(self, m):
    f = m / 0.3048
    return f
  def feetToMetres(self, f):
    m = f * 0.3048
    return m

converterObj = UnitConverter()

print(converterObj.metresToFeet(1.89))
print(converterObj.feetToMetres(6.0))
```

The definition of custom Python objects is specified using the `class` *keyword. Here we have defined a rudimentary unit converter class of object, which itself contains function definitions. The actual converter object is created by using the name of the class and from this object we can access the conversion functions.*

3 Python basics

Contents

Introducing the fundamentals

Python is a powerful, general-purpose computing language. It can be used for large and complicated tasks or for small and simple ones. Naturally, to get people started with its use, we begin with relatively straightforward examples and then afterwards increase the complexity. Hence, in the next two chapters we cover most of the day-to-day fundamentals of the language. You will need to be, at least a little, familiar with these ideas to appreciate the subsequent chapters. Much of what we illustrate here is called *scripting*, although there is no hard and fast rule about what is deemed to be a program and what is 'merely' a script. We will use the terminology interchangeably.

Here we describe most of the common operations and the basic types of data, but some aspects will be left to dedicated chapters. Initially the focus will be on the core data types handled by Python, which basically means numbers and text. With numbers we will of course describe doing arithmetic operations and how this can be moved from the specific into the abstract using *variables*. All the other kinds of data in Python can also be worked with in a similarly abstract manner, although the operations that are used to manipulate non-numeric data won't be mathematical. Moving on from simple numbers and text we will describe some of the other standard types of Python data. Most notable of these are the *collection* types that act as containers for other data, so, for example, we could have a *list* of words or a *set* of numbers, and the list or set is a named entity in itself; just another item that we can give a label to in our programs. Python also has the ability to let you describe your own types of data, by making an object specification called a *class*. However, this will be discussed in Chapter 7. We will end this chapter by introducing the idea of importing Python modules, which is a mechanism to allow a program to access extra functionality contained in separate files.

Finally, using Python is not only about the operation of programs, it is also important to consider what it means to the people who read it. Hopefully you will be writing clearly understandable code, with meaningful variable names and such like. Nonetheless, it is a good idea, when using any programming language, to get into the habit of adding human-readable comments to your programs, especially at points where the logic of what is happening is not so obvious. Such comments are simply textual descriptions that are separate from the functional part of the code. In Python, comments are usually introduced using the hash symbol[1] '#', whereupon all subsequent text on that line is for humans to read and not part of the program proper.

Getting started

Firstly, before using Python it must of course be installed on your computer. For Macintosh and Linux users Python is generally installed as standard, though you may wish to upgrade if the version is before Python 2.6. For Windows users Python is a simple double-click install, using the appropriate 'MSI' installer file. Windows, Macintosh, Linux and UNIX downloads are all available from http://www.python.org. However, for most flavours of Linux it is easiest to install or upgrade Python via the system's package manager. Later, in this book we will require that some extra Python modules are also installed, such as Numeric Python and Scientific Python packages. Because the websites for these may change in the future we list the up-to-date download locations for all extra packages on the Cambridge University Press website: http://www.cambridge.org/pythonforbiology.

There are two basic ways of running Python. The commands can be read and run from a file as a *script*. Alternatively, you can manually issue commands directly, one after the other, at what is called the Python prompt. To run a Python script on most computer systems you type within the command line interface, after the operating-system prompt, as illustrated in Figure 3.1. This will generally be set up as standard on Linux and Macintosh computers. However, for Windows care must be taken to set the PATH environment variable, as illustrated in Figure 3.2. Also for Windows systems, because the command line interface is less visible, it is common to execute a script by double-clicking on its icon. On-click execution can usually be set up on other systems too.

The file name for Python scripts traditionally ends in '.py', as illustrated in the example below, although strictly speaking it does not have to. By running the script we send the file containing lines of code to the Python interpreter, which reads it and acts on the contents. Also, if required, a script can have input values called *arguments* passed along when it is run, as illustrated for the Linux, UNIX or Mac command line:

```
> python scriptName.py argument1 argument2
```

This assumes that the executed 'python' command is on your *search path*, where the operating system knows to look for commands; otherwise you have to type the full path to its location.

The alternative to running Python from script files is to run Python alone, without a file, in an interactive mode. This mode gives you a prompt '>>>', where you can type manual input that is passed to the interpreter one line at a time. To start the interpreter with the Windows operating system you would click on the Python icon. To start from Mac OS X,

---

[1] The term *hash symbol* is used outside America; in America the term *pound symbol* or *number sign* is often used instead.

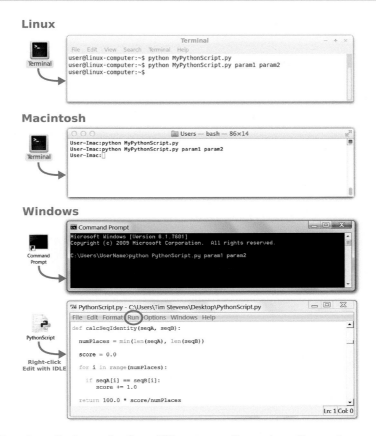

Figure 3.1. Running a Python script from different operating systems. For most computer systems it is commonplace to open a command-line interface and then type the 'python' command followed by the name, or file-system location, of the script, before pressing the *Return* or *Enter* key. For Windows systems the PATH environment variable will additionally need to be set up, as illustrated in Figure 3.2. In Windows Python scripts are also often run by double-clicking on the file icon or by using the Run menu from the IDLE editor (e.g. started via 'Edit with IDLE' in the right mouse menu for a Python file). IDLE is often included as part of Python installations for Windows.

Linux or UNIX this means opening a command-line shell and typing 'python', then pressing the *Return* or *Enter* key.

You can type commands at this prompt, pressing the *Return* or *Enter* key to issue each command and move on to the next line. Note that, by using the '-i' flag, it is also possible to run a Python script and then go into an interactive mode immediately afterwards. When the script is done it presents you with the prompt and awaits further instructions:

```
> python -i scriptName.py
[ Output of the script ]
>>>
```

When you are done, you quit by typing either the *Ctrl-d* key combination (so hold the 'Ctrl' key down and then tap the 'd' key) for Mac, Linux and UNIX systems or *Ctrl-z* for Windows systems. The Python prompt is convenient for testing out simple bits of code. More serious work is normally done with scripts, however. In this chapter you can work either way.

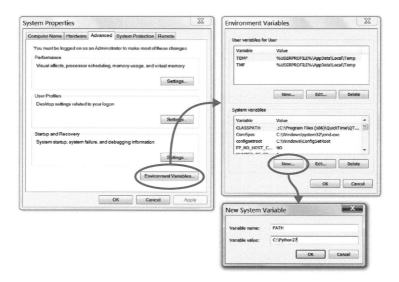

Figure 3.2. Setting the PATH environment variable in Windows. For the 'python' command to be recognised by Windows systems the PATH environment variable must include the location of the directory that contains the 'python.exe' file. An environment variable may be set in the Windows graphical interface via **Control Panel → System and Security → System → Advanced system settings**. If PATH is not already defined then the Python executable location may be specified via **New …**, for example, as '**C:\Python27**' or '**C:\Python33**', depending on the version. If the PATH is already defined then, after selection of this system variable in the lower table, using **Edit …** enables the addition of the Python location after any existing values, after a **semicolon**, for example, adding '**;C:\Python34**'. Note that the PATH specification has no spaces between entries (only '**;**') and **no** trailing slash '\'.

In Python 2 you can print a text message to the terminal window via the print command, for example:

```
print 'Hello world'
```

This automatically moves onto the next line because it prints a newline control character at the end. However, if you do not want to go to the next line put a comma at the end:

```
print 'Hello world',
```

In Python 3 the print statement changes to a *function*, which in simple terms means that it requires parentheses:

```
print('Hello world')
```

This function is also available in Python 2, although it doesn't print as nicely as in Python 3.[2] The Python 2 syntax for print is fairly ubiquitous in existing Python code. For new code it is probably best to use the Python 3 syntax, even when using Python 2. In this book we will use the print function rather than the print statement.

---

[2] In Python 2.6 and 2.7 you can get the Python 3 behaviour for print using a special statement: from __future__ import print_function.)

Figure 3.3. Starting the interactive Python interpreter on different operating systems. For most computer systems the interactive interpreter is started by issuing the 'python' command at an operating-system prompt. For Windows systems the PATH environment variable will need to be set first, however, as illustrated in Figure 3.2. Also, for Windows the interpreter may be started by double-clicking on the 'Python (command line)' icon (e.g. accessed via the start menu), or by double-clicking on the installed 'python.exe' file.

The print operation automatically converts anything that is not already a text string into text for display. Hence, for example, you can print numbers:

```
print(127)
```

Also, the value being printed might not be explicitly stated, coming instead from a calculation or other operation:

```
print(1 + 2 + 4 + 8 + 16)
```

And to print several things you can simply separate them with a comma, which as standard separates the printed items with a space:

```
print('The meaning of life is', 42)
```

If you want to see the value of some variable when running Python from a script file you need to explicitly use `print`. However, at the Python prompt just giving a variable name, and nothing else, on a line will print it out, albeit sometimes slightly differently. Note that `print` tidies things a little by rounding the last few decimal places, which is normally what you want:

```
>>> x = 0.333333333333333
>>> print(x)
0.333333333333
>>> x
0.333333333333333
>>>
```

Throughout our examples we will be adding comments to the Python commands. These describe what is going on at a given point in the code, and we encourage you to add helpful explanations to the relevant points in your programs. Even though such comments are passed into the Python interpreter, they are ignored and serve no purpose other than to inform anyone who reads the code. In Python, single-line comments are introduced with the '#' character, and everything after this character on that line constitutes the comment. Comments may go on a line of their own, but it is common practice to add them after a command. The example below illustrates the addition of a comment to the end of a line, which otherwise adds two numbers. As will be discussed below, do not be tempted to put a space at the beginning of the line.

```
>>> 2+2    # this is a comment
4
>>>
```

Whitespace matters

One of the fundamental design decisions Guido van Rossum made when designing the Python language was to designate different blocks of code using the space at the beginning of lines, rather than using curly braces/brackets as many other languages use (e.g. C, Perl, Java etc.). Thus the space at the beginning of a line matters in terms of how the following commands are interpreted. Space can be added by pressing the space bar or the *Tab* key, and moves the characters which follow to the right. Things like space and tab, which have no printed symbol, are often still considered as 'characters' in computing and are collectively referred to as 'whitespace'. If you have too little or too much whitespace at the beginning of a line you get a syntax error. The syntax error indicates that the Python interpreter was unable to process the input characters in a meaningful way.

```
>>>   2+2
4
>>>     2+2
 File "<stdin>", line 1
    2+2
    ^
IndentationError: unexpected indent
>>>
```

Note that whitespace after the beginning of a line does not matter (between *tokens* of the language):

```
>>> 2 + 2      # this is same as 2+2
4
>>>
```

Because whitespace at the beginning of a line is required to delineate blocks of code, as we illustrate later, it can be somewhat problematic if not used appropriately. In particular, you should never mix tabs with ordinary spaces (and from Python 3 this is illegal), because on some machines a tab may be equivalent to four spaces, but on others it may be eight, so code could break moving it from one machine to the next. In many text editors you can automatically convert tabs to ordinary spaces (sometimes this is called using 'soft tabs') and that is what we recommend.

Using variables

As we have already illustrated, we can create a named item, which here we call 'x' for simplicity, and assign it to a value:

```
>>> x = 17
```

Using the above example we can introduce some more jargon computing terms. On the left-hand side, before the equals sign, we have a *variable*. On the right-hand side we have a *literal*. The whole line here is a *statement* and specifies that the variable is set to have value equal to the literal. As you might expect from the term 'variable', the value of x may be changed by assigning a new value:

```
>>> print(x)
17
>>> x = 3
>>> print(x)
3
```

In Python the names we give variable data can contain only the usual 26 letters (upper and lower case), numbers and underscores ('_'), with the additional restriction that they cannot begin with a number. Note that names are case-sensitive, so the variables DNA, Dna and dna are all treated as distinct. In general, variables should have names that indicate what their purpose and/or type are, in order to make the code more understandable. For example, if you state freeEnergy = heat-entropy and not x = p-q, you can see at a glance what is intended and have your program more easily understood, including by yourself at a later date, without any additional comments.

We can use as many different variable names as we like and assign their value based on other variables. For example, in the following we assign a value to x and then assign a value for y based on x:

```
>>> x = 17
>>> y = x * 13
>>> print(y)
221
```

Unlike many computing languages, Python is not a language where you must initially specify, and then stick to, a given kind or *type* of data for a given variable. You could initially allocate a numeric value to 'x', without advance warning, and then later on change 'x' to some text. This differs from languages like C and Java, for example, where you would have to declare up front what type of data 'x' was to contain. In Python, the type of variable is specified by the type of whatever its value is set to. So if you redefine a variable its type may change. Although variables can change type, it is usually best to avoid that practice.

```
>>> x = 4       # x is set to the integer 4
>>> 3*x
12
>>> x = 7.1     # x now set to the floating point number 7.1
>>> 3*x
21.299999999999997
```

The above example reminds us that floating point calculations are not always precise. The answer could also depend on the Python implementation and version.

Simple data types

As with other computer languages, Python has various simple, inbuilt types of data. These are Boolean values, integers, floating point numbers, complex numbers, text strings and the null object.

Boolean values represent truth or falsehood, as used in logic operations. Not surprisingly, there are only two values, and in Python they are called True and False.[3] Example usage:

```
a = True
b = False
```

Integers represent whole numbers, as you would use when counting items, and can be positive or negative. In Python 2 there are two types of integers, plain integers and *long* integers.[4] Plain integers have a maximum size dependent on the specific Python implementation you are using. On a typical computer the largest plain integer would be $2^{31}-1$ or $2^{63}-1$ (for 32 bit and 64 bit respectively). There is no limit on long integers except for what can fit into available memory. In Python 3 there is only one type of integer, the long integer. Unless you are doing something unusual, there is no point worrying about this distinction or the difference between the two types, and in most situations in Python 2 the plain integers will suffice. Example usage:

```
x = -7
y = 123
```

Floating point numbers (in mathematics the *real* numbers), which are written with decimal points or exponential notation, are not always represented exactly, since a computer has only a finite amount of memory. This introduces issues to do with numerical errors,

---

[3] These existed in Python 2.2 but were then just synonyms for the integers 1 and 0. The Boolean type was properly introduced in Python 2.3.

[4] From Python 2.2 onwards, integers and long integers are treated uniformly, so most people can happily ignore the difference between the two types. Before that long integers had to have an 'L' at the end of the number, for example, `12345678901234L`.

and potential instability of numerical algorithms. However, such issues are common to all computer languages. Example usage:

```
z = 123.45
```

There is also an inbuilt data type to represent *complex numbers* which you would normally write in the form 'a+b*i*' (mathematical notation) or 'a+b*j*' (engineering notation). Although complex numbers occur quite naturally in mathematics, science and engineering, relatively few Python programs use them. The Python syntax follows the engineering style and the real and imaginary parts can themselves be integer or floating point:

```
x = 3+4j
y = 1.2-5.8j
```

Strings represent text, i.e. strings of characters. They can be delimited by single quotes (') or double quotes ("), but you have to use the same delimiter at both ends. Unlike some programming languages, such as Perl, there is no practical difference between the two types of quote, although using one type does allow the other type to appear inside the string as a regular character. Example usage:

```
r1 = 'Ala'
r2 = "Arg"
text = "It's a line with an apostrophe"
```

Python also allows multi-line strings, which start and end either with triple single quotes (''') or triple double quotes ("""). Example usage:

```
text = """Python also allows multi-line strings, which
start and end with a triple single quote or a triple
double quote."""
```

Note that the indentation inside the string does not have to align with the start of the statement. Any whitespace at the beginning or end of the internal lines, i.e. between the opening and closing triple quotes, does make a difference though. Hence, if the second line of text were indented, then those indentation spaces would be present in the string.

The last of the basic data types we cover here is a special built-in value called None, which can be thought of as representing nothingness or that something is undefined. For example, it can be used to indicate that a variable exists, but has not yet been set to anything specific. Example usage:

```
z = None
```

Finally, if you have a variable and want to know what its data type is then you can use the type() function. This actually generates a special object representing the type, though it prints out in an informative way:

```
print( type(x) )    #  'complex'
print( type(z) )    #  'NoneType'
```

Arithmetic

Python mostly uses a similar syntax to other computer languages for performing numerical arithmetic:

```
x + y        # addition
x - y        # subtraction
x * y        # multiplication
x / y        # division
x // y       # floored division
x % y        # remainder of x / y
x ** y       # x to power y
pow(x, y)    # x to power y
```

The variables x and y can be integers, floating point numbers or a mixture. If both are integers the result is also an integer, except in the case of division for Python version 3. Otherwise the result is a floating point number, even if it represents a whole number. Thus 4.6 + 2.4 is 7.0, not 7. This also includes the floored quotient, x//y, which gives the whole number part of the division of x and y as floating point. For example, 13.3//2.1 gives 6.0, not the integer equivalent.[5]

A non-programmer might wonder why x//y is useful at all. However, it turns out that it does come up in various contexts, but mostly when x and y are integers. This brings up an oddity, which Python, before version 3, shares in common with many computer languages, namely that for integers, the operation x/y is the same as x//y. A non-programmer might expect that 13/5 is equal to 2.6, but in fact it is equal to 2, the integer part of that. This is in contrast to doing division where at least one floating point number is involved like 13/5.0, 13.0/5 or 13.0/5.0, which are all indeed equal to 2.6. Hence in Python 2, if you have two integers and want to do the traditional non-integer division then you can explicitly convert one of them to a floating point number using the float() function, so, for example, float(13)/5. (There is also an int() function for converting floating point numbers to their integer part.)

It is a historic accident that integer division behaves this way, although the situation changes in Python 3, where integer division reverts to its more traditional 'human' meaning, so 13/5 now does equal 2.6. Accordingly, it is recommended that in Python 2 you avoid x/y if x and y are integers, but instead use x//y.

Example arithmetic results:

```
13 + 5          # 18
13.0 + 5        # 18.0

13 - 5          # 8
13 - 5.0        # 8.0

13 / 5          # 2 in Python 2; 2.6 in Python 3
float(13) / 5   # 2.6
13.0 / 5        # 2.6

13 // 5         # 2
13 // 5.0       # 2.0
```

As in most computer languages, multiplication and division have higher precedence than addition and subtraction, but arithmetic expressions can be grouped using parentheses to override the default precedence. So we have:

---

[5] This result is because 2.1 divides 13.3 six times with a remainder that is less than 2.1, or more precisely, 13.3 = 6.0*2.1 + 0.7.

```
13 * 2 + 5     # 31 since "*" has higher precedence than "+"
(13 * 2) + 5   # 31
13 * (2 + 5)   # 91
```

A common situation that arises is that a variable needs to be incremented by some value. For example, you could have:

```
x = x + 1
```

which increases the value of x by 1. Python allows a shorthand notation for this kind of statement:

```
x += 1
```

Also, it allows similar notation for the other arithmetic operations, for example:

```
x *= y
```

assigns x to be the product of x and y, or in other words x is redefined by being multiplied by y.

String manipulation

Text items in Python are called *strings*, referring to the fact that they are strings of characters. String functionality is an important part of the Python toolbox. For example, a file on disk (covered in Chapter 6) is read as a string or a list of strings; a file can be viewed as a collection of characters. Here, even if part of the loaded file represents a number, it is initially represented as a string of characters, not a proper Python numeric object. In Python, strings are not modifiable. This might seem like a limitation, but in fact it rarely is because it is easy enough to create a new, modified string from an existing string. And since strings are not modifiable it means that they can be placed in sets and used as keys in dictionaries, both of which are exceedingly useful.

In this section we will illustrate some basic manipulations on strings using the following example string:

```
text = 'hello world'    # same as double quoted "hello world"
```

In some ways a string can be thought of as a list of characters, although in Python a list of characters would be a different entity (see below for a discussion of lists). Note that when we refer to something in a string as being a *character*, we don't just mean the regular symbols for letters, numbers and punctuation; we also include spaces and formatting codes (tab stop, new line etc.). You can access the character at a specific position, or index, using square brackets:

```
text[1]              # 'e'
text[5]              # ' ' - a space
```

Note that the index for accessing the characters of a string starts counting from 0, not 1. Thus the first character of a string is index number 0. At first this can seem odd to non-programmers, but it is by far the most sensible convention, and is used in most modern computer languages.

Bear in mind that we cannot change the characters of a string. For example, we get an error if we try to change the first position to an 'H':

```
text[0] = 'H'            # Fails!
TypeError: 'str' object does not support item assignment
```

You can count backwards from the end of the string, where index -1 is the last character of the string:

```
text[-3]                 # 'r'
```

If a string has n characters, then the minimum value of the index is -n and the maximum value is n-1. If the index falls outside this range an error is generated; Python makes an *Exception* object which reports what the error was (see the next chapter for a description of these).

Python also has a very convenient slicing notation, to access a substring from within a string. The notation [start:stop] refers to the characters from position start up to but not including position stop. As with single indices, these positions can be negative. The fact that it is 'up to but not including' might seem odd, but as with the indices counting from 0, this turns out to be a sensible convention. In particular, if start and stop numbers in the slice notation are both non-negative then the number of characters in the resulting slice is just the difference between the two values (stop-start), or put another way [start:start+n] gives n characters.

As a further convenience, if you leave out the start entirely giving just [:stop], then the slice starts at the very beginning; the start point is taken to be 0. If you leave out the stop, so have [start:], then the slice continues to the very end; as if stop were taken to be the length of the string. Thus, for example, [:n] refers to the first n characters of the string.

```
text[1:3]               # 'el'
text[1:]                # 'ello world'
text[1:-1]              # 'ello worl'
text[:-1]               # 'hello worl'
```

This leads to the proper way to (effectively) change the first character of the example string. We can use a slice to access the characters we wish to keep and redefine text:

```
text = 'H' + text[1:]    # 'Hello world'
```

You can check if a substring is contained in a string:

```
'wor' in text           # True
'war' in text           # False
```

or is not contained in (is absent from) a string:

```
'wor' not in text       # False
'war' not in text       # True
```

There are two functions that let you determine the position of (the first occurrence of) a substring inside a string:

```
text.index('wor')       # 6
text.find('wor')        # 6
```

Note that the value returned is the index of the first character of the substring in the string. The difference between these functions is how they deal with the situation when the substring is not contained in the string. For the `index()` function an error is generated, but instead the `find()` function returns −1:

```
text.find('war')       # -1
```

It is a matter of taste which version you use. Nonetheless, it might have been better for `find()` to return `None` if the substring isn't present. You can search from the (right-hand) end of the string instead of the beginning:

```
text.index('l')        # 2
text.rindex('l')       # 9
text.find('l')         # 2
text.rfind('l')        # 9
```

When you read a file, you often end up with whitespace characters (newlines, carriage returns, tabs and spaces) that you want to get rid of, or deal with. There are various functions for this. Here we will consider a string with two leading spaces and two trailing spaces:

```
line = '  hello world  '
```

You can strip off the whitespace from both ends:

```
line.strip()           # 'hello world'
```

Note that since strings are not modifiable, this gives back a new string; it does not modify the original string. You can also strip whitespace from just the beginning (left) or end (right) of the string:

```
line.lstrip()          # 'hello world  '
line.rstrip()          # '  hello world'
```

There is no inbuilt function to remove all whitespace from everywhere in the string, including any in the middle. This is possible using the regular expression module, which we discuss in detail in Appendix 5.

You can split up your string into separate substrings according to the presence of whitespace. This creates a list of strings, where a 'list' is simply a container for the strings (here represented by square brackets). Lists are Python objects in their own right and are discussed further in the next section.

```
line.split() # ['hello', 'world'] - a list of two strings
```

Note that this automatically strips off the whitespace at the beginning and end before doing any splitting. You can also split on an arbitrary substring, noting that (quite sensibly) this does not strip off the whitespace at the beginning or end:

```
line.split('wor')      # ['  hello ', 'ld  ']
```

Given that you can split a string into parts, it is quite natural that you can also do the opposite and join a number of strings together into one long string. For example, given a variable that represents a list of strings, which we write inside square brackets and separate with commas:

```
myList = ['Homer', 'Marge', 'Maude', 'Ned']
```

you may want to create one long, combined string:

```
longText = 'Homer, Marge, Maude, Ned'
```

This is done using the `join()` function, where you connect the items from the list with some other connecting string (e.g. with commas and spaces). However, although you might expect the joining function to come from the list, it actually belongs to the connecting string. Thus, you do **not** do:

```
longText = myList.join(connectorString)    # Not used
```

Instead the correct Python way is:

```
longText = connectorString.join(myList)
```

The syntax can take a bit of time to become familiar, because the string that is linking things together might be defined on the same line where the joining occurs. Considering the following:

```
cities = ['London', 'Paris', 'Berlin']
connector = '->'
connector.join(cities)  # 'London->Paris->Berlin'
```

The last lines could be written as one, without an intermediate variable name:

```
'->'.join(cities)       # 'London->Paris->Berlin'
```

Thus, the connecting string is the thing that comes before the dot. A further point, which can catch you out, is that all the items that are to be joined together have to be strings; no other type will do. Also, the joining string is only added in-between the items of the list not at the beginning or end.

The `join()` function also allows you to concatenate items together without adding any extra characters, using an empty string. For example, suppose you have a list of one-letter codes for a DNA sequence (or protein or RNA) and want to create a string of all the letters joined together. Then you could do:

```
sequence = ['G', 'C', 'A', 'T']
seq = ''.join(sequence)            # 'GCAT'
```

You can also do string concatenation using the '+' operator, so an alternative to the above would be:

```
seq = sequence[0] + sequence[1] + sequence[2] + sequence[3]

# seq is 'GCAT'
```

This is generally not a good approach if the list is long, because it is much less efficient than using the `join()` method. And in any case you would usually not write out the list elements in full; you would use a loop to go through each item in turn (see the next chapter). On the other hand, for concatenating only a few strings together it is perfectly acceptable to do it this way. As another example, suppose you have some numbers and want to create a string with this information in it. Then you could do the following, converting the numbers to strings using `str()`:

```
x = 12
y = 5
text = "I have " + str(x) + " apples and " + str(y) + " oranges."
# the text is "I have 12 apples and 5 oranges."
```

Even here, though, Python offers an alternative, which is to use a *formatted* string. So we could write the above instead as:

```
text = "I have %d apples and %d oranges." % (x,y)
```

Here `%d` is a formatting code and represents the places in the text to insert the digits. The values for the digits are contained in the round-bracketed 'tuple' collection at the end (see below for discussion of tuples), after the bare `%` sign. Naturally, there should be as many formatting codes in the initial string as there are items to insert. If we were inserting other types of data then we would use different codes, for example, `%s` to insert a string and `%f` for a floating point value:

```
name = 'Barry'
weight = 82.173
text = "The weight of %s is %f kg" % (name, weight)

# Gives "The weight of Barry is 82.173000 kg"
```

We can optionally specify the number of decimal places to use for the floating point value by adjusting its formatting code. For example, `%.1f` can be used so that the weight is written out with one digit after the decimal place, rounding as appropriate:

```
text = "The weight of %s is %.1f kg" % (name, weight)

# Gives "The weight of Barry is 82.2 kg"
```

If you also wanted at least five total characters for the weight, padding with spaces, you would write `%5.1f`. It is notable that you can actually use `%s` for every type of data, because values will be automatically converted into a representative string, but if you want to fine-tune the appearance of floating point numbers then it is best to use the `%f` construct.

There are analogous options for the `%d` construct used with integers. So `%5d` means that at least five places are used to display the integer, and `%05d` means that you zero-pad the five places at the left, if necessary. For example, you could create a string with the time of day via:

```
hours = 12
minutes = 5
seconds = 43
t = "%02d:%02d:%02d" % (hours, minutes, seconds)
# t is "12:05:43"
```

Python has a notable tweak with string formatting: if the collection of values that is to be substituted only has one item then you can just use the item directly, rather than using brackets (which represents a tuple, see below). So

```
"My name is %s" % name
```

is equivalent to

```
"My name is %s" % (name,)
```

See Appendix 4 for a more complete table of formatting codes and a thorough description of the new-style formatting system specified with `string.format()` method.

Collection data types

As well as the simple data types, Python has several common collection data types, *tuples*, *lists*, *sets* and *dictionaries*, that provide a means of bringing multiple items together into a container.

The simplest collection type is a tuple. A tuple contains a fixed number of items and once it is created it cannot be modified. You can think of it as a fixed (immutable) and ordered collection of items. A tuple is defined using a left round parenthesis '(' at the start and a right round parenthesis ')' at the end. For example, we could have:

```
x = ()                  # empty tuple
x = ("Ala",)            # tuple with one item
x = (123, 54, 92, 54)  # tuple with four items
```

Note the peculiar-looking syntax for tuples with only one item inside; there is a comma (',') at the end. This is because otherwise Python would interpret the parentheses as an expression rather than a tuple. For example, (2+3) is a mathematical expression for the number 5, but (2+3,) is a tuple containing one item (again 5). So parentheses are used in both these contexts in Python, and the comma is a small irritation that results to avoid ambiguity.

The items inside a tuple can repeat themselves and be of different data types; you can mix numbers, strings or whatever. In common usage, however, the items tend to all be of the same type, as illustrated above. Also, an item inside a tuple does not have to be a simple data type, it can itself be a collection, or even a user-defined type (which we come to in Chapter 7). A nonsense example of a tuple with mixed types and repetition, where the last item is another tuple, inside the first, is:

```
x = ( 2, 2, 'banana', False, ('a','b') )
```

Like all the collection types, tuples may also be created using variables:

```
x = 1.2
y = -0.3
z = 0.9
t = (x, y, z)
```

The next simple collection type is a *list*. A list contains an arbitrary number of items, and new items can be added and existing ones removed. As with tuples, the items in a list remain in their specified order. The major difference between lists and tuples is that the contents of a list can be modified, whereas a tuple is fixed at the moment it is defined. A list is defined using a left square bracket '[' at the start and a right square bracket ']' at the end. For example, we could have:

```
x = []                  # empty list
x = ["Ala"]            # list with one item
x = [123, 54, 92, 54]  # list with four items
```

As with tuples, the items in a list can repeat and be of different data types, although in normal usage they tend to all have the same type. And again, an item does not have to be a simple data type. You can convert a tuple to a list with the inbuilt list() function, and you can convert a list to a tuple with the tuple() function:

```
t = (123, 54)
x = list(t)    # x is [123, 54], t is still (123, 54)
w = tuple(x)   # w is (123, 54), x is still [123, 54]
```

The next collection type is a *set*. A set contains an arbitrary number of items, and can be modified; new items can be added and existing ones removed. Unlike tuples and lists, however, the items in a set are not in any order. Also, an item can only appear once in a set; if you try and add the same item twice then the second time it will be ignored. Sets were introduced relatively late into Python[6], and so the syntax used a keyword, specifically `set(collection)` to get a filled set or `set()` to get an empty set. When we pass a collection (list, tuple or other set) to the construction the contents of the collection are used to define the contents of the set.

```
x = set()                   # Empty set

listData = [123, 54, 92, 54]
x = set(listData)           # Set with _three_ items
```

Note that the second set has three items, not four, because the `54` is repeated and so the second one is ignored. Because `set()` does not take more than one argument, extra brackets are often used to create an inner collection, for specifying multiple items directly:

```
x = set(1,4,9,16,25)     # Fails! - Multiple arguments
x = set([1,4,9,16,25])   # Works - Brackets make a single list
```

In Python 2.7 and in Python 3, although `set()` is still used for making sets with the contents of other collections, a new shorter notation can be used for directly defining non-empty sets.[7]

```
x = {123, 54, 92, 54}
```

Be aware that creating a set using an inner tuple requires an extra comma if the tuple contains only one item, otherwise the parenthesis will effectively be ignored. Using square brackets, to make an inner list instead, does not have this issue.

```
x = set(("Ala"))      # A set containing three letters!
x = set(("Ala",))     # Set containing one string item
x = set(["Ala"])      # Set containing one string item
```

As with tuples and lists, the items inside a set may represent a mixture of different kinds of data, so you could have a set containing both numbers and text if you wanted, although in normal usage they tend to all be of the same type. Additionally, an item does not have to be one of the simple Python types; you could place your own custom data objects in it.

There is a significant caveat with putting things in sets because it turns out that not all Python data types can be placed in one. Only items that can be described as *hashable* can go in. The concept of hashability[8] is perhaps too complex to describe at this point. However, the basic essence of the situation is that if an item is to be allowed within a set it cannot be modified internally, to take on a new value. If such value modifications were allowed then items inside a set could be changed so that they become indistinguishable, and this is inconsistent with sets not having repeats. The inbuilt simple types like integers and strings are not modifiable, because their values define what they are. Thus, these are hashable and

---

[6] It cannot be used for empty sets because that notation instead means an empty dictionary.
[7] They were allowed in Python 2.3 but were not done in the current way until Python 2.4.
[8] The value of an item is converted via a hash function into an index that allows the location of the item within a data structure to be determined efficiently, without the need to compare values with all the other items. Hence, for this look-up to work, the item's value must be fixed and thus also its index.

hence are allowed as set items. Modifiable collections like lists, dictionaries and other sets are not allowed as items, because when their content changes so does their value.

The final, main collection type is a *dictionary*. (Python dictionaries are equivalent to *hash maps* in Java and *hashes* in Perl and Ruby.) A dictionary is a mapping between a set of identifying keys and specific values, one for each key. An item in a dictionary is a `key:value` pair that represents one entry. A dictionary is modifiable; new entries can be added, existing entries can be removed and existing entries can be modified by changing the value associated with a key. A dictionary is defined using a left curly bracket '{' at the start and a right curly bracket '}' at the end. When written out in full the key and value of a pair is linked by a colon ':' and different pairs are separated with commas. For example, we could have:

```
x = {}                           # empty dictionary
x = {"Ala": 71.07}               # dictionary with one entry
x = {"Ala": 71.07, "Arg": 156.18}  # dictionary with two entries
```

The last dictionary above maps the string 'Ala' to the value 71.07 and the string 'Arg' gets mapped to the value 156.18. Although a dictionary value can be of any type, a key in a dictionary must be hashable (not internally modifiable), in the same way as an item in a set. Not only does this guarantee that the keys of the dictionary don't repeat, it also provides an efficient look-up mechanism. Neither the keys nor the values have to all be of the same data type, although in use they often are; it is quite common for keys to be strings. You can think of an English dictionary as being a map from words to definitions, in which case both the keys (the words) and the values (the definitions of each word) are text strings. Though, one difference is that an English dictionary is ordered (alphabetically) but a Python dictionary is not.

After having described the major collection types, a question arises as to why we need the tuple data type at all, since a list is everything that a tuple is and more. There are three basic reasons. The first reason is that tuples are more efficient computationally, although normally that is not a big issue. The second reason is that a program might want to return a tuple rather than a list to users because it does not want the collection to be modified; an example would be `(latitude, longitude)` coordinates where both items go together to give a meaningful outcome. The third reason is that, as mentioned above, lists can be modified and so are not hashable. Hence, lists cannot be used as items in a set or as keys in dictionaries, whereas tuples can. You can think of a tuple as a frozen list.

Although sets are modifiable and so not hashable, there is a variant of sets that is, and these are called frozen sets. Accordingly, a frozen set is an unordered collection of items, without repetition, and which cannot be modified. To create a frozen set you use the nomenclature `frozenset(...)` instead of `set(..)`. You can think of the relationship between sets and frozen sets as being the same as the relationship between lists and tuples. There are no inbuilt frozen dictionaries in Python, although you could add them yourself if you were really keen (by making a custom object which inherits properties of a dictionary, but where some of the innards are redefined).

List and tuple manipulation

In this section we will illustrate some basic manipulations on tuples and lists using the following tuple and list examples to do this:

```
t = (123, 54, 92, 87, 33)          # tuple
x = [123, 54, 92, 87, 33]          # list
```

When you need to fetch a value, from a tuple or list, the items are accessed by their index number: the position of the desired item relative to the start. Although Python uses round parentheses to define tuples and square brackets to define lists, when you want to access an item of a tuple or list, in both cases you use square brackets, i.e. myList[index], myTuple [index] will fetch the value at the index position of the list and tuple respectively. Note that the index for accessing items of tuples and lists starts counting from 0, not 1. Thus the first item of a tuple or list is the element with index 0. By way of example, here we obtain the first item of the above defined tuple and the third element of the list:

```
t[0]      # 123
x[2]      # 92
```

Python also lets you access items starting from the end of the tuple or list using negative indices, where the index -1 refers to the last item, -2 to the next-to-last etc.

```
t[-1]     # 33
x[-3]     # 92
```

If the tuple or list has n items, then the minimum value of the index is –n and the maximum value is n-1. If the index falls outside this range an error is generated; Python makes an *Exception* object which reports what the error was (see the next chapter for a description of these). As with strings, Python has a very convenient *slicing* notation, to access a range of items from within a tuple or list. The notation [start:stop] refers to the items from position start up to **but not including** position stop. As with single indices, these positions can be negative.

As a further convenience, if you leave out the start entirely giving just [:stop], then the slice starts at the very beginning; the start point is taken to be 0. If you leave out the stop, so have [start:], then the slice continues to the very end; as if stop were taken to be the length of the tuple or list. Thus, for example, [:n] refers to the first n items of the tuple or list. If you leave out both the start and stop points then you get back a copy of the original tuple or list, and this is a convenient way of doing that.

```
t[1:3]         # (54, 92)
t[1:]          # (54, 92, 87, 33)
x[1:-1]        # [54, 92, 87]
x[:-1]         # [123, 54, 92, 87]
x[:]           # [123, 54, 92, 87, 33] – A copy
```

You can check if an item is present inside a tuple or list by using the in operator, which gives you back a Boolean value (true or false):

```
92 in t        # True
93 in x        # False
```

Similarly you can also check if an item is not within a tuple or list:

```
92 not in t    # False
93 not in x    # True
```

Neither of these is a particularly efficient operation (computationally), so for long tuples and lists you may want to avoid doing too many of this kind of check if possible. With checks of this kind to be more efficient you would often use sets.

For lists and tuples you can count the number of occurrences of an item, which may be more useful than simply detecting its presence:

```
y = [3, 11, 7, 3]
y.count(3)          # 2
```

For lists you can check at which index a given item is located, noting that this gives the first occurrence if there is more than one:

```
x.index(87)         # 3
```

The `index()` function also exists for tuples in versions of Python from 2.6 onwards (it is slightly odd that it took Python so long to introduce it). Using `index()` is also not very computationally efficient, although it is not often needed. The `len()` function returns the number of items in the tuple or list; its length in other words:

```
len(t)              # 5
len(x)              # 5
```

So far we have been discussing how to access the items in a tuple or list. Tuples are not modifiable so the above is pretty much the end of the story for them. However, given that lists are modifiable there are more things we need to consider. You can change the item at a specific position in the list:

```
x[1] = 55           # x now [123, 55, 92, 87, 33]
```

If you mistakenly tried the same operation on a tuple you would get an error:

```
t[1] = 55               # Fails!
TypeError: 'tuple' object does not support item assignment
```

You can append new items to the end of a list:

```
x.append(17)        # x now [123, 55, 92, 87, 33, 17]
```

You can add a list's items on to the end of another:

```
y = [1, 2]
y.extend(x)         # y now [1, 2, 123, 55, 92, 87, 33, 17]
```

You can insert items at specific positions:

```
x.insert(1, 99)     # x now [123, 99, 55, 92, 87, 33, 17]
```

Here, for example, the index of 1 indicates what its final position will be after the insertion; it is inserted before the previous item at position 1. And if the index is the length of the list then the insertion is after the last existing item, which is equivalent to using `append`. Strangely, if the index is less than 0 it acts as if it were 0, inserting before the first item. Similarly, any index greater than the length of the list inserts after the last item. Either way, no error is triggered for having an out-of-range index.

You can remove a specific item from within the list, thus shortening it:

```
x.remove(92)          # x now [123, 99, 55, 87, 33, 17]
```

This will cause an error if the item is not in the list. You can also delete an item at a specific position with the `del` statement:

```
del x[4]              # x now [123, 99, 55, 87, 17]
```

This will cause an error if the index is out of range. Lists also have a `pop()` method which allows you to remove ('pop out') the item at a particular index, but this function will also return the value of the item, rather than discarding it entirely:

```
y = x.pop(2)          # x now [123, 99, 87, 17], y is 55
```

You can even change a range of items of a list to a new range of items using the slice notation, although this takes some getting used to:

```
x[1:3] = [19, 21, 5]  # x now [123, 19, 21, 5, 17]
```

This has replaced two items, at positions 1 and 2, with three new items.

You can reverse the order of the items in a list:

```
x.reverse()           # x now [17, 5, 21, 19, 123]
```

Also, lists can be sorted internally using the `sort()` function:

```
x.sort()              # x now [5, 17, 19, 21, 123]
```

If you want a new list to be generated as a result of the sorting, rather than the original list being altered, then you can use the `sorted()` function:

```
x = [123, 54, 92, 87, 33]
y = sorted(x)         # y = [33, 54, 87, 92, 123], x unchanged
```

Note the difference in syntax, so there is no `sort(x)` and also no `x.sorted()`. In fact the `sorted()` function can also be used on tuples and sets, but still returns a new list:

```
t = (123, 54, 92, 87, 33)
y = sorted(t)         # y = [33, 54, 87, 92, 123], t unchanged
```

The `sort()` and `sorted()` functions use the 'natural' sort order of the items; numeric and alphabetic. However, it is possible to specify another ordering method, for example, to get the reverse of the sort. These topics are covered in Chapter 5 on functions.

Set manipulation

In this section we will illustrate some basic manipulations on sets using the following set, containing three items:

```
s = {123, 54, 92}
```

The items in a set are not in any order and so it does not mean anything to access them by index. In fact, there is no way to inquire about a specific item in a set. However, you could put the same items in a tuple or a list and access that, but even then, because the items in a set are unordered, the order in the resulting tuple or list is arbitrary:

```
x = tuple(s)    # x now (123, 92, 54), s still a set
x = list(s)     # x now [123, 92, 54], s still a set
```

You can check whether an item is present in a set, which is a computationally efficient operation, like with dictionaries, but unlike lists and tuples:

```
54 in s         # True
55 in s         # False
```

You can also check whether an item is not in (absent from) a set:

```
54 not in s     # False
55 not in s     # True
```

The `len()` function returns the number of items in the set:

```
len(s)          # 3
```

You can add items to a set:

```
s.add(77)       # s now {123, 54, 92, 77}
```

If you try and add an item to a set that is already contained in the set then the operation will effectively be ignored:

```
s.add(123)      # s still {123, 54, 92, 77}
```

You can remove an item from a set:

```
s.remove(54)    # s now {123, 54, 92, 77}
```

If you try to remove an item that is not in the set then you get an error.

You can fetch and remove an arbitrary item directly from a set, noting that here the `pop()` does not take any arguments:

```
y = s.pop()     # s is one smaller, y is arbitrary
```

There are several functions that work with multiple sets, to do many of the operations associated with set theory (as are often illustrated with Venn diagrams), for example, to find unions and intersections.

```
s = set{1, 2, 3, 4, 5}
t = set{4, 5, 6, 7}

a = s & t    #  Intersection: {4, 5}
b = s | t    #  Union: {1, 2, 3, 4, 5, 6, 7}
```

More set operations are listed in the reference section of the Appendices.

Dictionary manipulation

In this section we will illustrate some basic manipulations on dictionaries using the following dictionary to do this:

```
d = {'Ala':71.07, 'Arg':156.18}
```

Dictionaries map keys to values. For the above dictionary we have two keys, `'Ala'` and `'Arg'`, where the key `'Ala'` gets mapped to the value `71.07` and the key `'Arg'` gets mapped to the value `156.18`. The elements of a dictionary are not stored in any particular order, thus it would be meaningless to access them by position. Instead they are accessed by key. Although Python uses curly brackets to define dictionaries, when you want to access the value of a key in a dictionary you use square brackets; `myDict[key]`.

```
x = d['Ala']              # 71.07
```

If you try to get the value for a key that does not exist, then an error is generated (an `Exception` object). There is an alternative syntax, however, which can be used to access a value and that does not generate an error if an unknown key is used. This is the dictionary's `get()` function, which returns the corresponding value if the key is known and `None` otherwise:

```
x = d.get('Ala')          # 71.07
x = d.get('Gly')          # None
```

There is an optional second argument to the `get()` function that specifies the default value to use if the key is not in the dictionary:

```
x = d.get('Ala', 57.05)   # 71.07
x = d.get('Gly', 57.05)   # 57.05, because 'Gly' is absent from d
```

Note that if the key is in the dictionary then this default value is ignored. Also, the dictionary is not changed by this call, so that after the second line above, the dictionary still does not have a key for `'Gly'`. If you want this kind of side effect to happen, then there is the slightly oddly named function `setdefault()` that you can use instead. This behaves exactly like `get()` (so the fact that the name starts with 'set' looks misleading) except that as well as returning a value it also adds the key to the dictionary with the default value if the key does not exist (which is why 'set' is not totally misleading):

```
x = d.setdefault('Ala', 57.05)  # 71.07
# d remains {'Ala': 71.07, 'Arg': 156.18}

x = d.setdefault('Gly', 57.05)   # 57.05
# d becomes {'Ala':71.07, 'Arg':156.18, 'Gly':57.05}
```

You can check whether a key is in a dictionary and, unlike with tuples and lists, checking containment in this way is an efficient computation:

```
'Ala' in d                # True
'Thr' in d                # False
```

An alternative, equivalent method for Python 2 is:

```
d.has_key('Ala')          # True
d.has_key('Thr')          # False
```

You can also check whether a key is not in a dictionary:

```
'Ala' not in d            # False
'Thr' not in d            # True
```

The `len()` function returns the number of (key, value) entries in the dictionary:

```
len(d)                    # 3
```

You can change the value for an existing key by reassigning it:

```
d['Ala'] = 71.04
# d now {'Ala':71.04, 'Arg':156.18, 'Gly':57.05}
```

You can introduce new entries in the dictionary by assigning a value with a new key:

```
d['Ser'] = 87.07
# d now {'Ala':71.04, 'Arg':156.18, 'Gly':57.05, 'Ser':87.07}
```

You can delete entries from the dictionary:

```
del d['Ala']
# d now {'Arg':156.18, 'Gly':57.05, 'Ser':87.07}
```

This will generate an error if the key is not in the dictionary. You can get hold of the keys in the dictionary:

```
x = d.keys()      # ['Gly', 'Arg', 'Ser']
```

The keys of a dictionary effectively form a set, but sets were introduced in Python long after dictionaries, so for historic reasons in Python 2 the `keys()` function gives back a list, rather than a set. Nonetheless it still makes no sense to talk about the order of the entries in a dictionary, and thus also the order of the list that is returned. Although the list of keys is nominally ordered, the order cannot be relied upon.

You can also get hold of the values associated with all of the keys from the dictionary:

```
x = d.values()    # [57.04, 156.18, 87.07]
```

Unlike keys, the values in a dictionary can be repeated. So strictly speaking what we should get is what mathematicians call a *multiset*, or a *bag*. But Python has no inbuilt data type for this, and we are given a list, although again it cannot be relied upon to have a specific order.

Finally, you can get hold of a list of (key, value) pairs as tuples:

```
x = d.items()     # [('Gly', 57.04), ('Arg', 156.18), ('Ser', 87.07)]
```

In Python 3 the `keys()`, `values()` and `items()` functions change to return what is called a *view* rather than a list, which is a subtle difference that means, for example, if you really want to get hold of a set or a list then in Python 3 you would have to do so via `set(d.keys())` or `list(d.values())` etc.

Importing modules

In Python much functionality is inbuilt and immediately available, as has been demonstrated in this chapter. However, one of the fundamentals of the Python language is the ability to import external modules (or libraries) into the current program. Naturally, we do this to make use of extra functionality that is available elsewhere. Such modules may be part of the standard library (see http://www.cambridge.org/pythonforbiology for links to the Python library documentation) that is automatically included with the Python installation, they may be extra libraries which you may have to install separately (such as NumPy or BioPython) or they may be other Python programs you have written yourself. Whatever the source of the module, they are imported into a program via an import command. For example, if we wish to access the mathematical constants π and *e* we can use the `import` keyword to get the module named `math` and access its contents with the dot notation:

```
import math
print(math.pi, math.e)
```

Also we can use the `as` keyword to give the module a different name in our code, which can be useful for brevity and avoiding name conflicts:

```
import math as m
print(m.pi, m.e)
```

Alternatively we can import the separate components using the `from ... import` keyword combination:

```
from math import pi, e
print(pi, e)
```

Because the `math` module used above was part of the standard Python library we didn't have to worry about installing it separately. The standard library includes a selection of modules that are useful in a variety of general situations. Many of these modules are listed in the Appendices at the end of the book.

If we import a module that is not part of the standard library, then we must make sure that it is installed on our computer system (and there are generally instructions at the download sites). If it is not installed we will get an `ImportError`. For example, if the Python Imaging Library (PIL) is not installed but we try to import its `Image` object we would see something like:

```
>>> from PIL import Image
Traceback (most recent call last):
  File "<stdin>", line 1, in <module>
    from PIL import Image
ImportError: No module named PIL
```

Lastly, if you want to write your own modules which can be imported into other programs then you need to be aware of how Python searches for module files. Essentially Python has a series of import directories (a 'search path') that it looks inside to find a file of the required name, starting with the inbuilt modules. If we try to `import moduleAbc` then Python will look for the module file named `moduleAbc.py` in its import directories. By default the import search path will include several directories, which contain the standard libraries, external installations and the current working directory. Within Python, `sys.path` gives the list of what is on the search path:

```
>>> import sys
>>> print(sys.path)
```

You will see that this list contains an empty string `''` for the current directory and various other standard '`lib`' and '`packages`' directories that are part of the Python installation. If you have a Python file that you want to import that is not in one of these directories, then you can add further directories to `sys.path` and then refer to modules inside that directory:

```
import sys
sys.path.append('/home/user/myModules/') # Contains userModule.py
```

```
from userModule import userFunction
```

Alternatively you can add entries to the PYTHONPATH environment variable via your computer's operating system, which will automatically be added to `sys.path` at run time. It is in this way that you can use all of the downloadable Python code that accompanies this book. Hence, if you download the material in its standard directory and put the full path (i.e. including any leading directories) to this in `sys.path` or the PYTHONPATH environment variable then you will be able to import any of the example code, such as:

```
from MachineLearning import neuralNetTrain, neuralNetPredict
```

Any sub-directories of the import directories listed in `sys.path` can also be treated as if they were modules, although before Python 3.3 this is provided they contain a file named '`__init__.py`', and which is typically blank or contains only '`pass`': a Python statement that does nothing. For example, if I have a personal module directory located at '`/home/user/myModules/`', which is on the module search path, I can then add a sub-folder called '`molecules/`' that contains the script '`anneal.py`' as well as any required '`__init__.py`' (for Python before version 3.3) so that I can do the following import:

```
from molecules import anneal
```

4   Program control and logic

Controlling command execution

On the whole, a program will normally run by executing the stated commands, one after the other in sequential order, like reading the lines of a book. Frequently, however, you will need the program to deviate from this, to jump to a line that is not the next line. There are three main ways of diverting from the line-by-line paradigm. The first way is through the use of *functions* (subroutines), where the program's execution jumps from a particular line of code to an entirely different spot, even in a different file or module, to do a task before (usually) jumping back again. You can even jump to a function's subroutine from inside another function, and do this repeatedly, so that there is a *stack* of jumps between the current line and the first jump point. Given their importance, functions will be discussed in a dedicated chapter that follows this one.

The second way of jumping between program lines is through the use of *conditional statements*. Here you can check if some statement or *expression* is true, and if it is then you continue on with the following block of code, otherwise you might skip it or execute a different bit of code. The third way is by performing repetitive loops through the same lines of code, where each time through the loop different values may be used for the variables. Usually such loops are either done a specific number of times or until something in particular happens.

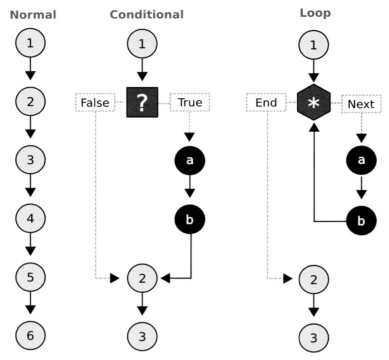

Figure 4.1. The changes to the flow of a program's line-by-line execution using conditionals and loops. The normal flow of a program's execution proceeds from one line to the next. The presence of a conditional test, using an 'if' statement, means a block of code is only executed if the test evaluates to be logically true, and otherwise the block is skipped. The presence of a loop, starting with a 'for' or 'while' statement, causes the execution of a block of code to be repeated a number of times from the start of the block and after the last repeat the program execution resumes after the block.

When you have a collection of items, like a list, a loop can be used to consider all of the items in turn; the loop *iterates* over the items of the collection. A significant number of people who are new to programming find this the hardest idea to get to grips with, although Python's syntax makes it about as easy as it can be.

Loops, conditional statements and functions are the three ways of controlling a program's flow that occur under ordinary circumstances. There is actually a fourth way that can cause a jump in program execution and that is if an error or *exception* occurs, i.e. something illegal has happened. In Python, like with Java and many other languages, when an error occurs inside a function, the exception propagates back up the stack of any function calls, until it finds the first of those functions. If the initial function does not specifically handle the error, then the program (or more precisely that specific *thread*) stops.

Code blocks

With all of the means by which Python code execution can jump about we naturally need to be aware of the boundaries of the *block* of code we jump into, so that it is clear at what point the job is done, and program execution can jump back again. In essence it is required that the end

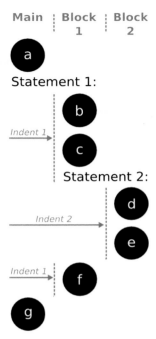

Figure 4.2. The indentation of code blocks within a Python program using spaces. In lines of Python code the different program blocks that relate to flow-control statements like 'if', 'for' and 'while' etc. are indented relative to the start of the line using whitespace. A flow-control statement in the main part of the program will have no indentation but will define the start of a program block that is indented to the first level. Any statements that are indented to this first level will be inside the block of the first statement. A further flow-control statement at the first level will define a new block at the second indentation level. Any lines of code that have less indentation than a preceding line are outside the previous block and are part of the existing block with the same indentation or are outside all blocks if there is no indentation.

of a function, loop or conditional statement be defined, so that we know the bounds of their respective code blocks.

One distinct and important difference between Python and many other computer languages is how it deals with the issue of defining execution bocks (that are jumped into). In many programming languages (such as Java, Perl, C, etc.), a block of code generally has a left curly brace ('{') at the start and a right curly brace ('}') at the end. In Python the extent of a block of code is completely determined by indentation, i.e. the addition of blank spaces at the start of the line. The start of a Python code block must be indented by some amount relative to whatever defines the block, such as a function declaration, a conditional statement or loop statement. The content of the block must all have the same indentation, except for when there are blocks inside the blocks, which are then indented further. And finally, the block ends when the indentation ends.

For people moving to Python from other languages the block indentation can take some getting used to. An advantage of the Python block syntax is that it removes clutter and makes the code look cleaner; there is never a confusion of closing brackets. However, there is one subtle issue: it is generally best to avoid using tab stops in Python code, unless your editor is set up to automatically convert tabs to spaces while you are typing. The reason for the potential problem is that in Python, tabs equate to giving space out to the column that is the next multiple of 8 counting from the beginning of the line. Many editors would show it instead out to the next multiple of 4 or 2, and so the code would look different, and this is particularly problematic if the code mixes tabs and spaces for indentation. The number of leading spaces used to indent each block is a matter of taste. The Python community generally recommends 4 spaces, but in this book we consistently use 2, allowing all our examples to fit within the width of the pages in a reasonable-size font.

Conditional execution

The 'if' statement

A conditional statement, or 'if' statement, is used to specify that some block of code should only be executed if some associated test is upheld; a conditional expression evaluates to True. This might also involve subsidiary checks using the 'elif' statement (the English equivalent of which might be 'or else if the following') to control an alternative block if the previous expression turns out to be False. There can even be a final 'else' statement (in English 'otherwise if all else fails') to do something if none of the checks are passed.

To give a solid example, the following uses statements that test whether a number is less than zero, greater than zero or otherwise equal to zero and will print out a different message in each case:

```python
if x > 0:
  print("Value is positive")

elif x < 0:
  print("Value is negative")

else:
  print("Value is zero")
```

The general form of writing out such combined conditional statements is as follows:

```python
if conditionalExpression1:
  # codeBlock1

elif conditionalExpression2:
  # codeBlock2
```

plus any number of additional elif statements, then finally:

```python
else:
  # codeBlockEnd
```

The elif statements are optional and independently the else statement is also optional, so you can just have an isolated if statement, which is a fairly common situation. Each block has to be indented relative to the if/elif/else. Note also the colon (':') that appears after the if/else/else statements. This is a mandatory part of the syntax, though it is easy to forget.

Each block must contain at least one line of code, but can have more. Python has a 'pass' statement that qualifies as a line of code but does nothing. So when you are first setting

if: elif: else:

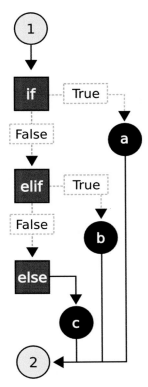

Figure 4.3. The flow of program execution with combined conditional statements. One of the blocks of code 'a', 'b' or 'c' will be executed depending on the truth or falsehood of statements that follow the 'if' and 'elif' keywords. If the conditional expression at the 'if' evaluates as true then code block 'a' is executed, otherwise if the conditional expression at 'elif' evaluates as true then instead code block 'b' is executed,. Finally, if none of the conditional expressions evaluates as true then code block 'c' is executed (so there is no additional conditional expression to evaluate for this, the block is just executed).

out your code, you can use this as a placeholder until you get around to adding the actual desired code:

```
if someExpression:
    pass                    # fill code in later
```

Comparisons and truth

The question naturally arises as to which expressions are deemed to be true and which false. Naturally, for the special Python logic items named True and False (often referred to as *Boolean* values) the answer is obvious. Also, the logical states of truth and falsehood that result from conditional checks like 'Is x greater than 5?' or 'Is y in this list?' are also clear. However, in Python even expressions that do not involve an obvious query can be assigned a status of conditional truthfulness; the value of an item itself can be forced to be considered as either `True` or `False` inside an `if` statement. For example, when tested in a conditional statement the number `1` will be interpreted as being `True` and the number `0` as `False`, although their values are otherwise numeric. For the Python built-in data types discussed in this chapter the following are deemed to be `False` in such a context:

```
None                  # nothingness
False                 # False Boolean
0                     # 0 integer
0.0                   # 0.0 floating point
""                    # empty string
()                    # empty tuple
[]                    # empty list
{}                    # empty dictionary
set(())               # empty set
frozenset(())         # empty frozen set
```

And everything else is deemed to be `True` in a conditional context.

Python has the standard comparison (or relational) operators:

```
x > y                 # x greater than y?
x < y                 # x less than y?
x >= y                # x greater than or equal to y?
x <= y                # x less than or equal to y?
x == y                # x equal to y?
x != y                # x not equal to y?
```

In terms of syntax, in Python 2 you can compare anything with anything, although if the expressions are not of the same data type the result is rather meaningless in some sense. And it is notable that in Python 3 you get an exception (error) if you try to use any of the comparison operators, other than `==` and `!=`, to compare expressions that do not have a natural comparison, like when comparing numbers with strings. A consequence of this is that in Python 3 you cannot sort the elements of a list, according to value, where the items have data types that are not comparable, because they cannot be ascribed as being greater than or less than one another.

Python has two additional comparison operators:

```
x is y                # rather than x == y
x is not y            # rather than x != y
```

The difference between the two is that `==` and `!=` compare the values while `is` and `not is` compare whether objects are identical. Everything in Python is an object, and you might be interested in whether two objects have the same value or are really the same object, a stronger condition. As an analogy in the real world, two people might have the same name but even if they have the same name it does not mean they are the same person. As an example in Python:

```
x = [123, 54, 92, 87, 33]   # x defined as a list
y = x[:]                    # y is a copy of x
y == x                      # True – same value
y is x                      # False – different object
```

Logic operations

Python also has standard logical connectives (`and`, `or`, `not`), although the resulting expressions are not necessarily Boolean (`True` or `False`). For example, consider the logical AND operation, also known as conjunction:

```
x and y
```

You might think that this should give the value `True`, one of Python's special Boolean logic objects, if `x` and `y` both evaluate to `True` in a conditional context (imagine each tested in an `if` statement), and the value `False` otherwise; the other logic object. However, the situation is more subtle than that, because the operation doesn't necessarily give back `True` or `False` logic objects. Rather, it gives back the `x` or `y` value, one of the values you put in, which only evaluate as being equivalent to `True` or `False` in a conditional test.

For example, given that we know that zero is considered to be `False` in a conditional test and other numbers are considered to be `True`:

```
z = 0 and 3        # z equals 0 rather than False
z = 1 and 2        # z equals 2 rather than True
z = 5 and 0        # z equals 0 rather than False
z = x and y        # z might equal x or y, depending on value
```

Considering the last line, if `x` evaluates to `True` in a conditional context then `z` evaluates to `y`; the truth is determined by the second item. And conversely, if `x` evaluates to `False` in a conditional context then `z` evaluates to `x`; which is known to be false. Naturally, this also holds when the values we are using in the logical operation are not so obviously `True` or `False` in comparisons:

```
[] and 5          # []   (False in conditional context)
3 and 5           # 5    (True in conditional context)
3 and ()          # ()   (False in conditional context)
```

Moving on from the AND operation there is also the OR operation, also known as disjunction:

```
x or y
```

Here, if `x` evaluates to `True` in a conditional context then the statement `x or y` immediately gives back `x`, and `y` is not even evaluated (given it doesn't matter whether it is true or false). If `x` evaluates to `False` in a conditional context then `x or y` evaluates to `y`.

There is also negation, the NOT operation:

```
not x
```

This does evaluate to a proper Boolean, so it evaluates to `True` if `x` evaluates to `False` in a conditional context, and vice versa.

If a special Boolean object is definitely required from an operation like AND or OR, rather than one of the values in the comparison, you can explicitly convert the result using `bool()`:

```
x = 3 and 5     # x is 5
y = bool(x)     # y is True
```

Although it may seem a little odd, you can do the above in one step and use the logic operation as if it were a single argument:

```
y = bool(3 and 5)   # y is True
```

In many circumstances explicitly converting the result of a logic operation to Boolean `True` or `False` is not required. Going further, it is sometimes positively useful to get back one of the actual values that was put into the operation. For example, suppose you have a variable `x`, which is sometimes true and sometimes not, and you wanted a text string that said 'Yes' if `x` was `True` and 'No' if `x` was `False`. You could do:

```
if x:
  text = "Yes"
else:
  text = "No"
```

But an alternative would be to write:

```
text = x and "Yes" or "No"
```

Here, if `x` is `True` then the AND operation evaluates to 'Yes' and the OR operation preserves this. If `x` is `False` then the AND operation evaluates to `False` and the OR operation evaluates to 'No'. This is perhaps a bit of trickery, based on the strings being true, but it can be convenient.

There is one more subtle point with these Boolean `True` operations. Considering the statement `x or y`, if `x` turns out to be true in a conditional context then the expression `y` would not be evaluated at all, given that the OR statement will be true no matter what `y` is. The same idea applies to `x and y`: this time `y` is not evaluated when `x` is false. This turns out to come in quite handy. For example, if `x` is a tuple or list, we could do:

```
x = [256, 128]

if x and x[0] > 10:
  # etc.
```

Here, if `x` is an empty list (so evaluates to `False` in a conditional context) then the second expression is not even evaluated, which is just as well since it would fail and generate an exception (error), because `x[0]` does not exist.

As an example of a conditional statement, we could have:

```
x = [123, 54, 92, 87, 33]

if len(x) > 10:
  x.append(999)
elif len(x) > 3:
  x.append(888)
else:
  x.append(777)

# x now [123, 54, 92, 87, 33, 888]
```

In this case, since `len(x)=5`, the `if` expression evaluates to `False` whereas the `elif` expression evaluates to `True` and so `888` is appended to `x`, and the `else` block is skipped.

There is another conditional statement allowed from Python 2.5 onwards:

```
expression1 if conditionalExpression else expression2
```

This reads a bit oddly because the condition is in the middle, but what this means is that if the `conditionalExpression` evaluates to `True` it evaluates to `expression1` and otherwise it evaluates to `expression2`. Thus instead of writing slightly tricky code like:

```
text = x and "Yes" or "No"
```

you can instead write:

```
text = "Yes" if x else "No"
```

Loops

When you have a collection (tuple, list, set, dictionary ...) it is quite natural to want to consider every element in the collection in turn, processing each in some way. This process is often called *iteration*. This is the job of the 'for' loop.

The 'for' loop

In this section we will illustrate the `for` loop using the following list:

```
values = [10, -5, 3, -1, 7]
```

Then an example of a `for` loop is:

```
for v in values:
  print(v) # block of code
```

What this means is that the variable v is set to each item from the list `values` in turn, and every time this happens the block of code is executed. So here v is first set to `10`, and the block of code is executed, then v is set to `-5`, and the block of code is executed again etc., until finally it is set to `7` and the block of code is executed for the last time. The situation shown for a list also applies if `values` is a tuple or set or dictionary. For a dictionary the variable v would be assigned to the keys of the dictionary, one after the other. Also, remember that sets and dictionaries are not ordered so the looping will not happen in any particular order.

Note that although `values` is defined before the `for` statement, the item variable v is introduced anew at this point. It does not matter if a variable named v already exists before the loop, but its original associated value will be overwritten. The item variable name v is arbitrary, and any other name could be used instead. Though it would be most confusing if it were called `values` ...

As with conditional statements, the block has to have at least one line of code in it, although it can naturally have more, and the colon (':') at the end of the `for` statement is mandatory. As another example, suppose you wanted to calculate the summation of the items in the list `values`. Then you could do:

```
total = 0
for v in values:
  total = total + v

# At the end total is 14
```

Following this in detail, doing `total + v` each time, we are doing the operations `0+10`, `10-5`, `5+3`, `8-1`, `7+7`, and updating `s` each time, so at the end `total` is `14`. It would not matter here whether `values` were a list, a tuple or a set, this loop would still result in the sum of the items in `values` being calculated. Note that `v` is not forgotten about after the loop ends, it holds the value of the last item it was set to, in this case `7`.

The 'while' loop

In addition to the `for` loop that operates on a collection of items, there is a `while` loop that simply repeats while some statement is `True`:

```
while conditionalExpression:
    # codeBlock
```

The `codeBlock` is repeated over and over until the `conditionalExpression` evaluates to `False`. Note that if the tested expression never evaluates to `False` then you have an 'infinite loop', which is naturally not helpful. Sometimes you can use a `while` loop to avoid having to initially generate a collection of items to then loop over. For example, you could generate a series of numbers by doubling a value at each iteration, until a limit is reached:

```
value = 1
while value < 32:
    # maybe do something with the value
    value *= 2

# value is 32 at the end
```

What happens here is that the value is doubled in each loop and once it gets to `32` the `while` test fails (32 is not less than 32) and that last value is preserved. Note that if the test were instead `value <= 32` then we would get one more doubling and the value would reach `64`.

Skipping and breaking loops

Python has two ways of affecting the flow of the `for` or `while` loop inside the block. The `continue` statement means that the rest of the code in the block is skipped for this particular item in the collection, i.e. jump to the next iteration.

```
values = [10, -5, 3, -1, 7]

total = 0
for v in values:
    if v < 0:
        continue        # skip the rest of the block for this v

    total += v

# At the end total is 20
```

So here the negative items are left out of the sum. An alternative way of writing the above without the `continue` statement is to only add the numbers we do want, rather than skipping those we don't:

```
total = 0
for v in values:
  if v >= 0:
    total += v

# At the end total is 20
```

It is a matter of taste which style you prefer. Note that either way the variable `v` at the end will still be the last value from the list, whether or not it is negative.

The second way of affecting the flow in a `for` or `while` block is the `break` statement. In contrast to the `continue` statement, this immediately causes all looping to finish, and execution is resumed at the next statement after the loop.

```
total = 0
for v in values:
  if v == 3:
    break

  total += v

# At the end total is 5, v is 3
```

Here the third item in the list is `3`, so at this point the loop terminates and the sum is only of the first two items in the list. Note that the loop variable, `v`, never makes it to the rest of the list, so stays `3` when the loop terminates.

If you have a `for` loop inside a `for` loop, then a `break` in the inner loop only results in a jump to the end of the inner loop, not the outer one. This can sometimes be a pain but the alternative choice would in general have been worse.

There is one extra twist with `for` loops when you have a `break` statement in the loop. Often you want to know if you have broken out of the loop, but otherwise execute some code if you have not, and Python has a special syntax for adding such code. For example, here we take values for `x` and test if they are greater than 5 and less than 10. If we find such a value `foundInRange` is defined to be `True` and the loop is stopped immediately with `break`. If no value passes the test, then the loop reaches the end and program execution reaches the `else:` block where `foundInRange` is alternatively defined as `False`.

```
for x in someList:
  if 5 < x < 10:          # Is x between 5 and 10?
    foundInRange = True
    break                 # Do not test any more
else:
  foundInRange = False     # Found nothing

print(foundInRange)
```

Thus this is a second context in which `else` can occur. Note that the `else` lines up with the `for`. The code in the `else` block is executed only if the `break` statement has not been executed.

Positional indices

In most loops you do not have to know the positional index of the associated variable in the list, you just need to know that the variable is set to one item after another in the list. But sometimes you do have to know the index. For example, in geometry a vector (e.g. x, y and z axis values to specify a position in three-dimensional space) can be implemented in Python as a list of floating point numbers. Suppose that you wanted to calculate the inner (or dot) product between two vectors, `vec1` and `vec2`.[1] The inner product is the summation of the products of the corresponding terms in each list. So in three dimensions, for example, it would be:

```
vec1 = (4, 7, 0)
vec2 = (1, 3, 5)

s = vec1[0]*vec2[0] + vec1[1]*vec2[1] + vec1[2]*vec2[2]    # 25
```

Here the calculation for the inner product is $(4 \times 1) + (7 \times 3) + (0 \times 5)$ yielding 25. We could also save writing out the whole calculation and use a loop to go through all of the required vector positions (indices) adding the product for each. Thus, instead we can write:

```
s = 0
for index in [0,1,2]:
  s += vec1[index] * vec2[index]
```

This is a good enough way of calculating the inner product if you know you are in three dimensions, but let's say you wanted to calculate this for vectors in an arbitrary number of dimensions, i.e. where the length of the vectors is not known beforehand. In this case the list of indices needs to match the size of the vectors. We can obtain such a list using `range()`.

The `range()` function takes one, two or three integer arguments. If it has one argument, n, then it generates a number sequence starting at `0` and finishing at `n-1`, in steps of `1`. If it has two arguments, `n1` and `n2`, then it generates a sequence starting at `n1` and finishing at `n2-1`, in steps of `1`.

```
range(7)            # [0, 1, 2, 3, 4, 5, 6] - show as list(range(7)) in Python 3
range(2, 7)         # [2, 3, 4, 5, 6]
```

If it has three arguments then the first two, `n1` and `n2`, are as before, and the third is the step size. There is one subtlety, however: if the step size is positive then the list stops at the greatest integer less than `n2` (which might not be `n2-1`) and if the step size is negative then the sequence stops at the least integer greater than `n2`. If you use a step size of `0` you get an error.

```
range(2, 13, 3)     # [2, 5, 8, 11]
range(7, 2, -2)     # [7, 5, 3]
```

Returning to the inner product calculation, we simply need to use the length of one of the vectors `len(vec1)` to get the required range of indices:

```
s = 0
for i in range(len(vec1)):
  s += vec1[i] * vec2[i]
```

---

[1] This would allow you to calculate the angle between two directions, for example. See Chapter 9.

Here `len(vec1)` gives the length of `vec1` and so `range(len(vec1))` gives a list which exactly contains the valid (non-negative) indices for `vec1` and `vec2`. Note, the code assumes that `vec1` and `vec2` are the same length, and that is something that should really be checked somewhere.

In Python 2 there is an alternative for `range()`, which is `xrange()`. The difference is that `range()` actually creates a whole list of numbers while `xrange()` does not; it just sets the relevant loop variable (here, n) to what it should be set to, one after the other. The difference between `range()` and `xrange()` is normally not an issue but could be if the length were large. In Python 3, `xrange()` no longer exists and `range()` behaves like the Python 2 `xrange()`.

There is another, arguably neater, way we could get at the indices and that is using the inbuilt `enumerate()` function, which gives both an index and the corresponding item. Specifically, it sets the loop variable to a two-tuple whose first item is the index and the second is the list item at that index:

```
s = 0
for i, x in enumerate(vec1):
  s += x * vec2[i]
```

It behaves like Python 3's `xrange()` in regard to not actually creating one big list (in this case of 2-tuples), so is reasonably efficient. It might seem possible to also use the `.index()` function to get the position of an item in the vector. However, this would only work if each term in `vec1` was unique, and there is no guarantee of that:

```
s = 0
for x in vec1:
  i = vec1.index(x)        # bad idea due to repeats
  s += vec1[i] * vec2[i]
```

Also, getting the index in this way is relatively slow. So overall it is much better to stick with the `range()` or `enumerate()` functions.

List comprehension

Sometimes loops are used to go through one collection of items simply to generate another collection of (different) items. For example, here the function `range(1,8)` gives a sequence `[1,2,3,4,5,6,7]`, which is used to make a list of square numbers:

```
squares = []
for x in range(1,8):
  squares.append(x*x)

# squares becomes [1, 4, 9, 16, 25, 36, 49]
```

However, when mapping one collection to another in this way, there is a neater alternative syntax called *list comprehension*. And often this will allow you to map one list of items onto another in a single line of code. In essence a list comprehension is a way of using a loop to build a list from the inside. Here is a simple example:

```
squares = [x*x for x in range(1,8)]   #  [1, 4, 9, 16, 25, 36, 49]
```

The first expression inside the square brackets (x*x) is what is placed in the resulting list, and the remainder, from the `for` to the closing bracket, is what generates the values for the loop variable x. Not only is this shorter to write than a conventional `for` loop, it is also computationally quicker.

Looping tips

As a word of warning, it is generally not a good idea to alter the number of items in a list while you are looping over that list, unless you really know what you are doing (and even then we typically only add to the end). The sequence of elements that the loop goes through will be hard to predict because the positional indices may not correspond to their original values. Suppose you have a list of values and want to remove any that are less than five. You could try the following:

```
values = [0, 1, 2, 3, 4, 5, 6, 7, 8, 9]

for val in values:
  if val < 5:
    values.remove(val)   # modify the loop list: bad idea
```

This does not work because the loop variable, val, gets confused about where it is in the list when the list is modified inside the loop. In this case some of the values less than five were skipped and not removed from the list:

```
print(values)   # [1, 3, 5, 6, 7, 8, 9]
```

Instead you could duplicate the list, here using the `list()` function, and this now gives the expected result:

```
values = [0, 1, 2, 3, 4, 5, 6, 7, 8, 9]

for val in list(values):
  if val < 5:
    values.remove(val)

print(values) # [5, 6, 7, 8, 9]
```

With a duplicate, the list that is iterated over is now not modified inside the loop. In the next example we show an alternative approach where the values are added to an initially empty list. In general, constructing a new list in this way is more efficient than making a copy and then removing internal items:

```
values = [0,1,2,3,4,5,6,7,8,9]
values2 = []

for val in values:
  if val >= 5:
    values2.append(val)

values = values2
```

The above is an example of filtering a collection according to a condition (and would also work with a set or tuple). For this task we could also use a list comprehension, which has no problems with modifying the list we are looping through and does the job on a single line of code. This combines a `for` loop with a conditional `if` check, although the latter is an optional part of the syntax.

```
values = [0,1,2,3,4,5,6,7,8,9]
values = [val for val in values if val >= 5]
```

If the code for the filter condition is complicated you may need to write the list comprehension over more than one line. Though, if the filtering is done by a function then it is easier to keep to a single line. You can also operate on the loop variable before adding it to the filtered list. For example, here we import a mathematical function to calculate the factorial of a number and then use that function to calculate the factorial value inside the list comprehension:

```
from math import factorial

values = [0,1,2,3,4,5,6,7,8,9]
result = [factorial(val) for val in values if factorial(val) >= 700]

print(result) # [720, 5040, 40320, 362880]
```

This approach is where a list comprehension might not be such a good idea, because although it is short, it calls the function (factorial()) in this case twice; once for the conditional test and once to construct the result list. Naturally this is best avoided if the function is slow to calculate. Hence, for efficiency reasons you might want to stick with the longer version:

```
values = [0,1,2,3,4,5,6,7,8,9]
result = []

for val in values:
  fac = factorial(val)

  if fac >= 700:
    result.append(fac)
```

As mentioned earlier, you can loop over the keys of a dictionary. Normally you are interested not just in the keys, but also in the value associated with the key in the dictionary. However, if you have a dictionary and its keys, you can easily look up the values. For example, suppose you have a dictionary, yearDict, for which the keys are the names of months, and the value corresponding to each key is the number of days in that month. If you want to calculate the total number of days from all the months it is a simple matter of looping through all of the keys of the dictionary (the months), then using each key in turn to get the value from the list, which is then added to the total.

```
yearDict = {'Jan':31,'Feb':28,'Mar':31,'Apr':30,
            'May':31,'Jun':30,'Jul':31,'Aug':31,
            'Sep':30,'Oct':31,'Nov':30,'Dec':31,}

total = 0
for month in yearDict:        # loop through keys
  total += yearDict[month]   # lookup value
```

Error exceptions

Python deals with errors caused by illegal circumstances, which are called *exceptions*, in a relatively graceful manner. Note that it is commonplace to say 'throw an exception' to mean that an exception was generated by some illegal program state. Here is a very simple example of something that will generate an exception object, indicating the type of error, in this case because it is illegal to divide a number by zero:

```
x = 1/0    # Throws a ZeroDivisionError
```

Catching exceptions

Rather than simply letting such things always break the program, Python allows you to 'try' something and then 'catch' any exceptions that might occur and do something about it, and potentially let the program carry on afterwards. The simplest version is:

```
try:
  # code block that might throw an exception
except:
  # exception handling code

# maybe let the program continue after
```

So if an exception occurs, the program immediately stops executing the code in the try block and instead executes the code in the except block. In the above case, every kind of exception error is trapped, so there is no way of knowing what kind of exception it was. However, you can specify what kind of exception is trapped. Python has many inbuilt exception types (classes). As illustrated above, if you try and divide by 0 then you get a ZeroDivisionError, and you can check specifically for that:

```
x = 1
y = 0
try:
  z = x + y
  w = x / y
  t = x * y
except ZeroDivisionError:
  print('divided by zero')

print('program did not stop')
```

In this specific example, an exception occurs in the calculation of w, so it is not set and, further, the calculation of t never happens and an error message is printed instead. Because the illegal division by zero was dealt with the program can then continue. You can even get hold of an exception object, which can be printed out to show the internal Python error message:

```
x = 1
y = 0
try:
  w = x / y
except ZeroDivisionError as errorObj:
  print(errorObj)    # prints 'integer division or modulo by zero'

print('program did not stop')
```

Before Python 2.6 the syntax for catching the exception object was different, so if you are using an older version of Python then you have to use a comma instead of 'as' (the comma syntax also works in Python 2.6 and 2.7 but not in Python 3):

```
try:
  w = x / y
except ZeroDivisionError, errorObj:
  print(errorObj)    # prints 'integer division or modulo by zero'
```

This reports the Python message about what happened, which you would get if the program failed, but naturally because we used the `try:` clause to catch the problem the program did not stop. There is a simple 'base' exception in Python called `Exception`, and that lets you trap all types of exception and get hold of the exception object:

```
try:
  # code block that might throw an exception
except Exception as errorObj:
  # exception handling code
```

Also, you can catch different types of exception by repeating the `except` statement. This allows different problems to be handled in different ways. For example, if you try to divide something by a text string (or a string by something) in Python you get a `TypeError`. So your division code could check for this and for zero division, printing an appropriate message for each situation:

```
try:
  w = x / y
except ZeroDivisionError as errorObj:
  print('divided by zero', errorObj)
except TypeError as errorObj:
  print('divided by something silly', errorObj)
```

Triggering exceptions

To cause your own exception you use the `raise` command. This does not have to be in response to an illegal Python state; you can force an exception to occur at any time. Often this is done to trap problems before an illegal state can occur. For example, you may wish to check that two vectors are of the same length before calculating the inner product (as above). Here it is better to have an informative error message that tells us in English that the vectors are of different lengths, rather than reach an illegal state and get an `IndexError`.

The `raise` command takes an argument which is an exception object, and normally you create this object right in the command itself. How to create an object is discussed in Chapter 7. But as an example of `raise`, suppose you are given a variable x and you want to check that it is greater than zero and throw an exception if that is not true. You could do:

```
if x <= 0:
  raise Exception('x must be > 0')
```

In the same way as when an illegal state triggers an exception, the `raise` causes Python to look up the stack of executed commands until it finds any code that catches the exception, but otherwise the program quits if nothing handles it. Note that in versions of Python before 2.6 you could use strings as error objects, instead of dedicated exceptions. However, you should not use strings as exceptions, even if you are using older versions of Python, since it will break if someone uses the code in a newer version.

The `raise` command is also sometimes used in an `except` block because you might want to re-throw the original exception after having done some of your own handling. In this case the program will stop because of the error, but you have the opportunity to do something else before the end. For example, this could be useful to do some clean-up, e.g. to delete temporary files, close an Internet connection etc.

```
try:
  w = x / y
except ZeroDivisionError as errorObj:
  print('Program error due to zero division')
  # do some clean-up (not shown)
  # then re-throw exception
  raise errorObj
```

In this case you don't have to create an exception object because one already exists. In fact, inside an `except` block Python allows you to not explicitly specify the exception object in the `raise` statement, in which case it will use the one that got you into that block in the first place:

```
try:
  w = x / y
except:
  print('Program error due to zero division')
  raise
```

Finally

You can also add code at the bottom of a `try`/`except` block that will be executed whether or not an exception was thrown, using `finally`:

```
try:
  w = x / y
finally:
  print('finished with division')
```

So here the final message is printed whether or not an exception occurred. This final block of code is a convenient location for cleaning up (for example, explicitly closing files that are open).

In older versions of Python you could not have both an `except` clause and a `finally` clause, but that was allowed starting in Python 2.5 (the syntax for the `except` clause here is only valid from Python 2.6):

```
try:
  w = x / y
except ZeroDivisionError as e:
  print('divided by zero')
finally:
  print('finished with division')
```

Further considerations

Stopping a program

We have mentioned how a program will stop when an error exception occurs, unless we specifically try to catch that error, and how we can manually trigger an exception using `raise`. However, it is also possible to simply end a program immediately at a particular point, perhaps the harshest way to control program flow. This kind of intervention is often not required. Programs in general will simply end after the Python interpreter reaches the last of the commands, when there is nothing more to execute. Also, we can conditionally execute most of a program inside an `if` block, and thus skip straight to the end of a program if a condition is not met. However, there are some circumstances where a program has an indefinite loop, perhaps waiting for specific events, which we might want to terminate in a deliberate manner. A good example of this is a graphical user interface, where the program does not simply end, but rather enters an event-driven loop whereby it waits to respond to the user providing input like typing at the keyboard or clicking/touching a graphical item on the screen. In this case the program will continue to run until the user specifically exits the program, maybe by selecting a 'Quit' option from a menu.

In general a Python program is stopped with the `exit()` function from the standard `sys` module, which naturally must be imported first. In the following example, although we have a loop which would normally carry on indefinitely, the `exit()` stops the program entirely at the first iteration:

```
import sys

while True:  # Loop indefinitely
  sys.exit() # Quit program

print('This never gets executed')
```

Technically using `sys.exit` raises a `SystemExit` exception, which in theory could be intercepted using `try` and `except`, for example, to run clean-up code (e.g. deleting temporary files or saving session data to disk) before re-raising the exception so the program actually stops. If you really do want a program to stop without doing anything else and with no chance of interception, then `os._exit()` can be used instead (taking note of the underscore and the zero which is an exit code):

```
import os, sys

try:
  sys.exit()
except SystemExit:
  print("Exit stopped")

os._exit(0) # Can't be stopped
```

Test code

When importing a Python file as a module we are generally interested in variable, object (class) and function definitions that do not immediately result in any significant program execution. However, the Python code of an imported module is nonetheless read line by line by the interpreter and any top-level commands[2] in that file will be executed in the usual manner. This might not be a problem, but we generally want to import just parts of a module without running it as if it were the main program. A Python module will often have a main program so that it can be tested and show examples for its use.

Accordingly, when writing your own Python modules it is commonplace to design them so that if they are executed directly they run as a main program and use any test code etc., but if they are only imported any code that is specific to their main program is not executed. Usually this is achieved by placing any test code at the bottom of the Python file and using the special internal variable, __name__. In Python, anything that begins and ends with a double underscore ('__') is part of the language's internal workings. The variable __name__ is set at runtime for each module that Python uses in a program. It is set to the string "__main__" in the main script (which itself may be used as a module in other programs) that you directly run, and it is set to the module name in all modules that are then imported, directly or indirectly, from that script.

Thus you often see the following in Python modules:

```
# implementation code

if __name__ == "__main__":
  # test code for implementation code
```

When the script is run directly then __name__ is set to "__main__" and the test code is executed, but if the code is imported as a module from somewhere else then __name__ is set to the module name and so the test code is not executed.

The sys library is often used in this context when your test code might require one or more arguments, such as a test file name to work with. Suppose you have a function, wordCount(), that calculates the number of words in a file, and you want to test it. You could put some test code at the bottom of the module:

```
def wordCount(fileName):
  # implementation
  # gives back the word count in fileName

if __name__ == "__main__":

  import sys

  if len(sys.argv) < 2:
    print('Error: no file name specified')
    sys.exit()

  fileNames = sys.argv[1:]
  for fileName in fileNames:
    print(fileName, 'word count =', wordCount(fileName))
```

Note that sys.argv[0] is the name of the Python script, so the file names are from position 1 onwards. See Chapter 5 for discussion of the function definition keyword def and Chapter 6 for discussion of command-line arguments sys.argv.

---

[2] Those not encapsulated in an abstract definition.

5 Functions

Contents

Function basics

Functions are the mainstay of most programming in Python. The point of a function is to group a specific set of operations in a specific, hopefully not too large, piece of code. A function in Python is what is sometimes referred to in other languages as a subroutine. A function is written as a block of code that is separate from the main program flow, but which can be called upon to operate in many separate places. Once defined, a function will not do anything in particular until instructed to act, often with a specific set of data. Many functions perform jobs that are general and can be used in more than one situation. The benefits from using functions in your programs are clear: they reduce the amount of code you have to write and maintain; they allow you to use existing code that is already tested; they improve the clarity of your code by separating mundane tasks from more important ones, and they allow you to recombine code in different ways to perform new tasks.

Below is a very simple example of a function being used: the function name is `abs` and it is used to find the absolute magnitude of a number, the positive value removing any minus sign. In this case we have used an inbuilt Python function but it is also possible to define new functions and import existing ones from other files.

```
x = abs(-3.0)
print(x)        # Result: 3.0
```

When discussing functions we will address two distinct issues. The first concerns the definition of the function; this is to state what operations it performs. The second concerns how the function is used within a working program; this is to say how a function is *called*, to do something in a given situation.

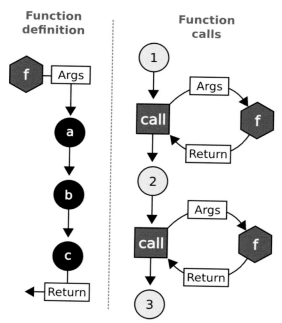

Figure 5.1. The definition of abstract functions and their execution within a program.
A function is an encapsulated block of code that is initially defined in an abstract way. A function generally accepts variable arguments, which would have different values in different situations, and can pass back or 'return' values to the program. A function can be called into use at any point in a program, once it is defined, and this often involves sending specific values for the functions' arguments. Upon calling, the flow of the program's execution is diverted to the function, but returns to the original point of the call when done, often sending back data as one or more return values.

Function definitions and calls

A function carries a name that identifies it so that it can be used thereafter in the program. Thus, when making new functions it is advisable to give functions informative names, usually so that you instantly know what its purpose is. It is possible to define two functions with the same name, but the one defined secondly will overwrite the first one. Here we define a simple function named 'sayHello' that prints a line of text to the screen:

```
def sayHello():

  print('Hello world')
```

The 'def' keyword tells the Python interpreter that a function definition follows, whose name is 'sayHello'. Note that the brackets are a required part of the definition and that the code block that the function contains (just a single print line in this case) is indented relative to the 'def'. The above just defines the function in an abstract way and nothing will be printed when the definition is made. To actually use a function you need to invoke it (*call* it) by using its name and a pair of round parentheses:

```
sayHello()  # Prints 'Hello world'
```

If required, a function may be written so it accepts input, which can be any kind of Python object. In the next example we specify a variable called 'name' in the brackets of the function definition and this variable is then used by the function. Although the input variable is referred to inside the function the variable does not represent any particular value. It only takes a value if the function is actually used in context.

```
def sayHello(name):

  print('Hello ' + name)
```

When we call (invoke) this function we specify a specific value for the input. Here we pass in the value 'Marge', so the name variable takes that value and uses it to print a message, as defined in the function.

```
sayHello('Marge')   # Prints 'Hello Marge'
```

When we call the function again with a different input value we naturally get a different message. Here we illustrate that the input value can also be passed in as a variable (text in this case).

```
text = 'Homer'
sayHello(text)      # Prints 'Hello Homer'
```

A function may also generate output that is passed back or *returned* to the program at the point at which the function was called. For example, here we define a function to do a simple calculation using input (x) to create an output (y):

```
def calcFunc(x):

  y = 2*x*x + 4*x + 1

  return y
```

Once the return statement is reached the operation of the function will end, and anything on the return line will be passed back as output. Here we call the function on an input number and catch the output value as result. You can visualise the invocation of the function as transforming the input into the output. Notice how the names of the variables used inside the function definition are separate from any variable names we may choose to use when calling the function.

```
number = 7
result = calcFunc(number)

print(result)   #    127
```

Strictly, all Python functions give back a return value. In cases where no specific return statement is given, like with sayHello() above, the value passed back from the function is implicitly the None object. This alludes to the fact that we are not obliged to catch the output of a function as a variable; even if a function returns a value the caller can choose to ignore it. So, for example, we could just do:

```
calcFunc(5)
```

In this case the call is rather pointless, since the only reason to use this function is to generate the value that is passed back. But it's possible that a function can return a value that is not of interest in all circumstances.

A function can accept multiple input values, otherwise known as *arguments*. These are separated by commas inside the brackets of the function definition. Here we define a function that takes two arguments and performs a calculation on both, before sending back the result.

```
def calcFunc(x, y):

  z = x*x + y*y

  return z

result = calcFunc(1.414, 2.0)

print(result)   #  5.999396
```

Note that this function does not check that x and y are valid forms of input. For the function to work properly we assume they are numbers. Depending on how this function is going to be used, appropriate checks could be added.

Return values

There can be more than one `return` statement in a function, although typically there is only one, at the bottom. Consider the following function to get some text to say whether a number is positive or negative. It has three return statements: the first two return statements pass back text strings but the last, which would be reached if the input value were zero, has no explicit return value and thus passes back the Python `None` object. Any function code after this final `return` is ignored.

```
def getSign(value):

  if value > 0:
    return 'Positive'
  elif value < 0:
    return 'Negative'

  return

  print('Hello world')  # This line is ignored

print getSign(33.6) # Result: 'Positive'
print getSign(-7)    # Result: 'Negative'
print getSign(0)     # Result: None
```

All of the examples of functions so far have returned only single values; however, it is possible to pass back more than one value via the `return` statement. In the following example we define a function that takes two arguments and passes back two values. The return values are really passed back inside a single tuple, which can be caught as a single collection of values. However, if required the returned values can be extracted immediately to give a clean syntax.

```
def calcFunc(x, y):

  w = x*x - y*y
  z = x*x + y*y

  return w, z

values = calcFunc(1.414, 2.0) # Grab output as a whole tuple

print(values)  # Result: (-2.000604, 5.999396)

a, b = calcFunc(1.414, 2.0) # Grab individual values

print a  # Result: -2.000604
print b  # Result: 5.999396
```

Input arguments

In the previous section we saw that a function can take a specified number of input arguments. However, compared to many programming languages Python is somewhat flexible in the way it accepts arguments.

Optional arguments

In a function definition Python allows default values to be specified for arguments. For example, suppose we have a function that will run a simulation, and the number of steps in the simulation is one argument that we allow. Also suppose that 1000 steps is a sensible default value; the value that will be used in the absence of any other information. Then we could define the first line of the function with:

```
def runSimulation(numberSteps=1000):
```

We can call the function and optionally specify or not specify the value for `numberSteps`. Thus we can do either

```
runSimulation(500)
```

or

```
runSimulation()
```

In the first case the function is run with the variable `numberSteps` set to `500`; the value that we specified. In the second case no value is specified, so `numberSteps` is set to the default value of `1000`. Of course we could explicitly pass in `1000` to the function if we wanted to.

When using defaulting arguments you have to be cautious when your default value is a Python object that is *mutable*, i.e. can be changed internally. This applies to Python lists and dictionaries, but not things like numbers, text strings or tuples. A potential problem arises where a default value is mutable because each time the function is called the same Python object is used; it will not make a new one each time. Consider the following:

```
def myFunction(parameters=[]):

  parameters.append(100)
  print(parameters)

myFunction()  # Result: [100]
myFunction()  # Result: [100, 100]
myFunction()  # Result: [100, 100, 100]
```

Each time the function is run the same `parameters` list remains and is added to each time. To avoid this kind of problem, rather than having an empty list as a default value we use the `None` object, which is of course immutable, and then define a new empty list as required.

```
def myFunction(parameters=None):

  if parameters is None:
    parameters = []

  parameters.append(100)
  print(parameters)

myFunction()  # Result: [100]
myFunction()  # Result: [100]
```

A function can have any number of non-defaulting (mandatory) and defaulting arguments, as long as the defaulting ones come last in the definition; so that when we pass values to the function the mandatory arguments are filled first. Accordingly, if we add another argument to the `runSimulation` function which represents, say, the initial temperature of the simulation, we would do:

```
def runSimulation(initialTemperature, numberSteps=1000):
```

In this case `initialTemperature` does not have a default and must be specified when the function is called. If the new argument is specified with a default value then it is allowed to follow the other defaulting arguments, so, for example, we could have:

```
def runSimulation(numberSteps=1000, initialTemperature=300.0):
```

You could call the second form of the function above by explicitly stating the values of `numberSteps` as `500` and `initialTemperature` as `400.0`:

```
runSimulation(500, 400.0)
```

or by stating only the first argument explicitly, i.e. `numberSteps` as `500`, leaving `initialTemperature` at the default vale of `300.0`:

```
runSimulation(500)
```

or just use both default values.

```
runSimulation()
```

The question then naturally arises how you can use the default value for `numberSteps` but not for `initialTemperature`, given that the latter is the second argument in the function. Fortunately, Python allows names to be used for arguments when calling functions, so you can identify exactly which argument is being passed in. Thus the following is allowed

```
runSimulation(initialTemperature=400.0)
```

Here `numberSteps` is set to the default `1000` and `initialTemperature` to `400.0`.

You can use names for function arguments even if they do not have default values or even if there is no particular need to; it is a matter of taste, but should be considered if it adds clarity. Remember that if you use the name for any argument when calling a function then all subsequent arguments must also use the named syntax. Accordingly, one could use a named argument at the end:

```
runSimulation(500, initialTemperature=400.0)
```

or name both arguments:

```
runSimulation(numberSteps=500, initialTemperature=400.0)
```

or use named arguments in reverse order:

```
runSimulation(initialTemperature=400.0, numberSteps=500)
```

which are all equivalent to

```
runSimulation(500, 400.0)
```

The ability to use any order for named arguments is particularly useful so that you don't have to remember the order in which they were defined in the function.

It is generally a good idea as a programmer that if you add arguments to a function definition that has already been used in various applications then you should add the new arguments to the end of the function definition and specify default values for all of them. This way any existing code employing the function continues to work without modification. (Of course it is another matter if the function has changed what it actually does.)

Example: reverse transcribe a DNA sequence

To illustrate what we have described with a more realistic situation, next we define a function which will take a string of letters representing a DNA or RNA sequence as an input argument and create the *reverse complement* sequence, which is the sequence of an opposing strand that will form a tight interaction through hydrogen-bonding base pairs (see Chapter 11 for further explanation). Naturally we give the function an informative name, and specify two input arguments, one which is mandatory and represents the sequence and the other which is optional to indicate whether we have a DNA sequence (A, C, G and T letters) or an RNA sequence (A, C, G and U letters). This optional argument is named `isDna` and defaults to `True`, i.e. that we have a DNA sequence, not RNA.

```
def reverseComplement(sequence, isDna=True):

  from string import maketrans

  if isDna:
    sequence = sequence.replace('U','T')
    transTable = maketrans('ATGC', 'TACG')
```

```
else:
  sequence = sequence.replace('T','U')
  transTable = maketrans('AUGC', 'UACG')

complement = sequence.translate(transTable)
reverseComp = complement[::-1]

return reverseComp
```

Internally this function relies on the `translate()` function which is built into Python strings (like the input sequence) and the `maketrans` function that is imported from the string module; this makes a character substitution table between equivalently positioned letters from two strings. Also, it is notable that we use the `replace()` function of strings to guard against having the wrong kinds of letter (i.e. T versus U) in the input compared to the `isDna` argument. The upshot of all of this is that the input sequence has letters swapped according to the pairs G ↔ C and A ↔ T for DNA or A ↔ U for RNA to create complement. The reverse of this `reverseComp` is generated using the handy slice notation with a negative step (`[::-1]`). This final string is what we want to pass back from the function, and thus we use it with `return` at the end. The function is readily tested with some example sequence strings:

```
seq1 = 'GATTACA'
seq2 = "AUGGUG"

print(reverseComplement(seq1))                  # TGTAATC
print(reverseComplement(seq1, isDna=False))     # UGUAAUC
print(reverseComplement(seq2, False))           # CACCAU
```

Anonymous arguments

There is one extra subtlety with function arguments (which is beyond what many novice programmers would need to know, so feel free to skip this section). In a function definition, Python has the ability to specify arbitrary anonymous arguments at the end of the list of ordinary arguments. It uses a special syntax for these extra arguments, which can be confusing when you first see it. Consider the following illustrative function:

```
def testFunc(item, *args, **kw):
  print('Mandatory argument:', item)
  print('Unnamed arguments:', args)
  print('Keyword dictionary:', kw)
```

Here we have one mandatory argument, `item`, and at the end there are two special arguments, `*args` (short hand for 'arguments') and `**kw` (short hand for 'keywords') which collect any number of extra arguments that may be passed in. The asterisks (*) have to be used as shown; one for `args` and two for `kw`, but the actual argument names can be different. When the function is called with more arguments than just a single `item` value, these extra arguments are placed inside the `args` and `kw` variables. Simple listed arguments, i.e. those that do not use the `name=value` syntax, are put in the `args` tuple and those that do use the `name=value` syntax are put in the `kw` dictionary, with the keys being the argument names and the values being what these are mapped to. Inside the function `args` and `kw` are used without the asterisks (except possibly when passed into another function). Note that in a function definition you don't have to collect both kinds of arbitrary arguments; you can have `*args` without `**kw` or vice versa.

The following examples illustrate how using this arbitrary argument syntax adds a large degree of flexibility to function calls, although this is sometimes at the expense of reducing clarity.

```
testFunc('Hello', 1, 99, valueA="abc", valueB=7.0)
```

Here the variable `item` is set to the value `'Hello'`, the variable `args` is set to the tuple `(1, 99)` and the variable `kw` is set to the dictionary `{'valueA':"abc", 'valueB':7.0}`.

```
testFunc('Hello', valueA="abc", valueA=7.0)
```

Here `args` is empty and `kw` is set to the dictionary `{'valueA':"abc", 'valueB':7.0}`.

```
testFunc('Hello')
```

Here only the mandatory `item` is specified; `args` and `kw` are both empty.

You may be thinking why you would ever want to have anonymous arguments in a function. Indeed, in general they are often avoided, because it means it is not particularly obvious what the arguments are expected to be. However, there are some situations where the use of anonymous arguments does make sense. For example, you may have a function whose input arguments are dictated by some other consideration, e.g. it might be a function which will be called by an application that someone else wrote, and which you can't change, but where the arguments are not needed to make the function work. In that case one might write the function definition as follows, to indicate that the only argument that matters is the first one:

```
def myPublicFunc(item, *irrelevantArgs, **irrelevantKw):
  print('Ignoring all arguments except', item)
```

Also, you sometimes have a function that passes forward most of its information to another function without using it directly, and that second function may have lots of arguments. Rather than listing all the arguments in the first function, it is quite common to collect anonymous arguments and just pass them all onto the second function. So, for example, suppose there is a function, `setStyle`, that a `draw` function calls to set lots of things like the colour, line style, shading etc. The `draw` function might then be defined as follows, passing the anonymous arguments to another function that is called inside:

```
def draw(points, *args, **kw):
  # some code
  setStyle(*args, **kw)
  # some more code
```

This would work if the definition of `setStyle` accepts anonymous arguments like

```
def setStyle(*args, **kw):
```

or named arguments

```
def setStyle(fgColor="red", bgColor="black", linestyle="plain"):
```

Consider the following call to the `draw` function:

```
draw(points, bgColor="green")
```

In the first case where `setStyle` accepts anonymous arguments the `kw` dictionary is set to `{'bgColor': "green"}`, and the `args` tuple is empty. In the second case where there are named arguments `bgColor` is 'green' but the other arguments have default values; i.e. `fgColor` is 'red' and `linestyle` is 'plain'.

When calling functions with anonymous arguments the asterisk syntax should be used with care, given that

```
setStyle(*args, **kw)
```

is different to

```
setStyle(args, kw)
```

In the former the function can have any number of unspecified arguments which are filled in by the elements of the tuple and the dictionary. For the latter the function would be expecting **exactly** two arguments; here the first argument is a tuple and the second a dictionary.

Lasty, the anonymous arguments notation is sometimes convenient if the number of arguments being passed into a function is large. You can define the argument tuple and keyword argument dictionary before the function call, thus reducing clutter:

```
tupleArgs = (value1, value2, value3)

dictKw = {'color':'blue', 'depth':3, 'gamma':0.271728}

myFunction(*tupleArgs, **dictKw)
```

Variable scope

Next we move on to consider the *scope* of variables inside and outside functions. The scope of a variable is the context within a program where the variable's name has meaning. It is often convenient to think about this in terms of whether a variable operates locally or not; i.e. whether a variable only exists inside a block of code.

Module and function variables

If a variable is defined outside a function then its name has meaning for the whole Python file or *module* in which it resides. Such a variable is usually visible to other modules, so can be imported from one module to another, unless it is specifically protected from being exported by having a name beginning with double underscore '__'. If a variable is defined inside a function then it is visible only inside that function.

```
def mathFunction(x, y):
    z = (x+y)*(x-y)
    return z

answer = mathFunction(4, 7)

print(answer) # Fine

print(z) # Fails; z does not exist outside function definition
```

In general most Python code will be defined as functions and thus most variables should be defined inside those functions. Keeping variables inside functions, and thus their scope local, has several distinct advantages: the value of the variable cannot be altered outside the function, so there is no 'pollution' from elsewhere; you don't have to worry whether your variables are being used by other bits of code[1] and the logic is usually more obvious, since it is contained.

An occasion where a variable might be defined outside a function is when it is intended to be a constant, so not modified after its original definition. For example, consider the following two functions to convert masses between units of pounds and kilograms:

```python
KILOS_PER_POUND = 0.45359237

def poundsToKilos(pounds):

  kilos = pounds *  KILOS_PER_POUND
  return  kilos

def kilosToPounds(kilos):

  pounds = kilos /  KILOS_PER_POUND
  return  pounds
```

The constant POUNDS_PER_KILO could of course also be defined inside the functions, but it is useful to both, and does not change. Thus in this case it makes more sense to define the constant at the module level, outside the functions.

Global variables

If a variable is defined at a module level and if it is desired to change it inside a function, then Python requires a 'global' statement in the function to get hold of the variable from outside. For example, suppose we have a counter that is incremented inside a function, then we could do

```python
counter = 0

def someFunction(argument):

  global counter
  counter += 1
  performOperation(argument)
```

In some ways it is good that Python has this arduous way of modifying variables defined at module level, because it is generally not a good idea to code this way. Often you would avoid using global entirely by passing variables in as arguments and collecting them as output:

---

[1] If a variable is defined outside a function then code that changes it is not 'thread safe', without extra code to manage that.

```
counter = 0

def someFunction(argument, counterVal):

  counterVal += 1
  performOperation(argument)

  return counterVal

counter = someFunction(input, counter)
```

Further considerations

Nesting functions

Python allows functions to be defined inside functions; one function is said to be *nested* inside the other. Typically you would do this only if the inner function is only of use in the very limited context of the outer function. As with variables, this is one way to prevent functions from being used in another module. However, it can make code harder to read, so it is generally best to use this technique sparingly.

Nonetheless, sometimes a function inside a function makes sense. For example, in Python, lists can be sorted with the sort() function, and it has an optional argument which is a function that determines the sort order (otherwise the 'natural' sort order is used). Suppose we have a function that draws atoms, and we want to draw atoms with smaller 'z' coordinate first (assuming that 'z' is the direction out of the screen). Then we could do

```
def drawAtoms(atoms):

  def atomKey(atom):
    return atom.z

  atoms.sort(key=atomKey)
```

Here the sort() function takes an argument, key (allowed since Python 2.4), that specifies a function which takes as argument an item of the list and returns a value which will be used to compare different items in the list. Here the value returned is the z value of the atom, so sorting will put atoms with smaller z values first.

Lambda: anonymous functions

So far we have been discussing the normal named Python functions. However, Python also has some seemingly strange functionality to define anonymous functions. Here's how you can define an anonymous function using the lambda keyword; note how it is immediately assigned to a variable and that a colon ':' is used to separate the inputs from what is returned.

```
cube = lambda x: x*x*x

print(cube(3))   # Result: 27
```

This is equivalent to:

```
def cube(x):
  return x*x*x

print(cube(3))  # Result: 27
```

A lambda function can only be simple; it cannot contain control statements like if or for. So its use is restricted. There might seem little point in defining such limited functions that you cannot name, but there are a few situations where they are very handy. Consider that you have a function that accepts another function as an argument; in this case the argument function is called if there is an error:

```
def jobFunc(arg1, errorFunc):
```

Now imagine that you want to pass some arguments to errorFunc to say how the error should be displayed:

```
def jobFunc(arg1, errorFunc('Warning', color='Red')): # Wrong
```

The above is not what we want because errorFunc is always called when jobFunc is defined. What we really want to do is only call errorFunc with the warning message somewhere inside jobFunc, when and where it is needed. Accordingly we can use lambda to wrap errorFunc without it being called.

```
def jobFunc(arg1, lambda: errorFunc('Warning', color='Red')):
```

As another demonstration, consider the sorting example for nested functions in the previous section. We could use lambda instead:

```
def drawAtoms(atoms):
  atoms.sort(key=lambda atom: atom.z)
```

Decorators

From Python 2.4 the language was extended with a new syntax that allows the convenient wrapping of one function with another; to modify a function so that its output passes directly to another for further processing. Doing this was actually possible before the new syntax was invented, but the specification of how one function wrapped another had to come at the end, after the definition of the function. This meant that it was not always obvious at first glance that a function had been wrapped, and consequently that its output might be modified. Consider, for example, the following trivial example function to do a simple mathematical operation:

```
def getSumSquares(n):
  result = sum([i*i for i in range(n)])
  return result
```

Next we consider a wrapping function `addEmphasis` that is designed to work in a variety of situations (including for the function defined above). It takes an `inputFunc` function and defines an inner `modifyFunc` function that runs the first `inputFunc` but modifies the result, in this case to make text with added '*' symbols. Note that the main function passes back only a reference to the inner `modifyFunc`; it only defines a new function based on the input one and doesn't actually run anything.

```
def addEmphasis(inputFunc):
  def modifyFunc(*args):
    result = inputFunc(*args)
    result = '****%d****' % result
    return result
  return modifyFunc
```

The function addEmphasis is designed so that an input function can be redefined to include the modification. For example, `getSumSquares` can be changed from a numeric output to produce text with asterisk symbols:

```
print(getSumSquares(127))   # Gives: 674751

getSumSquares = addEmphasis(getSumSquares) # Redefine

print(getSumSquares(127))   # Gives: ****674751****
```

With the decorator syntax, rather than redefining a function at the end, we 'decorate' it at the start with a wrapper function specification using the '@' symbol:

```
@addEmphasis
def getSumSquares(n):
  result = sum([i*i for i in range(n)])
  return result
```

This '@' syntax can be used to add any number of *decorator* functions. Take, for example, another general wrapping function, which in this case times how long something takes to run:

```
def usingTimer(function):
  from time import time

  def timer(*args, **kw):
    start = time()
    output = function(*args, **kw)
    end = time()
    print('Function call took %f seconds' % (end-start))

    return output

  return timer
```

To add this wrapping to the specification we only need to add one line to the start:

```
@usingTimer
@addEmphasis
def getSumSquares(n):
  result = sum([i*i for i in range(n)])
  return result

print(getSumSquares(127))
```

And so when the same test is applied again the result is further modified to print out the timing information.

```
Function call took 0.000074 seconds
****674751****
```

6 Files

Computer files

A computer file is a means by which data is stored on a permanent basis, or at least until it is deleted. It is held in a place such as a hard disk drive or removable storage device that is separate from the active, temporary memory of a computer. While the active memory may hold the current program and an amount of data, files represent a larger archive of stored data and the general idea is that this should survive when the computer is switched off. Parts of this saved data may be copied into the active memory as required. Loading data from files (which may be stored locally or transmitted via a network) places data into the active memory so that it can be worked upon efficiently. This data might be the code for a computer program which can then be executed to do a job. Naturally we save program instruction code as a file so that it may be used as many times as desired, without having to rewrite anything.

This chapter will focus on data files that store information for programs to work with, rather than the program files themselves, given that we can trust the Python interpreter to handle the loading and running of Python code. We will show how data can be read into a program and written out from a program, e.g. to and from files stored on disk. Such data files come in a large variety of shapes, sizes and forms (unlike Python files, which conform to a single, precise standard). Information can be stored in an endless number of ways, sometimes at the whim of the programmer, but fortunately in the spirit of cooperation (including mutual financial interest) particular types of data are often stored in a standardised way, with a known specification.

Stored data is represented as a series of binary numbers, i.e. zeros and ones (merely the absence or presence of a signal) and a connected series of data that goes together, as a named unit, is what we mean by a file. However, there are two distinct types of file: *binary* and *plain text*. The difference between these is that plain text files only use a limited series of binary codes to describe data as character symbols, like digits and letters. Binary files are not restricted in this sense, but their interpretation is dependent upon having the right kind of computer system and/or program to load them. Plain text files are much more universal and keep to a standard set of binary codes, so that they will be interpretable in the same way whatever the computer system or programming language. In this chapter we concentrate on data stored as plain text files, and this will cover most of the file standards used in biology.

File formats

When data is stored in a plain text file, although interpreting the individual component characters is trivial, if we are to understand what the contents of the file actually mean we have to understand the way in which the data inside the file is structured. This is just like written language where knowing the alphabet is not enough, we also need to understand concepts like words, sentences and punctuation. Ultimately it is the decision of the computer programmer as to how the data in saved files is structured. However, where it is important that files should be understood by a variety of programs the data will be represented in a standardised, and hopefully documented, way. The data standard for a file is often referred to as a file *format*. Virtually all plain text formats consider the stored data according to lines; they are subdivided by special end-of-line control characters[1] at the end of each line.

A very common file structure is to have one record of data per line, often with a single header line at the top of the file, to describe the contents of the lines. One possibility is that the file represents a table with each line describing one row. The fields (or cells) in each row, one for each column of the table, may be demarked in various ways: this could be according to the position within the row, i.e. the position of a character relative to the start, or specified by special separating characters, like commas or whitespace (tabs, blank spaces etc.). An alternative to a fixed order of fields is that the lines consist of pairs of named keys (identifiers) and corresponding values. Here the keys will generally come from a fixed, allowed set of keys and in some instances the data values that are addressed by a single key may span many lines.

Another common file structure is to have tags that identify and specify the beginning and end of a record. Often these tags can be nested, one inside the other, thus denoting containment. For example, the XML (e**X**tensible **M**arkup **L**anguage) data standard uses tags where the record starts with text like '<NAME>' and ends with the text '</NAME>', where here 'NAME' is the identifier for the element.

Sometimes a programming language will have its own, inbuilt formats for representing data structures created in that language. This process is referred to as *serialisation*. In general the serialisation format will be specific to the language in question, and may require special modules to be installed. However, if data is only going to be used in a single language

---

[1] Not a regular printable symbol.

```
ATOM    473  N    SER A   92      10.056   6.423   5.078  1.00  0.73           N
ATOM    474  CA   SER A   92      10.391   7.707   5.748  1.00  0.87           C
ATOM    475  C    SER A   92      11.422   7.439   6.842  1.00  0.94           C
ATOM    476  O    SER A   92      12.423   8.118   6.941  1.00  1.07           O
```

Figure 6.1. An example of position (column) formatted data. This extract is from a Protein Data Bank file which is used to represent data about molecules and their three-dimensional structures.

```
_struct_ref.db_name                     UNP
_struct_ref.db_code                     GLPA_HUMAN
_struct_ref.entity_id                   1
_struct_ref.pdbx_db_accession           P02724
_struct_ref.pdbx_align_begin            1
_struct_ref.pdbx_seq_one_letter_code
;MYGKIIFVLLLSAIVSISASSTTGVAMHTSTSSSVTKSYISSQTNDTHKRDTYAATPRAH
EVSEISVRTVYPPEEETGERVQLAHHFSEPEITLIIFGVMAGVIGTILLISYGIRRLIKK
SPSDVKPLPSPDTDVPLSSVEIENPETSDQ
```

Figure 6.2. An example of a plain text format that uses keys and values. Here the last value extends over multiple lines. This extract is a fragment of a file format called mmCIF. As with the PDB format mmCIF is used to represent data about biological molecules and their structure.

```
<PDBx:atom_site id="16809">
    <PDBx:Cartn_x>-5.882</PDBx:Cartn_x>
    <PDBx:Cartn_y>-7.295</PDBx:Cartn_y>
    <PDBx:Cartn_z>0.189</PDBx:Cartn_z>
    <PDBx:label_asym_id>B</PDBx:label_asym_id>
    <PDBx:label_atom_id>N</PDBx:label_atom_id>
    <PDBx:label_comp_id>GLY</PDBx:label_comp_id>
    <PDBx:label_entity_id>1</PDBx:label_entity_id>
</PDBx:atom_site>
```

Figure 6.3. An example of a file format containing tagged elements. A truncated fragment of an XML file from the Protein Data Bank (PDBML).

environment, using serialisation can offer an efficient means to store the active data, in terms of both speed and programming ease. Python has a serialisation method which is referred to as *pickling*. Such 'pickle' files are usually textual, but they are not so easy for a person to interpret. For example, the Python list data structure:

```
x = [1, 2, 'a', 'b', True, None]
```

is saved by the pickling method as:

```
(lp1
I1
aI2
aS'a'
aS'b'
aI01
aNa.
```

Given most file formats, it is normally easier to write files than it is to read them; it is easier to extract data from a controlled and standardised in-memory representation (e.g. Python structures) than it is to interpret someone else's text, which may not be standardised or even fully understood. So when writing a program to read a file you need to parse the file (determine its syntactic structure) and also confirm that the content is valid, however that may be defined. When writing a file you just have to make sure that you are following the rules for the file layout. A common programming paradigm is to first read one or more files, do some processing and then write out one or more files. Although the processing is normally the major objective of the program, it is not uncommon to have situations where most effort needs to be spent creating the code to do the reading and writing of the files, especially for simple programs. If there is already an existing piece of tested code, such as the importable BioPython module, which can be used to read and write files, then this is often used in preference to spending time writing something new. See Chapter 11 for examples that use BioPython to read and write files.

Reading files

Using 'open'

When reading a file within a Python program, you obviously need to specify the location and the name of the file, collectively termed the *path* to the file. The file's path is then used by the inbuilt `open()` function, or similar, to get access to the file's contents. The path can be *absolute*, which is to say that it starts at the top or *root* of the operating-system file hierarchy, or the path can be relative to the current working directory (folder). Initially the current working directory is the directory where the Python interpreter starts, although it can be changed from inside the code if needed. In Python the file path is just a string, so, for example, you could have:

```
path = 'examples/dataFile.txt'
fileObj = open(path)
```

The open function passes back what is termed a *file handle*, a Python object used to access and represent the open file. Here the file handle is stored in a variable called `fileObj`. It might be tempting to use the more natural variable name 'file', but in Python the word 'file' is already used to describe the type (class) of such objects. Hence, it is best to avoid overriding the internal variable name. An alternative to the descriptive but verbose 'fileObj' is the shorter 'fh', meaning 'file handle'.

There is the possibility that a program may fail to open the file with the specified path. For example, the stated file may simply not exist (e.g. the name was wrong), or you may not have permission to read it. Under such circumstances, when you try to open it the function will throw an error, a standard Python exception (`IOError`). Otherwise, if all goes well, when you are done reading a file then you can explicitly say you are finished, by using the `close()` function, a method that is known to the file object class. A listing of methods associated with file objects is given in Appendix 2.

```
fileObj.close()
```

Usually you do not have to explicitly close a file, because when the variable name `fileObj` goes out of scope and is no longer meaningful, such as at the end of a loop or function, Python will automatically close it.[2] However, it is generally a good habit to explicitly close a file handle when you know you are done, just in case it is important. Also, this lets someone who reads your code know that you have finished using the file.

The `open()` function can take an optional second argument: a 'mode', specifying how the file should be opened. These modes include 'r' to read from a file and 'w' to write to a file, but, as you might have spotted in the above examples, the 'r' is optional, given that it is the default. Although we could still have explicitly stated reading mode:

```
fileObj = open(path, 'r')
```

Reading lines of data

The file object has certain functionalities associated with it, which allow the underlying data in the file to be read. The most commonly used functions are: `read()`, `readline()` and `readlines()`. The `read()` function is used if you want to read an entire file in one go, into one long string.

```
data = fileObj.read()
```

The `read()` function has an optional argument that specifies the required number of bytes (addressable units of information) to load, but reading the entire file in one go is more common for this function. Of course, if the file is huge this might not be a good idea, because of memory limitations. Accordingly, the `readline()` function reads only one line from the file, and is often placed in a loop to process multiple lines, without having to load everything at once.

```
line = fileObj.readline()
```

As far as `readline()` and `readlines()` are concerned, each line of the file is defined as a text string ending in the *newline* character, or a string that stops at the end of the file without a newline. In other words the newline character separates the lines. Note that the newline character at the end of the line is not removed when using this function, i.e. it is included as the last character of the returned string.

For Unix-derived computer systems (e.g. Linux, OS X) the '\n' newline character is the normal convention. However, for Windows computers the normal convention is that a line ends with the two characters '\r\n', but that also works for the above example because the last character is still '\n'. In some situations a file might have lines that only end with '\r' and, given the way we have opened the file here, this would not automatically be recognised by 'readline' as the end of the lines, even though it is intended to be.

Python has provided a convenient way to deal with the '\r' versus '\n' end-of-line issue. (The newline '\n' and carriage return '\r' concepts are originally from the humble mechanical typewriter.) The mode argument to the `open()` function can include the character 'U', to specify universal line interpretation, so, for example:

```
fileObj = open(path, "rU")
```

This means that when the file is read every occurrence of '\r\n' is replaced with '\n' and every occurrence of just '\r' is replaced with '\n'. This is the recommended method of

---

[2] When the variable is garbage collected; its memory released.

opening a text file when you are not sure of its line endings, unless of course the '\r' characters (singly or in combination with '\n') are required and mean something specific for the file being considered.

Every time you read part of a file, for example, using readline(), a register of which line is next to be read, the *file pointer*, advances in the file. Hence, the next time you read some more of the file, you read from where the previous read ended. When the file pointer reaches the end of the file then the next readline() gives back an empty string; a conveniently False value. Thus if you want to process one line at a time in a file you could do the following where the loop continues as long as the line is True:

```
fileObj = open(path, "rU")
line = fileObj.readline()
while line:
  # process line
  line = fileObj.readline()
fileObj.close()
```

However, there is a more elegant alternative to using readline() repeatedly: from Python 2.2 onwards you don't have to manage the lines yourself. Rather, the open file acts as an iterable object which leads to much simpler code, i.e. so you can loop through the file as if it was a list, yielding the lines inside the loop:

```
fileObj = open(path, "rU")
for line in fileObj:
  pass # process line

fileObj.close()
```

The function readlines() reads all the lines in the file in one go, and returns a list of the lines; a list of strings. Accordingly, an alternative way to process an entire file would be to do:

```
fileObj = open(path, "rU")
lines = fileObj.readlines()
fileObj.close()
for line in lines:
  pass # process line
```

Again, as with the read() function, this is a reasonable approach if the file is not too large. There is also an optional argument for readlines() giving a number of bytes to read, whereupon that amount of data will be read, including any extra bit required to complete a final, otherwise partial line. Another option, which is slicker, but arguably less clear, is to open and read the file in a single statement:

```
for line in open(path, "rU"):
  pass # process line
```

Here the file is closed implicitly, because it was not assigned to a variable, and this is a case where that is acceptable coding style. It is obvious, given that no explicit variable name is stated, that the file is no longer used once the loop has finished.

Another alternative to manually closing a file object is to use the `with ... as` statement, which was introduced on Python 2.5. For example, we could write:

```
with open(path, "rU") as fileObj:
  for line in fileObj:
    pass # process line
```

Here the `with` statement assigns the opened file object to the `fileObj` variable in a special way. We won't go into the precise details of what is happening, but the basic principle is that a `file` class of object has inbuilt methods (`_enter_` and `_exit_`) to deal with its setup and release. In this case the result is that the file is closed at the end of the `with` code block. Note the `with` and `as` keywords are a general part of Python, and not specifically related to files.

Working with variable file names

When writing a program we often want it to be able work with different files on different occasions, each with a different name or file path. Naturally it is possible to sidestep this issue by simply writing file paths into the program code and change the program whenever it needs to work on a different file. However, it is often far easier to accept the name of the file (or files) as variable arguments when we run the Python program. To do this we could either use a graphical interface (see Chapter 26) to select files using a file browser, or if we are running a program using the `python` command we can put the name of the file argument after the name of the Python script. So, for example, here we refer to a file called 'inputFile.txt' in a directory called 'data/' which we include after the script name when running the program:

```
> python programFile.py data/inputFile.txt
```

Note that the directory separator '/' would be a backslash '\' on Windows systems. Inside the `programFile.py` script to access the name of the file from the command line we simply need to import `sys.argv`. This is a list containing all of the text fields that were filled in on the command line, starting with the name of the Python script itself, when the script was run:

```
import sys

pyScriptName = sys.argv[0] # 'programFile.py'
dataFileName = sys.argv[1] # 'data/inputFile.txt'

workFunction(dataFileName) # Use the file name for something...
```

Thus by accessing `sys.argv[1]`, and higher numeric indices for further files, we can define file paths (or any other parameters we want to pass to the script) in a dynamic way when we run the program.

File reading examples

Reading whitespace-separated files

For our first practical example we will begin with reading a simple yet commonly used kind of file, one where each line has several fields that are separated with whitespace. By 'whitespace' we mean tab stops ('\t') or one or more spaces. An example of such a file would be the following, where we first have a descriptive header line and then subsequent lines with three text fields; the first is the name of a chromosome, the second is a base-pair position in the chromosome and the last is a value representing an experimentally determined value for that position:

```
chromosome    position       value
chr1     3417953       0.74634
chrX     152662801     0.50036
chr7     55281536      0.82376
chr4     9168943       0.73375
chr1     13170641      0.42181
```

For the purposes of our example we will assume that the above lines are in a file called 'chromoData.tsv' which lies in the 'examples' sub-directory of the current working directory, where '.tsv' gives a hint that the format is tab-separated values. In order to process this file we will first read the separate header line with .readline(), given that it doesn't contain data we are interested in. Then we will loop through the remainder of the lines, by iterating over the file object, and for each line we will use the string function split() to separate the line into a list of substrings. Without any arguments split() will separate the fields according to whitespace, which is what we want. For a different file format we could specify a different separator, so, for example, for comma-separated fields we would use split(',') or for tab-separated fields split('\t'), both of which can accommodate data items with internal spaces.

```python
fileObj = open('examples/chromoData.tsv')
values = []
header = fileObj.readline() # Don't need this first line

for line in fileObj:
  data = line.split()

  chromosome, position, value = data
  position = int(position)
  value = float(value)

  values.append(value)

mean = sum(values)/len(values)
print('Mean value', mean)
```

For each line we obtain a list with three items and these are extracted into separate chromosome, position and value variables. Initially these will be text strings, given that they were just read from the file, but in the case of the position and value we generally want to convert them from strings into integer and floating point number data types respectively (though in this simple example we have not used the position). Accordingly we use the int() and float() functions to do the conversion. Once a variable is a numeric data type we can then perform mathematical operations, like finding the mean value as illustrated.

We will consider field-delimited formats again in the readListFile() function below, where we handle things in a more general way, allowing different data type conversion functions and field separators to be specified as function arguments.

```
>swissprot|P02724|GLPA_HUMAN Glycophorin-A;
MYGKIIFVLLLSEIVSISASSTTGVAMHTSTSSSVTKSYISSQTNDTHKRDTYAATPRAH
EVSEISVRTVYPPEEETGERVQLAHHFSEPEITLIIFGVMAGVIGTILLISYGIRRLIKK
SPSDVKPLPSPDTDVPLSSVEIENPETSDQ
```

Figure 6.4. An example of a sequence entry from a FASTA-format file. Here the data comprises an annotation line, which contains the database codes and name of a protein, followed by the amino acid sequence of the protein, represented as one-letter codes.

Reading FASTA files

The FASTA text format is used to store nucleotide or protein sequences, using one-letter codes (as illustrated in Figure 6.4). Each file can store one or more sequence entries. An individual sequence entry spans two or more lines: the first line is a textual comment or identifier for the sequence (e.g. a gene or protein name with database codes), and the remaining lines contain the sequence of residue codes. The first comment line for each sequence begins with the '>' character. (Originally the semicolon character ';' was also allowed at the beginning of a line to indicate a comment, but that is no longer commonly used.) Each line subsequent to the comment line contains part of the sequence, until either the end of the file is reached or the next comment line occurs, indicating the start of the next sequence. The lines of residue codes are normally no more than 60 characters long, each character representing one nucleotide or amino acid in the sequence.

Here we will define a function to read a FASTA file, and return a list of the sequences, with each element of the list representing one sequence as a string containing one-letter codes. It would be a relatively simple modification to also collect the annotation (name) data. In this function we choose to read one line at a time because the length and number of sequences can become very large, even a whole genome. Each sequence is potentially specified across multiple lines so we need to keep track of that, and it is normally only clear that the end of a sequence record is reached when the next comment line is found, or the end of the file is reached. The example below does this by creating a list `seqFragments` and appending each part of the sequence as it finds it, and then at the end joining all the parts together using the `join()` function.

The function accepts a single argument, which is the name of the file to open (the full path if not in the current directory). Within the function the file name is used to create a file handle object, opened for reading in universal mode '`rU`', and two empty lists are initialised: one to collect complete sequences and one to store fragments of sequences as they are extracted from separate lines. The line of the opened file is read by using a `for` loop, to iterate through the file data as it is extracted from the `fileObj`. The loop naturally yields lines until the end of the file is reached.

```
def readFastaFile(fileName):

  fileObj = open(fileName, 'rU')
  sequences = []
  seqFragments = []

  for line in fileObj:
    if line.startswith('>'):
```

```
   # found start of next sequence
   if seqFragments:
     sequence = ''.join(seqFragments)
     sequences.append(sequence)
   seqFragments = []

  else:
   # found more of existing sequence
   seq = line.rstrip() # remove newline character
   seqFragments.append(seq)

if seqFragments:
# should be the case if file is not empty
sequence = ''.join(seqFragments)
sequences.append(sequence)

fileObj.close()

return sequences
```

Inside the loop we check whether the line begins with the comment identifier '>', and if it does the line is either at the first sequence record or it has found the start of a new record. In the latter case the complete one-letter sequence of the previous record is defined, by joining all of the fragments from separate lines, and added to the list of sequences. After joining, each list of fragments is then reset for the next sequence record. If the line does not begin with a comment identifier we must be on a sequence line, in which case the trailing '\n' character is removed and the line is stored in the list of sequence fragments (to be joined at the end of the record). After the loop ends, and any remaining sequence is added, the list of sequences is passed back.

An alternative, shorter and perhaps more understandable version would be to just concatenate strings together:

```
def readFastaFile(fileName):

  fileObj = open(fileName, 'rU')
  sequences = []
  seq = ''

  for line in fileObj:
    if line.startswith('>'):
      if seq:
        sequences.append(seq)
      seq = ''

    else:
      seq += line.rstrip()

  if seq:
    sequences.append(seq)

  fileObj.close()

  return sequences
```

In Python it is sometimes recommended to avoid too much string concatenation, given that it can be less efficient than other methods. For short files it would not matter, but for longer ones the `join()` method works slightly faster. This is an example where it pays to be a bit more careful and write slightly longer, and perhaps more opaque, code. Nonetheless, it is up to the programmer to decide what to optimise and what not to optimise.

We shall pause to consider what might go wrong with the above code. Someone might pass in the name of a file that does not exist, or for which the user does not have read permission, in which case the `open()` function will throw an exception, indicating the error. There are various functions in the `os` module (see Appendix 3) that can help avoid such problems. For example, to check whether a file exists you can do:

```
import os

fileName = 'examples/chromoData.tsv'

if os.path.exists(fileName):
  print('File exists')
  # ...
```

Alternatively, someone might use a file that exists and for which the user has read permission, but which is not actually a FASTA-format file, or is not a recent FASTA file where comment lines start with '>'. This will lead to junk output, rather than an error. You could check that the first line starts with the character '>', and throw an exception if it does not. Of course it's possible there is a non-FASTA file that happens to start with '>'. You could check that all the other lines have valid nucleotide or protein one-letter codes. It is up to the programmer to decide how much to check for. Though, the more you want your code to be used by other people, the more checks you should have. It should be noted that the BioPython module that can read FASTA format will do some of these checks for you: see Chapter 11 for examples.

Reading PDB files

PDB (Protein Data Bank) files were invented in the 1970s to describe the three-dimensional coordinates of biological macromolecules. As the name suggests this was initially designed for proteins, but the same system is now commonly used to represent DNA, RNA, carbohydrates, lipids, small molecules and any other biologically important molecule. PDB files can contain the description of multiple molecules and multiple structures, and can hold lots of other descriptive information. However, in this section we ignore all the complexities and concentrate only on the parts that specify the spatial coordinates. A PDB file is both key/value and line oriented, with the key at the start of each line giving context to the data in the remainder of the line. The coordinates we are interested in are in records where the line starts with the six characters 'ATOM ' (with two spaces at the end), which can be thought of as the key. The x coordinate is given in columns 30 to 37, y in columns 38 to 45 and z in columns 46 to 53, assuming that the first column is column 0.

The following example reads a PDB file to calculate the *centroid* of a structure, the average position of the atoms. Strictly speaking, this should be biased by the weight of each atom, but we ignore that issue here (and in practice it does not make much of a difference). In a drawing application, if you rotate a molecule on the screen, it is generally desired to rotate it about the centroid, otherwise the rotation looks odd.

The function takes the name of the PDB file as an argument, and returns the number of atoms found as well as the average x, y and z positions. As a PDB reader the function is very simple and naïve, and in any serious program you would do best to use an existing and tested function, like the one in the BioPython module. Nonetheless, the function will serve to illustrate the principles involved.

Initially we open the file object, read all of the lines and then immediately close it. Next, variables representing the numbers of atoms and the totals for the x, y and z coordinates are initialised to zero, before looping though each of the lines. If a line begins with the desired 'ATOM ' key the atom count is increased, the coordinates are extracted and the coordinate totals are increased. The coordinate data is initially just text characters from the file and needs to be converted to Python numbers (which can be added numerically). The Python float() performs the conversion from test string to floating point number. So, for example, the string '12.572' would be converted to the number 12.572.

```
def calcCentroid(pdbFile):

  fileObj = open(pdbFile, 'rU')

  natoms = 0
  xsum = ysum = zsum = 0

  for line in fileObj:
    if line[:6] == 'ATOM  ':
      natoms = + = 1
      x = float(line[30:38])
      y = float(line[38:46])
      z = float(line[46:54])
      xsum += x
      ysum += y
      zsum += z

  fileObj.close()

  if natoms == 0:
    xavg = yavg = zavg = 0
  else:
    xavg = xsum / natoms
    yavg = ysum / natoms
    zavg = zsum / natoms

  return (natoms, xavg, yavg, zavg)
```

Once the looping is done and the additions are complete, the averages are defined by dividing the summation of each coordinate type by the total number of atoms. Note that if the PDB file has no atom records the averages are simply set to zero, and we cannot divide by zero in any case. The function is then readily tested:

```
print(calcCentroid('examples/protein.pdb'))
```

Of course it's possible that someone calls the calcCentroid() function with an argument that is not a PDB file, or even a file that does not exist. If the file does not exist, or you do not have permission to read it, then the function will throw a standard Python exception

(`IOError`) when it tries to open it. If the file exists but is not a PDB file then most likely there will be no lines starting with the text 'ATOM ' and so the function will just return the tuple (0, 0, 0, 0). It's also possible in this case that there is a line starting with 'ATOM ' (by coincidence) but it does not have three floating point numbers in columns 30 through 53, in which case a standard Python exception (`ValueError`) will be thrown when the `float()` function is called.

There is always a question as to how you deal with bad input to a function. There is no perfect answer. Sometimes you might want to throw standard Python exceptions. In other cases you might want to check for conditions that might lead to an exception and instead return some sensible default. Alternatively you might want to throw your own exception to give a more informative warning to the user, rather than the standard Python one. It is a matter of taste and circumstance.

Reading XML files

Extensible Markup Language (XML) is a way of storing information in files in a standard, textual way. Although it is rather verbose, it is very popular and there are many tools for parsing XML files, which makes it relatively easy to use. An XML file is ordered like a tree, with a containment hierarchy, so in some sense like the directory structure of a file system. At the outermost level is the 'root' of the data tree. Each node in the tree is called an XML element. Each element has a tag defining what kind of element it is, and may also have any number of attributes, some text and can contain any number of other (child) elements. Each element, except for the root element, has a unique parent element. The XML tools let you navigate this tree.

An XML file needs to be syntactically well formed, and the parsing tools will automatically check for this. An XML file may also be required to be valid, in the sense of satisfying some 'schema', which defines what the hierarchy can be, including a specification of the tags, attributes and parent/child relationships. The parsing tools will also automatically check for validity, if the XML file specifies a schema. Schemas can be defined either with a DTD (Document Type Definition) or with an XML schema. We do not consider this issue further here. Note that just because an XML file is well formed and valid does not mean that the data it contains is correct or meaningful.

Python, from version 2.5, includes an XML parser called ElementTree, which provides a very convenient way of reading XML files. It can also be used to write XML files. Because ElementTree does all the tricky parsing and validation work, in some sense it is easier to read XML files than it is to write them. So when you read an XML file you only need to pay attention to the information you want, but when you write an XML file you have to include all the information that the schema is expecting.

ElementTree includes a quick C-language implementation of the parser (hidden underneath the Python), and it is recommended that this is how you use it:

```
from xml.etree import cElementTree as ElementTree
```

The first step when using ElementTree to read an XML file is to parse it using the module's `parse()` function, which provides a handle to the XML tree object:

```
xmlFile = 'examples/protein.xml'
tree = ElementTree.parse(xmlFile)
root = tree.getroot()
```

The `parse()` function accepts either the path to the file or a file handle object. The `getroot()` function on the tree handle then returns the root (top) object. From the root object you can navigate down the tree hierarchy, extracting the information you need using several functions that ElementTree provides for every node element.

Given a `node` element, you can access any text that may be associated with it via `node.text`. Access to the attributes is obtained by treating a `node` almost as if it were a dictionary. So `node.keys()` returns the attribute names, and `node.get(name)` returns the value of the attribute with the given `name`, or `None` if there is no attribute with that `name`. However you cannot use the syntax `node[name]`. This is because instead `node[n]` returns the nth child of the node.

The `node.find(pattern)` function lets you find the first descendant of a `node` that matches the `pattern`, or `None` if there are none matching. At its simplest, the `pattern` can just be a tag, which would then find the first child of the `node` that has that tag. But you can get further down the tree by using a Unix-style file-system path syntax, so, for example, `find('PubDate/Year')` would find the first grandchild where the tag is `'Year'` and its parent (so the child of the original node) has tag `'PubDate'`.

You can even use wildcards for any of the tags on the path, so `find('*/Year')` would match all children and find the first grandchild where the tag is `'Year'`, i.e. the intermediate tag does not matter. However, you unfortunately cannot use wildcards to match part of a tag, so `find('Pub*/Year')` would not work.

The `findall(pattern)` function works in the same way as `find(pattern)` except that it returns all matching elements, instead of just the first one. Also, the `findtext(pattern)` function returns the text of the first element that matches the pattern, or `None` if there is no match. This is convenient shorthand, so

```
text = node.findtext(pattern)
```

is the same as

```
element = node.find(pattern)
if element:
  text = element.text
else:
  text = None
```

Reading PubMed XML files

PubMed[3] is a search engine that lets you access the MEDLINE database of citations for articles in the life sciences. You can download the citations in various formats. In this section we consider the XML format.

---

[3] http://www.ncbi.nlm.nih.gov/pubmed.

As with all XML formats, although you can escape from the pain of parsing using ElementTree, you still have to understand what the schema (or 'data model') is. The schema for PubMed XML is defined by a DTD (Document Type Definition), although reading this is not very enjoyable. And as with all schemas it is quite possible that it will change in future, so application code can break. Here we use the DTD that was valid on 1 January 2009. We will show how to extract and print the journal year and title and the article title and abstract, from a collection of PubMed XML files (which, for example, can be downloaded from the PubMed website).

The root object has the tag 'PubmedArticleSet' and underneath that are one or more children with the tag 'PubmedArticle', although here we will just look at the first child. Underneath that, there is either a child with tag 'NCBIArticle' or 'MedlineCitation' and we will assume the latter. Continuing down the hierarchy we eventually get to the information we want:

```python
def printPubmedAbstracts(xmlFiles):

  for xmlFile in xmlFiles:
    tree = ElementTree.parse(xmlFile)
    root = tree.getroot()

    citationElem = root.find('PubmedArticle/MedlineCitation')
    pmid = citationElem.findtext('PMID')
    articleElem = citationElem.find('Article')
    journalElem = articleElem.find('Journal')
    journalTitle = journalElem.findtext('Title')
    journalYear = journalElem.findtext('JournalIssue/PubDate/Year')
    articleTitle = articleElem.findtext('ArticleTitle')
    articleAbstract = articleElem.findtext('Abstract/AbstractText')

    print('PMID = %s' % pmid)
    print('journalYear = %s' % journalYear)
    print('journalTitle = "%s"' % journalTitle)
    print('articleTitle = "%s"' % articleTitle)
    print('articleAbstract = "%s"' % articleAbstract)
    print('')
```

The PMID is the PubMed ID of the citation. With variants of this code you could create your own kind of short summary of MEDLINE citations.

Writing files

As with reading a file, in order to write a file in a Python program, you need to specify the path to the file and use the 'open' function to define an object and get access to it. The difference is that the file mode here is 'w', which stands for 'writing', instead of 'r' or 'rU':

```python
path = 'output.txt'
fileObj = open(path, "w")
```

This creates an (initially) empty file. Note that if a file with that path already exists then it will first be deleted before the new file is created, so this can be a dangerous operation. You can check whether the file exists using `os.path.exists(path)` and then ask the user whether they really want to overwrite what is there. If you want to append to the end of a file, rather than create a new one, then you can use the append mode:

```
fileObj = open(path, "a")
```

The most commonly used writing functions are `write()` and `writelines()`. There is no function corresponding to `readline` because `write` serves that purpose. Thus `write(data)` writes out the string `data` and it doesn't care whether or not there is a newline character (or carriage return) in the string. The `writelines(lines)` function writes out a list of strings, and again is indifferent as to whether or not the strings have a newline character. Given the fact that these functions are somewhat free-form with regard to what is written out and how the lines of the file are defined (i.e. with '\n' or '\r\n'), you need to be sure that the end-of-line characters are added in the right way.

As with reading, when you are done with a file-handling object for writing then you should generally close it explicitly:

```
fileObj.close()
```

Example FASTA format writer

Suppose we have a sequence as either a list or string of one-letter codes. Then we can create a FASTA file with that sequence in it quite easily. A function is defined which takes arguments for the textual comment, the residue sequence and file location:

```
def writeSingleFastaSequence(comment, sequence, fastaFile):

  fileObj = open(fastaFile, "w")

  fileObj.write('> %s\n' % comment)

  for (n, code) in enumerate(sequence):
    if n > 0 and n % 60 == 0:
      fileObj.write('\n')

    fileObj.write(code)

  if n % 60 != 0:
    fileObj.write('\n')

  fileObj.close()
```

Notice how much simpler this is than reading a FASTA file, although getting the newline characters correct is still a bit fiddly. The function outputs 60 one-letter codes on each line. The `enumerate` function is useful in circumstances like this because it not only gives the residue `code` but it also gives its position, `n`, in the sequence. If `n` is a multiple of 60 (which is what 'n % 60 == 0' checks) and if n > 0 then a carriage return should be added to the output. And a final carriage return also needs adding at the end if n is not a multiple of 60 (if it is then one has already been added).

To expand upon the above example and write multiple sequence records to a FASTA-format file the next function takes multiple sequence strings and a corresponding list of comments. Also, we accept an optional argument to state the width of the sequence lines. To increase the efficiency of the process we avoid looping though all of the characters (and writing them separately); the sequences have the newline characters inserted at the appropriate places and are written in their entirety. The number of sequence lines is calculated using the width, so that by default a sequence of 60 residues or fewer needs one line but 61 residues need two. The number of lines is used to generate a list of sub-sequences by taking the appropriate slice out of the `seq` for each line, i.e. from `width*x` to `width*(x+1)` for each. The sub-sequences are joined with newlines to generate a long string of lines.

```
def writeFastaSeqs(comments, sequences, fastaFile, width=60):

  fileObj = open(fastaFile, 'w')

  for i, seq in enumerate(sequences):

  numLines = 1 + (len(seq)-1)//width
  seqLines = [seq[width*x:width*(x+1)] for x in range(numLines)]

  seq = '\n'.join(seqLines)
  fileObj.write('> %s\n%s\n' % (comments[i], seq))

  fileObj.close()
```

Note how the format '`> %s\n%s\n`' inserts both the comment and the sequence text at the same time, thus avoiding calling `file.write()` again. The `join()` function adds '\n' in between all the sequence lines but we still need that extra '\n' at the end. See Chapter 11 for examples that illustrate how to write FASTA files using BioPython modules.

Column-delimited formats

Next in this chapter we will look at making new file formats to work with your programs. In general, however, you might consider avoiding this entirely. If there is already a well-defined standard that is used for a particular kind of data, like FASTA for sequences or PDB (or more recently mmCIF) for molecular structures, then that should be the first choice, especially if you want other people or programs to understand your data. Also, for an arbitrary set of data you could use an existing standardised system like XML and benefit from the large number of available Python modules to deal with it.

Nonetheless there are occasional situations where there is a pertinent need to read and write data in a custom format, especially where the data is fairly simple and the files will only be used in a limited, perhaps internal, set of situations. As an example we will choose a simple file format, which is easy to read, write and for a human being to understand. This will consist of an initial header line that states what the various items of data represent and then subsequent lines, one for each of the data elements in a list. On each data line we will use a piece of text (commonly a single character like a space, tab stop or comma) to separate or delimit the various items on that line. Note that we should choose the separator string carefully so that it is not something that will be contained in the data and disrupt the delineation of different items.

In the function below, to write out the data we first create a header line, to indicate what each of the fields represents, and then loop through the list of data to create the remaining lines of the file. There is a check to make sure the heading list is the same size as the first item of data and the heading line is formed using the `.join()` function of the separator string. This combines all the elements into one text string and is then written out to the file, combining it with a newline character. For the data lines the separator joins the `formats` variable to create a single one-line `format`, which will be the template to say how to convert each row of data into the appropriate line of text, where each item has the correct numerical precision and padding etc. The actual data lines are created from a tuple of each `row` via the '`%`' formatting operator and are written out with a newline character.

```
def writeListFile(fileName, data, headings, formats, separator='\t'):

  if len(data[0]) != len(headings):
    print("Headings length does not match input list")
    return

  fileObj = open(fileName, 'w')

  line = separator.join(headings)
  fileObj.write('%s\n' % line)

  format = separator.join(formats)
  for row in data:
    line = format % tuple(row)
    fileObj.write('%s\n' % line)

  fileObj.close()
```

To create a specific type of file using this general function the headings, format and separator can be specified, i.e. so they are invariant for the function. For example, here is a file format specification which uses four items (a string, two integer numbers and floating point number) on a line separated by tabs:

```
def writeChromosomeRegions(fileName, data):

  headings = ['chromo', 'start', 'end', 'value']
  formats = ['%s', '%d', '%d', '%.3f']
  writeListFile(fileName, data, headings, formats, ' ')
```

Which could produce something like:

```
chromo start end value
chr1 195612601 196518584 0.379
chr1 52408393 196590488 0.361
chr1 193237929 196783789 0.473
chr1 181373059 6104731 0.104
chr2 7015693 7539562 0.508
chr2 9097449 9108209 0.302
```

The equivalent functions for reading our files are fairly simple. We just need to skip the first line, assuming of course we already know what the data represents, and then loop through the remainder of the lines. For the data lines we remove the last, newline character with `.rstrip()` and split them according to the specified separator, again defaulting to a tab space, and put the resulting list as an entry in the larger list, `dataList`, which is returned at the end of the function. Note that because the values read from the files are just text characters we need to appropriately convert anything which should not remain a Python string, like numbers or `True`/`False` values. This is illustrated below by the use of the `converters` argument, which contains a list of functions (`int`, `float` etc.) to transform the text from the file in the appropriate way. If a conversion is not required for an item then the list simply contains `None`.

```python
def readListFile(fileName, converters, separator='\t'):

  dataList = []
  fileObj = open(fileName, 'rU')
  header = fileObj.readline()       # Extract first line

  for line in fileObj:              # Loop through remaining lines
    line = line.rstrip()

    data = line.split(separator)

    for index, datum in enumerate(data):
      convertFunc = converters[index]

      if convertFunc:
        data[index] = convertFunc(datum)

    dataList.append(data)

  return dataList
```

We can then use this general file-reading function to make something more specific, i.e. by defining a separator and conversion functions appropriate to a particular job. In the example below we use a space as a separator and leave the first value as text, convert the second and third values to integers and convert the fourth to a floating point value, i.e. so we could read files made with `writeChromosomeRegions()`.

```python
def readChromosomeRegions(fileName):

  converters = [None, int, int, float]
  dataList = readListFile(fileName, converters, ' ')

  return dataList
```

There is a standard Python module, called 'csv' (after Comma Separated Value), which will do most of the above handling of delimited text files. Unfortunately, it uses different methods to open files in Python 2 and Python 3. In Python 2 the binary, `'b'`, flag is used to open the file, and in Python 3 an extra `newline` argument is used instead. The underlying reason for the complication is because the `csv` module is designed to cope with new lines being present in the middle of an item of data. Hence, the module does not read or write the data with the standard line-by-line method and makes a separate assessment about how to split the data into rows.

To deal with all of this we have created a small function that can distinguish between Python 2 and Python 3 using `sys.version_info.major` (which gives the value 2 or 3 for the respective versions) and use the `csv` module in the correct way. The construction of the function is similar to `writeListFile`, where we write a header and then the data lines, but here the actual writing is done using the `writerow()` method of a `csv.writer` object. Also, it is notable that what we have called the data `separator`, the `csv` functions call the `delimiter`.

```
import csv
import sys

def writeCsvFile(fileName, data, headings, separator='\t'):

  if sys.version_info.major > 2:
    fileObj = open(fileName, 'w', newline='')
  else:
    fileObj = open(fileName, 'wb')

  writer = csv.writer(fileObj, delimiter=separator)
  writer.writerow(headings)

  for row in data:
    writer.writerow(row)
  fileObj.close()
```

There are also complications due to the Python version in our `csv` reader function `read-CsvFile()`, but the file reading itself is fairly straightforward. We simply create a `csv.reader` object that can be looped through, row by row. We ignore the first (index zero) header row and convert the text to the required data types in the same way we did in `readListFile()`.

```
def readCsvFile(fileName, converters, separator='\t'):

  dataList = []

  if sys.version_info.major > 2:
    fileObj = open(fileName, 'r', newline='')
  else:
    fileObj = open(fileName, 'rb')

  reader = csv.reader(fileObj, delimiter=separator)

  for n, row in enumerate(reader):
    if n > 0:  # n = 0 is the header, which we ignore
      for index, datum in enumerate(row):
        convertFunc = converters[index]

        if convertFunc:
          row[index] = convertFunc(datum)

      dataList.append(row)

  fileObj.close()

  return dataList
```

Further considerations

File operations with the standard library

Python provides a library, or module, called 'os' and a sub-module called 'os.path' that together provide a lot of functionality which is useful when dealing with files. These two modules provide functionality to easily navigate around the file system.

Here is a list of some of the more useful functions. For more details and fuller descriptions, see the Appendices at the end of this book or documentation on the Python website http://www.python.org.

`os`:

`chdir(path)`	change the current working directory to be `path`
`getcwd()`	return the current working directory
`listdir(path)`	returns a list of files/directories in the directory `path`
`mkdir(path)`	create the directory `path`
`rmdir(path)`	remove the directory `path`
`remove(path)`	remove the file `path`
`rename(src, dst)`	move the file/directory from `src` to `dst`

`os.path`:

`exists(path)`	returns whether `path` exists
`isfile(path)`	returns whether `path` is a "regular" file (as opposed to a directory)
`isdir(path)`	returns whether `path` is a directory
`islink(path)`	returns whether `path` is a symbolic link
`join(*paths)`	joins the `paths` together into one long path
`dirname(path)`	returns directory containing the `path`
`basename(path)`	returns the path minus the `dirname(path)` in front
`split(path)`	returns `(dirname(path), basename(path))`

One reason the `path.join` function is provided is to abstract away details about the operating system, such as whether the directory separator is '/' (Unix, Linux and OS X) or '\' (Windows). For example,

```
os.path.join('home', 'test', 'mydoc.txt')
```

returns '`home/test/mydoc.txt`' on Unix and '`home\test\mydoc.txt`' on Windows.

Suppose you want to find all files ending with a specified suffix (e.g. '.txt') in a directory, and recursively include all sub-directories. One way to do this using the functions listed above in Python is:

```
import os

def findFiles(directory, suffix):

  files = []
  dirfiles = os.listdir(directory)

  for dirfile in dirfiles:
    fullfile = os.path.join(directory, dirfile)

    if os.path.isdir(fullfile):
      # fullfile is a directory, so recurse into that
      files.extend(findFiles(fullfile, suffix))

    elif dirfile.endswith(suffix):
      # fullfile is a normal file, and with correct suffix
      files.append(fullfile)

  return files
```

Note that `listdir(directory)` returns only the simple, base name of each file, i.e. without the names of the directories in which it is located, so `os.path.join()` is used to create the full file/directory name. Also, make sure you notice how this function calls itself, to add more files for each sub-directory.

Suppose instead you wanted to remove all files ending with a specific suffix in some directory, and recursively including all sub-directories. The code would be very similar:

```python
import os

def removeFiles(directory, suffix):

  dirfiles = os.listdir(directory)

  for dirfile in dirfiles:
    fullfile = os.path.join(directory, dirfile)

    if os.path.isdir(fullfile):
      # fullfile is a directory, so recurse into that
      removeFiles(fullfile, suffix)

    elif dirfile.endswith(suffix):
      # fullfile is a normal file, and with correct suffix
      os.remove(fullfile)
```

Pickling data

If you do not need a quick format that is human-readable, then one option is to leave the reading and writing of files entirely to Python, by using its *pickling serialisation* modules, to convert in-memory Python objects to a stream of characters that can be saved. Note that if you are loading non-standard Python objects from a pickle file the reading routine must be aware of the definitions of these objects. To make a text representation of a data structure (list, dictionary or whatever) and then save it to file you could do:

```python
import pickle

fileObj = open('saveFile.pickle', 'wb')

pickle.dump(data, fileObj)

fileObj.close()
```

And to get the data back again:

```python
import pickle

fileObj = open('saveFile.pickle', 'rb')

data = pickle.load(fileObj)

fileObj.close()
```

In Python 2 there is a faster implementation of the `pickle` module called `cPickle`, but the latter might not be available in all implementations. In Python 3 the `pickle` module automatically uses the faster implementation if it is available, so there is no `cPickle`. Also, in Python 3 the file objects must be opened in binary (`'b'`) mode; in Python 2 they do not have to be.

7 Object orientation

Creating classes

For simple tasks involving short programs, you can survive perfectly well with the standard Python data types for holding information, such as lists and dictionaries. However, for more complicated tasks involving long programs, this often becomes unwieldy. There are various ways to deal with this issue, but one of the most fruitful is the ability to define your own data types: objects built to your own specification, organised in the way that is convenient to you. Modern computer languages do this via the introduction of bespoke object definitions that are known as *classes* and this kind of thinking is generally termed *object-oriented programming*.

When creating your own custom data types, the *class* is the definition of a particular kind of object in terms of its component features and how it is constructed or *implemented* in code. The term *object*, however, refers to a specific instance, or occurrence, of the thing which has been made according to the class definition. The making of an object of a given class is what is usually termed *instantiation*. A convenient analogy is to think of the blueprint for a house being like a class, but the actual, solid house being the object. Also, given a single blueprint one may build many instances of different house objects, all to the same design. It is quite common to use the words 'class' and 'object' interchangeably, even in the same context, although they mean different things, and it is important to understand the difference. As it happens, everything that is brought into existence in Python is an object, so even integer and floating point numbers are objects, although most of the time you can work without noticing that.

There are various definitions of what constitutes object-oriented programming, and here the exact details do not matter much, since we are using what Python has provided in this regard. One common principle seen in many programming languages is that the class definition should make available certain useful functionality, and that internal information, about how a specific class is implemented, should be hidden from the outside. This is called *information hiding* and *encapsulation*. In this regard, however, Python does not take a particularly strict view and you can prod and probe virtually every part of an object, if you

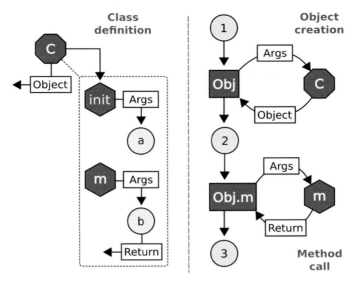

Figure 7.1. The definition of an abstract class and the use of objects made with the class. A class definition is an abstract prototype that contains the specification for creating a custom kind of object data structure. The class is a container for various named values, which may be simple attributes or functions, which will then belong to any object created using the class definition. The initialisation function '__init__()' will be called whenever a new object is made, and this can accept arguments and set the initial state of the object. After the class specification has been defined, a particular instance can be created using the class name. Such an object is generally associated with a variable name so that it may be referred to later, including so that methods (functions bound to the object) can be called using the dot notation.

know how. Nonetheless, you can work as though there were encapsulation, and we will use this practically in the next chapter. As an example, suppose you have a `Molecule` class that models how biological molecules are constructed. This might have defined a function called `getName()`, which gives back the name of the molecule as a text string. For the person who is using this function, there should be no requirement to know how the molecule's name is stored internally or how the function is implemented, just that the function exists and provides the name of the molecule.

Another principle of object-oriented programming is that, in general, a class should be capable of being extended by you, or someone else, through the introduction of what is known as a *subclass*. In essence, the subclass *inherits* all the functionality of the original class (which is called its *superclass*) and then adds something extra. To continue the above example, someone might decide to extend the `Molecule` class by introducing a subclass called `Protein`, which has an extra function `getSequence()` that returns the sequence of the protein. Because of the inheritance from the `Molecule` superclass, the `Protein` class still has the function `getName()` available to it.

Classes can have links to other classes, aside from any subclass-superclass hierarchy. For example, the `Protein` class might have a link to a quite different class called `AminoAcid` that models the amino acid residues that make up the protein molecule. Often when linking different classes it is common to provide functionality that allows you to interrogate the links between objects. Accordingly, the `Protein` class could have a function `getAminoAcids()` that returns a list of objects, each of which would be an `AminoAcid`. If the `Protein` class is used to make an actual object named `myProtein`, then calling the function on this object will give back specific objects that describe its component residues.

The superclass-subclass mechanism can generally be thought of as an 'is a' relationship, whereas the links between classes can generally be thought of as 'has a' or 'has' relationships. So, a `Protein` 'is a' `Molecule`, whereas a `Protein` 'has' `AminoAcid` residues.

Class definition

The definition of a class specification is begun using the `class` keyword, followed by a name used to identify the class:

```
class Molecule:
  # contents of the implementation
```

The implementation code, which defines how the class is constructed and operates, is indented relative to the start of the `class` keyword, in keeping with the style of other separated code blocks elsewhere in Python. In particular, functions are usually defined within the indented code block of a class, such that these functions 'belong' to that class definition.

For defining a subclass, which is based on the definition of some other class, we have to include the superclass in parentheses:

```
class Protein(Molecule):
  # extra contents for subclass implementation
```

Unlike many computer languages (e.g. Java), Python allows *multiple inheritance*, which is to say that a class can have more than one superclass, and inherit properties from all of these. In this case the names of the superclasses are listed, separated by commas, within the parentheses. Multiple inheritance, however, is generally rare; it is much more common to have only one superclass, like the `Protein` above, or no superclass at all like `Molecule`.

In Python 2 there is a slight complication with Python classes, which stems from the fact that 'new-style' classes were introduced, for technical reasons.[1] In most circumstances you can define things in the 'classic' way as we have illustrated, but if you are using a version of Python from 2.2 to 2.7 you can use the new style by making the class definition a subclass of the basal Python class, which uses the keyword `object`.[2]

```
class Molecule(object):  # New style
  # contents of the implementation
```

In Python 3 all classes inherit from `object` and you do not have to specify this explicitly (although you can).

When writing your own class code, it is common practice, but certainly not mandatory, to save each class definition within a single, dedicated file on disk and to call that file after the name of the class. For example, we might have a class called `Molecule` implemented in a file named 'Molecule.py' and `Protein` implemented in one named 'Protein.py'. Generally

---

[1] In order to unify data types and object classes, which until Python 2.2 had remained separate.

[2] This is a confusing name since it is a class, and not an object, which is an instance of a class.

this goes along with the notion that your class definitions may be useful in different situations, whereupon each definition can be imported, to be used within any Python program as needed. To access such classes, saved in other files (modules), one has to import the class, for example, here via:

```
from Molecule import Molecule
```

and

```
from Protein import Protein
```

Above, the first name in each line is the name of the module, which relates to the name of the file ('`Molecule.py`' or '`Protein.py`') from which to load data. Note that the '`.py`' suffix of the file name is not included. The second name, after the `import` statement, is the name of the class (`Molecule`, `Protein`). It is only our convention that the module names and class names are the same, but it is a common one which you should be familiar with.

There is an alternative approach, especially for closely related classes, where multiple class definitions are contained in a single file. For example, we could decide to put both `Molecule` and `Protein` into the same file, here somewhat arbitrarily called '`Molecules.py`'. In this case the import, assuming you need access to both classes in your code, would instead look like:

```
from Molecules import Molecule, Protein
```

How exactly object class definitions are arranged, in terms of which module they reside in and what the module names are, are both matters of taste and convenience.

Class functions

Class functions are functions (subroutines) that are defined within the construction of a class, so that the function's capabilities are available to any object made according to that class specification. They are defined in the same way as ordinary functions, albeit indented within the class code block, but there is one extra twist: the function is aware of the object to which it belongs. Class functions are accessed from the variable representing the object via 'dot' syntax, linking the object's name to the function name. For example, using a function we define below, for a `Molecule` object we might do:

```
name = myMolecule.getName()
```

Here the `getName()` function knows which `Molecule` object to use when fetching the name, without any additional information. When a class function is written, the first argument in its definition is special and represents the object that the function would be called from.[3] In this way you have a handle on the object that you can use in the function code. As with all arguments to functions, you can call this argument for the object itself anything you want. However, the universal and unwavering convention in Python is to call it '`self`', and nothing else, ever. If you call it anything else, and a Python programmer spots it, don't be surprised if annoyance results.

---

[3] In Java, as a comparison, this handle is not included in the definition of the function, rather it is implicitly made available via the '`this`' variable.

Here we define a class and include a class function within:

```
class Molecule:

  def getName(self):
    # function implementation
```

Because the class function takes `self` as the first argument, the function can be used (called) any time you have made an object using the class.

Notice that when you call the function (not forgetting the brackets, see above) you do not include the `self` argument inside the parentheses, since it is automatically known that it refers to the object that is calling the function. This oddity can initially cause people confusion, but a convenient way to think about it is that the object, in this case `molecule`, substitutes for `self` inside the class and inside the actual function. Indeed the `self` argument really is set as the molecule when the program is run.

There is an alternative way to call class functions, which explicitly passes in the object. This method turns out to be useful in certain circumstances when we are dealing with superclasses and subclasses. So instead of the above, you could use the name of the class and pass the object as an argument, which perhaps makes it easier to see how `self` is filled:

```
name = Molecule.getName(molecule)
```

Here `getName()` is the general function definition in the `Molecule` class, rather than one bound to a particular instance of an object. Hence, the specific object to be operated on must be explicitly passed in to the function call, because otherwise the object, in this case `molecule`, is not known. Note, however, that this method still uses the 'dot' syntax. The first `object.function()` way of calling reads better, and is shorter, than the second way. So, unless you definitely need to use the second version, which can happen when subclasses come into play, then the first version is preferred.

From the context of the implementation, inside the code that constructs the class, you write function calls with `self` as the object, which of course is filled in when a real object instance is made. For example, suppose we have a function that provides the name of the molecule with the first letter capitalised. To implement this function we first get hold of the molecule's name, without capitalisation, using a call to the other function, before we do the job of changing the text:

```
class Molecule:

  def getName(self):
    # function implementation

  def getCapitalisedName(self):
    name = self.getName()
    return name.capitalize()
```

Notice that in `getCapitalisedName()` we are assuming that `name` is set to a Python string and not `None`, otherwise `capitalize()` would cause an error, and generate an `AttributeError` exception because the `capitalize()` function is only guaranteed to be present for string objects. You could alternatively protect against this by not calling the capitalisation function unless the name is definitely a string:

```
class Molecule:

  def getName(self):
    # function implementation

  def getCapitalisedName(self):
    name = self.getName()
    if name:
      return name.capitalize()
    else:
      return name
```

The order of the function definitions inside the class implementation code does not matter, so here `getCapitalisedName()` could have been listed before `getName()`. Although, be warned that if by mistake you actually specify a function definition more than once inside a class then the last occurrence replaces the previous one.

Moving on to consider subclasses, which build upon some other class definition, we could include new functions:

```
class Protein(Molecule):

  def getSequence(self):
    # function implementation

  def getAminoAcids(self):
    # function implementation
```

You would call these functions in the expected way, for example:

```
sequence = protein.getSequence()
```

However, it is especially notable that there is no need to repeat the code for `getName()` or `getCapitalisedName()`, which are already defined in the `Molecule` superclass; this is a major point of using class inheritance. The `Protein` class automatically inherits these functions from the other class on which it is based, so, for example, one can do:

```
name = protein.getName()
```

The `Molecule` class, however, knows nothing about the extra functions in the `Protein` subclass. Accordingly `.getSequence()` cannot be used for an object made from the `Molecule` definition but only from an object made from the `Protein` definition.

Class and object attributes

Object classes serve to bring together information and class functions that act on that information. To hold this information, Python classes have simple *attributes*; plain variables tied to the object. A very simple example would be to associate a variable called `description` to the `Molecule` class so that for a given instance called `molecule`, we can do the following to get its value:

```
print(molecule.description)
```

There are two ways in which such variables can be associated with a class. They can be an *object attribute* and belong to a specific object, or a *class attribute* and belong to the class as a whole. Class attributes are defined outside all function blocks and will have the same value for all objects. Object attributes are defined inside class functions and must be set explicitly for each object.

For example, the `AminoAcid` class might have a dictionary containing the molecular masses of amino acids, with the key being the one-letter code of the amino acid.[4] The weight of amino acids is independent of which particular protein molecule we may be considering,[5] so in this case we could define a dictionary as a class attribute.

```
class AminoAcid:

  massDict = {'A': 71.07, 'R': 156.18, ... }
```

Such class attributes can be accessed either via an object instance or via the class name. As with functions, attribute access is via the 'dot' syntax, but note that there are no parentheses when you access an attribute; that is how you can tell the difference between a function call and an attribute access. Here, if you have an object `lysine` made from the `AminoAcid` class, then you can get to the dictionary attribute via:

```
massDict = lysine.massDict
```

However, because it is a class attribute you could also access it via the name of the class:

```
massDict = AminoAcid.massDict
```

Class attributes are often used for variables which do not change, because they are defined for the class as a whole and their values are made available to all instances of that kind of object.

The `AminoAcid` class might have another class attribute called `acceptableCodes`, which lists the codes that are deemed to be valid for an `aminoAcid`. In fact, here these could just be the keys from the `massDict` dictionary, so we could do:

```
class AminoAcid:

  massDict = {'A': 71.07, 'R': 156.18, ... }
  acceptableCodes = set(massDict.keys())
```

In Python 2, the above `massDict.keys()` call gives back a list of keys for mostly historic reasons, because lists and dictionaries were introduced into Python long before sets. Although we could leave it as a list, we turn it into a set for efficiency reasons; determining whether something is in a set is faster than determining whether it is in a list. In Python 3 `massDict.keys()` returns a *view* onto the keys, and the `set()` converts this explicitly into a set.

To access the `massDict` class variable inside a class function we write '`self.massDict`'. But it is notable that outside a class function, in the context of the main code

---

[4] Note that this does not work very well if you have non-standard amino acids in your protein.
[5] OK, in reality isotope abundances and protonation states will affect the mass a bit, but we'll ignore that sort of thing here.

block for a class, there is no `self` so above we would not write '`self.massDict.keys()`'. Also, when the Python interpreter sees this usage of `massDict` it is not yet finished with reading the class definition, so technically the class does not yet exist. Thus, the syntax '`AminoAcid.massDict.keys()`' is also not allowed inside the class code block. Instead, a class attribute that refers to a previous class attribute just uses the name directly, as illustrated here.

Interestingly, bare function names, without the parentheses, are also class attributes, so you can get hold of them and then call them later. For example, with the function `getSequence` in the `Protein` class you could do:

```
getSequenceFunc = Protein.getSequence
getSequenceFunc(protein)        # same as protein.getSequence()
```

The `AminoAcid` class might have an attribute for the one-letter code of the amino acid. The code letter depends on the particular amino acid in question, and not on the class as a whole. Thus, this provides an example where it is more natural to have an object attribute, rather than a class attribute. Accordingly, we might have a function `setCode()` where this attribute, which we imaginatively call `code`, can be set (based upon the input value) by using the 'dot' syntax to tie the variable to `self`:

```
class AminoAcid:

  def setCode(self, code):
    self.code = code
```

Then for the object `lysine`, built from this class, you can get at its `code`, once it has been set:

```
lysine = AminoAcid()
lysine.setCode('Lys')

code = lysine.code
```

It is relatively easy to see the particular object instance called `lysine` becomes the `self` in the function definition. Taken together, you might have the following implementation of the `getSequence()` function to fetch residue codes from a list of `AminoAcid` objects:

```
class Protein(Molecule):

  def getSequence(self):
    return [aminoAcid.code for aminoAcid in self.aminoAcids]
```

This assumes that the code has indeed been set for all the amino acids making up the protein; and that the attribute `self.aminoAcids` is defined somehow for the `Protein` class. As it happens, in the situation described here we would probably not introduce the function `setCode()` at all. Instead, we would set the important `code` attribute when we create the object in what is called the *constructor*, a special function called whenever a new object is made, as discussed in the next section. Nonetheless other, less fundamental attributes might use setter functions.

Lastly we must come clean and admit that Python actually has another way to set attributes, namely:

```
aminoAcid.code = 'A'
```

This syntax allows you to set any attribute to any value, without having to write setter functions, or even having to define the attribute elsewhere. Because it is so simple, this syntax is very popular among Python programmers. However, we have chosen to introduce the more formal way first, because classes depend on knowing which attributes are present, what the values of the attributes are and possibly how those values relate. For example, in `setCode()` we can easily introduce a check to ensure that an amino acid code is valid. When designing a new class, setting all kinds of attributes in an unregulated way is not the best way to start. However, as we will go on to discuss in the Properties section at the end of Chapter 8, there is actually a way to get the best of both worlds so that you can use the above assignment syntax while under-the-hood it will really be using a special setter function.

Class constructors

All objects have a life cycle; at the beginning you have to create a new object, and at the end the object is removed. The creation of an object is handled in a special function that is called a *constructor*, and its removal is handled in a function that is called a *destructor*. Python has automatic *garbage collection*, which means that objects are (eventually) deleted from memory when they are no longer accessible to the program, so normally you do not have to define a destructor. Even when you do want to clean up deliberately, when you are done with an object, it is normal to define a separate function and call it explicitly, rather than use a destructor, since the latter is only called when Python gets around to doing its garbage collection.

A constructor function, when present, is always called whenever the corresponding object is created. The constructor's definition always has the special name __init__; double underscore followed by 'init' followed by double underscore.[6] As with all Python class functions, the first argument is the object handle, i.e. 'self'. After the `self` you can have any number of other arguments that might be useful to set the object up, or maybe no arguments at all. Many classes have a *key*, an attribute that uniquely identifies objects of a given class. Usually this key is either passed into the constructor function as one of the arguments or it is deduced from the arguments that are passed in. For example, suppose you decided that for a `Molecule` there is a name that identifies it. We could then do:

```
class Molecule:

  def __init__(self, name):
    # contents of constructor function
```

Unlike a normal class function, which you would call using the same name as it was defined with, here you do not utter the constructor name '__init__' at all to call it. Instead, the class name is used directly with round parentheses. Thus, to create a `Molecule` you would do:

```
molecule = Molecule(name)
```

The next question is what to do in the constructor with the information that is passed in. Naturally, that is entirely up to the person writing the code, according to what the

---

[6] Anything in Python that begins and ends with a double underscore is 'magic' relating to the internal workings.

requirements of the class are. A very common practice is to keep a reference to the arguments that are passed to the constructor, so that they can be referred to later via the object:

```
class Molecule:

  def __init__(self, name):
    self.name = name
```

What this syntax means is that any object of the `Molecule` class now has an attribute that is called `name`; the variable is bound to the `self` that represents the particular instance of an object. You can thus do something like:

```
molecule = Molecule(name)                    # Make new object
print('molecule name = %s' % molecule.name)  # Use the object
```

Note that the attribute `self.name` is created on-the-fly inside the constructor. You do not have to otherwise specify that you intend to create this, or any other, attribute in Python. Here the value is set directly from an argument that is passed into the constructor when the new object is made, but it is possible that the value is somehow determined indirectly.

As with any function, often you should check that the value passed into an object constructor makes sense. So here, for `Molecule`, perhaps we want the name to be set and not be the empty string or `None`. Accordingly, we could check that the name is defined (semantically true) and if it is not we can trigger an error and go no further:

```
class Molecule:

  def __init__(self, name):
    if not name:
      raise Exception('name must be set to something')
    self.name = name
```

You can create attributes in any class function (or directly on the object), but it is normal to create most of them in the constructor, either directly in the constructor function or indirectly via another function that is itself called from within the constructor. Sometimes an attribute might not yet have a known value at the moment when the object is created, and in this case it is normal practice to set it to a default value, for example, `None`, if there is no other means of determining it. The value can then be set later when the information is available.

As another example, the `AminoAcid` constructor could be implemented as follows, giving an informative error message about what should have been done:

```
class AminoAcid:

  def __init__(self, code):
    if code not in self.acceptableCodes:
      text = 'code = "%s", must be in list %s'
      raise Exception(text % (code, sorted(self.acceptableCodes)))
    self.code = code
```

For a subclass, such as the above `Protein`, you might decide that the superclass constructor is all that you need; remembering that a subclass inherits all of the functions and attributes of its superclass. In this case the subclass would have no extra constructor defined, and when such an object is created it will automatically use the superclass constructor. To create a subclass object you still use its class name, even if it doesn't explicitly have its own version of the constructor function:

```
protein = Protein(name)
```

Typically, however, a subclass would define its own `__init__` function and thus override the superclass constructor, because there are usually more setup operations that need to be done for this kind of object, compared to the class on which it is based.

It is very common that, if you do overwrite the constructor, you also want to call the superclass constructor from inside the subclass constructor; the setup for the superclass is also useful for the subclass and you want to implement both pieces of code. To illustrate this, for the `Protein` class it's possible we would pass in the sequence of one-letter codes, which determine the amino acid components of the protein, as part of its constructor, and so have:

```
class Protein(Molecule):

  def __init__(self, name, sequence):
    Molecule.__init__(self, name)
    # now the Protein specific initialisation
```

Note the syntax: inside the class the constructor is called via `__init__`, not via the class name; we are not making a new object yet, just calling the setup function. Also note that `Molecule.__init__` is the constructor for the `Molecule` superclass. We cannot use the more normal syntax `self.__init__` in this context because that refers to the Protein `__init__` function, not the Molecule `__init__` function.

Given that we are passing in the molecular sequence to the constructor, it would be natural to create the `AminoAcids` objects there:

```
class Protein(Molecule):

  def __init__(self, name, sequence):
    Molecule.__init__(self, name)
    self.aminoAcids = []
    for code in sequence:
      aminoAcid = AminoAcid(code)
      self.aminoAcids.append(aminoAcid)
```

Note that the `sequence` could be a list of one-letter strings each of which represents one code, or it could be a string with each position representing one one-letter code, it does not matter which because you iterate over them (consider each element in turn) in the same way in Python.

Example: molecule, protein and amino acid classes

Putting the code for the class definitions all together in one place, we have:

```
class Molecule:

  def __init__(self, name):
```

```
      if not name:
        raise Exception('name must be set to something')
      self.name = name

    def getName(self):
      return self.name

    def getCapitalisedName(self):
      name = self.getName()
      return name.capitalize()

class Protein(Molecule):

    def __init__(self, name, sequence):
      Molecule.__init__(self, name)
      self.aminoAcids = []

      for code in sequence:
        aminoAcid = AminoAcid(code)
        self.aminoAcids.append(aminoAcid)

    def getAminoAcids(self):
      return self.aminoAcids

    def getSequence(self):
      return [aminoAcid.code for aminoAcid in self.aminoAcids]

    def getMass(self):
      mass = 18.02  # N-terminus H and C-terminus OH
      aminoAcids = self.getAminoAcids()
      for aminoAcid in aminoAcids:
        mass += aminoAcid.getMass()
      return mass

class AminoAcid:

    massDict = { "A": 71.07, "R":156.18, "N":114.08, "D":115.08,
                 "C":103.10, "Q":128.13, "E":129.11, "G": 57.05,
                 "H":137.14, "I":113.15, "L":113.15, "K":128.17,
                 "M":131.19, "F":147.17, "P": 97.11, "S": 87.07,
                 "T":101.10, "W":186.20, "Y":163.17, "V": 99.13 }

    acceptableCodes = set(massDict.keys())

    def __init__(self, code):
      if code not in self.acceptableCodes:
        text = 'code = "%s", must be in list %s'
        raise Exception(text % (code, sorted(self.acceptableCodes)))
      self.code = code

    def getMass(self):
      return self.massDict[self.code]
```

And here is an example of these objects being used, with some self-explanatory `print` statements:

```
water = Molecule('Aqua')
print('molecule attributes')
```

```
print('molecule name =', water.name)

print('molecule function calls')
print('molecule name =', water.getName())
print('molecule capitalisedName =', water.getCapitalisedName())

myProtein = Protein('Fictitious', 'MPKAILV')
print('protein attributes')
print('protein name =', myProtein.name)
print('protein amino acids =', myProtein.aminoAcids)

print('protein function calls')
print('protein name =', myProtein.getName())
print('protein amino acids =', myProtein.getAminoAcids())
print('protein sequence =', myProtein.getSequence())
print('protein mass = ', myProtein.getMass())
```

We will expand on the idea of using classes to describe molecules in Chapter 8, where we create an object hierarchy to describe molecular structures and the coordinates of their component atoms.

Further details

This next section discusses some of the finer technical details of Python classes. If you are a newcomer to object orientation then this section can be skipped on a first reading, as all of the subsequent examples in this book that involve Python objects can be understood using what we have discussed above.

Class and object dictionary

Python classes have a special inbuilt attribute called __dict__, which, as its name implies, is a Python dictionary. This is used to store class attributes, including functions. The key to access an entry in the dictionary is the name of the attribute (or function), and the dictionary value is the value of the corresponding attribute (or a function reference). For example, if we have:

```
class AminoAcid:

  massDict = { ... }
  acceptableCodes = set(massDict.keys())
```

then `AminoAcid.__dict__` would have an entry with key `'massDict'` and one with key `'acceptableCodes'`. Note that the dictionary keys are Python strings.

Instances of objects also have an attribute called __dict__ , and this is used to store the object's own attributes. For example, if we have a class definition:

```
class Molecule:

  def __init__(self, name):
    # ...
    self.name = name
```

and we create a corresponding object via:

```
molecule = Molecule('myMoleculeName')
```

then `molecule.__dict__` has an entry with key `'name'` and value `'myMoleculeName'`.

When Python needs to access an attribute for an object it first checks in the object `__dict__` for a key equal to the attribute name, and if that does not exist it then checks in the associated class `__dict__`. This can be used to set a default value using a class attribute that can then optionally be overridden by an object attribute of the same name. Class attributes take up less memory than an equivalent object attribute because the latter has space allocated in the object `__dict__` for every single object instance of the class.

Python's inbuilt `dir()` function, which lists available attributes for a given object or class, is related to the `__dict__` attribute. Acting on a class it gives a list of the keys in the class `__dict__` and acting on an object it returns a merger of the keys for the class `__dict__` and for the object `__dict__`:

```
dir(Molecule)
# list that includes ['__init__', 'getName', 'getCapitalisedName']
```

Although it is not a common practice, a class attribute or object attribute (including a function) can be removed completely using a `del` statement. Naturally, class attributes must use the `del` on the class and object attributes use the `del` on an object:

```
del AminoAcid.massDict   # removes massDict from class
del molecule.name        # removes name from molecule
del aminoAcid.massDict   # AttributeError exception
del Molecule.name        # AttributeError exception
```

In effect, the `del` operation is working on the corresponding `__dict__`:

```
del molecule.name        # same as: del molecule.__dict__['name']
```

String attribute access

Sometimes it is useful to access an attribute using a string that contains the name of the attribute, rather than directly using the attribute itself. The following examples will hopefully make the distinction clear. Firstly, there is the inbuilt Python function, `hasattr()`, which says whether a given object has a given attribute:

```
hasattr(molecule, 'name')    # True
hasattr(molecule, 'style')   # False
```

Note that the second argument is indeed a string that describes the name of the attribute. This function also works for class attributes and in that case the first argument can also be a class rather than an object:

```
hasattr(AminoAcid, 'massDict')   # True
```

There are two accompanying functions, `getattr()` and `setattr()`, also inbuilt. For example, we can get an attribute value using a string containing the attribute's name via:

```
name = getattr(molecule, 'name')
# same as: name = molecule.name
```

and we can set an attribute value in a similar manner, noting that we pass in three arguments, because we need the value to assign:

```
setattr(molecule, 'name', 'newMoleculeName')
# same as: molecule.name = 'newMoleculeName'
```

The `getattr()` and `setattr()` functions might seem redundant. After all, we would usually just do:

```
name = molecule.name
```

In the above examples that would be a fair point, because the attribute of interest is a specific one. However, sometimes in code we more generally access different attributes using a variable name, rather than by directly referring to a particular name known to Python, and that is where these additional functions come in very handy. It is not unusual to see code like the following, which is general for any attribute; in this case a check to make sure a named attribute is not accessed unless it is defined:

```
if hasattr(molecule, attr):
  value = getattr(molecule, attr)
  # ...
```

For example, `attr` might be `'aminoAcids'`, which in our example code exists if `molecule` is a `Protein` but not if it is just a plain `Molecule`. In the next section we see a better way to check if an object is an instance of a specific class.

To complete the suite of inbuilt attribute functions, there is also a `delattr()` function, which is equivalent in functionality to the `del` statement:

```
delattr(molecule, 'name')   # same as: del molecule.name
```

Class information

Normally when an object is manipulated it is just the attributes and functions associated with it that are of interest. However, sometimes the class itself is of interest. There is a special object attribute called `__class__`[7] that gives a handle on the class definition of an object, although it should be noted that this exists only for user-defined classes:

```
molecule.__class__         # same as: Molecule
```

It might not seem obvious why you would need this, given that the class definition was required to make the object in the first place. Nevertheless, it is surprisingly useful in more complex situations where you are manipulating objects from a number of different classes, and you need to check that the right thing happens to the right kind of object.

The class definition itself has a special attribute called `__name__` that gives the class name as a string:

```
Molecule.__name__          # 'Molecule'
```

These two special attributes do not appear in the internal attribute dictionary `__dict__`.

---

[7] Remember that double underscores indicates internal Python workings or 'magic'.

Classes have another special attribute called __doc__, which does appear in the __dict__. This is used for adding textual comments to help document the class, in the same way that the corresponding attribute (also named __doc__) is also used to document a function. It is filled in from the triple-quoted Python string that appears immediately after the class definition, if such a string exists:

```
class Molecule:
  '''This class describes a biological molecule'''
  # ...


Molecule.__doc__    # 'This class describes a biological molecule'
```

As well as attributes accessed from objects and classes, there are also inbuilt Python functions that operate on classes and objects to say something about them. There is a function, isinstance(), that checks whether a specified object is an instance of a class (was made with that definition). The class might not be the direct class of the object but could also be a superclass:

```
molecule = Molecule('moleculeName')
isinstance(molecule, Molecule)     # True
isinstance(molecule, Protein)      # False


molecule = Protein('proteinName', 'QWERTY')
isinstance(molecule, Molecule)     # True
isinstance(molecule, Protein)      # True
```

This can be useful when you have objects that derive from a known class, but which include a mixture of subclass and superclass versions, where you only want to operate in a given way if it is actually a member of the subclass, with its extra bits:

```
if isinstance(molecule, Protein):
  aminoAcids = molecule.aminoAcids  # exists since it's Protein
  # ...
```

There is an associated function, issubclass(), that works on two classes instead of an object and a class:

```
issubclass(Molecule, Molecule)    # True
issubclass(Molecule, Protein)     # False
issubclass(Protein, Molecule)     # True
issubclass(Protein, Protein)      # True
```

Related to detecting which kind of object you have, there is another concept called the *type* of an object. The type of an object is given by the inbuilt type() function. In Python 3 the type of an object gives its class, and the same is true in Python 2 for 'new-style' classes, which inherit from object. In Python 2 all user-defined 'old-style' classes have the same type.

```
type(3)        # <class 'int'>   in v3, <type 'int'> in v2
type(3.14)     # <class 'float'> in v3, <type 'float'> in v2
type('red')    # <class 'str'>   in v3, <type 'str'> in v2
type(())       # <class 'tuple'> in v3, <type 'tuple'> in v2
type([])       # <class 'list'>  in v3, <type 'list'> in v2
```

```
type(set())     # <class 'set'>   in v3, <type 'set'> in v2
type({})        # <class 'dict'>  in v3, <type 'dict'> in v2
type(molecule)  # <class 'Molecules.Molecule'> in v3
                # and in v2, new style class
                # <type 'instance'> in v2, old style class
```

The `type()` function actually gives you back another Python object, and the type of such an object is the seemingly odd `<class 'type'>` in Python 3, or `<type 'type'>` in Python 2. The type objects can also be obtained otherwise:

```
int         # <class 'int'>   in v3, <type 'int'> in v2
float       # <class 'float'> in v3, <type 'float'> in v2
str         # <class 'str'>   in v3, <type 'str'> in v2
tuple       # <class 'tuple'> in v3, <type 'tuple'> in v2
list        # <class 'list'>  in v3, <type 'list'> in v2
set         # <class 'set'>   in v3, <type 'set'> in v2
dict        # <class 'dict'>  in v3, <type 'dict'> in v2
Molecule    # <class 'Molecule'> in v3 and v2
```

In Python 2 for 'old style' classes, to get at the type of `Molecule` without using the `type()` function we need to import the `types` module and then we have:

```
import types
types.InstanceType  # <type 'instance'> in v2 old style class
```

Checking the object type comes in handy when some variable could have one of a few types, and how it gets processed depends on the type. For example, a function might have an argument that could be a tuple or a dictionary, where the following code could be used:

```
def f(x):
  if isinstance(x, tuple):
    # do something
  elif isinstance(x, dict):
    # do something else
  else:
    # error: raise an exception
```

8 Object data modelling

Data models

This chapter delves more deeply into the topic of creating custom Python objects using class definitions. Given that we have discussed the basics of object-oriented programming in Chapter 7, we now move on to illustrate how such mechanisms can be used in a practical, scientific sense. If you are interested in only a light introduction to Python, you might consider skipping this chapter on a first reading. However, the objects we discuss here will underpin many of the examples given later on in this book, in Chapters 15 and 20, so you may like to look back to see how such things are constructed.

In the previous chapter we saw how to introduce our own types of data object into Python, using classes. Here we move on to look at how to use a number of different, but connected, classes to construct what is often known as a *data model*. A data model is an abstract description of concepts that can be used to build a computational version of some topic or real-world situation that you are interested in. Essentially, you examine the kind of information you wish to describe and divide it up into conceptual parcels. Each of these will become one kind of computer object (a class with attributes, functions and links to other classes), which then allows you to create a synthetic analogue of the thing you are interested in. Virtually all programs, irrespective of size, rely on some kind of underlying model to organise data, although this may not use object-oriented programming and is often not formalised in any way. No data model can be expected to be a perfect computer representation of what it describes, but the idea is to make it good enough to serve a useful purpose, by having some of the properties of the things being modelled.

Designing a molecular structure data model

In this chapter we construct and implement an example data model which represents the three-dimensional structures of large biological molecules. If you are unfamiliar with the basic principles of biological molecules and their structures, see the introductions to Chapters 11 and 15, which aim to be suitable for non-biologists. Specifically the data model will be for linear polymers, such as DNA, RNA and protein, where a longer molecule is built of smaller components linked together into a chain. It is a relatively simple data model, and it could certainly be extended, but we will avoid adding complications and keep things as clear as possible for this book. As such, we will make various simplifying assumptions about molecules and biology, but that is the case with all data models, it is all just a matter of degree. Specifically, we will ignore issues such as how the molecules might have a few extra or a few absent atoms (mostly hydrogen ions and small modifications) or how the molecules might have extra links, which are not part of the main linear chain, like the disulphide links found in some proteins. We will not use any formal computer methods to describe the construction of the data model. Instead, we will rely upon relatively plain English. There are formal modelling techniques, like UML (Unified Modeling Language), for example, but such things are well beyond the scope of this book.

Our model will describe the identities and the relative three-dimensional positions of all of the atoms which collectively can be considered a macromolecular structure; the precise shape of large biological molecules. This structure may be composed of any number of polymer molecules that come together, but is frequently used to describe just one molecule. Each molecular chain will have a distinct biological type, i.e. DNA, RNA or protein, and we can mix polymer types however we like. For example, we might want to consider the structure of a protein bound to a section of DNA.

We will sometimes expect more than one set of three-dimensional coordinates for a given molecule, which means that for the same set of atoms we can describe alternative arrangements or *conformations*. Describing multiple conformations is useful to indicate situations where the precise structure is uncertain and to describe the outcome of dynamical simulations of the molecule, where each set of coordinates could represent a different point in time or a different outcome. By allowing discrete collections of coordinates for a given molecule, we generate what is sometimes referred to as a *structural ensemble*. This term is used to emphasise the 'togetherness' of a bundle of related conformations.

In our model we will identify a given structure by a name, which will be a textual identifier, and we will also include a non-mandatory property, the Protein Data Bank identifier, to indicate when the data has come from an entry in the main biological coordinate database. The Worldwide Protein Data Bank[1] is a publicly available database that stores the structures of molecules. These were mostly determined by X-ray crystallography but many have been determined by other techniques such as nuclear magnetic resonance (NMR). Despite the name suggesting that the PDB database is only for proteins, these days it contains coordinate data for DNA and RNA too, although the protein structures vastly outnumber the other types. The structures that we are modelling might have been entered into this database, and we want to keep track of that. Accordingly, we use the textual PDB identifier that is unique to each entry in the PDB. Naturally, the PDB has its own data model to describe biological structures and their associated data, and it is far more extensive and complicated than the one we are using here. In their data model the PDB identifier is mandatory, but in our data model we will make it optional; the data doesn't have to come from this database in every case.

---

[1] http://www.wwpdb.org.

There are many design decisions in our example data model, about which things to describe, which things we ignore and what rules we apply. We will discuss the aspects of our particular model as we go through the example. However, which precise details we have chosen is not the most important thing; the idea is to empower you to create your own data models to do exactly what you want.

Implementing a data model

Once you have designed a data model, you can implement it in terms of code. Things are not necessarily fixed, however; in practice it is very common to modify a data model and the corresponding code in a continuous process of improvement and adaptation. It is somewhat unusual to get the model 'correct' the first time around. Also, it is also almost inevitable that more functionality is added over time, and this often requires a change to the data model.

We have stated that we will use a data model to allow an ensemble of different conformations for the same molecule. There are various ways that this aspect of the model could be implemented, but there are two appealing choices. We could have Python classes that describe the molecular composition once and have multiple sets of coordinates which emanate from this reference; each atom would have alternative coordinates for the different conformations. Alternatively, we could have multiple descriptions of the molecular composition, each of which holds a single set of coordinates. Here we choose the second option, and although this approach has advantages and disadvantages, it is mostly chosen here in view of coding simplicity.

Our molecular structure data model will have a *top* class called Structure, which will group all the objects belonging to other classes. There will be three additional classes: Chain, Residue and Atom, and the overall hierarchy is illustrated in Figure 8.1. The Chain[2] class will represent one polymer molecule. The class called Residue[3] will model the individual chemical compounds or *residues* that have been linked together into a chain to form the molecule; for proteins these will be amino acids and for DNA and RNA these will be nucleotides. Lastly, the Atom class will, somewhat unsurprisingly, represent the atoms that are found in the linked chemical components. The Atom class will contain a single set of three-dimensional coordinates for that atom. The classes in our data model will be linked together to form a *containment hierarchy*; structures contain chains, chains contain residues and residues contain atoms, thus going from the largest entity and subdividing it into progressively smaller, but still meaningful, parts.

A typical life cycle of a Python object involves first creating it, then possibly modifying it, and finally deleting it (or leaving it to cease to exist when a program stops). Our model will explicitly handle some deletion cases, like removing a Chain from a Structure. However, it is very tempting to not worry at all about deleting objects, and often you can get away with this. Sometimes there is simply no need to add deletion functions, because you are only ever going to explicitly create and modify objects, never remove them. To this end it

---

[2] We decided against using the name Molecule to emphasise the fact that this class is a container for a series of smaller chemical entities.

[3] See Chapter 11 for a description of why these are called residues.

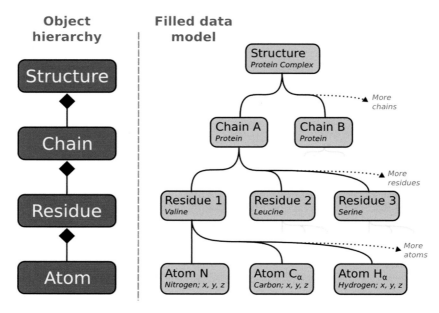

Figure 8.1. A simple hierarchical data model for organising macromolecular structure data. The example data model for containing the structures of biological polymer molecules is composed of four object classes. The top of the hierarchy is a Structure, which represents the 3D conformation of a group of molecules. Each Structure contains Chain objects, each of which represents a different molecule of a specific type (DNA, RNA or protein). A Chain contains Residue objects that represent smaller chemical components that are linked together into a chain, i.e. amino acids or nucleotides, thus representing the protein or nucleic acid sequence. Each Residue object contains Atom objects, which are of a specific chemical element type and carry the actual spatial coordinates.

is handy that Python has a 'garbage collection' mechanism, which means that once you no longer have a handle to an object, it will (eventually) automatically be cleared away.

Structure

We start with the construction of the `Structure` class. As mentioned, we require a name and optionally provide a PDB identifier code. More than one `Structure` object with the same name will be allowed, each with its own set of coordinates. Hence, we introduce another mandatory attribute called `conformation`,[4] which is a number that specifies which set of coordinates within an ensemble we are considering. In many circumstances we will only have one conformation, so we issue a default value of `0`, even though it is mandatory.

This naturally leads to the following first attempt at the class definition and constructor code:

```
class Structure:
  def __init__(self, name, conformation=0, pdbId=None):
    if not name:
```

---

[4] Several other molecular structure data models use the term 'model' instead of 'conformation', but we avoided it here because in a chapter about data models, talking about other kinds of 'model' would be needlessly confusing.

```
    raise Exception('name must be set to non-empty string')
self.name = name
self.conformation = conformation
self.pdbId = pdbId
```

Remembering that the __init__ function, the constructor, is called each time an instance of this class of object is made, we store the name, conformation and pdbId as attributes by binding their values onto variables that are linked to self, which provides a handle to any actual object instance made using this class. Note that we have used a convention whereby attribute names are lower case except when a new 'word' starts, and then the first character of that is capitalised, thus here giving pdbId. A popular alternative is to keep attribute names all lower case but use underscores to separate the words, which would give pdb_id. No doubt there are other conventions, and it mostly doesn't much matter what you do, as long as you are consistent as an aid to readability.

In the class constructor we check whether name is defined. This is done using the clause 'if not name' to check if the value is logically false, e.g. None or an empty string, and in these cases we deem the name to be undefined. An undefined name means something is wrong, so we cause an error by raising an exception object. However, we have not checked that name is actually a text string. Someone could try to create a Structure by passing in any Python object that evaluates as true (a non-zero number, for example) and it would pass the check but violate our intention about what the name should be. Hence, if you were being cautious you would check the type of name before using it. Similarly, checks can be made for the other input arguments, and so in effect introduce *run-time type checking* into the constructor. Here, for the sake of brevity, we will avoid such caution.

Another thing we have not checked here is whether values are meaningful, even if they are of the correct Python data type. For example, we do not know whether the pdbId, if set, is actually a valid identifier. To determine if the pdbId is a valid PDB identifier code is not trivial, but the example at the start of Chapter 15 will give you a hint at a solution if you are really keen.[5] We ignore such issues here, but it illustrates that no matter how many checks you make, there are almost certainly some checks that you have not made. Also, part of the solution is to not pass junk into your data model in the first place (despite the fact that users may try).

For the pdbId we have set the default to None rather than an empty string. This is a matter of taste, but generally in such situations we use None because this pretty much always means 'not set'. For pdbId an empty string could be taken to mean the same thing, since real PDB identifiers are never empty strings, but in other situations an empty string might be a legitimate setting.

In data modelling there is the notion of an object's *key*; this is something that uniquely identifies an object amongst other objects of the same class. Here we intend that the name and the conformation uniquely identify Structure objects, so these two attributes taken together are a natural key for this class. If we were diligent, and really wanted to enforce this to be a key, then we should add a check in the constructor that (name, conformation) has

---

[5] The wwPDB offers a web service to fetch structural data, which can be accessed using Python.

not already been used by an existing `Structure`. Again, for reasons of simplicity we ignore that issue here, but if we wanted to worry about it then we would have to keep track either of all the `Structure` objects that we created or of all the associated names and conformations, for example, using a set or list of `(name, conformation)` pairs.

This brings up another design decision: a `Structure` object has a `name` and `conformation`, but we have not stated whether we are allowed to change them. This depends on how we intend to use them. For example, if we have an application where the `name` is intended to be a friendly way of identifying a `Structure` to the user then we might want to allow the user to change it to something they prefer. In contrast, the `conformation` is effectively just an index number into the coordinate elements of a structural ensemble, and so there is no reason to allow that to be changed. Indeed if it could be modified then that might create more trouble than it was worth. If we allow an attribute to change we call it *changeable* and otherwise we call it *frozen*. When an attribute is frozen it can only ever be set once, and normally that would mean in the constructor (when the object is made). In Python you have to take some extra steps to make attributes frozen, and we will discuss this later. For now we will in effect assume that everything is changeable.

Another issue with attributes is the matter of how many items they are allowed to represent, according to the data model, which is termed their *cardinality*. Specifically, the cardinality is represented with whole numbers where the *low cardinality* represents the minimum number of items that can be represented, while the *high cardinality* represents the maximum number. Because we have stated that the `name` is mandatory it always represents exactly one thing, thus the low cardinality is 1 and the high cardinality is also 1. We can write the overall cardinality of this attribute, minimum to maximum, as being '1..1'. Similarly, the cardinality of `conformation` is also '1..1'. Conversely, because the `pdbId` is optional there might be none or one, so for this the cardinality is '0..1'.

Perhaps at some point we decide that we are going to allow references to more than one PDB identifier in a given `Structure` object. This would fundamentally change the data model, and the constructor then might become, noting the plural name for the last attribute:

```
class Structure:
  def __init__(self, name, conformation=0, pdbIds=None):
    # etc.
```

Here we might intend that `pdbIds` is specified as a list or tuple, containing strings representing PDB identifier codes, or otherwise left undefined as `None`. The low cardinality is still 0, because there might be no PDB identifiers, but we now have no upper limit, so the high cardinality is effectively unbounded, which we label as '*'. This case gives an overall cardinality of '0..*'. Obviously the high cardinality for any attribute has to be greater than 0, otherwise it can never exist. If it is 1 then the attribute is normally spelled in the singular (`pdbId`) and if it is greater than 1 then the attribute is normally spelled as a plural (`pdbIds`).

When the high cardinality is greater than 1 another issue comes into play. In this case we have a collection and there is the question of whether the items in the collection are in any particular order, or not. For `pdbIds` we have stated that we intended it to be defined by a list or tuple, collections that do have ordered items. Consequently, it is natural to assume that the attribute is also ordered. Alternatively, we might have allowed it to be defined by a set, in which case it is natural to assume it to be unordered. Deciding whether something is ordered or unordered can be critical in some contexts. In any case, from here on we stick with the singlular `pdbId` attribute, rather than `pdbIds`.

Changing the high cardinality of an attribute changes the data model fairly dramatically (it has to be specifically coded in the classes) so it is a good idea to think carefully about the situation being modelled. It might be tempting to always assume that the high cardinality is unbounded ('*') because it is more general, but this is a bad idea if it really ought to be 1. For one thing it means dealing with a collection containing a single object instead of just the single object itself, which can make for confusing and error-prone code.

Finally, we create a `Structure` object in the usual way, by using the name of the class and passing in values for the attributes:

```
structure = Structure('Chromosome Regulator', 0, "1A12")
```

or we could write it using named input attributes:

```
structure = Structure(name='Chromosome Regulator', pdbId="1A12")
```

As another example we could avoid passing in a PDB identifier, given that this attribute is not mandatory and will take the default value of `None`.

```
structure = Structure(name='Chromosome Regulator')
```

Chain

As we mentioned previously, a structure may comprise more than one molecule. Each molecule, because it is a chain of linked amino acids or nucleotides, will be described using the `Chain` class. Our data model is made with the assumption that each chain belongs to a unique structure, so is effectively contained by that structure. This is an important design decision and has all kinds of ramifications. What we are describing here, when one kind of object is said to contain another, is what is known in data modelling as a *parent-child relationship*. The `Structure` object is the parent and the `Chain` object is the child. Also, we often talk in terms of classes rather than objects and say that `Structure` is the parent class of `Chain`.

If a parent object contains children, then each child object must belong to that parent; a child object is only meaningful within the context of its parent. A consequence of this is that the parent object must be created before its children. Also, if a parent object is deleted then all of its children must also be deleted. An alternative here would be to make a `Chain` free-standing, meaning it could appear in more than one `Structure`. That would be a perfectly plausible scenario, but it is not especially helpful here and not the way we will model it. A parent class can have many different kinds of child classes. Thus, we could have a `Technique` class, which represents how the structure was determined. This could also be a child of `Structure`. When a `Structure` is deleted both `Technique` children and `Chain` children would need to be deleted. To keep things simple, here we just consider `Chain`. Note that `Structure` itself has no parent. Had we decided to model things differently, we might have introduced a class `Ensemble`, as the parent of `Structure`, and then `Ensemble` would have no parent.

A parent-child relationship is an example of a 'link' between objects. Links are generally harder to manage than simple attributes (like name and pdbId), because there are two ends to consider, one for each object, and they need to be made consistent with each other. One way to manage links is to have one of the classes keep track of everything. For example, in the previous chapter we had a parent class, Protein, and a child class, AminoAcid, and the parent class managed everything. So a Protein object had a list self.aminoAcids. But an AminoAcid object had no reference to the Protein being its parent. It's quite possible that this is not a problem; for example, an AminoAcid might only appear in contexts where the Protein also appears. Having all the information at one end of a link makes it easier to manage. On the other hand, it usually makes it harder to use.

Here we choose to manage object to object links from both ends. Accordingly, a Structure object will have the attribute self.chains to access its children and the Chain class will have self.structure to access its parent. We will need to keep both of these attributes synchronised at creation and deletion of a Chain object. Both ends of this link have cardinality. A Chain object has to have a Structure parent and there can be only one of them, so that cardinality is '1..1'. A Structure object can have any number of Chain children (zero or more), and so that cardinality is '0..*'. We will assume that Chain children are ordered for a given parent, and that this order is the order in which they were created.

We will model the Chain class as having an identifying code, which is a string, and a descriptive molType, where the latter can be 'protein' or 'DNA' or 'RNA'. Relative to the parent Structure, we will assume that the code uniquely identifies the Chain. Or to put it another way, a given Structure object has only one Chain child with a given code. Thus, the object key for a Chain, relative to its parent, is code and the full key is (structure, code). This is a typical situation for a child class; it has a key that identifies it relative to its parent and then together with the parent itself this is the full key for the child. This then leads to the following proposals for the implementation of Structure and Chain respectively:

```
class Structure:
  def __init__(self, name, conformation=0, pdbId=None):
    if not name:
      raise Exception('name must be set to non-empty string')
    self.name = name
    self.conformation = conformation
    self.pdbId = pdbId

    self.chains = [] # For the children

  def delete(self):
    for chain in self.chains:
      chain.delete()

  def getChain(self, code):
    for chain in self.chains:
      if chain.code == code:
        return chain

    return None
```

The attribute that links the `Structure` parent to children is `self.chains`, and this is initialised as an empty list, to be filled in as the child objects are made. The `delete()` function is fairly straightforward: when the structure is deleted its chains disappear. It is notable that there is no specific deletion of the structure object itself, i.e. we don't do `del self`, because it is at the top of the hierarchy and will simply disappear when it is no longer associated with any Python variables (it will be garbage collected). We have also included a function to help get hold of a specific `Chain` object, whereby a code string is accepted as an argument. Here the function loops through the list of children (`self.chains`) in order to find a `Chain` with a matching attribute, and, because this is the unique key to identify the `Chain` within its parent, there will never be more than one possible match.

The `Chain` class is constructed by accepting the parent structure, the `code` value as its key and a molecule type. The `__init__()` performs some checks to make sure the arguments are reasonably sensible, and this includes determining whether the parent structure already contains a `Chain` with the input `code`, which needs to be unique. If there are no errors, the attributes are associated with the `self.` variables. Lastly, the constructor adds the `Chain` (represented here by `self`, but filled in with an actual object at run time) onto the `structure.chains` list, the link from the structure parent to its children. Note that we do not modify the parent's link to its child until after we have checked that everything is ok.

```
class Chain:
  allowedMolTypes = ('protein', 'DNA', 'RNA')

  def __init__(self, structure, code, molType='protein'):
    if not code:
      raise Exception('code must be set to non-empty string')
    if molType not in self.allowedMolTypes:
      raise Exception('molType="%s" must be one of %s' %
                      (molType, self.allowedMolTypes))

    # check that key code is not already used
    chain = structure.getChain(code)
    if chain:
      raise Exception('code="%s" already used' % code)

    self.structure = structure
    self.code = code
    self.molType = molType

    structure.chains.append(self)

  def delete(self):
    self.structure.chains.remove(self)
```

The `delete()` function for the `Chain` is a simple matter of removing the object (again represented by `self` in the class definition) from its parent's list of children. Also, unless we have a specific handle on the object, all notion of it will be lost and it will eventually be removed from memory when Python performs garbage collection.

Note that in the above example we have chosen to store the chains of a structure using a list. That means that the `structure.getChain()` function is not particularly efficient, because it has to loop over a list looking for a matching item. If the `chains` had been unordered (relative to the `structure`) then an alternative would have been to have a dictionary, `chainDict`, where the key is `code` and the value is the `chain`. The `chains` could be obtained from `chainDict.values()`. If we desire both efficiency and ordered children we could have both the dictionary, `chainDict`, and the list, `chains`, although we would have to keep them synchronised:[6]

```
# Alternative Chain implementation

class Structure:
  def __init__(self, name, conformation=0, pdbId=None):
    # ... initial part as before
    self.chainDict = {}
    self.chains = []

  def getChain(self, code):
    return self.chainDict.get(code)

class Chain:
  def __init__(self, structure, code, molType='protein'):
    # ... initial part as before
    structure.chainDict[code] = self
    structure.chains.append(self)

  def delete(self):
    del self.structure.chainDict[self.code]
    self.structure.chains.remove(self)
```

Practically, the inefficiency with lists isn't an issue here because we expect that a given structure will at most have only a few chains, so we will stick with the simpler list implementation.

As a final point on the `Chain` class, although we have only used a single value as the identifying key, it would also be possible to have a key consisting of two values, e.g. `(code, molType)`, where the key is passed around as a tuple.

Residue and atom

The same issues come up all over again with `Residue`, a child class of `Chain`, and `Atom`, a child class of `Residue`. For `Residue` we will assume that it has an identifying key (relative to its parent) called `seqId`, and an optional attribute called `code`. The `Chain` class will get an extra function, `getResidue()`, which returns the child `Residue` with a given `seqId` (or `None` if there isn't one). We will assume that the `Residue` children of a `Chain` are ordered, according to the order of their creation. However, this time we will use both a dictionary and a list in the implementation because a `chain` can have many (so hundreds of) `residues` and we want an efficient implementation for `chain.getResidue()` and a list of the `residues` in sequential order. We will also add another function, `getAtoms()`, into the `Chain` class, which will return all the atoms of all the residues in the chain.

---

[6] From Python 2.7 we also have the option of using an ordered dictionary, which is a special collection type in the `collections` module.

For `Atom` we will assume that it has a key to identify it, within its parent `Residue`, called `name`, and an additional mandatory attribute `coords` giving the three-dimensional (X, Y and Z) coordinates of the atom. It is a design decision to make coordinates mandatory, and it would be perfectly valid to instead make them optional, so we could represent 'no 3D information'. We will assume that the `Atom` children of a `Residue` are ordered, by order of creation. We are going to add a `getAtom()` function into the `Residue` class and so we will again use a dictionary to make this efficient.

This leads to the following proposal for the implementation of `Residue` and `Atom`, and a modified implementation of `Chain`, noting that there is nothing especially tricky here and the `class` construction uses the concepts already described:

```python
class Chain:
  allowedMolTypes = ('protein', 'DNA', 'RNA')

  def __init__(self, structure, code, molType='protein'):
    # ... initial part as before

    self.resDict = {}                # Children
    self.residues = []               # Children
    structure.chains.append(self)    # Parent's link

  def delete(self):
    for residue in self.residues:
      residue.delete()
    self.structure.chains.remove(self)

  def getResidue(self, seqId):
    return self.resDict.get(seqId)

  def getAtoms(self):
    atoms = []
    for residue in self.residues:
      atoms.extend(residue.atoms)
    return atoms
```

For the `Residue` class remember that the unique key to identify it from its parent `Chain` is the `seqId`, so this is what is used in the `getChain()` look-up to make sure we don't have any repeats. When we construct a `Residue` it goes in its parent's `chain.resDict`, for quick look-up (with the `seqId`), and in the `chain.residues` list, to have the objects in order (an alternative would be a single ordered dictionary from the `collections` module; available from Python 2.7). When a `Residue` is deleted both of these operations are reversed; we remove its reference from both the list and dictionary.

```python
class Residue:
  def __init__(self, chain, seqId, code=None):
    if not seqId:
      raise Exception('seqId must be set to non-empty string')
    residue = chain.getResidue(seqId)
    if residue:
      raise Exception('seqId="%s" already used' % seqId)
```

```
    self.chain = chain
    self.seqId = seqId
    self.code = code

    self.atomDict = {}              # Children
    self.atoms = []                 # Children
    chain.resDict[seqId] = self     # Parent's link
    chain.residues.append(self)     # Parent's link

  def delete(self):
    for atom in self.atoms:
      atom.delete()
    del self.chain.resDict[self.seqId]
    self.chain.residues.remove(self)

  def getAtom(self, name):
    return self.atomDict.get(name)
```

Lastly for the Atom, it is the same approach again. In this case the key to identify an atom is its name, so this is used to check for repeats and in the Residue's dictionary to look up its children. Because this is the final class in our data model there are no children of Atom. The atom record naturally holds the important coordinate information, which defines the three-dimensional structure as coords, a NumPy array containing x, y and z axis positions.[7] We are using an array for this to make geometric manipulations easier. Note that we check the coords is a collection of three items, although we could be more rigorous and check the data type etc. Also, we have an attribute to state what chemical element the atom is, which may not be obvious from the name.

```
from numpy import array

class Atom:
  def __init__(self, residue, name, coords, element):
    if not name:
      raise Exception('name must be set to non-empty string')
    atom = residue.getAtom(name)
    if atom:
      raise Exception('name="%s" already used' % name)
    if len(coords) != 3:
      raise Exception('Coordinates must contain three values')

    self.residue = residue
    self.name = name
    self.coords = array(coords)
    self.element = element

    residue.atomDict[name] = self   # Parent's link
    residue.atoms.append(self)      # Parent's link

  def delete(self):
    del self.residue.atomDict[self.name]
    self.residue.atoms.remove(self)
```

At each level, in the constructor and delete() functions, you need to look upwards to the parent and downwards to the children.

---

[7] See Chapter 9 for discussion of NumPy.

Populating the model: reading PDB files

In Chapter 6 we made a quick and dirty attempt at reading structural coordinate information from a PDB-format file, when we wrote the function `calcCentroid()`. In that function we used the information as we read it, and did not bother storing it for later use. In contrast, here we will create one or more `Structure` objects, so we can perform any number of manipulations and interrogations on the data, after it has been loaded. There are various aspects of a PDB-format file which we will consider when loading its data into our Python objects:

- A PDB file can have multiple structure conformations (models) in it, so our function will return a list of structures.
- A PDB file is record oriented and most records are one line long. The first six characters of the record determine the record type.
- The 'HEADER' record gives a short description and a date but most importantly for us it also provides the PDB identifier.
- The 'TITLE' record gives a further description of the molecular assembly. We will use this to specify the `name` attribute of the `structure`.
- A 'MODEL' record indicates when a new structure is starting, and gives numbers for the conformation. It is optional, so in a PDB file with only one set of coordinates it can be missing. If there is a 'MODEL' record then there is a matching 'ENDMDL' record when the description of that conformation finishes.
- An 'ATOM' record specifies the chain, residue and atom information, including the coordinates, for one atom. This means that the chain and residue information is specified over and over again, once for each atom in the residue.

A fuller description of the (current) PDB file format can be found via the wwPDB website (http://www.wwpdb.org). There are many more record types, but we will only use the ones mentioned above and ignore the remainder. Also, it should be noted that, when this book was being written, a decision was taken to phase out the (now quite old) PDB file format and use the mmCIF format instead. However, we stick with the PDB format, which is still widespread, as an example for teaching Python.

Below we define the `getStructuresFromFile()` function, which takes the location of a file and gives back a list of `Structure` objects. At the top of the function the variable `structure`, representing the current working `Structure` object, is initialised to `None`. This is done so that later in the code, when we come across an 'ATOM' record for the first time we know that we need to create a new `Structure` object. The variable `structure` is again set to `None` when an 'ENDMDL' record is found, because that indicates that this is the end of the current conformation; the next 'ATOM' record, if there is another, is for the next conformation where a new `structure` would then need to be created. The `conformation` is initialised to `0` because there might not be any 'MODEL' record. However, if there is a 'MODEL' record then instead the `conformation` is taken from that. The `name` and `pdbId` are initialised to default values. In theory these do not have to be initialised because they will all be set up in due course, when the 'TITLE' and 'HEADER' records have been read, but it might be the case (for non-standard PDB files) that those records are missing. After the initialisation the remainder of the function involves looping through the lines from the file to interrogate the various records, making the required Python objects as we go along:

```python
def getStructuresFromFile(fileName):

  fileObject = open(fileName)
  structure = None
  name = 'unknown'
  conformation = 0
  pdbId = None
  structures = []

  for line in fileObject:

    record = line[0:6].strip()
    if record == 'HEADER':
      pdbId = line.split()[-1]

    elif record == 'TITLE':
      name = line[10:].strip()

    elif record == 'MODEL':
      conformation = int(line[10:14])

    elif record == 'ENDMDL':
      structure = None

    elif record == 'ATOM':
      serial    = int(line[6:11])     # not used here
      atomName  = line[12:16].strip()
      resName   = line[17:20].strip()
      chainCode = line[21:22].strip()
      seqId     = int(line[22:26])
      x         = float(line[30:38])
      y         = float(line[38:46])
      z         = float(line[46:54])
      segment   = line[72:76].strip()
      element   = line[76:78].strip()

      if chainCode == '':
        if segment:
          chainCode = segment
        else:
          chainCode = 'A'

      if not structure:
        structure = Structure(name, conformation, pdbId)
        structures.append(structure)

      chain = structure.getChain(chainCode)
      if not chain:
        chain = Chain(structure, chainCode)

      residue = chain.getResidue(seqId)
      if not residue:
        residue = Residue(chain, seqId, resName)

      if not element:
        element = name[0] # Have to guess
```

```
    coords = (x,y,z)
    atom = Atom(residue, atomName, coords, element)

  fileObject.close()

  return structures
```

Values for the various attributes are extracted from the lines of the file, according to the kind of record present. When dealing with numeric data like the `seqId` (an integer) or the x, y and z coordinates (floating point numbers) the data extracted from the line will initially be just a string of characters, i.e. not a real Python number object. Hence, for these values we need to explicitly convert the text with `int()` or `float()` as appropriate. The textual values receive a little processing to remove spaces from their ends, using the `strip()` method of strings.

Note that as well as the `structure`, the `chain` and `residue` are also created as and when they are needed; each time we come across an 'ATOM' line we definitely need to make a new `atom`, but a new `chain` or `residue` is only needed when we come across its first atom. Here we detect when a new `chain` is needed when the `chainCode` changes and a new `residue` when the `seqId` number changes. Naturally, when we start a new `chain` we will also make a new `residue`, given that it doesn't yet have any children to put atoms into. For the `chainCode` we do a little checking to make sure it wasn't blank[8] and if it is empty we try to substitute with the `segment` letter, and if this fails a plain 'A'. Also, if the chemical element is missing we take the first character of the atom name as a guess. These are just two examples of the kind of checking that should be done if you are accepting coordinate data from a variety of sources and are being rigorous. Here we consider only using official files from the PDB, so we are confident that they are well formed and conform to all the standards.

In the above example it may seem odd that we define the `atom` variable, since in fact it is not used for anything, and we could just have written

```
    Atom(residue, atomName, coords, element)
```

Either way, it looks like we are immediately throwing away the object we have just created, leaving it to be garbage collected. However, the parent `residue` has a handle to all its `atoms`, via both a dictionary and a list, so this does not happen. In turn `chain` has a handle to all its `residues`, and `structure` has a handle to all its `chains`. So it all works out nicely and all we need to pass back at the end of the function is a list of structures, which contain everything else. The code is then easily tested on an appropriate PDB-format file:

```
testStructs = getStructuresFromFile('examples/glycophorin.pdb')

structure = testStructs[0]
chain = structure.getChain('A')
for residue in chain.residues:
  print(residue.seqId, residue.code)
```

There is a notable deficiency in the code; the `molType` of the `chain` is not specified explicitly, so it is set to the default value, `'protein'`. This is not very satisfactory if the `chain` is DNA or

---

[8] Technically a single space, in some older files.

RNA. A cleverer approach would be to try and determine, perhaps from the `residue.code` and/or `atom` names, what the `molType` is. This raises the issue that it could take a few 'ATOM' records before the `molType` might be determined accurately, by which time the `chain` has long since been created, perhaps with the incorrect `molType`. That would be a problem if the `molType` was not supposed to change, i.e. was frozen. If we do allow it to change then we could set the `molType` accurately by looking at what kinds of atom are present. A function to do this, which looks at a few characteristic atoms in a `residue`, might be as follows:

```
def guessResidueMolType(residue):

  if residue.getAtom("CA") and residue.getAtom("N"):
    return 'protein'

  elif residue.getAtom("C5'") and residue.getAtom("C3'"): # DNA/RNA

    if residue.getAtom("O2'"):
      return 'RNA'
    else:             # This is "2'-deoxy"
      return 'DNA'

  return 'other'
```

If we insisted that the molecule type could not be changed, an alternative method would be to process the entire PDB file, deduce the `molType` and then afterwards create the `structure` and associated `chains` etc. We would still need to store the information as we process the PDB file, and that means some kind of intermediate data structure, though this could be a simple one involving dictionaries and lists.

Refined implementation

Getter and setter functions

Given the way we have implemented the classes, we can directly access and manipulate the class attributes. For example, if we have a structure object we can get the attribute values:

```
name = structure.name
conf = structure.conformation
```

We can also change these values:

```
structure.name = 'new name'
structure.conformation = 235
```

Although the former seems harmless enough, the latter might be a bad idea. It's possible the application might rely on these two attributes not changing. Or even if the information is allowed to change, it's possible that the values being set might be illegal in some way. For example, in the constructor we checked that `name` was not an empty string and that is not done here, so it is perfectly possible that we could do:

```
structure.name = ''
```

The standard way to deal with situations like this in most modern computer languages (e.g. Java) is to design the class so that attributes are private and thus not accessible in this way. Instead we would have *getters* and possibly *setters*, which are functions that allow access to query and redefine the relevant information, but in a guarded way. For example, here we could have:

```
class Structure:
  # ...
  def getName(self):
    return self.name

  def setName(self, name):
    if not name:
      raise Exception('name must be set to non-empty string')
    self.name = name
```

We see that the setter function `setName()` has some validity checking in it. Access then becomes:

```
name = structure.getName()
structure.setName('new name')
```

If for some reason we thought that the `name` should not be changeable, we would simply not provide the setter function, so only the getter function would exist. This approach still has a couple of problems. First of all, this access is rather verbose to use. For example, instead of:

```
residue.chain.structure.name = 'new name'
```

we would have to do:

```
residue.getChain().getStructure().setName('new name')
```

Secondly, as it stands, the attribute name is still accessible, so someone could still use direct access, albeit deliberately or by mistake. In Python, nothing can be totally hidden from the user in the implementation of a class. However, there are ways to signal the clear intent to disallow access. Instead of using `name` for the attribute we could use `_name`, so starting with an underscore. Then the getter and setter become:

```
class Structure:
  # ...
  def getName(self):
    return self._name

  def setName(self, name):
    if not name:
      raise Exception('name must be set to non-empty string')
    self._name = name
```

With this change `structure.name` no longer works, but `structure._name` does, although the underscore in front is a warning that this is not intended to be used. Another alternative is to use `__name`, starting with two underscores. In this case `structure.__name` does not work, and trying to use it results in an `AttributeError` exception. This is a bit of Python 'magic' to make the attribute somewhat private. However, it turns out that a determined person could still get at the attribute, just using another bit of Python magic: an underscore and the class name have to be joined to the beginning of the attribute name. Thus, here the access would be via `structure._Structure__name`. There is no real privacy in Python.

Setter functions often have validity checking in them, so they seem to serve a useful purpose. On the other hand getter functions are often very boring, and just give back an attribute, so they often seem to serve little purpose. Nevertheless, there are a few clear situations when they actually do something useful. The first case is when the high cardinality of the attribute is greater than 1; we have a collection of items. Returning to an example of this from earlier in the chapter, we could have decided that the `Structure` class has `pdbIds`, instead of just one `pdbId`. The naïve getter function would simply be:

```
class Structure:
  # ...
  def getPdbIds(self):
    return self.pdbIds
```

Unlike strings and integers, collections are modifiable; their contents can be changed. Thus, returning the value in this way would allow the person using the function to directly manipulate the collection, which may cause problems for the objects:

```
pdbIds = structure.getPdbIds()
del pdbIds[0]  # delete first one; changes structure.pdbIds
```

An alternative implementation could return a copy of the collection instead of the internal attribute:

```
class Structure:
  # ...
  def getPdbIds(self):
    return list(self.pdbIds)
```

Here the list that is returned from the function can be manipulated without any harm:

```
pdbIds = structure.getPdbIds()
del pdbIds[0]  # does not change structure.pdbIds
```

Another case when a getter function is useful is when the associated value is not directly stored in the object, but instead is calculated on-the-fly. For example, suppose we wanted to have a function, `structure.getMass()`, which returns the molecular mass of the structure. We might choose not to store the mass at all, but instead do something like:

```
class Structure:
  # ...
  def getMass(self):
    mass = 0
    for chain in self.chains:
      mass += chain.getMass()
    return mass
```

and in turn the `chain.getMass()` could use `residue.getMass()` and that could use `atom.getMass()`. Even the `Atom` class might not store the mass but instead calculate it on-the-fly using a dictionary based on the `element`.

Our final example of useful getter functions illustrates that the underlying implementation of a class can change, but any code that uses a getter function instead of direct attribute access does not itself have to change. Here, we might decide that calculating the mass on-the-fly is a slow procedure, so we want to calculate it once and then cache the result:[9]

```
class Structure:
  # ...
  def getMass(self):
    if hasattr(self, 'mass'):
      # already been calculated so just return value
      return self.mass

    # not been calculated yet so do it
    mass = 0
    for chain in self.chains:
      mass += chain.getMass()
    self.mass = mass # cache for next time
    return mass
```

This has changed how the class is implemented, but all code that uses the `getMass()` function is unaffected.

Properties

Python has a really handy mechanism for allowing the syntax of direct attribute access but with the protection of using getter and setter functions. This is called a Python *property* and is made available via the inbuilt `property()` function. In Python 2 it is only (properly) available for 'new-style' classes; those that inherit from `object`. In Python 3, all classes are 'new style' so they all support this mechanism. It is best to illustrate with an example, so consider the code:

```
class Structure(object):
  # ...
  def getName(self):
    return self._name

  def setName(self, name):
    if not name:
      raise Exception('name must be set to non-empty string')
    self._name = name

  name = property(getName, setName)
```

What this means is that use of the `name` attribute automatically calls `getName()` or `setName()`, rather than accessing a simple attribute. For example:

```
name = structure.name
# equivalent to: name = structure.getName()
```

[9] Strictly, such a cache should be reset, as part of the model implementation, when any atom is added or removed.

```
structure.name = name
# equivalent to: structure.setName(name)
```

Note that the information is stored internally using the variable `self._name`. Here the variable could not instead be `self.name` because that would result in `getName()` or `setName()` being called recursively, and so end up with an 'infinite' loop. In the property definition the setter function is optional, and this provides a good way to implement frozen attributes. As an example, if we suppose `chain.molType` was frozen we could have:

```
class Chain(object):
  # ...
  def getMolType(self):
    return self._molType

  molType = property(getMolType)
```

In this case, trying to assign the attribute's value would give rise to an `AttributeError` exception:

```
chain.molType = 'DNA'  # --> AttributeError exception
```

The `property()` function has a third optional argument to specify a delete function, which is called when the attribute is deleted using 'del', and a fourth optional argument to specify documentation for the property.

From Python 2.6, there is an alternative syntax for specifying a property, using decorators (see Chapter 5). Again these only apply to 'new-style' classes.

```
class Structure(object):
  # ...
  @property
  def name(self):
    return self._name

  @name.setter
  def name(self, value):
    if not value:
      raise Exception('value must be set to non-empty string')
    self._name = value
```

This is not any shorter than the non-decorator form, and it also looks odd that both the get function and the set function are called `name`.

9 Mathematics

Contents

Using Python for mathematics

Given that Python is an interpreted programming language, rather than a fast compiled language, many people do not consider it for writing programs that involve extensive numerical work. While Python programs are certainly slower to execute than the equivalent written in something like C or FORTRAN, mathematical functionality certainly exists in Python and has the inherent advantages of the language; it is easy for people to use and conveniently links to other helpful data structures. Of course speed of calculation may not be so important, for a scientific investigation it may not matter if something takes 1 second or 0.1 second to run. Fortunately, computers get faster and the Python interpreter becomes improved, so you can do quite a bit of numerical work without concern. However, if calculation speed really is important in a given situation then there are a few things you can do to make things faster while still keeping the convenience of Python. For example, you can write code in C, a very efficient numerical language, and use it from within Python (this is called a *C extension*), effectively extending the vocabulary of the interpreted language with speedy subroutines. More recently the language Cython has helped make C extensions very easy to write. Cython is a Python-like language, and virtually all Python programs can be interpreted by it, without alteration, but the language ultimately generates C code that can be compiled. Cython can be used to call fast library code written in pure C, and can incorporate a mixture of Python and C data structures in the same code; although less flexible, the C data structures are very efficient. Writing C extensions and Cython modules is discussed in Chapter 27.

Python includes standard arithmetic operations as part of the core functionality: add, multiply etc. There is an additional module, `math`, which always comes packaged with Python and which provides further numerical functionality: logarithms, trigonometry etc. For numerical calculations that are not especially intensive, the core functionality and the `math` module will often suffice. There has been a history of trying to provide modules for quick numerical algorithms in Python. The first attempt, begun in 1995, was called Numeric, and the second attempt was called Numarray. These two are now deemed to be obsolete, but the third attempt, begun in 2005, is called NumPy (http://numpy.scipy.org/), incorporates elements from the earlier attempts and will hopefully last longer.

NumPy provides support for basic numerical operations, with an emphasis on specifying calculations that operate on a whole array of numbers at once. As will be discussed below, its operations include functionality for random numbers, linear algebra and Fourier transforms. It is implemented in C underneath,[1] and thus is quick to run, but can naturally be accessed in Python. NumPy is relatively easy to use because you are still working with Python commands, but the way that some things work, especially how to think about numeric array operations, can take some learning. For serious linear algebra work in Python, NumPy is the method of choice. There is another closely related package, called SciPy[2] (Scientific Python), which adds some higher-level numerical capabilities, such as integration, optimisation and signal processing. NumPy and SciPy are not part of the standard Python software release, so require a separate download and installation, although modern download managers ought to make this fairly easy to do. See http://www.cambridge.org/pythonforbiology for details of where to download SciPy and NumPy. In some sense these packages could be deemed to be the Pythonic answer to the analogous capabilities in, for example, the MATLAB system.[3]

The Python '`math`' module

The `math` module is most useful for trigonometric, exponential and logarithmic functionality. All of the examples in this section assume that the `math` module functionality is imported, usually in its entirety with the following statement:

```
import math
```

So that, for example, you could to the following, and access the logarithm, square root and exponential functions using the 'dot' notation:

```
a = math.log(x)        # logarithm, base e, of x
b = math.sqrt(x)       # square root of x
c = math.exp(x)        # e to the power of x
```

Alternatively, the functions of the `math` module may be imported individually:

```
from math import log, sqrt, exp
```

---

[1] NumPy relies on fast calculation routines from a library called LAPACK (http://www.netlib.org/lapack/).
[2] http://www.scipy.org/.
[3] The core linear algebra part of LAPACK is called BLAS. For some operations it is important to have a very efficient implementation of BLAS, such as is provided by ATLAS (http://www.math-atlas.sourceforge.net).

This method avoids the 'dot' notation, and consequently is faster to execute, though you now have to remember where the function was imported from:

```
a = log(x)
b = sqrt(x)
c = exp(x)
```

See Appendix 3 for a full list of the `math` module functionality.

As a slightly more complex example, the following takes a collection of angles in degrees and makes a list containing their sines. This involves two `math` operations: `math.sin()` and `math.radians()`. Unsurprisingly, the former calculates the sine of an angle that is passed in, but the latter is also required because the sine calculation requires that the angles are in units of radians (i.e. 2π for a full circle), not degrees. Hence, if we pass angles in degrees, `math.radians()` is used to convert the angle values to radians.

```
angles = [0, 30, 45, 60, 90, −90]
sines = [math.sin(math.radians(angle)) for angle in angles]

# [0.000, 0.500, 0.707, 0.866, 1.000, −1.000] (rounded to 3 places)
```

Note that in the above code we use a list comprehension to build a list (`sines`) from the inside, according to another list (`angles`); this is more compact and quicker than using a conventional `for` loop. Going a little further, if the inverse operation (valid in the range −90 to 90 degrees) were needed then the following implementation would do that, using `math.degrees()` to convert from radians and `math.asin()` for the inverse sine operation (arcsine):

```
angles = [math.degrees(math.asin(sine)) for sine in sines]
```

Example: mean angle

Below we give a more involved example that uses the `math` module inside a function definition. This function will be used later on in this book, where it is used to calculate the average angle value from a list of input angles. Later, in context, it will be used to look at the angles of chemical bonds that come from alternative molecular shapes (conformations) to give the average bond angle. Because angles are a circular measure (i.e. 360° equals 0°) we cannot take the simple mean value of the numbers. Instead we find the average sine and cosine of the angles, which oscillate between plus and minus one and so don't have a problem when angles go past a full turn or become negative. Both the sine and cosine are required because each function on its own is not unique to a single angle, e.g. the sine of 45° equals the sine of 135°. Accordingly, we use the averages of both sine and cosine to generate the average angle via a special inverse tangent function, and naturally the mathematical operations come from the `math` module:

```
from math import sin, cos, degrees, radians, atan2

def meanAngle(angles, inDegrees=True):
  sumCos = 0.0
  sumSin = 0.0

  for angle in angles:
    if inDegrees:
      angle = radians(angle)
```

```
    sumCos += cos(angle)
    sumSin += sin(angle)

  N = len(angles)
  meanAngle = atan2(sumSin/N, sumCos/N)

  if inDegrees:
    meanAngle = degrees(meanAngle)

  return meanAngle
```

The inverse tangent function used is specifically `atan2`; while a normal inverse tangent (`atan`) operates on a single number and gives an angle between $-90°$ and $+90°$, this function uses both sine and cosine values to calculate the inverse tangent and also which quadrant of the circle it lies in. Thus the result is an angle between $-180°$ and $+180°$. Finally, note how all calculations are done in radians, so we must explicitly convert the input and output when working with values specified in units of degrees, signalled by the input argument `inDegrees` being `True`.

Rounding

The math module also has a couple of handy functions that convert floating point numbers to integral floating point numbers, i.e. round up or down to the nearest whole number, but give a result that is still floating point. The function `floor(x)` converts x to the largest whole number (integral) value less than or equal to x, and similarly `ceil(x)` converts x to the smallest integral value greater than or equal to x:

```
math.floor(5.25)        # 5.0
math.ceil(5.25)         # 6.0
```

Even though these functions remove any fractional part, after the decimal point, in both cases the data type returned is floating point. Hence these could be converted to actual Python integers using the `int()` function:

```
int(math.floor(5.25))   # 5
int(math.ceil(5.25))    # 6
```

Note that `int(x)` is the integer part of x and so `int(math.floor(x))` is not the same as `int(x)` when x is negative:

```
int(math.floor(-5.25))  # -6; the integer less than the value
int(-5.25)              # -5; the integer part of the value
```

The more usual kind of rounding, to the nearest whole number or decimal place, would be done with the inbuilt `round()` function; this is not in the `math` module.

```
round(8.49)             # 8.0     ; rounded down
round(8.51)             # 9.0     ; rounded up
round(3.141592, 1)      # 3.1     ; to one decimal place
round(3.141592, 3)      # 3.142   ; to three decimal places
round(9621, -2)         # 9600.0 ; to nearest hundred
```

As illustrated above, the second argument in round() specifies how many decimal places the rounding is done to, and this number can be negative to round off at positions above 1.0; to the nearest ten, hundred, thousand etc. Note that even rounded numbers will still be subject to floating point errors.

In Python 3 the behaviour of round() changed slightly. Consider the case when the second argument takes its default value, 0. In Python 2, the function rounds to the number with the larger magnitude if the fractional part of the argument is 0.5, but in Python 3 this only happens if that number is even, otherwise it rounds to the number with the smaller magnitude.

```
round(7.5)          # 8.0 in both Python 2 and Python 3
round(8.5)          # 9.0 in Python 2, 8.0 in Python 3
```

Thus in Python 3 you always get an even number in this circumstance.

Plotting

To go along with mathematical operations it is naturally often handy to plot the numerical values as a graph or chart. To illustrate this we will use the popular Matplotlib module which is often included with SciPy and NumPy packages. Naturally, the following examples assume that we have installed the Matplotlib, otherwise you will get an error from the import command. From the matplotlib module we will import pyplot, which has lots of helpful functions that can be used to display information as charts and graphs.

```
from matplotlib import pyplot
```

To use pyplot we first create some example values in a list and then call the .plot() function, passing the data in as an argument. This will create a line graph from the values list, but doesn't immediately display anything on screen.

```
values = [x*x for x in range(10)]
pyplot.plot(values)
```

To actually display the line graph on screen we call the .show() function, which should cause a pop-up window to appear.

```
pyplot.show()
```

In order to create a graph with multiple data lines we can repeatedly use plot() with different data lists, which will all be added to the same graph, before finally invoking show().

```
valuesA = [x*x for x in range(1,10)]
valuesB = [100.0/x for x in range(1,10)]

pyplot.plot(valuesA)
pyplot.plot(valuesB)
pyplot.show()
```

Note that once we call show() the current graph data is completely cleared, i.e. we will get a blank plot if we repeat show() immediately after. So far the graphs have all been plotted by supplying height (y axis) information for the values, which are plotted in order. However, by passing two data lists to each plot we can also specify the values for the x axis:

```
xVals = range(21,30)
yVals = [100.0/x for x in range(1,10)]

pyplot.plot(xVals, yVals)
pyplot.show()
```

As standard the `plot()` function will use relatively sensible defaults to determine the look of the graphs. However, there are many extra options that can be specified to control the drawing of the lines, axes, legends etc. Here we create a thicker purple line by setting the `color` and `linewidth` attributes. Also, we give the line plot a textual label which will appear if `legend()` is used. For the axes we control the range of displayed values for the y axis with `ylim()`, here making it wider than the default, which would only go up to the extremes of the data range, and manually specify the values for the tick marks on the axis using `yticks()` (naturally `xlim()` and `xticks()` are also available).

```
pyplot.plot(xVals, yVals, color='purple',
            linewidth=3.0, label='DataName')
pyplot.legend()
pyplot.ylim(0, 101)
pyplot.yticks([0, 25, 50, 75, 100])
```

As an alternative to simply showing the graph on screen we could also write an image out to a file. Accordingly we use `savefig()`, though note that we do this before we call `show()`, otherwise the latter would clear the current graph of data. Here we export to PNG format by using a file name with a '`.png`' extension and state that we require 72 dots per inch output resolution:

```
pyplot.savefig("TestGraph.png", dpi=72)
pyplot.show()
```

As well as simple line graphs Matplotlib can easily create many other types of data display. For example, here we create a scatter plot by using `scatter()` (rather than `plot()`), where we set the style for the marker symbol and its size (here `s=40`). Note that, for illustrative purposes, the `valsB` list is generated by using `random.gauss()` to take random samples from a normal distribution with a mean of `0.0` and standard deviation `1.0`.

```
from random import gauss

valsA = range(100)
valsB = [gauss(0.0, 1.0) for x in valsA]
pyplot.scatter(valsA, valsB, s=40, marker='*')
pyplot.show()
```

We could also show the data as a bar chart, where `valsB` will represent the heights of the bars:

```
pyplot.bar(valsA, valsB, color='green')
pyplot.show()
```

Although we could use `bar()` to show histogram data, i.e. where initial values have been grouped into different regional ranges (bins), we could instead let the `hist()` function do the work of binning values (counting the number of data points in each range). Here, we create a

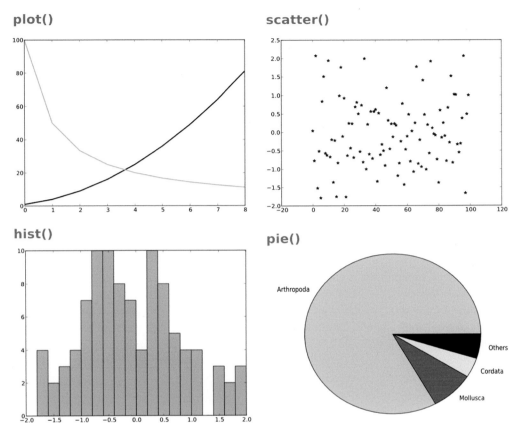

Figure 9.1. Example graphs and charts generated using Matplotlib. The examples relate to the named line graph, scatter plot, histogram and pie chart functions that are available in the 'pyplot' module, e.g. via `pyplot.hist(data)`.

histogram of the random (normally distributed) sample of points by stating we want the data grouped into 20 bin regions between the limits of −2.0 and 2.0:

```
pyplot.hist(valsB, bins=20, range=(-2.0, 2.0))
pyplot.show()
```

As a final example, the `pie()` function is, as you might guess, handy for making pie charts and naturally this has options to specify the colours and textual labels for each of the segments:

```
sizes  = [83, 8, 4, 5]
labels = ['Arthropoda', 'Mollusca', 'Cordata', 'Others']
colors = ['#B00000', '#D0D000', '#008000', '#4040FF']
pyplot.pie(sizes, labels=labels, colors=colors)
pyplot.show()
```

There's a lot more functionality and graph types in `pyplot` (contour plots, colour matrices, box-and-whisker plots etc.) that we don't have room to cover here. However, Matplotlib is well documented at the Matplotlib website, which we link at http://www.cambridge.org/pythonforbiology.

Linear algebra

In this section we give a short introduction to *linear algebra* for non-mathematicians. A familiarity with some of the concepts involved will be assumed when discussing various parts of subsequent chapters. In particular for this book, linear algebra is helpful in the understanding and manipulation of three-dimensional coordinates. Here, rather than discussing how separate positions on each spatial axis (x, y and z) are used, we can describe a 3D point in its entirety as a single *vector* that groups the axis positions. Also, changes of position (*transformations*) may be described with *matrices*, which specify how vector locations are relocated. An example of an easily visualised transformation is rotation; a shape, defined by a collection of coordinate positions, is moved to a new set of coordinates to change the orientation of the shape.

The way that data is represented in linear algebra is in terms of vectors, which represent positions in space. Although the 'space' we mention in this book is usually the three-dimensional in-out, up-down and left-right kind we all recognise it can also be used for more abstract spaces where the 'axes' are merely independent qualities. An example of this might be *colour space* where you can define a colour (or rather a colour vector) by its red, green and blue components. In the language of Python a vector is represented by an ordered collection of numbers, an *array*, where each number specifies the location of the point along each axis (dimension). You can think of these just as a list of floating point numbers; this will have a known length and will not contain any other kind of Python object. To take a biological example, an atom in a molecule has a location (relative to the other atoms), which consists of three coordinates (x, y and z) and it is normal to place these values in an array; the three coordinates correspond to what is normally described as a point in space. If you were studying the dynamics of a molecule then in addition there would be a fourth coordinate, time, and you would be dealing with space-time, although normally you would only consider one time value after the other, rather than all in one go.

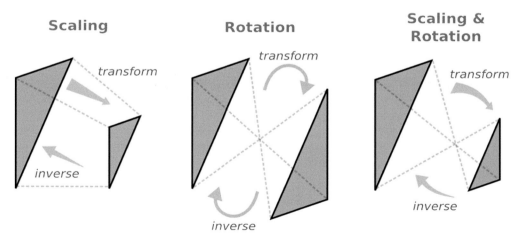

Figure 9.2. Matrix transformations of spatial coordinates, their inverses and combinations. With reference to a triangle of points, various simple linear transformations of coordinates are illustrated. Each transformation may be specified as a matrix and applied to coordinates, specified as vectors, using matrix multiplication. For linear transformations the application of two subsequent transformations can be represented by a single (combined) transformation. All of these simple transformations have inverse transformations which will restore the coordinates of the points to their original values.

The 'linear' part of linear algebra refers to the fact that the fundamental characteristic of a coordinate transformation does not change with scale (its magnitude). For example, if you break down a transformation into two smaller parts, then it does not matter if you do the full transformation or instead do the two smaller parts one after the other. If you rotate a shape by 30° around some axis and then again by a further 45° around the same axis, you get the same result as if you had instead just rotated the shape in one step by 75° (30+45).

The 'algebra' part of linear algebra refers to the fact that different types of transformation may be melded together to define a single transformation. For example, if you rotate a shape about one axis by some angle, and then around another axis by a different angle, then there is a single equivalent rotation that could do the whole job, potentially about a third axis. The converse of combining transformations is also true because a single transformation can be described by combinations of other transformations. This point leads to the idea that transformations may be reversed; a forward and corresponding reverse transformation results in no change. For example, a rotation about a given axis of a given angle has an opposite rotation: around the same axis but in the opposite direction, so that if you combine both rotations it is equivalent to not doing any rotation at all.

If you imagine a vector to represent a point in space, it is convenient to think of what are called *unit vectors*, vectors of length 1, as merely representing directions. The way linear transformations are normally described is in terms of what happens to the unit vectors that point directly along the coordinate axes (in a positive sense). For example, for 3D space we would consider the transformation of the three vectors (1,0,0), (0,1,0) and (0,0,1), which respectively represent the unit vectors along the positive x, y and z axes. Because the transformation is linear, once you know how it acts on these unit vectors, you know how it acts on any vector.

Suppose for a given transformation operation (1,0,0) gets mapped (transformed) to (a,b,c), and (0,0,1) gets mapped to (d,e,f), and (0,0,1) gets mapped to (g,h,i). Then an arbitrary input vector (x,y,z) gets mapped to the final vector (x',y',z') via:

$$(x', y', z') = x(a, b, c) + y(d, e, f) + z(g, h, i)$$
$$= (xa + yd + zg, xb + ye + zh, xc + yf + zi)$$
$$= (ax + dy + gz, bx + ey + hz, cx + fy + iz)$$

This looks pretty horrible, but fortunately computers usually do the calculations. A way that you might think about what is happening above is that the x component of the input vector is multiplied by elements a, b and c (which define the transformation), but each product contributes to a different axis of the output vector; xa is added to the new x', xb is added to the new y', and xc the new z'. Similarly, the y and z components have their own multiplicative elements to define contributions to the new vector. Thus, overall the elements of the transformation (a to i) say how to combine (multiply and add) the input coordinates to make the output.

Matrix transformations

Another way to write transformations mathematically is in terms of a *matrix*, which is a rectangular encoding of the transformation. You can think of this as merging the mapping vectors for each axis into a single block. When we do this we normally write the vectors in column, rather than row, fashion. For example, for the transformation above, the two vectors would have 3 elements and the matrix would have 3×3 elements:

$$\begin{pmatrix} x' \\ y' \\ z' \end{pmatrix} = \begin{pmatrix} a & d & g \\ b & e & h \\ c & f & i \end{pmatrix} \begin{pmatrix} x \\ y \\ z \end{pmatrix}$$

The way to read this is that the first element (x') of the output vector on the left is found by adding up the element-by-element multiplication of the first row of the matrix with the input vector on the right; $ax + dy + gz$, and so on for the other two elements y', z'.

There is a special transformation, the *identity matrix*, which transforms vectors to themselves (i.e. causes no change at all):

$$\begin{pmatrix} x' \\ y' \\ z' \end{pmatrix} = \begin{pmatrix} 1 & 0 & 0 \\ 0 & 1 & 0 \\ 0 & 0 & 1 \end{pmatrix} \begin{pmatrix} x \\ y \\ z \end{pmatrix} = \begin{pmatrix} x \\ y \\ z \end{pmatrix}$$

The advantage of this matrix notation is that the *composition* of two transformations (applying one after the other) becomes easier to manage. For example, if we have a second transformation:

$$\begin{pmatrix} x'' \\ y'' \\ z'' \end{pmatrix} = \begin{pmatrix} a' & d' & g' \\ b' & e' & h' \\ c' & f' & i' \end{pmatrix} \begin{pmatrix} x' \\ y' \\ z' \end{pmatrix}$$

then the composition of both transformations is:

$$\begin{pmatrix} x'' \\ y'' \\ z'' \end{pmatrix} = \begin{pmatrix} a' & d' & g' \\ b' & e' & h' \\ c' & f' & i' \end{pmatrix} \begin{pmatrix} a & d & g \\ b & e & h \\ c & f & i \end{pmatrix} \begin{pmatrix} x \\ y \\ z \end{pmatrix}$$

$$= \begin{pmatrix} a'a + d'b + g'c & a'd + d'e + g'f & a'g + d'h + g'i \\ b'a + e'b + h'c & b'd + e'e + h'f & b'g + e'h + h'i \\ c'a + f'b + i'c & c'd + f'e + i'f & c'g + f'h + i'i \end{pmatrix} \begin{pmatrix} x \\ y \\ z \end{pmatrix}$$

The final matrix is said to be the *product* of the two initial matrices. The element at an arbitrary cell location (m,n), the cell that lies in both row m and in column n, in the final matrix is the sum of the element-wise product of row m in the first matrix and column n in the second matrix.[4] Note that when applying two transformations it is important to apply them in the correct order; usually you cannot swap matrices. For example, a rotation of 90° about the x axis then a rotation of 90° along the y axis is not the same as doing the rotations in the opposite order.

Mathematicians have introduced a succinct notation for vectors and matrices, by hiding all of their internal elements, which makes the product formula look clearer. So instead of writing (x,y,z) we write this with a single letter, say w:

---

[4] It follows that you can find the product of matrices that have different overall sizes, but only if the size of the rows in the first is the same as the size of the columns of the second.

$$w = \begin{pmatrix} x \\ y \\ z \end{pmatrix}$$

and similarly for w' and w''. In Python this could be implemented as a list (or tuple) and we would have x = w[0], y = w[1] and z = w[2]. The matrices would also be written with single letters, for example, T:

$$T = \begin{pmatrix} a & d & g \\ b & e & h \\ c & f & i \end{pmatrix}$$

The transformation from $w=(x,y,z)$ to $w'=(x',y',z')$ is then written simply, with implied matrix multiplication, as:

$$w' = T\,w$$

and the composition of two transformations, T and T', is then:

$$w'' = T'\,T\,w$$

which can also be described as the outcome of the first transformation (w') being applied to the second:

$$w'' = T'\,w'$$

or in terms of a new transformation (T'') that combines both:

$$w'' = T''\,w$$

Here T'' is defined as the product of the matrices[5] T' and T. This means that the value of T'' at cell location (m,n) is the sum of the element-wise product of row m of T' and column n of T. In Python, these matrices could be implemented as lists of lists.

The identity matrix is commonly denoted by the letter I:

$$I = \begin{pmatrix} 1 & 0 & 0 \\ 0 & 1 & 0 \\ 0 & 0 & 1 \end{pmatrix}$$

If we have a transformation T, then the inverse transformation T^{-1} is the one such that the product matrix is the identity matrix:

$$T\,T^{-1} = I$$

An inverse does not necessarily exist for a transformation matrix (e.g. projecting a volume on to a line), but if it does then it is unique and it also acts as an inverse from the other side:

$$T^{-1}\,T = I$$

---

[5] Mathematically we would have: $T'[m][n] = \sum_{k=1}^{3} T'[m][k] \times T[k][n]$.

The example we have used here has been in terms of vectors of size 3 and 3×3 matrices. However, the whole thing generalises and indeed the matrices do not even have to be square, although if they are not square then they do not have a (proper) inverse. So if we wanted to project a 3D shape, say, from its (x,y,z) coordinates to a representation in terms of the coordinates (x,y) on a computer screen, then we could use a 2×3 matrix:

$$\begin{pmatrix} x' \\ y' \end{pmatrix} = \begin{pmatrix} a & b & c \\ d & e & f \end{pmatrix} \begin{pmatrix} x \\ y \\ z \end{pmatrix}$$

Multi-dimensional arrays

The term *dimension* is often used in popular parlance to describe the number of independent axes that can be used to span a given space. So, for example, the coordinate (x,y,z) of an atom location is often said to be a point in three-dimensional space, and if you add a time coordinate, t, then you have a point, (x,y,z,t), in four-dimensional space-time. Theoretical physicists, or at least some of them, are happy to work even in 10 and 24 dimensions.

In Python, especially when using the NumPy module, you will often come across the word 'dimension' used in a different way, as an indication of the *rank*; how many axes are needed to describe an array of numbers, rather than the size of the space they operate on. The dimension of an array can be defined as the number of indices that are required to specify an element of the array. For example, with a vector, w, we specify an element using one index, for example, w[m]. With a matrix we use two indices, for example, T[m][n]. Thus vectors have dimension one and matrices dimension two, even if both work in real 3D space, and it is possible to have arrays of higher dimension. The vector (x, y, z) is said to have size 3. When using Python's numpy module we can define a matrix as a single array object, rather than a list of lists. In this case we would use combined indices like T[m,n] to access elements and the size of the array is referred to, perhaps confusingly, as its *shape*; so a 2×3 matrix is said to have shape (2, 3), and total size $2\times3 = 6$. Note that an array in NumPy can be of any size, and thus may describe vectors and matrices.

We will mostly use the NumPy terminology in this chapter when we consider multi-dimensional arrays. However, as we illustrate below, we can implement the functionality of multi-dimensional arrays in Python ourselves, using lists of lists etc., if we cannot or do not want to import the numpy module.

Python multi-dimensional lists

This section illustrates how multi-dimension array functionality can be used in Python without working with NumPy. This would only be recommended for simple linear algebra tasks where efficiency is not important. Python defines multi-dimensional arrays using collections of collections of collections etc. Here the collection may be a Python list or, if the associated data does not need modifying, a tuple. In particular, matrices can be implemented using collections of collections. For example, we could define a 2×3 matrix using a list of lists:

```
x = [[1,2,3],[4,5,6]]
```

The elements are accessed in the normal Python way:

```
y = x[1][2]            # Value in row 1 column 2 (y equals 6)
x[0][1] = 7            # Value in row 0 column 1 set to 7
                       # x becomes [[1,7,3],[4,5,6]]
```

The rows of a matrix can be accessed easily:

```
row = x[0]             # Gives [1,2,3]
```

and the columns slightly less easily:

```
col = [y[2] for y in x]   # Gives [3,6]
```

Note that `len(x)` gives the number of rows and `len(x[0])` (or any element) gives the number of columns of the matrix, assuming the matrix is not empty:

```
len(x)                 # 2
len(x[0])              # 3
```

Strictly speaking, for any function that manipulates a matrix, it should be checked that we actually have a matrix, and in particular that it has a consistent number of entries in each row. In the interest of brevity, here we leave out most checks. Without such caution, if someone tried to use nonsense data it would raise an exception,[6] but the error message might not make it clear what the problem is.

Standard Python does not provide any inbuilt functionality for matrix (or higher multi-dimensional array) manipulation, so it all has to be implemented by the programmer. As an example we illustrate a function that creates the *transpose* of a matrix. The transpose of a matrix is a new matrix with the first row of the original matrix becoming the first column in the new matrix, the second row becoming the second column etc. It is relatively simple to implement, here using list comprehension:

```
def transposeMatrix(x):
  nrows = len(x)
  ncols = len(x[0])

  return [[x[n][m] for n in range(nrows)] for m in range(ncols)]
```

This just goes through all the elements in the transposed order.

```
x = [[1,7,3], [4,5,6]]
y = transposeMatrix(x)      # y = [[1,4],[7,5],[3,6]]
```

Matrix multiplication is a bit harder:

```
def multiplyMatrices(x, y):
  rowsX = len(x)
  colsX = len(x[0])
  rowsY = len(y)
  colsY = len(y[0])

  if colsX != rowsY:
    message = 'x is %d x %d; inconsistent with y which is %d x %d'
    raise Exception(message % (rowsX, colsX, rowsY, colsY))
```

---

[6] Make an error object.

```
z = rowsX * [0]    # Constructs a list of zeros, of size rowsX

for i in range(rowsX):
  z[i] = colsY * [0]
  for j in range(colsX):
    for k in range(colsY):
      z[i][k] += x[i][j] * y[j][k]

return z
```

Here it is checked that the number of columns in x is the same as the number of rows in y, because that is an easy mistake to make when using the function. There is a triple loop to do the sums, which is why matrix multiplication is slow, at least when it is done this way.

It might be tempting to initialise z using:

```
z = rowsX * [colsY*[0]]
```

but that does not work because it creates a matrix with literally the same list in each row, so the additions would affect all the rows, not just the given row. So the initialisation is done in two steps. First a list of the correct length is created:

```
z = rowsX * [0]
```

Here it does not matter what is used as the repeated item, because that is immediately replaced by a list of the correct length in the loop and the summation is done.

```
for i in range(m):
  z[i] = colsY * [0]
```

If this is as sophisticated as the code is required to get in terms of linear algebra, and efficiency is not an issue, then it is acceptable to do things this way. However, for anything more sophisticated, e.g. matrix inversion, it makes most sense to use the NumPy module.

NumPy package

For the examples in this section we will assume that the numpy module has been installed, given that it is not a standard part of Python, and that when not explicitly stated it is imported as follows:

```
import numpy
```

Array objects

NumPy has its own version of multi-dimensional arrays, called the N-dimensional array,[7] which allows efficient manipulation. A basic Python multi-dimensional list can be converted to a NumPy array and vice versa. For example, a 2×3 matrix can be defined in NumPy using:

```
x = numpy.array([[1,2,3],[4,5,6]])
```

---

[7] The data type is ndarray in NumPy documentation.

And to get back standard Python lists we use the `tolist()` method of the array:

```
listOfLists = x.tolist()
```

The elements in a NumPy array can be accessed in the same way that ordinary Python matrices (like lists of lists) are accessed, although there is an alternative syntax that is quicker and avoids some of the brackets:

```
x[1][1]                        # 5
x[1,2]                         # 6
```

NumPy will determine the data type of matrices from the data types of its elements. For example, the array x above will be of type `int` because its contents are integers. If instead you wanted x to be floating point you could either make one of the numbers explicitly floating point or you could specify the data type of the array at construction:

```
x = numpy.array([[1,2,3],[4,5,6]], dtype=float)
```

Note that NumPy has its own data types, so you can be even more specific about which kind of number you require, beyond the regular Python types. Here we specify 32-bit precision numbers:

```
x = numpy.array([[1,2,3],[4,5,6]], dtype=numpy.float32)
```

These special types are usually not needed in regular Python, but can be very handy when interfacing array data with C code (see Chapter 27).

The shape of a NumPy array (i.e. number of rows, columns) is determined via:

```
x.shape                   # (2, 3)
```

and its total size via:

```
x.size                    # 6 (= 2 x 3)
```

What NumPy calls the dimension of an array is the same as the length of the shape:

```
x.ndim                    # 2
len(x.shape)              # 2
```

There are various ways of making arrays of a standard kind, without having to convert other Python data structures. For example, we can create arrays of specified size consisting of all zeros, all ones or an identity matrix (zeros but ones on the diagonal):

```
x = numpy.zeros((2,3))             # 2 x 3 matrix full of 0.0
x = numpy.ones((3,2))              # 3 x 2 matrix full of 1.0
x = numpy.identity(3)              # 3 x 3 identity; floating point
x = numpy.identity(3, numpy.int)   # 3 x 3 identity; integer
```

The regular arithmetic operations work on NumPy arrays, and operate in an element-by-element manner:

```
x = numpy.array([1.0, 2.0, 3.0])
y = numpy.array([3.0, 4.0, 5.0])
x + y      # array([4.0, 6.0, 8.0])       i.e. 1+3, 2+4, 3+5
x * y      # array([3.0, 8.0, 15.0])
x - y      # array([-2.0, -2.0, -2.0])
x / y      # array([0.33333333, 0.5, 0.6])
```

Also, arithmetic can involve single numbers, whereupon all elements of the array are operated on with that number:

```
x + 1.0  # array([2.0, 3.0, 4.0])
y * 5.0  # array([15.0, 20.0, 25.0])
```

To perform other mathematical operations on arrays, NumPy has the array equivalent of most of the functions found in the `math` module and these work efficiently to perform the operation for each element of the array. As well as NumPy arrays the functions will accept regular Python lists or tuples as input, but an array is returned:

```
angles = numpy.array([30.0, 60.0, 90.0, 135.0])
radians = numpy.radians(angles)
cosines = numpy.cos(radians)    # array([0.866, 0.50, 0.0, -0.707])
numpy.log([10.0, 2.71828, 1.0]) # array([2.302585, 1.0, 0.0])
numpy.exp([2.302585, 1.0, 0.0]) # array([10.0, 2.71828, 1.0])
```

Array operations and methods

As well as accessing arrays with explicit index numbers, NumPy supports a slice notation that is very similar to lists and tuples in standard Python. However, because an array can have a rank greater than one, like a matrix which has both rows and columns, then a slice expression may be specified for each of the array dimensions, and a comma is used to separate the expressions for the different dimensions:

```
x = numpy.array([[1,2,3], [4,5,6]])

x[0]        # array([1, 2, 3])          - row zero
x[0,:]      # array([1, 2, 3])          - row zero, as above
x[:,2]      # array([3, 6])             - column two
x[-1,:]     # array([4, 5, 6])          - last row
x[:,1:]     # array([[2, 3],[5, 6]])   - column one onwards
x[1,0:2]    # array([4, 5])             - row one, first two columns
x[::-1,:]   # array([[4, 5, 6],[1, 2, 3]]) - reversed rows
x[:,(2,1,0)] # array([[3, 1, 2],[6, 4, 5]]) - new column order
```

Note that the last example listed above, where the columns of the matrix are shuffled into a new order by specifying a tuple of indices, provides a way of sorting a matrix so that its rows or columns appear in numerical order, comparing values at selected index. Here we use `numpy.argsort()` to get an array of indices (`idx`) that represents the order of the numerical values in column one of x. These indices are then used to make a new matrix with sorted rows:

```
x = numpy.array([[4,4], [5,1], [8,3], [7,2]])
idx = numpy.argsort(x[:,1]) # array([1, 3, 2, 0]) - column one order

x[idx,:] # array([[5,1], [7,2], [8,3], [4,4]])
         # re-ordered rows, by column one value
```

We can use the array index and slice notation not only to extract values, but also to assign values:

```
x = numpy.array([[1,1,1], [1,1,1], [1,1,1]])

x[1]    = (2,3,4)  #  x; array([[1,1,1], [2,3,4], [1,1,1]])
                   #  new row one

x[:,2] = (5,6,7)   #  x; array([[1,1,5], [2,3,6], [1,1,7]])
                   #  new column two

y = numpy.zeros((2,2))

x[:2,:2] = y       #  x; array([[0,0,5], [0,0,6], [1,1,7]])
                   #  replace 2 x 2 elements with 0

x[:,:] = 3         #  x; array([[3,3,3], [3,3,3], [3,3,3]])
                   #  replace all elements with 3
```

NumPy arrays have a number of inbuilt functions (methods) which can be accessed from them using the dot notation. Where appropriate we can often specify which axis (e.g. rows or columns for a matrix) to operate on:

```
x = numpy.array([[3,6],
                 [2,1],
                 [5,4]])

x.min()          # 1 ; minimum value
x.max()          # 6 ; maximum value
x.max(0)         # array([5,6]) ; maximum value row
x.max(axis=0)    # same as above
x.sum()          # 21 ; summation of all elements
x.sum(0)         # array([10, 11])  ; add rows together
x.sum(1)         # array([9, 3, 9]) ; add columns together
x.mean()         # 3.5  ; the mean value of the elements
x.mean(1)        # array([4.5, 1.5, 4.5]) # mean of each row
```

Note that the specification of the axis argument can be a little confusing until you are used to the way things work. Thus, for example, although axis 1 refers to columns, x.sum(1) will add up the elements within each row; it is as if all the columns have been combined into one.

NumPy cleverly lets you create a new array by changing the shape of an existing array. For example, to create a 2×3 matrix you can first create a vector of size 6, here using arange() (the array equivalent of range()), and then just reshape it:

```
x = numpy.arange(1,7)      # array([1, 2, 3, 4, 5, 6])
x = x.reshape((2, 3))      # array([[1, 2, 3], [4, 5, 6]])
```

Of course the reshaping only works if the total size matches. You can even, for example, reshape a 2×3 matrix into a 3×2 matrix:

```
y = x.reshape((3, 2))      # array([[1, 2], [3, 4], [5, 6]])
```

Note that this does not reshape x itself but creates a new array with the new shape. Also, the reshaping we have just done here is not the same as the transpose of the matrix, where rows and columns are switched. The transpose of the matrix is given by:

```
y = x.T                    # array([[1, 4], [2, 5], [3, 6]])
```

or equivalently

```
y = x.transpose()          # array([[1, 4], [2, 5], [3, 6]])
```

Matrix multiplication is exceedingly simple in NumPy. If you have two matrices x and y then their matrix product is obtained using the `dot()` function:[8]

```
x = numpy.array(((1,1),(1,0)))
y = numpy.array(((0,1),(1,1)))
z = numpy.dot(x, y)          # array([[1, 2], [0, 1]])
```

What the above is saying in terms of matrices is that

$$x\,y = \begin{pmatrix} 1 & 1 \\ 1 & 0 \end{pmatrix} \begin{pmatrix} 0 & 1 \\ 1 & 1 \end{pmatrix} = \begin{pmatrix} 1 & 2 \\ 0 & 1 \end{pmatrix}$$

It might seem tempting to just use '*' for multiplication, and although this is a valid operation in NumPy, it just multiplies the two matrices together element by element, so is not the same as matrix multiplication:

```
z = x * y          # array([[0, 1], [1, 0]])
```

There is actually a specific `matrix` data type in NumPy that does allow '*' to be used for matrix multiplication, but it is not used very often and it is generally best to stick with the commonly used `array`. NumPy also has functions that do some of the trickier operations involved in linear algebra; for example, there is a function to calculate the inverse of a matrix. This is accessed via the `linalg` sub-module:

```
x = numpy.array(((1,1),(1,0)))
y = numpy.linalg.inv(x)          # array([[0., 1.], [1., -1.]])
```

Note that the inverse is floating point even if the original matrix is integer. What the above is saying in terms of matrices is that

$$x\,y = \begin{pmatrix} 1 & 1 \\ 1 & 0 \end{pmatrix} \begin{pmatrix} 0 & 1 \\ 1 & -1 \end{pmatrix} = \begin{pmatrix} 1 & 0 \\ 0 & 1 \end{pmatrix}$$

There is much more to the `linalg` module, as described in the NumPy documentation,[9] including various decompositions and eigenvector calculation.

Linear algebra examples

For the last part of this chapter we will work through two practical examples which use some of the ideas discussed above.

Rotation matrices

Often we want to consider transformations that move and reorient coordinates but still preserve their shape, i.e. the positions of the coordinates relative to one another. An example of this, which will be discussed further in later chapters, is transformations on complete

---

[8] The function name derives from the *dot product* operation commonly applied to vectors.
[9] See link at http://www.cambridge.org/pythonforbiology.

molecular structures, in which case we are interested in the coordinate positions of the atoms that make up the molecule. Although in some situations we may want to move specific atoms to generate new molecular shapes, often we don't want to distort the precious (experimentally determined) data and merely wish to reposition the molecule. There can be many reasons for moving a molecule's coordinates, a few of which include: creating a view to make a graphical representation; setting up a system for energy calculations and dynamic simulations; superimposing structures to find where conformations differ. Whatever the reason, a commonly required operation is rotation, a kind of coordinate transformation that can be described by a *rotation matrix*.

The simplest rotations to consider are those that rotate about one of the three coordinate axes. First, consider rotation by an angle A (specified in radians) about the z axis. The unit vector along the z axis, *(0, 0, 1)* would not be affected by this rotation because it lies exactly along the direction which we rotate around; effectively it gets transformed to itself. For the same rotation, the unit vector along the x axis, *(1, 0, 0)*, would naturally be altered according to the sine and cosine of the angle; it moves away from a pure x direction to gain a y component, specifically transformed to *(cos A, sin A, 0)*.[10] Similarly, the unit vector along the y axis, *(0, 1, 0)*, moves to gain an x component and is transformed to *(−sin A, cos A, 0)*. Thus, combining the transformations for the individual axis vectors, the rotation by an angle A about the z axis is given by the matrix:

$$R_{z,A} = \begin{pmatrix} \cos A & -\sin A & 0 \\ \sin A & \cos A & 0 \\ 0 & 0 & 1 \end{pmatrix}$$

Note that this indeed transforms things correctly, for example:

$$R_{z,A} \begin{pmatrix} 1 \\ 0 \\ 0 \end{pmatrix} = \begin{pmatrix} \cos A & -\sin A & 0 \\ \sin A & \cos A & 0 \\ 0 & 0 & 1 \end{pmatrix} \begin{pmatrix} 1 \\ 0 \\ 0 \end{pmatrix} = \begin{pmatrix} \cos A \\ \sin A \\ 0 \end{pmatrix}$$

A rotation by an angle A about the x axis is similarly given by the matrix

$$R_{x,A} = \begin{pmatrix} 1 & 0 & 0 \\ 0 & \cos A & -\sin A \\ 0 & \sin A & \cos A \end{pmatrix}$$

And a rotation by an angle A about the y axis is given by the matrix

$$R_{y,A} = \begin{pmatrix} \cos A & 0 & \sin A \\ 0 & 1 & 0 \\ -\sin A & 0 & \cos A \end{pmatrix}$$

It turns out that all rotations, about any direction, can be composed of products of rotations about the main axes, although this does not necessarily help. The formula for the rotation about an arbitrary direction, n (a unit length axis of rotation), by an angle A is much more

---

[10] This is assuming that the rotation follows the convention known as the right-hand rule.

complicated, and we won't describe it in fine detail. It turns out that it is easiest to define it by how it acts on an arbitrary vector w:

$$R_{n,A}w = (w \cdot n)n + (\cos A)(w - (w \cdot n)n) + (\sin A)n \wedge w$$

Here $w \cdot n$ is the sum of the element-wise product (also called the *inner product* or *dot product*) of the two vectors, and $n \wedge w$ is what is known as the *cross-product*.[11] The term $w - (w \cdot n)n$ can be thought of as representing the projection of the vector w along the axis of rotation n. This component of the vector is not affected by the rotation, but the remaining component is, hence the sine and cosine terms. Thinking in terms of two vectors we can explicitly represent the calculations involved in generating the dot product (a single number) and cross-product (another vector), although if you were doing this in earnest you would use the `dot()` and `cross()` functions in NumPy:

```
vec1 = (x1, y1, z1)
vec2 = (x2, y2, z2)

dotProduct = x1*x2 + y1*y2 + z1*z2
crossProduct = (y1*z2-z1*y2, z1*x2-x1*z2, x1*y2-y1*x2)
```

Combining the expression for rotation about an arbitrary axis with the knowledge of how to calculate dot and cross-products we can derive the following function to generate a rotation matrix in Python. It takes an axis direction (specified as a vector, not necessarily of unit length) and an angle (in radians) to define a rotation matrix that will perform the required rotation operation, via matrix multiplication.

Note how the example makes use of the `sin` and `cos` trigonometric functions and the square root function from the `math` module. We will not go through the mathematical details, but you can see what the construction of the matrix involves: dividing the input axis by its length to generate an axis vector of unit length; calculating variables for sine and cosine of the angle, to avoid repeated calculation; and construction of the final rotation matrix using the required expressions, involving the angle-derived variables and axis coordinates.

```
import math

def getRotationMatrix(axis, angle):
  vLen = math.sqrt( sum([xyz*xyz for xyz in axis]) )
  x, y, z = [xyz/vLen for xyz in axis]

  c = math.cos(angle)
  d = 1-c
  s = math.sin(angle)

  R = [[c+d*x*x,    d*x*y-s*z,  d*x*z+s*y],
       [d*y*x+s*z,  c+d*y*y,    d*y*z-s*x],
       [d*z*x-s*y,  d*z*y+s*x,  c+d*z*z  ]]

  return R
```

---

[11] Generating a vector at right angles to both of the vectors operated on.

Note that this does not use NumPy and what is returned is an ordinary Python list of lists, but it can be converted using `numpy.array()`, if desired. For example, in the following we get a NumPy array of the rotation matrix representing rotation by 60 degrees around the axis (1, 1, 1) so that its transformation can be applied by matrix multiplication using the `dot()` function:

```
import math

axis = (1, 1, 1)
angle = math.radians(60)   # convert from degrees to radians
rotMatrix = numpy.array(getRotationMatrix(axis, angle))

vector1 = numpy.array([2, -1, -1])   # A test vector
vector2  = rotMatrix.dot(vector1)   # [1, 1, -2]
```

The axis could be passed in as a NumPy vector rather than an ordinary Python vector; the function works in either case.

Torsion angle

The following example combines use of both the `math` module and NumPy. The objective is to make a function that is able to calculate what is known as a *torsion angle*.[12] While the standard kind of angle involves three points, which we can imagine as a 'V' with the angle being the amount of turn between the two ends about the connecting point, a torsion angle is defined by four points. You can imagine these four points as forming a 'Z' shape made of three lines, where the torsion angle is the twist between the first and last lines. Our illustration of a 'Z' on a flat page is a torsion angle of 180°, but if one of the end lines came directly out of the page then the torsion angle would be 90°.

Measuring torsion angles will be helpful in later chapters where the four defining points are atoms of a molecule and the lines connecting them are chemical bonds. For protein molecules especially, the torsion angles of atoms along the backbone provide quite a bit of useful information. In this context the torsion angle is defined to be the angle between chemical bonds and we can define these bond vectors as the difference vectors between atom positions: what we have to add to one atom position to get to the other. The central chemical bond (the middle line of 'Z') is the axis about which the rotation between the other bonds is measured.

We now provide the function that calculates the torsion angle. Firstly, we import the required NumPy and mathematical functions. Notice that we have not used the NumPy `cross()` function before; this calculates the *cross-product* between two vectors, which was mentioned briefly in the context of rotations. If you imagine two directional lines in three-dimensional space, emanating from the same point, the cross-product will be a new line at right angles to both the other lines.[13]

```
from numpy import cross, dot, array
from math import sqrt, acos
```

---

[12] This is also known as *dihedral angle*.

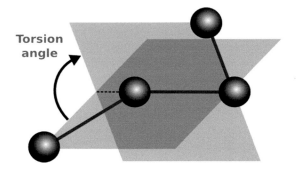

Figure 9.3. A cartoon visualisation of the torsion angle defined by a sequence of four points.
A torsion angle, also known as a dihedral angle, is the measurement of the amount of twist between two planes about the axis of their intersection. In three dimensions a torsion angle may be defined by four distinct points, as is commonly found in molecular structures, where the points represent atoms, and where the central two atoms define the intersection axis and each end atom defines the direction of a plane.

A function is defined which takes a list of four coordinate positions to calculate the torsion angle between them. Here the order of the input coordinates is critical; we will measure the twist about the axis defined by the central two coordinates (coord2 and coord3). We will not describe the mathematics behind the calculations in great detail, but we will describe the basic operation performed at each step.

Inside the function the first procedure is to define three vectors, which may represent the bonds between three pairs of atom coordinates, by subtracting coordinates from one another; the resulting difference vector represents how to move from one coordinate to the other. Two perpendicular cross-product vectors are then calculated: for the first and second vectors and for the second and third vectors. The basic idea here is that even though the cross-product vectors are perpendicular to the second (central) vector they will have the same angle of twist (relative to the central vector) between them as the first and third vectors do. However, because they are perpendicular to the central vector the angle between them is the angle we want: the angle of twist around the axis of the central vector. Hence, the perpendicular vectors are then used to calculate three dot products. The first represents the 'shadow' projection of one vector on the other, which naturally depends on the angle between them, and the other two dot products will simply be the lengths of the vectors squared. From these the cosine of the torsion angle is obtained.

```
def calcTorsionAngle(coord1, coord2, coord3, coord4):

  bondVec12 = coord1 - coord2
  bondVec32 = coord3 - coord2
  bondVec43 = coord4 - coord3

  perpVec13 = cross(bondVec12, bondVec32)
  perpVec24 = cross(bondVec43, bondVec32)
```

---

[13] By convention this direction is right-handed; if you consider a typical graph, with the x axis going across the bottom and the y axis going up on the left side, then the cross-product will come out of the page. The length of the cross-product is also important because it depends on the sine of the angle between the two directions.

```
projection   = dot(perpVec13, perpVec24)
squareDist13 = dot(perpVec13, perpVec13)
squareDist24 = dot(perpVec24, perpVec24)

cosine = projection / sqrt(squareDist13*squareDist24)
cosine = min(1.0, max(-1.0, cosine))
angle = acos(cosine)

if dot(perpVec13, cross(perpVec24, bondVec32)) < 0:
  angle = -angle

return angle
```

Note that we check to make sure the cosine is between -1 and $+1$; we might have a small error in the calculation from the use of floating point numbers so we use the `min()` and `max()` functions to reinforce the limits. The angle is calculated by using the inverse cosine function `acos()` (meaning arccosine). Finally, we use a dot and cross-product to check whether we actually calculated a negative angle; because the inverse cosine will only give a value between zero and π radians (180°) we don't otherwise know whether the first coordinate is 'above' or 'below' the plane defined by the other three. We can then test the function:

```
from numpy import array, degrees

p1 = array([2,1,1])
p2 = array([2,0,0])
p3 = array([3,0,0])
p4 = array([3,1,-1])

angle = calcTorsionAngle(p1, p2, p3, p4)
print( degrees(angle) )     # -90.0
```

10 Coding tips

Contents

Improving Python code

This chapter is concerned with improving Python code and we will illustrate, using short code snippets, various tips that help with speed, memory use and coding clarity. There may be several aspects of a program that we seek to improve, but we can't necessarily expect to improve all of them all of the time. Often optimisation of a Python program is about compromise; you may make a program run faster at the expense of using more memory. Clarity is an especially important aspect that we will be mindful of when making suggestions, and in general we recommend making code more easily understood over mild improvements in performance. Finding and correcting errors in code can take a long time, sometimes longer than the program took to write in the first place, so keeping the code easy to understand is especially important.

A basic programming approach that the authors often follow, and which may be helpful for others, is a three-point plan:

1. Firstly, make the code work: an inelegant program is better than one that doesn't work.
2. Next do it properly: with a working reference, you can take a step back and criticise your approach.
3. Then make it better: only once your program is working, and the general approach won't change, is it worth optimising.

During this chapter we will assume that the first point is covered and think about the last two. However, it is worth noting upfront that time spent on optimising (point 3) has the tendency for ever diminishing returns, so we will try not to be too obsessive.

Code clarity

When it comes to writing clear Python code or generally being 'elegant' we wouldn't want to give fixed rules about something so subjective. However, we can show you some extremes to avoid and highlight some of the more common conventions that other programmers will be using. Considering the following versions of an example loop in Python, can you tell what it is supposed to be doing scientifically?

```
m = []
for a, b in l:
  m.append(rf(a, b, 54.7))
```

How about this:

```
transResult = []
for globalCoordPosX, globalCoordPosY in mainCoordinateList():
    transResult.append(angleTransformationFunc(globalCoordPosX,
                                        globalCoordPosY, 54.7))
```

Or maybe:

```
newCoords = []
for x, y in coords:
  newCoords.append( rotate(x, y, 54.7) )
```

It is often tempting to be lazy and use minimalist names for variables, but in the long run, especially if you have to maintain the code, and understand what it means at a later date, this approach is often counterproductive. Also, if you write clear code in the first place there is much less need to document your programs. In contrast, great verbosity is not always what is called for. Some verbosity is good, but too much obscures what you are trying to do, so keep it simple enough to be obvious. Single-letter variables can be absolutely fine in the right spot: usually where they are used immediately (e.g. in a loop) or used for simple counters etc. The verbose example above also hints at something else that can be very important, which is the choice of the variable names per se. The meaning in English of a variable's name should give a huge clue to what it does or represents, so call a variable after what it is. And if it is a collection, like a list or set, consider using the plural like `coords` or using a hint like `coord-List` or `coordSet`, so you never have to think what you have.

Some readers may already have noticed the frequent use of *camel case* in this book: giving names humps with capital letters like `camelCase`. This is merely convention and not a requirement of the language. While you will see it frequently in Python and Java, other languages more frequently have underscores, i.e. `under_score`, in order to join words into variable names. There are several official Python style recommendations,[1] which the authors admittedly don't always stick to in the book, like indenting with four spaces (that would mean running out of room or a really small font), but adhering to the spirit of these and not being too silly with your placement of whitespace etc. really helps in the long run.

An important thing about coding conventions is that, as long as there is consistency, you can impart extra information. For example, for variable names the authors follow the following conventions. If a variable is global to a module, i.e. outside all functions and classes, as is common for constants, then it is written in upper case:

```
AVOGADRO_NUMBER = 6.02214e23
GYROMAG_RATIOS = {'1H':267.513, '2H':41.065, '13C':67.262, '15N':19.331,}
```

---

[1] PEP8: https://www.python.org/dev/peps/pep-0008/.

Function and regular variable names begin with lower case, but class names begin with upper case, so the following visually represent two completely different things in our convention:

```
snack = BananaSplit()        # Object instance
twoHalves = bananaSplit()    # function call
```

Naturally, functions are usually distinguished from regular variables by the presence of parentheses.

Improving speed

When it comes to making Python programs faster there are a few general points to consider before going through specifics. We can think of program execution time in terms of the number of operations performed and how long individual operations take. Hence, to improve speed we try to reduce the number of operations and/or perform faster individual operations. Also, an important aspect of this is how the number of operations scales with the size of the problem. For example, the time taken could stay roughly constant, increase linearly or increase in an exponential manner (to name only a few possibilities) as the number of data items increases. Naturally the scaling depends on the algorithm in question. Hence, if things seem too slow it is worth questioning whether a particular algorithmic approach is a smart one, especially if the number of data items is large and the speed scales poorly with the size of this data. As a very simple example, albeit somewhat contrived, to add all the integer numbers between a and b you could do:

```
s = 0
for x in range(a, b+1):
  s += x
```

or

```
s = sum(range(a, b+1))
```

or

```
s = (b+1-a) * (a+b) / 2.0
```

The first two examples involve an operation for each item in a list (albeit explicitly or from sum()), so their execution time is roughly proportional to the length of the list. The last formula involves a small, fixed number of operations, which is much quicker for large ranges, although you could argue that the code is not as easy for a novice to understand. Whether analyses of this kind are worthwhile will depend on the context of the operation within the program as a whole. For example, if you were doing this kind of summation many times, in a long loop, the overall gains from optimisation would be significant, but if it is only done occasionally or if you're only adding a few values the difference is probably negligible. On the whole we advise opting for clarity if an operation is not a performance bottleneck. Sometimes it is easy to spot where the bottlenecks are in your code and you can simply time how long something takes, as illustrated below. However, if it is not easy, there are a number of profiler programs, such as the inbuilt Python module cProfile,[2] that can be used to inspect how long the various operations take and thus show you where to focus attention.

---

[2] From Python 2.5.

Even when keeping the same basic Python algorithm, some surprisingly simple programmatic changes or rearrangements can yield speed differences. This chapter will subsequently go through many more specific examples of these, but common things to look out for are: avoiding unnecessary function calls (which are relatively slow in Python), taking care to remove unnecessary operations from long loops and making good use of Python collections. For the examples below we are using `usingTimer()`, as described in Chapter 5, as a *decorator* to wrap the test code in a function that times how long the operation takes to run. Firstly, we have an example which uses the `math` module to take the square root for a range of numbers, the result of which is added to a total:

```
import math

@usingTimer
def testFunc(x):
  y = 0
  for i in range(x):
    y += math.sqrt(i)
  return y
```

Testing using a large number the result takes just over 2 seconds:[3]

```
testFunc(10000000)
>>> Function call took 2.046932 seconds
>>> 21081849486.439312
```

Now a small modification is done to obtain `sqrt` at the start,[4] rather than accessing it repeatedly from the `math` module inside the loop, using dot notation:

```
from math import sqrt

@usingTimer
def testFunc2(x):
  y = 0
  for i in range(x):
    y += sqrt(i)
  return y
```

Testing reveals that this new version is about a third of a second quicker:

```
testFunc2(10000000)
>>> Function call took 1.672865 seconds
>>> 21081849486.439312
```

Rewriting the function to use a list comprehension, thus avoiding explicit Python loops, is quicker still. Note that in this version we have also avoided constructing large lists. In Python 2 we would use the iterable `xrange()` instead of the list-creating `range()`, but in Python 3 the `range()` function has the same behaviour as `xrange()` in Python 2. And by using round parentheses for the list comprehension we make a list *generator* object, rather than an actual large, filled list.

---

[3] With a 3.2-GHz 64-bit AMD processor.
[4] If we use the original `math` import we could do `sqrt = math.sqrt`.

```
@usingTimer
def testFunc3(x):
    yList = (sqrt(i) for i in range(x))     # in Python use xrange(x)
    return sum(yList)

testFunc3(10000000)
>>> Function call took 1.360537 seconds
>>> 21081849486.439312
```

Another common tactic to make Python run faster is to investigate whether there are any fast, pre-constructed libraries that you can import (as Python modules) to do the job. An obvious example would be to use NumPy for mathematical work involving arrays. Here the imported code is written in a fast compiled language underneath and avoids having to go through Python loops. An example of the gain from using `numpy.array` objects is illustrated below, and the improvement is impressive, over nine times faster than the original:

```
from numpy import arange, sqrt

@usingTimer
def testFunc4(x):
    yArray = sqrt(arange(x))
    return yArray.sum()

testFunc4(10000000)

>>> Function call took 0.221356 seconds
>>> 21081849486.439312
```

If there is a requirement to make a Python program even faster, above and beyond what we can achieve with changes to coding style, then we might consider parallelisation (using multiple processors) or write a module in a fast compiled language like C. We cover these more advanced topics in Chapter 27.

A compendium of tips

After more than enough moralising about coding style, we get to some simple tips of things in Python that might not be so obvious, especially for newcomers to Python. Some of the points have been covered in earlier chapters, but we reproduce them here to give a single resource. This really is just a big list, although we have attempted to split things up into vague subject categories.

Many of the examples we give here are simply aspects of Python that are commonly overlooked. This may be because non-standard options are easily overlooked, because of recent additions to the language or because the code looks a bit scary. An example of the latter would be the following, where `zip(*)` is used to extract separate lists of the first and second elements from a list of 2-tuples. So instead of something like:

```
data = [(2,5),(6,7),(8,7),(8,4),(2,5),(6,4),(3,2)]
xList = [pair[0] for pair in data]
yList = [pair[1] for pair in data]
```

You might see:

```
xList, yList = zip(*data)
```

Without wishing to suggest anything about whether the `zip(*)` approach is better, it is a proper part of the language and is something you see in others' code, so it is good to be aware of it.

Python has many features and it is easy to overlook some of them, even though they might be useful. A very simple example of this is the following for calculating logarithms in any base, which might be done naturally as soon as one learns about the `log()` function:

```
from math import log
y = 1e7
x = log(y) / log(10)
```

As it happens the base for the logarithm is the second argument to the `log()` function, so instead we can do:

```
from math import log
y = 1e7
x = log(y, 10)
```

This shows the value of carefully reading the Python documentation.

Simple operations

When taking powers of numbers the `pow()` function is slower than the `**` operator or simply using multiplication. So instead of:

```
v = pow(w, 1.5)
x = pow(y, 3)
```

try:

```
v = w**1.5        # Faster
x = y*y*y         # Faster still
```

When it comes to taking square roots, sometimes you can avoid the operation entirely, e.g. by using the square, which is quicker to calculate. A common application of this is in finding distances; the smallest square distance will still be the smallest overall. Thus, instead of doing:

```
from math import sqrt

for x, y in data:
  if sqrt(x) < y:
    doSomething()
```

consider:

```
for x, y in data:
  if x < y*y:
    doSomething()
```

The values of two variables can be swapped cleanly in one line by assigning the elements in two implicit tuples. Hence the three lines involving a temporary variable:

```
temp = x
y = x
x = temp
```

become one neat swap:

```
x, y = y, x
```

Different variables can be set to the same value with a concatenated assignment, so instead of:

```
x = 0
y = 0
z = 0
```

you can do:

```
x = y = z = 0
```

This does not work for variables that represent collections. Thus:

```
u = []
v = []
```

means something different to:

```
u = v = []
```

because in the first example we have two separate lists, and in the second example they both represent the same list.

Comparison operators can be concatenated, e.g. to check a value is within an upper and lower limit. So instead of:

```
if (lower < x) and (x <= upper):
  doSomething()
```

the following is simpler:

```
if lower < x <= upper:
  doSomething()
```

The Boolean and ternary (one-line `if` clause) operators can avoid simple multi-line clauses like:

```
aList = [1, 0, 0, 1, 1, 0]

for x in aList:
  if x:
    print('Yes')
  else:
    print('No')
```

This first alternative using Boolean logic works but is a little dangerous, because it depends on both the alternatives being logically true (which is fortunately the case here for the strings `'Yes'` and `'No'`):

```
aList = [1, 0, 0, 1, 1, 0]

for x in aList:
  print(x and 'Yes' or 'No')
```

A better alternative is to use the ternary operator:

```
aList = [1, 0, 0, 1, 1, 0]

for x in aList:
  print('Yes' if x else 'No')
```

It is sometimes convenient to use `or` to provide a default value, so in the following example the variable is assigned to `None` if the function call returns zero:

```
x = sum(myList) or None
```

Strings

When joining strings into longer text it is faster to use a `separtor.join()` call than redefining the string each time. Thus for big lists avoid:

```
seqList = ['C','A','T','G','G','C','T','C','T','C']

seqString = ''
for letter in seqList:
  seqString += letter
```

when you can do:

```
seqList = ['C','A','T','G','G','C','T','C','T','C']
seqString = ''.join(seqList)
```

though in later versions of Python the latter is only about 10% quicker.

Similarly, using a formatted string is faster than string concatenation:

```
line = 'First:' + s1 + ' Second:' + s2 + ' Third:' + s3     # Slower
line = 'First:%s Second:%s Third:%s' % (s1, s2, s3)          #Faster
```

If you need to get sequential letters you can use the handy `ord()` and `chr()` functions, which respectively fetch and decode character numbers (according to the Unicode scheme).

```
startNum = ord('A')
tenLetters = [chr(startNum+i) for i in range(10)]

# Gives ['A', 'B', 'C', 'D', 'E', 'F', 'G', 'H', 'I', 'J']
```

Removing characters from a string can be done with the often forgotten `.translate()` method, though this is used slightly differently in Python 3 compared to Python 2. Here we remove 'A', 'T' and ',' characters:

```
seq = 'A,T,G,A,C,A,T,C,A,T,G,G,C,T,C,T,C'

# Python 2
seq = seq.translate(None, 'AT,')

# Python 3
transTable = str.maketrans('', '', 'AT,')
seq = seq.translate(transTable)

# Both give 'GCCGGCCC'
```

Collections

To see if a collection (list, set, tuple, dictionary) is empty just test whether it is logically true. So rather than

```
if len(myList) == 0:
  doSomething()
```

instead do:

```
if myList:
  doSomething()
```

To copy a list you can use the `list()` keyword or use the `[:]` slice notation:

```
duplicateList = list(firstList)
duplicateList = firstList[:]
```

A slice notation can also be used to get a reversed copy of a list (remembering that the last element of the slice notation is the step):

```
revList = firstList[::-1]
```

This is more compact than copying and then using `reverse()`:

```
revList = list(firstList)
revList.reverse()
```

or using the `reversed()` iterator, which is handy when going through loops in reverse order (for example, `for x in reversed(a List):`), but needs an explicit conversion to make a duplicate list:

```
revList = list(reversed(firstList))
```

For dictionaries don't forget the `.get()` and `.setdefault()` methods. So:

```
if x in myDict:
  y = myDict[x]
else:
  y = defaultValue
```

becomes:

```
y = myDict.get(x, defaultValue)
```

or if the default value is `None` simply:

```
y = myDict.get(x)
```

If the default value should be actually put in the dictionary then you can do:

```
y = myDict.setdefault(x, defaultValue)
```

In Python 2 if you want to simply enquire whether something is present in a dictionary it is simpler, and slightly faster, to use `in` rather than call `has_key()`.

```
if myDict.has_key(key):
  doSomething()
```

becomes:

```
if key in myDict:
  doSomething()
```

In Python 3 dictionaries no longer have the `has_key()` method.

It may sometimes be helpful to construct a dictionary from a list. Rather than going through a loop, a list of 2-tuples with `(key,value)` pairs can be used:

```
listData = [(1,'Apples'), (2, 'Bananas'), (3, 'Cherries')]
dictData = dict(listData)
print(dictData[2])
```

In Python 2 to do the reverse you can use `.items()` to get a list of all pairs, or `.iteritems()` to get an iterator object, which can be looped though like a list but which yields one item at a time, and so saves memory by not making the complete list:

```
for k, v in dictData.items():      # Makes a list
  print('Key: %d, Value: %s' % (k,v))

for k, v in dictData.iteritems(): # Uses an efficient iterator
  print('Key: %d, Value: %s' % (k,v))
```

In Python 3 there is no `.iteritems()` method and `.items()` returns an iterable *view* on the items in the dictionary, rather than a list.

The `zip` keyword can be used to combine corresponding elements from multiple lists, which is handy for dictionaries when you initially have separate lists for the keys and values:

```
keys = [1, 2, 3]
values = ['Apples', 'Bananas', 'Cherries']
listData = zip(keys, values) #  [(1,'Apples'), (2,'Bananas'), (3,'Cherries')]
dictData = dict(listData)
```

The next tip was mentioned before, but we repeat it in the compendium, and it can reverse the above operation (although for dictionaries, `.keys()` and `.values()` also do the job). If you already have data in a list of lists (or tuples) then `zip` can neatly extract the elements which share the same index.

```
listData = [(1,'Apples'), (2,'Bananas'), (3,'Cherries')]
numbers,  fruits = zip(*listData)
```

The way to imagine this one is that the call is actually `zip((1,'Apples'), (2, 'Bananas'), (3, 'Cherries'))`, with the * extracting the items in the list as separate arguments. The `zip` then combines the first elements and the second elements together, exactly as above. This is neater than using the equivalent list comprehension:

```
listData = [(1,'Apples'), (2,'Bananas'), (3,'Cherries')]
numbers = [x[0] for x in listData]
fruits = [x[1] for x in listData]
```

The `zip` can also come in handy as a compact notation for looping through two lists, although in Python 2 it does make a new list, so is not so space efficient. Accordingly, something like:

```
for i, aValue in enumerate(aList):
  bValue = bList[i]
    print(aValue, bValue)
```

could become:

```
for aValue, bValue in zip(aList, bList):
  print(aValue, bValue)
```

In Python the set data type is sometimes overlooked, especially by those who started with early versions of Python. Nonetheless, it is exceedingly useful and can avoid the need to do looping with lists, as long as order is not important (or can be reconstructed). There is a caveat to such set operations, however: the elements must be *hashable*, which means they cannot be internally modifiable, a requirement to keep things unique. In essence, sets can contain most objects, numbers, strings, tuples and frozen sets but cannot contain other sets, lists or dictionaries.

Looking up elements in a set is fast, so where you have lots of look-ups to do, instead of:

```
for x in firstList:
  if x in veryLongList:
    doSomething()
```

you can make things quicker with:

```
bigSet = set(veryLongList)
for x in firstList:
  if x in bigSet:
    doSomething()
```

Note this assumes that the speed gained using `bigSet` for look-up makes up for the time spent creating the set in the first place.

Sets provide a neat way of removing duplicates from a list, as long as you don't want to preserve order, you just convert to a set and back to a list again:

```
myList = ['apple', 'banana', 'lemon', 'apple', 'lemon', 'lemon']
uniqueList = list( set(myList) )   # ['lemon', 'apple', 'banana']
```

To get the common elements of several lists using set operations is neat and efficient, although it may be prudent to simply work with sets in the first place:

```
a = ['G','S','T','P','A']
b = ['A','V','I','L','P']
intersection =  set(a) & set(b)
commonList = list(intersection)   #  ['A', 'P']
```

Likewise to find elements that are present in either list:

```
a = ['G','S','T','P','A']
b = ['A','V','I','L','P']
union = set(a) | set(b)
combinedList = list(union) # ['A', 'G', 'I', 'L', 'P', 'S', 'T', 'V']
```

When constructing lists it can be quicker and more compact to use list comprehensions than loops. For example:

```
squares = []
for x in range(1001):   # in Python 2 use xrange(1001)
  squares.append( x * x )
```

is slower than:

```
squares = [x*x for x in range(1001)]
```

Also, if we don't need the whole loop, but just need to iterate though it, we can use round parentheses to make a *generator* object (which has no length as such and does not have indices).

```
squares = (x*x for x in range(1001))  # Using () not []
for y in squares:
  doSomething()
```

```
squares[3]  # Fail: This will not work on () generators.
```

It is sometimes overlooked that list comprehensions can be concatenated, although it is easy to take this sort of thing too far:

```
[(x,y) for x in range(3) for y in range(3)]

# Gives [(0, 0), (0, 1), (0, 2),
#        (1, 0), (1, 1), (1, 2),
#        (2, 0), (2, 1), (2, 2)]


[(x,y) for x in range(3) for y in range(x,3) if x+y >1]

# Gives [(0,2), (1,1), (1,2), (2,2)]
```

Sometimes you may wish to construct a list of blank lists, to put items into later. For this it is tempting to do:

```
data = [[]] * 3

print(data)          # Gives [[], [], []]
```

but here the same list object was repeated three times internally:

```
data[1].append(True)

print(data)          # Gives [[True], [True], [True]]
```

so try a list comprehension instead:

```
data = [[] for x in range(3)]

data[1].append(True)
print(data)          # Gives [[], [True], []]
```

Although perhaps not such common operations, the any and all keywords can be used to find whether any or all elements in a list hold a certain condition. Accordingly:

```
for x in myList:
  if x < 2:
    doSomething()
    break
```

becomes:

```
if any(x < 2 for x in myList):
  doSomething()
```

Likewise:

```
if len(myList) == len([x<2 for x in myList]):
  doSomething()
```

is the same as:

```
if all(x<2 for x in myList]):
  doSomething()
```

For obtaining a sorted list, the inbuilt `sorted` function is useful when you don't want to modify the original list. So instead of :

```
b = list(a)
b.sort()
```

you can do:

```
b = sorted(a)
```

If you want to sort a list on something other than the items' innate value you can construct a list of 2-tuples which will be sorted on the first item (which contains the values to sort on). Here we sort according to the length of the strings:

```
aList = ['homer', 'bart', 'maggie', 'lisa', 'marge']
bList = [(len(x), x) for x in aList]
bList.sort()
aList = [x for (lenX, x) in bList]

# Gives ['bart', 'lisa', 'homer', 'marge', 'maggie']
```

However. the `key` option of `sort()` is much more nifty and allows you to pass in the function that is used to generate the sort key:

```
aList = ['homer', 'bart', 'maggie', 'lisa', 'marge']
aList.sort(key=len)
```

Sometimes when dealing with objects we would like to sort on the value of a particular attribute. You can readily write a function to fetch that attribute (for any object in the list, as required by the sort operation), and thus generate a key for the sort. So, for example:

```
def getSortAttr(obj):
  return obj.something

objList = [objA, objB, objC]
objList.sort(key=getSortAttr)
```

However, you can also use the `key` option in combination with the `operator` module. The function `operator.attrgetter()` uses the name of an attribute to create a separate on-the-fly function[5] which sends back the value of an attribute, which in this case is the value to sort with. So an alternative to the above is:

---

[5] A callable object.

```
from operator import attrgetter

objList = [objA, objB, objC]
objList.sort(key=attrgetter('something')) # Name of attribute as a string
```

The functions `operator.itemgetter` (for selecting items in a collection) and `operator.methodcaller` (for invoking class functions) can also be used in a similar manner.

Loops

We've been using `enumerate()` throughout the book, but it is still something novices occasionally overlook. So instead of:

```
myList = ['e', 'f', 'g']
for i in range(len(myList)):
  print(i, myList[i])
```

do:

```
myList = ['e', 'f', 'g']
for i, val in enumerate(myList):
  print(i, val)
```

And from Python 2.6 you can use a second argument to specify the start point for the index:

```
for i, val in enumerate(myList, 5):
  print(i, val)

# Gives:
# 5 e
# 6 f
# 7 g
```

In Python 2 when looping though sequential numbers, such as indices, consider using `xrange()` rather than `range()`. This saves space because it only yields numbers on demand. Helpfully an `xrange` still has a length and can be indexed. In Python 3 `xrange` is effectively renamed `range` and replaces the old list constructor.

```
for x in xrange(100, 1000000): # Doesn't make all the numbers (Python 2)
  doSomething()
```

To make an indefinite[6] loop, use a `while` loop that tests something that is logically true, although don't forget to break out of the loop eventually:

```
while 1:
  test = doSomething()
  if test:
    break
```

Because loops are constructs that allow you to repeat operations many times, when thinking about speed a general principle is to put as few operations into the loop as possible. For example, when doing function calls in a loop to construct a list using `.append()`, a speed

---

[6] Some might say infinite.

improvement can be made if the dot notation call is done only once outside the loop. For example:

```
aList = []
for x in someBigList:
  if testFunc(x):
    aList.append(x)
```

becomes the faster:

```
aList = []
addToList = aList.append
for x in someBigList:
  if testFunc(x):
    addToList(x)
```

Related to the above, if you know how long a list will be it is faster to pre-construct it in a quick manner and curate it using indices, rather than appending repeatedly.

```
aList = [0] * n
bList = [0] * n
for i in range(n):
  aList[i] = someCall(i)
  bList[i] = anotherCall(i)
```

If you need two loops and have to break out of both of them, cunning use of else, continue and break can do the job without having to set any flags:

```
for a in oneList:
  for b in anotherList:
    if discoverSomething(a,b):
      # Quit inner loop and subsequently the outer too
      break
  else:
    # Without a break we get here at the end of the inner loop
    # Continuing the outer loop the next break is skipped
    continue
  # Only get here due to the first break
  break
```

If you have a loop that may cause an error (throw an exception) then it may be tempting to do a precautionary check to stop errors before they occur. However, it is generally quicker to let the exception happen and then catch it in a safe way. This is because with try: there is no repeated checking and extra time is taken only if an error is encountered. So, for example:

```
for x in bigList:
  if rareEvent(x):
    rareEventOccurred(x)
  else:
    commonTask(x)
```

can be modified into:

```
for x in bigList:
  try:
    commonTask(x)
  except SpecialException:
    rareEventOccurred(x)
```

Functions

For keeping code tidy, depending on circumstances, you can consider collecting a function's arguments into a dictionary before it is called:

```
animal = longFunctionName(name='Benny', age=9, furColor='black',
                          legs=4, tail=True)
```

The above can, if it helps, be changed to:

```
kwds = dict(name='Benny',
            age=9,
            furColor='black',
            legs=4,
            tail=True)
x = longFunctionName(**kwds)
```

If you need to collect function arguments for calling some other function internally, but don't want to have to state them explicitly, you can collect them with '*' notation and pass them on untouched.

```
def func1(a, b, c, *args):
  d = func2(*args)
  ...
```

To call a different function depending upon a value you could use conditional statements:

```
if x = 1:
  funcA()
elif x = 2:
  funcB()
elif x ==3:
  funcC()
```

or perhaps more elegantly do a dictionary look-up:

```
funcDict = {1:funcA, 2:funcB, 3:funcC}
funcDict[x]()
```

Memoization is a handy technique if you have a slow operation, but the operation may be repeated, so you can pass back the result from last time (given the same inputs) rather than calculating again. In Python we can implement this in a function by using a default argument such as a dictionary. It sometimes causes trouble if a dictionary or list is used for a default function argument, because the collection is not automatically renewed each time the function is called (the same object is always used). However, for memoization we can exploit this to make a function cache.

```
def verySlowCalcOnSimpleData(data, cache={}):

  key = getCacheKey(data) # This could be as simple as making a tuple

  answer = cache.get(key)

  if answer is not None:  # Assumes the answer is never None
    return answer # Quick response

  answer = someVerySlowCalculation(data)

  cache[key] = answer # Store the result for next time

  return answer
```

Note that the input data itself can sometimes be used as a key to the cache dictionary, but in general a key may need generating so that it can be used in a dictionary.

Comparisons

Although we have covered comparison operators in earlier chapters, there are a few occasions where it is easy to get unexpected results with what may seem like fairly simple tests. For example, you might do the following to detect an undefined value:

```
if not x:
  # Do something if x is logically false
```

This will work in all situations where the variable has a value that is logically false, like `0.0`, `None` or an empty list. However, sometimes you actually want the test to be more specific, for example, to catch `None` when the variable is undefined but not catch `0.0` when it is a useful numeric value. In this case the test should be:

```
if x is None:
  # Does not get here if x is zero
```

Clearly, sometimes you do actually want the more general test '`if not x`' to catch all logically false values, depending on the situation.

When comparing whether two things are the same we can use either '`==`' or '`is`'. Remember that the double equal sign checks whether the two items have the same value, but the `is` keyword checks whether they are the same Python object. This can be important where there are two different objects but they hold the same value. For example, when comparing integer and floating point numbers:

```
x = 5.0  # Floating point
y = 5    # Integer

if x == y:
  # Succeeds; same values

if x is y:
  # Fails; different objects
```

Hence when working with numbers '`==`' is usually what we want. Conversely, when working with more complicated Python objects it is often best to use the `is` keyword, so you know you really have the same object, because classes can actually be written to yield the same value in a comparison, even if they are not the same entity.

In many situations it is useful to act according to what the data type of an object is. Remembering that we can get the data type of a Python object with the inbuilt `type()`, we could check to see if a variable is of a specific type:

```
if type(x) is list:
  # Do something only if x is a list
```

The above example relies on the fact that the keyword `list` represents a data type, even though we often use it as a function, to create lists. Even if there is not a convenient keyword, in Python 2 we can get hold of all inbuilt Python data types via the `types` module, e.g.

```
import types
print(dir(types)) # List things in the module
```

In Python 3 the `types` module exists but does not include data types with built-in keywords (like `int` or `list`). This module enables you to do:

```
if type(x) is types.FunctionType:
  # Do something only if x is a function
```

The comparison data type could also be generated within the statement using the right kind of object (here a list, albeit an empty one):

```
if type(x) is type([]):
    # Do something only if x is a list
```

Type checking can also be done using the inbuilt `isinstance()` function:

```
if isinstance(x, list):
    # Do something only if x is a list
```

And this is especially handy if there are several different data types that you want to catch. For example, we may wish to do something if the variable is a list or a tuple or a set.

```
if isinstance(x, (list, tuple, set)):
    # Do something if x is a list, tuple or set
```

Alternatively, this could be achieved in a more verbose manner by combining multiple tests with the OR operator. Using `isinstance` is helpful when working with object classes and their hierarchies, especially considering that with old-style Python version 2 classes[7] using `type()` doesn't help:

```
class Something:        # Python: old style
  def func(self):
    print("First class")
```

---

[7] A new-style class in Python 2 would be defined using `class Something(object):`. In Python 3 all classes are new style and do not need to subclass `object`.

```
class Different:
  def otherFunc(self):
    print("Second class")

a = Something()
b = Different()

type(a) is type(b)       # True; not what you might expect! Python only
                         # These are both type 'instance'
isinstance(b, Something) # False
isinstance(a, Different) # False
```

As mentioned before, `isinstance` can also be used when one object is a subclass of another. Taking an example from Chapter 7 where a `Protein` is a subclass of `Molecule`:

```
myProtein = Protein('Enzyme A', sequence)  # a pre-specified sequence

if isinstance(myProtein, Molecule):
  # Succeeds : Classes are different, but Protein is a subclass of Molecule
```

In Python version 2 the comparison operations (such as `==`, `!=`, `<`, `>`, `<=`, `>=` etc.) will work with all different data types, and while we might expect to compare floating point numbers with integers we can actually compare all types, whether it makes much sense or not:

```
a = 'Banana' # A text string
b = 5        # An integer

a == 5       # False; as you might expect
a > b        # True in Python 2; arbitrary, but consistent
```

Where a value comparison doesn't make too much sense Python arbitrarily deems one value to be greater than the other, including for custom classes, although it always does this in a consistent way. This may seem odd, but such comparisons are handy in various situations. For example, you can sort a list containing items of mixed data types and get the same result each time. However, it can be argued that allowing meaningless comparisons allows errors to go unnoticed. It is for perhaps this reason that in Python version 3 comparisons cannot be made between incompatible types, so here comparing integer with floating point works, but integer and string does not.

When dealing with comparisons and arrays created using the NumPy module, we have to be especially careful. It might be expected that the comparison operations with a `numpy.array` object give back a single `True` or `False`, as is generally the case in Python. However, they actually give back arrays of `True` and/or `False`:

```
from numpy import array

a = array([1, 0, 9, 1])
b = array([2, 0, 3, 1])

print(a == b)   # Gives array([False, True, False, True])
```

As you can see, the array comparisons are element-wise comparisons. While this kind of result can be handy in a NumPy context it has some important consequences. For example, unlike with lists, sets or tuples, a NumPy array comparison cannot be placed directly in a logic test:

```
if a == b:
   # Never gets here! Raises a ValueError
```

Also, although you can put such arrays in a list, the list cannot be sorted (because sorting needs plain `True` or `False` to come from value comparison).

```
myList = [a, b]
myList.sort()    # Fails! Raises a ValueError
```

Miscellaneous

Lastly, we move on to the last few tips that don't seem to fit in the other categories.

To write some test code in Python that is not run when the file is imported as a module, but which is run when the file is used directly, the `__name__` attribute for the module can be inspected. This attribute belongs to the local Python file and is only equal to the text string `'__main__'` when called directly.

```
if __name__ == '__main__':
   testSomething()
```

When performing imports it is good practice to issue them at the top of a file where possible, so you can see at a glance what is being used. Also, doing imports outside functions is generally quicker. However, imports inside function definitions may be needed to avoid circular imports (i.e. `a` imports `b` which imports `a` which imports `b` ...). So avoid:

```
def func(x, y):
   from SomeModule import someFunction
   someFunction(c)
```

when you can do:

```
from SomeModule import someFunction

def func(x, y):
   someFunction(x)
```

Importing everything from a module with the '`*`' notation slows down execution (including because it interferes with the optimisation of the Python compiler). It may also lead to other problems:

```
from library import *        # Generally bad
from somethingElse import *   # Even worse to do it twice in one module
```

The problem is that you may not be sure what you've imported, and you might have imported something with a name that clashes with one of the internal names or one of the names from another import. Also, if you do this for multiple modules it won't be at all clear which thing comes from which module. If you need lots from the imported modules it is better to consider using dot notation, if you don't want to make imports explicit. So, for example:

```
import library
import somethingElse
```

```
x = library.magic()
y = somethingElse.magic()
```

If you need to know what attributes and methods an object has, try the `dir()` keyword:

```
dir(obj)
```

Or maybe try the `pydoc` graphical interface and have a browse:

```
pydoc -g # Python 2
pydoc3 -b # Python 3
```

11 Biological sequences

Bio-molecules for non-biologists

This section is aimed at programmers who do not have much biological training, to explain a little about where biological sequences come from. Naturally we must omit a large amount of detail if we're going to keep things short enough for this book. The emphasis will be about how information is stored, transferred and interpreted in biological systems to ultimately give the chemistry of life. We leave details of the current understanding of the precise mechanisms to your enthusiasm and further reading.

      Life can be thought of as a set of controlled chemical reactions and interactions that build and maintain organisms. When there is no control over biochemistry the raw materials of an organism soon succumb to decay; complex biological molecules turn into much simpler, more stable forms. The specific set of chemical reactions and interactions that allow life to live and reproduce are mostly directed by protein molecules with occasional roles for RNA molecules.[1]

Proteins

The different kinds of protein molecules that direct the processes needed for life are different because they are made up of different sequences of smaller entities, amino acids (see

---

[1] RNA stands for ribonucleic acid and it is present in several fundamentally important biochemical situations.

One-letter sequence

MYGKIIFVLLLSEIVSISASSTTGVAMHTSTSSSVTKSYISSQTNDTHKRDTYAATPRAH
EVSEISVRTVYPPEEETGERVQLAHHFSEPEITLIIFGVMAGVIGTILLISYGIRRLIKK
SPSDVKPLPSPDTDVPLSSVEIENPETSDQ

Three-letter sequence

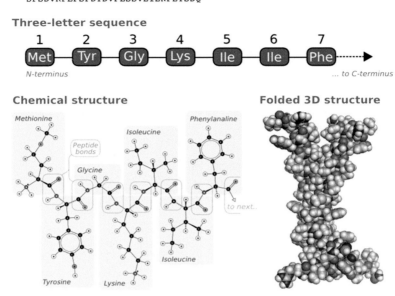

Figure 11.1. **A protein amino acid sequence and the polymer molecule it represents.** A protein is a polypeptide chain consisting of a sequence of amino acid residues, and this may be represented in several different ways. In the simplest form, one-letter codes are listed sequentially, where each letter represents a different kind of amino acid. The sequence may also be represented by three-letter amino acid codes. For both kinds of sequence the amino acids are listed in order starting from the N-terminus, which has an unlinked amine chemical group. Such sequences are really just simplifications of the underlying chemical structure, which in most biological situations adopts a particular compact three-dimensional folded structure.

Figure 11.1). This sequence specifies their shape, physical properties, movement and chemical activity. There are 20 common amino acid types that are joined together into chains of varying length to make the various proteins. The amino acids are joined together into a linear sequence by chemical bonds that are referred to as peptide links, thus protein chains are frequently referred to as *polypeptides*. Most proteins adopt a particular three-dimensional arrangement as segments of the amino acid entities within the chain come together into one or more compact globules. The final shape of a protein is usually vitally important for its function, and this shape is governed by the combination and order of amino acids in the polypeptide chain. However, it should be noted that the relationship between protein sequence and structure is exceedingly complex, such that we cannot generally predict a protein's structure directly from its sequence alone.

The amino acids that are linked into a protein chain are often referred to as *residues*. The origin of this term is somewhat archaic; it stems from the early days of biochemistry. When the sequence of amino acids in a protein was first discovered, it was done by carefully removing only the amino acid at the start of the protein chain using chemical cleavage. This removed one kind of amino acid, separating it from the remainder of the protein, and the amino acid was left as a chemical residue (i.e. the leftover) from the cleavage reaction. Successive rounds of amino acid removal, on a shortening protein chain, gives successive chemical residues, each of which corresponds to a particular kind of amino acid. Thus the order of the kinds of chemical residue reveals the order of the amino acids that made up the protein chain. Today you will see the term *residue* used when one wants to refer to a particular amino acid in a particular position of a protein chain. The term is also frequently used in the same way for the entities that make up chains of DNA and RNA, the other types of biological molecules that have a linear sequence.

When a protein is constructed it is made inside a living cell by joining amino acids together via peptide links, in the correct order for that type of protein[2] in a process called *translation*. The information about which one of the 20 types of amino acid is joined to the previous one in the sequence, at the growing end of a protein chain, is determined by a different kind of molecule; an RNA. RNA molecules are also made up of chains of smaller entities, which in this case are called nucleotides (completely different to amino acids that are found in proteins). RNA molecules in this instance can be thought of as messages, because they are relaying the information to create proteins. The origin of the sequence information that RNA transfers to protein ultimately comes from DNA, arguably the most famous of the biological molecules. It should be noted that not all RNA molecules are used to make proteins; the non-coding RNAs have various other roles in a cell.

The sequence of components in RNA is essentially a short-lived copy of the information that is stored in molecules of DNA. So even though the actual chemical reactions of life mostly happen because of proteins, the blueprint of how to make the proteins comes from the DNA. DNA is the permanent store of information present in every cell. There is a little caveat to this point because some cells, like red blood cells in human beings, lose their DNA. For the red blood cell this gives it more space to fulfil its role of carrying oxygen around the body, at the cost of having a short lifespan: its RNA messages will eventually run out and it will no longer be able to make new protein (which all cells must do to survive).

DNA

DNA is present in a cell because it was passed from parent to offspring. Half of your DNA sequence will come from your mother and half from your father. Of the total DNA inside a cell, only part of it will be used to make RNA messages, and thus ultimately proteins. The regions of DNA that are used to make RNA, by specifying its sequence, are called *genes*. The remainder of the DNA that is not part of any gene may have a biological role or it may be junk. Junk DNA does not have any specific function, but it is perhaps useful in providing space around genes so that life can evolve by shuffling genes without damaging them.

The parts of DNA that are neither junk nor genes are critically important. Included in such regions are DNA sequences that determine which genes are actually used on a given occasion. For example, consider a brain cell and a muscle cell inside a human; both cells have the same DNA but one helps you think and the other helps you move. The different jobs that the different cells do are only possible because they make different kinds of protein molecules.

---

[2] Cells have a specific kind of molecular particle, containing protein and RNA chains, called the *ribosome*, which does this job.

One-letter sequence

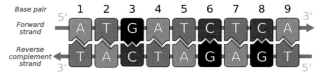

ATGATCTCAGGATGTATGGAAAAATAATCTTTGTATTACTATTGTCAGGTAAGTGATTTTATTTCATCTT
GGTTCTGTTATATTGGGTATGAGATCATAGAATAAAATATGAACTACCCTATTTTAGTTCTATCTTATTT
AAATCAATAAATGAGTAGTATTTCCTCTTCCAGTCTGGTGGATGGATTTTACTGGAACTCAGCTACCAAT
GTGGGGGAAATGGCACAAGGGAGCCCAGTATTTATGGCCAAATCCAGTTTTCTAGTATGAGAAGCTTACT
TCAATTCTAAGTCTAGCTAGAATTAAAATAATTTT

Double-stranded base pairs

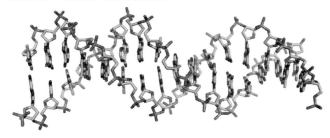

Double helical structure

Figure 11.2. A DNA sequence, the double-stranded base pairs and the molecule it represents.
A DNA chain is commonly represented as a one-letter sequence where each letter represents a different type of nucleotide chemical component. By convention the letters are listed by starting from the 5′ end of the DNA strand (a numbering that relates to the position of a free hydroxyl chemical group on a ribose sugar ring). In most cases the DNA sequence is really a representation of a double-stranded molecule where, even though only the sequence of one strand is specified, the opposite strand is deducible as the reverse complement, following the normal G:C, A:T base-pair rules. The normal three-dimensional structure of the double-stranded DNA is a double helix.

They make different protein molecules because different sets of genes are active. In each type of cell some genes will be switched off and some will be switched on. It is DNA that lies outside a gene that provides these on/off switches (often near the starts of genes). We will forego detailed discussion about how these gene switches are controlled, but suffice it to say that in the case of muscle cells and brain cells in humans the initial difference in gene activation is made early in development, when a baby is just a tiny embryo.

Chemically DNA is composed of a chain of four different kinds of smaller component called nucleotides.[3] People often refer to DNA as having four bases, which refers to the part of the DNA residues that differs between the four types. The common part of the residues is a sugar scaffold joined by phosphate groups. Because DNA is the permanent store of information in a cell it is packaged up in a protected form, as a double helix with many special proteins bound to it. This double helix structure means that DNA is really composed of two nucleotide chains aligned along their length, commonly called *strands*. There are strict rules about how the two strands are aligned and joined to one another; each base pairs up with only

---

[3] Strictly 2′-deoxynucleotides if you know your ribose sugar chemistry.

one other type (A with T and G with C) on the adjacent strand. Such sequences of nucleotides that are able to pair up according to these rules are said to be *complementary*. Thus although there are two chains, one strand always mirrors the information of the other. Together, one long pair of DNA strands inside a cell is called a *chromosome*, although in higher organisms these strands are never naked DNA; they have lots of proteins bound to them to compact the chromosome and control the reading of genes.

Transcription

When the DNA information is read in a process called *transcription* its double helix is unwound,[4] in a small area, so that its bases, on what is called the template strand, are exposed. These exposed nucleotides specify how an RNA molecule is made. RNA is a polymer composed of four different types of nucleotide residue just like DNA. RNA chains do not form stable double helices like DNA, but they can associate with a length of DNA following the same base-pair rules (the exception being that the T base in DNA is replaced by the similar U in RNA, which also pairs with A). Each exposed nucleotide in the DNA chain will only bind one complementary kind of RNA nucleotide, which is put on the end of the growing RNA chain; thus the DNA sequence dictates the RNA sequence in a predictable way.

Which physical DNA strand of the two acts as the template to make RNA varies; it could be either. In other words, regions of both DNA strands are used as RNA templates, but a specific gene will only use one strand. Accordingly, when an RNA molecule is made, its sequence mirrors that of the DNA; its nucleotides are joined into a chain by physically binding the template DNA strand. Given that RNA uses the same base-pair rules as DNA, it will be complementary to the template. Because the other DNA strand is also complementary to the template (usually it pairs up to form a helix), so the RNA and the other DNA strand have the **same** sequence. The DNA strand which has the same sequence as the RNA is called the *coding* strand. When dealing with gene sequences in computing it is usually the case, for example, when looking in a bioinformatics database, that you will be working with the sequence of the coding DNA strand, which is the same as the RNA sequence. Also, even though RNAs really have U bases instead of T bases, in bioinformatics an RNA sequence will often be presented with Ts, as if they were U; certainly this is programming laziness, but it does mean that most programs don't care whether the sequence came from RNA or DNA, and after all they are often representations of the same information.

Translation

Most RNA molecules go on to specify protein amino acid sequences in a process called *translation*; these are called messenger RNAs (mRNA).[5] Because there are 20 (common) types of protein amino acids and only four RNA nucleotides, a combination of RNA nucleotides is required to specify each amino acid. By a mechanism which we will not get into, at a point within an mRNA (starting with the sequence 'AUG') each subsequent group of **three** bases, called a *codon*, directs one of the 20 common protein amino acids to be joined

---

[4] By special proteins called helicases.
[5] The RNAs that don't act as messengers often have a direct physical role in biochemistry, just like proteins.

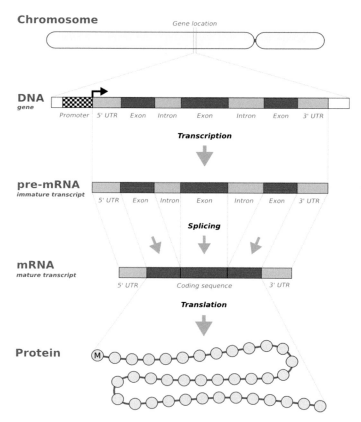

Chromosome

Gene location

DNA
gene

Promoter | 5' UTR | Exon | Intron | Exon | Intron | Exon | 3' UTR

Transcription

pre-mRNA
immature transcript

5' UTR | Exon | Intron | Exon | Intron | Exon | 3' UTR

Splicing

mRNA
mature transcript

5' UTR | Coding sequence | 3' UTR

Translation

Protein

Figure 11.3. An overview of transcription and translation processes. When a protein is made in a cell its sequence is determined by the coding sequence contained in a messenger RNA molecule in a process called translation. This RNA is created from a gene region in the DNA, which is itself part of a chromosome, in a process called transcription. In most cases an initial RNA transcript, effectively a copy of part of one of the DNA strands, must be processed by splicing, to remove non-coding introns.

onto a growing protein chain. Because DNA can be copied into RNA from either of its two strands and because on each strand there are three possible ways to group the nucleotides into codons, DNA has six *reading frames*, i.e. six possible ways for the same region to be used to make a protein sequence. Of course one gene only uses one reading frame, but different genes exploit all of the six possibilities.

With four types of base in RNA there are 64 (4×4×4) possible combinations of triplet nucleotide sequence that can make a codon. A few of the codons will cause the production of a protein to stop, but most will specify the inclusion of one specific amino acid. Because there are more codons than amino acid types, an amino acid is often specified by several different codons. The relation between the triplet of bases in codons and the amino acid that is added to a protein is known as the *genetic code* (see Table 11.1). This genetic code may be different in different kinds of organism, but only slightly so.

An important complication to the way in which RNA transmits its sequence message comes from the fact that it usually has large non-coding sections removed before it becomes a mature mRNA and its sequence is translated into protein. The regions of an RNA chain that

Table 11.1. *A table of the standard genetic code.* *Messenger RNA molecules are translated into protein sequences at a large molecular assembly called a ribosome. This takes subsequent groups of three RNA nucleotides, called codons (starting from an 'AUG') and joins different amino acids into a growing polypeptide chain, according to which codon sequence is present. The mapping between the three-letter RNA codons and each amino acid is known as a genetic code. In most cases multiple codons (of which there are 64 in total) correspond to a single amino acid (of which there are 20 in total). Some of the codons do not correspond to any amino acid and cause the protein chain synthesis to stop. The table shown is the standard genetic code, which is valid for the genes found in the cell nucleus of most of the most commonly studied organisms. There are small variations compared to this code in bacteria and organelle genomes (e.g. mitochondria and chloroplasts).*

Codon positions			Amino acid translation or stop			Codon positions			Amino acid translation or stop		
1st	2nd	3rd	One-letter code	Three-letter code	Full name	1st	2nd	3rd	One-letter code	Three-letter code	Full name
U	U	U	F	Phe	Phenylalanine	A	U	U	I	Ile	Isoleucine
U	U	C				A	U	C			
U	U	A	L	Leu	Leucine	A	U	A			
U	U	G				A	U	G	M	Met	Methionine
U	C	U	S	Ser	Serine	A	C	U	T	Thr	Threonine
U	C	C				A	C	C			
U	C	A				A	C	A			
U	C	G				A	C	G			
U	A	U	Y	Tyr	Tyrosine	A	A	U	N	Asn	Asparagine
U	A	C				A	A	C			
U	A	A			**STOP**	A	A	A	K	Lys	Lysine
U	A	G				A	A	G			
U	G	U	C	Cys	Cysteine	A	G	U	S	Ser	Serine
U	G	C				A	G	C			
U	G	A			**STOP**	A	G	A	R	Arg	Arginine
U	G	G	W	Trp	Tryptophan	A	G	G			
C	U	U	L	Leu	Leucine	G	U	U	V	Val	Valine
C	U	C				G	U	C			
C	U	A				G	U	A			
C	U	G				G	U	G			
C	C	U	P	Pro	Proline	G	C	U	A	Ala	Alanine
C	C	C				G	C	C			
C	C	A				G	C	A			
C	C	G				G	C	G			
C	A	U	H	His	Histidine	G	A	U	D	Asp	Aspartic acid
C	A	C				G	A	C			
C	A	A	Q	Gln	Glutamine	G	A	A	E	Glu	Glutamic acid
C	A	G				G	A	G			
C	G	U	R	Arg	Arginine	G	G	U	G	Gly	Glycine
C	G	C				G	G	C			
C	G	A				G	G	A			
C	G	G				G	G	G			

are removed are called *introns* and those that remain are called *exons*. The RNA is said to be *spliced*: the ends of the exons are joined as the introns are lost. Introns are very common in the human genome and their presence makes it significantly more difficult to detect which bits of a gene are actually used to make protein sequences.

Even though DNA, RNA and protein really are sequences of chemical compounds, linked together into a chain, it is often sufficient to represent them simply as a sequence of letters or residue codes. You can perform many useful analyses simply by knowing what the order of amino acids or nucleotides is, without having to consider all of the atoms that are present in the real molecule.

DNA sequencing

Today the majority of the sequence information for DNA, RNA and protein in various organisms comes from the sequencing of just DNA. Because of the rules of nucleotide pairing and because of the genetic code (three nucleotides give one amino acid) it is easy to determine an RNA and protein sequence once you know the gene-coding regions in the DNA. It may be difficult to work out where the coding regions of a gene start and end in a large section of DNA, but the conversion to the different types of sequence is trivial.

DNA is sequenced with a special kind of chemical reaction, which these days is often performed by a computerised machine. In essence many copies of a DNA strand are made using an enzyme (a protein that catalyses the required chemical reaction), and the nucleotides that are added to the end of the growing strands are detected. A common way (used in Sanger and Illumina sequencing methods) of detecting the nucleotide added is to have the reaction occasionally stop, when an inhibiting compound is incorporated at the growing end. Here there are four different inhibitors that take the place of each of the DNA nucleotides. The aim is to get some of the copied DNA strands to stop growing at every single nucleotide position. The sequence is revealed by detecting which inhibitor stopped the chain growing at each position; i.e. which nucleotide is at the end of each length of strand. The different inhibitors at the end of the DNA strands are designed to glow with different colours to make them easy to identify. The reaction can happen in distinct cycles (e.g. Illumina method) to give subsequent nucleotide reads, or the DNA strands can be sorted by size so that the end nucleotide can be detected afterwards (the Sanger method).

The actual DNA that is used in the sequencing reaction commonly comes from an organism's set of chromosomes (which collectively are referred to as a *genome*), but it is also possible to have DNA which comes from the amplification of a small section of a genome (i.e. a small quantity is copied to give a large amount) or to use DNA that has been copied from RNA (i.e. opposite to the usual flow of information) using a special enzyme.

Using biological sequences in computing

Whatever the origins of a biological sequence, before writing programs to work with biological sequence information one must first have the sequences represented in some data structure; ideally this should suit the purpose of any subsequent analyses. There are various ways in which people store sequences, ranging from the simplistic to the exceedingly complex, and each will have its own advantages and disadvantages.

The commonest and simplest method is to store sequences as text; i.e. as strings of letters, where each letter represents a different kind of residue. Thus for DNA and RNA we will be working with alphabets of four letters, representing nucleotides, and for proteins an alphabet of 20, representing amino acids. For the standard set of residues that make up the majority of biological polymers, this representation is sufficient as we have more than enough letters on a standard keyboard. However, a simple one-letter representation is not good enough if we need to describe unusual amino acids (both naturally modified and artificially created). In such circumstances people usually resort to three-letter code strings for amino acids: for example one can distinguish between proline 'PRO' and hydroxyproline 'HYP'.[6] One can go further still and define a biological sequence as a series of purpose-made object data structures, rather than a series of text codes. While using lists of complex objects will be cumbersome and unnecessary for many tasks, they are certainly a good choice if you need to work with the underlying atoms within a residue, as is the case in structural biochemistry.

In Python, a sequence of one-letter residue codes will usually be represented as a string data type and three-letter codes as a list, although other arrangements are of course possible. Also, if we are being cautious with our sequences then we may like to check that our data structures only contain valid codes. A biological sequence can also be included in a larger data structure if it needs to be annotated with further information. Although Python dictionaries can be used for this purpose when you need something quick, we sometimes advocate defining a custom object that can link your sequences to other data.

For testing and demonstration purposes, like the examples in this book, sequence data can be entered directly into the code of your programs. Of course for real-world applications of programs we would want to have our programs work on arbitrary sequences that we read in from a file or database. These could be sequences that have been output from another program, something you have obtained by searching a large sequence database or even an entire genome sequence that you have downloaded. Interacting with files and databases directly is dealt with in Chapters 6 and 20, and for the moment we will simply demonstrate with short sequences.

Once you have your sequences in some kind of data structure, it is time to start analysis. While we cannot hope to anticipate all that you might need to do, we can at least give some idea of what is possible. At the same time we aim to show how some of the things that are commonly done with sequences can be readily achieved with Python. The following examples are simple scripts that all deduce some property of an input DNA, RNA or protein sequence that gives some real-world information or prediction about the sequence. Note that in all of the examples we will forego checking that the sequences we used are valid: that they are the right kind of object and that they contain only the known types of residue code or letter. In an important real-world application you would clearly make such checks before you try to run any analyses and the BioPython modules that we demonstrate at the end can help you do this.

---

[6] This example uses the Protein Data Bank naming system.

Translate a DNA sequence into protein

The first example script is designed to determine the sequence of amino acids in a protein, starting from a DNA or RNA sequence by using the genetic code, stored in a Python dictionary, to perform the translation. The situation is generally more complicated, because precisely which section (or sections) of a nucleotide sequence end up being used is not always clear; there can be the issue of finding a gene amongst a large amount of DNA and working out how the RNA that is made (*transcribed*) from the gene is processed by splicing to give a mature messenger RNA. For now we leave such problems aside.

Firstly, we define a dictionary that contains our genetic code. Here we use strings containing three nucleotide letters as the dictionary's keys; these are the codons. The value associated with each codon is the three-letter code of the appropriate amino acid or the None object if it is a stop codon.

```
STANDARD_GENETIC_CODE = {
          'UUU':'Phe', 'UUC':'Phe', 'UCU':'Ser', 'UCC':'Ser',
          'UAU':'Tyr', 'UAC':'Tyr', 'UGU':'Cys', 'UGC':'Cys',
          'UUA':'Leu', 'UCA':'Ser', 'UAA':None,  'UGA':None,
          'UUG':'Leu', 'UCG':'Ser', 'UAG':None,  'UGG':'Trp',
          'CUU':'Leu', 'CUC':'Leu', 'CCU':'Pro', 'CCC':'Pro',
          'CAU':'His', 'CAC':'His', 'CGU':'Arg', 'CGC':'Arg',
          'CUA':'Leu', 'CUG':'Leu', 'CCA':'Pro', 'CCG':'Pro',
          'CAA':'Gln', 'CAG':'Gln', 'CGA':'Arg', 'CGG':'Arg',
          'AUU':'Ile', 'AUC':'Ile', 'ACU':'Thr', 'ACC':'Thr',
          'AAU':'Asn', 'AAC':'Asn', 'AGU':'Ser', 'AGC':'Ser',
          'AUA':'Ile', 'ACA':'Thr', 'AAA':'Lys', 'AGA':'Arg',
          'AUG':'Met', 'ACG':'Thr', 'AAG':'Lys', 'AGG':'Arg',
          'GUU':'Val', 'GUC':'Val', 'GCU':'Ala', 'GCC':'Ala',
          'GAU':'Asp', 'GAC':'Asp', 'GGU':'Gly', 'GGC':'Gly',
          'GUA':'Val', 'GUG':'Val', 'GCA':'Ala', 'GCG':'Ala',
          'GAA':'Glu', 'GAG':'Glu', 'GGA':'Gly', 'GGG':'Gly'}
```

Now we define a sequence. This initial sequence is really only for testing purposes. In reality of course we want to accept a variety of different sequences from files and databases.

```
dnaSeq = 'ATGGTGCATCTGACTCCTGAGGAGAAGTCTGCCGTTACTGCCCTGTGGGGCAAGGTG'
```

Given the nucleotide sequence, we take each group of three nucleotide letters and use the group as a key to look up the corresponding amino acid code, remembering of course that we must convert any DNA T residues into RNA U residues (which our genetic code dictionary requires). Assuming we find an amino acid code we add it to the list which represents the protein sequence. If we cannot find an amino acid for a codon, then we have reached a stop codon, whereupon our protein sequence is complete and we can immediately stop the translation. Note that we define the coding as a three-letter sub-sequence using the slice notation `seq[i:i+3]`, remembering that this will take letters from position `i`, up to but not including `i+3`. At the end we pass back the list of amino acid codes. This operation is put into a Python function, so that we can repeat the operation with any sequence and genetic code.

```
def proteinTranslation(seq, geneticCode):
  """ This function translates a nucleic acid sequence into a
      protein sequence, until the end or until it comes across
      a stop codon """

  seq = seq.replace('T','U') # Make sure we have RNA sequence
  proteinSeq = []

  i = 0
  while i+2 < len(seq):
    codon = seq[i:i+3]
    aminoAcid = geneticCode[codon]

    if aminoAcid is None: # Found stop codon
      break

    proteinSeq.append(aminoAcid)
    i += 3

  return proteinSeq
```

Note that there are many ways in which we could have extracted the groups of three letters from the input sequence. In this instance we used a `while` loop, and the loop continues as long as there are still at least three letters remaining, i.e. that the index plus two $i+2$ is still within the length of the sequence (and also unless the `break` is triggered by a stop codon). Here index `i` will be the position of the first letter in the codon and $i+2$ will be the last letter. Getting these 'boundary conditions' correct (so it is $i+2$ not $i+1$ or $i+3$) is one of the tricky bits of computer programming. Of course at the end of the loop we increase the index by three for the next round.

To actually run the function on our test sequence call the function by using its name in association with the variable for the test sequence and the variable that holds the genetic code: these get passed to the function as an argument. The resulting protein sequence is passed back to fill in the value of the `proteinSeq` variable.

```
protein3LetterSeq = proteinTranslation(dnaSeq, STANDARD_GENETIC_CODE)
```

Converting a DNA sequence to an RNA sequence is much easier, because all we have to do is replace T letters with U letters, as we already had to do when using the genetic code dictionary, and we can use the inbuilt Python functionality (assuming the sequence is stored as a text string) to do this.

```
rnaSeq = dnaSeq.replace('T','U')
```

Estimate molecular mass

This next script estimates the mass of a DNA, RNA or protein molecule (in units of daltons). This is only an estimate because various residues reversibly bind hydrogen ions under different conditions (i.e. pH affects whether H^+ ions are joined to the acidic and basic sites) and we are assuming standard proportions of the various isotopes.[7] Nonetheless this estimate

---

[7] The versions of a chemical element with different atomic weights, caused by different numbers of neutrons in the nucleus.

will be useful enough to say where we expect DNA or protein to lie on an electrophoresis gel[8] or mass spectrometer trace.

Firstly, we define a function, hopefully with a sensible and informative name, and specify that it takes one argument `seq`, which is a sequence, and one argument `molType`, which states whether we are using a protein sequence, a DNA sequence or an RNA sequence. Note that we set a default value for `molType` to be `'protein'`, so that we can work with protein sequences without having to explicitly specify the value.

Inside the function we define a dictionary that stores the average molecular weights of the different kinds of residue. Internally this dictionary contains three inner sub-dictionaries, one for each of the different molecule types. We access the correct inner dictionary using the `molType` as a key. The one-letter residue codes then act as the keys to the inner dictionary to extract the appropriate molecular masses.

Next we define a variable to hold the total for the molecular mass. This is initially defined with a value equal to that of the molecular mass of water, because the average residue masses in the dictionary do not take account of the end residues that have extra atoms (OH at one end and H at the other) because they are only linked on one side, instead of both sides.

```
def estimateMolMass(seq, molType='protein'):
  """Calculate the molecular weight of a biological sequence assuming
     normal isotopic ratios and protonation/modification states
  """

  residueMasses = {
      "DNA": {"G":329.21, "C":289.18, "A":323.21, "T":304.19},
      "RNA": {"G":345.21, "C":305.18, "A":329.21, "U":302.16},
      "protein": {"A": 71.07, "R":156.18, "N":114.08, "D":115.08,
                  "C":103.10, "Q":128.13, "E":129.11, "G": 57.05,
                  "H":137.14, "I":113.15, "L":113.15, "K":128.17,
                  "M":131.19, "F":147.17, "P": 97.11, "S": 87.07,
                  "T":101.10, "W":186.20, "Y":163.17, "V": 99.13}}

  massDict = residueMasses[molType]

  # Begin with mass of extra end atoms H + OH
  molMass = 18.02

  for letter in seq:
    molMass += massDict.get(letter, 0.0)

  return molMass
```

The `for` loop extracts each element of the sequence in turn, which will be a single nucleotide or amino acid letter. This letter is then used to look up the appropriate value of molecular mass in the dictionary. The `.get()` function of the dictionary is used so that a default value for the mass can be specified, just in case we have a letter in the sequence that is not in the dictionary. In such a circumstance using a guess for an average mass of an unrecognised residue, rather than `0.0`, may be appropriate under some circumstances. The molecular mass of the current

---

[8] A means of separating molecules in a sample according to size, by passing an electric current through a gelatinous substance in which differently sized chains move at different speeds.

residue is then added to the total, and the `for` loop moves onto the next letter in the sequence. Finally the `return` statement is used so that the value of the total molecular mass is passed back to the point in the program where the function was called from. To test this function we could do something like:

```
proteinSeq = 'IRTNGTHMQPLLKLMKFQKFLLELFTLQKRKPEKGYNLPIISLNQ'
proteinMass = estimateMolMass(proteinSeq)
```

or for DNA, noting that we have to specify the molecule type:

```
dnaMass = estimateMolMass(dnaSeq, molType='DNA')
```

Simple sub-sequence properties

The following examples move from considering a whole sequence to looking at the properties of regions or overlapping sub-sequences from within a larger sequence. Although we have only room to give a few simple examples, this kind of analysis is very important in trying to ascertain the biological function of the different parts of a sequence and in some cases to help guide experimentation.

Finding a sequence motif

The next example script is designed to find a particular smaller sub-sequence within a larger sequence. This kind of operation is useful because specific small sequences, called *motifs*, often have important biological roles. Examples of this in DNA include specific sequences which indicate the start of genes, specify where the coding regions of genes are spliced (introns are removed) or allow a protein to bind. There are also examples in proteins including where specific sequences are used to interact with other proteins, bind to DNA or direct that the protein should be modified (e.g. by joining sugars to it).

The following is a simple example of how to find a fixed sub-sequence within a larger sequence:

```
seq = 'AGCTCGCTCGCTGCGTATAAAATCGCATCGCGCGCAGC'
position1 = seq.find('TATAAA')
position2 = seq.find('GAGGAG')
```

This will give the values for `position1` as 15 and `position2` as −1 (because it cannot be found). The negative number that is given back from .find() when a substring is not found is a bit of an oddity. The authors feel that giving back the `None` object would have been more appropriate.

In many cases, however, it is not just one single well-defined sub-sequence that corresponds to a motif with a biological function. More usually there are a range of similar sequences that are all found to fulfil the same role. Accordingly, next we will not be defining a sequence motif as a single stretch of residues, but rather as a residue *profile*.

A residue profile states, for each position in the motif, what range of residues are found and how often one kind of residue is present compared to the others. Obviously, the residue profile that we use to try to detect a particular motif will have been determined beforehand, almost always by comparing sequences that we know (e.g. by experimentation) will function in the desired way.

This particular example attempts to find the region of a DNA sequence called the 'TATA box'. This is a biologically important region which, as the name suggests, often contains a particular sequence. The biological role of this sequence is to help define where the start of a gene is. Note that only some genes use the TATA box system.

Firstly, we define the profile that encodes the preferred sequence letters for each position in a section of DNA. The profile is stored as a Python dictionary, where the keys to the dictionary are the nucleotide letters and the values are lists of scores. Each score list contains a number for each position that indicates how likely it is to find the given nucleotide at that location.

```
profile = {
  'A':[ 61, 16,352,  3,354,268,360,222,155, 56, 83, 82, 82, 68, 77],
  'C':[145, 46,  0, 10,  0,  0,  3,  2, 44,135,147,127,118,107,101],
  'G':[152, 18,  2,  2,  5,  0, 10, 44,157,150,128,128,128,139,140],
  'T':[ 31,309, 35,374, 30,121,  6,121, 33, 48, 31, 52, 61, 75, 71]}
```

Next we define a function that takes a DNA sequence and a profile to determine where the profile best matches. A `for` loop is used to define the start point `i` of each sub-sequence that we wish to test against the profile. Note that we have ensured that the sub-sequence does not fall off the end of the larger input sequence because we stop the loop at the end of the sequence **minus** the profile length, for which we use the variable `width`.

We calculate the score for each sub-sequence by looking up the scores for the component letters using the profile dictionary: we use the nucleotide letter to first get the correct sores list and then the profile position `j`, to extract the right score. The position we interrogate from the input sequence is defined as `i+j`; i.e. we add the position within the sub-sequence to the absolute start of the sub-sequence within the larger sequence. The best score that has been found so far is recorded, together with the position where that score occurred.

```
def matchDnaProfile(seq, profile):
  """ Find the best-matching position and score when comparing a DNA
      sequence with a DNA sequence profile """

  bestScore = 0
  bestPosition = None # Just to start with

  width = len(profile['A'])

  for i in range(len(seq)-width):
    score = 0

    for j in range(width):
      letter = seq[i+j]
      score += profile[letter][j]

    if score > bestScore:
      bestScore = score
      bestPosition = i

  return bestScore, bestPosition
```

Rather than finding the best score and best position, we could alternatively have reported the scores at all positions, or even a list of positions that all give a score above a specified threshold. To test our function with our profile and DNA sequence, and report back which section of the sequence gave the best hit, we do the following:

```
score, position = matchDnaProfile(dnaSeq, profile)
print(score, position, dnaSeq[position:position+15])
```

GC content of DNA

The next example investigates a DNA sequence by measuring its GC content: i.e. the percentage of the total base pairs that are G:C (rather than A:T). All we need to do for this is to take the sequence of one strand of DNA and simply count how many of the nucleotides are G or C. Measuring the GC content of DNA is biologically relevant because regions of a chromosome that are rich in G and C give a hint that they might be coding for genes.

To make things more interesting, rather than just report the final GC content for the whole of a sequence, we will measure the GC content for every possible 10 residue sub-sequence and then plot the values along the length of the sequence as a graph. In other words we will perform the calculation on a sliding window of residues.

Firstly, we define a function that takes a DNA sequence and a window size (optionally) as input and gives a list of numerical GC content values as output. We will take the output data and use it to draw a graph using an external Python module called Matplotlib, which is very useful for plotting numerical data (see Chapter 9 for more details). As with the profile search above, we will use a `for` loop to scan through the sequence, while taking care to avoid falling off the end. However, this time, because we don't need the position of each nucleotide within the search window (we only needed this before to get the right position in a profile), we can find the number of G and C letters by using the `.count()` method that is built into Python strings and lists.

```
def calcGcContent(seq, winSize=10):
  gcValues = []

  for i in range(len(seq)-winSize):
    subSeq = seq[i:i+winSize]
    numGc = subSeq.count('G') + subSeq.count('C')
    value = numGc/float(winSize)
    gcValues.append(value)

  return gcValues
```

Each of the measurements for each sliding position are added to the output list, which we can then plot as a graph as follows. Note that this example assumes that we have installed the Matplotlib module (see http://www.cambridge.org/pythonforbiology for download links), otherwise you will get an error from the `import` command.

```
from matplotlib import pyplot

gcResults = calcGcContent(dnaSeq)

pyplot.plot(gcResults)
pyplot.show()
```

Protein hydrophobicity plot

Now we will move on to another example which produces data which we can display as a graph, but this time it will be for a protein sequence. The task here is to generate a plot of how water-hating, or to use the proper term *hydrophobic*, a given stretch of residues is. An amino acid may be hydrophobic if the atoms in its side chain[9] represent an arrangement that does not favour interactions with water molecules; typically this means they don't carry a charge or chemical groups that can form any hydrogen bonds. It is often useful to find such hydrophobic regions because by shunning water they make important interactions inside the folded core of proteins or allow a protein to be inserted into a cellular membrane. It is in the context of cell membranes that this example is based.

A cellular membrane is a double layer (*bilayer*) of hydrophobic lipid[10] molecules into which specific proteins are embedded by virtue of a hydrophobic anchor. A membrane defines the outer extent of each cell, and various internal compartments, with special functions, inside it. Biologically the lipid component of a membrane creates a barrier to most molecules and the protein component allows selective passage for some molecules, in line with the requirements of the cell.

The next example function aims to predict whether a protein possesses a sufficiently hydrophobic segment of residues (which will fold into a helix) that will allow it to be inserted into a cell's system of membranes. This is a simplistic prediction, as in reality there are other factors that govern whether a segment is used, but nonetheless it is sufficiently accurate to find over 70% of membrane spans.

Initially we define a hydrophobicity scale: a number associated with each amino acid letter that says how water-hating it is. For this example we will use the GES scale,[11] but there are several others to choose from.

```
GES_SCALE = {'F':-3.7,'M':-3.4,'I':-3.1,'L':-2.8,'V':-2.6,
             'C':-2.0,'W':-1.9,'A':-1.6,'T':-1.2,'G':-1.0,
             'S':-0.6,'P': 0.2,'Y': 0.7,'H': 3.0,'Q': 4.1,
             'N': 4.8,'E': 8.2,'K': 8.8,'D': 9.2,'R':12.3}
```

We define the function that will perform the search so that it accepts a protein sequence and hydrophobicity scale dictionary as mandatory inputs, and an optional input to specify a search window size. The philosophy of this function differs a little from those above because it includes an optimisation to calculate quickly; i.e. minimising the number of operations performed.

An index i is defined to loop through the sequence and, because it is useful in several spots, we define j to be i plus the search width. The adding up of the hydrophobicity score for each segment can take place inside one of two separate sections, depending on the result of an if statement. This statement is set up such that the first time we add up scores (detected by the score being at its start value of None) we consider all of the positions from i up to j. After this first summation, rather than repeating the summation for the whole of the next section, we use the fact that the next section only differs from the previous one at its first and last positions.

---

[9] The part that differs between each kind of amino acid.
[10] A kind of oily fat.
[11] Engelman, D.M., Steitz, T.A., and Goldman, A. (1986). Identifying nonpolar transbilayer helices in amino acid sequences of membrane proteins. *Annual Review of Biophysics and Biophysical Chemistry* 15: 321–353.

Accordingly, to get the score for the next section we take the existing score and take away the score of the residue we have just left behind (i-1) and add the score of the new end residue (j-1: we go up to but do not include position j). This is a speed optimisation because overall fewer operations are performed, but it will be prone to the accumulation of small floating point errors: however, such errors will not grow to anything significant for something as short as a protein sequence.

```
def hydrophobicitySearch(seq, scale, winSize=15):
  """Scan a protein sequence for hydrophobic regions using the GES
     hydrophobicity scale.
  """

  score = None
  scoreList = []

  for i in range(len(seq)- winSize):
    j = i + winSize

    if score is None:
      score = 0
      for k in range(i,j):
        score += scale[seq[k]]

    else:
      score += scale[seq[j-1]]
      score -= scale[seq[i-1]]

    scoreList.append(score)

  return scoreList
```

As before we can execute the function with an example sequence and plot the results with Matplotlib.

```
from matplotlib import pyplot

scores = hydrophobicitySearch(proteinSeq, GES_SCALE)

pyplot.plot(scores)
pyplot.show()
```

Measuring repetitiveness

It is often useful to measure how repetitive a section of DNA or protein sequence is. For DNA regions of repetitive sequence are often associated with non-coding, and especially 'junk' DNA; hence if we are searching for genes and their regulatory regions we look for a sequence that is not repetitive. For protein, repetitive amino acid sequences are associated with the parts that are unstructured: the bits that don't fold into a compact globule. This can be especially useful for multi-domain proteins (several independent compact globules along the length of the chain), where repetitive sequence regions can identify the flexible linkers between the compact, inflexible, folded domains.

For this example we will measure the repetitiveness along a sequence by counting how many different kinds of residue are present within a given search window. If a given sub-sequence only uses a limited number of residue types, rather than a varied mixture, we deem that sequence to be repetitive or have low *sequence complexity*. For a more formal mathematical definition we turn to information theory and use a measure which derived from the theory of *information entropy,* as initially described by Shannon.[12]

Considering that DNA and protein have different numbers of possible residue types (4 versus 20), it is clear that for these different kinds of molecule the expectation of how many *types* of residue we will see on average in any given segment will be markedly different. For example, in a given 12-residue segment, for DNA we would expect to see every kind of nucleotide, and if all nucleotides are equally likely there would be on average three of each kind. For 12-residue protein segments, it is impossible to have all amino acids present, and we might reasonably expect the *average* occurrence of any kind to be less than 1.

Accordingly, because the expectation of how likely it is to find a given type of residue varies according to the situation, we will measure sequence repetitiveness by comparing the measured occurrences with what we would expect given a random selection. Here we will refer to such a random selection as the *null hypothesis*: what we would expect in the absence of extra information (see also Chapter 22 for further discussion). In reality there are many null hypotheses that you could choose from, if you were searching within gene sequences and already know that G:C base pairs are more common than A:T. However, for this exercise we will assume an unbiased null hypothesis, where each residue type is equally likely: the baseline for nucleotides is 25% and the baseline for amino acids is 5%. We will refer to the formulation we use for this comparative measure of repetitiveness as the *relative entropy,* also known as the Kullback-Leibler divergence.[13]

The actual example code will be broken up into two separate functions; one will calculate the relative entropy and the other will scan through a sequence compiling the results. The first function takes a sequence, a list of possible residue types, which of course will differ for DNA and protein, and does the appropriate mathematics (see the last column of Table 11.2) to give back a single value H (a traditional letter to use for entropy). Note that this function does not contain any code specific to a given context. This means that the function will be useful in other situations, one of the main reasons to write it in a separate block of code. A separate function also helps to keep the Python code readable.

The workings of the function involve calculating the expected base-level proportion base, then filling the dictionary prop, which stores the abundance of each kind of residue we have in the input sequence. The residue counts are converted to a proportion (i.e. between 0 and 1) by dividing by the sequence length, making sure that we do floating point division (dividing two integers gives an integer in Python before version 3). Armed with the base level and the proportions we then loop through each residue type and add the local repetitiveness measure for that type to the relative entropy total. The occurrence of base-2 logarithms throughout is a means to ensure that the answer is output in units of *bits* (common in information theory).

```python
def calcRelativeEntropy(seq, resCodes):
  """Calculate a relative entropy value for the residues in a
     sequence compared to a uniform null hypothesis.
  """
```

---

[12] Shannon, C.E. (1948). A mathematical theory of communication. *Bell System Technical Journal* 27(3): 379–423.
[13] Kullback, S., and Leibler, R.A. (1951). On information and sufficiency. *Annals of Mathematical Statistics* 22(1): 79–86.

```
from math import log

N = float(len(seq))

base = 1.0/len(resCodes)

prop = {}
for r in resCodes:
  prop[r] = 0

for r in seq:
  prop[r] += 1

for r in resCodes:
  prop[r] /= N

H = 0
for r in resCodes:
  if prop[r] != 0.0:
    h = prop[r] * log(prop[r]/base, 2.0)
    H += h

H /= log(base, 2.0)

return H
```

The second function will use the `calcRelativeEntropy()` function as it scans through an input sequence. In this example, and unlike the previous ones in this chapter, rather than having our search window stop before it falls off the end of the input sequence we will use a cheat that will enable the search to go right to the last residue. This involves pretending that the sequence is circular; rather than falling off the end the search will jump back to the beginning of the sequence. Whether it is appropriate to cheat is at the programmer's discretion, but in this instance it gives results that appear reasonable, due to the fact that we are repeating a real segment of sequence.

Table 11.2. *The relative entropy formulation to measure sequence repetitiveness.* *The repetitiveness of a biological sequence may be formulated mathematically by calculating a relative entropy value D_{KL} (the Kullback-Leibler divergence). This is simply the summation, considering all the residue (amino acid or nucleotide, depending on the molecule) types, of the observed proportion of each type (P_i) multiplied by the log-ratio of the observed proportion divided by the proportion expected in random sequences (Q_i, which is always 0.25 in the above example). The relative entropy is illustrated for various degrees of sequence repetition, showing that the measure represents the variety of different residue types, in a sample of fixed size.*

Sequence	Residue proportions				Relative entropy (bits) $D_{KL} = \sum_i P_i log_2(\frac{P_i}{Q_i})$
	G	C	A	T	
GGGGGGGG	1.000	0.000	0.000	0.000	2.00
TCTCTCTC	0.000	0.500	0.000	0.500	1.00
GGGGCCCC	0.500	0.500	0.000	0.000	1.00
GAAGACGA	0.375	0.125	0.500	0.000	0.59
GCATTACG	0.250	0.250	0.250	0.250	0.00

The makeup of the function is quite simple: it takes the input sequence and search window size, and also an optional variable to flip between using DNA and protein sequences. Note that the length of the input sequence `lenSeq` is recorded before we perform our cheat of copying residue codes from the beginning of the sequence to make an artificial extension. Before entering the loop we define the residue codes that we will use for the relative entropy calculation, i.e. the four nucleotides for DNA or 20 amino acids for protein. Within the loop, rather than using the `.append()` method to add each score to the list we first make a blank list of the right size by repeating zeros (`[0.0] * lenSeq`), to which the individual scores are then added. This technique can often be used to make your code execute more quickly. The index `i` in the `for` loop goes all the way up to `lenSeq`, because we have added the extra sequence and the entropy score is calculated from the sub-sequence that is sliced out of the input sequence.

```
def relativeEntropySearch(seq, winSize, isProtein=False):
  """Scan a sequence for repetitiveness by calculating relative
     information entropy.
  """

  lenSeq = len(seq)
  scores = [0.0] * lenSeq

  extraSeq = seq[:winSize]
  seq += extraSeq

  if isProtein:
    resCodes = 'ACDEFGHIKLMNPQRSTVWY'
  else:
    resCodes = 'GCAT'

  for i in range(lenSeq):
    subSeq = seq[i:i+winSize]
    scores[i] = calcRelativeEntropy(subSeq, resCodes)

  return scores
```

We can then test the function in the usual manner, and make a graph of the results:

```
from matplotlib import pyplot

dnaScores = relativeEntropySearch(dnaSeq, 6)

proteinScores = relativeEntropySearch(proteinSeq, 10, isProtein=True)

pyplot.plot(dnaScores)
pyplot.plot(proteinScores)
pyplot.show()
```

Protein isoelectric point

The last of the examples in this chapter is an example that involves an *optimisation*. So far all of the numerical values that we have calculated have been deduced analytically by applying some equation: there is a direct method to get from the data to a precise answer. However, it is commonplace to come across problems where the values we are interested in are not directly accessible. In reality such problems may range from those for which it is genuinely difficult to imagine any formulaic method (hard problems: see Chapter 25) to those where a direct formulation is merely inconvenient or slow.

The topic of this example is the estimation of the *isoelectric point* of a protein, which we will call the *pI*. This is a measurable property of a protein: it is the pH[14] at which the protein carries no overall electric charge. This is something that is often used to characterise and isolate proteins, for example, by performing electrophoresis (moving particles through a porous substance with an electric current) in a gel with a pH gradient.

Proteins have electric charges because certain kinds of amino acids, together with the chain termini (the unlinked ends), are capable of accepting or losing a hydrogen ion (H^+). The groups that are capable of gaining a hydrogen ion, and thus a positive charge, are called *basic*: this includes the residues arginine, lysine and histidine[15] and also the N-terminus of the protein, the start of the chain where there is a free amine group. The groups which lose a hydrogen ion gain a negative charge; they are neutral before the loss. These groups are called *acidic* and include the residues aspartic acid, glutamic acid, cysteine, tyrosine and the C-terminus of the protein: the end of the protein chain where there is a free carboxylic acid group.

In any given situation whether or not these basic and acidic groups carry a charge depends on the hydrogen ion concentration of the environment: the pH. In a solution with a *low* pH the concentration of H^+ is *high*,[16] and so there are lots of free ions to bind to the basic groups, giving them a positive charge, and also lots of free ions to bind to the acidic groups, removing the negative charge and making them neutral. Conversely, with a *high* pH, the concentration of H^+ is *low*, whereupon the ions are lost from the protein; basic groups become neutral and acidic groups are left with a negative charge.

The different basic and acidic groups do not bind to hydrogen ions equally strongly. For example, aspartic acid very easily loses H^+; at neutral pH 7.0 they are almost all lost, but for tyrosine at a neutral pH hardly any are lost. The strength of any acid or base can be described by the *acid dissociation constant*, referred to as the pK_a. This has a formal mathematical definition using the concentrations of hydrogen-bound and unbound components,[17] but is most easily remembered as the pH at which on average *half* of the groups will be bound with H^+. Any one specific group can of course only be bound to a whole hydrogen ion or no hydrogen ion, so these constants represent the average over time as H^+ is dynamically lost and gained. The pK_a value for aspartic acid is 4.4, so at pH 4.4 it will have H^+ half of the time, and thus its average electric charge will be -0.5: half negative because the free half is negative. For aspartic acid, as pH *increases* it will become increasingly negatively charged as it will be bound to H^+ less of the time. Conversely the pK_a value for lysine is 10.0, thus at pH 10.0 it will be half bound by H^+, but because this residue is basic the ions add a positive charge, rather than neutralise a negative one. So for lysine, as the pH *decreases* more H^+ binds and it becomes more positively charged.

---

[14] i.e. hydrogen ion concentration in solution.

[15] See Table 11.1 for the three- and one-letter amino acid codes.

[16] pH is the negative $\log_{10}$ of the H^+ ion concentration.

[17] $pK_a = -\log_{10} K_a$, where K_a is the concentration of unbound component times the concentration of free hydrogen ions, divided by the concentration of hydrogen-bound component.

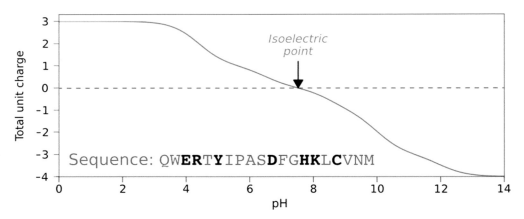

Figure 11.4. A graph of how overall peptide charge varies with pH. The estimated overall charge for an example protein sequence is shown for various pH values, which correspond to hydrogen ion concentrations. At different pH values the various acidic and basic chemical groups on some of the amino acids release or accept hydrogen ions, thus changing the average (considering many molecules over time) electric charge. The pI or isoelectric point is the pH value where the individual charges balance to give zero overall charge.

To calculate the pI we must find the pH where we think the positive and negative charges in the protein balance, hence must first have a method for estimating the charge of a protein chain at a given pH. We then use this method to test different values of pH, until we home in on the value where the overall charge is zero. It is possible to use an exhaustive method whereby we systematically test lots of pH values which differ by a very small amount until we find one that gives the charge closest to zero. However, here we will use a more intelligent method that will find a good answer in only a few steps. Although the example problem is not so challenging and we could have used the exhaustive method, for many other, larger problems using an intelligent optimisation is essential to get a reasonable answer in a reasonable time.

The optimisation algorithm we will use employs a divide-and-conquer strategy. We test various pH values by stepping between test points and for a given pH value and whether the resulting charge is above or below zero (positive or negative) tells us in which direction we must search next for a better answer. Also, if we come across a better guess for the pI (i.e. a pH that predicts a charge closer to zero) then we reduce the step size (how far to go for the next guess) by half so that we get increasingly close to the optimum value and don't overshoot far. Note that we are only able to use this strategy because we know how the problem works; it is well behaved because we know that there is only one solution and we know how far ahead to look for a better answer. Not all problems will be so simple, and for the more difficult situations we can employ the methods discussed in Chapter 25.

The function `estimateCharge` is designed to estimate the charge of a given sequence at a given pH. The basis of the calculation involves estimating the proportion of dissociated acidic and H[+]-bound[18] basic amino acids from reference pK$_a$ values. This procedure does not

---

[18] Often called *protonated* because an H[+] ion, without any electrons, is really just a bare proton.

take account of the effect that the sequence and folding of a protein has upon the dissociation constants of its component residues, i.e. the pK_a values in any real situation vary according to how charged residues interact with each other. Also, this calculation assumes we are in water at a standard temperature.

The function takes two input variables which are helpfully named and returns a single charge value. We define a dictionary of pK_a values for basic and acidic amino acids, keyed by their code letters. Note that we also have values for '+' and '-', which are symbols that will be used to represent the charge-carrying groups that arise from the free N- and C- termini. We define another dictionary, isAcid, so that we can look up whether each charge group acts as an acid or not.

For each amino acid letter in the sequence we find the pK_a value of the amino acid from the pK_a dictionary. If an amino acid is neither acidic nor basic (uncharged), and thus not present in this dictionary, the .get() function will helpfully give a value of None. If we do get a pK_a value we do the mathematics with the pK_a and the input pH[19] to calculate how much of the group will be dissociated (free from H^+).

If the residue is acidic we add a negative charge for the dissociated proportion of the residue. Otherwise, if the amino acid is basic, we have positive charge for the proportion of the amino acid that remains *associated* with hydrogen ions. The associated proportion is what remains after we subtract the dissociated proportion, hence 1-proportion, and we do not bother to multiply by +1 for a positive charge. The estimated charge of the individual amino acid is added to the running total. And at the end we return the total charge from the function to be used elsewhere.

```
def estimateCharge(sequence, pH):
  """Using pKa values estimate the charge of a sequence of
     amino acids at a given pH"""

  pKaDict = {'+': 8.0,'-': 3.1,'K':10.0,'R':12.0,
             'H': 6.5,'E': 4.4,'D': 4.4,'Y':10.0,'C': 8.5}

  isAcid = {'+':False,'-':True,'K':False,'R':False,
            'H':False,'E':True,'D':True,'Y':True,'C':True}

  total = 0.0

  for aminoAcid in sequence:
    pKa = pKaDict.get(aminoAcid)

    if pKa is not None:
      r = 10.0 ** (pH-pKa)
      dissociated = r/(r+1.0)

      if isAcid[aminoAcid]:
        charge = -1.0 * dissociated
```

---

[19] When we have a pK_a value we calculate the ratio between the hydrogen ion concentration (obtained from the input pH) and the dissociation constant K_a, because according to the dissociation equation $K_a = [H^+][A^-]/[HA]$, the ratio $K_a/[H^+]$ is the ratio between dissociated and associated concentrations. We then use the ratio, r, of dissociated $[A^-]$ to associated $[HA]$ to calculate the proportion of the dissociated component relative to the total, i.e. dissociated $= [A^-]/([HA] + [A^{-}]) = r[HA]/([HA]+r[HA]) = r/(1+r)$.

```
    else:
      charge = 1.0 - dissociated

    total += charge

  return total
```

The `estimateIsoelectric` function uses the `estimateCharge` function defined above to estimate the pH at which a protein sequence will be neutrally charged. To the input sequence of letters we add the + and - symbols to represent the charge groups at the N and C termini (strictly speaking these don't have to be at the ends because order is unimportant). We define an initial pI guess `bestValue` of zero before starting our search for the point of neutrality, as we know that the pI is not going to be less than this. Also, the charge at this starting pH is estimated from this initial value and an increment size of 7.0 is defined (somewhat arbitrarily) to determine the next pH value along the scale that will be tested.

Now we set up a `while` loop to search for the pH at neutrality, but we do not aim to calculate the pH at which the charge is exactly zero, we just want to get close; the result, just like the pK_a values, will only be an estimate, so a very precise value is not necessary. Thus, rather than performing the loop until the charge is exactly zero we only continue until it is less than an acceptable small value (`0.001` in this case). We test to see if the absolute value[20] of charge for the best pH found so far is greater than the threshold. Otherwise the loop will stop and the last value of the best pH, close to the pI, will be recorded.

If the test charge is smaller than the smallest found so far we record the best pH from the value tested and we record the smallest charge found thus far. Otherwise, if the test pH gives no improvement to the smallest charge, we reduce the step size variable `increment` to half its value, to narrow-in on a better value. Also, if the tested charge was less than zero we know that we should step in the reverse direction (multiply by minus one) to get closer to zero. Finally, when the `while` loop exits the last pH recorded will be one corresponding to neutrality, so we return this value.

```
def estimateIsoelectric(sequence):
  """Estimate the charge neutral pH of a protein sequence.
     This is just a guess as pKa values will vary according to
     protein sequence, conformation and conditions.
  """

  sequence = '+' + sequence + '-' # assumes seq is a string
  bestValue = 0.0
  minCharge = estimateCharge(sequence, bestValue)
  increment = 7.0

  while abs(minCharge) > 0.001:
    pHtest = bestValue + increment
    charge = estimateCharge(sequence, pHtest)

    if abs(charge) < abs(minCharge):
      minCharge = charge
      bestValue = pHtest
```

---

[20] The magnitude of a value regardless of sign.

```
    else:
      increment = abs(increment)/2.0
      if minCharge < 0.0:
        increment *= -1

  return bestValue
```

To run this we simply call the function with a protein one-letter sequence. Also, to see how quickly the test pH value homes-in on the pI, you may like to insert `print(pHtest)` inside the `while` loop.

```
pI = estimateIsoelectric(proteinSeq)
```

Obtaining sequences with BioPython

If you wish to use any of the above examples in a real situation you will naturally want to get your sequences from a database or file where they are stored, rather than having to type sequence letters into a Python file. With a working knowledge of Python you would hopefully be able to extract sequence information from the various text file formats. However, for the most part this is a solved problem and in many instances you can rely upon existing modules to extract sequences from files or databases. A good source of such modules is the BioPython library, and below we illustrate how to use BioPython to get sequence data from a few common sources. These examples assume that BioPython is installed correctly on your system and available to Python as the `Bio` module. See http://www.cambridge.org/pythonfor-biology for links to download BioPython.

Reading and writing FASTA files

To read a FASTA-format file using BioPython we use the `SeqIO` module, which in this case takes an open file object and extracts each sequence of the file, in turn creating a special object for each record. These objects have the attributes `.id` and `.seq` that respectively allow us to get hold of Python strings representing the identifier and one-letter sequence.

Quite simply we import the `SeqIO` BioPython module and create an open file object for the sequence file (in FASTA format) that we wish to read.

```
from Bio import SeqIO

fileObj = open("examples/demoSequences.fasta", "rU")
```

The `SeqIO` module has a `parse()` function that takes the file object and a file format string to yield sequence record objects. Here the records are extracted in a `for` loop and allocated to the `protein` variable, which we can then interrogate. In this example we send the sequence to the `estimateIsoelectric` function defined above.

```
for protein in SeqIO.parse(fileObj, 'fasta'):
  print(protein.id)
  print(protein.seq)
  print(estimateIsoelectric(protein.seq))

fileObj.close()
```

Writing a FASTA file using BioPython is slightly trickier because we have to first create the right type of BioPython objects (`SeqRecord`), which we then pass into a function for writing. Despite the complication of making these objects there is the added benefit that the sequence will be checked, e.g. that it has the right set of letters, before it is written.

We make several more imports from the BioPython library. The `SeqRecord` is the final object we wish to make, and which will be written out. The `Seq` object is needed internally to make a `SeqRecord` and `IUPAC` is needed to check the sequence letters according to some (the IUPAC) standard.

```
from Bio.SeqRecord import SeqRecord
from Bio.Seq import Seq
from Bio.Alphabet import IUPAC
```

An open file object is created in writing mode, with the desired output file name. If we were prudent we would check that we are not overwriting an existing file (using `os.path.exists()`).

```
fileObj = open("output.fasta", "w")
```

Next a `Seq` class of object is made, which accepts a one-letter sequence in its construction, and a sequence validation alphabet specification, which in this case is the IUPAC protein codes. This sequence object is in turn used to make a `SeqRecord` which associates the sequence object with an identifier (and potentially other kinds of annotation).

```
seqObj = Seq(proteinSeq, IUPAC.protein)
proteinObj = SeqRecord(seqObj, id="TEST")
```

The `proteinObj` (a `SeqRecord` class object) is then written to file using the `SeqIO.write()` function. Note that this takes a list of sequence records, as we can have many sequences in one file, hence we put `proteinObj` in a list of one, using square brackets. The other arguments to this function are naturally the open file object to write to and the format type of the file.

```
SeqIO.write([proteinObj,], fileObj, 'fasta')

fileObj.close()
```

Accessing public databases

If, rather than getting a sequence record from a file, we wish to get data directly from a database then there are a few helper functions in BioPython that allow easy access to some large sequence databases via Internet-based services, rather than having to talk to the database directly. For example, if we wish to read a FASTA-format file from NCBI GenBank we can do the following, although we naturally have to know the identifier for the sequence we want.

We import the `Entrez` module, set the email address attribute (to identify ourselves, as encouraged by the database) and then call a function to fetch a given entry based on a given database type `"protein"`, return format type `"fasta"` and sequence identifier number.

```
from Bio import Entrez

Entrez.email = 'mickey@disney.com'
socketObj = Entrez.efetch(db="protein", rettype="fasta",
                          id="71066805")
```

The above function call gives back an open socket object (assuming the connection works) that can be used in the same way as a normal open file object, although it actually represents a connection via the Internet. Hence, the reading of the sequence is done in the same way as for the FASTA file:

```
dnaObj = SeqIO.read(socketObj, "fasta")
socketObj.close()

print(dnaObj.description)
print(dnaObj.seq)
```

In a similar way we can read a SWISSPROT record using the ExPASy module, although it should be noted that the function to find the sequence and get an open socket (get_sprot_raw) is different to before and requires different arguments, given that such specifications depend on the exact details of the Internet service that the database provides.

```
from Bio import ExPASy

socketObj = ExPASy.get_sprot_raw('HBB_HUMAN')
proteinObj = SeqIO.read(socketObj, "swiss")
socketObj.close()

print(proteinObj.description)
print(proteinObj.seq)
```

12 Pairwise sequence alignments

Contents

Sequence alignment

The alignment of biological sequences is probably the most widely used operation in bioinformatics. In essence sequences are aligned so that we can determine how similar they are, and from this all sorts of useful information can come, such as whether two sequences are related by evolution (they have a common ancestor) or whether they have a similar biological function. The process of comparison is called *alignment* because the trickiest part of the process is to say which bits of two sequences are equivalent to one another; how residues of the different sequences can be paired up. Usually when we align sequences we seek to determine the best alignment out of the vast number of possible comparisons by finding the combination of residue pairs, one from each sequence, which gives the highest overall score for similarity.

Once a sequence alignment has been achieved, and assuming you trust the results, you can treat the aligned regions as having a degree of equivalency. If the alignment is good

```
--ALIGN-MENTS        ACGCTCGCT----TCGCGACG
ANALIGDPVENTS        ACGATCGCTCGCATCGCGACG
```

Figure 12.1. Example alignments of protein and DNA one-letter sequences. Dashes are used to represent positions where one sequence has a residue with no equivalent in the other; i.e. a gap is required to pair up the residues properly. Note that with the protein sequence (left) we can match positions not only by the residues being identical, but also by the residues being similar.

enough you might be able to say, for example, that two DNA sequences relate to the same kind of gene, despite the nucleotides not being exactly the same. It should always be remembered, however, that a sequence alignment can only give a limited amount of information about the underlying biology, but it is often an excellent starting point. Even where the knowledge gained is distinctly incomplete, a sequence alignment is quick to perform and often helpful to guide experiments. You might significantly narrow down the number of possibilities of what a section of DNA or protein could be, or say what it definitely is not, with one simple database search, i.e. doing alignments against a database of well-studied sequences. Sequence alignments are also done in a laboratory setting to guide procedures, for example to determine which part of a protein to investigate.

As you might expect for the cornerstone technique of bioinformatics, sequence alignment comes in many different varieties. Which kind of sequence alignment is used will largely depend upon what kind of biological question is being asked.[1] The following are examples of situations that sequence alignments may be used for, with attention paid to what is different about the bioinformatics of each case.

Sequence classification

You will often find this task referred to as *sequence annotation*, i.e. to associate extra information with a sequence to say what it is or does, or perhaps at least say what it might do. The principle behind this is to compare a query sequence (maybe newly discovered) with other sequences that we have some information about. If the query can be aligned with some of these other sequences then we can often infer that our query sequence has some of the same properties as those that it matches, and the degree of similarity says something about the confidence of the inference. This can be done with all the different kinds of biological polymer molecule: DNA, RNA and protein. An illustrative example would be when the elephant genome sequence is being processed, because there is a massive amount of information about the genomes of other mammals (human, mouse, rat, cat, dog etc. . . .) then there's a good chance that any given gene will align well with other genes that we already have knowledge of. Accordingly, if we find the elephant gene that aligns best with mammal genes labelled as *beta-globin* then it is highly likely that this gene is the elephant's version of beta-globin and that it functions in the same way, making up the haemoglobin protein to carry oxygen in red blood cells. On this occasion we could align either DNA sequences or protein sequences, as we are working with relatively closely related species. In situations where the degree of similarity is not so good protein sequence alignments can yield better results than DNA; because, for reasons that will be discussed later, protein sequence similarity tends to be better preserved during evolution than for DNA.

Conservation analysis

Once we have some sequences that we are sure are related to one another, we can begin to look at how the sequences differ, despite the common connection. Such sequences are often different versions of a gene, which function in the same way, from different organisms.

---

[1] And sometimes regrettably which software is available.

The basic principle of this type of analysis is that when the biological role of a particular set of DNA sequences (and thus also any protein produced) is conserved, the residues in the sequence that are important for this function are also conserved, but those that are not so important are more free to vary. DNA sequences naturally change when cells divide (and the changes are passed on to future generations when organisms reproduce) because of the error-prone nature of DNA replication. If a sequence change occurs that is detrimental to the function of the cell or organism, then the change will tend not to be passed on; the cell may die, offspring may not survive or the descendants will not be as successful as those within the population that are unaltered. Conversely, changes that are of little or no detriment will be tolerated. These could be at unimportant genetic locations, for example, the last position of a codon is often irrelevant for determining which amino acid is produced; or they could be variations that do cause a noticeable change but which function just as well, like when one amino acid changes for another that can act in the same way.

Simply by aligning sequences and discovering positions that significantly preserve residue type we can tell that those positions are important, even if we do not yet know why they are important. Also, if we can classify sequences that we know act differently despite being similar, then the individual changes in the sequence can often explain why the sequences as a whole act differently. To take an example from the study of genetic diseases, if you look at the beta-globin gene in people who have sickle-cell anaemia and compare it to those who do not have the disease, it is very easy to generate a sequence alignment to see that there is a change in the DNA, and hence protein sequence, of the seventh codon which is only present in those with the disease.[2] Further investigation shows that this change really is the underlying cause of the disease; it causes haemoglobin to stick together aberrantly.

When you look in detail at positions in a protein sequence and measure how well the residues are preserved, then the reasons and effects are often best understood by considering the folded structure of the protein; i.e. by considering the three-dimensional locations of the atoms. Amino acid residues that are involved in a specific chemical reaction that is catalysed by the protein, at its *active site*, are usually very well preserved. Other residues, for example, in the folded core of the protein, may be well conserved because of their importance in determining the shape of the protein, although some variation will be tolerated in the amino acids if they are replaced by similar types that fit together in a similar way. Positions that are not so important for the shape of a protein, generally the residues on the surface and those in flexible regions, will tend to vary the most. However, even in such locations there are some constraints on which amino acids are tolerated for normal function; for example, a change could make a necessary flexible region inflexible.

If we step backwards from the scale of an individual gene or protein and look at the context of lots of genes on the chromosomes which make up a whole genome, then we can observe trends that show how the genome as a whole is evolving. A good example of this is that when the human genome is compared to the chimpanzee genome it becomes apparent that the human chromosome 2[3] has no single chimpanzee equivalent; indeed there are two chimp chromosomes that correspond to the human one. We are certain of this because the relative location and identity of equivalent human genes is preserved, even if the length of

---

[2] There are actually at least two types of amino acid difference that are known to cause the disease.
[3] The numbering system for chromosomes simply goes from the largest to the smallest, excluding the sex chromosomes (X and Y).

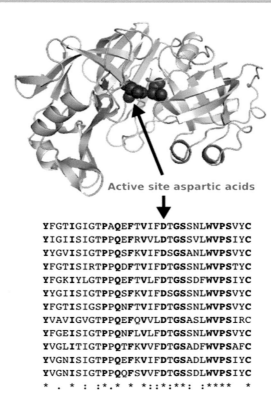

Active site aspartic acids

```
YFGTIGIGTPAQEFTVIFDTGSSNLWVPSVYC
YIGIISIGTPPQEFRVVLDTGSSVLWVPSIYC
YYGVISIGTPPQSFKVIFDSGSANLWVPSVYC
YFGTISIRTPPQDFTVIFDTGSSNLWVPSTYC
YFGKIYLGTPPQEFTVLFDTGSSDFWVPSIYC
YYGIISIGTPPQSFKVIFDSGSSNLWVPSVYC
YFGTISIGSPPQNFTVIFDTGSSNLWVPSVYC
YVAVIGVGTPPQEFQVVLDTGSASLWVPSIRC
YFGEISIGTPPQNFLVLFDTGSSNLWVPSVYC
YVGLITIGTPPQTFKVVFDTGSADFWVPSAFC
YVGNISIGTPPQEFKVIFDTGSADLWVPSIYC
YVGNISIGTPPQQFSVVFDTGSSDLWVPSIYC
* . *  :  :*.* * *::*:**: :****  *
```

Figure 12.2. The conservation of residues in a protease enzyme's active site. The active-site residues are those that are involved in the chemistry of the enzyme's reaction, and because of this they tend to vary little during evolution, as long as the function of the enzyme is preserved. Invariant sites in this alignment are marked with '*'. Such important residues may be discovered by doing multiple alignments of evolutionarily divergent sequences, e.g. from different species, to see which amino acid residues are most conserved. If an active site is not preserved this may tell us something important about the function of the protein.

chromosome differs. Going on from this, further analysis shows that the human chromosome has been created from the merging of two smaller ones; other monkeys and apes have two rather than one, so we are sure that two chromosomes is the ancestral situation.

Genetic trees

On the whole, groups of biologically important sequences are similar to one another because they are generated from common ancestors during the process of evolution. Such sequences that can be linked by their ancestry are often loosely described as a *family*. A family of sequences will arise as sections of chromosomes are duplicated and as different groups of organisms evolve into separate species. As sequences diverge from one another, their functions will also diverge, i.e. for different roles and different situations. However, as long as there is sufficient similarity to link one sequence to another we can still infer that sequences belong to a family.

Going a step further we can say not only that sequences are related, but also how they are related. This is to say in what order they diverged from, or on occasion merged with, one another and hence build a family tree for the sequences. For example, by using the most

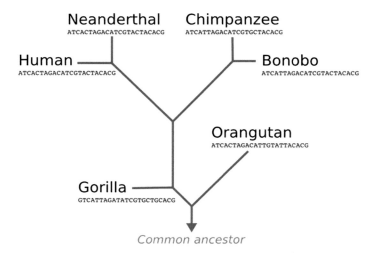

Neanderthal
ATCACTAGACATCGTACTACACG

Chimpanzee
ATCATTAGACATCGTGCTACACG

Human
ATCACTAGACATCGTACTACACG

Bonobo
ATCATTAGACATCGTACTACACG

Orangutan
ATCACTAGACATTGTATTACACG

Gorilla
GTCATTAGATATCGTGCTGCACG

Common ancestor

Figure 12.3. A cartoon phylogenetic tree, illustrating the evolutionary relationship of a few ape species. We often build phylogenetic trees using biological sequences by finding the arrangement of taxa (which here represent ape species) which would involve the simplest genetic changes, thus predicting the order of evolutionary divergence and grouping the most similar taxa together. The illustrated tree is a rooted tree, to show the relationship of the taxa to their last common ancestor. In this stylised diagram the line lengths do not convey any special meaning, but in general branch lengths are used to represent evolutionary distance (dissimilarity). The example DNA sequences are the same part of the mitochondrial cytochrome c oxidase gene COX1 from each of the species.

variable features in the non-coding regions of DNA we can build trees of closely related human families. In order to build a genetic tree from a group of sequences the basic procedure is to generate sequence alignments, and hence similarity scores, for all pairs of sequences. The task then is to use the scores to arrange the sequences into the most likely tree-like arrangement; for example, by considering which arrangement is most frugal with regard to the implied residue changes. Building an accurate tree is often very difficult, given that the number of possible ways of arranging a tree can be immense.

Often the family trees of one or more genes will be used to construct a family tree for a group of whole, distantly related, organisms. Accordingly, by studying slowly changing features, like the genes for ribosomal RNA and large DNA insertion events,[4] we can determine how most of the major groups of living organisms are related to one another. Also, if we use a multitude of genes to build a consensus as to how a group of organisms is related to one another, then we can determine where in the tree a particular change occurred. It is also possible to seek genes that do not follow the main evolutionary trend. Such rogue genes are often the result of genetic material being transmitted by some method other than normal reproduction; examples of this include transfer via viruses and during symbiosis.[5]

Although genetic trees of sequences are made possible by doing alignment, they are special enough to get their own section. Hence, for more detailed information and Python programming examples of how simple genetic trees may be constructed see Chapter 14.

[4] e.g. virus DNA that gets left behind in a genome.
[5] Where different organisms live closely with one another, to the advantage of one or both.

Protein structure prediction

We cannot yet automatically determine the three-dimensional structure of a protein just from its amino acid sequences,[6] but looking at the variations and conservation in protein structures as a whole it is apparent that structural features are more preserved than the protein sequences. Sequences may vary considerably between proteins (that really are related) and yet have very similar three-dimensional structures; so much so that we are often unable to produce a convincing alignment using the residue sequence alone. Hence, if we actually can make a good sequence alignment between two protein sequences then we can generally be confident that their structures are similar, i.e. structure is better preserved than sequence. In effect we are saying that the two proteins had a common ancestor that was structurally similar to both. This observation of structural preservation being better than our ability to detect related sequences is at the heart of a process called *comparative modelling*. Comparative modelling allows us to make a good guess at the structure of one protein if we already know the structure of a related one, and as you might expect the closer the relationship between the two proteins (evolutionarily speaking) then the better our guess. This structural guess is usually called a *model*. In essence, an alignment is made with our query sequence and various potential *template* sequences that have a known structure. If the degree of sequence similarity is suitably large then we may attempt to build a model of our query protein structure using information from the template. Such information might include the position of sub-structures like helices and extended strands[7] or the distances and angles between atoms.

Whether to align protein or DNA

If you're looking at the coding regions of genes and their resulting proteins, when approaching a particular sequence alignment problem, there is often the option of aligning nucleotide sequences or protein sequences. In general, which kind of alignment to do will be governed by the specific purpose of the investigation. For example, if you are studying the spread of an influenza virus through human populations, then because the rates of nucleotide change in the virus genome are high, and because you wish to study every genetic change in fine detail, aligning nucleotide sequences would be the best option. Conversely, if you are interested in protein structure, maybe by doing comparative modelling, then using nucleotide sequences would introduce unnecessary error and you would always use the protein sequence, albeit possibly translated from the DNA. In such circumstances, because the nucleotides can have one of four bases, the chances of a random match are quite high compared to an amino acid which is one of 20; i.e. the chances of spurious matches in protein sequences are much smaller. Also worth noting is the fact that some DNA changes have no effect on the protein at all, and thus are irrelevant for many questions arising from protein sequence. Lastly, when dealing with amino acids, we almost always calculate a score of an alignment based upon the degree to which they match, rather than just saying how many are identical, which is often the case for DNA.[8] In general, this is possible because the chemical structure of amino acids

---

[6] Current computer methods are not accurate and/or fast enough to determine the native protein conformation from the vast total number of possibilities; there is no known short-cut from a single sequence to structure. Recently significant progress has been made, however, using large numbers of aligned sequences: http://evfold.org.

[7] See Chapter 15 for discussion.

[8] Although strictly speaking not all DNA substitutions are equally likely.

allows you to say how similar they are, for example, in terms of size and charge or in the ability to form a loop. This can reveal relationships, e.g. conservation caused by the protein's structure, which would otherwise not be visible by considering simple matches alone.

. . . .

Calculating an alignment score

Our first Python examples dealing with sequence alignments are simple subroutines to give a measure of how good an alignment is. This is important because although there are many possible ways of arranging letters to get an alignment the one we almost always want is the best one: the one with the highest score. The easiest way to calculate a score for two aligned sequences is to calculate how many residue pairs are identical; this is to say we can measure the *sequence identity*.

Sequence identity

The Python function to measure sequence identity is fairly simple. It accepts two sequences and compares them under the assumption that these sequences are aligned, i.e. that the first position in one is equivalent to the first position in the other. The procedure is then to start with a score of zero and each time we have a position in both sequences where the letter is the same we add one to the score. At the end we give back the score divided by how many places we compared, so that we have an average. We also multiply this by 100 so that we get a percentage figure, which is the conventional representation.

Note that in the function we define the number of places for comparison (numPlaces) as the length of the shortest sequence, i.e. we take the minimum of the two sequence lengths to guarantee that we won't overshoot the smallest of the pair.

```
def calcSeqIdentity(seqA, seqB):

  numPlaces = min(len(seqA), len(seqB))

  score = 0.0

  for i in range(numPlaces):

    if seqA[i] == seqB[i]:
      score += 1.0

  return 100.0 * score/numPlaces
```

We can now test this function by defining some short protein sequences, and then applying the function to pairs of sequences

```
seq1 = 'ALIGNMENTS'
seq2 = 'ALIGDVENTS'
seq3 = 'ALIGDPVENTS'
seq4 = 'ALIGN-MENTS'

print(calcSeqIdentity(seq1, seq2)) # 80.0%
print(calcSeqIdentity(seq1, seq3)) # 40.0%
print(calcSeqIdentity(seq4, seq3)) # 72.7%
```

Note that, as expected when seq1 is compared to seq2 the result returned is 80%. However, for seq1 and seq3, which as you will note only differ by one extra 'P' residue, the value falls to 40%. This is because although the first few residues align well, once the extra residue is reached the sequences are no longer in step. The situation is resolved by inserting a dash (representing a gap) in the shorter sequence, so that 'P' is paired with '-' and the ends of the sequences align nicely. Accordingly seq4 and seq3 give a score of 72.7% (eight out of eleven). As you can see inserting a gap in the right place can be crucial. We will illustrate later how gaps can be inserted automatically to give an optimal alignment without human intervention.

Substitutability

Next we will move on from measuring a simple sequence identity to the more subtle measure of *sequence similarity*; this is to say that sequence pairs in an alignment can have a score even when they are not the same. The notion of similarity in this case is somewhat subjective and ultimately depends on the kind of biology you are working with. Nevertheless the general idea when scoring how similar two residues are, when aligned as a pair, is to consider how *substitutable* one residue type is for another; in other words, how likely they are to have been swapped or exchanged for one another. Residues that commonly swap for one another are deemed to be similar and give high scores, while those that rarely swap are dissimilar and give low scores. High similarity in this instance doesn't necessarily mean that two residues are always chemically similar, although they often are. Strictly speaking the substitutability of one residue type for another depends on the exact context of the residue (where it is in a chromosome or protein etc.) but we can ignore this complication for now and consider just an average value for swap-ability.

The substitutability of one residue for another is stored as a two-dimensional array, commonly called a *substitution matrix* or *similarity matrix*. The idea is that each score value in the matrix represents the substitutability of two residue types, e.g. 'A' to 'G' in DNA or 'V' to 'L' in proteins. The two residue types can be thought of as indicating the row and column of an element in a matrix, although in our Python examples we will encode matrices as dictionaries of dictionaries. Using dictionaries we can look up the score for two residue types by using the residue letters directly as keys, without having to work out the numbers for the matrix row and column. With a substitution matrix dictionary the first key (residue letter) identifies a sub-dictionary from inside the main dictionary and the second key gets the final value from inside the sub-dictionary.

Below is an example of a very simple substitution matrix that would give the same scores as if you were measuring sequence identity. i.e. a score of one where residues are identical and zero elsewhere.

```
DNA_1 = {'G': { 'G':1, 'C':0, 'A':0, 'T':0 },
         'C': { 'G':0, 'C':1, 'A':0, 'T':0 },
         'A': { 'G':0, 'C':0, 'A':1, 'T':0 },
         'T': { 'G':0, 'C':0, 'A':0, 'T':1 }}
```

Remembering that two keys are needed to extract a value (one for the main dictionary and one for the sub-dictionaries) we would get 1 for identical residue look-ups like DNA_1['G']['G'] and 0 for non-identical keys like DNA_1['G']['A'].

Changing track slightly, rather than scoring DNA sequences for matches we could also score for complementarity (i.e. using Crick and Watson's pairing rules), with 1 for A:T or G:C matches and -1 for mismatches. Expressed as a Python dictionary this would be:

```
REV_COMP = {'G': { 'G':-1, 'C': 1, 'A':-1, 'T':-1 },
            'C': { 'G': 1, 'C':-1, 'A':-1, 'T':-1 },
            'A': { 'G':-1, 'C':-1, 'A':-1, 'T': 1 },
            'T': { 'G':-1, 'C':-1, 'A': 1, 'T':-1 }}
```

Moving on to a more sophisticated matrix, as illustrated above, you will note that substitution scores can have negative values (mismatch) and that a score of zero is often used to indicate indifference. In the DNA_2 matrix below note that identical residue keys give a score of 1 but non-identical -3. In other words the mismatches are strongly penalised; in an alignment three identical residues are required to balance one mismatch. Also note that the example uses the residue code 'N', which in this instance for DNA means any[9] unidentified residue, which is indifferent in an alignment, given that we cannot tell if it is good or bad and so scores zero with everything.

```
DNA_2 = {'G': { 'G': 1, 'C':-3, 'A':-3, 'T':-3, 'N':0 },
         'C': { 'G':-3, 'C': 1, 'A':-3, 'T':-3, 'N':0 },
         'A': { 'G':-3, 'C':-3, 'A': 1, 'T':-3, 'N':0 },
         'T': { 'G':-3, 'C':-3, 'A':-3, 'T': 1, 'N':0 },
         'N': { 'G': 0, 'C': 0, 'A': 0, 'T': 0, 'N':0 }}
```

The next example is part of a substitution matrix for protein sequences. It is a fairly famous one called BLOSUM62 (often the default in many programs). You will of course note that the matrix is much larger than for DNA because we have 20 regular amino acids, plus 'x' for unknown type. We have only shown the first four sub-dictionaries here, but the full matrix can be found in the on-line material (available via http://www.cambridge.org/pythonforbiology). There are usually many variants of a given substitution matrix type. Here we specifically use the '62'[10] version of BLOSUM series because it is a good general-purpose one. You would commonly consider using different matrix versions to tune your alignment for more closely related or distantly related sequences for which substitution preferences are known to differ.

```
BLOSUM62 = {'A':{'A': 4,'R':-1,'N':-2,'D':-2,'C': 0,'Q':-1,
                 'E':-1,'G': 0,'H':-2,'I':-1,'L':-1,'K':-1,
                 'M':-1,'F':-2,'P':-1,'S': 1,'T': 0,'W':-3,
                 'Y':-2,'V': 0,'X':0},
            'R':{'A':-1,'R': 5,'N': 0,'D':-2,'C':-3,'Q': 1,
                 'E': 0,'G':-2,'H': 0,'I':-3,'L':-2,'K': 2,
                 'M':-1,'F':-3,'P':-2,'S':-1,'T':-1,'W':-3,
                 'Y':-2,'V':-3,'X':0},
            'N':{'A':-2,'R': 0,'N': 6,'D': 1,'C':-3,'Q': 0,
                 'E': 0,'G': 0,'H': 1,'I':-3,'L':-3,'K': 0,
                 'M':-2,'F':-3,'P':-2,'S': 1,'T': 0,'W':-4,
                 'Y':-2,'V':-3,'X':0},
```

---

[9] A bad, but probably intentional, pun; 'N-y' for 'any'. Such ambiguities are often caused by uncertainties in the recording of sequences, especially at the ends of a DNA read.

[10] Sixty-two is the percentage sequence identity limit of the blocks of multiply aligned sequences that are used to count the substitutions.

```
'D':{'A':-2,'R':-2,'N': 1,'D': 6,'C':-3,'Q': 0,
     'E': 2,'G':-1,'H':-1,'I':-3,'L':-4,'K':-1,
     'M':-3,'F':-3,'P':-1,'S': 0,'T':-1,'W':-4,
     'Y':-3,'V':-3,'X':0}}
```

SNIP: THE FULL MATRIX CARRIES ON FOR 17 MORE SUB-DICTIONARIES

As with the DNA matrices we use two keys to get the substitution score and have positive, zero and negative values. Note that the matrix, like the DNA matrix examples, is symmetric,[11] i.e. BLOSUM62['A']['R'] equals BLOSUM62['R']['A']. Unlike the DNA examples the diagonal of the matrix is not uniform, which is to say that the score for residue types being the same in an alignment differs. For example, BLOSUM62['A']['A'], meaning an exact alanine match, gives a score of 4, but an exact asparagine match BLOSUM62['N']['N'] gives a higher score of 6. Thus 'A' is less well conserved (more swappable for something else) than 'N'.

We will not go into fine detail about how substitution matrices are calculated until Chapter 14. In essence the idea is that you first generate good, well-curated alignments of multiple sequences using as much information as humanly possible from structure and function etc. and you then count how many times one residue type is substituted for another within the alignment. Then in various ways these counts are converted into whole-number scores, relative to some baseline value. If you are really interested, we recommend the early papers on the PAM[12] and BLOSUM[13] matrices. These two protein matrices are calculated in slightly different ways, but together they give a good idea of the underlying principles.

Calculating sequence similarity

The next example of a Python function will consider a substitution matrix like the ones discussed and use it to calculate an overall similarity score for two aligned sequences. The inputs to the function are two strings of sequence letters and the similarity matrix.

```
def calcSeqSimilarity(seqA, seqB, simMatrix):

  numPlaces = min(len(seqA), len(seqB))

  totalScore = 0.0

  for i in range(numPlaces):

    residueA = seqA[i]
    residueB = seqB[i]

    totalScore += simMatrix[residueA][residueB]

  return totalScore
```

---

[11] This isn't universally true for all substitution matrices though.
[12] Dayhoff, M.O., Schwartz, R., and Orcutt, B.C. (1978). A model of evolutionary change in proteins. *Atlas of Protein Sequence and Structure* (volume 5, supplement 3 ed.). Washington DC: National Biomedical Research Foundation. pp. 345–352.
[13] Henikoff, S., and Henikoff, J.G. (1992). Amino acid substitution matrices from protein blocks. *PNAS* 89(22): 10915–10919.

```
# Test with pre-defined substitution matrices
# DNA example
print(calcSeqSimilarity('AGCATCGCTCT', 'AGCATCGTTTT', DNA_2))

# Protein example
print(calcSeqSimilarity('ALIGNMENT', 'AYIPNVENT', BLOSUM62))
```

The `calcSeqSimilarity()` function is very similar in construction to the previous `calcSeqIdentity()` function, except that this time rather than seeing if two residue letters are equal, we use them as keys to look up a similarity score in the substitution matrix. Note that this function has one big deficiency: it cannot deal with gaps ('-'). To address this problem we could put entries for gaps into the similarity matrix. However, a simpler solution is to introduce a separate gap penalty; gaps are generally undesirable but are tolerable if the subsequent alignment matches well. If we complicate things slightly more we can have different gap penalties depending on whether we are inserting a new gap or extending an existing one. Generally extending an existing gap (i.e. putting one dash after another) has the smaller penalty. This is equivalent to saying that we score alignments more highly if they use fewer, longer gapped regions. The following modified function uses gap penalties `insert` and `extend`, which both carry default values. Note that we have a different name for the new function and its input sequences (`alignA` and `alignB`) to reinforce the fact that it is working on a pair or aligned sequences, including any gaps, rather than just plain sequences:

```
def pairAlignScore(alignA, alignB, simMatrix, insert=8, extend=4):

  totalScore = 0.0

  n = min( len(alignA), len(alignB) )

  for i in range(n):

    residueA = alignA[i]
    residueB = alignB[i]

    if '-' not in (residueA, residueB):
      simScore = simMatrix[residueA][residueB]
    elif (i > 0) and ('-' in (alignA[i-1], alignB[i-1])):
      simScore = -extend
    else:
      simScore = -insert

    totalScore += simScore

  return totalScore

# Test
print(pairAlignScore('ALIGDPPVENTS', 'ALIGN--MENTS', BLOSUM62)) # 28
print(pairAlignScore('ALIGDPPVENTS', '--ALIGNMENTS', BLOSUM62)) # -3
```

Pay special attention to the logic above in the `if`/`elif`/`else` statement. If a gap is not amongst the two residue codes then the score for the pair is obtained as before from the similarity matrix dictionary. Otherwise, we do have a gap and thus carry on to check the other two conditions. If the position in the sequence `i` is not at the very start (`i > 0`) and one of the previous positions was a gap we subtract the `extend` penalty. And if all else fails we have a gap that starts anew, so we subtract the `insert` penalty.

Optimising pairwise alignment

Given that we have discussed the principle of how we can measure the match quality of an aligned pair of sequences, we now turn to the problem of determining which alignment out of all the possible combinations is the best (highest scoring). Consider for a moment the following three examples. The last alignment is the best, with the middle one a close second.

```
ALIGNMENTS---        A--LIGN-MENTS        --ALIGN-MENTS
ANALIGDPVENTS        ANALIGDPVENTS        ANALIGDPVENTS
```

We can represent these same alignments as a comparison matrix, where each element represents the pairing of a residue from one sequence with a residue from the other. In this case we have indicated the aligned (paired) residues for a given row and column with 'x'. Each alternative alignment can be viewed as a different route through the comparison matrix and gaps are simply jumps in one sequence or another; down rows or across columns.

```
  ALIGNMENTS            ALIGNMENTS            ALIGNMENTS
Ax.........          Ax.........          A.........
N.x.......           N.........           N.........
A..x......           A.........           Ax........
L...x.....           L.x.......           L.x.......
I....x....           I..x......           I..x......
G.....x...           G...x.....           G...x.....
D......x..           D....x....           D....x....
P.......x..          P.........           P.........
V........x.          V.....x....          V.....x....
E........x           E......x...          E......x...
N.........           N.......x..          N.......x..
T.........           T........x.          T........x.
S.........           S.........x          S.........x
```

It is clear that each separate alignment possibility is a different way of placing the dashes, which represent the gaps. With a total alignment length of 13 and placing three gaps inside the shorter sequence we have $(13 \times 12 \times 11)/(3 \times 2 \times 1) = 286$ ways of arranging the gaps. If we extend this calculation to the not unreasonable and biologically typical scenario of five gaps in 100 positions then the result is $(100 \times 99 \times 98 \times 97 \times 96)/(5 \times 4 \times 3 \times 2 \times 1) = 75,287,520$. And this still does not consider another class of alignment possibilities with gaps being present in both sequences like:

```
------ALIGNMENTS
ANALIGDPVENTS---
```

As you can see the number of possible combinations grows very rapidly with the length of the sequence. Indeed, for almost all situations it is impractical to check them all when doing an alignment. Nevertheless we can still find the best alignment of a pair of sequences by using a clever trick, which allows us to neglect checking the vast majority of alignments. The principle behind this is commonly referred to *dynamic programming*. This is perhaps a misleading name,

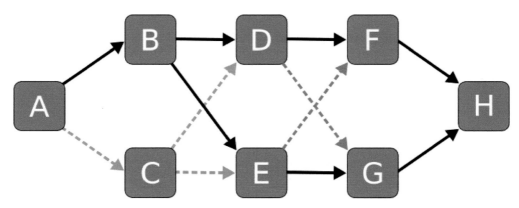

Figure 12.4. A simple feed-forward network to illustrate the principles of dynamic programming. The objective of the dynamic programming algorithm is to find the best route through the network without having to consider all possible routes. By finding the best route to each intermediate location the algorithm excludes many route possibilities.

because the idea doesn't actually involve anything especially dynamic, and isn't anything novel in terms of computer programming. However, the idea is really very cunning.

Dynamic programming

The dynamic programming algorithm, which we commonly use to align pairs of sequences, is based on the idea that if your big problem can be broken down into local sub-problems that occur repeatedly, then the solution to each sub-problem only needs to be calculated once. In terms of sequence alignment the big problem is to find the highest-scoring arrangement and the local sub-problems involve smaller sub-alignments. For a simple abstract example imagine a problem where you wanted to find the quickest overall route from point A to point H, where you had to travel via various intermediate points (you could imagine that each point represents a town and travelling represents a train ride). The arrangement of the points is as shown in Figure 12.4, such that at each intermediate stage you have to go through one of two points.

As illustrated, starting at A you must go via B or C then through D or E, then F or G before finally getting to H. Each leg of the journey takes a different time and all of these times are added to get the total travel time. If I want to find the quickest journey from A to H then I have to potentially consider all of the possible routes, A, B, D, F, H, A, C, E, F, H etc. . . . It might seem that you have to calculate the time for all possible journeys before you know which way is best. However, this problem can be broken into smaller sub-problems which allow us to discard certain routes at an early stage. For example, we can calculate the shortest journey time from A to D and A to E; potentially these could go through either B or C, but if both of the shortest routes go through B we need never consider point C ever again. When extending the routes through F and G we only have to extend our best routes from D and E, which we have already determined only use point B. In general, if we add up journey times as we go, and given that at each point there are only two previous directions that we could have come from, we can choose the best direction to get to that point and ignore all routes that come from the other direction. Thus in the end when we reach H there are only two routes to choose between, one from F and one from G. At the start we couldn't know which of F or G would be optimal in the end, but in calculating the routes (sub-problems) to these final points we have only been following the best paths for each stage, which is much more efficient than following them all. Sequence alignment is analogous to this kind of journey analysis, the difference being that we want the route of maximum alignment score (not minimum time) and we have three routes into each point rather than two; the three routes are (i) to put a gap in one sequence, (ii) put a gap in the other sequence or (iii) align two residues.

If we start from the beginning of two sequences and consider the alignment growing by inserting a gap in one sequence or the other, or by putting the next two residues together, at each point the number of possibilities grows by a multiple of three each time. However, to get to a particular residue pairing there are three routes that it could have come from; and like in the above example only the best one to this point needs to be continued. Specifically, what sequence alignment does is to extend the (sub-total) scores of these three routes with gap penalties or a similarity score for the next residue pair. By taking the maximum of these scores the other two are discarded and their routes cut; two of the sub-alignments that get to this point do not need to be extended any further, because there is already a better alignment to get to the same place. Every time the alignment is extended, multiple new possibilities are generated, but at the same time previous sub-alignments can be disregarded. By repeatedly extending the alignment, and pruning sub-optimal alternatives along the way, the end of the sequences is eventually reached. Initially as the sequences are compared, there is no way to know which sub-alignments are discarded and which win out, but the decisions that are taken (whether to add gaps or pair up residues) at each step along the way can be remembered. So, starting from the end point the winning decisions can be followed, backwards, to find what gave the best score at each point and hence the best alignment.

Pairwise alignment with Python

The following Python function is an example of how dynamic programming is used in practice to generate an optimal alignment for a pair of sequences. The method is very similar to the one described by Needleman and Wunsch.[14] Another popular dynamic programming method for sequence alignment was introduced by Smith and Waterman, which focuses more on local solutions.[15] However, in this case the underlying principle is the same, and the only major difference is that a Smith-Waterman alignment doesn't allow the intermediate alignment scores to dip below zero, so that local (off diagonal) alignments are more prominent.

The following discussion will split the alignment function into more easily understood sections but the code, in its concatenated entirety, can be viewed in the on-line material at http://www.cambridge.org/pythonforbiology. First we begin with the function definition

[14] Needleman, S.B., and Wunsch, C.D. (1970). A general method applicable to the search for similarities in the amino acid sequence of two proteins. *Journal of Molecular Biology* 48(3): 443–453.

[15] Smith, T.F., and Waterman, M.S. (1981). Identification of common molecular subsequences. *Journal of Molecular Biology* 147: 195–197.

that takes two one-letter sequences, a substitution/similarity matrix and two different gap penalties. Note that you would normally check that the input values were sensible (e.g. sequences were really one-letter and upper case) before using the function

```
def sequenceAlign(seqA, seqB, simMatrix=DNA_2, insert=8, extend=4):
```

Next we define two numbers which represent the size of the comparison grid. As illustrated above, the comparison grid is simply a matrix of one sequence versus the other, so that each point is a different possible residue pair. The size of the grid is one larger than the lengths of the sequences because we require an extra row and column at the start of the matrix (left and top) to contain starting values to begin growing the alignments from.

```
numI = len(seqA) + 1
numJ = len(seqB) + 1
```

Next we create the matrices (actually lists of lists here, but we could use NumPy matrices) which represent the comparison grid. We will actually use two matrices of equal size; one is used to contain the sub-totals of the alignment scores and the other is used to record which route (sub-alignment) was taken, i.e. whether the best route to this point involved a gap or a residue comparison. In the route matrix we will use a code to say which of the three directions was taken, where 0 represents the pairing of two residues, 1 is for a gap in seqB (but a residue in seqA) and 2 is for a gap in seqA (but a residue in seqB). We build the preliminary blank matrices of the right size initially with zeros:[16]

```
scoreMatrix = [[0] * numJ for x in range(numI)]
routeMatrix = [[0] * numJ for x in range(numI)]
```

Then for the route matrix we adjust the top and left edges, except the first element, so that they contain route codes that represent gaps. This means that for alignments where one sequence is indented relative to the other the only route to get to that point is by having a series of leading gaps in the indented sequence. We will leave the edges of the score matrix at zero, so as not to penalise alignments that begin with a gap. However, in some situations you might want to have a penalty for this.

```
for i in range(1, numI):
  routeMatrix[i][0] = 1

for j in range(1, numJ):
  routeMatrix[0][j] = 2
```

Given that the edges of the route and score matrix are now defined, we now fill in the remainder of the matrix values by looping through i and j, the indices of the rows and columns, starting at 1, rather than 0, due to the extra edges which were added.

```
for i in range(1, numI):
  for j in range(1, numJ):
```

We need to decide which gap penalty is relevant for each matrix element. So we use some logic which means that the insertion penalty is used, except if the previous position has a gap, in which case we use the extension penalty. Note that we make this decision twice and define

---

[16] See the description of list comprehensions in Chapter 4 for an explanation of the Python.

two separate gap penalty values, one for each sequence. Detecting whether the previous position has a gap is a simple matter of checking the route matrix in the previous row (i-1) or column (j-1) to see if it has a gap code (1 or 2).

```
penalty1 = insert
penalty2 = insert

if routeMatrix[i-1][j] == 1:
  penalty1 = extend

elif routeMatrix[i][j-1] == 2:
  penalty2 = extend
```

Next we look up the similarity score for comparing the residue codes in this row and column. Note that we get the position for the residues within the two sequences using i-1 and j-1; this is because the row and column in the comparison matrix is one larger than the equivalent position in the sequences, due to the extra initialisation values put at the start.

```
similarity = simMatrix[ seqA[i-1] ][ seqB[j-1] ]
```

With the gap penalties and the similarity scores defined, we make a list of the three possible path options for extending the alignment at this position. These paths are the sub-total scores of three neighbouring matrix elements (representing the three shorter sub-alignments that meet at this point) with either the similarity score added or a gap penalty subtracted. The first path (route code 0) means adding two residues and using their similarity score, effectively going diagonally in the comparison matrix i-1, j-1 to i,j). The second path (route code 1) means having a gap in seqB and subtracting a penalty, going down a row i-1, j to i,j. Likewise the third path (route code 2) subtracts a penalty for a gap in seqA and goes across a column i, j-1 to i,j.

```
paths = [scoreMatrix[i-1][j-1] + similarity, # Route 0
         scoreMatrix[i-1][j] - penalty1, # Route 1
         scoreMatrix[i][j-1] - penalty2 # Route 2
```

The best sub-total score is simply the maximum value in the paths list, and the route code is the index (position starting at zero) of this score. Note how the paths.index is the reason for using the numeric route codes; they are arbitrary, but come naturally from the order of the above list.

```
best = max(paths)
route = paths.index(best)
```

Finally for the loop we store the best score as the sub-total for this matrix element, and the best route to get to this point in the route matrix. In a subsequent loop these values will be considered to extend the alignment further. The route might not survive, but at least it will be considered.

```
scoreMatrix[i][j] = best
routeMatrix[i][j] = route
```

Now that the matrix of scores and matrix of routes are filled in we have reached our destination; routes have met at the ends of the two sequences. The task now is to work back from the end of the paths (bottom right of the comparison matrix), following the winning routes backwards to the start of the sequences, at each point adding the appropriate gap or paired residues. Thus, we define initially blank alignment lists, one for each input sequence that we will fill with residue codes and dashes in reverse order:

```
alignA = []
alignB = []
```

Next we get the row and column for the end of the alignment, which are one less than the corresponding sizes, given that list indices start at zero. We also record the score for this final position, which we can give back as the overall score for the overall alignment.

```
i = numI-1
j = numJ-1
score = scoreMatrix[i][j]
```

The next loop fills in the alignment strings by going back along the rows and columns (decreasing values of i and j), following the winning route at each step. Naturally we continue while we still have sequence alignment to fill in. Note how the while loop continues if either the row or column is bigger than zero; it only stops when both are zero, when at the very start of the alignment. Each time the operations go through the loop there will be a new row and column combination, and thus a new route code is defined for this point.

```
while i > 0 or j > 0:
    route = routeMatrix[i][j]
```

Depending on the route code, i.e. which of the paths won out to get to this point, we grow the alignment strings in one of three ways. Route 0 indicates that the maximum score came from comparing two residues, so in this case we add residue letters to both alignment strings and decrease both the row and column by one. Routes 1 and 2 indicate that the best score came from having a penalty and inserting a gap, in which case we add a dash to one sequence and a residue to the other, decreasing either the row or the column as appropriate. Again, note that the indices for getting residue letters from the sequences are one less than the row and column indices.

```
    if route == 0: # Diagonal
      alignA.append( seqA[i-1] )
      alignB.append( seqB[j-1] )
      i -= 1
      j -= 1

    elif route == 1: # Gap in seqB
      alignA.append( seqA[i-1] )
      alignB.append( '-' )
      i -= 1

    elif route == 2: # Gap in seqA
      alignA.append( '-' )
      alignB.append( seqB[j-1] )
      j -= 1
```

Lastly we reverse the alignment lists, given that we were working backwards from the end, and convert them to text strings which we send back with the overall score. Note that working with lists of text characters in this way, which we only join into a string at the end, is quicker than repeatedly extending strings, especially for long sequences.

```
alignA.reverse()
alignB.reverse()
alignA = ''.join(alignA)
alignB = ''.join(alignB)

return score, alignA, alignB
```

And the function can be tested accordingly:

```
seqA = 'WFSEPEIST'
seqB = 'FSRPAVVIST'

score, alignA, alignB = sequenceAlign(seqA, seqB, BLOSUM62)

print(score)   # 17
print(alignA) # WFSEPE--IST
print(alignB) # -FSRPAVVIST
```

Quick database searches

As a final thought to this chapter we acknowledge another kind of alignment that is routinely performed. This is to have a query sequence and compare it with a large database of other sequences to find alignment matches from a potentially vast number of targets.

What we have been discussing so far are rigorous alignments of limited numbers of sequence pairs. The dynamic programming method and its various derivatives will give you the best alignments, but it has the disadvantage that it is prohibitively slow if you are aligning a query sequence with a database containing thousands (e.g. the human genome) or even millions of sequences (e.g. the whole EMBL or NCBI database). When searching such databases, the approach is to use query methods that can take clever short-cuts, i.e. it uses rules that are generally true in order to quickly remove a large number of sequences from consideration, without having to do full sequence alignments with every member of the database. The most famous of these methods is the BLAST routine.[17] Although such heuristic methods have opened up a vast array of convenient sequence information, strictly speaking they can miss things that a more rigorous approach would not.

Below we give an example of how we can run BLAST searches on sequence databases using Python. The objective here is not to rewrite the BLAST program itself, but rather to wrap the external program in code and use it as if it were a Python module which can interface with all of our other code. This naturally has the advantage of using an efficient, tried and tested program but still being able to work with Python. Thus what we will do is create a function that makes the required input, runs the external executable (i.e. the actual BLAST program, which must already be installed on the system) and interprets the output, collecting the data back into the Python data structures.

---

[17] Altschul, S., Gish, W., Miller, W., Myers, E., and Lipman, D. (1990). Basic local alignment search tool. *Journal of Molecular Biology* 215(3): 403–410.

Using BLAST from Python

Before we get started we need to consider how to create a sequence database that the BLAST program can search through. This is not a normal sequence file or database, but rather it is a special, optimised series of files that is special to the BLAST system and which makes the comparison of sequences efficient. Fortunately there is a program called 'makeblastdb', which is available in the BLAST+ package,[18] that can be used to make the special database from sequences stored in common text-based formats like FASTA.

Before the examples we import all of the external functions and modules that will be required. Specifically we need the substitution function (`sub`) from the regular expression[19] module to help format the sequences for the input files. The `call` function from the `subprocess` is a convenient way to run programs that are external to Python. The `ElementTree` module will allow us to read the XML[20] formatted BLAST output.

```
from re import sub
from subprocess import call
from xml.etree import ElementTree
```

Now a function is defined to make the special sequence database for BLAST. This is not a conventional relational database, e.g. where small amounts of data are accessed and modified, but rather a specially formatted version of sequence data to be queried, which allows BLAST to efficiently find matches. The function takes a FASTA-formatted file as input, a distinguishing name for the database, `formatdbExe`, the location of the external 'makeblastdb' program to run and `isProtein`, a switch to state whether the sequence is protein and not nucleic acid (RNA or DNA). Note that the input FASTA sequence file is often very large, representing a whole database (e.g. UniProt) or even a genome. Obtaining genome sequence information will be discussed later in Chapter 17.

```
def makeBlastDatabase(fastaFile, databaseName,
                      formatdbExe=None, isProtein=True):
```

If a value for `formatdbExe` is not passed in we will assume that the database creation program is available to the local system as a command called `makeblastdb`. If the program name is unavailable, or not set correctly, then an error exception will be triggered when we come to run the program via `call()`.

```
  if not formatdbExe:
    formatdbExe = 'makeblastdb'
```

Next we define a variable which will indicate the database type. The `makeblastdb` program accepts a value `'prot'` (protein) or `'nucl'` (nucleic acid) after its `-dbtype` option to indicate whether the database is protein or not.

```
if isProtein:
  molType = 'prot'
else:
  molType = 'nucl'
```

---

[18] The equivalent in the older BLAST2 package is 'formatdb'
[19] Regular expressions are discussed more fully in Appendix 5.
[20] eXtensible Markup Language: a text-based general data storage format, somewhat similar to the HTML used in web pages.

Then all of the options and arguments that will be used when running the `makeblastdb` program are collected in a list. These would be the values that one would type at a command line prompt to run the program.

```
cmdArgs = [formatdbExe,
           '-dbtype', molType,
           '-in', fastaFile,
           '-out', databaseName]
```

This list is then passed to the `call()` function to actually run the program, noting that we catch any exceptions that may occur. If an exception is triggered we print the command that was tried, together with the original Python error (`err`) so we can determine what the problem was.

```
print('Making BLAST database %s...' % databaseName)

try:
  call(cmdArgs)

except Exception as err:
  print('BLAST database creation failed')
  print('Command used: "%s"' % ' '.join(cmdArgs))
  print(err)
  return

print(' ...done')
```

With the function defined we can now run the `makeblastdb` program on a FASTA-format file as if it were Python. Naturally we only need to do this once for each database, unless it changes.

```
fileName = 'EcoliGenome.fasta'

makeBlastDatabase(fileName, 'ECOLI_PROT')
```

However, we may have to specifically state the location of the `makeblastdb` program if the system doesn't know that location of the executable file:

```
makeBlastDatabase(fileName, 'ECOLI_PROT', '/usr/bin/makeblastdb')
```

Next we come to a similar Python wrapper function that will run the actual BLAST program, given a query sequence and a database, created as described above. For the sake of simplicity, several of the BLAST input options will be ignored and only part of the output will be read. However, this function can be expanded to use as much detail as is required.[21]

The Python function that controls the BLAST search takes arguments for the query sequence (as one-letter codes) and the location of the formatted database to search. There are optional arguments to specify the BLAST program location (otherwise the command `'blastp'` will be tried), the maximum cut-off expectation value (the E-value score), the type of substitution matrix and the number of processing cores to use. It is important to note that different BLAST programs are used to specify which molecule type (protein or nucleic acid) is used in the query and independently in the database. Accordingly, the following programs are available to specify which type of BLAST search will be run:

---

[21] For full details of all the options available see the BLAST+ package documentation in the help section at the NCBI: http://blast.ncbi.nlm.nih.gov.

'blastp': protein query against a protein database.

'blastn': nucleic acid query against a nucleotide database.

'tblastn': protein query against a translated nucleotide database.

'blastx': translated nucleic acid query against a protein database.

'tblastx': translated nucleic acid query against a translated nucleotide database.

Note that the 'tblastx' method translates nucleotide sequence to protein sequence in all reading frames, so may yield amino acid matches even if the base sequences are quite different, but naturally only one of the reading frames encodes protein.

The blastSearch() function makes use of the subprocess module again, but this time we also import Popen and PIPE. The Popen class is a more general way of controlling external programs than a simple call and we use it here because we are directly controlling the standard input to and output from BLAST using Python, rather than via the operating system (an alternative to this approach would be to read and write temporary files). The PIPE import is a value that allows us to achieve this by setting the stdin and stdout attributes of Popen so that data will be communicated or 'piped' to and from Python.

```
from subprocess import call, Popen, PIPE

def blastSearch(seq, database, blastExe=None, eCutoff=10,
                matrix='BLOSUM62', cpuCores=None):
```

As with the previous function we set a default name for the program, here assuming 'blastp' is available and executable on the system, if the name was not specifically set by the user. As discussed above, the choice of the specific BLAST program dictates what type of search is being performed and what molecule type (protein or nucleic acid) the query and database represent.

```
if not blastExe:
  blastExe = 'blastp'
```

If the number of processor (CPU) cores was not specified we take the maximum available as specified by the handy multiprocessing module. Using multiple cores will allow part of the BLAST search to run more quickly, because certain tasks can be computed in parallel.

```
if not cpuCores:
  import multiprocessing
  cpuCores = multiprocessing.cpu_count()
```

The query sequence is tidied, making it upper case and inserting a newline '\n' every 60 characters to make regular FASTA-format sequence strings.

```
querySeq = seq.upper()
querySeq = sub('(.{60})(.)',r'\1\n\2', querySeq)
```

The sequence is then placed into a string formatted as a FASTA entry by adding an annotation line (with arbitrary name) before a sequence line. These two lines represent the data that would be contained in a FASTA file containing a single sequence, which we will send as input to BLAST.

```
querySeq = '>UserQuery\n%s\n' % querySeq
```

The program command arguments are collected in a list as before, noting that we don't need to set any input or output files because by default BLAST will use standard input and output:

```
cmdArgs = [blastExe,
           '-outfmt', '5',        # => XML output
           '-num_threads', str(cpuCores),
           '-db', database,
           '-evalue', str(eCutoff),
           '-matrix', matrix]

print(' '.join(cmdArgs))
```

The `Popen` class is used to make an object that encapsulates the BLAST process, setting attributes so that the standard input and output data is sent via Python. If there are any problems with this, e.g. if the BLAST executable is not found, then we catch the exception and report the error that occurred.

```
try:
  proc = Popen(cmdArgs, stdin=PIPE, stdout=PIPE)

except Exception as err:
  print('BLAST command failed')
  print('Command used: "%s"' % ' '.join(cmdArgs))
  print(err)
  return []
```

Nothing will be run by BLAST until we send it some input data and we do this using the `communicate()` method of the process object. This function call gives back two values, representing standard output and error data respectively. If we hadn't set `stdout=PIPE` above this data would simply be sent to the screen.

```
stdOutData, stdErrData = proc.communicate(querySeq)
```

The results are collected by reading through the XML-formatted text contained in the `stdOutData` string using the `ElementTree` module. The `fromstring()` function will give back an object representing the tree-like hierarchy of the XML data. To navigate this tree we start at the base, which is named `root` here, from which we can follow any node (like branches) to get to the data within.

```
results = []
root = ElementTree.fromstring(stdOutData)
```

From this start point, to collect the required data we use the `.find()` function, going from point to point within the structure defined by the XML file. What this function does is to look for specific elements in the tree-like structure of the XML file which match a particular text string. Naturally, knowing which things to look for depends on knowing the schema of the XML file, i.e. what the names of the elements are and how they relate to one another. See Chapter 6 for further discussion of XML files. Note that the format of these XML files could change in future versions of BLAST, so if this example does not work directly look at the actual output and adapt the following code for the actual XML elements in the file.

To get to `iteration`, which will store all the sequences matched in the database (there is only one iteration in a regular BLAST search) we go to elements named `Blast Output_iterations` and then `Iteration`.

```
iteration = root.find('BlastOutput_iterations').find('Iteration')
```

The matches to the query that BLAST finds (if any) will be stored in the XML inside the `'Iteration_hits'` section as `'Hit'` elements. If there are no `'Iteration_hits'` then we simply return an empty list for the results.

```
ihit = iteration.find('Iteration_hits')

if ihit:
  hits = ihit.findall('Hit')
else:
  return []
```

Otherwise if there are `Hit` elements representing the sequence matches to the query in the database we loop through them to find the `Hsp` elements (high scoring pairs).

```
for hit in hits:
  hsp = hit.find('Hit_hsps').find('Hsp')
```

Then the remainder of the database match information is stored in a dictionary using the name as a key. Note that this is just one example of how to store the results in Python data structures.

```
hitDict = {}
```

The name of the match (the hit) is extracted from the `'Hit_def'` attribute using `hit.findtext()` which we then store in the dictionary with the `'def'` key. Similarly, its sequence length is stored as the `'len'` after conversion of the XML string to a Python integer:

```
hitDict['def'] = hit.findtext('Hit_def')
hitDict['len'] = int(hit.findtext('Hit_len'))
```

For all the values associated with each high-scoring pair we take the tag name for the value (as used in the XML file) and remove the first four characters to make a simpler `key`, thus avoiding all the keys beginning with `'Hsp_'`, e.g. `'Hsp_score'` becomes `'score'`. The actual value is then obtained with `.findtext()` and converted to a Python data type, with `int()` or `float()` as needed. First we collect the floating point values:

```
for tag in ('Hsp_bit-score', 'Hsp_evalue'):
  key = tag[4:]
  hitDict[key] = float(hsp.findtext(tag))
```

Then the integer values:

```
for tag in ('Hsp_score', 'Hsp_query-from',
            'Hsp_query-to', 'Hsp_hit-from',
            'Hsp_hit-to', 'Hsp_query-frame',
            'Hsp_hit-frame', 'Hsp_identity',
            'Hsp_positive', 'Hsp_gaps',
            'Hsp_align-len'):
  key = tag[4:]
  hitDict[key] = int(hsp.findtext(tag, '0'))
```

And finally the plain text values:

```
for tag in ('Hsp_qseq', 'Hsp_hseq', 'Hsp_midline'):
  key = tag[4:]
  hitDict[key] = hsp.findtext(tag)
```

Once the hit is processed we store the dictionary, with all the extracted values in the `results` list. A dictionary could have been used to store the results, but here we wish to preserve the order of the matches.

```
  results.append(hitDict)
return results
```

Finally, the Python BLAST search wrapper function can be tested with a query sequence and a BLAST sequence database created previously.

```
seq = 'NGTISYTNEAGKIYQLKPNPAVLICRVRGLHLPEKHVTWRGEAIPGSLFDFA' \
      'LYFFHNYQALLAKGSGPYFYLPKTQSWQEAAWWSEVFSYAEDRFNLPRGTIK' \
      'ATLLIETLPAVFQMDEILHALRDHIVGLNCGRWDYIFSYIKTLKNYPDRVLP'
```

The location of the database is the file path to the BLAST database files, which includes the database name, but excludes any file extensions:

```
database = 'ECOLI_PROT'
```

The results that come back after running the specified BLAST program (the path for which should be set appropriately) are a list of names and data dictionaries for any sequence matches.

```
results = blastSearch(seq, database'/usr/bin/blastp',
                      eCutoff=1.0, cpuCores=4)
```

The various sequences and scores are obtained from the results dictionary for each match (hit) by using the appropriate keyword:

```
for hitDict in results:
  print(hitDict['def'],  hitDict['score'], hitDict['evalue'])
  print(hitDict['qseq'])
  print(hitDict['midline'])
  print(hitDict['hseq'])
```

13 Multiple-sequence alignments

Multiple alignments

Expanding from an alignment of just two sequences, the more sequences you can align together then the more information you have and the more accurate your alignment will be. The caveat to this is that the closeness of the relationship between sequences is important and should also be taken into account. Consider that for two very closely related sequences, the differences are significant and similarities less significant, because you expect similarity. In contrast, for distantly related sequences residue differences are positively expected, so the similarities are more significant and differences less so.

    Given that we have described how alignments can be made for pairs of sequences, the next topic is to show how we can include more than just two sequences in an alignment to make a *multiple-sequence alignment*, and how the overall or average properties of such an alignment can be measured. As a naïve example of the benefits of multiple-sequence alignment, consider the following alternative alignments for two sequences:

```
GCGCATG--GCGCAT              GCGCAT--GGCGCAT
GGGCATGCGGCGCAT              GGGCATGCGGCGCAT
```

There is no way to know which alignment is best; the gap appears equally good in either position. However, if there is a third sequence which supports one scenario over the other then we can make a better judgement, in this instance supporting the first scenario.

```
GCGCATG--GCGCAT
GGGCATGCGGCGCAT
GCGCATGCCCCGCAT
```

When aligning pairs of sequences, which you can imagine as a two-dimensional problem, we can use the dynamic programming trick. However, as the number of sequences in a multiple alignment increases the complexity of the problem increases; in effect we get a new dimension of possibilities for each extra sequence. If the comparison of two sequences required a grid of points then three requires a volume of points, four a 4D hypervolume etc. In essence the alignment problem grows very complex very quickly with extra sequences. The objective is to find the gap placements in the multiple-sequence alignment to give the optimum alignment score for all of the sequences at the same time. Overall, there is no general fast method for guaranteeing the generation of optimal multiple-sequence alignments.

Progressive pairing

The general multiple-alignment problem might not be solvable in an optimal mathematical sense, but there are nonetheless means for getting very good (and hence scientifically useful) solutions quickly. The way that this is done considers that the sequences being aligned have arisen from a divergent, branching evolutionary process. Thus, rather than optimising the alignment of each sequence with all of the others, many methods attempt to join sequences into an alignment in the order of their similarity to one another; this is the *progressive pairing* approach. The alignment process can thus be broken down into discrete stages: the building of a family tree for the input sequences and then the creation of the overall multiple alignment, by progressively pairing alignments of smaller numbers of sequences in the same order as the branches of the family tree join (often done in a weighted manner depending on the lengths of the tree branches).

The tricky problem of building (phylogenetic) trees is discussed later in Chapter 14. Initially we will make only crude multiple-sequence alignments by joining sequences in the input (arbitrary) order. Nevertheless you can use later Python examples to make family trees and thus employ a better order for combining alignments. There are of course other approaches that can be taken to generate multiple alignments, aside or on top of the progressive pairing approach. However, many of these are slower and more specialised for increased accuracy. Usually these more rigorous approaches will not give answers too dissimilar to solutions from progressive pairing; indeed one common speed-up trick for more exhaustive searches is to restrict the search for optimal solutions to only routes that lie near the initial, quick solution.

Alignment consensus and profiles

Before we begin to look at routines to generate multiple alignments we will look at some routines to analyse multiple alignments, given that this will be useful in their construction. Firstly, we will consider the *consensus* of a sequence alignment. This means the compression of multiple aligned sequences into one representative or average sequence. This will typically be a completely unnatural sequence, but it will have many of the overall properties of the alignment. Consider the following multiple alignment, with the consensus line below. You can see that the residues in the consensus sequence are simply formed by taking a poll of the most popular residue type at each position. Note that in this particular example we have not considered gaps; a residue letter always wins out over a dash, but this needn't always be the case.

```
Seq1:  SRPAPVVIILIILCVMAGVIGTILLISYGIRLLIK
Seq2:  TVPAPVVIILIILCVMAGIIGTILLISYTIRRLIK
Seq3:  HHFSEPEITLIIFGVMAGVIGTILLISYGIRRLIK
Seq4:  HEFSELVIALIIFGVMAGVIGTILFISYGSRRLIK
       -----------------------------------
Cons:  HVPSPVVIILIILGVMAGVIGTILLISYGIRRLIK
```

Generating a consensus sequence

The following Python example takes a (multiple) alignment and produces a representative consensus sequence. The input alignment is assumed to be a list of one-letter sequences with gaps, with each list assumed to be of the same length. The second input to the function is a threshold value so that if no single residue type exceeds this fraction, at any of the alignment positions, then the position is deemed to be undefined and the residue code 'x' will be used. With the default threshold, 0.25, if we have an alignment containing ten sequences, at least three residues of the same kind must be present to give something in the consensus.

Initially, the function is defined along with some initial values, one for the length of the sequences, n (we happen to measure the first sequence of the alignment), one for the number of sequences in the alignment, nSeq, and an empty consensus string that will be filled in and then passed back at the end. Note that we convert the number of sequences into a floating point number because later on we will be using it to do some division, and we want to ensure that this operation produces a floating point number.[1]

```
def consensus(alignment, threshold=0.25):

  n = len(alignment[0])
  nSeq = float(len(alignment))
  consensus = ''
```

We loop though an index representing all of the positions (columns) within the alignment. Each time we define an empty counts dictionary that will hold the number of each type of residue observed at this position.

```
  for i in range(n):
    counts = {}
```

Another loop is defined inside the first one to iterate over each sequence. We get the appropriate letter for the current sequence using the position index, and skip the remainder of the loop if the letter is actually a dash. Finally in the loop we increase the count for this kind of residue letter by one, using the dictionary trick involving .get(), so that a starting value of zero is obtained if we have not seen a particular letter before during this cycle.

```
    for seq in alignment:
      letter = seq[i]
      if letter == '-':
        continue

      counts[letter] = counts.get(letter, 0) + 1
```

Now that the counts for the residue letters have been filled for this position we next calculate the fractions of the total that each letter represents. This is a simple matter of dividing the count for a letter by the number of aligned sequences and then putting the fraction and the corresponding letter in a list. Note that we use small, two-element sub-lists to store the fractions, rather than use a dictionary as we did for the counts, so that we can sort the values.

---

[1] Remember in Python 2 dividing two whole numbers gives a whole number.

```
fractions = []
for letter in counts:
  frac = counts[letter]/nSeq
  fractions.append([frac, letter])
```

The list of fractions is sorted and because the numeric value is the first element in the sub-lists, the fractions will be put in value order of the number, rather than alphabetic order of the letter. However, the letters will still go along for the ride. The largest fraction (`bestFraction`) and corresponding letter are the ones we want to use to build the consensus sequence with, thus we take the last element (`[-1]`) of the sorted `fractions` list, remembering that the values are sorted low to high. Although here we represent the proportion of each letter using a floating number between zero and one, it would also be possible to use integers, i.e. between zero and the number of sequences.

```
fractions.sort()
bestFraction, bestLetter = fractions[-1]
```

If the winning fraction is below our significance threshold then we extend the consensus sequence with an 'x', rather than use an infrequent residue. Otherwise the most popular residue is added to the end of the consensus, and the consensus string is returned from the function.

```
if bestFraction < threshold:
  consensus += 'X'

else:
  consensus += bestLetter

return consensus
```

Note that we simply concatenated the consensus string to the next letter using '`+=`', but we could also have put the letters in a list and used `''.join()`. Finally, the function can be tested with an alignment represented as a list of sequences:

```
alignment = ['SRPAPVVIILIILCVMAGVIGTILLISYGIRLLIK',
             'TVPAPVVIILIILCVMAGIIGTILLISYTIRRLIK',
             'HHFSEPEITLIIFGVMAGVIGTILLISYGIRRLIK',
             'HEFSELVIALIIFGVMAGVIGTILFISYGSRRLIK']

print(consensus(alignment)) # HVPSPVVIILIILGVMAGVIGTILLISYGIRRLIK
```

Generating an alignment profile

The second alignment analysis we will do is to generate a *profile*. A profile is a per-position statistic saying how much of each kind of residue there is at a given location. Because a profile has positional information it can still be aligned just like a simple sequence alignment, only this time we are dealing with fractions of residue types, rather than single residues. The Python function to generate a profile from an alignment is given below. It is very similar to the consensus generation, excepting that we don't have to choose a winning residue. The profile that is generated at the end is represented as a list of dictionaries; we locate an alignment position with a list index, then the sub-dictionary for this index gives the fractions of each residue type present. Note how we use the `counts` dictionary initially to do as the name suggests, and store counts, but later it is used to store the fractions; there is no reason to introduce another variable.

```python
def profile(alignment):

  n = len(alignment[0])
  nSeq = float(len(alignment))
  prof = []

  for i in range(n):
    counts = {}

    for seq in alignment:
      letter = seq[i]
      if letter == '-':
        continue

      counts[letter] = counts.get(letter, 0) + 1

    for letter in counts:
      counts[letter] /= nSeq

    prof.append(counts)

  return prof

alignment = ['SRPAPVVIILIILCVMAGVIGTILLISYGIRLLIK',
             'TVPAPVVIILIILCVMAGIIGTILLISYTIRRLIK',
             'HHFSEPEITLIIFGVMAGVIGTILLISYGIRRLIK',
             'HEFSELVIALIIFGVMAGVIGTILFISYGSRRLIK']

print(profile(alignment))

# First sub-dict: {'H': 0.5, 'S': 0.25, 'T': 0.25}
```

Profiles of this kind are often used as *position-specific scoring matrices*, for example, in programs like PSI-BLAST.[2] The idea here is that you build a profile for a family of sequences which have something of interest in common, and then search other sequences with the whole profile. This allows you to find sequences that share the properties of your aligned family as a whole, rather than finding ones that are similar to the individual members; this increases the sensitivity of similarity searches. A family profile conveys family-specific information, like the presence of a highly conserved (or invariant) site, which would not be recorded if you looked for sequences with a general substitution table.

Profile alignments

Once we have calculated a profile for an alignment we can align it with something else, just as we do in a pairwise sequence alignment. We actually require the ability to align pairs of profiles for the demonstration function to create multiple alignments. When we create a

---

[2] Altschul, S.F., Madden, T.L., Schäffer, A.A., Zhang, J., Zhang, Z., Miller, W., and Lipman, D.J. (1997). Gapped BLAST and PSI-BLAST: a new generation of protein database search programs. *Nucleic Acids Research* 25: 3389–3402.

multiple-sequence alignment by progressively adding sequences, we have to combine profiles for one alignment (one branch of the tree) with another alignment (a different branch); we cannot do a plain sequence alignment.

We won't go through the profile alignment function in great detail because it is so similar to the regular sequence alignment function discussed in Chapter 12. Nonetheless, there are still a few key differences. Firstly and most obviously, we are not passing-in sequences (strings of letters), but rather profiles (lists of residue fraction dictionaries). Secondly, when we calculate the residue similarity scores, rather than comparing one residue with another, we compare all of the residue fractions with each other. Accordingly, say that at a given position we have 50% 'A' and 50% 'G' in one profile and 25% 'C' and 75% 'G' in the other profile, then the final similarity score comes from the addition of four values; the combinations A:C, A:G, G:C and G:G each using a different similarity value from the substitution matrix, multiplied by the weight for that pair. You can imagine the weights for each combination being the area within a square defined by edges that have lengths defined by the individual profile weights. For example, the A:C combination is 50% times 25%, giving 12.5% to the score. Thirdly, the function also modifies the gap penalties because the input profiles are generated from alignments that may carry gaps. The modification is to multiply the penalties by the total weight for the profiles at each point; any missing weight will be due to gaps so the total is not necessarily 100%. This multiplication reduces the gap penalties, thus gaps are penalised less when gaps were present in the input profiles.

```
def profileAlign(profileA, profileB, simMatrix, insert=8, extend=4):

  numI = len(profileA) + 1
  numJ = len(profileB) + 1

  scoreMatrix = [[0] * numJ for x in range(numI)]
  routeMatrix = [[0] * numJ for x in range(numI)]

  for i in range(1, numI):
    routeMatrix[i][0] = 1

  for j in range(1, numJ):
    routeMatrix[0][j] = 2

  for i in range(1, numI):
    for j in range(1, numJ):

      penalty1 = insert
      penalty2 = insert

      if routeMatrix[i-1][j] == 1:
        penalty1 = extend

      elif routeMatrix[i][j-1] == 2:
        penalty2 = extend

      fractionsA = profileA[i-1]
      fractionsB = profileB[j-1]

      similarity = 0.0
      totalWeight = 0.0
      for residueA in fractionsA:
```

```
            for residueB in fractionsB:
                weight = fractionsA[residueA] * fractionsB[residueB]
                totalWeight += weight
                similarity += weight * simMatrix[residueA][residueB]

        penalty1 *= totalWeight
        penalty2 *= totalWeight

        paths = [scoreMatrix[i-1][j-1] + similarity, # Route 0
                 scoreMatrix[i-1][j]   - penalty1,    # Route 1
                 scoreMatrix[i][j-1]   - penalty2]    # Route 2

        best = max(paths)
        route = paths.index(best)

        scoreMatrix[i][j] = best
        routeMatrix[i][j] = route

  profileOutA = []
  profileOutB = []

  i = numI-1
  j = numJ-1
  score = scoreMatrix[i][j]

  while i > 0 or j > 0:
    route = routeMatrix[i][j]

    if route == 0: # Diagonal
      profileOutA.append(profileA[i-1])
      profileOutB.append(profileB[j-1])
      i -= 1
      j -= 1

    elif route == 1: # Gap in profile B
      profileOutA.append(profileA[i-1])
      profileOutB.append(None)
      i -= 1

    elif route == 2: # Gap in profile A
      profileOutA.append(None)
      profileOutB.append(profileB[j-1])
      j -= 1

  profileOutA.reverse()
  profileOutB.reverse()

  return score, profileOutA, profileOutB

alignA = ['SRPAPVV--LII', 'TVPAPVVIILII']
alignB = ['HHFSEPEITLIIF', 'H-FSELVIALIIF']

print(profileAlign(profile(alignA), profile(alignB), BLOSUM62))
```

The output of the function is a numeric score and two aligned, gapped profiles. Note that the output profiles were made by using the `.append()` function to add to the end of the lists, even though we went through the sequence positions in reverse order (decreasing i and j). This results in the profile lists being created in the reverse order to what we require, hence reverse() is used to flip the order at the end.[3] A gap is indicated in an output profile by using the None object; this is arbitrary but something that evaluates as false is handy.

Generating simple multiple alignments in Python

With a function that can align profiles, we can now consider building multiple alignments. Ideally, as previously mentioned you would make a genetic tree of the sequences and join them into a multiple alignment in the same manner as the tree branches. However, because we'll leave tree generation until later, we will only do a crude multiple-sequence assembly and add sequences one at a time to a growing multiple alignment, which will be enough to illustrate the procedure. Because sequence alignment works with a pair of inputs, we will have a pair of profiles that are to be aligned, each (potentially) representing smaller alignments that have already been combined. Repeating this procedure the multiple alignment grows, in terms of the number of sequences, as more profiles are added. Note that it would also be possible to generate multiple alignments using consensus sequences, rather than profiles. A function to do this, consensusMultipleAlign, is given in the downloadable material that accompanies this book (http://www.cambridge.org/pythonforbiology). You will note that because a consensus is a regular sequence representation only regular sequence alignment is required within the function, rather than having to do profile alignments.

Profile-based multiple alignment

Before we define the next function we make the required imports to get a substitution matrix (used for testing) and the regular pairwise alignment function sequenceAlign(). The module we are importing from relates to Chapter 12 and is available in the on-line material for this book (http://www.cambridge.org/pythonforbiology). The profile-based multiple-alignment function takes a list of unaligned sequences and a substitution matrix as input. Then we initialise a value for the number of input sequences.

```
from Alignments import BLOSUM62, sequenceAlign

def simpleProfileMultipleAlign(seqs, simMatrix):

  n = len(seqs)
```

We begin the multiple alignment by aligning the first two sequences and placing the gapped output in the multipleAlign list. Obviously we would need to check that we had at least two sequences before using this function. Note that if we were doing this properly we would need to start with the edges of a genetic tree.

```
  score, alignA, alignB = sequenceAlign(seqs[0], seqs[1], simMatrix)
  multipleAlign = [alignA, alignB]
```

Next the function loops through an index, one for each input sequence, but starting at 2. We start here because we have already used the first two sequences to get the multiple alignment started. Then in the loop we generate two profiles, one for the existing alignment and one for

---

[3] We could alternatively have used insert(0) to add to the start of the lists, though the way we have shown is quicker.

the sequence to be added, `seqs[i]`, noting that this is placed inside a list as required by the profile-generating function.

```
for i in range(2,n):
  profA = profile(multipleAlign)
  toAdd = [seqs[i],]
  profB = profile(toAdd)
```

Now we do the alignment of the two profiles, and get two gapped profile alignments back, as well as a score which in this instance we ignore.

```
score, alignA, alignB = profileAlign(profA, profB, simMatrix)
```

With the two gapped and aligned profiles, we have to use the positions of gap insertions to combine the new sequence into the existing multiple alignment. Note how the output profiles are just a guide to place insertions; the deepening alignment always uses the original sequences. We repeat this operation twice, once to collect and insert gaps for the starting multiple alignment, and once for the newly added sequences. Accordingly, we loop through the alignment using `enumerate()` so that we extract both an index number and a fractions dictionary (which is what a profile is composed of). If the fractions dictionary is `None`, i.e. missing, then we have a gap in that profile and we place the current index in a list of gap positions.

```
gaps = []
for j, fractions in enumerate(alignA):
  if fractions is None:
    gaps.append(j)
```

Once the gap locations are collected, all the sequences in that part of the alignment are looped through and are redefined with extra dashes, i.e. we add a column of '-' at the gap location. In this round we are putting dashes into the original `multipleAlign` list.

```
for j, seq in enumerate(multipleAlign):
  for gap in gaps:
    seq = seq[:gap] + '-' + seq[gap:]

  multipleAlign[j] = seq
```

Then a second round of gap insertion is done, this time collecting locations from the second align profile, `alignB`, and placing the gaps in the `toAdd` list.

```
gaps = []
for j, fractions in enumerate(alignB):
  if fractions is None:
    gaps.append(j)

for j, seq in enumerate(toAdd):
  for gap in gaps:
    seq = seq[:gap] + '-' + seq[gap:]

  toAdd[j] = seq
```

With all the gaps placed, the `toAdd` list is then added to the `multipleAlign` list to form a new, deeper alignment. The `.extend()` function is used because we are joining two lists.

```
multipleAlign.extend(toAdd)
```

Finally, outside the loop, having combined all of the sequences, we pass back the completed multiple alignment and test the function.

```
  return multipleAlign

# To test and print out the result
seqs = ['SRPAPVVLIILCVMAGVIGTILLISYGIRLLIK',
        'TVPAPVVIILIILCVMAGIIGTILLLIISYTIRRLIK',
        'HHFSEPEITLIIFGVMAGVIGTILLLIISYGIRLIK',
        'HFSELVIALIIFGVMAGVIGTILFISYGSRLIK']
align = simpleProfileMultipleAlign(seqs, BLOSUM62)
for k, seq in enumerate(align):
  print(k, seq)

# Result:
# 0 -SRPAPVV--LIILCVMAGVIGTI--LLISYGIRLLIK
# 1 -TVPAPVVIILIILCVMAGIIGTILLLIISYTIRRLIK
# 2 HHFSEPEI-TLIIFGVMAGVIGTILLLIISYGIR-LIK
# 3 -HFSELVI-ALIIFGVMAGVIGTI--LFISYGSR-LIK
```

Note that in the above test, after we have obtained the alignment we loop through the list of sequences of which it is composed using the inbuilt `enumerate()` function. This is so we efficiently get back an index number at the same time as the one-letter sequences.

Interfacing multiple-alignment programs

The above examples show you the basic mechanisms of how multiple alignments can be made. However, unless you wish to develop your own routines, when doing multiple alignments in most laboratory situations the best solution is to use existing, tried, tested and documented programs. Common programs like ClustalW,[4] MUSCLE[5] or T-Coffee[6] are easily interfaced with other routines by using Python to write input files and read output files. With experience you could write your own functions to achieve this; however, if you are using common file formats then it is easiest, and safer, to use the BioPython modules. The following example shows how you can make input for, execute and read the output of ClustalW. ClustalW is arguably the most widely known multiple-alignment program, and one that offers a good compromise between speed and accuracy. Using the same principles you can easily adapt the approach to use other programs.

Using ClustalW from Python

We will use input to the ClustalW program in the simple FASTA format, which you are hopefully already familiar with. Of course, if the data is not already in a FASTA file we will

[4] Thompson, J.D., Higgins, D.G., and Gibson, T.J. (1994). CLUSTAL W: improving the sensitivity of progressive multiple sequence alignment through sequence weighting, position specific gap penalties and weight matrix choice. *Nucleic Acids Research* 22(22): 4673–4680. Download via http://www.clustal.org/.
[5] Edgar, R.C. (2004). MUSCLE: multiple sequence alignment with high accuracy and high throughput. *Nucleic Acids Research* 32(5): 1792–1797.
[6] Notredame, C., Higgins, D.G., and Heringa, J. (2000). T-Coffee: a novel method for fast and accurate multiple sequence alignment. *Journal of Molecular Biology* 302(1): 205–217.

have to make one; indeed the following example will make the files for you starting from a simple list of one-letter sequences as we have been using in the examples. Although plain sequence strings are fine for basic analyses you may wish to consider more detailed sequence representations, like the `Bio.Seq` module, when situations are more complicated.

Firstly, in our example we import all of the external modules we will need upfront. It is generally a good idea to do this because you can quickly look at the start of a file and see all of the modules you are depending upon. Most of the modules are BioPython[7] libraries, so it is assumed that this is installed on your system.

```
import os
from Bio.Seq import Seq
from Bio.SeqRecord import SeqRecord
from Bio.Alphabet import IUPAC
from Bio import SeqIO, AlignIO
```

Two file names are defined; one for the input to the alignment program (`.fasta`) and one for the output from the program (`.aln`).

```
fastaFileName = "test2.fasta"
alignFileName = "test2.aln"
```

Next we loop though the sequences, `seqs`, defined as needed, to make a list of sequence record objects. For each one-letter sequence we first make a BioPython `Seq` object using the standard IUPAC[8] protein sequence alphabet. If there were an invalid protein sequence we would get an error. The basic sequence object is then combined with a name and a description to make a fully fledged sequence record object. The `SeqRecord` object is required for writing a FASTA file.

```
records = []
for i, seq in enumerate(seqs):
  seqObj = Seq(seq, IUPAC.protein)
  name = 'test%d' % i
  recordObj = SeqRecord(seqObj, id=name, description='demo only')
  records.append(recordObj)
```

We use the `open()` function to create a file handle object into which the sequence records are written. The `SeqIO` module of BioPython has a `.write()` function that can output various formats, so naturally we use the desired `'fasta'` option. With the file written the file handle is closed; this is not essential here but a good habit to get into.

```
outFileObj = open(fastaFileName, "w")
SeqIO.write(records, outFileObj, "fasta")
outFileObj.close()
```

With the input file made, the next job is to run the external alignment program; in this case ClustalW. This requires we have installed the alignment program and that it can be run as the command `'clustalw'` from the operating system. We will invoke the command using the `subprocess.call()` function, as we illustrated for BLAST in Chapter 12. Also, in this

---

[7] http://biopython.org.

[8] International Union of Pure and Applied Chemistry; an organisation that standardises chemical nomenclature.

instance we specify -INFILE and -OUTFILE options, to define the input and output file names respectively, formatted in the form *–OPTION=VALUE* (unlike with BLAST where the option name and associated value are separated with a space). If desired, we could read the ClustalW manual and include other options to control other aspects like gap penalties and the substitution matrix.

```
from subprocess import call
cmdArgs = ['clustalw',
           '-INFILE=' + fastaFileName,
           '-OUTFILE=' + alignFileName]
call(cmdArgs)
```

A perhaps better, but more complicated, alternative to the call() is to use the subprocess. Popen class, which amongst other things provides a means of executing jobs in parallel; very handy to do if you have lots of alignments and a multi-core computer.

After the system execution we then open and read the output alignments file. We use the AlignIO module from BioPython to do the interpretation of the file, specifying the read format as 'clustal'. Note that we use the .read() function because we only have one alignment in the file, if there were several we would use .parse() instead, to get back a list of alignments. The reading function makes an alignment object from which we can print out attributes, like the length, and loop through to get at the individual records. Note that the record variables are BioPython objects and thus come with .seq and .id attributes from which you can easily get at the one-letter sequence and name.

```
fileObj = open(alignFileName)
alignment = AlignIO.read(fileObj, "clustal")

print("Alignment length %i" % alignment.get_alignment_length())
for record in alignment:
  print(record.seq, record.id)
```

Finally, we will illustrate how you can write the alignment object out into a different format (other than the Clustal-format file we already have). This is a simple matter of making a list of alignments and an output file handle object, with a new name, which we pass on to the AlignIO.write() function. In this example we are using the 'phylip' format corresponding to the PYHLIP suite[9] of programs that makes and analyses phylogenetic trees.

```
alignments = [alignment,]
outputHandle = open("test2.phylip", "w")
AlignIO.write(alignments, outputHandle, "phylip")
```

---

[9] Felsenstein, J. (1993). PHYLIP (Phylogeny Inference Package) version 3.5c. Distributed by the author. Department of Genetics, University of Washington, Seattle.

14 Sequence variation and evolution

Contents

A basic introduction to sequence variation

Naturally the genetic codes of different kinds of organism differ to support their different construction, habits and biological requirements. Also, genetic codes vary between individuals of the same species, despite the large degree of similarity that binds them as a species. It is such variation within species that provides the opportunity for offspring to differ from their parents and potentially gain an improvement or specialisation, which in time may give rise to a new species. In a more modern context, sequence variations have a vital role in our understanding of genetic diseases and are becoming increasingly important for the development of pharmaceuticals, where the effectiveness and side effects of drugs may vary significantly according to the genotype of a person.

The variety of species and individuals is all down to the variety of genome sequences, but in order to discover as much as we can about the consequences of and reasons for this variation we should understand something about the mechanism by which sequences can change. In this chapter we will not go in to immense detail about the underlying mechanics, but simply cover the main principles and, importantly as far as bioinformatics is concerned, describe what kinds of change are detectable in the biological sequences. It should be remembered that it is only changes that are passed on to offspring which will influence evolution directly, but other variations may be important, for example, in the study of cancer-causing mutations.

Mechanisms of genetic change

You can think of DNA changes, and hence for RNA and protein too, as arising from one of four general ways: from recombination events where DNA strands exchange; as a consequence of damage to DNA; from errors in the replication of DNA at cell division; and by the action of mobile genetic elements, like viruses and transposons. We will introduce these points separately.

Recombination is the shuffling of large sections of DNA (i.e. many bases at a time) as a result of crossover between two different sections of DNA double helix. Two regions of DNA come together and, in a controlled way, the two double helices are broken and joined back together, in an exchanged manner so that the new DNA molecules are made of two regions from different origins. There are two notable situations where this occurs, for deliberate biological reasons. The first is during *meiosis*, the cell division that gives rise to gametes (egg and sperm cells). This kind of recombination usually occurs between sister chromosomes (i.e. the two copies of a given kind) where they share significant similarity, just before the chromosomes separate to form eggs or sperm, which carry only one copy of each kind of chromosome. The end result is that offspring have chromosomes that are not identical to their parents', but rather versions that are a spliced combination of the originals. This is an important means of generating genetic variation within a species and, because recombination occasionally occurs at the wrong spot with an offset between the chromosomes, is a means by which entire genes get duplicated. The second notable occurrence of recombination involves genes of antibodies, i.e. for the immune system. In this instance the recombination is used to form a diverse array of immune cells each producing different antibodies. This is part of the way that the immune response adapts to the potentially limitless variety of invading organisms. The antibody genes contain many alternative coding regions (i.e. exons) in different groups and the splicing brought about by recombination effectively selects a different coding region from within each group, to create different final exon combinations in each cell, so that it makes antibodies that bind a different target.

DNA is a relatively inert biological molecule, which is important for its role as the store and transmitter of inherited information. Nevertheless, there are still means by which the chemical structure of DNA can be disrupted. This can be as a result of various things including: high energy radiation (X-rays, gamma rays); ultraviolet light; highly reactive free-radical compounds, including those generated as a natural consequence of breathing oxygen; high temperatures and chemical toxins. DNA damage is a constant part of life and as such many repair mechanisms have evolved to fix things. Usually the damage can be fixed directly by repair enzymes, but if it gets too bad a cell will often commit suicide. Sometimes, however, the repair may not reproduce the original chemical structure or the damage may escape being fixed, so that when the DNA is replicated the base-pair matching at that position goes awry and the sequence changes. Because DNA damage is a somewhat random process and localised to small areas the sequence changes it creates mostly involve only a single base pair; a *single-nucleotide polymorphism*. However, larger changes are possible, for example, when there are double-stranded DNA breaks that are joined back together in the wrong way, as can be seen in some cancer cell lines.

As hinted at in the discussion of DNA damage, the replication of DNA strands is a time when variations become consolidated. However, it is also a time when variations can initially occur because DNA replication itself is slightly error-prone. This is important because it allows the changes to feed evolutionary processes, but they are mistakes nonetheless; DNA is produced where the occasional base pair doesn't match. The reason for this kind of mistake is because of the chemical structure of the DNA bases themselves. The bases are in a state of

Nucleotide Polymorphism

GCATCGTTCAC**T**CAGCCATCGGA**C**TACG
GCATCGTTCAC**G**CAGCCATCGGA**A**TACG

Insertion / Deletion

GCATCGTTCACT**CAGCCA**TCGGACTACG
GCATCGTTCACT------TCGGACTACG

Variable Copy Number

GCATC**GTTCCA**------------TCAGC
GCATC**GTTCCAGTTCCA**------TCAGC
GCATC**GTTCCAGTTCCAGTTCCA**TCAGC

Figure 14.1. An overview of the basic types of small-scale genetic change. Changes in DNA sequence often arise through somewhat random processes, like errors in replication or recombination, but in general variation is accepted non-randomly by natural selection. The simplest change, shown as an alignment mismatch, is the substitution of one type of nucleotide for another, called a single-nucleotide polymorphism (SNP), and there may be many substitutions in a given length of DNA. Insertions and deletions are apparent where a section of a sequence does not have an equivalent alignment match in another sequence. Knowing the ancestor sequence will reveal whether the change was actually an insertion or a deletion, though in general such changes are called 'indels' without disclosing the mechanism. Insertions and deletions are more likely to occur where the DNA sequence is repetitive, which in turn results in variation in the numbers of a repeated sub-sequence.

structural flux; there is an exchange between the normal form and another chemical structure. In chemical structure terms there is *tautomerism*, an equilibrium between different double-bonded forms (the double bond can switch from C=C to C=N by the movement of a hydrogen). While the standard chemical form is far more common, the occasional brief occurrence of the alternative form version results in a structure where different base pairs can form, compared to the normal Crick-Watson pairing (G:C, A:T). If the alternative form appears during replication the wrong base may be incorporated into the new strand, thus giving a base-pair mismatch. In many organisms, including humans, once the newly added base reverts to its normal form the pair mismatch can be detected and immediately removed by the proof-reading apparatus of the DNA replication machinery.[1] Occasionally, however, the mismatch still escapes, and although this is a very rare event (maybe of the order of 1 in 10^{10} for mammal genomes), given a large number of total bases (6×10^9 in a human cell), a large number of cell division events and all the individuals of a population it will undoubtedly happen from time to time. An escaped base-pair mismatch may still be repaired by enzymes, but as either of the two bases could be replaced, to give a matching pair, the fix may either regenerate the original sequence or consolidate a change, to give a single-nucleotide polymorphism.

The other general way of DNA changing is via the action of viruses and transposons, which can both be thought of as mobile genetic regions. Some viruses hide their own genome inside that of their host, using enzymes they bring along, thus evading host defences to become virulent at a later time. Some of these are *retroviruses*,[2] which have RNA viral

---

[1] Many virus replication enzymes lack proof-reading ability so that although error-prone their genome can mutate, adapt and evolve at a very high rate.
[2] HIV, which causes AIDS, is an example.

genomes, but make a DNA copy using a reverse transcriptase enzyme, i.e. the reverse of the normal DNA to RNA transcription. Virus DNA that is inserted into its host's genome naturally causes a change in the DNA sequence, but it may not stay there. Often viral DNA is cut out or *excised* from its host, sometimes leaving parts of its DNA behind. For virus DNA that remains, it can sometimes change so that the viral sequence remains in an inactive form, but is carried from that time onwards as its host reproduces. Transposons are like viruses in that they can be considered as genetic elements that can move and they often have similar means of inserting and excising their genetic material. However, unlike viruses, transposons lack the means to escape a cell: coat proteins, infection receptors and the like. They are sometimes colloquially called 'jumping genes'. Transposons and their remnants are responsible for large proportions of many genomes, forming repetitive, non-functional sequences (at least in the cellular sense; they arguably have a role in evolution).

Whatever the cause of a change in DNA sequence, its persistence relies upon it being accepted. If a change is detrimental and kills a cell or organism or carries a distinct disadvantage, then that change will tend not to be passed on to future generations. If a change is neutral or carries an advantage, then it will tend to be accepted. Thus, although the causes of sequence variation are highly random, their acceptance by evolution is not at all random. Indeed, when we observe variation in biological sequences most interest is in determining what the consequences are for the biology, for individuals and species, rather than the causal mechanics.

Conservation and variation

In earlier chapters we have analysed sequences to detect their similarities and thus to form alignments. The purpose of alignment is not only to group sequences, but also to say in what precise ways the sequences differ, or are preserved. If we look down a column of letters along the various positions of a multiple alignment some locations will use more residue types than others, and if we are considering protein sequences we can see places where the *chemical character* of the amino acids may remain the same despite the precise residues being different. Accordingly, we can measure how *conserved* or *variable* a given position is. Combining many positions we can say how variable a whole region or whole gene is. We can analyse sequence variability and find changes of biological or medical importance, and also learn something about the evolution and origins of the sequences.

As discussed, when organisms reproduce, sequence changes naturally occur. However, not all changes in the DNA are of consequence and many have no effect at all on an organism. This is because not all DNA has an immediate biological role, and even within the regions that do there can often be several sequences that perform a job equally well. Generally, changes in DNA which are not important for biological function occur more frequently; there is no reason for them not to be passed on.

In the genomes of many organisms there is a high proportion of non-functional, often repetitive junk DNA between genes. This is not to say that all DNA between genes is useless, given that such intergenic regions must contain control elements to regulate gene expression, promoters and enhancers, and also structural DNA to maintain chromosomes, like telomeres to protect chromosome ends and centromeres to allow replication. Some regions of non-functional DNA tend to show the highest rate of change during evolution and thus the largest variation between individuals of a species. Human DNA fingerprinting, for example, which may be used to identify criminals or detect family members, works by looking at *hypervari-able* regions that are different in almost every person. Such fingerprinting would not work nearly so well if gene-coding regions were used; there would be far fewer differences and finding two individuals with the same sequence (i.e. not the real criminal) would be much more likely, and in some cases positively expected.

The task of some analyses, rather than to just detect variations, is to measure the relative **rate of change** of variation. If we can find sites where the rate of change of the sequence is above or below the normal expected value, then this tells us something about the process of evolution at a fine scale. A common rate measure for variations in the coding regions of genes that go on to make protein is to look at the number of DNA substitutions that do change the amino acid sequence, compared to those that do not; the synonymous, *silent substitutions*. Remember that the number of three-base codons (64) is larger than the number of amino acids, and there are usually different ways of coding for the same amino acid.

In regions that have more silent changes than active ones the acceptance of the sequences during evolution indicates a *purifying selection*; this sequence is important and there is a reason why the protein sequence is preserved. Where there are proportionately more active changes than silent ones, compared to the average, then this can indicate a region where there is *positive selection*. Such regions indicate that the rate of evolution at these sites is greater than

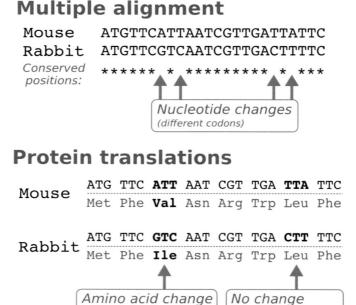

Figure 14.2. How changes in a protein sequence depend on codons. When DNA sequences that correspond to the protein-coding region of a gene are aligned we can observe mismatches that arise from nucleotide substitutions. However, in order to determine what effect, if any, there will be in the protein sequence we must consider what the differences are in the three-letter codons. Because an amino acid is usually encoded by more than one three-letter codon a nucleotide change may have no effect; the change can be synonymous. Whether there is a difference will depend on the genetic code and the precise codon change.

normal and that continuous change and adaptation is advantageous. Regions of positive selection in the human genome include genes involved in the immune system, which are ever changing to cope with the continuous appearance of new harmful bacteria, viruses and parasites.

Homologues and species

When comparing different versions of the same gene or protein, say in a multiple alignment, the sequences analysed are all related to one another. Saying that they are the same kind of gene is not only to say that they do the same job, but also that they have a common ancestor. Considering all the species of mammals on Earth, they all use haemoglobin to transport oxygen via blood and they all have globin genes to make the protein part of this. Thus we also know that the common ancestor[3] of mammals had haemoglobin particles and globin genes. The origins of globin undoubtedly go back even further than this to the time when backboned animals were new to our planet. The globin genes have diverged as the various species have split from one another, with any sequence change being carried on to descendants of that line. Genes or proteins that are known to be related to one another by the fact that they share a common ancestor are said to be *homologous*. It is a common mistake to mix up sequence similarity with homology; it may be stated that you can 'measure sequence homology', when strictly speaking what is meant is that the sequences are sufficiently similar that we can **infer** homology: a common ancestry.

Considering again the globin gene, as you may already know, there are even different kinds of globin gene **within** a single genome. Normal haemoglobins are a combination of alpha and beta versions of globin; two copies of each protein make the final particle. If we consider globins that are used in an embryo and fetus even more globin versions are present: gamma, delta, epsilon, zeta. Each version comes from a different gene and because they are so similar we know that they all have a common ancestor and were generated by *gene duplication* within a genome. So there are two basic means by which homologues are generated: when species separate or when genes duplicate. Accordingly for a pair of homologous genes or proteins we can say whether they are *orthologues* or *paralogues*.

Orthologues are different versions of the same gene in different species, generated by the fact that there was a common ancestor which also had the gene. For closely related species this is usually a straightforward concept, but for more distantly related species the definition becomes fuzzier, given that functions can diverge and genes can be copied within an organism. For example, the human PAX6 gene (involved in formation of the iris of the eye) has two orthologues in fruit fly,[4] named *eyeless* (ey) and *twin of eyeless* (toy).

Paralogues are genes that are related by the fact that they arose from gene duplication within a genome. The *eyeless* and *twin of eyeless* genes already mentioned are good examples; there was one ancestor gene and a duplication event generated the homologues. This is not to say that the duplication occurred in the fruit fly we see today, but rather in some ancestor that gave rise to many species, including the common laboratory fruit fly *Drosophila*. Looking at the globin genes where we have six close paralogues it is obvious that here there must have been multiple gene duplication events.

---

[3] Strictly speaking a population rather than an individual.
[4] *Drosphila melanogaster.*

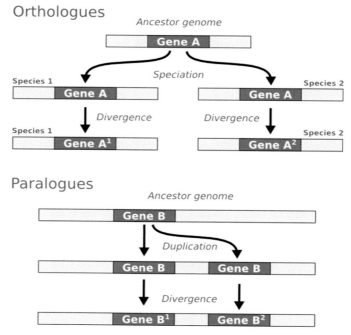

Figure 14.3. The classification of homologous sequences as orthologues or paralogues. DNA sequences in different contexts may be similar to one another because they share a common ancestor (they are homologous) and any differences have arisen by evolutionary divergence. There are two general ways in which such homologues can arise: an original sequence may follow two different evolutionary routes because different species separated from one another, thus creating orthologues, with different versions of the sequence in each genome. Alternatively, within a single genome a sequence may be duplicated and the different copies may then take on different roles, diverging from one another, creating paralogues.

When looking through sequences to detect homologues it is often the case that we find only sections of a gene or protein that are similar, while the remainder of the sequence is distinctly dissimilar. The reason for this is that the limits of genes are not static in evolution; via sequence changes they can expand and contract and be recombined in new ways, so that only parts of their common ancestry remain. It is a common occurrence to have only part of a gene duplicated, and in some instances duplication can merge genes or parts of genes. The units that are most commonly shuffled around in this way correspond to exons, i.e. the protein-coding parts of genes between the introns. By shuffling whole exons it is more likely that a sensible coding sequence will be maintained; breaking up exons is more likely to result in codon frame-shifts and nonsense protein code. At a protein level it is clear that the shuffling of exons gives rise to shuffling, duplication and recombination of entire domains, the functional units of proteins which are typically autonomously folding and globular. Multi-domain proteins, where each part of the gene has a potentially different ancestry, are very common in many genomes, including human. Accordingly if you search for protein homologues you are typically searching for domains that share a common ancestor, rather than whole genes. It often does not matter whether protein homologues comprise whole coding regions or are just part of a larger gene as far as the analysis of the family is concerned. So long as they do have a common ancestor you can tell something about conservation, protein structure and function.

Often we seek to link sequences as orthologues and paralogues. This helps us understand the process of evolution that has given rise to the observed sequences. Also, if we have some knowledge, experimental or otherwise, about one sequence then we can also say something about the related sequences; often they have a similar function. We can group them into a family where we can see general trends more clearly; multiple-sequence alignment is improved and we can identify conservation to identify features important to function (e.g. binding sites and catalytic sites) and protein structure. Indeed the presence of a common ancestor for proteins, and the preservation of 3D coordinates between homologues, is at the heart of a process called comparative modelling, which will be discussed more in Chapter 15. Often there is a choice between studying DNA sequence or protein sequence, and which you do on a given occasion depends on what you are interested in. Protein structure is far more conserved than amino acid sequence, which is in turn more conserved than nucleotide sequence, so for detecting the conservation in remote protein homologues, with similar 3D folds, you would use protein sequences. However, given that DNA changes are the underlying mechanism, and thus the most sensitive measure, if precise changes are studied nucleotide sequences are used.

Phylogenetic trees

Given a group of homologous sequences we can often go beyond saying that they are related and build a *phylogenetic tree* to say how they are related to one another. The idea with such a tree is to reconstruct the way that sequences have diverged during evolution. This can be used to reconstruct the events of how genes, non-coding regions and even protein domains arose. Given enough information we can look at a large scale to say how whole species are related, and if we look at the fine details how individuals within a family are related. Of course on some occasions we already know the inheritance tree, by using knowledge of parentage. This enables us to follow traits including physical differences, biochemical differences (e.g. blood groups) and inherited disease symptoms. However, it is only if we study the inherited differences at the biological sequence level that we can understand the molecular reasons, which in turn improves medicine and biology.

In history, evolutionary and family trees were built according to observable charac-teristics. If two species shared certain anatomical characteristics they would be deemed to be more closely related. This works well in some cases, but not in others (such as knowing where to place the elephant, whale and duck-billed platypus in the evolution of mammals). The reason for this difficulty is that people were only following a few subjective measurements. DNA sequencing allows us to place evolutionary lineages with much more confidence, because the detection of sequence is a precise thing and there are vastly more data points to follow: potentially every base pair, gene and transposon. Nevertheless, we sometimes still have to resort to anatomical comparisons when DNA is unavailable, as with dinosaurs, but the more bones the better.

When constructing a phylogenetic tree of sequences the basic principle is to think of the most similar sequences being the most closely related, analogous to the anatomical means of grouping organisms. When looking at sequence evolution we often think in terms of the most frugal explanation or *parsimony*; it is reasonable to assume that minimal changes are the most likely, so we would think that a nucleotide is less likely to change from say T to G to

C than it is to go directly from T to C. Accordingly when we build a phylogenetic tree we assume that the correct one is, or is close to, the one that involves the minimum amount of overall sequence change. Absolute parsimony isn't always a good idea in all situations: with distantly related sequences, and those with a high rate of change, the chances of having intermediate residue changes is significant, so it is better to think in terms of the long-term equilibrium of sequence. Also some things may be similar by chance and not because of a common ancestor, although this becomes increasingly unlikely overall if we consider increasingly more sequence data. However, there may simply not be enough data to form a firm opinion, even if building some sort of optimised tree is computationally possible.

When trying to work out real inheritance and evolutionary relationships more information will yield better results. Thus when we look at the relationships between species it is best to consider as much sequence and as many sequences as possible, although given the choice it is better to have sequences that sample a tree widely and evenly. Tree-building becomes more inaccurate, with regard to the underlying truth, the longer the branch, so it is best to have lots of linking sequences and hence shorter branches. Also, some sequences (genes, proteins or whatever) may be better than others at uncovering the relationships, particularly if the rate of sequence change is the right magnitude; too many changes and the assumption of parsimony is weaker, but too few changes and there isn't enough evidence to support a hypothesis. Accordingly, when we study fast-moving things, like the mutation of viruses, we look at rapidly changing genes, and for slow things like speciation we look at slowly changing things: ribosomal RNA genes, mitochondrial 'housekeeping' genes and rare transposon and duplication events.

When we have confidently built a phylogenetic tree, analyses of sequence variation gives us more information than can be obtained from alignments. We will be able to spot which changes occurred first and whether the same change has occurred more than once.

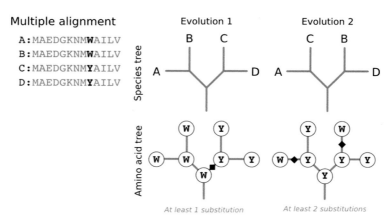

Figure 14.4. How phylogenetic trees may be used to count substitution events. A multiple alignment can be used to find sequence differences and thus describe conservation (or variation) as the proportions of different residues present at a given position. However, if we consider the evolutionary divergence of the sequences from one another, as illustrated by a phylogenetic tree, we can get more information about the evolutionary rate of change. The illustrated trees place the aligned sequences in two different hypothetical situations. For the first tree only one W↔Y amino acid substitution is required to generate the observed sequences, whereas for the second two substitutions are required. Naturally, the substitutions shown are the simplest scenarios, but overall the most frugal changes will be the most likely.

As illustrated in Figure 14.4, consider for example four sequences A, B, C and D, two of which, A and B, have residue **W** at a position and two of which, C and D, have residue **Y** at the same position. If we know that the pairs A and B and C and D are more closely related as a whole, then we know that one residue substitution was enough to make the observed situation; the ancestor of the sequences might have had **W** or **Y** but one substitution is enough to generate the A and B (**W**) branch or C and D (**Y**) branch. Conversely if the most closely related pairs **overall** are A and C and B and D then each pair contains a mix of **W** and **Y** residues. In this case there must have been at least two substitution events, one on each branch from the ancestor, which could only have one of the two residues. Accordingly, by considering the overall relationship between sequences we can make much better measurements of the rate of change than we can from just a multiple alignment.

Similarity measures

The first Python examples in this chapter involve measuring values for the similarity and hence conservation of sequences in a multiple alignment. This is achieved by looking at the various positions of the alignment and how similar the residues are down each column. This is not necessarily the best way to measure conservation (and consequently the opposite measure of variability), because the choice of sequences can be subjective and treating all sequences equally does not account for the expectation in the amount of change; the more distantly related sequences are as a whole the more change is expected. Nevertheless if we choose sequences carefully and know that the evolutionary distance between them is appropriate for the question we wish to ask then this type of analysis is pertinent, and quicker to perform than many more complex methods.

Measuring conservation

As the first example we define the `getConservation` function, which expects input arguments consisting of an alignment and a substitution matrix, so that we can score the similarity between sequence elements. Note how initially, after defining a conservation list which we will fill and pass back at the end, we use the `profile()` function defined in the previous chapter to convert the alignment into a profile: a list of dictionaries containing the composition of each position. In order to use the profile-generating function we must either have defined it in the same Python file or use an `import` command like the one below to access it from another location. The module import we illustrate is what you would need if importing from `MultipleAlign.py` which has been placed on the current Python import path (e.g. current working directory). This file is downloadable from the on-line archive which accompanies this book (see http://www.cambridge.org/pythonforbiology).

```
from MultipleAlign import profile

def getConservation(align, simMatrix):

  conservation = []
  prof = profile(align)
```

So armed with a compositional profile generated from the input alignment we next have to convert the compositional data for each alignment position into a single value representing the degree of residue conservation, according to some input matrix of similarity scores. Thus we loop through all of the dictionaries of the profile. Within the loop, for each composition dictionary we call the .items() function, which is an inbuilt function or *method* of standard Python dictionaries. In Python 2 this generates a list of (key, value) pairs which serves two purposes: firstly we will need both the residue letter (key) and composition (value) when we look through all the residue combinations, and secondly having a list allows us to sort the data. In Python 3, the .items() function returns a *view* onto the items and a list() function is applied to this to get an actual list.

```
for compDict in prof:

    items = list(compDict.items())   # list() not needed in Python 2
```

We sort the items list according to the composition value, the second element of each pair, so that we get a ranking from which we can find the highest-scoring residue code. Note that if we sorted according to the first element of the pairs we would get the list in alphabetical order, not the required scoring order. The sort is performed using the specified key function, which takes as its one argument an element of the list, and returns the desired comparison value determined by that element. Here that is the second element, x[1], which is used for the comparison. See Chapter 5 for a more detailed description of lambda. If we don't pass a key function then the list would be sorted according to x itself, which means the first element, x[0].

```
    items.sort( key=lambda x: x[1] )
```

Next we define an initial score of zero and then make two loops through the items containing residue codes and composition values, so that we generate all possible combinations of two residue types. The score for each combination is simply the multiplication of the compositional fractions and the similarity score from the substitution matrix. You can imagine the two composition values defining lengths on the side of a unit square, and thus their product represents an area of the whole.

```
    score = 0.0

    for resA, compA in items:
      for resB, compB in items:
        score += compA * compB * simMatrix[resA][resB]
```

With the score total for this position defined we now scale its value so that we can easily compare different positions. Accordingly, we find the maximum possible value for the score at this position and divide by that, so that all scores will be between zero and one. We define the best possible score as being the diagonal element of the substitution matrix for the residue with the highest composition at this location. Accordingly, we get the best residue code from the end of our sorted list of items (index -1 to get the highest value) and use it as both of the keys in the matrix dictionary look-up.

```
    bestLetter = items[-1][0]
    maxScore = simMatrix[bestLetter][bestLetter]
```

The score is scaled by dividing it by the maximum value and added to the conservation list, which once filled is then passed back at the very end, outside the profile loop.

```
    score /= maxScore
    conservation.append(score)

  return conservation
```

We can test the function by defining some alignments and calling the function with a substitution matrix appropriate to DNA or protein. Note that we can import the previously defined matrices, which are part of the file `Alignments.py`.

```
from Alignments import DNA_1, BLOSUM62

align1 = ['AAGCCGCACACAGACCCTGAG',
          'AAGCTGCACGCAGACCCTGAG',
          'AGGCTGCACGCAGACCCTGAG',
          'AAGCTGCACGTGGACCCTGAG',
          'AGGCTGCACGTGGACCCTGAG',
          'AGGCTGCACGTGGACCCTGAG',
          'AAGCTGCATGTGGACCCTGAG']

print(getConservation(align1, DNA_1))

# [1.0, 0.51, 1.0, 1.0, 0.76, 1.0, ...]

align2 = ['QPVHPFSRPAPVVIILIILCVMAGVIGTILLISYGIRLLIK-------------',
          'QLVHRFTVPAPVVIILIILCVMAGIIGTILLISYTIRRLIK-------------',
          'QLAHHFSEPE---ITLIIFGVMAGVIGTILLISYGIRRLIKKSPSDVKPLPSPD',
          'QLVHEFSELV---IALIIFGVMAGVIGTILFISYGSRRLIKKSESDVQPLPPPD',
          'MLEHEFSAPV---AILIILGVMAGIIGIILLISYSIGQIIKKRSVDIQPPEDED',
          'PIQHDFPALV---MILIILGVMAGIIGTILLISYCISRMTKKSSVDIQSPEGGD',
          'QLVHIFSEPV---IIGIIYAVMLGIIITILSIAFCIGQLTKKSSLPAQVASPED',
          '-LAHDFSQPV---ITVIILGVMAGIIGIILLLAYVSRRLRKRP-----PADVP-',
          'SYHQDFSHAE---ITGIIFAVMAGLLLIIFLIAYLIRRMIKKPLPVPKPQDSPD']

print(getConservation(align2, BLOSUM62))

# [0.3, 0.38, 0.09, 0.8, 0.08, 1.0, 0.64, 0.1, 0.32, 0.27, ...]
```

Conservation display

In the next example we will use the conservation values to make a very simple display, consisting of symbols that can be placed under the alignment to indicate which positions are conserved and which vary. Such a display, often involving '*', ':' and '.' characters to represent perfect, high and moderate conservation respectively, is very commonly seen in the output of programs that generate multiple alignments. A more sophisticated graphical display might involve rendering text with different colours. We will not give such an example here, but the same conservation values could be used.

The function is defined as accepting an alignment, a similarity matrix and a list of threshold values as input arguments. The list of thresholds will define the values that separate the conservation into different categories, and hence dictate which symbol is used. After the initial function definition we define an empty string which will be filled with symbols to indicate the similarity at each position. We use the previously defined `getConservation()`

function to extract the conservation values from the input alignment, and we extract the list of thresholds as three singular values.

```
def makeSimilarityString(align, simMatrix, thresholds):

    simString = ''
    conservation = getConservation(align, simMatrix)
    t1, t2, t3 = thresholds

    for score in conservation:

        if score >= t1:
            symbol = '*'
        elif score >= t2:
            symbol = ':'
        elif score >= t3:
            symbol = '.'
        else:
            symbol = ' '

        simString += symbol

    return simString
```

The remainder of the function simply involves looping through each score and comparing it with the threshold values, which are assumed to be in descending order. If the score is at or above the first threshold, the symbol is defined as '*'. Otherwise if it's at or above the second it is a ':', and at or above the third value a '.'. Finally, if the score was below the smallest threshold the symbol is set to a space, representing little or no conservation. At the end of each loop the symbol is added to the growing text string, which is then passed back at the `return` statement.

We test this function by printing out the sequences in the alignment with the line of newly generated conservation symbols underneath.

```
symbols = makeSimilarityString(align2, BLOSUM62, (1.0, 0.5, 0.3))

for seq in align2:
    print(seq)

print(symbols)
```

And the result in this case is:

```
QPVHPFSRPAPVVIILIILCVMAGVIGTILLISYGIRLLIK-------------
QLVHRFTVPAPVVIILIILCVMAGIIGTILLISYTIRRLIK-------------
QLAHHFSEPE---ITLIIFGVMAGVIGTILLISYGIRRLIKKSPSDVKPLPSPD
QLVHEFSELV---IALIIFGVMAGVIGTILFISYGSRRLIKKSESDVQPLPPPD
MLEHEFSAPV---AILIILGVMAGIIGIILLISYSIGQIIKKRSVDIQPPEDED
PIQHDFPALV---MILIILGVMAGIIGTILLISYCISRMTKKSSVDIQSPEGGD
QLVHIFSEPV---IIGIIYAVMLGIIITILSIAFCIGQLTKKSSLPAQVASPED
-LAHDFSQPV---ITVIILGVMAGIIGIILLLAYVSRRLRKRP-----PADVP-
SYHQDFSHAE---ITGIIFAVMAGLLLIIFLIAYLIRRMIKKPLPVPKPQDSPD
.. : *: .     :. **: **:*::..*::::: ....:.*:          .
```

In the above examples we have been thinking of conservation as being an amount of something. However, we can also say in what way the residues are being conserved, especially for amino acids that have chemical characteristics which may be preserved, even if the exact residue types are not. An example would be a position where we have histidine, lysine and arginine residues ('H', 'K' and 'R'), where although the amino acid type does change, the position is one that can be categorised as having a positive electric charge, a property of all three types.

Here we illustrate a Python list that defines some amino acid categories, based upon their properties. Each category is represented by a pair of items in a tuple. The first item is the name of the category and the second is a string containing all of the residue codes for that category. Note that we have simple one-letter categories for the individual amino acids, so that if an alignment only has one residue type in a given position the pure amino acid is stated, rather than giving a larger, less precise category.

```python
AA_CATEGORIES = [('G','G'), ('A','A'), ('I','I'), ('V','V'),
                 ('L','L'), ('M','M'), ('F','F'), ('Y','Y'),
                 ('W','W'), ('H','H'), ('C','C'), ('P','P'),
                 ('K','K'), ('R','R'), ('D','D'), ('E','E'),
                 ('Q','Q'), ('N','N'), ('S','S'), ('T','T'),
                 ('-','-'),
                 ('acidic',     'DE'),
                 ('hydroxyl',    'ST'),
                 ('aliphatic',   'VIL'),
                 ('basic',       'KHR'),
                 ('tiny',        'GAS'),
                 ('aromatic',    'FYWH'),
                 ('charged',     'KHRDE'),
                 ('small',       'AGSVTDNPC'),
                 ('polar',       'KHRDEQNSTC'),
                 ('hydrophobic','IVLFYWHAGMC'),
                 ('turnlike',    'GASHKRDEQNSTC'),
                 ('undef',       'AGVILMFYWHCPKRDEQNST-')]
```

The function example that employs amino acid categorisation naturally takes this list as an argument, as well as an alignment to work on. Inside the function a list is defined that will be filled with the residue categories, one for each position, then the next command makes a profile from the alignment using the previously discussed profile() function.

```python
from MultipleAlign import profile

def getAlignProperties(align, categories):

  properties = []

  prof = profile(align)
```

The bulk of the function involves looping through the dictionaries containing composition fractions, defined for the positions of profile, and getting the list of residue letters. Here the actual composition values are not used, but the dictionary keys give us the list of residue types found at that location.

```
for fracDict in prof:

  letters = fracDict.keys()
```

Given the residue letters the task is now to go through each category, which is listed in order of size from smallest to largest, to find the first, and hence smallest, category that contains all of the residue letters for this position. Accordingly for each category in turn we make a loop through each letter and determine if it is within the string of letters for that category (group). If the letter is not in the category, then the current category cannot possibly be right, so we use the break keyword to quit the current loop through the letters immediately, thus going on to test the next category. If there is no residue letter that causes the break to be triggered, i.e. all letters were present for the group, then the program flow will reach the else: statement, whereupon the name of the current category is added to the list to represent the current alignment position. With that done we don't need to test any more categories for this position so a different break is issued, this time to quit the loop through categories, thus going on to the next position. At the very end of the function the list of properties is returned as you would expect.

```
for name, group in categories:
  for letter in letters:
    if letter not in group:
      break                        # quit inner loop

  else:                            # all letters are in group
    properties.append(name)
    break                          # quit outer loop

return properties
```

The function is tested with an alignment and the categories list for the amino acids. In this instance we print the results by looping through the category names and their index numbers simultaneously, courtesy of Python's enumerate() function

```
catNames = getAlignProperties(align2, AA_CATEGORIES)

for i, category in enumerate(catNames):
  print(i, category)
```

Calculating substitution matrices

Next we will move on from conservation to a means of measuring the variations, or more specifically the kinds of residue substitution, present in a multiple-sequence alignment. We will use the analysis of substitutions to illustrate how an empirically derived substitution matrix, like the BLOSUM variant we have been using, can be measured. Of course it may seem that there is a somewhat circular argument in what we will be doing; an initial substitution matrix is needed to make the alignments from which we calculate a new substitution matrix. Clearly the composition of the new matrix will be influenced by the choice of the first. However, this is not all folly. Firstly, there are many ways of making and improving alignments which don't rely on a substitution rule. For proteins this would typically involve using alignments of three-dimensional structures. If homologous proteins of known structure can be aligned spatially then we have an alternative means to detect equivalent residue pairs. Secondly, even if we use one matrix to generate the input alignments the choice of sequences in the alignment has a vast influence on the outcome. So, for example, we might use a general matrix to perform alignments of a very narrow class of sequences, but the measured substitutions will reflect the choice of input. In this way a substitution matrix becomes more specialised, and thus better able to detect substitutions of the specialist class.

The function to calculate substitution naturally accepts a list of alignments to analyse, and various other arguments, but it is notable that these include a number that is used to smooth the data. The idea of data smoothing might seem dubious at first but it has an important function here. What the function will do in the end is compare the observed number of substitutions, given the alignments, and the expected number of substitutions for each possible pair of residue types. If the number of observations is very small then such a comparison is much less meaningful. The addition or removal of an observation might cause a large proportional change, but this isn't significant. Consider a substitution type where the number of observations is 4, but the expected value is 2. The observation is double what you expect, but it is likely that an extra couple of observations arose by pure chance. If the number of observations were 40 and the expected 20, then this is much more meaningful. Accordingly, the smoothing procedure used to deal with this is only significant when the number of observations is small (i.e. about the same size as the smoothing value), and all it does is make the data move towards the expectation, such that a particular point is not unduly significant. Small numbers of observations are of course much less problematic if the total amount of alignment data that you can analyse is large, but you don't always have the luxury of this.

We will compare observed and expected substitutions by calculating the logarithm of a probability ratio: a log-odds score. Hence, we import the `log` function from Python's standard mathematical library. Logarithms are convenient for substitution tables so that addition can be used to get an overall score for alignments etc. If raw probabilities were used then we would need to multiply values instead, which would often result in floating point numbers that are far too small for a computer to quickly and accurately deal with. The function name is then defined with its arguments and two variables are initialised: empty dictionaries that will hold the statistical data. The `alphabet` argument is a list of all possible residue letters, whether or not they are observed in the alignments, and the `maxVal` argument is used to scale the substitution matrix at the end.

```
from math import log

def calcSubstitutionMatrix(alignments, alphabet, maxVal, smooth=5):

  matrix = {}
  counts = {}
```

The blank dictionary that will contain the substitution matrix is filled with smaller sub-dictionaries, which in turn are filled with zero values. The residue letters are the keys in both dictionaries so two letters are required to extract each value: the substitutability of one residue for another. The dictionary containing the residue counts is also filled, with an initial value of zero for each letter.

```
for letterA in alphabet:
  subDict = {}

  for letterB in alphabet:
    subDict[letterB] = 0

  matrix[letterA] = subDict
  counts[letterA] = 0
```

Initially the function will measure the observed number of residues of each kind and the number of substitutions of one residue type to another. However, in the end we need to calculate simple probabilities as the fraction of total possible events we observe. Accordingly, variables for two totals are initialised to zero, one for the total number of pair substitutions and one for the total number of residue letters across all sequences. Note that these numbers have floating point representation (0.0 not 0) because we will eventually use them for division and wish to avoid integer rounding in Python 2.

```
totalRes = 0.0
totalSub = 0.0
```

The main part of the function involves looping through the input alignments to collect the substitution data. It would have been possible to call upon the `profile()` function that we have used previously, but on this occasion we would have had to loop through all of the sequences and positions in any case, to calculate the totals and residue counts. Thus in this instance it is simpler to do everything within the same set of loops. After the initial loop we record the number of positions in the particular alignment, and then loop through these positions, initialising a list which will collect the residue letters observed at this position, eventually doing yet another loop though the residue letters, `seq[i]`, available from all the sequences at this location (i). If the letter represents a gap it is skipped, otherwise it is added to the `letters` list.

```
for align in alignments:

  numPos = len(align[0])

  for i in range(numPos):

    letters = []

    for seq in align:

      letter = seq[i]
      if letter == '-':
        continue

      letters.append(letter)
```

With the letters collected we go through the list in two loops to get all of the possible residue pairs. The count for each letter is increased inside the first loop and inside the second loop we add each pair observation to our matrix. The matrix is thus filled with the observed substitutions, although since the letter does not change we really mean *preservation*. The overall residue count is increased by `numLetters`, the length of the `letters` list (so not including gaps), and the substitution total is increased by the square of `numLetters`, because we compare all against all to fill the matrix.

```
for letterA in letters:
  counts[letterA] += 1

  for letterB in letters:
    matrix[letterA][letterB] += 1

numLetters = len(letters)
totalRes += numLetters
totalSub += numLetters * numLetters
```

Once the residue and substation counts have been collected the next task is to calculate the various probabilities and the log-odds scores for the final matrix. To calculate the expected substitution probability we first calculate the average composition of each residue type by considering the number of counts over all alignments, residue positions and sequences.

```
averageComp = {}
for letter in alphabet:
  averageComp[letter] = counts[letter]/totalRes
```

A variable for the maximum score is initialised and then we loop through all residue pairs, given the possibilities defined in our alphabet of residue letters:

```
maxScore = None
for resA in alphabet:
  for resB in alphabet:
```

For each pair of residue types we calculate the expected substitution probability by multiplying the average compositions of each type. This is equivalent to saying that the expectation for a given pair, in the absence of any other information, depends only upon the probability of observing each residue type independently. With the expectation calculated we do a check to make sure that it is not a zero value (False in the if statement); if it is zero then one of the residue types is not observed at all in the data, thus we have nothing to do and so skip the loop.

```
expected = averageComp[resA] * averageComp[resB]

if not expected:
  continue
```

The observed substitution count is simply the pair count of this combination, as stored in matrix. This is then used to calculate a weighting value that is used to smooth the data in the event of the number of observations being low. Note that if observed is much larger than the smooth value the weight is close to zero, if they are equal the weight is one half, and if observed is comparatively small the weight is about one.

```
observed = matrix[resA][resB]
weight = 1.0 / (1.0+(observed/smooth))
```

The observed substitution counts for the pair are converted into a probability by dividing by the total number of substitution pairs. Then the smoothing is applied by redefining the observed probability as a combination of the original observed probability and the expected probability. With a low weight (lots of observations) the observation probability is barely altered, but with a large weight (few observations) the probability is closer to the expectation.

```
observed /= totalSub
observed = weight*expected + (1-weight)*observed
```

Lastly in the loops the log-odds score is calculated as the logarithm of the probability ratio; we check whether this score is the new maximum score value and then put the score into the matrix for the current residue letters.

```
logOdds = log(observed/expected)

if (maxScore is None) or (logOdds>maxScore):
  maxScore = logOdds

matrix[resA][resB] = logOdds
```

Once the loops are done it simply remains to scale the log-odds values in the substitution matrix and convert to integers. The integer conversion is somewhat traditional, but does mean that calculations with the substitution matrix are quicker. Note how the maximum score is converted to a positive value using `abs()` because the score logarithm could be negative. The maximum score is used to divide the matrix values to get a relative figure. However, we also multiply the log-odds score by the `maxVal` argument passed in at the start, which means we generate a range of integer values up this figure. Without `maxVal` using `int()` here would be pointless; we would only get zero or one.

```
maxScore = abs(maxScore)

for resA in alphabet:
  for resB in alphabet:
    matrix[resA][resB] = int(maxVal*matrix[resA][resB]/maxScore)

return matrix
```

The function is tested with a list of alignments, a list of residue letters and a scaling, maximum value for the matrix. Note how we can cheekily get the residue letters from the keys of an existing substitution matrix.

```
aminoAcids = BLOSUM62.keys()
print(calcSubstitutionMatrix([align2,], aminoAcids, 10))
```

Phylogenetic trees

Moving on from the more basic ways of considering substitutions and alignments, the next examples relate to the generation of phylogenetic trees. These are branching structures which in general aim, however falteringly, to represent the order in which different *taxa* (the sequences, species, strains etc. being compared) diverged from one another during the process of evolution. The aim is to give the order of the divergence and the evolutionary distance between the branch points and observed sequence data. Naturally, the more information is available the better the chances are of a computer-generated tree having accurate evolutionary distances and the real, historical divergence order. However, it should be noted that you can always generate a tree, however bad or uninformative the data, so caution is always advised in estimating the true, underlying relationship.

Phylogenetic trees using neighbour-joining

Tree-building can be a very difficult task. This is basically because the number of branch combinations increases very rapidly with the number of input taxa. The number of combinations can be so vast that we cannot routinely test all combinations to give the optimum tree arrangement. Hence what is often done, and which we will do here, is to generate a quite good, but not optimal, tree using a computationally fast method. The method used here is the one introduced by Saitou and Nei; the neighbour-joining method.[5] This is a quick way of generating a tree that is usually not so far from a global optimum and which is also relatively easy to understand mechanistically. The generated tree will not necessarily be at the global optimum because the algorithm is 'greedy' in that it only optimises the local solution for pairing up sequences, i.e. it never considers the tree as a whole. Nonetheless it is still a speedy and useful example to be used in subsequent analyses, and it can also provide a starting point for more globally aware optimising methods, if more accuracy is required. As with most of the examples in this book, the aim of the tree-generating example is not to give a high-performance, gold-standard program, but rather to lead you through a relatively simple piece of code that can be readily understood and which illustrates the major points of the topic.

The neighbour-joining algorithm involves repeatedly finding the closest pair from amongst the input sequences (and sub-trees after the first cycle) and joining these to form a new larger sub-tree until all sequences have been considered and only one, fully joined tree remains. Thus the first step for the tree-building example is to calculate what is termed a *distance matrix*, where the rows and columns of the matrix represent the different input sequences and the matrix elements represent the evolutionary distance between those sequences. Here we will only compare isolated input sequences; however, the method could be adapted to consider multiple sequences for each taxon (e.g. each input is a group of sequences representing a different species) so that the evolutionary distance between organisms is better represented as a whole. The tree that will result in the end is what is called an *unrooted tree*. This means that the tree has no indication of which taxon came first, evolutionarily speaking, and hence where the common (but unseen) ancestor fits in. By making an unrooted tree we are only showing the branch relationships between the input taxa. We recommend further reading if you wish to build *rooted* trees: those with an ancestor start point.

The evolutionary distance that separates two sequences in this instance will be estimated using the scores that are generated by performing pairwise sequence alignment. Thus the values will depend on the similarity of paired residues as defined by a substitution matrix. With DNA sequences it is common to use a more complex evolutionary model, to say how each change relates to distance; there are many more sophisticated ways of calculating the distance between sequences, not all of which can be used with the relatively simple neighbour-joining method used here. However, by using the substitution matrices we will keep things very simple for illustrative purposes and be able to use existing Python functions. For further reading on distance matrix generation, and tree generation in general, the PHYLIP[6] software package is a good starting point.

---

[5] Saitou, N., and Nei, M. (1987). The neighbor-joining method: a new method for reconstructing phylogenetic trees. *Molecular Biology and Evolution* 4(4): 406–425.

[6] Felsenstein, J. (1989). PHYLIP – Phylogeny Inference Package (Version 3.2). *Cladistics* 5: 164–166.

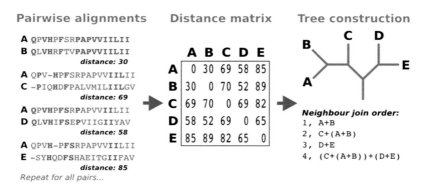

Figure 14.5. The basic construction of a phylogenetic tree using neighbour-joining. Although many more rigorous methods exist, phylogenetic trees can be constructed from sequences using the relatively simple neighbour-joining method. This involves creating pairwise alignments between all sequences to determine their similarity, and thus also rough evolutionary distance from one another. The resulting distance matrix is then used to construct the tree. This is done in a cycle where the most similar pair each time is combined to make a branch point and reduce the size of the matrix, until all sequences are joined.

Before we get to the main tree-generating Python function we will construct some ancillary functions that will be used inside the main routine. This will help understanding of the process by breaking it into operations that can be thought about separately. Also, the function that makes the tree will be neater and easier to understand. The downside to this way of doing things is the relative speed penalty incurred with calling Python functions. Thus if speed was an issue you could forego the separate functions and embed their code directly in the main section. If even more speed is required we would recommend a Python/C hybrid language approach, for example, using Cython, as discussed in Chapter 27.

To start with we make a simple distance matrix generation function that accepts a list of sequences and a substitution matrix as input arguments. It defines an initially blank matrix[7] of zeros with a row and column for each sequence (n sequences), such that the matrix elements will eventually be filled with the distance between all possible sequence pairs. Then we calculate a list of maximum possible scores for sequence comparisons, by comparing each sequence with itself using the `calcSeqSimilarity()` function defined in Chapter 12. These values will give a baseline for the other scores. The row and column indices in the matrix are then looped through (`i` and `j`) and for each index the corresponding sequence is obtained. Note how two `for` loops are used but we don't go from index 0 to n for these; instead the first loop goes up to n-1, and the second starts from i+1. This means we only consider one half of the matrix, excluding the diagonal. We can reduce the number of loops this way, and thus save some time, because the matrix is symmetric, i.e. `matrix[i][j]` is the same as `matrix[j][i]`.

```
from Alignments import sequenceAlign, calcSeqSimilarity
from math import exp
```

---

[7] Here we are using a list of lists, but we could use NumPy arrays.

```
def getDistanceMatrix(seqs, simMatrix):

  n = len(seqs)
  matrix = [[0.0] * n for x in range(n)]
  maxScores = [calcSeqSimilarity(x, x, simMatrix) for x in seqs]

  for i in range(n-1):
    seqA = seqs[i]

    for j in range(i+1,n):
      seqB = seqs[j]

      score, alignA, alignB = sequenceAlign(seqA, seqB, simMatrix)
      maxScore = max(maxScores[i],maxScores[j])
      dist = maxScore - score

      matrix[i][j] = dist
      matrix[j][i] = dist

  return matrix
```

The sequences for each matrix element are aligned using the input substitution matrix. The alignment thus generated is passed back with a score, which is then converted into a distance. The maximum possible score is calculated as the maximum from amongst the two maxScores list values for the sequences compared in this iteration. The distance is then calculated as the difference between this maximum possible score and the observed alignment score. Accordingly, if the sequences are identical the alignment score value is equal to the maximum and the distance is zero, and the more dissimilar the sequences, the larger the distance between them. If we were to use the simple DNA substitution matrix DNA_1, the distance would represent the number of non-identical positions. The distance value is then inserted into the matrix at the given row and column.

Here we test the function using previously defined lists of sequences (without gaps) and substitution matrices. The output is looped through row by row and the values in each row are displayed in a formatted manner (to two decimal places using '%.2f') to keep things neat on screen.

```
seqs = ['QPVHPFSRPAPVVIILIILCVMAGVIGTILLISYGIRLLIK',
        'QLVHRFTVPAPVVIILIILCVMAGIIGTILLISYTIRRLIK',
        'QLAHHFSEPEITLIIFGVMAGVIGTILLISYGIRRLIKKSPSDVKPLPSPD',
        'QLVHEFSELVIALIIFGVMAGVIGTILFISYGSRRLIKKSESDVQPLPPPD',
        'MLEHEFSAPVAILIILGVMAGIIGIILLISYSIGQIIKKRSVDIQPPEDED',
        'PIQHDFPALVMILIILGVMAGIIGTILLISYCISRMTKKSSVDIQSPEGGD',
        'QLVHIFSEPVIIGIIYAVMLGIIITILSIAFCIGQLTKKSSLPAQVASPED',
        'LAHDFSQPVITVIILGVMAGIIGIILLLAYVSRRLRKRPPADVP',
        'SYHQDFSHAEITGIIFAVMAGLLLIIFLIAYLIRRMIKKPLPVPKPQDSPD']

distMatrix = getDistanceMatrix(seqs, BLOSUM62)
distMatrix = getDistanceMatrix(align1, DNA_1)

for row in distMatrix:
  print(['%.2f' % x for x in row])

# ['0.00', '2.00', '3.00', '4.00', '5.00', '5.00', '5.00']
# ...
```

Next we define another function to find out which element of a distance matrix represents the closest sequences, and thus which sequences should be joined next when making a tree. Note that this is not as simple as finding the smallest distance, rather we calculate the value q (which is often described in the form of the Q-matrix) as the distance scaled by the number of branch points for the tree ($n-2$) minus the sum of the row and column at that point. For each row and column (again only needing to consider half of the matrix) we determine whether the q value is smaller than the best so far. If so (and for the initial loop, where $minQ$ is $None$) then we redefine the best q, and record the row and column for this element as joinPair.

```
def getJoinPair(distMatrix):

  n = len(distMatrix)

  minQ = None
  joinPair = None

  for i in range(n-1):
    sumRow = sum(distMatrix[i])

    for j in range(i+1, n):
      sumCol = sum(distMatrix[j])

      dist = distMatrix[i][j]
      q = (n-2)*dist - sumRow - sumCol

      if (minQ is None) or (q < minQ):
        minQ = q
        joinPair = [i,j]

  return joinPair
```

At the end of the function the indices for the closest pair of sequences are returned. Because of the `for` loops these indices are in a specific order, which helps later when the tree is built. We can test the function using a previously defined distance matrix:

```
print(getJoinPair(distMatrix))
```

The final function we require before we get to the main tree construction is one to determine the distance from a point on the tree (sequence or branch point) to the branch point that is created when joining up with another part of the tree. This function is required because the neighbour-joining algorithm works using a distance matrix, and each time we join sequences we need to know the distances to this new join point (node). Effectively the joined parts of the tree are replaced by a new single node; i.e. two sequences are replaced with an ancestor. When the distance matrix is recalculated we need to know how far each of the joined sequences is from the new point.

The function takes a distance matrix and two indices, i and j. The first index is the row/column for the sequence we want to find the distance from, and the second index is the sequence we are joining with. The row and column for the two indices are extracted directly, remembering that the distance matrix is symmetric, so the j column is the same as the j row.

```
def getDistToJunction(distMatrix, i, j):

  n = len(distMatrix)
  row = distMatrix[i]
  column = distMatrix[j]

  dist = distMatrix[i][j] + (sum(row)-sum(column))/(n-2)
  dist *= 0.5

  return dist
```

The distance from sequence index `i` to the new node is the average distance (hence multiplying by `0.5`) between the two joined sequences (`distMatrix[i][j]`) and the distance to the whole tree, i.e. for `n-2` nodes. The distance to the rest of the tree is calculated by taking the total (whole tree) distance for one sequence away from the total distance for the other and then dividing by the number of nodes to get an average. If the total tree were not considered the branch point would be exactly halfway between the two sequences. However, with the rest of the tree potentially being closer to one of the sequences than the other the branch point is pulled to one side.

Finally we get to the function that actually builds the tree. This takes an abstract distance matrix as input, so that it can be used in many different situations (not just for sequences), but which can be generated here using `getDistanceMatrix()` using a list of sequences and a residue similarity matrix as arguments. As mentioned previously this could be modified to take multiple sequences for each taxon, rather than just one. Initially in the function a list `joinOrder` is initialised, which will record the order in which sequences are joined to build the tree, and we use the function defined above to make the initial matrix of distances between sequences. The `tree` variable is initialised as a list containing index numbers, one for each sequence in the distance matrix. An alternative to using simple numbers here would be to use names for the sequences if we had them. What is actually used for the tree construction is not important; it just needs to be a tag to identify each taxon. As the main part of the function proceeds the `tree` variable will be modified, and each time a new branch is formed the tags for the two joined sections will be combined (into a tuple). In the end the nesting of pair sequence tags in this list will represent the structure of the tree.

```
def neighbourJoinTree(distMatrix):

  joinOrder = []
  n = len(distMatrix)
  tree = list(range(n))  # do not need list() in Python 2
```

Next we perform an iterative loop using the `while` statement. The loop continues while the length of the distance matrix is greater than two, i.e. while we still have to work out which possible bits of tree to combine. Each time we go through the loop a new branch point is defined and the length of the distance matrix will decrease by one (two sequences disappear and one new branch point appears). When there are only two tree parts to join there are no more choices to be made; the last remaining join is made and the tree is complete. Inside the loop we use the `getJoinPair()` function defined above to determine which indices, stored as `x` and `y`, represent the closest pair, and thus should be joined during this round.

```
  while n > 2:

    x, y = getJoinPair(distMatrix)
```

Just as the distance matrix gets smaller each time we join sequences so too does the `tree` list. When we combine branches we remove the representation of those branches from the list but add a new value for the larger, joined branch. Accordingly, the rows and columns of the distance matrix have a direct correspondence to the positions in `tree`. The variable `node` is defined as a tuple containing the two parts of the tree that are to be combined. These two parts correspond to the positions `x` and `y` (in the distance matrix and `tree` list) and can represent single, unjoined sequence tags or tags that are already in a branching structure. The new branch node is added to the `joinOrder` list and the `tree` list. To see the tree grow you could issue '`print(tree)`' during the iteration.

```
node = (tree[x], tree[y])
joinOrder.append(node)
tree.append(node)
```

Then we delete the `x` and `y` elements from the `tree` list because these parts have now been combined, and added as the new node to the end. Note that the `y` element is deleted before the `x` because we want to delete the largest list index first. Deleting the smaller index `x` first would shuffle the remainder of the list containing the `y` position along by one, complicating matters.

```
del tree[y]
del tree[x]
```

Next we have to adjust the distance matrix in light of the newly joined positions. First we calculate the distance from the joined `x` and `y` positions to the new branch point using the function `getDistToJunction()` explained in detail above.

```
distX = getDistToJunction(distMatrix, x, y)
distY = getDistToJunction(distMatrix, y, x)
```

A new row, containing zeros, is added to the distance matrix. This will later be filled with distances to the new branch point. Note that this new row has one more element than the existing rows because it won't be extended like the others in the following loop.

```
distMatrix.append([0] * (n+1))
```

Then we loop through all positions in the distance matrix, except the newly added one, and select those that have not been joined this round (not `x` or `y`). The distance from each of these positions to the new node is then defined as the average of the distances to the `x`, `y` points minus the respective distances of `x` and `y` to their joining node.

```
for i in range(n):
  if i not in (x,y):

    dist = (distMatrix[x][i]-distX) + (distMatrix[y][i]-distY)
    dist *= 0.5
```

This distance is added to the end of the distance matrix row. Also, the new row (`n`) at the bottom has the appropriate element filled in. Thus we have filled in the new distance values for the ends of row and column `i`.

```
    distMatrix[i].append(dist)
    distMatrix[n][i] = dist
```

Once the loop through matrix positions is complete we will have a new row and column in the distance matrix for the newly joined branch point. So next we delete the old `x` and `y` positions from the matrix. Firstly we delete two rows, and then for the remaining rows we loop through and delete the appropriate two columns. At the end of the `while` loop we decrease `n` by one, because after the merge the distance matrix is now one row/column smaller.

```
del distMatrix[y]
del distMatrix[x]

for row in distMatrix:
  del row[y]
  del row[x]

n -= 1
```

At the end of the function we convert the tree list to a tuple, effectively saying that the two remaining parts are joined (remember tuples are immutable). We add the whole tree to the list of join events. Finally the function passes back the tree and branch combination order, each of which is useful in different ways.

```
tree = tuple(tree)
joinOrder.append(tree)

return tree, joinOrder
```

For the above function the output tree data just uses sequence identifiers, it doesn't contain any distance information. However, it is trivial to make modifications that will allow the function to also pass back `distX` and `distY`, the distance to the branch points, along with the identifiers, e.g. `node = ((tree[x], distX), (tree[y], distY))`. Finally we can test the tree generation, first generating the distance matrix with a list of protein sequences and the BLOSUM residue substitution scores:

```
distMatrix = getDistanceMatrix(seqs, BLOSUM62)
tree, treeJoinOrder = neighbourJoinTree(distMatrix)

print(tree) # Result : (((7, (0, 1)), (4, 5)), ((2, 3), (6, 8)))
```

Tree-guided multiple-sequence alignments

Although we may just want to admire the evolutionary relationship suggested by a phylogenetic tree we can also use trees for many subsequent bioinformatics exercises. One important example is in the generation of multiple-sequence alignments. When making a tree, pairs of sequences are compared by alignment, but we can combine the sequences in the same order as they are liked into the tree to build a deeper multiple alignment. The example of a multiple alignment we gave earlier in Chapter 13 simply combined sequences in an arbitrary input order. Now we aim to do much better and link them in a way that better represents how they diverged. The process of multiple alignment in our example will be much the same; profiles are aligned and gaps are inserted so that sequences can be stacked; it is just the order of comparing the profiles that differs (which now represents the branching nodes of the tree). We could make further improvements to the multiple alignments by weighting the different branches when we combine the sequences. However, we will leave such complexities to programs like ClustalW etc.

The multiple-alignment function naturally takes a list of one-letter sequences and a substitution matrix as input arguments. Then inside the function we create a dictionary `multipleAlign` which will hold the sequences as they are combined. Because we will be combining groups of sequences that correspond to the branches of a phylogenetic tree, this dictionary will hold all of the partial alignments as they are joined together. The keys of this dictionary represent the indices of the sequences that were used to make each sub-alignment. Initially nothing has been combined, so the dictionary is filled with isolated sequences alone in a list (`[seq,]`), keyed by a single index. As the multiple alignment builds new indices will be added to indicate which sequences have been combined. These new indices will be of the tuple form `(x,y)` where `x` and `y` could be plain index numbers or other, already joined keys. For example, the key '`(3,(1,2))`' will be for an alignment that first combined sequences 1 and 2 which was then added to 3 – of course here the keys '`(1,2)`' and '`3`' will already exist because we are following the branches of the tree from outside inwards.

```
from MultipleAlign import profile, profileAlign

def treeProfileMultipleAlign(seqs, simMatrix):

  multipleAlign = {}
  for i, seq in enumerate(seqs):
    multipleAlign[i] = [seq,]
```

With the alignment dictionary initialised, the `neighbourJoinTree()` function is called using a distance matrix generated with the same input sequences and substitution matrix. This will return not only the structure of the tree, in terms of indices, but also the order in which sequences were combined. It is this second value that we will use to combine the sequence alignment. A copy is made of the tree combination order, which can be altered as we make the alignment. The original unmodified `treeOrder` list will be returned from the function as output.

```
  distMatrix = getDistanceMatrix(seqs, simMatrix)
  tree, treeOrder = neighbourJoinTree(distMatrix)
  joinStack = treeOrder[:]
```

Next a `while` loop is set up to continue combining sequences while we still have tree branches to consider. Note how the first (index zero) element is removed from the `joinStack` list, thus shortening it while at the same time providing the keys (`keyA` and `keyB`) to extract the next sub-alignments to be joined. Eventually all branch points will have been considered and the `joinStack` list will be empty, whereupon the loop will stop.

```
  while joinStack:
    keyA, keyB = joinStack.pop(0)
```

These keys are then used to access the relevant sub-alignments from the larger dictionary. These sub-alignments will then be combined, once gaps have been inserted, to make a new, larger multiple alignment. This is equivalent to joining two branches of the phylogenetic tree.

```
    subAlignA = multipleAlign[keyA]
    subAlignB = multipleAlign[keyB]
```

The profiles (lists of residue fraction dictionaries) are calculated using the previously defined `profile()` function for each sub-alignment. These are then passed into the profile alignment function, again previously defined, to generate a score (`s`, which will be completely ignored here) and two aligned profiles, i.e. with gaps.

```
profA = profile(subAlignA)
profB = profile(subAlignB)

s, profAlignA, profAlignB = profileAlign(profA, profB, simMatrix)
```

We insert the gaps in the previously combined multiple alignments (or single sequence at the beginning) by looking for gaps in the newly aligned profile. Accordingly, we loop though the profile alignment to find positions (index `i`) where there is no fraction dictionary, but rather a `None` object, indicating a gap.

```
gaps = []
for i, fractions in enumerate(profAlignA):
    if fractions is None:
        gaps.append(i)
```

With the gaps collected, another loop is performed, this time through each of the sequences in the sub-alignment. Each sequence is redefined by the insertion of dashes '-' at the positions specified by the `gaps` list. Note that as the gaps are inserted the sequence string will grow, so although all gap positions after the first are meaningless, with regard to the initial contents of `seq`, once all preceding gaps have been inserted the next gap location will be in-step with the longer redefined `seq`. After the gap insertion the sequence is re-inserted in the sub-alignment at the correct index `[j]`.

```
for j, seq in enumerate(subAlignA):
    for gap in gaps:
        seq = seq[:gap] + '-' + seq[gap:]

    subAlignA[j] = seq
```

The gapping procedure is repeated using the gaps from the second of the aligned profiles (`profAlignB`) to insert dashes into the second of the sub-alignments (`subAlignB`).

```
gaps = []
for i, fractions in enumerate(profAlignB):
    if fractions is None:
        gaps.append(i)

for j, seq in enumerate(subAlignB):
    for gap in gaps:
        seq = seq[:gap] + '-' + seq[gap:]

    subAlignB[j] = seq
```

Now that the sequences have had the appropriate gaps inserted the two smaller alignments are combined and placed in an expanded `multipleAlign` dictionary. The new key for this alignment is simply a tuple containing the two keys of the smaller sub-alignments that were combined in this iteration through the loop. The old sub-alignments are no longer needed and can now be removed from the dictionary by deleting the entries for the old, now combined, keys.

```
newKey = (keyA, keyB)
multipleAlign[newKey] = subAlignA + subAlignB

del multipleAlign[keyA]
del multipleAlign[keyB]
```

Finally we return the deepest alignment in the dictionary, which is the one that was stored using the final value of `newKey` from the loop, and the tree data. This represents the last joining in the `treeOrder` list, and thus represents the tree as a whole.

```
return multipleAlign[newKey], tree, treeOrder
```

We test the function with a list of input sequences and then print out the gapped output sequences from the returned alignment.

```
align, tree, order = treeProfileMultipleAlign(seqs, BLOSUM62)

for seq in align:
  print(seq)
```

And this gives:

```
-LAH-DFSQ--PVITV-IILGVMAGIIGIILLLAYVSRRLR-KR-----P-PADVP
QPVH-PFSRPAPVVIILIILCVMAGVIGTILLISYGIRLL-------------IK
QLVH-RFTVPAPVVIILIILCVMAGIIGTILLISYTIRRL-------------IK
MLEH-EFS--APVAIL-IILGVMAGIIGIILLISYSIGQIIKKRSVDIQP-PEDED
PIQH-DFP--ALVMIL-IILGVMAGIIGTILLISYCISRMTKKSSVDIQS-PEGGD
QLAH-HFSE--PEITL-IIFGVMAGVIGTILLISYGIRRLIKKSPSDVKPLPSP-D
QLVH-EFSE--LVIAL-IIFGVMAGVIGTILFISYGSRRLIKKSESDVQPLPPP-D
QLVH-IFSE--PVIIG-IIYAVMLGIIITILSIAFCIGQLTKKS-SLPAQVASPED
-SYHQDFSH--AEITG-IIFAVMAGLLLIIFLIAYLIRRMIKKPLPVPKPQDSP-D
```

Calculating substitution rates using trees

Another way to use sequences while following a phylogenetic tree structure is to estimate the rate of residue substitution. This will not be a rate in a specific time sense, but a rate in terms of the evolutionary distance between sequences. Strictly speaking the calculations will be of the minimum rate of substitution because we might not see all intermediate residue changes, i.e. although it is possible for a residue to change from X to Z to Y, unless we observe Z we will assume a direct transition from X to Y, the most parsimonious solution. There are two example Python functions that will measure substitution rates: the first is just to give a simple numeric rate value for the positions along a set of aligned sequences, and the second is to work out the relative rates of passive and active DNA substitutions in protein-coding regions.

In the following rate calculation functions we will be following residue letters back along the branches of a tree. In order to determine whether there has been a substitution event at a particular tree branch, at a particular sequence position, we estimate what the ancestral residue might be. This is a very simple guess, and other more complex probabilistic estimates can be made, but it is good enough for illustrative purposes. In essence the function works by taking two sets for residues, one for each of the branches of a tree, and determining whether these sets have any residue codes in common. If they do it is assumed that these are the ancestral residue codes. When the outer branches of the trees are used then there will only be a single residue letter in each set. However, for inner branch points where substitution events may have occurred (and there is no good guess at when the ancestor might be) each set can contain several residue codes.

The function is defined to take two sets of codes. Because we are using the Python set objects we can use their inbuilt `.intersection()` function to find residue letters that are present in both input sets. Then we perform some tests to determine what the ancestor residue codes might be and whether we think a substitution event has occurred.

```
def ancestorResidue(codesA, codesB):

  common = codesA.intersection(codesB)
```

If there are some residue codes in common this common set is passed back as the ancestor set. Also, it is assumed that a substitution has not occurred if the number of common residues is smaller than one or both inputs: then the ancestor represents a clarification, rather than a divergence; any input set that is larger than the common set would have already had its substitutions counted earlier, when the ambiguous set was defined.

```
  if common:
    return common, False
```

If there are no residue codes in common, the ancestor is set to the union of the two input sets. In this way ambiguous sets are generated. If the inputs are both full a substitution event has occurred, otherwise we have one or more gaps and no substitution is counted.

```
  else:
    union = codesA.union(codesB)

    if codesA and codesB:
      return union, True
    else:
      return union, False
```

Next we define a function that follows a tree of sequences and uses the ancestor residue estimation to calculate minimum substitution rates. This function takes an alignment and tree joining order as inputs; obviously these relate to the same sequences. We determine the number of alignment positions n and the number of tree nodes. Two lists of zeros are then initialised to be filled in with the relative and absolute substitution rates for each position. The relative rates will be compared to the average for all positions.

```
def calcSubstitutionRates(align, treeOrder):

  n = len(align[0])
  numNodes = float(len(treeOrder))
  absRates = [0.0] * n
  relRates = [0.0] * n
```

Next a dictionary is defined and initially filled with the gapped sequences from the input alignment. The sequences are converted from a string of letters into a list of Python sets, each containing a single letter. Note that gaps are replaced by empty sets. As we compare sequences in tree order, to count substitutions, the elements of this dictionary will be replaced with new sets of residues representing ancestral sequences (i.e. at tree branch points). The initial dictionary keys are simply indices within the alignment.

```
treeSeqs = {}
for i, seq in enumerate(align):
  sets = []

  for letter in seq:
    if letter == '-':
      sets.append(set())
    else:
      sets.append(set([letter])) # Could use sets.append({letter})
                                      from Python 2.7

  treeSeqs[i] = sets
```

Then we loop though the tree generation order, and in each iteration we extract the indices of the combined branches (a and b) from the start of the list. Note that by using .pop() the list is shortened, and when the list is empty the `while` loop will stop. The indices are then used to extract the relevant sequences, and a new list is defined which will be filled with a guess at the ancestor sequence.

```
while treeOrder:

  a, b = treeOrder.pop(0)
  seqA = treeSeqs[a]
  seqB = treeSeqs[b]
  seqC = []
```

To fill the ancestor sequence, all the alignment positions are looped through and the corresponding pair of branching sequence elements extracted (seqA[i], seqB[i]). The ancestor residue data is predicted with the function defined above and we catch the output values for the ancestral residue codes and a Boolean value (`True` or `False`) that tells us whether a substitution has occurred. If it has, the list of counts for the absolute rate calculation is incremented by one at this position.

```
  for i in range(n):

    residueSet, swapped = ancestorResidue(seqA[i], seqB[i])
    seqC.append(residueSet)

    if swapped:
      absRates[i] += 1.0
```

With this tree branch point considered, we store the ancestor sequence in the dictionary of sequences, combining the old keys in a tuple to make a new key for the ancestor. This is the same way that the node identifiers were made in the `neighbourJoinTree()` function.

```
  treeSeqs[(a, b)] = seqC
  del treeSeqs[a]
  del treeSeqs[b]
```

After the `while` loop the average substitution rate is calculated as the total number of residue substitutions, across all positions, divided by the maximum possible number of substitutions: one for each branch node at each alignment position.

```
meanRate = sum(absRates)/(numNodes*n)
```

Finally, we loop though the alignment positions and calculate the per-site rate values. The absolute rate is simply the count in the `absRates` list divided by the number of branch points. The relative rate is the difference of the absolute rate from the mean, as a proportion of the mean rate, and thus gives a positive or negative value depending on whether the local rate is larger or smaller than average. At the end the lists of absolute and relative rates are passed back as output.

```
for i in range(n):
  rate = absRates[i] / numNodes
  absRates[i] = rate
  relRates[i] = (rate-meanRate)/meanRate

return absRates, relRates
```

To test the function we use our multiple-alignment function to generate the alignment and tree data for some sequences. The alignment and order of tree construction are then passed as arguments to calculate the substitution rates, which we then loop through and print.

```
align, tree, joinOrder = treeProfileMultipleAlign(seqs, BLOSUM62)
absRates, relRates = calcSubstitutionRates(align, joinOrder)

for i, absRate in enumerate(absRates):
  print('%2d %6.2f %6.2f' % (i, absRate, relRates[i]))
```

The final example function in this chapter will detect sequence substitutions in the same manner as the substitution rate calculation, by following a guiding tree. However, instead of treating all substitutions equally we will subdivide them into two categories: those that change a codon to encode a different amino acid (active) and those that do not (passive). As mentioned above, regions of high active substitution may indicate positive selection in evolution, and more passive regions may indicate preservation of function. Naturally this analysis only works for DNA sequences corresponding to the protein-coding region of genes. We will assume that the input alignment is of the coding DNA strand (so not reverse complement of the genetic code) and starts exactly at the start of the first codon to be considered (so no offset).

The function takes an alignment, tree data and a genetic code as arguments. As before the number of positions and number of nodes in the tree are defined. Then counters for the active and passive substitutions are initialised and the dictionary to contain the sequences, and ancestor sets, is defined. Much of the function's logic is the same as described previously in `calcSubstitutionRates()`. However, once the ancestor residues are predicted things begin to differ.

```
def calcActivePassive(align, treeOrder, geneticCode):

  n = len(align[0])
  numNodes = float(len(treeOrder))
  active = 0
  passive = 0
```

```python
treeSeqs = {}
for i, seq in enumerate(align):
  sets = []

  for letter in seq:
    if letter == '-':
      sets.append(set())
    else:
      sets.append(set([letter]))

  treeSeqs[i] = sets

while treeOrder:

  a, b = treeOrder.pop(0)
  seqA = treeSeqs[a]
  seqB = treeSeqs[b]
  seqC = []

  for i in range(n):

    residues, swapped = ancestorResidue(seqA[i], seqB[i])
    seqC.append(residues)
```

If a substitution event is detected then we have to consider the codons in which the current DNA residues reside. The codon start position is defined as the multiple of three at or before the current position: if we divide `i` by three, round to the integer and then multiply by three, we will round down to a multiple of three. The sub-sequences for the codons are defined for each of the sequences by taking a slice of each from the codon start position to a position three residues along.

```python
    if swapped:
      codonStart = (i//3)*3 # // is integer division

      subSeqA = seqA[codonStart:codonStart+3]
      subSeqB = seqB[codonStart:codonStart+3]
```

Because the variables containing the sub-sequences are actually lists of Python sets (generated with the `ancestorResidue()` function) the actual triple-letter codons are generated by looping through all the residue possibilities for each of the three positions and joining the residue combinations (x+y+z). For each combination the amino acid code is then looked up using the DNA codon, according to the input genetic code dictionary. Note that because our genetic code dictionary uses RNA sequence keys, we first have to replace T with U in the DNA codon. The amino acid code is added to the set of amino acid codes `aminoAcidsA` using the `.add()` function of Python sets.

```python
      aminoAcidsA = set()
      for x in subSeqA[0]:
        for y in subSeqA[1]:
          for z in subSeqA[2]:
            codon = x+y+z
            codon.replace('T','U')
            aminoAcidsA.add( geneticCode.get(codon) )
```

The process is repeated for the second sub-sequence, i.e. the other branch of the tree.

```
aminoAcidsB = set()
for x in subSeqB[0]:
  for y in subSeqB[1]:
    for z in subSeqB[2]:
      codon = x+y+z
      codon.replace('T','U')
      aminoAcidsB.add( geneticCode.get(codon) )
```

If the detected substitution leaves any of the amino acid codes the same, comparing the two branches of the tree, we increase the count for the passive substitutions, otherwise if the code does change then the active count is increased. Note how the intersection of the two sets of amino acid codes is used to find whether there are codes in common, in which case we assume a passive change.

```
if aminoAcidsA.intersection(aminoAcidsB):
  passive += 1
else:
  active += 1
```

Finally, in the `while` loop the new ancestor sequence is stored and the already considered parts of the tree are deleted. And at the end of the function the counts for the active and passive DNA substitutions are passed back at the `return` statement.

```
treeSeqs[(a, b)] = seqC
del treeSeqs[a]
del treeSeqs[b]

return active, passive
```

To test we first generate an alignment and tree data and then make the active and passive counts, using an imported genetic code dictionary.

```
seqs = ['AAAGTGGATGAAGTTGGTGCTGAGGCCCTGGGCAGGCTG',
        'AAAGTGGATGATGTTGGTGCTGAGGCCCTGGGCAGGCTG',
        'AAAGTGGATGAAGTTGGTGCTGAGGCCCTGGGCAGGCTG',
        'AAAGTGGATGAAGTTGGTGCTGAAGCCCTGGGCAGGCTG',
        'AAAGTGGATGAAGTTGGTGCTGAGGCCCTGGGCAGGCTG',
        'CATGTGGATGAAGTTGGTGGTGAGGCCCTGGGCAGGCTG',
        'AAAGTGGACGAAGTTGGTGCTGAGGCCCTGGGCAGGCTG',
        'CATGTGGATGAAATTAGTGGTGAGGTCCTGGGCAGGCTG',
        'AACGTGGATGAAGTTGGTGGTGAGGCCCTGGGCAGGCTG']

from Sequences import STANDARD_GENETIC_CODE as SGC

align, tree, joinOrder = treeProfileMultipleAlign(seqs, DNA_1)
active, passive = calcActivePassive (align, joinOrder, SGC)

print('Active: %d Passive: %d' % (active, passive))
```

15 Macromolecular structures

An introduction to 3D structures of bio-molecules

So far in this book the more biological chapters have focussed on sequences: a linear and effectively one-dimensional representation of biological macromolecules. Studying sequences allows us to study the flow of biological information from the genome and how DNA, RNA and protein macromolecules evolve. However, this representation is somewhat removed from the physical reality of the biochemical soup of life, which of course occurs in three-dimensional space. We can even think in terms of four dimensions, if you consider time and how things change. Naturally, change in biological molecules is at the core of all life processes; nothing stands still. Here we will keep things relatively simple, however, and will not delve into the time-dependent, dynamic aspects. Hence, this chapter simply relates to the three-dimensional arrangements of biological molecules.

Here our primary focus is on the structure of proteins and RNA. This is not to say that the structure of DNA is not important, it is of course vital, but the difference is that for proteins (and directly functional, untranslated RNA) our understanding of the way biology works is so much more dependent on a precise three-dimensional structure. DNA, with its double helix, is necessarily an inert and repetitive structure. Things happen to cause deviations from this regularity when DNA is activated and deactivated (for reading), transcribed into mRNA, replicated, repaired etc., but it is the proteins of the cell that are the causal agents for these specific events. The way that proteins interact with DNA is just one of a plethora of different actions they perform to create the life-sustaining processes within organisms. The ability of an organism's proteins to do a multitude of, usually very precise, jobs stems from the fact that different proteins, encoded in different gene transcripts,[1] have different sequences of amino acid residues. The combinations of amino acids cause the different protein chains, initially made in a linear way, to *fold* into different three-dimensional structures. It is the precision of the various protein structures, i.e. that the same amino acid sequence virtually always gives the same three-dimensional arrangement of atoms, which allows proteins to perform a task and evolve according to this task, albeit catalysing a chemical reaction, interacting with another biological molecule or whatever.

Studying the structures of proteins, and the occasional non-translated RNA, allows us to work out how they operate; what their molecular mechanics are. This not only improves our understanding for its own sake, but also allows us to intervene in biology at an atomic level, as we do when we make new medicines and pesticides etc. For medicine in particular, the ability to say why things happen at this very small scale has allowed us to design new compounds to affect biology in a knowledgeable way, to cure an ailment or disease. Before we had such precise atomic knowledge the best we could do was test a vast array of existing compounds, just in case one of them had a desirable effect.

Protein structure

When we determine the amino acid sequence of a protein we gain the knowledge of which types of residue have been linked into a polypeptide chain. Because we know the chemical structure of the individual amino acid components, and because protein chains are formed in a regular and predictable way, we therefore know virtually all of the atoms and *covalent bond* connections[2] that are present in the entire protein molecule. In general the only deviations from this overall chemical structure will occur where small parts of specific chemical groups are not static, as when hydrogen ions hop on and off acidic residues, or if the protein is subsequently modified by enzymes. Such *post-translational* modifications include the formation of cross-links (between cysteine residues), cutting of the peptide backbone and the addition of other moieties like sugars, fats and phosphate groups. While modifications complicate the affair, if we are unable to do specific experiments to determine what has happened (e.g. mass spectrometry) then we can often discover what has occurred once we determine the overall three-dimensional structure.

We will now consider why proteins fold into their respective shapes. Protein folding is a deep topic because the number of potential conformations for a typical polypeptide chain is vast and the relationship between protein sequence and structure is generally not predictable. Even where it is possible to investigate a sufficient number of hypothetical three-dimensional arrangements, knowing which arrangement is correct, the *native conformation*

---

[1] Recall that due to splicing variations a single gene can give multiple RNA transcripts.
[2] The tight and somewhat stable kind of bond commonly seen between the biologically abundant elements: carbon, nitrogen, hydrogen and oxygen.

observed in nature, from purely theoretical considerations requires exceedingly long computational calculations. Fortunately, in molecular biology we generally don't have to make such tricky predictions because we can determine protein structure by performing experiments and making observations. Because protein folding is an exceedingly complex topic, most discussions about its mechanisms are well beyond the remit of this programming book. Nevertheless we will describe some of the basic principles, specifically what kind of forces are involved in holding a protein structure together, because this helps us understand the features we observe in structure data.

Overall the folding of molecules can be thought of in terms of energy. The atoms of a molecule, because they are in constant thermal motion,[3] are able to change relative position so that the overall conformation moves towards the lowest, most stable energy. Generally you can think of this as the three-dimensional arrangement that forms the most stabilising interactions between atoms. Strictly speaking a molecule will not be static, at its energy minimum, because it will move about due to temperature (it has kinetic energy). Accordingly, we often think of a molecule's native state as being a set of similar conformations that are close to the energy minimum, albeit bumbling about. It should always be remembered that the higher the temperature the wider are a molecule's motions and the further it can stray.

Proteins fold into compact, globular structures because of the way amino acids interact with one another and whether they interact (or do not interact) with water molecules, the primary biological solvent that surrounds them. Sometimes a protein will have cysteine residues that form covalent *disulphide* links (under oxidising conditions) that tie different parts of the protein together, but most of the compactness and precision of folding is due to weaker, non-covalent interactions, including those with water molecules. In simple terms the residues that can form stabilising interactions with water lie on the outside and those that cannot lie on the inside (in the core). Admittedly there are some kinds of proteins that aren't really dissolved in water directly, including those that are embedded in lipid bilayers (the fatty membranes that surround cells and their internal compartments). However, even here it is the ability of particular amino acids to interact with or avoid water that is behind the formation of a compact structure.

The atoms around the peptide links, which form the backbone of a protein's amino acid chain, are capable of interacting in a stabilising way with water and amongst themselves; the amide (N-H) and carboxyl (C=O) groups form polar *hydrogen bonds*. All things being equal the interaction with water is stronger, but the other parts of the amino acids that stick out from their backbone, the *side chains*, tip the balance so that the protein backbone is mostly stabilised by the backbone atoms hydrogen bonding with each other, and not water. The different amino acids have chemical structures that govern whether their side chain can make a significant interaction with water. Side chains containing atomic groups that can form relatively strong hydrogen bonds (O-H, N-H, C=O) and those that carry an electric charge are said to be *hydrophilic* (water-loving), because they can make stabilising interactions with water. Those that do not are described as *hydrophobic* (water-hating). Strictly speaking there is not a set dividing line between hydrophobic and hydrophilic; it is more a matter of degree. In an aqueous (water) environment, the hydrophobic and hydrophilic residues segregate when a protein folds, to form a hydrophobic core and hydrophilic exterior, i.e. a globule. This is just

---

[3] Except at absolute zero: -273.15 degrees Celsius or -459.67 degrees Fahrenheit.

a general trend though; the protein globule is stabilised further by the hydrogen bonds along the backbone, which tend to form regular patterns of hydrogen-bonding networks, called *secondary structure*. Also, the electric charges and polarities will push and pull the structure into the final shape. This final conformation is one where the core residues (mostly hydrophobic) come together and give rise to another weaker, but widespread, kind of interaction described as the *van der Waals* force, and thus the core packs tightly. This weak non-bonding interaction is actually present all the time between close atoms, including those from water, but in many situations it is swamped by other, stronger interactions. As a final point on protein folding, it should be noted that some large sections of amino acid sequences do not have a significant hydrophobic component. These regions will typically not form a single stable, folded structure because they don't have the ability to form a hydrophobic core. Usually this results in the region being highly dynamic or *unstructured* and is commonly seen at the ends of protein chains and as flexible linkers between folded *domains*, which are compact and globular.

Protein structure is often described in terms of a structural hierarchy, which helps us understand the final form as a combination of smaller elements; which is to say nothing about the actual mechanism of folding. This hierarchy is roughly described as follows:

Primary structure: this is the sequence of amino acids in the polypeptide chain and traditionally also includes any disulphide links and post-translational modifications. Essentially, this describes the covalent bond connectivity of the residues. It should be noted that although amino acids have two mirror-image (chiral) forms, unless explicitly stated, it will always be the usual, biologically abundant left-handed form that is present.

Secondary structure: this is a regular arrangement of hydrogen bonding along the protein backbone giving characteristic twist angles (dihedral/torsion) to the polypeptide chain. The common secondary-structure categories are alpha-helix, beta-sheet, turn and random coil (no regular structure). Secondary structure is often displayed in terms of the Ramachandran angles: the twist of the backbone either side of the alpha carbon atoms (the atom where the amino acid side chain branches). The twist about the peptide bond is not so descriptive here because it is almost always very flat.

Tertiary structure: this is how secondary-structure elements of one polypeptide chain come together to form a compact structure. In essence this is the folded, three-dimensional structure of one protein molecule in isolation.

Quaternary structure: this is the structure that results when multiple proteins and/or proteins and other molecules (RNA, DNA, small molecules) come together to form a larger composite structure that is commonly termed a *complex*. Some protein complexes consist of multiple copies of the same kind of protein coming together, often in a highly symmetric way.

Membrane proteins

As already hinted at, there are some proteins that do not reside wholly in water, instead they lie within lipid membranes[4] of cells. In some sense the lipid membrane can be viewed as an

---

[4] Fatty soap-like molecules that form double layers.

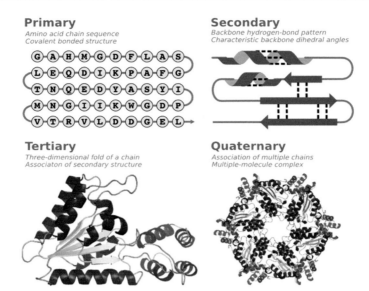

Primary
Amino acid chain sequence
Covalent bonded structure

Secondary
Backbone hydrogen-bond pattern
Characteristic backbone dihedral angles

Tertiary
Three-dimensional fold of a chain
Associaton of secondary structure

Quaternary
Association of multiple chains
Multiple-molecule complex

Figure 15.1. The hierarchical levels at which protein structures are commonly described. Proteins may be described in various ways, from a simple sequence of amino acids to a full folded, three-dimensional structure. The primary structure of a protein is the covalently bound chain of amino acid residues. The secondary structure represents the formation of characteristic hydrogen-bonding patterns (and thus also angles of twist) between its backbone peptide groups. The two most common secondary-structure types are the α-helix and the β-sheet (formed of β-strands, represented here as arrows). The tertiary structure represents the three-dimensional shape of one protein chain and the tertiary structure is how multiple three-dimensional protein chains combine to form larger assemblies. The quaternary structure represents the association of multiple molecules, which may or may not be of different types, into a larger complex.

alternative solvent to water, and thus it is a somewhat different environment for protein folding. It is a very hydrophobic (water-hating) environment, without polar or hydrogen bonding groups that could interact with water, and thus excludes water molecules. This exclusion of water is part of the reason why membranes exist in cells: to define different compartments to isolate different biochemical environments, including the limits of the cell itself. Although cell membranes are distinct layers, made of fatty molecules, they are not entirely static; their component lipid molecules, and other embedded molecules such as membrane proteins, can move about (diffuse), albeit in a mostly two-dimensional manner. While the lipids form a barrier, membrane proteins are inserted to give the membrane biological function, e.g. to transport specific compounds across the barrier.

The membrane-spanning part of a membrane protein is commonly referred to as a *transmembrane* domain. The transmembrane part may only be a small part of a larger protein, where the other parts are in water, but in some cases it may comprise almost all of the protein. The folding of a transmembrane protein domain is naturally somewhat different to that of aqueous domains. Indeed, most membrane proteins are inserted through a hole in a membrane as they are made. However, it is not the insertion into a membrane that makes a protein a membrane protein, but rather its amino acid composition. Firstly, a special signal (a special kind of amino acid sequence at the start of the protein chain) is needed to get them to the

Transmembrane proteins

Figure 15.2 (Plate 1). The general form of transmembrane proteins that reside in a lipid bilayer. Transmembrane proteins are embedded within the plane of a hydrophobic cellular membrane, which is composed of a double layer of lipid molecules and other membrane proteins. In contrast to aqueous proteins that reside in water, a membrane protein adopts a structure so that its hydrophobic (water-hating) parts lie within the membrane, often with hydrophilic (water-loving) parts protruding into the aqueous regions either side of the membrane. There are two common types of structure that transmembrane proteins adopt to form a hydrophobic membrane domain: an α-helical bundle or a β-barrel. A black and white version of this figure will appear in some formats. For the colour version, please refer to the plate section.

membrane when they are made. Secondly, in order to persist in a membrane the protein needs to have a large number of hydrophobic amino acids (those that cannot form especially favourable interactions with water) to form the transmembrane domain. Such protein domains are much more hydrophobic than those that sit in water; if they were not so they would simply dissolve in the water, leaving the membrane. Accordingly, you can often spot a transmembrane span by looking at the amino acid sequence, because it has so many hydrophobic residues in a concentrated region.

Within a membrane the polypeptide backbone of a protein forms regular hydrogen bonds with itself. Indeed, there is no water to interact with in any case. And because the domain crosses the membrane, only interacting with the aqueous environment at the edges, there are only two basic ways that a protein can pack into a compact, functional shape. The most common of these is the alpha-helical bundle, where the rods of the helices (i.e. the hydrogen-bonded secondary-structure part) cut across the plane of the membrane and lie roughly parallel with one another. The other transmembrane form is the beta-barrel, where a beta-sheet secondary structure zigzags its way across the width of the membrane, each strand hydrogen-bonding with the next to eventually form a closed ring. Although these structural strategies are different they have a similarity in that they both present a very hydrophobic surface of amino acid side chains, to remain stable in the membrane. The inside of such folds is pretty hydrophobic too, otherwise they would tend to migrate to the water surface. Overall, transmembrane structures are notably different from the aqueous proteins, and they are important to consider because they are quite common; about 30% of all proteins have a transmembrane domain. Unfortunately, however, you will see proportionately few three-dimensional structures for them, because they are notoriously difficult to do the normal experiments with, given their need to be surrounded by lipids (or similar).

RNA structure

Unlike its DNA cousin, RNA doesn't form long, continuous double-helical structures. RNA, once it is synthesised from a DNA template during transcription, is a single strand of nucleotide polymer which may then fold into a three-dimensional structure. Any structure might not be of much importance if the RNA is only acting as a messenger for protein synthesis. However, for RNA molecules that are not translated, and have other roles in biology, their three-dimensional conformation is often essential for their function, just as it is for proteins. Unlike proteins, however, which form globules due to the presence of hydrophobic residues, RNA structures tend to form initially because of the hydrogen bonding that occurs between pairs of bases, i.e. the same kind of complementary interaction that occurs between DNA strands. Such base pairing describes the secondary structure of the RNA. In this way an RNA can fold back on itself, or interact with another RNA molecule, to form a small region of double-stranded *duplex*. A common feature seen in RNA structures is *stem-loops*, where the strand is pinched together because there are two relatively close complementary sections (the stem), leaving a loop at the end. The full conformation (tertiary structure) is a three-dimensional arrangement of the loops and duplex regions relative to one another. Because RNA has only a few, chemically similar kinds of residue its structural diversity is somewhat limited compared to that of proteins. Nonetheless, there are some RNAs that are complex enough to form biological catalysts, just like protein enzymes; these are called *ribozymes*.

Determining macromolecular structures

There are a variety of methods that can be used to determine the three-dimensional structures of large biological molecules, to various degrees of precision or *resolution*. However, the vast majority of high-resolution structures of biological molecules were determined by just two experimental techniques: X-ray crystallography and nuclear magnetic resonance.

X-ray crystallography, as you might guess, works by forming crystals made of the molecule of interest, and is the most common form of macromolecular structure determination.

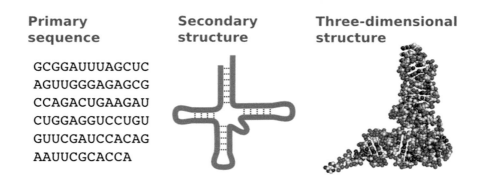

Primary sequence

GCGGAUUUAGCUC
AGUUGGGAGAGCG
CCAGACUGAAGAU
CUGGAGGUCCUGU
GUUCGAUCCACAG
AAUUCGCACCA

Secondary structure

Three-dimensional structure

Figure 15.3. The sequence, secondary structure and folded, three-dimensional structure of an RNA molecule. The illustrated RNA sequence is from a tRNA molecule, which forms a particular three-dimensional structure to perform its biological role (transferring a particular amino acid to the ribosome particle during transcription). The RNA molecule contains short internal regions of complementary base pairs and folds by looping to form hydrogen bonds within itself, as illustrated by the secondary structure.

Growing the crystals to start with is often the most difficult part of the process. However, if good crystals can be made, then because the molecules are held together in a lattice, the regular array of atom positions acts to diffract a beam of X-ray radiation,[5] thus creating a *diffraction pattern* consisting of spots called *reflections,* once the beam has passed though the crystal. Several diffraction patterns are collected by rotating the sample in a precisely known way, to view the structure from a variety of angles. The diffraction patterns can be combined, by mathematical means that we will not go into, to derive a three-dimensional map of the *electron density* of the atoms. The structure of the macromolecule is then determined by fitting its covalent chemical structure into the electron density to find the best fit. For proteins this usually involves a computer program that first finds the backbone route (fairly easy because of the characteristic spacing and angles along the polypeptide) and then fits the side chains, although this last part often needs manual assistance.

Nuclear magnetic resonance (NMR) is the second most common form of macro-molecular structure determination. Although overall many more structures have been determined by crystallography, NMR is particularly common for smaller structures, where it works best; indeed there is generally a limit to the size of the molecule that can be studied this way. NMR typically uses a concentrated solution of the subject (although solid-state samples are becoming more common), which is then placed in an extremely powerful electromagnet. In such a strong magnetic field some of the atomic nuclei, those that have *spin-active* isotopes, will align with the magnetic field. While the spin-active hydrogen-1 and phosphorus-31 isotopes are very abundant in nature, the carbon-13 and nitrogen-15 isotopes are rare and thus biological samples are usually artificially enriched with these. The magnetic alignments of spin-active atomic nuclei are then detected by passing a pulse of radio waves though the sample, so that they resonate. Because the resonance frequencies of the atoms depend on their chemical and structural environment, individual atoms often have distinct resonant frequencies. By the use of specially designed radio pulses the magnetisation can be moved from one atom to another, selectively to covalently bound atoms or those that are close in space. Such experiments correlate the observed resonances so that, often after a lengthy analysis, they can be connected together to identify which resonance goes with which atom and then how the resonances are arranged in space, thus yielding a three-dimensional structure.

Comparative modelling

When we do not have direct experimental evidence for the structure of a protein, we can sometimes still come up with a good guess called a *model* if we know the structure of a closely related protein. This method is known as *comparative modelling* or *homology modelling*. Strictly speaking even the structures of proteins determined by X-ray crystallography and NMR can also be thought of as models, as prior information about normal molecular geometries is used and there is always some uncertainty. However, direct experimental data constrains the models much more (and generally crystallography more so than NMR) and the more data you have the closer the model will be to the native conformation.

Comparative modelling relies on the observation that when proteins evolve their structures change more slowly than their amino acid sequence does. Hence, if we can detect

---

[5] X-rays are chosen because their wavelength is on the same scale as the separation between atoms.

two proteins that have sufficiently similar amino acid sequences, and thus infer a common ancestry, then we can be confident that they have structural similarities. Also, the closer the sequence similarity between two such homologous proteins, the closer their structural similarity will be. There are two basic steps when building a structural model based upon the structure of a protein homologue: find a homologue of known structure and then use the homologue's structure to guide the building of a model.

For the query protein of unknown structure we use its sequence to find potential homologues which do have a known structure, to act as the structural *template*. Template detection uses a special kind of sequence alignment, which is especially sensitive and accurate. Rather than using the regular, general substitution matrices like BLOSUM or PAM, comparative modelling tends to use family-specific scoring matrices or, for even better homologue detection, substitution tables that are specific to the *structural environment*. Using environment-specific substitution data allows an alignment to be sensitive to the way that amino acid changes in evolution depend on structure. For example, serine swapping for proline is more common in turns than in alpha-helices, because proline tends to disrupt helices. We know the structural environment for each position in a sequence alignment because we know the structure of the template, and the best guess is that the query sequence has the same structural environment, even if the residues differ. Such structural environments are typically defined by combining side-chain hydrogen bonding, solvent exposure[6] and secondary-structure categories. Thus, for example, you would have substitution matrices for exposed alpha-helix; buried, side-chain hydrogen-bonding alpha-helix; exposed beta-sheet, to name but a few.

Once a template is selected, and the sequence-structure alignment tells us which query residues are equivalent to which template residues, the next step is to build the computer model. Generally the backbone of the model is built first, then the side chains, given that these may vary significantly between the query and template, and finally loops are modelled in the regions that were not aligned, i.e. where there were gaps. The initial model may be built by assembling fragments of the query structure using the conformations borrowed from one or more templates. Alternatively, it may be built by the application of *spatial restraints*, derived from the templates, on to the model of the query polypeptide. The model is then subjected to a minimisation procedure to find the conformation that best satisfies these restraints. A popular program for such restraint-based modelling is MODELLER.[7]

- -

Using Python for macromolecular structures

In the following Python examples we will mostly examine and manipulate existing structural data, i.e. the coordinates of atoms. The idea is that you should become familiar with how to handle structural information. We deliberately avoid going into the computational aspects of how to determine structures in the first place. We will leave such vast and specialist topics to your future diligence.

---

[6] Buried or on the surface.

[7] Sali, A., and Blundell, T.L. (1993). Comparative protein modelling by satisfaction of spatial restraints. *Journal of Molecular Biology* 234: 779–815.

Obtaining structure data

Before we can begin to manipulate macromolecular structure data we must initially get hold of the coordinate information. Firstly, if you are using the downloadable material that goes with this book[8] there will be a few example files of structures saved in Protein Data Bank (PDB) file format. Alternatively, we could use the power of Python to download data directly from the PDB website's download service. The following code achieves this by making use of the urllib module (in Python 3; urllib2 in Python 2), which is a standard part of a Python installation. This module will do all the hard work and we will use it to send a request to the PDB web service, the response to which will be a plain text file containing the required structural data, and all we need to do is specify the identifier (pdbId) of the entry that we wish to download.

Initially, we import the web-handling urlopen() function. The module this resides in changed from Python 2 to Python 3 so we first try the Python 3 form and if that does not work then try the Python 2 form, using a try / except (you could also check whether sys.version [0] is '3').

```
try:
  # Python 3
  from urllib.request import urlopen
except ImportError:
  # Python 2
  from urllib2 import urlopen
```

Then define a Python string that contains the URL where the PDB data can be downloaded from, noting that it is a formatted string template with %s indicating where the database identifier will be inserted.

```
PDB_URL = 'http://www.rcsb.org/pdb/cgi/export.cgi/' \
          '%s.pdb?format=PDB&compression=None'
```

The function is then defined, and accepts an identifier and an optional file name, where the PDB data will be saved, as arguments. If no file name is specified (or conditionally evaluates to False, like an empty string) then the file name is specified by adding '.pdb' to the database identifier.

```
def downloadPDB(pdbId, fileName=None):

  if not fileName:
    fileName = '%s.pdb' % pdbId

  response = urlopen(PDB_URL % pdbId)
  data = response.read().decode('utf-8')

  fileObj = open(fileName, 'w')
  fileObj.write(data)
  fileObj.close()

  return fileName
```

---

[8] http://www.cambridge.org/pythonforbiology.

We use the web-reading `urlopen()` function to generate what is called a response object. This object is then used to fetch the PDB file into a string using the `read()` function. In Python 3 this comes back as bytes, not as a string, and in order to be able to write it to a file it needs to be converted to a string via a decoding, here using UTF-8. This extra decoding step is not needed in Python 2. This string is then simply written to file. The file name that was used is then returned at the end. Note that if you were to use this function regularly it would be advisable to add a few checks, just in case things go wrong; check that the URL query really worked and maybe warn the user if attempting to overwrite an existing file. The function is easily tested, in this case to generate a file with a defaulted name of `'1A12.pdb'`.

```
fileName = downloadPDB('1A12')
```

For most of the subsequent examples we will be working with the simple structure data model that was described in Chapter 8 and which is available with the web material in the `Modelling.py` file. Hence, to be able to test these functions, you will need to load the PDB file data into our `Structure` class of objects as illustrated below:

```
from Modelling import getStructuresFromFile

strucObjs = getStructuresFromFile(fileName)
```

Of course our data model and object classes for macromolecular structure are fairly simple, so they can be used as examples in this book. If you require a more complex but comprehensive set of objects, the `Bio.PDB` modules in BioPython can be used as an alternative. Some of the basics of these modules are described briefly towards the end of this chapter.

Simple geometric manipulations

With some structure data in hand the job now is to get used to working with its various components. Given that the data we are working with consists of three-dimensional coordinates, many of the manipulations involve thinking about geometry. Accordingly, the first example is designed to find the centre of mass (centroid) of a structure. Although similar functionality has been illustrated earlier, in Chapter 5, this time we are using the objects in our structure data model. This is done so that the loading and saving of data are entirely separate from the operations we want to focus on, thus making things clearer and more general.

Firstly, a dictionary is defined to give the atomic numbers of the chemical elements commonly found in biological molecules. If precise results were really important to you, the dictionary would contain many more elements, and list atomic masses rather than the atomic number (number of protons); however, this dictionary is sufficient for most purposes; the most important thing is that hydrogen has less weight. There is also an import for the NumPy function `zeros`, so the centre of mass can be initialised as a `numpy.array` object. Importing from the `numpy` module will simplify our code by allowing us to perform operations, quickly and with a minimal number of commands, on entire arrays of numbers at once (see Chapter 9). The `numpy` module is often not part of the standard Python installation, but its use is exceedingly widespread and it is available to download via the Scientific Python website (see http://www.scipy.org).

The function `getCenterOfMass` is defined, accepting a `Structure` object (as we describe in Chapter 8) to work on as an argument. Inside the function the variables to represent the central position and the total mass are set to zero. Then we have three nested loops, which iterate over all of the chains, residues and atoms, all according to the structure of our data model objects, thus accessing each of the atoms in the structure in turn.

```python
from numpy import zeros

ATOMIC_NUMS = {'H':1, 'C':12, 'N':14, 'O':16, 'P':31, 'S':32}

def getCenterOfMass(structure):

  centerOfMass = zeros(3, float)
  totalMass = 0.0

  for chain in structure.chains:
    for residue in chain.residues:
      for atom in residue.atoms:
        mass = ATOMIC_NUMS.get(atom.element, 12.0)
        centerOfMass += mass * atom.coords
        totalMass += mass

  centerOfMass /= totalMass

  return centerOfMass
```

The nested loops get to the individual atoms, whereupon we use the symbol for the atom's chemical element (`atom.element`) as key in the dictionary of atomic numbers to extract an approximate relative mass for the atom. Note that if the key is not present, we get a mass equivalent to carbon. The mass is then multiplied with the atomic coordinates, in an element-wise fashion given that `centerOfMass` and `atom.coords` are both NumPy arrays, and also added to the totals. In this way heavier atoms will contribute more to the total compared to lighter ones, e.g. carbon will contribute 12 times as much as hydrogen. The mass is also added to a separate total, to record the combined mass of all atoms. Once the loops are finished the array for the coordinate totals is then divided by the total mass to get the per-atom average. Of course if we were not using a weighted summation, we could divide by the number of atoms.

We can test the function on data loaded as a `Structure` object:

```python
struc = getStructuresFromFile('examples/1A12.pdb')[0]
print(getCenterOfMass(struc))
```

In the next examples we will actually change structural coordinates. Usually we don't want to mess about with the coordinates too much; however, it is very common to move (translate) and rotate a structure, to reposition it while still leaving the relative position of the coordinates unaffected. The Python example for rotating a structure involves using what is often called the *rotation matrix*. For three-dimensional space a rotation matrix is a three by three square array of numbers which represent the transformation of Cartesian (x, y and z) coordinates. In essence, to use the matrix we take a vector containing a positional coordinate and multiply it (matrix style) to get a new vector, which is now rotated relative to the origin (the zero point on all axes).

The function that generates the rotation matrix is the one described in Chapter 9, and can be imported from the corresponding example module and is used inside the function below, which takes a given structure and rotates it by a given angle about a given axis.

Naturally the molecular structure, axis vector and angle are passed in as arguments, although we have defaults set to rotation about the x axis (1,0,0) and zero angle. The `getRotationMatrix` function operates on the axis and angle to produce the matrix, and we convert it to an `array` so that we can use the `dot()` function, which calculates the dot product between arrays; and if we use it on matrices this is the same thing as matrix multiplication. With the rotation matrix defined the task is then to loop through all of the chains, residues and atoms, to get each atom in turn so that we can rotate its coordinates.

```
from numpy import array, dot
from Maths import getRotationMatrix
from math import pi

def rotateStructure(structure, axis=(1,0,0), angle=0):

  rMatrix = array(getRotationMatrix(axis, angle))

  for chain in structure.chains:
    for residue in chain.residues:
      for atom in residue.atoms:
        newCoords = dot(rMatrix, atom.coords)
        atom.coords = newCoords

rotateStructure(struc, (0,1,0), pi/2)
```

For each atom we calculate the dot product of the rotation matrix and the original coordinate array; this is performing the matrix multiplication operation. Once the new, rotated, coordinates are fully defined, the attribute of the atom object is updated accordingly (remembering that it is stored as an array). Note that because we are manipulating the structure internally we don't need to return anything from the function. Given that we passed the structure into the function in the first place it will be available after the function is run, albeit now with rotated atom coordinates.

Representing structures as NumPy arrays

The examples above used standard Python loops and didn't make full use of the NumPy array objects that store the atom coordinates. Accordingly, given that each atom position is described as a single NumPy array (a vector of x, y and z coordinates) each atomic coordinate can be placed inside another larger array, to form a two-dimensional matrix which represents the whole structure. To be able to work with whole 2D arrays of coordinates we must first transfer each atom's coordinates from the data model into the 2D array. Once defined, the 2D array data can be manipulated without having to repeatedly loop through each atom in turn; most operations can work on all atoms at the same time. Internally there will actually be looping going on, but we do not have to think about that and can trust that it will occur in an efficient and expected manner.

The following function loops through all of the atoms in a structure and places the coordinates in a NumPy `array` object. This function is thus general and can be used any time we wish to move from the hierarchical molecular data model to a numeric array system. Firstly, we import the required functions from the `numpy` module: `array` which will convert a Python list or tuple and `dot` for matrix multiplication.

```
from numpy import array, dot

def getAtomCoords(structure):

  coords = []
  atoms = []

  for chain in structure.chains:
    for residue in chain.residues:
      for atom in residue.atoms:
        coords.append(atom.coords)
        atoms.append(atom)

  return atoms, array(coords)
```

Inside the function we simply loop through the chains, residue and atoms and append the atom coordinates to a larger list of coordinates. We also collect a list of the atom objects at the same time, so that we will know which coordinate goes with which atom; obviously useful if we want to update the atom positions after changing the coordinates. Finally, we return the list of atoms and the coordinates, taking special notice of the fact that the coordinates are converted from the normal Python list into a NumPy `array` object (which here will be two-dimensional).

We will now use this array-generating function within another function that rotates an input structure. This does exactly the same job as the rotation function described earlier, but the innards are much simpler to write because we use one 2D array for all the coordinates and avoid writing loops. As before, the function takes an input angle and axis to define a rotation matrix which is converted to an `array`. The `getAtomCoords()` function just described is used to get the array object that represents all of the coordinates.

```
def rotateStructureNumPy(structure, axis=array([1,0,0]), angle=0):

  rMatrix = array(getRotationMatrix(axis, angle))

  atoms, coords = getAtomCoords(structure)

  coords = dot(coords, rMatrix.T)

  for index, atom in enumerate(atoms):
    atom.coords = list(coords[index])
```

The actual rotation operation is exceedingly simple: we just use the NumPy `dot()` operation to multiply the coordinate array by the rotation matrix;[9] this can be described as applying a coordinate *transformation*. Finally, the updated coordinates are used to update the coordinates of the atom objects. Note that we loop through the atom objects with the `enumerate()` function, to get not only the atom but also an index position. This index is then used to identify the correct location in the coordinate array, given that the order of the atom list exactly matches the order of coordinates; the `getAtomCoords()` function guarantees this. The position for each atom is updated in one operation by assigning one coordinate array (containing three numbers) to the three atom attributes, i.e. the array is unpacked into separate components.

---

[9] The transposed rotation matrix is the last argument so it operates on the (X, Y, Z) spatial dimension of the data, rather than the (long) atom dimension.

Here we test the function by rotating the previously loaded structure's coordinates $\pi/2$ radians (90°), about the y axis.[10] Because the coordinates are modified internally on the atom objects, there is no value to catch when the function is called. Note that the numeric constant `pi` is not known to Python until we import it from the `math` module.

```
from math import pi

rotateStructureNumPy(struc, (0,1,0), pi/2)
```

The next example function moves on from rotation to do an arbitrary transformation of the structural coordinates. Such a change is called an *affine transformation* in maths-speak, and involves a transformation matrix, which may or may not be a rotation, and a translation: a vector to specify movement from one location to another. The transformation matrix may be used to stretch, reflect, shrink, enlarge and rotate the structure.

Firstly, we import the `identity` function from the NumPy module, which is used to create an identity matrix, where the diagonal elements are one and the other elements are zero. The identity matrix represents no transformation at all when used in multiplication, just as normal multiplication by 1 doesn't change a number. We use an identity matrix of size three[11] as the default value of the `transform` variable, so that unless we state otherwise no matrix transformation occurs. The default for the translation is a vector of zeros; so no movement occurs.

```
from numpy import identity

def affineTransformStructure(structure, transform=identity(3),
                             translate=(0,0,0)):

  atoms, coords = getAtomCoords(structure)

  coords = dot(coords, transform.T)

  coords = coords + translate

  for index, atom in enumerate(atoms):
    atom.coords = coords[index]
```

As with the rotation example we use the `getAtomCoords()` to get the list of atom objects and the corresponding array of coordinates. The `dot()` call is used to apply the transformation, i.e. by matrix multiplication. The translation, to move the coordinates, is achieved by simply adding the translate vector, containing the numbers to add to the x, y and z locations, to the coordinate array. Although the `coords` variable is a two-dimensional `numpy.array`, and the `translate` vector is only one-dimensional (just three values, one for each axis), the array addition repeats over all of the rows in the larger array. This addition does a simple element-wise addition for the x, y and z numbers in each atom row; numbers at equivalent columns in the two arrays are added together to make a new array of the same size as the coordinate array.

Below, the function is tested on a loaded structure with a transformation that creates a mirror image (the coordinates take on the opposite sign; negative to positive, positive to negative) and a translation which is a movement of 10 units along the x and y axes. Using the basic PDB writer function `writeStructureToFile()` that is available in the on-line material

---

[10] The y axis has the first and last numbers, x and z, set to 0 and the middle number, for y, varies.

[11] As a Python list of lists this would be `[[1,0,0],[0,1,0],[0,0,1]]`.

we can write the modified structure, with transformed coordinates, to a new file so that we can view the structure in graphics software to see the result of our handiwork. Using graphical software to view molecular structures will be discussed later in this chapter.

```
mirrorTransform = array([[-1,0,0], [0,-1,0], [0,0,-1]])

translate = array([10.0,10.0,0.0])

affineTransformStructure(struc, mirrorTransform, translate)

from Structures import writeStructureToFile
writeStructureToFile(struc, 'testTransform.pdb')
```

Distances and angles

The next examples involve looking inside structures to extract particular measurements, like atomic distances and angles. Specifically, the angles we will be looking at are *torsion* (or *dihedral*) angles, which represent the angle of twist about an axis; here the axis will be along a chemical bond and the twist will be the relative orientation of two neighbouring bonds.

Firstly, to get the distances between two atoms is very easy: we simply get the coordinates of each atom, get the difference between the coordinates (stored in the variable deltas), add together the squares of these differences, and finally take the square root of the total.[12] Naturally because we have NumPy arrays most of this is done in an element-wise manner. It should be noted that the sum of squares can be calculated as the dot product of deltas with itself (though we could also do (deltas*deltas).sum(axis=1)). Because we are using dot on one-dimensional arrays (vectors) in this instance we get a single number out, i.e. we are not using it for matrix multiplication. So given two of our Structure.Atom objects, named atomA and atomB, we do:

```
from math import sqrt

deltas = atomA.coords - atomB.coords
sumSquares = dot(deltas, deltas)
distance = sqrt( sumSquares )
```

We move on from this toy example to a function that takes a structure and gives back a list of atoms that are within a certain distance from a specified point. This might be useful if you wanted to work out which other atoms a given atom may be interacting with. The function takes a structure object, position and a distance limit, to specify the catchment radius relative to the position. Initially within the function a blank list is created to collect the close atoms, then we convert the input xyz location into a NumPy array and we extract the coordinate array from the structure object as illustrated previously. Next we define limit2 as the square of the limiting search radius. This is done so that when we are looking though the coordinates we do not have to repeatedly calculate square roots, which would needlessly slow things down. Given that distances are always positive numbers, we can say whether one distance is larger than another (the search limit in this case) by comparing the squares of the distances, rather than the actual distances.

---

[12] This is simply applying Pythagoras' rule for calculating the length of the hypotenuse of a right-angle triangle in three dimensions.

```
def findCloseAtoms(structure, xyz, limit=5.0):

  closeAtoms = []
  xyz = array(xyz)
  atoms, coords = getAtomCoords(structure)
  limit2 = limit * limit

  deltas = coords - xyz
  squares = deltas * deltas
  sumSquares = squares.sum(axis=1)

  boolArray = sumSquares < limit2
  indices = boolArray.nonzero()[0] # Array for first and only axis

  closeAtoms = [atoms[i] for i in indices]

  return closeAtoms
```

Next in the function `deltas` is calculated as the difference from the atom coordinates to the `xyz` position. The square distance for each atom's separation is calculated by multiplying the differences with themselves and then summing along the spatial dimensions (`axis=1`) to give `sumSquares`. These are then compared to the square of the limiting radius. Using comparisons on NumPy arrays will act in an element-wise manner to generate an array of Boolean values. Hence `boolArray` will have true elements where the square distance is less than the limit. The indices of the non-zero (true) elements are extracted with `.nonzero()`, taking note that this gives a tuple with separate indices for each array axis. Thus, because `boolArray` is only one-dimensional, here we take the first and only array of indices (`[0]`). These indices can finally be used to select the appropriate group of close atoms from the larger list via a list comprehension. This list of close atoms is then passed back at the end of the function. We can test the function with a `Structure` object, here finding atoms that are close to the centre of mass coordinates that were determined earlier:

```
atoms = findCloseAtoms(struc, (18.89, 0.36, 1.24), 5.0)
for atom in atoms:
  print(atom.residue.code, atom.name)
```

Now we will move on to calculate the twist angle between atoms, imagined as a group of four atoms in a rough rectangle, connected by three bonds so that three edges are connected and one edge is not. The angle we will be measuring is the amount of twist between the first and last bonds, around the central bond. This can be visualised by looking directly along the central bond, so that the other two bonds make a 'V' shape; the *torsion* angle is the angle made by this shape.

We can use the function described in Chapter 9 that calculates torsion angles to measure the ϕ and ψ angles for an amino acid residue we find in a protein structure. These angles represent the twist of the polypeptide chain backbone either side of the residue's alpha-carbon atom position (where the side chain comes out). The input to the function is naturally a `Residue` object, from our structure data model, and an option to specify whether we want the angle to be represented in units of degrees (the default) or in radians. Because of the choice of angle units, we also import a standard function which allows us to do the conversion between degrees and radians.[13]

---

[13] 180° equals π radians.

```
from Maths import calcTorsionAngle

from math import degrees

def getPhiPsi(residue, inDegrees=True):
```

Initially the ϕ and ψ angles are initialised as `None`, so that if we fail to find certain backbone atoms (this happens for the first and last residue of a chain) there is an indication that the angles were not definable. Next we get a list of all of the residues in the input residue's chain, so that we can get hold of the residues either side. Given that we are using our data model objects, getting the list of residues is a simple matter of first finding the `Chain` object to which the input `Residue` object belongs (this is the parent link), and then following the `.residues` link to get the list of all residues (children) for the chain.

```
phi = None
psi = None

chain = residue.chain
residues = chain.residues
```

For the input residue we find three `Atom` objects representing the polypeptide backbone, by fetching atoms with the appropriate names. For each atom we collect the array of their coordinates.

```
atomN  = residue.getAtom('N')
atomCa = residue.getAtom('CA')
atomC  = residue.getAtom('C')

coordsN  = atomN.coords
coordsCa = atomCa.coords
coordsC  = atomC.coords
```

A positional index is defined to represent the location of the current residue in the polypeptide chain.

```
index = residues.index(residue)

if index > 0:
```

An index greater than zero means we are not at the start and thus can find the previous residue in the chain, by taking one from the index and fetching the residue from the list. We use the previous residue to fetch the atom, and then the coordinates, for the preceding carbonyl carbon ('C') atom, which is the last of the positions required to calculate the `phi` angle. Armed with the four coordinates, and using them in the correct order, the angle is calculated using the function described previously. Then the angle is converted to units of degrees if required.

```
residuePrev = residues[index-1]

atomC0 = residuePrev.getAtom('C')
coordsC0  = atomC0.coords

phi = calcTorsionAngle(coordsC0, coordsN, coordsCa, coordsC)

if inDegrees:
  phi = degrees(phi)
```

If the index is less than the last possible position, then we are able to find the next residue in the chain, by adding one to the index, and thus also are able to calculate the `psi` angle in the same manner. This torsion angle requires the coordinates of the backbone nitrogen atom in the next residue, i.e. it is on the opposite side of the input residue, compared to `phi`.

```
if index < ( len(residues)-1 ):
  residueNext = residues[index+1]

  atomN2 = residueNext.getAtom('N')
  coordsN2  = atomN2.coords

  psi = calcTorsionAngle(coordsN, coordsCa, coordsC, coordsN2)

  if inDegrees:
    psi = degrees(psi)

return phi, psi
```

Finally, the pair of angles is returned from the function. We can test the code by calling the function on all of the residues in a structure (although strictly speaking we ought to check that it is a protein first). Note that we exclude the first and last residues by using the slice notation `[1:-1]`. The measured ϕ and ψ angles are placed into separate lists, which are then passed as input to a function from the `pyplot` module to generate a scatter plot which allows us to visualise the data (see http://www.cambridge.org/pythonforbiology for Matplotlib download sites). This kind of display is called a *Ramachandran plot*, after one of its inventors, and will often have background colours indicating the likelihood of finding different angles. Positions of the ϕ and ψ angles on the Ramachandran plot are indicative of the secondary structure of the residues. Although each secondary-structure type (defined in terms of hydrogen bonding) will cover a range of angle values, a typical alpha helix will be at ($\phi = -60°$, $\psi = -45°$) and a typical beta-strand at ($\phi = -135°$, $\psi = 135°$).

```
from matplotlib import pyplot

phiList = []
psiList = []
for chain in struc.chains:
  for residue in chain.residues[1:-1]:
    phi, psi = getPhiPsi(residue)
    phiList.append(phi)
    psiList.append(psi)

pyplot.scatter(phiList, psiList)  # Scatter plot
pyplot.axis([-180,180,-180,180])  # Set bounds
pyplot.show()
```

Note that if you have a number of structural models, for example from an ensemble calculated by NMR, then you may wish to calculate the average of torsion angles. There is a function described in Chapter 9 which does this, given that angles are a cyclic measure (e.g. $-180°$, $180°$ and $540°$ mean the same thing) and a simple numerical average is no good.

Structural subsets

Next we will consider dissection of molecular structures into smaller parts. This sort of thing is done in many instances. You may, for example, want to remove a flexible region from the analysis of your molecule. Alternatively, you might want to select only a certain kind of residue or certain kinds of atoms. The latter may be done to define the backbone path of the molecular chain, which is useful when comparing structures with dissimilar sequences.

The example Python function we describe makes a subset of a structure by making a restricted copy of another structure, including only the atoms which are required. Alternative methodologies might be to remove atoms from an existing structure, or only load certain atoms in the first place, and these approaches may save a bit of computer memory. Firstly we import the definitions of the classes of structural objects we wish to make:

```
from Modelling import Structure, Chain, Residue, Atom
```

A function is then defined which takes an input structure and three other, optional, arguments that specify which chains, residues and atoms to consider. If any of these arguments is not specified (so defaults to None), it is taken to mean that no filtering is done for that kind of component and all are included. The chainCodes argument is assumed to be a collection of letter codes, e.g. ['A', 'B'], the residueIds is assumed to be a collection of residue numbers and atomNames, as you might expect, a collection of atom names. You can use any of the common Python collection types here, list, tuple or set, although these will be converted to sets using set() to remove repeats and give the best speed performance.

```
def filterSubStructure(structure, chainCodes=None,
                       residueIds=None, atomNames=None):
```

Within the function we determine a name for the new Structure object we are going to make by using the template Structure object's name, and then adding '_filter' plus other strings that list which chain codes, residue numbers (converted to strings) and atom names we have selected. Note how we first check to see if a chain, residue or atom specification was defined (not None, and hence true) before the name is extended.

```
name = structure.name + '_filter'

if chainCodes:
  name += ' ' + ','.join(chainCodes)
  chainCodes = set(chainCodes)

if residueIds:
  name += ' ' + ','.join([str(x) for x in residueIds])
  residueIds = set(residueIds)

if atomNames:
  name += ' ' + ','.join(atomNames)
  atomNames = set(atomNames)
```

Next the class definition for Structure is used to make a new instance of that kind of object, which we refer to as filterStruc. Although we defined a new name for this new object, we keep the conformation number and PDB identifier from the original; these indicate the origin of the data, and have not changed.

```
conf = structure.conformation
pdbId = structure.pdbId

filterStruc = Structure(name=name, conformation=conf, pdbId=pdbId)
```

The main body of the function is to loop through all of the chains, residues and atoms of the input selecting only those we wish to duplicate. Thus first we go through each `Chain` object and, if we have specified a filtering list for its code (`chainCodes`), we exclude any that are not mentioned; the loop, and hence chain, is skipped by using the `continue` command. If a chain is not excluded then we initialise a list that will contain residues to copy:

```
for chain in structure.chains:
  if chainCodes and (chain.code not in chainCodes):
    continue

  includeResidues = []
```

For each included chain we loop through its `Residue` objects and perform a similar check to see if the residue should be included. If the `residueIds` argument was filled but the residue number is not present then that residue is skipped. Otherwise, we go on to collect a list of atoms.

```
for residue in chain.residues:
  if residueIds and (residue.seqId not in residueIds):
    continue

  includeAtoms = []
```

Again, in the same sort of way we check to see if each atom's name is in our list of things to include, and if successful the list of template `Atom` objects is expanded.

```
for atom in residue.atoms:
  if atomNames and (atom.name not in atomNames):
    continue

  includeAtoms.append(atom)
```

If we have notionally decided to include a particular residue but that residue does not contain any of the required atom types, then there is no need to copy this residue at all.[14] When there are some atoms to copy for this residue, i.e. `includeAtoms` is not empty, both the list of atoms and the `Residue` object are placed in the `includeResidues` list. We could have placed the atoms in a big list on their own, but it is convenient to keep them with the corresponding residue, given that we need to specify the `Residue` (parent object) when making an `Atom` (child object).

```
  if includeAtoms:
    includeResidues.append( (residue, includeAtoms) )
```

If the residue list is not empty, we can make a new chain in the new `Structure` object, which is passed in at `Chain` creation to specify the parent link. With the chain now made we loop through the list of residues and corresponding atoms to make new `Residue` and `Atom` objects in the new structure. Notice that we use the attributes of the original objects when making the new ones. Thus, the residue copies will have the same number and code, and the new atoms

---

[14] This may occur while selecting amide hydrogens in amino acids, given that proline residues have none.

will have the same names and coordinates (albeit in a new array). Also, remember when making these objects within our structure we always have to specify the parent object, going up the data model hierarchy.

```
if includeResidues:
    filterChain = Chain(filterStruc, chain.code, chain.molType)

    for residue, atoms in includeResidues:
        filterResidue = Residue(filterChain, residue.seqId,
                                residue.code)

        for atom in atoms:
            coords = array(atom.coords)
            Atom(filterResidue, name=atom.name, coords=coords)
```

Finally in the function the new `Structure` object, with selectively copied components, is passed back:

```
return filterStruc
```

The function can be tested by specifying the chain, residue and atom selection. Here we select chain `'A'`, all residues (so the filter is `None`) and the backbone heavy atoms `['N', 'CA', 'C']`.

```
chainCodes = set(['A'])
residueIds = None                       # No residue filter: all of them
atomNames  = set(['N','CA','C']) # Heavy backbone atoms (not H)

chain_A_backbone = filterSubStructure(struc, chainCodes,
                                      residueIds, atomNames)
```

We could make a dedicated, streamlined function to make a complete copy of a structure. However, using the above `filterSubStructure()` function without passing any chain, residue or atom selection results in a full copy of the input structure. Thus we could be cheeky and do the following to pretend we had a dedicated copy function:

```
def copyStructure(structure):

  return filterSubStructure(structure, None, None, None)
```

Coordinate superimposition

In accordance with earlier examples that show how sequences can be aligned, this section describes how coordinates in three-dimensional space may be aligned by superimposing atom positions. Naturally this is a geometric operation that allows us to compare the shapes of structures, rather than a per-residue analysis. Essentially the coordinate alignment involves moving the centres of the structures to the same point (translation to the centre) and then finding the optimal rotation which gives the minimal deviation between corresponding pairs of coordinates; we superimpose equivalent atoms to place them as close together as possible without distorting the shape of the structures. Overall this procedure is commonly used to superimpose structures with identical atoms, i.e. different structural conformations (models). However, we can also attempt to superimpose structures of molecules with different atoms (different residue sequences). Below there is an example of this, where we find a common set of equivalent atoms, which can be used to guide two different molecules. There may be better ways to do this, but it illustrates how it can be done with relatively simple Python.

Note that similar functionality is also available via BioPython, by use of the `Bio.PDB.Superimposer` object. However, we define our own functionality here to explain the basics of what is happening and to give experience with Python. The following Python examples work with the structure data model we have described and are split into several separate functions, so that we can more clearly explain what is going on at each step. Although these procedures are usually applied to structures of proteins, it is possible to do the same thing to other kinds of molecule, including RNA and DNA.

Centring coordinates

Firstly in this section we will need to import some mathematical functions from the `numpy` module to allow us to efficiently work with arrays of numbers. Note especially that the `sqrt` and `exp` functions have the same names as the equivalents in the `math` module, but work on both NumPy arrays and single numbers.

```
from numpy import zeros, ones, cross, sqrt, linalg, exp, identity
```

Before we align structural coordinates by optimising rotations, we first need to move the input structures so that their centres are aligned. Effectively this means calculating the centre of mass of each structure and moving its coordinates so that it is repositioned with the centre at the zero points of all the axes (x, y and z). The centre needs to be the zero point because the rotational transformations we will consider are all rotations about this centre. The function that moves an array of coordinates naturally takes the array as an input argument, together with an array of weights. Note that both arguments must be NumPy `array` objects, which allows us to perform operations on whole arrays of numbers at once, rather than using loops; this is quicker to run and arguably easier to read (once you understand NumPy).

The weights array allows different kinds of atom to have different degrees of influence, typically derived from the atom's mass. Inside the function we first multiply the elements of the coordinates array by the weights array, taking special notice that we use the transpose of `coords` array where rows are switched with columns. This transposition is required because element-by-element multiplication of a one-dimensional array (`weights`) with a two-dimensional array (`coords`) works on a per-row basis. If we switch rows with columns then we get three rows that correspond to all of the x, y and z values respectively, and what we require is to multiply all of these rows separately with the weights. Given the weighted coordinates (`wCoords`) we then find the summation of the x, y, and z coordinates independently, i.e. we add up along the rows (specified with `sum(axis=1)`), the result of which is an array of three numbers. This array of totals is then divided by the total weight to get an average that represents the centre of the coordinates (`center`). Finally, all the coordinates are moved to the new centre (0, 0, 0) by taking away the old centre position; another example of a per-row `numpy.array` operation. The new coordinates and the old centre position (which represents a translation operation) are passed back from the function.

```
def centerCoords(coords, weights):

  wCoords = coords.transpose() * weights

  xyzTotals = wCoords.sum(axis=1)

  center = xyzTotals/sum(weights)

  coords -= center

  return coords, center
```

Aligning coordinates

The next function we describe is the clever mathematical part of the whole coordinate alignment operation. It takes two arrays of coordinates, and an optional array of weights, then calculates the rotation that best superimposes equivalent pairs of coordinates (corresponding atoms from each structure). The underlying mathematical theory is somewhat involved, and we will not discuss this in great detail, but by reading through the Python code it should be clear as to what the overall effect of the commands is.

The function is defined, and if we do not pass in an array of weights a new one is created using ones(), to give a series of 1's the same length as the coordinates (one for each atom).

```
def alignCoords(coordsA, coordsB, weights=None):

  n = len(coordsA)
  if weights is None:
    weights = ones(n)
```

The alignment routine is basically a minimisation of the differences in positions between corresponding coordinates, which we can achieve by finding an optimum rotation transformation. We define an initial matrix as the weighted dot product (matrix multiplication) of the one coordinate array and the transpose of the other. We need to make the transpose (switch rows for columns) to align the long, atom axes of the arrays so we end up with a 3×3 matrix. You can imagine each coordinate array as a transformation between atom number and spatial position. Accordingly, by applying one array to the other we get the transformation of positions, through a common set of atoms, to new positions, thus defining a spatial transformation.

```
  rMat = dot(coordsB.transpose()*weights, coordsA)
```

The mathematical magic that performs the actual minimisation is an operation called *singular value decomposition* or *SVD* for short. Luckily the linear algebra module of NumPy (linalg) has a convenient svd() function that will do all the hard work for us. Effectively this takes the 3×3 transformation matrix defined above and splits (factorises) it into three components, the combination of which would perform the same transformation. These three components are as follows: a rotation matrix, an array of linear scaling factors and an opposing rotation matrix.[15]

---

[15] More formally a matrix of eigenvectors of rMat multiplied by its transpose; an array of single numbers, each of which is the square root of the corresponding eigenvalues; another matrix of eigenvectors for the transpose of rMat multiplied by rMat. If you want to know more details about what is going on, see the following reference: Kabsch, W. (1978). A discussion of the solution for the best rotation to relate two sets of vectors. *Acta Crystallographica* 34A: 827–828.

The detailed explanation of why this is done is probably very confusing if you don't already have a good understanding of what are known as eigenvectors. However, it is sufficient to say that the matrices generated represent very special directions, which allows us to get directly at the rotation that we need to apply to align the coordinates.

```
rMat1, scales, rMat2 = linalg.svd(rMat)
```

Before using the extracted rotation matrices to align the coordinates, we must first check whether the SVD has given transformations that would cause a mirror image; the decomposition does not distinguish between normal and reflected solutions. To address this problem we calculate what is known as the *determinant* of the matrices: a single number that represents the overall scaling factor for the transformation. Helpfully, the linalg.det() function easily calculates the determinant, which will be -1 or $+1$ for our matrices. Given that the SVD operation has factored out a change in size, all that remains is a change in sign. The two determinants are multiplied, so that if they have opposite sign the result is -1.

```
sign = linalg.det(rMat1) * linalg.det(rMat2)
```

We check whether the sign is negative, and if so we know to remove the reflection. This is done by flipping the sign of the numbers in the last column in the matrix.

```
if sign < 0:
  rMat1[:,2] *= -1
```

Then the optimised rotation matrix, which will allow us to transform the coordinates to do the superimposition, is calculated by multiplying the two matrices obtained by the linalg.svd decomposition. Effectively this is reconstructing most of the original rMat transformation, except that the scaling, extracted by the SVD factorisation, has been eliminated so that only a pure rotation remains.

```
rotation = dot(rMat1, rMat2)
```

We use the final rotation matrix to transform one set of coordinates, so that it is repositioned closer to the other set. Naturally we only move one coordinate array, and the other is left alone. The dot() function is used to perform the matrix multiplication on the arrays: apply the coordinate transformation. Lastly, we return the calculated rotation matrix and the updated coordinates:

```
coordsB = dot(coordsB, rotation)

return rotation, coordsB
```

Calculating root-mean-square deviation

In order to help us interpret the coordinate superimposition, and because it is required in subsequent examples, we define a function that will calculate the variation in the coordinates across the atom positions represented in our arrays. Strictly this is the mathematical definition often called *root-mean-square deviation* or *RMSD* for short. As the name suggests the RMSD value is calculated by taking the differences in coordinate positions, squaring them, finding the average value and then the square root of this. This is effectively the average distance of coordinate spread, although it should be noted that it is an average of squares, not the distances themselves, and so is biased more towards the larger deviations having more influence. (Note that the `alignCoords()` function can easily be adapted to calculate RMSD for the overall transformation, but we have deliberately avoided complicating the function further.)

The function that calculates the RMSDs takes an array of reference coordinates, a list of the other coordinate array to compare with and a list of weights, so that each atom position can be biased separately. Inside the function we initialise a list to hold the RMSD values for each structure, find the total of all the input weights (using the handy, inbuilt `sum()` function) and initialise an empty array of zeros that is the same size as the coordinate arrays, which will hold the summation of positional differences.

```
def calcRmsds(refCoords, allCoords, weights):

  rmsds = []
  totalWeight = sum(weights)
  totalSquares = zeros(refCoords.shape)

  for coords in allCoords:
    delta = coords-refCoords
    squares = delta * delta
    totalSquares += squares
    sumSquares = weights*squares.sum(axis=1)
    rmsds.append( sqrt(sum(sumSquares)/totalWeight) )

  nStruct = len(allCoords)
  atomRmsds = sqrt(totalSquares.sum(axis=1)/nStruct)

  return rmsds, atomRmsds
```

The bulk of the function involves looping through the list of coordinate arrays and comparing them to the reference. The operations in this loop all involve whole array objects, and so when we add, subtract, multiply and divide the operations are applied to all elements on the arrays at the same time. This is the advantage of using the NumPy arrays: it simplifies the code and avoids having to write more loops. Accordingly `delta` is the array of all coordinate differences, and the elements of this whole array are squared to give `squares`. The square coordinate differences are added to the array of totals for use later. The squared deviation for each atom is calculated as the sum of the square values along the spatial axis (i.e. $x^2 + y^2 + z^2$). Here this is done using the `squares.sum(axis=1)` operation, thus we get the total for each atom separately and form another array. This is then multiplied by the weights for the atoms to give `sumSquares`, which represents the contribution of each atom to the coordinate 'deviation'. Lastly, the `sumSquares` is summed over all atoms to give a single value, which is divided by the total weight to find the average atomic square deviation for each structure. The square root of this (hence root-mean-square deviation) is placed in the RMSD list.

Once the loop is complete the RMSD values for the individual atoms are calculated. Given that the square differences for the atoms were added to `totalSquares` for all of the coordinate arrays (all structures) the average of these is then used to calculate each atom's RMSD over the whole set of conformations. As before, this is all done with NumPy operations, to work on whole arrays without loops.

Aligning a structure ensemble

Given that we now have a mathematical function that can find the optimum rotation to superimpose two arrays of coordinates, we now require a function that will superimpose more than two coordinate arrays, i.e. so we can superpose all the conformations of a whole *structure ensemble*. We will do this by repeating pairwise superimpositions relative to some reference. Repeating the superimposition of structure pairs is not actually the optimal way to align structures; it would be better to have a method that compared all against all at the same time. However, more complicated methods are beyond the scope of this book, and the result we get here will be very good in most circumstances. Indeed we try to be a little clever in the way that we perform pair-superimposition in the following examples. Initially, we arbitrarily align all coordinates with the first set, from which we calculate an average (albeit not real) structure. We then find the real structure which is closest to the average to use as a reference for a second round of superimposing alignment; we assume the reference is somewhere near the 'middle' of the spread of coordinates. Also, in the second round we change the weights, which may initially be set according to atom mass, to values that represent how variable each atom position was during the first round. Thus, the final coordinate superimposition will be biased towards the atom positions that were most similar in the first round and dissimilar regions (flexible parts of an ensemble, for example) will not have much influence. This frees us from having to specify which parts of the structures are most invariant and gives better overall results. It should be noted that the whole superimposition procedure may be repeated more than once to get the coordinates to converge more, although the convergence is usually very good with only one pass.

```
def superimposeCoords(allCoords, weights, threshold=5.0):

  nStruct = len(allCoords)
```

The reference coordinate array, that will remain stationary during the alignment, is arbitrarily set to the first from the list of all arrays. We then create an array of zeros of the same size to which we will subsequently add all of the coordinates (element by element inside the array) after they are aligned to the reference, so that we can find the average. The list of rotation matrices, which we pass back at the end of the function, starts empty and will be filled as we do the alignments.

```
  refCoords = allCoords[0]
  meanCoords = zeros(refCoords.shape)
  rotations = []
```

The first round of coordinate superimposition (alignment) is achieved by looping through all the coordinate arrays and aligning them to the reference set. Note that because the reference array is simply the first array in the list we don't need to align the first coordinates (with themselves). Accordingly, when the loop index (generated by `enumerate()`) is zero we can skip the alignment and set the rotation matrix as the 3×3 identity matrix, `identity(3)` (representing no rotation). Otherwise, the coordinate alignment is done with the appropriate function and the optimised rotation matrix and updated coordinates are obtained. These new coordinates are put back into the `allCoords` array at the current index, replacing the previous coordinates (the alignment makes a new coordinate array and doesn't affect the originals). Then the rotation matrix is collected into the `rotations` list. Finally in the loop, the coordinate array is added to the total held in `meanCoords`, which is divided by the number of structures, after the loop, to give the average positions.

```
for index, coords in enumerate(allCoords):
  if index == 0:
    rotation = identity(3)
  else:
    rotation, coords = alignCoords(refCoords, coords, weights)
    allCoords[index] = coords # Update to aligned

  rotations.append(rotation)
  meanCoords += coords

meanCoords /= nStruct
```

We calculate the RMSD values for the structures and atoms using the above defined function. Note that these values will be relative to the coordinate average (passed in as the first argument) and will respect the weights. From the resulting RMSDs it is a simple matter to find the best (smallest) value. The index of this value in the `rmsds` list then allows the selection of the appropriate coordinate array within the list of all coordinates. This best set of coordinates will then become the reference structure for a second round of superimposition, with adjusted weights. Effectively this reference array is the structure closest to the mean of the aligned coordinates.

```
rmsds, atomRmsds = calcRmsds(meanCoords, allCoords, weights)

bestRmsd = min(rmsds)
bestIndex = rmsds.index(bestRmsd) # Closest to mean
bestCoords = allCoords[bestIndex]
```

Adjusted weights are defined for a second round of coordinate alignment. The weights that are used are derived from the RMSD values for the corresponding atoms. We scale the RMDS value by a threshold value, which effectively defines the sensitivity to structural variation. Then the new weights are calculated as the negative exponent of the scale values squared. In this way atoms with small RMSD values will have the largest weights and as the variance increases the weighting will diminish exponentially. Note that the below is done on NumPy `array` objects (and the `exp()` function is the `numpy` version) so that the operations are performed on all of the elements in the array at once.

```
weightScale = atomRmsds/threshold
weights *= exp(-weightScale*weightScale)
```

We define an array, which will contain the average coordinates, by making a copy of the coordinate array that was used as the superimposition reference: the set that was closest to the mean. A loop is then used to go through all of the coordinate arrays, noting that if the index matches our reference coordinates (`indexBest`) then we skip that loop; we don't have to add the coordinates to the average as they are already in the required array, and because it is the reference there is no need to do the superimposition. For the other coordinate arrays, the superimposing alignment is performed, optimising the rotation to the reference. Each new rotation matrix is replaced in the list of rotations and naturally the rotation matrix for the reference is left as it was before. Updated coordinates are inserted back into `allCoords` at the appropriate index and then added, element-wise, to the array used to calculate the coordinate average.

```
meanCoords = bestCoords.copy()

for index, coords in enumerate(allCoords):
  if index != bestIndex:
    rotation, coords = alignCoords(bestCoords, coords, weights)
    rotations[index] = rotation
    allCoords[index] = coords # Update to aligned
    meanCoords += coords

meanCoords /= nStruct
```

The summation of the coordinate arrays is then divided by the number of structures to get the revised average (mean) coordinate set. Finally, a separate function is used to calculate the RMSD values of the structures and individual atoms. Note that these values compare the new coordinates with the average set of coordinates[16] and also that we use redefined, unbiased weights (set to 1.0 courtesy of NumPy's handy `ones()` function) so that we report the observed distance variation for each atom equally. Then the final coordinate array, the RMSDs and the list of rotation matrices are passed back at the `return` statement.

```
weights = ones(len(weights))
rmsds, atomRmsds = calcRmsds(meanCoords, allCoords, weights)

return allCoords, rmsds, atomRmsds, rotations
```

Given that we now have a function that can perform an optimised superimposition of coordinate arrays, another function is required that will apply the procedure to `Structure` objects. Accordingly, we define a function that takes a list of such objects, from which the coordinate arrays can be extracted, perform the coordinate superimposition, and then update the original `Atom` objects with the new coordinates. This function requires a way of getting the weightings of different types of atoms, so we use the previously defined `ATOMIC_NUMS` to give that information, although it is likely that in practical situations such a dictionary will be defined in another, more general module.

```
def superimposeStructures(structures):
```

In the function the `weights` variable is initially set to `None`, although it will be set to a list of numbers once we come across the first structure object. Then we define an empty list to contain all of the coordinates for all of the structures; this will be a list of two-dimensional arrays.

```
weights = None
allCoords = []
```

The structures are inspected in turn inside a loop, and for each we use the existing `getAtom-Coords()` function to get the coordinate array and the corresponding list of `Atom` objects. In essence, we convert from our data model to NumPy arrays. Also in the loop we define the list of weights, if it is not already defined, by using the atomic numbers of the atoms, which we look up in a dictionary. This of course assumes that the weights for one structure will do for all structures, a reasonable assumption because we are only superimposing the same kinds of

---

[16] It would also be possible to find a new medioid (closest to mean) structure and use that for the RMSD calculation.

atoms. Next the coordinates are redefined by moving them to the centre (zero on all axes), using the `centerCoords()` function and respecting the atom weights, which are now in an `array` object. Finally at the end of the loop the centred coordinate array is put in the list containing all coordinates.

```
for structure in structures:
  atoms, coords = getAtomCoords(structure)

  if weights is None:
    weights = [ATOMIC_NUMS[atom.element] for atom in atoms]
    weights = array(weights)

  coords, center = centerCoords(coords, weights)
  allCoords.append(coords)
```

We then run the function that actually performs the coordinate superimposition on the arrays of coordinates. The results we get back are: the array of modified coordinates, a list of RMSD values for each structure, a list of RMSD values for each atom (across all structures) and a list of the rotations that were applied to each of the structures to do the superimposition. Although the rotation data is not useful here, it is used in other contexts, as illustrated in later examples.

```
results = superimposeCoords(allCoords, weights)
allCoords, rmsds, atomRmsds, rotations = results
```

Lastly in the function we go through each of the structures, using `enumerate()` to get a list index as well as a `Structure` object. The `getAtomCoords()` function is used to get a list of atoms and an `array` of the original coordinates (which we are not actually interested in). Then a second loop is performed, going through each atom index (`j`) and `atom` object. The coordinates of the atom are then updated from the array that contains the new, superimposed locations. Note how we use the two indices to get the correct group of coordinates from the `allCoords` array, and that the group is immediately unpacked into three variables. With the `Atom` objects updated with new positions, the loops and then the function end, passing back the RMSD values that specify how good the coordinate superimposition was for each structure, and the individual atoms.

```
for i, structure in enumerate(structures):
  atoms, oldCoords = getAtomCoords(structure)

  for j, atom in enumerate(atoms):
    atom.coords = allCoords[i][j]

return rmsds, atomRmsds
```

We can test the function with two or more structures, as long as they represent the same atoms. Here we load the same structure from file twice, into two separate objects, then rotate one of them before trying the coordinate superimposition. This is a reasonable test because the structures are the same, aside from the transformation, and thus the superimposition should be near perfect, with RMSD values of zero.

```
from Modelling import getStructuresFromFile

strucA = getStructuresFromFile('examples/1A12.pdb')[0]
strucB = getStructuresFromFile('examples/1A12.pdb')[0]
```

```
rotateStructureNumPy(strucA, (1,0,0), pi/3)

rmsds, atomRmsds  = superimposeStructures([strucA, strucB])

print(rmsds)  # Hopefully all close to zero
```

If we get a whole ensemble of structural models, for example by downloading data from an NMR-derived PDB entry, then we can test the alignment on the corresponding list of structures.

```
fileName = downloadPDB('1UST')
strucObjs = getStructuresFromFile(fileName)

coords, rmsds, atomRmsds, rotation = superimposeStructures(strucObjs)

print(rmsds)
```

Homologous structure alignment

In this section we will link the structural examples described in this chapter with the sequence alignment routine described in Chapter 12. The objective here is to superimpose two homologous structures (i.e. they have a common ancestor) which have similarly shaped structures, but which have different residue sequences, and hence different atoms. Because our structural superimposition function only optimises the transformation between two lists of coordinates that are of the same size, we first need to determine a subset of atoms that are common to both structures. This is achieved by first doing a sequence alignment to gauge which residues are equivalent (those that pair up in the alignment), and then selecting backbone atoms that are present on both residues of each pair. Despite there being potentially different side-chain atoms for the aligned residues the backbone atoms will be common and we can use their coordinates alone to do a structural alignment.

The first thing we need before doing the structural alignment is to convert the sequences, of two molecular chains we wish to align, from the structure data model representation of `Residue` objects into one-letter residue codes. Because the data model uses three-letter codes, we define a dictionary that gives the equivalent one-letter codes, the required input to our previous sequence alignment function. Note that we have residue codes for DNA and RNA as well as protein residues, although the nucleic acid codes are not truly three-letter.[17] We need such unaltered codes in our dictionary so that we can detect an unknown code. An alternative would be to have a new sequence alignment function that works with lists of three-letter codes, but that is more work.

```
THREE_LETTER_TO_ONE = {'ALA':'A','CYS':'C','ASP':'D','GLU':'E',
                       'PHE':'F','GLY':'G','HIS':'H','ILE':'I',
                       'LYS':'K','LEU':'L','MET':'M','ASN':'N',
                       'PRO':'P','GLN':'Q','ARG':'R','SER':'S',
                       'THR':'T','VAL':'V','TRP':'W','TYR':'Y',
                       'G':'G','C':'C','A':'A','T':'T','U':'U'}
```

---

[17] In a PDB file the codes in the strictest sense will have three characters, i.e. 'A ' or 'G ', but our reader function removes the excess spaces.

Extracting the sequence of one-letter residue codes from a chain is a fairly simple matter of looping through each of the residues that belong to the input chain. For each residue the above dictionary is used to convert the (mostly three-letter) PDB residue codes, which are used in the structure data model, into a string of one-letter codes that can be used in the alignment routines. The one-letter codes are added to a list, which is then joined into a long string after the loop is done. Note that the dictionary look-up uses a `.get()` function call, which supplies an 'x' character for an unknown residue type that has no key in the dictionary.

```
def getChainSequence(chain):

  letters = []

  for residue in chain.residues:
    code =  residue.code
    letter = THREE_LETTER_TO_ONE.get(code, 'X')
    letters.append(letter)

  seq = ''.join(letters)

  return seq
```

Now we define the function that does the sequence alignment to get the pairs of corresponding residues. These pairs are then used to define a subset of common backbone atoms which are used to perform the coordinate superimposition. Although this function is fairly long and may look complicated at first glance, it is mostly connecting together existing functionality to do the job. Firstly, we make sure that we have imported the required substitution matrix (the default for the function) and the function that will perform the pairwise sequence alignment.

```
from Alignments import sequenceAlign, BLOSUM62
```

The function is defined and takes two `Chain` objects as mandatory input arguments and has optional arguments for the names of the atoms to align and for the substitution matrix to use in the alignment. The atom names default to the heavy polypeptide backbone atoms, and so will be present in all amino acids (including proline). The substitution matrix defaults to a fairly general one, although a better one for structural purposes may of course be passed in. Both defaulting arguments naturally assume that our molecular chains are proteins. If we were using this function on nucleic acid structures we could need to input a different set of atom names, for the ribose sugar phosphate backbone, and a different substitution matrix.

```
def seqStructureBackboneAlign(chainA, chainB,
                    atomNames=set(['CA','C','N']),
                    simMatrix=BLOSUM62):
```

Inside the function we initially find the two `Structure` objects that contain the input chains (following the parent link). These objects will be used later on in selecting structural subsets, and when applying the transformations for the final coordinate superimposition. Then we use the chains to get a complete list of all `Residue` objects that we are going to align.

```
  structureA = chainA.structure
  structureB = chainB.structure
  residuesA = chainA.residues
  residuesB = chainB.residues
```

The `getChainSequence()` function defined earlier is used to obtain the one-letter sequence strings from the two `Chain` objects, so that we can pass input to the function that performs the pairwise sequence alignment (see Chapter 12). The alignment strings that are passed back from the function will contain gaps ('-' symbol) to indicate how one sequence is aligned with the other. Gaps will indicate that a residue has no equivalent in the other sequence, but otherwise we will have pairs of residue letters that indicate which positions are deemed to be equivalent when comparing the two molecules. Although in this instance we are ignoring the alignment score that is generated, we could be more prudent and use it to check whether the sequence alignment is sufficiently good to proceed further. Note that in order to align the two sequences we also pass in the similarity matrix as an argument.

```
seqA = getChainSequence(chainA)
seqB = getChainSequence(chainB)

score, alignA, alignB = sequenceAlign(seqA, seqB, simMatrix)
```

Next a few variables are initialised: two lists to contain the locations of the aligned residues (relative to their position in the chain, not the alignment) and positional counters that are used to track these locations as we look through the sequence alignment.

```
pairedPosA = []
pairedPosB = []
posA = 0
posB = 0
```

Then the output strings from the alignment (`alignA` and `alignB`) are interrogated to find the residue pairs. This is achieved by defining a loop that goes through the complete range of index positions in the alignment string. For each index `i`, we look at the residue codes at that alignment position and work out the position of each residue relative to the start of its chain (`posA`, `posB`). In essence the sequence position of each residue in the molecule is out of step with the position in the alignment because of the gaps that are inserted as padding, to make the alignment possible. If at a given position there is a gap symbol in one of the alignment strings then we know that this represents a point where a residue on one sequence has no equivalent in the other sequence. Accordingly, we increase a counter (`+= 1`) for the sequence that does not have the gap, because for this sequence we still go one position along the chain. Hence, if `alignA[i]` is a gap, we increment `posB`, and vice versa. Otherwise, if there are no gaps, both residue chain positions are added to the lists that record the aligned residue pair and both positions increment for the next loop. Note that we add the old residue positions before the increment, so that we start the positions from zero; handy for Python lists.

```
for i in range(len(alignA)):
  # No dashes in both at same location

  if alignA[i] == '-':
    posB += 1

  elif alignB[i] == '-':
    posA += 1

  else:
    pairedPosA.append(posA)
    pairedPosB.append(posB)
```

```
posA += 1
posB += 1
```

Then we simply use the list of paired alignment positions, which are the indices to the residues in their respective lists, to get hold of the required `Residue` object and place its number (`seqId` attribute) in a list. In essence, the positions in the alignment are not the same thing as the residue number in the structures, so we need a conversion between the two.

```
filterIdsA = [residuesA[p].seqId for p in pairedPosA]
filterIdsB = [residuesB[p].seqId for p in pairedPosB]
```

The `filterSubStructure()` function defined above, which takes a structure and copies specified parts, is used to define new `Structure` objects representing only backbone atoms. Note that we make a filter to select the required coordinates using the chain's code, a list of residue numbers obtained via the sequence alignment, and the names of the backbone atoms which the structures will have in common. The resulting sub-structures will have exactly the same number of atoms, with a one-to-one correspondence that can be used for structure superimposition.

```
backboneStrucA = filterSubStructure(structureA, [chainA.code],
                                    filterIdsA, atomNames)
backboneStrucB = filterSubStructure(structureB, [chainB.code],
                                    filterIdsB, atomNames)
```

For each of the backbone-only sub-structures we collect a list of atoms and the array of coordinates (we move from the data model to the NumPy array). These atom lists are not used in the end, but are generated by the function in any case. The weights that are used to perform the superimposition are all set to `1.0` for each atom, so all coordinates carry equal worth in the initial instance.

```
atomsA, coordsA = getAtomCoords(backboneStrucA)
atomsB, coordsB = getAtomCoords(backboneStrucB)
weights = ones(len(atomsA))
```

The two sets of coordinates are moved to the centre (zero on all axes) using the previously defined function. It is important to do this so that the remainder of the structural superimposition, described by a rotation, can be estimated. Note that the translations that were applied to move the structures are recorded, i.e. in the `centerA` and `centerB` vectors, so that we can use them again when we align the full structures.

```
coordsA, centerA = centerCoords(coordsA, weights)
coordsB, centerB = centerCoords(coordsB, weights)
```

With the coordinates extracted, for those atoms we wish to superimpose, we use the superimposition function we described above, noting that the input argument is a list of coordinates. In this instance we are not actually interested in the relocated coordinates, but rather in the rotational transformation that was applied. Also, we collect the RMSD values to report at the end.

```
coords = [coordsA, coordsB]
c, rmsds, atomRmsds, rotations = superimposeCoords(coords, weights)
```

The rotations that were recorded, from the coordinate superimposition, and the locations of the coordinate centres are used to transform the original structures. Thus, although this transformation data is calculated only using the common, sequence-aligned, subset of back-bone atoms, the transformations are applied to all of the atoms from both structures. Finally, at the end of the function the RMSD values for the chains and individual atoms are returned, to see how well the structures are superimposed.

```
affineTransformStructure(structureA, rotations[0], -centerA)
affineTransformStructure(structureB, rotations[1], -centerB)

return rmsds, atomRmsds
```

We can test the function with two known homologous proteins. Here we download the PDB file data and extract the file data to define `Structure` objects. Note that in this example we only use the first available coordinate set (hence the `[0]` to select the first). For the first structure this means the first conformation of an NMR-derived ensemble of structures. The second structure only has one set of coordinates in any case, the usual situation for crystal structures. From both structures we select the chains with code `'A'` and perform the sequence-coordinate alignment.

```
struc1 = getStructuresFromFile(downloadPDB('1UST'))[0]
struc2 = getStructuresFromFile(downloadPDB('1HST'))[0]

chain1 = struc1.getChain('A')
chain2 = struc2.getChain('A')

rmsds, atomRmsds = seqStructureBackboneAlign(chain1, chain2)

print(rmsds)
```

External macromolecular structure modules

We will now look briefly at alternative ways of using macromolecular data in Python. The first of these is the PDB module of BioPython, which provides a more capable and tested alternative to our very simple structure data model. The second is the graphical program PyMol, which can be imported as a Python module so that coordinate data can be rendered as pretty pictures. (See http://www.cambridge.org/pythonforbiology for download links to BioPython and PyMol.)

Structures in BioPython

The PDB sub-module of BioPython contains functionality to read, write, manipulate and investigate macromolecular structure data. In the following example, after loading the PDB module, we make a parser object that can make a `PDB.Structure` object using the data from a file. Although this method may seem a little clunky compared to our earlier examples of a single function that can be used to import the data, having an intermediate object does allow the programmer to have more line-by-line control of the file reading.

```
from Bio import PDB

fileName = 'examples/1UST.pdb'
parser = PDB.PDBParser()
struc  = parser.get_structure('Name', fileName)
```

Once furnished with the main structure object, we can then extract the first set of coordinates (first conformation) from the structure and loop through all of the chains, residues and atoms to get at the coordinates in a convenient manner, noting that the names of object attributes differ from the previous data model:

```
conformation = struc[0]

for chain in conformation:
  for residue in chain:
    atomNames = [a.name for a in residue]
    print(chain.id, residue.id[1], residue.resname, atomNames)
    caAtom = residue['CA']
    print(caAtom.name, caAtom.coord, caAtom.bfactor)
```

Writing the data to disk requires calling a function to define a special object (called `writer` here) which can be used to save the file:

```
outFileName = 'test.pdb'
writer = PDB.PDBIO()
writer.set_structure(struc)
writer.save(outFileName)
```

Referring back to one of the early examples we gave in this chapter, to find the atoms within a given distance of a specified point, we can write an equivalent function that uses the BioPython `PDB.Structure.Structure` object, rather than the simple `Structure` object we defined in Chapter 8. The mathematics of the function are the same as in `findCloseAtoms()`, but the collection of coordinates and atom objects is naturally different because of the different data model. Note that in this example we have been a little more rigorous and check, using the inbuilt `isinstance()` function, that the input object is of the required type: `PDB.Structure.Structure`. Looping through the chains, residues and atoms is fairly straightforward, although the data model is slightly different. For example, here we use a conformation (coordinate model) number to specify which coordinate set to use within the structure object; our bespoke model is simpler and loads alternative conformations as entirely separate objects.

```
def findCloseAtomsBioPy(structure, xyz, conf=0, limit=5.0):

  if not isinstance(structure, PDB.Structure.Structure):
    raise Exception('Structure must be Bio.PDB.Structure class')

  closeAtoms = []
  xyz = array(xyz)
  limit2 = limit * limit

  coords = []
  atoms = []
  confModel = structure[conf]

  for chain in confModel:
    for residue in chain:
      for atom in residue:
        coords.append(atom.coord)
        atoms.append(atom)
```

```
deltas = coords - xyz
squares = deltas * deltas
sumSquares = squares.sum(axis=1)

boolArray = sumSquares < limit2
indices = boolArray.nonzero()[0]

closeAtoms = [atoms[i] for i in indices]

return closeAtoms
```

Structures in PyMol

To view structures graphically with the PyMol[18] program, assuming that it is installed, it is a simple matter to import its functionality into a Python script.[19]

```
import pymol
```

With PyMol imported we call a function which causes the graphical environment to appear, so that we can see step-by-step what effects our instructions have.

```
pymol.finish_launching()
```

Using the name of a PDB file we have previously downloaded we load the structure data for a protein called glycophorin[20] into PyMol and then issue several commands to render the structure in a particular graphical form. It is notable that PyMol doesn't have a rich data model for this data. Thus, rather than working with objects that represent each of the structural entities, we make copious use of the `pymol.cmd` module to perform most of the operations using special instruction strings, which unfortunately we must learn. Here we load the structure and give it a name, which can be used to identify the whole data set in later commands.

```
fileName = downloadPDB('examples/1AFO')
strucName = 'Glycophorin'
pymol.cmd.load(fileName, strucName)
```

Two subsets of the structure are defined: one called `'bb'` to represent the protein backbone, where the 'magic' string to select the backbone heavy atoms is `'name n+c+o+ca'`; and another called `'tmd'` that corresponds to the transmembrane[21] part of the structure, from residue number 71 to 100.

```
pymol.cmd.select('bb','name n+c+o+ca')
pymol.cmd.select('tmd', 'resi 71-100 and name n+c+o+ca ')
```

---

[18] The PyMOL Molecular Graphics System, Version 1.5.0.4 Schrödinger, LLC.

[19] Note that during testing of this code on some Linux computer platforms, we discovered that in order for the PyMol program to work properly from a Python script we also had to define the `'PYMOL_PATH'` environment variable, which can be done from within Python by importing the os module and then setting `os.environ['PYMOL_PATH'] = os.path.dirname(pymol.__file__)`: the location where PyMol's Python modules are installed.

[20] This protein is found in the membranes of red blood cells.

[21] i.e. the hydrophobic helices that anchor this protein in a lipid membrane.

Next various commands are issued to change the way that various bits of the structure are displayed. In order, the operations do the following: colour the whole protein grey, colour the backbone selection red, display the backbone path as a 'cartoon' ribbon, colour the trans-membrane residues blue and finally hide the lines to the hydrogen atoms.

```
pymol.cmd.color('gray', strucName)
pymol.cmd.color('red', 'bb')
pymol.cmd.show('cartoon', 'bb')
pymol.cmd.color('blue', 'tmd')
pymol.cmd.hide('lines','hydro')

outFileName = strucName + '.pdb'
pymol.cmd.save(outFileName, strucName, 0, 'pdb')
```

And the graphical display may be output as a picture file (here in PNG format):

```
pymol.cmd.png(strucName+'.png')
```

Finally we can quit the display program, but otherwise leave any Python script running:

```
pymol.cmd.quit()
```

16 Array data

Contents

Multiplexed experiments

In many areas of biological and medical science, as new techniques and machinery are developed there is a tendency to record ever increasing amounts of data. A notable example of this is comes with 'next-generation' DNA sequencing, which we discuss further in Chapter 17. In general though, with high-throughput methods the idea is to perform many small experiments, of the same design, in parallel. When we simultaneously detect the outcome of many assays the procedure can be described as being multiplexed. This not only has speed advantages but can also reduce costs and improve consistency between experiments. And naturally, to handle large numbers of experimental assays it is important to use computers for the processing and analysis of data.

A multitude of modern techniques involve parallel experiments, including the detection of potential drug compounds, RNA molecules, antibodies and protein crystals, to name only a few. However, in this chapter we do not have space to cover the informatics of lots of specific techniques, so instead we cover general themes, such as data organisation, normalisation and comparison. Also, all of the examples will be based on the notion of the experimental data being arranged as a rectangular array, which in turn is often a consequence of the physical manner in which the assays were performed and detected, on some form of regular grid.

Although there has been a recent trend to use the R programming language for working with array-based assays, Python together with its NumPy and SciPy libraries is naturally more than capable. In a change from much of this book, where we describe code that is simply based on Python functions, here we will use an object-oriented framework. Hence we create a 'Microarray' class that will tie together experimental data and various functions

a) Red channel   b) Green channel   c) Both channels

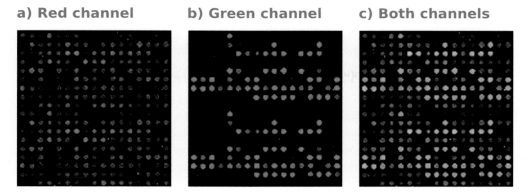

Figure 16.1 (Plate 2). A microarray image composed of red and green colour channels. Here each spot represents a different probe molecule and the colour intensity reflects the binding of DNA within two samples, each labelled with a different fluorescent marker. Original image courtesy of Paul Edwards and Karen Howarth, University of Cambridge. A black and white version of this figure will appear in some formats. For the colour version, please refer to the plate section.

that operate on it. The ideas behind this approach are discussed in Chapters 7, 8 and 15. Though unlike Chapter 15, where we have a relatively complicated hierarchical data model involving several classes, here we describe only a single class to organise things into a helpful construct.

Much of what we describe in this chapter relates to data management and many of the underlying programming principles are the same as things that have been discussed before. Hence the programming examples, which relate to array analyses, will focus on the organisation and analysis of the data rather than its acquisition. We will assume that the experiments have been done correctly and with due diligence, although some emphasis will be put on the handling of errors or noise in the data. These Python examples will start fairly simply, to illustrate the fundamental principles of what the data means and how we represent it in computational data structures. Examples will grow more complex by the incorporation of analytical methods, many of which are discussed in other chapters, e.g. clustering data.

Microarrays

A microarray is a means of performing many small-scale experiments on a sample at the same time. These experiments will all be of the same kind, i.e. have the same experimental design, but individual experiments will have different conditions or components. On the whole these experiments will be physically arranged as spots in a rectangular grid on a solid surface (the *matrix*) and have their test components immobilised on that surface, so they cannot mix. The basic reason for doing things in this manner is to make things quicker and easier. Lots of small experiments are performed at the same time, requiring a proportionately small amount of sample and providing the same set of conditions for each test (or at least very similar; there can be inhomogeneities across the array). Naturally, the outcome of the experiments has to be detected at the end and the final state of the microarray is generally measured using optical methods. Most microarrays are designed to detect the binding of components from a sample, to the different targets in the array, by using fluorescence. Here the binding causes an element of the array to glow when irradiated with UV light. In terms of computing, what is important is that we know what distinguishes the components of the different miniature experiments within the array, and then at the end how much signal, e.g. fluorescence, is detected from each.

The actual solid support for the array of experiments is typically made of glass, plastic or silicon and the experimental components are chemically bonded to its surface in a small regular array (placed there by machine). The components are generally bio-molecules, such as DNA, protein or even glycans (poly-sugars), but could also be samples of cells (i.e. a tissue microarray) or even small molecules. In the case of DNA microarrays the DNA strands of differing sequences are immobilised, with one sequence to each spot, on the solid matrix and bind to complementary single-stranded nucleotides, i.e. they hybridise through base-pair interactions. The samples that are applied to such an array will contain mixtures of fluorescent-labelled DNA strands, so that those with sequences that are complementary to the spots hybridise, to cause that part of the array to emit a certain colour of light visible when illuminated with UV light. Naturally, to know which DNA sequences have been detected in this way requires that the sequence of each spot in the array is known. For protein microarrays the situation is similar but the array spots are immobilised proteins, commonly antibodies, which detect other molecules in a specific manner via non-covalent interactions.

Whatever the type of array component (be it DNA, protein or whatever) and however it is detected we will use the same basic Python data structure to represent all kinds of microarray; they all have an array of spot elements in a rectangular grid and they all are detected by means of some kind of scalar signal. Although this abstract description can be applied in several situations, it could naturally be customised or extended for more specialised purposes. It should be noted that we have chosen to associate the array with parallel data layers, e.g. for red and green fluorescence *channels*, which is commonplace for microarrays. Accordingly an array element may be associated with several different signal values and the system is flexible, so that we can describe anything from a single array of values to multiple layers representing different kinds of both processed and unprocessed data.

What we gain from microarrays is a measure of interaction or reaction for each of the spots in the array. An array will tell us how strongly a given sample interacts with each spot component. One of the Python examples, to do hierarchical clustering, will analyse this further to show similarities within a microarray. This is a common process to visually indicate similarities between rows and columns. Here we can look back to some of the phylogenetic tree-building code and borrow a function, illustrating the usefulness of keeping Python functions abstract and general. We also show how you can look for similarities and differences in array data, for example by comparing different colour channels. In such circumstances it can be important to use controls and normalise the data, to test whether the detection worked equally well in each case and to remove systematic error. With this in mind several of the following examples are based on normalisation techniques which will allow the comparison of different data arrays even if the overall levels of signal differ, although which particular technique is used will vary for each situation.

Handling array data

The Python examples for microarrays that we describe involve creating our own kind of object. We are not forced to take this approach, and could work with isolated numeric arrays or use Python data structures like lists or dictionaries. However, we want to show how object orientation can organise data in a helpful way, improving consistency and convenience. In this case we will define one class of custom object called a `Microarray`, which as the name suggests will house all the data relating to one microarray experiment (which in turn represents lots of spots of miniature sub-experiments). This class of object will contain all the microarray data in terms of what each array spot element represents (in terms of conditions or components) as well as the detected signals for each spot element. We will store the arrays of signal information as two-dimensional NumPy arrays, which will make doing mathematical operations easier, but the rest of the data model will be regular Python for ease of bookkeeping.

The `Microarray` class will also contain its own functions (i.e. *methods* for that class) that will operate on the data contained within the data structure. This is a convenient way to write a program because you do not have to import the functions or pass the microarray object into the operation, rather the class gains methods that are innately bound to the data, and thus they are immediately accessible to it. The downside of doing this is that the programming is a little trickier, which is why for teaching purposes we initially avoid classes in the book, but it is generally good practice to use this style when you have a distinct concept that pulls all the data together, albeit a microarray or a molecular structure (as illustrated in Chapter 8). So although there may be more programming in the initial instance things will hopefully be easier in the long run.

Reading array data

We will define the `Microarray` class a bit later, but first we will look at what kind of raw, experimental data it will contain. This will be illustrated by writing functions that extract data for a `Microarray` object from the contents of files, specifically text files and image files. An alternative approach would be to make a blank `Microarray` object and then have that load data into itself, but here we aim to first show the kind of actual underlying data we are dealing with.

Importing text matrices

The load function will assume a simple file format where we have one line of text for each array element and the columns of data in the file are identifiers for the array coordinates (these could be the row and column of the array) and the actual data signal value. For example, this could be something like:

```
A          B          1.230
A          C          4.510
A          D          0.075
B          C          4.999
B          D          0.258
C          D          2.312
```

We will not assume that all elements of the array are represented and we will not assume that the array row and column identifiers (i.e. A, B, C and D in the above example) are either continuous or in any particular order.

The experimental microarray data will be represented as two-dimensional NumPy arrays. Accordingly, we make various imports from the `numpy` library for the mathematical array operations that will need to be done:

```
from numpy import array, dot, log, sqrt, uint8, zeros
```

The function to load data from a file and create a `Microarray` object takes a file name to load from as the first argument, an identifying name for the array as the second argument and an optional third argument to state what the default signal value is for an array element that we have no data for. This last argument may or may not be used depending on what data we read in, but we should at least be aware of incomplete or failed data points. The default here is zero, rather than `None`, given that we are dealing with NumPy arrays that don't mix data types.

```
def loadDataMatrix(fileName, sampleName, default=0.0):

  fileObj = open(fileName, 'r')
```

Empty sets are created which will contain the identifiers for the rows and columns in the microarray data. In the end these could be filled with just numbers representing the array coordinates, or they could be text labels. The only rule about these identifiers is that because they are in sets they cannot be internally modifiable items like lists (they must be *hashable*[1]). The values in these sets will be keys to access the numeric data stored in the `dataDict` dictionary.

```
  rows = set()
  cols = set()
  dataDict = {}
```

Next we loop though all the lines in the open file object and split each line according to internal whitespace to give values for the `row`, `column` and the numeric `value`. Naturally this operation would be different if the format of the file was different.

```
  for line in fileObj:
    row, col, value = line.split()
```

A check is made to ensure that each row identifier that comes from the file has an entry in `dataDict`. If it does not then we make a new inner dictionary within the main one using the `row` identifier as the key. Note that we cannot fill this dictionary, or any other data collection, in advance because we do not know what rows or columns we have until the file has been read.

```
    if row not in dataDict:
      dataDict[row] = {}
```

The actual signal `value`, which is a floating point number, is then added to the data by using the row and column identifiers as keys for the main and inner dictionaries, although we first check the dictionary to guard against using repeats for the same element. The `row` and `col` that were just used are then added to the set of `rows` and `cols`. Because they are `set` data types it does not matter if we have seen them before, given that a set automatically ignores repeated items. Note that the value is converted using `float()` because initially it is just a text string loaded from a file, and not a Python number object.

---

[1] See discussion in Chapter 3.

```
    if col in dataDict[row]:
      print('Repeat entry found for element %d, %d' % (row, col))
      continue

    dataDict[row][col] = float(value)

    rows.add(row)
    cols.add(col)
```

After all the lines have been processed we then convert the set of row and column identifiers to sorted, ordered lists. Only now that the total range of these identifiers has been collected can they be used to create the axes of the NumPy 2D value array. The sizes of this array will naturally be based on the number of row and column identifiers, which are recorded as nRows and nCols.

```
  rows = sorted(rows)
  cols = sorted(cols)

  nRows = len(rows)
  nCols = len(cols)
```

The NumPy array to store the values is initialised as an array of zeros of the required size, with the data type (the last argument) set to be floating point numbers:

```
  dataMatrix = zeros((nRows, nCols), float)
```

The rectangular dataMatrix array is then filled by extracting values from the dataDict, although there is provision to replace missing values with the default value (hence we use the .get() dictionary call). Note how we use enumerate() to extract index numbers (i and j) as we loop through row and column identifiers. These indices are then used to fill the correct position in the array. At the end the ordered list of rows and cols will be used so that the indices in the NumPy array can be used to look up the original data labels that they refer to.

```
  for i, row in enumerate(rows):
    for j, col in enumerate(cols):
      value = dataDict[row].get(col, default)
      dataMatrix[i,j] = value

  fileObj.close()
```

With the data collected we use it to construct a Microarray class of object, made with the definition described below. This is then passed back from the function.

```
  return Microarray(sampleName, dataMatrix, rows, cols)
```

Extracting array image data

The next example reads raw array data from a pixelated image, i.e. a picture of the whole array, which contains separate layers of data recorded, or at least stored, as separate colour channels. Each colour channel records a separate signal for the same spots, given that two samples were labelled with different fluorescent dyes that can be assayed independently.

The input file is read as a pixmap image that contains red, green and blue colour channels (RGB) using functionality that is discussed more fully in Chapter 18. Because we are dealing with image data we import from the Python Imaging Library (PIL),[2] as discussed in Chapter 18 (so you may like to skip ahead to learn more about images) to handle all the tricky tasks of making image pixmaps and saving the image data to a file. It should be noted that this is not part of the standard Python library and must be installed separately. Also, we import a function imageToPixmapRGB from the Images module (part of the downloadable data that accompanies this book) that will convert the image data into a NumPy array. And as you may expect we import some NumPy functions to manipulate numeric array data.

```
from PIL import Image
from Images import imageToPixmapRGB
from numpy import array, dstack, transpose, uint8, zeros
```

The array import function itself takes the name of the file to load and a human-readable name for the data. Also we specify the number of rows and columns (optionally if different from the rows) that the image represents. While it is certainly possible to do image processing to guess where the circular spots in the array image are located it is far easier to specify the grid size upfront and then simply subdivide the image into equally sized rectangles, corresponding to the rows and columns. Here we will simply take the signal for each spot as the total amount of signal within each grid cell, though this could be refined by fitting circles and removing noise etc.

```
def loadArrayImage(fileName, sampleName, nRows, nCols=None):
```

If the number of data columns was not specified when the function is called we set it to be equal to the number of rows. The numeric matrix that will contain the signal information dataMatrix is then constructed initially as an array for zeros of the required size, noting that the first axis has three layers before we specify rows and columns (3, nRows, nCols), which will be used to store the separate colour components. It is a matter of taste whether the different layers use the first or last axis of the array, but here we put it first because it makes the code slightly simpler overall, even though this is the opposite of how the data is stored in the image.

```
  if not nCols:
    nCols = nRows

  dataMatrix = zeros((3, nRows, nCols), float)
```

Using the imported modules, an object representing the image is generated from the input file with the Image.open() method, and this is them converted to a numeric array with the function from Chapter 18.

```
  img = Image.open(fileName) # Automatic file type
  pixmap = imageToPixmapRGB(img)
```

The size of the pixel data along each of its axes is easily determined from the numeric array. By dividing the total image width and height by the number of columns and rows respectively we get a measure of the grid size, which we will use to subdivide the image data. We calculate both floating point grid sizes (dx, dy) and integer sizes (xSize, ySize) because we need precise values to define the grid start points but a fixed number of pixels to find the end points,

---

[2] See http://www.cambridge.org/pythonforbiology for information about how to download and install PIL.

and thus give blocks of equal area. Note the integer size calculation involves adding one pixel because we will be taking a slice out of the image array up to, but not including, the end value, but that this also means we subtract one prior to division to avoid overshooting the edge of the image.

```
height, width, depth = pixmap.shape

dx = width/float(nCols)  # float() not needed in Python 3
dy = height/float(nRows)
xSize = 1 + (width-1)//nCols
ySize = 1 + (height-1)//nRows
```

Looping through each microarray row the first pixel position for that image section (yStart) is calculated by multiplying the row number by the row depth in the image (dy) and converting to an integer. The last pixel position will be just inside the limit (yEnd), which is calculated as the start plus the integer grid width (ySize).

```
for row in range(nRows):
  yStart = int(row*dy)
  yEnd   = yStart + ySize
```

Similarly, within each row we calculate the range of pixels to select a column of data from the image.

```
for col in range(nCols):
  xStart = int(col*dx)
  xEnd   = xStart + xSize
```

The data corresponding to an individual microarray grid element (i.e. spot) is a rectangular region of pixels sliced from the image pixmap, using the row and column bounds just calculated. The data from this sub-region is summed along both the width and height axes of the array (but not colour axis), hence we use .sum(axis=(0,1)) to give the total signal for the grid element. This is then stored in dataMatrix at the required row and column, noting that the ':' specification for the first axis of the array means that we are setting all the colour channels at the same time.

```
elementData = pixmap[yStart:yEnd,xStart:xEnd]
dataMatrix[:,row, col] = elementData.sum(axis=(0,1))
```

Note that if width is not a multiple of nCols then the last column has fewer pixels in the sum, and similarly for the last row, if height is not a multiple of nRows. Finally at the end of the function we create a Microarray object, as described below, with its name and data array.

```
return Microarray(sampleName, dataMatrix)
```

The 'Microarray' class

Now we come to define the Microarray class of object. This is specified according to the principles discussed in Chapter 8. The __init__ constructor, which is called each time an object of this type is made, requires a sample name and a NumPy array of data. We can also pass in values for the row and column data labels. These are not mandatory, and if not set they will be filled with sequential integer numbers for each row and column. Note that the def keyword is indented relative to the `class` keyword, to specify that the function definition is inside, and thus part of, the `class` specification.

```
class Microarray(object):

  def __init__(self, name, data, rowData=None, colData=None):
```

Firstly, inside the constructor function we store the name for the microarray data by assigning it to `self.name`. Recall that `self.` arguments are used so that variables are tied to the `Microarray` object, so that when a particular instance of a microarray object is made the `self` will represent that actual object, rather than the abstract specification stated within the `class` construction. Then we make a copy of the input data (using `array()` will make a copy as a NumPy array), which converts any input lists or tuples and means the original input won't be changed if it is used elsewhere.

```
  self.name = name
  data = array(data)
```

Next the sizes of axes in the array data are extracted with the `.shape` attribute. If there are three axes in the data we assume these represent the respective data channels (e.g. colours), rows and columns. Otherwise if there are two data axes we assume there is only one channel and we re-cast the `data` array as a single element within a larger array, so that it is forced to have three axes (i.e. shape is `(1, nCols, nRows)`) even though there is only one layer of data.

```
  shape = data.shape

  if len(shape) == 3:
    self.nChannels, self.nRows, self.nCols = shape

  elif len(shape) == 2:
    self.nRows, self.nCols = shape
    self.nChannels = 1
    data = array([data]) # or data.reshape((1, self.nRows, self.nCols))
```

If the number of data axes doesn't fit what we require an error exception is triggered so the program will stop (if the exception is not handled).

```
  else:
    raise Exception('Array data must have either 2 or 3 axes.')
```

With the `data` now potentially adjusted, to ensure that it has three axes, we associate it with a `self.` variable so that it is tied to the object. Also we take a copy of the data which will be left in its original form, so that at any point in the future we can revert to the original state if we wish,

```
  self.data = data
  self.origData = array(data)
```

Lastly in the constructor function the row and column labels are associated with the object, noting that the `or` keyword is used so that if either list of labels is `None` (empty or otherwise logically false) then they are defined as a sequential range of integer numbers, one for each row or column.

```
  self.rowData = rowData or range(self.nRows)
  self.colData = colData or range(self.nCols)
```

This example is somewhat lazy, given that we have not made any checks to ensure that the data is of the correct type or that the rowData or colData is the correct size (if set). Naturally these checks should be made in real-world applications.

As a very simple example in the toolkit of functions for Microarray we create a method so we can revert the self.data array back to the original values, if we want. The function takes the self as its argument, which at run time will be filled with a particular occurrence of the object, so that it can then access all of the attributes (and other functions) linked to that object; in other words we can use self. inside this function. The function simply works by assigning self.data to a copy of the original data and resetting self.nChannels, in case that had changed.

```
def resetData(self):

  self.data = array(self.origData)
  self.nChannels = len(self.data)
```

Exporting array data

With the constructor function complete we know that the Microarray objects can be made with the required set of attributes, for the data, rows and columns etc. The next task is to create other functions within the class definition that provide objects of that class with any special functionality that we need. After having discussed constructing and loading data into the Microarray objects we next turn to getting data out. As with the import functions we will consider both text files and images, the latter of which will be handy to indicate the changes that occur when we process and analyse the data.

We define the writeData function inside the above class definition, hence all of the code for the function is indented. Internally the function works by opening a file object (fileObj) to write out and loops through the rows and columns of the array, converting the row and column identifiers to strings with str(), just in case they are stored as numbers.

```
def writeData(self, fileName, separator=' '):

  fileObj = open(fileName, 'w')

  for i in range(self.nRows):
    rowName = str(self.rowData[i])

    for j in range(self.nCols):
      colName = str(self.colData[j])
```

For each row and column combination we use the indices (i,j) to get the data from the array for all array channels:

```
      values = self.data[:,i,j]
```

The line of text that will be written to the file is constructed using a list of data that has the name of the row and column at the start and then string representations of the numeric data in values. We convert the floating point numbers to strings with three decimal places with the format '%.3f', though we could increase the number of places if needed (see Appendix for detailed discussion of string formatting codes).

```
lineData = [rowName, colName]
lineData += ['%.3f' % (v,) for v in values]
```

The actual line to write is created by using the separator string (by default a space) and the
.join() method to combine the separate lineData strings into one. The line is finally written
to the file object with a trailing newline character, before the loops move on to the next item.

```
line = separator.join(lineData)
fileObj.write(line + '\n')
```

When we call this function we will do so from an instance of a Microarray object (here called
rgArray). Using example data that accompanies this book as a test[3] we can load the array data
from an image and export it as a text file:

```
imgFile = 'examples/RedGreenArray.png'
rgArray = loadArrayImage(imgFile, 'TwoChannel', 18, 17)
rgArray.writeData('RedGreenArrayData.txt')
```

For the next export example we will define an internal class function (a method) that creates a
picture representing the microarray data. This will be very useful to users and programmers to
get a visual representation of the experimental values in the array. The second argument after
self is a number that determines how large a square to use to represent each element of the
microarray, i.e. we are aiming to make a picture of the array using coloured squares. The
channels argument can be used to specify which layers of the array data will be used to create
the red, green and blue components of the image, bearing in mind that the Microarray could
have many layers of data. It will be specified as a list (or tuple) of integer indices to select the
layers and may contain None to specify that a colour channel should be blank (zeros).

```
def makeImage(self, squareSize=20, channels=None):
```

Because we will be making an image file that uses eight data bits to store each colour
component the numeric values are adjusted so they fit the integer range 0 to 255 (2^8-1).
Accordingly the extreme values present in the array data are found using the handy functions
built into NumPy arrays and the overall range is calculated.

```
minVal = self.data.min()
maxVal = self.data.max()
dataRange = maxVal - minVal
```

The adjusted array adjData contains pixel colour intensities and is a copy of the self.data
value array that has its lower limit subtracted (so that the pixmap has a minimum value of
zero, corresponding to black here) and which is then scaled so that the upper limit is set to
have the value 255 (the brightest colour). The array is then converted into an unsigned 8-bit
(uint8) version of itself; this way of storing numbers is the way that they are represented in
our image data.

```
adjData = (self.data - minVal) * 255 / dataRange
adjData = array(adjData, uint8)
```

Next, if the array channels (layers) to take for image construction were not passed in then we
decide on some sensible defaults. If there is only one channel in the data, the red, green and

---

[3] See http://www.cambridge.org/pythonforbiology.

blue components of the image (which will end up grey) will all come from the only data layer (index 0). Otherwise we will simply take the first layers of the array up to a maximum of three (we will fill missing RBG channels with zeros later).

```
if not channels:
  if self.nChannels == 1:
    channels = (0,0,0) # Greyscale
  else:
    channels = list(range(self.nChannels))[:3]
```

In the next step we will allow for blank colour channels. For example, if we want to specify that an image should use red only we could set channels as (0, None, None), so that the first array Microarray.data layer makes the red colour but there is no green or blue. Hence if a None is found in the channels we append an array of zeros of the required size to the pixmap list. Otherwise we add the required layer from the adjusted data. Using channels could also result in a different colour order to the original, e.g. by specifying channels as (2, 1, 0) the layers that usually represent red and blue would be swapped.

```
pixmap = []
for i in channels:
  if i is None:
    pixmap.append(zeros((self.nRows, self.nCols), uint8))
  else:
    pixmap.append(adjData[i])
```

We will also allow for the channels to be shorter than three, in which case we simply add missing channels as zero arrays to pixmap.

```
while len(pixmap) < 3:
  pixmap.append(zeros((self.nRows, self.nCols), uint8))
```

The three-dimensional image pixmap array is created by stacking the three colour layers along the depth axis (hence dstack()) and this is used with the PIL module to make an Image object called img. Given that we usually don't want the array elements only represented by single pixels, which would be too small to distinguish, the whole image is resized so we have squareSize pixels in each row and column, and hence much larger colour blocks. The final image object is then passed back at the end of the function.

```
pixmap = dstack(pixmap)
img = Image.fromarray(pixmap, 'RGB')

width = self.nCols * squareSize
height = self.nRows * squareSize
img = img.resize((width, height))

return img
```

To test the function we will again load the red and green example image as a Microarray.

```
imgFile = 'examples/RedGreenArray.png'
rgArray = loadArrayImage(imgFile, 'TwoChannel', 18, 17)
```

The image generation function can be used to make a picture with 20×20 pixel squares, so we can see whether the data is faithfully reproduced, albeit not in its original spotty form. Note

a) Red-green output   **b) Yellow-blue output**   **c) Normalised and clipped**

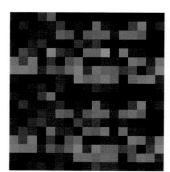 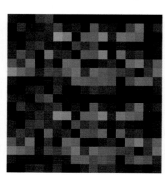 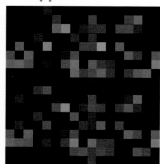

Figure 16.2 (Plate 3). Image pixmaps constructed from array data stored in the Microarray class.
Shown from left to right are: a red and green representation of the original, imported data; the original
data displayed as yellow and blue colours (the original red-channel data becomes red and green in the
image and the original green data becomes blue); clipped and normalised data where values less than half
the maximum are set to zero. A black and white version of this figure will appear in some formats. For
the colour version, please refer to the plate section.

that we display the `Image` object generated immediately using its inbuilt `.show()` method (see
Figure 16.2a).

```
rgArray.makeImage(20).show()
```

To make an image only containing red by taking data from layer `0` in the array we can specify
`channels` as (`0,None,None`).

```
rgArray.makeImage(20, channels=(0,None,None)).show()
```

And if we want to change the order of the layers so that the originally red and green data
(indices `0` and `1`) are mapped to yellow and blue respectively we do the following, illustrating
the result in Figure 16.2b:

```
rgArray.makeImage(20, channels=(0, 0, 1)).show()
```

Value normalisation

Now the examples move onto methods that adjust and normalise our data. The aim here is
generally to preserve the features and correlations within the data, while adjusting the data
values so that they have a standard range or fit better to some kind of distribution. The reasons
for doing this may simply be mathematical, e.g. to calculate scores or probabilities, but
normalisation has a very important role in making different experiments (both experiment
spots within an array and whole arrays) comparable with one another.

The first function of this kind is used to clip the lowest, base value of the data so that
it does not drop below a specified `threshold`; elements that have smaller values will be set to
this limit. Note that if the absolute threshold is not specified it is taken to be some proportion
of the maximum value, which in this case arbitrarily defaults to `0.2`. This function would be
handy to eliminate erroneous negative values, for example, or to disregard microarray
elements that are deemed to be insignificant because they are below some noise level. Also,

as with the makeImage() function we allow the channels to be specified to state which layers of the array data should be considered. If this is not specified it defaults to the indices for all layers range(self.nChannels). Note that channels is deliberately converted to a tuple and then placed in a list so that it can be used to index a subset from the self.data array (this is a consequence of the way NumPy array indices work).

```
def clipBaseline(self, threshold=None, channels=None, defaultProp=0.2):

  if not channels:
    channels = range(self.nChannels)

  channels = [tuple(channels)]
```

The maximum value is found from the required channels of the array data, and if the threshold (clipping) value is not specified it is calculated with the default proportion.

```
maxVal = self.data[channels].max()
if threshold is None:
  limit = maxVal * defaultProp
else:
  limit = threshold
```

By comparing the whole array self.data with the limit we generate an array of Boolean values (True/False) that state whether each element was less than the threshold value. The indices of the positions where this is True are provided by the nonzero() function of NumPy arrays. These array elements corresponding to these indices are then set to the lower limit.

```
boolArray = self.data[channels] < limit
indices = boolArray.nonzero()

self.data[indices] = limit
```

After clipping, the data is then centred by subtracting the baseline value, to give a new base value of zero, and finally scaled (effectively stretched) to restore the original maximum value.

```
self.data[channels] -= limit
self.data[channels] *= maxVal / (maxVal-limit)
```

We now consider various simple normalisation and adjustment methods. The normaliseSd function will scale the data values according to the standard deviation in the measurement, thus we divide all signal values by the standard deviation. We include the scale argument so that the data can also be arbitrarily scaled at the same time. This kind of adjustment is useful when comparing different instances of microarrays, where the actual range of signals (e.g. detected fluorescence values) probably ought to be the same on different occasions, but where there is variation in magnitude simply by the way the microarray is constructed. For example, one microarray might have a systematically larger amount of substrate printed on it compared to another. If there is confidence that different arrays are showing the same range of underlying data then this normalisation, according to standard deviation, is reasonable. The operation is done separately on all of the data layers (one for each channel), though we could extend the function to accept only a limited number of channels. The data is multiplied and divided appropriately in an element-by-element manner in the NumPy array. The standard deviation is obtained using the .std() method inbuilt into the NumPy array objects, as we describe in Chapter 22.

```
def normaliseSd(self, scale=1.0):

  for i in range(self.nChannels):
    self.data[i] = self.data[i] * scale / self.data[i].std()
```

Another similar kind of normalisation is simply to make sure that microarray values are scaled relative to their mean value. This is done for similar reasons as above, but rather than saying the variation in values is the same across different arrays, we assume that the mean values should be the same, or nearly so. This is indeed a reasonable assumption in many situations.

```
def normaliseMean(self, scale=1.0):

  for i in range(self.nChannels):
    self.data[i] = self.data[i] * scale / self.data[i].mean()
```

If we do one of the above normalisations then it is often handy to represent the data values as positive and negative numbers either side of the mean value, so we can see what is above or below the average, rather than as a merely positive intensity. This is readily achieved by subtracting the mean value from all the data:

```
def centerMean(self):

  for i in range(self.nChannels):
    self.data[i] -= self.data[i].mean()
```

Combining the centring of the data and scaling by the standard deviation we get a commonly used operation which is called Z-score normalisation,[4] as we discuss in Chapter 22. All this really means is that we move the data values so that they are centred at zero and scaled to put the standard deviation at the values ± 1.0.

```
def normaliseZscore(self):

  self.centerMean()
  self.normaliseSd()
```

Another kind of normalisation is to scale the values to some upper limit, e.g. so they are at most 1.0. This is done by dividing by the maximum value. This operation is useful if you know what the maximum value in the array is, or tends towards. For example, this could be a strong reference signal for an element acting as a positive control. Here we add the option to consider either each data layer separately (perChannel=True) or the maximum value from all the data (perChannel=False).

```
def normaliseMax(self, scale=1.0, perChannel=True):

  if perChannel:
    for i in range(self.nChannels):
      self.data[i] = self.data[i] * scale / self.data[i].max()

  else:
    self.data = self.data * scale / self.data.max()
```

Likewise we could normalise the rows separately, relative to the maximum values in each row. Here the data values are divided by an array representing the maximum value along only one

---

[4] Synonymous with standard score or Z-value normalisation.

axis, hence `axis=1`, to get the sum of the values (over the column positions) for each row. The slice notation `[:,None]` is a convenient way of changing what would otherwise be a one-dimensional array of maxima, in one long row vector, into a two-dimensional array with several rows (one number in each) and one long column. This means the division then scaling of each row of values by a different number.

```
def normaliseRowMax(self, scale=1.0):

    for i in range(self.nChannels):
        self.data[i] = self.data[i] * scale / self.data[i].max(axis=1)[:,None]
```

Row normalisation is useful where each row represents a different kind of data. An example would be where each column corresponds to a different nucleotide (or amino acid) sequence change in one molecule and each row represents a set of target molecules that are being bound. By normalising by row we will get the relative signal that illustrates which sequence changes give the best binding to each target. This may be useful in finding the change that leads to optimal binding for several targets, but naturally we lose information about the comparative strength of binding between targets.

Alternatively we can normalise the rows so they are scaled according to their mean value.

```
def normaliseRowMean(self, scale=1.0):

    for i in range(self.nChannels):
        self.data[i] = self.data[i] * scale / self.data[i].mean(axis=1)[:,None]
```

The same operations can be done for the columns in the data, it just depends on how the array is arranged. Note that here we use `axis=0` and so don't have to do convert the array of maxima into a column vector.

```
def normaliseColMax(self, scale=1.0):

    for i in range(self.nChannels):
        self.data[i] = self.data[i] * scale / self.data[i].max(axis=0)

def normaliseColMean(self, scale=1.0):

    for i in range(self.nChannels):
        self.data[i] = self.data[i] * scale / self.data[i].mean(axis=0)
```

If the microarray contains elements that represent control values, i.e. where you know what the expected results for these are, then you can scale the whole array relative to the known signal for these reference points. Here the reference values are specified by passing in the lists of the rows and columns they are found in, which are used to extract the corresponding values from the data. Taking a slice from `self.data` using separate rows and column specifications `[rows, cols]`, rather than finding specific indices, may seem odd, but is often convenient in NumPy (e.g. this is what `.nonzero()` gives). The `[rows, cols]` notation will affect all of the elements where all of the indices coincide, so can actually be more efficient than stating separate coordinates. The mean of the reference values is then used to divide the whole array. To add a bit of diversity we revert to allowing the `channels` to be specified to state which data layers to operate on.

```
def normaliseRefs(self, rows, cols, scale=1.0, channels=None):

    if not channels:
      channels = range(self.nChannels)

    channels = tuple(channels)

    refValues = self.data[channels, rows, cols]

    for i in channels:
      self.data[i] = self.data[i] * scale / refValues[i].mean()
```

A different way of normalising data values, especially for fluorescence intensity data, is to convert it into a logarithmic scale, effectively compressing its dynamic range. Note that we clip the baseline to remove any negative values and add 1.0 so we don't take the logarithm of any zero values. After conversion to a log scale we can then apply other normalisation techniques to compare different microarrays or rows etc. (Obviously we cannot do center-Mean() before the log conversion, because we want all values to be positive.)

```
def normaliseLogMean(self):

    self.clipBaseline(threshold=0.0)
    for i in range(self.nChannels):
      self.data[i] = log( 1.0 + self.data[i] / self.data[i].mean() )
```

We can test all of the above by creating a Microarray object named testArray that uses some example data from a text file. We illustrate the result of the various normalisation methods by writing out an image after each point.

```
testArray = loadDataMatrix('examples/microarrayData.txt', 'Test')
testArray.makeImage(25).save('testArray.png')

# Log normalise

testArray.normaliseLogMean()
testArray.makeImage(25).save('normaliseLogMean.png')

# Normalise to max and clip

testArray.resetData()
testArray.normaliseMax()
testArray.clipBaseline(0.5)
testArray.makeImage(25).save('clipBaseline.png')

# Normalise to standard deviation

testArray.resetData()
print("Initial SD:", testArray.data.std())
testArray.normaliseSd()
print("Final SD:", testArray.data.std())
```

Another handy way to do normalisation, albeit in a less scientific way, is to perform *quantile normalisation* and this is commonly used in DNA microarrays where consistency can be an issue. The process here is to make the distribution of data values in the array match some other, external distribution. This other distribution could be different microarray data or a mathematical distribution like a normal distribution (Gaussian). The matching of distributions is achieved by replacing each real microarray data value with the value from the reference distribution that has equal rank, so the highest value is replaced by the highest reference value, the second highest with the second highest reference and so on. While this may seem a little like cheating, quantile normalisation is especially useful if you suspect that the distribution of values in the microarray has been distorted or skewed, but at least the order of values conveys information.

The quantile normalisation procedure can be done by using NumPy as we illustrate below. The objective is to replace items in `values` by selecting items with the equivalent rank from `refData`. Note that we don't just sort replacement values because we want the ranks of these numbers in the original data order. First the data array is flattened into a one-dimensional vector and the indices of the values are extracted in size order (`.argsort()` does this). Hence, `order` represents the selection that sorts `values`. To take an example, if the flattened data is `[2.5, 7.1, 0.0, 5.9]` then the indices `order` is `[2, 0, 3, 1]` (2 is the position of the smallest value, 0 the position of the next smallest etc.).

```
def normaliseQuantile(self, refData, channel=0):
  # could be to a different channel

  values = self.data[channel].flatten()
  order = values.argsort()
```

Similarly the reference `refData` distribution is flattened into `refValues` (assumed to be an array of the same size as `self.data`) into a vector. Then `refValues` is sorted, putting its elements into size (and hence rank) order, so that we obtain an array of replacement values. The array of indices in original value order (`order`) is itself subject to `.argsort()`. This may seem confusing but what you get is an array of the **ranks** of each value, and thus a mapping from the original values to the replacement reference values. For example, if `values` is `[2.5, 7.1, 0.0, 5.9]` then the `refSelection` is `[1, 3, 0, 2]`, where each number is the size rank (starting at zero) of the equivalent data value. Once defined, `refSelection` allows us to redefine `values` by taking the reference values in the original rank order. Finally a new `self.data` is made by arranging `values` into the original shape.

```
  refValues = refData.flatten()
  refValues.sort()

  refSelection = order.argsort()
  values = refValues[refSelection]
  self.data[channel] = values.reshape((self.nRows, self.nCols))
```

And we can do a similar thing to quantile normalise each row separately. However, here we can use an internal reference distribution, which is the average for all the rows. We do not flatten the data arrays into a vector as each row is a vector and is dealt with separately. Accordingly we determine the order of elements of increasing value in each row (`orders`). The `refValues` is defined by sorting the values in each row and taking the average for each column (so each is the average of values with equivalent rank from each row). The `self.data` rows are then replaced with those of matching rank from the `refValues` averages.

```
def normaliseRowQuantile(self, channel=0):

  channelData = self.data[channel]
```

```
orders = channelData.argsort(axis=1)
sortedRows = array(channelData)
sortedRows.sort(axis=1)
refValues = sortedRows.mean(axis=0) # average over columns

rows = range(self.nRows)
self.data[channel,rows,:] = refValues[orders[rows,:].argsort()]
```

We can test the quantile normalisation using example data loaded from an image. For the reference we will use the data in layer 1 (green) to normalise layer 0 (red).

```
imgFile = 'examples/RedGreenArray.png'
rgArray = loadArrayImage(imgFile, 'TwoChannel', 18, 17)

rgArray.normaliseQuantile(rgArray.data[1], 0)
rgArray.makeImage(25).show()
```

Changing array channels

Moving on from making adjustments to the values within the array data we next consider operations that can add and remove whole signal layers. Naturally it is possible to create one `Microarray` object with different channels based on another, manipulating the signal data outside the class. Hence, for example, we could take the red and green (index 0 and 1) layers from one object and construct another, here making an array which visualises as yellow and blue:

```
red = rgArray.data[0]
green = rgArray.data[1]

yellowBlue = Microarray('yellowBlue', [red, red, green])
yellowBlue.makeImage(20).show()
```

However, in keeping with the object-oriented approach, we can add any general functionality to the `Microarray` class as methods. As examples we will add functions that replace, add and remove complete layers of array data (i.e. corresponding to one colour channel). Though, in the situations where we are supplying new values the input data must be of the correct size to be added to the array. In order to guarantee this we first describe the `checkDataSize()` function, which will trigger an error if the input for a layer is not of the same size as the existing array data. This function first makes an `array()` copy of the data, in case it was input as lists or tuples, and then the input size (accessed with `.shape`) must naturally match the number of rows and columns.

```
  def checkDataSize(self, channelData):

    channelData = array(channelData)
    if channelData.shape != (self.nRows, self.nCols):
      msg = 'Attempt to use data of wrong size'
      raise Exception(msg)

    return channelData
```

With the above function available to check any input we now add the functions to change the array layers. First `setChannel()` is constructed to replace all the data for an array layer specified at a given (existing) `index`.

```
def setChannel(self, channelData, index=0):

    channelData = self.checkDataSize(channelData)
    self.data[index] = channelData
```

The second function adds an entirely new layer of array data after the existing data. This involves using the NumPy function `append()` with the setting `axis=0` to create a new larger array (along the first axis). The `self.nChannels` attribute that records the number of individual layers is naturally incremented to keep consistency.

```
def addChannel(self, channelData):

    from numpy import append
    channelData = self.checkDataSize(channelData)

    self.data = append(self.data, channelData, axis=0)
    self.nChannels += 1
```

The function to swap channels takes two indices (which we really ought to check are valid) and uses tuples to index subsets of the NumPy arrays, i.e. assign the layer values which were at `indexA` to `indexB`, and vice versa.

```
def swapChannels(self, indexA, indexB):

    self.data[(indexB, indexA)] = self.data[(indexA, indexB)]
```

The function to remove a layer uses the NumPy `delete()` function, noting that this does not change the input arrays, but rather gives back a new array with the required part removed, which we then assign to `self.data`.

```
def removeChannel(self, index):

    from numpy import delete
    if index < self.nChannels:
        self.data = delete(self.data, index, axis=0)
        self.nChannels -= 1
```

Lastly there is a function to combine two channels specified via indices. This has a `combFunc` option to specify how the layers should be combined, which otherwise defaults to addition. If a special `combFunc` is passed in it must be the name of a function that accepts two equal-size NumPy arrays to perform the combination operation and creates another array of the same size. The `replace` option states which layer the new, combined data should be put into. This could be an entirely new layer, if the value is left as `None`, or it replaces an existing layer given an index. The required addition or replacement operation is easily achieved by using the existing methods we discussed above.

```
def combineChannels(self, indexA, indexB, combFunc=None, replace=None):

    if not combFunc:
        import operator
        combFunc= operator.add

    channelData = combFunc(self.data[indexA], self.data[indexB])

    if replace is None:
        self.addChannel(channelData)
```

```
    else:
      self.setChannel(channelData, replace)
```

We can test the above function using the example red and green data:

```
imgFile = 'examples/RedGreenArray.png'
testArray = loadArrayImage(imgFile, 'TwoChannel', 18, 17)

# Red and green added to channel 2
testArray.combineChannels(0, 1, replace=2)

# Display channel 2 as yellow
testArray.makeImage(20, channels=(2, 2, None)).show()
```

Array analysis

This final section continues the construction of the `Microarray` class (so note the relative indentation to the above Python code) and moves from simple manipulations to a few practical analyses. In general we will look at how we can find differences and similarities in the data. Only a few basic examples will be given, but analyses may be taken further using ideas discussed in other chapters. Hence we could use microarray data for principal component analysis (Chapter 23) or with machine learning methods (Chapter 24).

Differences and similarities

A very simple way of comparing two signal-intensity arrays is to take one away from the other and display the resulting positive or negative values as red and green colour channels. Hence we calculate the array of differences `diff`:

```
imgFile = 'examples/RedGreenArray.png'
rgArray = loadArrayImage(imgFile, 'TwoChannel', 18, 17)
diff = rgArray.data[0]-rgArray.data[1]
```

Then the differences are stored in the first two colour channels, flipping the sign for the green channel, and the values are clipped at `0.0` to remove any negative values.

```
rgArray.setChannel(diff, 0)
rgArray.setChannel(-diff, 1)
rgArray.clipBaseline(threshold=0.0, channels=(0,1))
rgArray.makeImage(20).show()
```

Alternatively we could use the `Microarray.combineChannels()` function to perform the comparison. For example, to illustrate where values on two channels are similar we can multiply the values, so the result is largest where the values from both channels coincide.

```
from operator import mul # Multiply

rgArray = loadArrayImage(imgFile, 'TwoChannel', 18, 17)
rgArray.combineChannels(0, 1, combFunc=mul, replace=2)
rgArray.makeImage(20, channels=(2,2,None)).show()
```

a) Product; similarities b) G-score; differences c) Hierarchical cluster

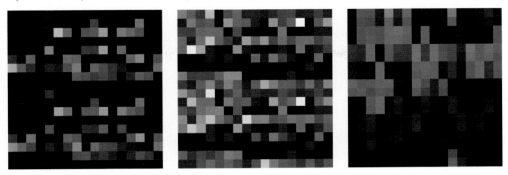

Figure 16.3 (Plate 4). The results of various array comparison procedures. Shown from left to right are: the product of red and green channels, displayed as yellow, to illustrate the coincidence between channels; the G-test scores of red $\times$ log$_2$(red/green), which is designed to show where the values in the two channels are different and of significant value; hierarchical clustering to produce shuffled rows and columns in the test data, which here shows that there are replicates for each row. A black and white version of this figure will appear in some formats. For the colour version, please refer to the plate section.

As another example, next we find the logarithm (here in base 2) of the ratio of the red and green channels. Combining two-channel red and green microarray data in this way is commonplace (for example, when making 'MA' plots). To achieve this we first define a small helper function `log2Ratio` which will accept two input data arrays and give back the combined, comparison array, noting that we take a copy of the original data and add a small amount to each array, to ensure that we do not divide by zeros or take logarithms of zero:

```
from numpy import log2

def log2Ratio(data1, data2):
  data1 = array(data1) + 1e-3
  data2 = array(data2) + 1e-3

  return log2(data1/data2)
```

We can use this function to combine the red and green channels (index 0 and 1), placing the result in the blue channel (index 2).

```
rgArray = loadArrayImage(imgFile, 'TwoChannel', 18, 17)
rgArray.combineChannels(0, 1, combFunc=log2Ratio, replace=2)
```

The result can be visualised by selecting only the blue channel, here making a greyscale image.

```
rgArray.makeImage(20, channels=(2,2,2)).show()
```

A further alternative to show differences between the two channel intensities is to use logarithms to calculate `x*log(x/y)`, where `x` and `y` are two colour channels, which is a convenient way of showing the information content[5] of one distribution over another and which is used in the G-test (see Chapter 22). Similar to the previous example red and green

---

[5] For probability distributions, the sum of such values gives the Shannon relative entropy.

channel arrays are both shifted away from zero by a small amount, so that zeros do not occur in the division that will follow.

```
from numpy import log2

def gScore(data1, data2):
  data1 = array(data1) + 1e-3
  data2 = array(data2) + 1e-3

  return data1 * log2(data1/data2)
```

This can be tested as before, though we normalise the values so the logarithms are scaled into the same range as the other channels:

```
rgArray = loadArrayImage(imgFile, 'TwoChannel', 18, 17)
rgArray.combineChannels(0, 1, combFunc=gScore, replace=2)
rgArray.normaliseMax(perChannel=True)

rgArray.makeImage(20, channels=(2,2,2)).show()
```

With values calculated in this way we can perform significance tests, as described in Chapter 22, given that the random expectation is that values will be chi-square distributed. Here the comparative term has been calculated from the perspective of the red channel, but we could also do it for green (i.e. `green * log(green/red)`).

Hierarchical clustering

The next example moves on to rearranging the data (swapping rows and columns) so that we can better see similarities or correlations in the elements. This rearrangement will involve hierarchical clustering, and uses the same idea as building a phylogenetic tree (see Chapter 14), so that the most similar rows and columns are placed next to one another. The aim is that any rows or columns that show similar patterns will be placed together in easily visualised sub-groups. The image-generating method described earlier can then be used to make a visual representation (see Figure 16.3).

The function `__hierarchicalRowCluster` takes an array of data as a two-dimensional matrix and uses this to build a matrix of the distances between each row, so the element `distanceMatrix[i][j]` is the distance between row `i` and row `j`. By 'distance' here we mean the similarity between different rows, and measure this here as the Euclidean distance between row vectors (root of sum of differences squared). In different situations you could consider other means of estimating the similarity between rows.[6] It should be noted that the function name begins with a double underscore '__'. In Python a function starting, and not ending, with a double underscore is effectively private,[7] so it can't be called directly from an instance of a class. A function is kept private like this if you don't want to expose it as a normal method (a function linked to a class), and only use it internally within the class definition. Hence, here we make the clustering function private and use it in the inner

---

[6] Hamming distance or Shannon mutual information entropy would be examples.

[7] In Python no function is truly private given that it can be accessed from an object if you know what its internal, 'mangled' name is.

workings of `Microarray` but do not allow method calls from an object like `microrray1`. `__hierarchicalRowCluster(data)`.

```
def __hierarchicalRowCluster(self, dataMatrix):

  from SeqVariation import neighbourJoinTree

  n = len(dataMatrix[0])
  distanceMatrix = zeros((n, n), float)
```

We loop through each layer and row in the `dataMatrix` and take this row away from the whole array, so we get a matrix of differences to this row. The differences are squared, added up along the row and the square root is taken, giving a distance from `row` to all other rows. These distances are then placed at position `i` for this row within the `distanceMatrix`.

```
  for channelData in dataMatrix:
    for i, row in enumerate(channelData):
      diffs = channelData - row
      sqDiffs = diffs * diffs
      sqDists = sqDiffs.sum(axis=1)
      distanceMatrix[i,:] += sqDists
```

The function `neighbourJoinTree` which was described previously in Chapter 14 is reused here to create a hierarchical tree from the distance matrix. Note that we use `.tolist()` to make a list of lists from the array of distances because the tree-generating function did not make use of NumPy arrays earlier in the book.

```
  tree, joinOrder = neighbourJoinTree(distanceMatrix.tolist())
```

The hierarchical tree structure is then interrogated by following its branches, effectively flattening it into a single list of node indices, which represents the order of rows for the shuffled matrix. The list `rowOrder` is initially defined as a copy of the tree, but is then processed to insert the contents of any sub-lists (or tuples) directly into the main list. The use of a `while` loop here is because the size of the `rowOrder` list will grow as the tree branches are flattened. We record an index `i`, which represents the position in the list that is being processed, and continue until the end of the list. The inner `while` loop checks whether the list item at position `i` is an integer (the end of a tree branch), and if it is not the contents of the sub-list (`rowOrder[i]`) are inserted into the main list using the slice notation `[i:i+1]`, replacing the original range covering the sub-list with its contents. The use of the second `while` loop is needed because the new element at position `i` may itself be another sub-list.

```
  rowOrder = list(tree)

  i = 0
  while i < len(rowOrder):

    while not isinstance(rowOrder[i], int):
      rowOrder[i:i+1] = rowOrder[i]

    i += 1

  return rowOrder
```

The row clustering routine is actually used by the `hierarchicalCluster` function, which is not private and so provides the method `microarray.hierarchicalCluster()`. This clusters the rows in `self.data` and then reorders the array according to the row hierarchy. The resulting `data` array is then transposed (flip rows for columns) and clustered again, effectively clustering the columns, which forms the new column order.

```
def hierarchicalCluster(self):

    rows = self.__hierarchicalRowCluster(self.data)

    swapped = transpose(self.data, axes=(0,2,1))
    cols = self.__hierarchicalRowCluster(swapped)
```

The reordered `rows` and `cols` are used as indices into the NumPy arrays to shuffle the data, according to the hierarchical clustering, noting that we don't affect `self.data`.[8] The data is then used to make an entirely new `Microarray` object, with a different order for its rows and columns.

```
    data = self.data[:,rows] # Rearrange
    data = data[:,:,cols]

    # data = array(data.tolist()) # to fix PIL.Image bug

    name = self.name + '-Sorted'
    rowData = [self.rowData[i] for i in rows]
    colData = [self.colData[j] for j in cols]

    sortedArray = Microarray(name, data, rowData, colData)

    return sortedArray
```

When testing the hierarchical clustering we get back a new `Microarray` object with a different order of rows and columns:

```
sortedArray = rgArray.hierarchicalCluster()
sortedArray.makeImage(20).show()

print(rgArray.rowData)
print(sortedArray.rowData)

# [0, 1, 2, 3, 4, 5, 6, 7, 8, 9, 10, 11, 12, 13, 14, 15, 16, 17] - Original rows
# [4, 13, 2, 11, 6, 15, 5, 14, 7, 16, 1, 10, 3, 12, 8, 17, 0, 9] - Shuffled rows
```

---

[8] Note that due to a bug in PIL at the time of writing, an extra line `data = array(data.tolist())` is required for proper function.

17 High-throughput sequence analyses

Contents

High-throughput sequencing

Given the decreasing cost required to determine the sequence of nucleic acids, sequencing is used in increasingly wider contexts. Rather than only determining the genome sequence of an organism, high-throughput techniques allow researchers to investigate much more, such as the variation within individuals of a population, the amount of expression of individual genes in a given sample (e.g. by detecting RNAs) and the sequences which are bound to particular protein components. A sequencing run on one of the latest-generation sequencing machines may generate many gigabases ($>10^9$ bp) of data and so much of the task for bioinformatics is to make sense of the raw sequence data: to put it into a genomic, biological context. For organisms with a known genomic sequence the primary task when processing high-throughput sequence data is to simply map relatively short bits of sequence called 'reads' that come from the sequencing machine to a reference genome. Only then can the detected sequences be understood. By mapping newly acquired sequences on to the known chromosomes the whole database of information that annotates the genome, such as the position of genes and regulatory sequences, indicates which DNA features were detected. In this chapter we will give an introduction to various basic computational procedures involving high-throughput sequence data which can be achieved, or at least handled, using Python. Because this is a vast and rapidly expanding subject we can only lightly touch on the core concepts here, though hopefully we have provided solid starting points for further development.

Sequencing for biochemical analysis

Although the advances in fast and relatively low-cost DNA sequencing technology were initially driven by the desire to acquire and assemble genome sequences, there are an increasing number of methods that rely upon mapping many short DNA reads to a complete, or nearly complete, genomic sequence. This might be done to find differences in sequences compared to the reference genome, i.e. finding sequence polymorphisms. However, it is also common to use high-throughput methods for the biochemical analysis of cells. For example, the DNA sequences could be a large complement of cDNAs, which are prepared using reverse transcription from messenger RNA, to show which genes were actually being read in a given sample or cells; this method is called *RNA-seq*. Alternatively the DNA could be fragments of genomic origin that have been specially selected in some way. Chromatin immumoprecipitation sequencing or *ChIP-seq* is such an example, where genomic DNA is chemically cross-linked to proteins that are bound to the chromosomes before it is cut into small fragments. Here the cross-linking is not specific for any particular kind of DNA-binding protein, but the fragments of DNA with their associated proteins are purified using antibodies that bind to and select only one kind of protein. The end result is to produce a sample of DNA sequences that were in close contact with a specific type of protein when it was functioning inside cells. Reversing the DNA-protein cross-links then allows the associated DNA to be sequenced, thus indicating where in the chromosomes the protein was originally present. ChIP-seq is frequently used to see which DNA sequences are associated with modified histone[1] proteins (usually methylate or acetylate), and this in turn indicates which sections of the DNA are active or inactive for transcription.

Short-read mapping

Irrespective of whichever technique was actually used to generate the DNA segments their sequence must be mapped to a reference genome to find from where in the chromosomes the sequence originated. Effectively, mapping short sequence reads back to a pre-assembled genome sequence that allows the reads to be annotated with all the known genomic information. This will include aspects such as: whether the sequence is from a gene, is a regulatory region, is a structural region[2] or is non-functional; which gene, if any, the sequence is from (or near to) and whether the sequence is an intron or an exon. Often the actual base-pair sequence of a read is not the point of main interest; the location within the genome is. Naturally, to find where DNA fragments come from requires an alignment of the read sequences to the reference genome sequence to find where they match. Usually only the two ends of the fragments are read for the first 100 or so base pairs,[3] but this is generally enough to locate the sequence within the genome. Also in this case, the pairing of the sequences from the two fragment ends can help the mapping: if you know the range of lengths of the DNA fragments (for example, using information from *gel electrophoresis*) then you know how far apart the *paired-end reads* could be, and thus restrict alignments to only genome positions where the reads are relatively close together. Unfortunately there may still be more

---

[1] Histones are proteins that wrap around and package DNA to form *chromatin*, and thus control access to the DNA.
[2] E.g. telomere, centromere.
[3] For example, Illumina machines typically sequence 100 bases, at the time or writing.

Transmembrane proteins

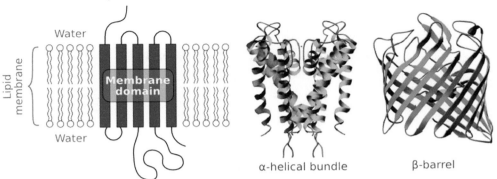

α-helical bundle   β-barrel

Plate 1 (Figure 15.2): The general form of transmembrane proteins that reside in a lipid bilayer. Transmembrane proteins are embedded within the plane of a hydrophobic cellular membrane, which is composed of a double layer of lipid molecules and other membrane proteins. In contrast to aqueous proteins that reside in water, a membrane protein adopts a structure so that its hydrophobic (water-hating) parts lie within the membrane, often with hydrophilic (water-loving) parts protruding into the aqueous regions either side of the membrane. There are two common types of structure that transmembrane proteins adopt to form a hydrophobic membrane domain: an α-helical bundle or a β-barrel.

a) Red channel   **b) Green channel**   **c) Both channels**

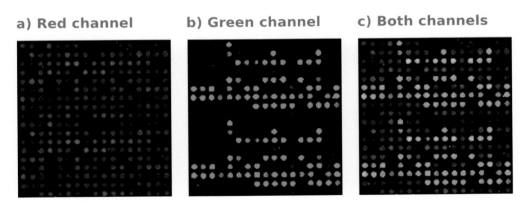

Plate 2 (Figure 16.1): A microarray image composed of red and green colour channels. Here each spot represents a different probe molecule and the colour intensity reflects the binding of DNA within two samples, each labelled with a different fluorescent marker. Original image courtesy of Paul Edwards and Karen Howarth, University of Cambridge.

a) Red-green output   b) Yellow-blue output   c) Normalised and clipped

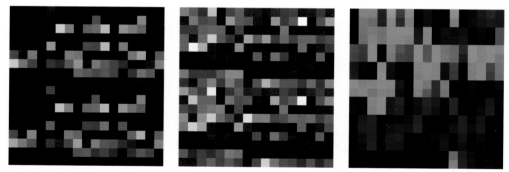

Plate 3 (Figure 16.2): Image pixmaps constructed from array data stored in the `Microarray` class. Shown from left to right are: a red and green representation of the original, imported data; the original data displayed as yellow and blue colours (the original red-channel data becomes red and green in the image and the original green data becomes blue); clipped and normalised data where values less than half the maximum are set to zero.

a) Product; similarities   b) G-score; differences   c) Hierarchical cluster

Plate 4 (Figure 16.3): The results of various array comparison procedures. Shown from left to right are: the product of red and green channels, displayed as yellow, to illustrate the coincidence between channels; the G-test scores of red $\times \log_2$(red/green), which is designed to show where the values in the two channels are different and of significant value; hierarchical clustering to produce shuffled rows and columns in the test data, which here shows that there are replicates for each row.

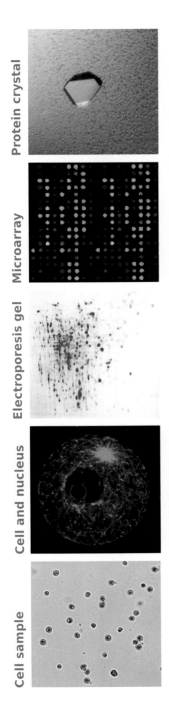

Plate 5 (Figure 18.1): Examples of a variety of different kinds of images used in biology. Shown from left to right are: a microscope image of a mammalian cell culture (courtesy Dr. Anja Winter, University of Leicester); a red-green fluorescence microscope image of an oocyte and its nucleus (courtesy Dr. Melina Schuh, MRC Laboratory of Molecular Biology); a two-dimensional electrophoresis gel of a plant proteome (courtesy Prof. Paul Dupree, University of Cambridge); an image of a DNA microarray (courtesy Karen Howarth, University of Cambridge); a protein crystal that has been grown for structure determination by X-ray crystallography (courtesy Dr. Aleksandra Watson, University of Cambridge).

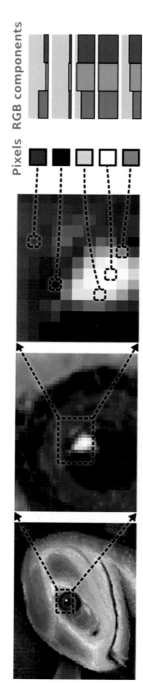

Plate 6 (Figure 18.2): An image, its component pixels and their RGB colour-space values. A section of the book cover picture is shown magnified at different levels to reveal the array of square pixels that the digital image is composed of. For the highest magnification, example pixels with different colours are selected and the component red, green and blue (RBG) values that constitute each colour are shown as histograms.

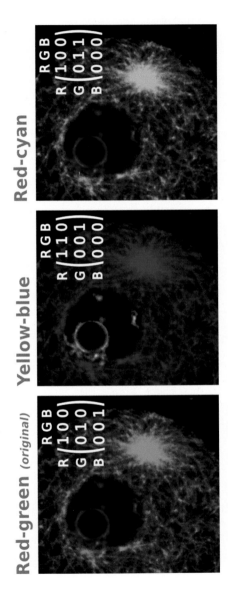

Plate 7 (Figure 18.3): Matrix transformations of pixmap colours. A red-green coloured (i.e. two-channel) fluorescence microscope image of a cell is shown alongside colour-adjusted yellow-blue and red-cyan versions. Inset in each image is the RGB colour transformation matrix relative to the red-green image.

Original image   **Edge detected**   **Feature codes**

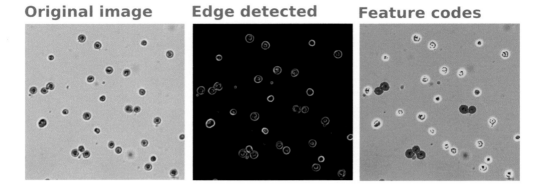

Plate 8 (Figure 18.5): Images from micrograph cell-counting procedure. An original microscope image of mammalian cells is shown alongside the results of the edge detection and a grey-scale version where the different cell-edge features are labelled with colours after blob analysis. Isolated cells are yellow, overlapping cells are blue and small fragments are red.

Plate 9 (Figure 19.5): Example of 2D peak picking for a protein gel. An image of a two-dimensional polyacrylamide gel, with the peak picked maxima shown as small red crosses. The gel has been stained with Coomassie Brilliant Blue dye to show the abundance of different proteins that have been separated according their size and isoelectric value.

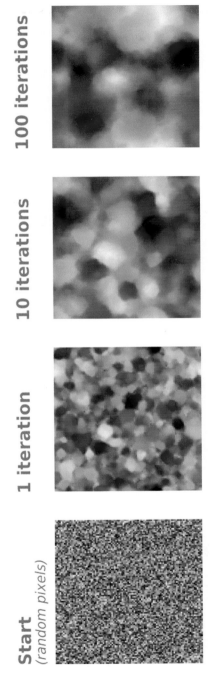

Start
(random pixels)

1 iteration

10 iterations

100 iterations

Plate 10 (Figure 24.4): Example self-organising map output. Results from of an initially random 100×100 colour pixel map (left) and the effect of the self-organising map on the colours after 1, 10 and 100 iterations.

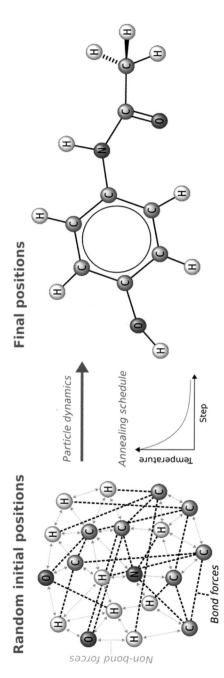

Random initial positions

Non-bond forces

Bond forces

Particle dynamics

Annealing schedule

Temperature

Step

Final positions

Plate 11 (Figure 25.4): An overview of the simulated annealing procedure, as applied to a molecular structure. The method of simulated annealing can be applied to the problem of computing a molecular structure given information relating to the bonding and non-bonding forces between atoms. Starting from random positions the forces that result from a particular arrangement of atoms can be estimated using knowledge of the forces' energy potentials. In turn this leads to a dynamics simulation where the forces, momenta and thus changes of atomic position can be modelled at discrete time steps. By applying an annealing schedule with a decreasing effective temperature, the simulated kinetic energy of the atoms will initially be large, so that large differences in conformation can be explored, but end up making only small adjustments, to better home in on a precise solution.

than one genome match for a particular sequence read, especially if the region has a repetitive sequence. Here the ambiguity can sometimes be resolved by sequencing for longer, e.g. reading the DNA fragments for more than 100 base pairs, but this gives diminishing returns as longer reads become more error-prone. Sometimes a sequence read may not match at all, if there has been a genuine substitution or an error in the sequencing (common at the end of reads). Fortunately in situations where sequences differ slightly we can do the sequence alignments in a permissive way, to accept small changes where the expectation is that the quality of the reads decreases with length; i.e. the chances of a mismatch increase with length.

The alignment of short, high-throughput DNA sequence reads to a genome is not done using the types of sequence alignment discussed previously, i.e. not using dynamic programming or programs like BLAST. Such methods would be too slow. Instead genome mapping methods pre-index the genome sequence for a quick look-up, and commonly use the Burrows-Wheeler transform for data compression.[4] The genome index means that a small query sequence can be mapped to the genome by extracting the known genome positions for its constituent sequence(s) that have been previously located; this is somewhat similar to finding data in a Python dictionary using its key. The general idea is to avoid having to align a query sequence to the large number of possible short sequences in the whole genome each time. Rather, significant matches are found with a quick look-up which can eliminate the vast majority of the genome sequence. This strategy is optimised for large numbers of reads being mapped to the same target (the genome sequence). This would be impractical for general pairwise alignments of arbitrary sequence databases because the indexing process, required before using a new target sequence, is designed for large contiguous sequences. Indexing is memory-intensive and is proportionately slow but, given the target is fixed for a given genome sequence,[5] the cost is returned many times over for the mapping of large numbers of small sequences to a single target.

Python examples

For some of the Python examples relating to high-throughput sequence analysis we will be using the *HTSeq* library (see http://www.cambridge.org/pythonforbiology for download and installation instructions) and will not attempt to write our own classes of object. This is a fast library providing objects that deal with genomes, sequences and annotations etc. It is especially helpful in simplifying the reading and writing of the various data file formats that are used to store sequence and sequence-related data. For example, we will use it to load FASTQ[6] files that contain the sequencing information that comes from a sequencing machine and GFF[7] files containing genome annotation information. Initially we will illustrate how to obtain genome sequences and map sequence reads, albeit controlling an external program to index the genome and do the actual Burrows-Wheeler alignment. Next we will load the

---

[4] See: Burrows, M., and Wheeler, D. (1994). A block sorting lossless data compression algorithm. Technical Report 124, Digital Equipment Corporation.
[5] Actually occasional new, refined *genome builds* will be released.
[6] Cock, P.J., Fields, C.J., Goto, N., Heuer, M.L., and Rice, P.M. (2010). The Sanger FASTQ file format for sequences with quality scores, and the Solexa/Illumina FASTQ variants. *Nucleic Acids Research* 38(6): 1767–1771.
[7] http://gmod.org/wiki/GFF3.

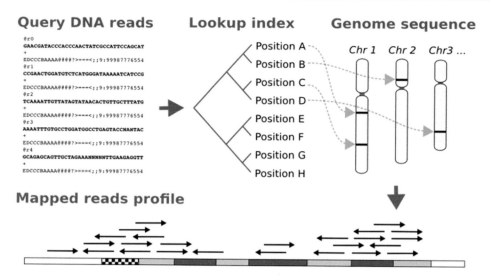

Figure 17.1. High-throughput mapping of sequence reads to an indexed genome. Large numbers of short DNA sequence reads, such as those that come from a ChIP-seq experiment, may be mapped to a complete genome sequence to identify which positions they represent. Rather than comparing each short read with the whole genome, many high-throughput methods map the sequences by using a pre-constructed index, to quickly connect a large set of sub-sequences with their genome locations. The result of the mapping is a genomic profile, illustrating any hotspots in the chromosome sequences that have multiple reads.

resulting alignment information to show how the results can be accessed in Python, including making graphs. Finally we illustrate how to link the alignment results, and thus the DNA sequence reads, to genomic annotations that describe genes, exons, introns and such like.

Mapping sequences to a genome

In this next section we illustrate how we can use Python to control the process of mapping large numbers of relatively short DNA sequence reads to a known genome sequence. For the intensive calculations we will be relying on an existing external program that is quick and well tested.

Obtaining genome sequences

While we can obtain genome sequences using a web browser, we may sometimes also wish to automate the process, which can naturally be done with Python scripts. For this example we will be illustrating how to get data from the NCBI,[8] but hopefully it is clear how the procedure would be adapted to use other databases. The following downloads genome sequence information, in the simple FASTA format, in a similar way to the example used earlier to download Protein Data Bank (PDB) files containing macromolecular structure data.

---

[8] National Center for Biotechnology Information in the United States.

On-line resources change, so naturally the examples have been tuned to work with the way that genomes can be accessed at the time of writing. In the future, however, the various Internet addresses we use will probably require adjustment. Here we use the File Transfer Protocol (FTP) service available at the following URL:

```
FTP_ROOT = 'ftp://ftp.ncbi.nlm.nih.gov/genomes'
```

And as an example we illustrate the downloading of the following *E. coli* bacterium genome data (and naturally we checked with a web browser to see the file names of the sequences that were available):

```
GENOME = '/Bacteria/Escherichia_coli_536_uid58531/NC_008253.fna'
```

The equivalent for a human chromosome sequence would use something like the following, noting that there is a sub-directory to specify the chromosome:

```
GENOME = '/H_sapiens/CHR_10/hs_ref_GRCh8_chr10.fa.gz'
```

We will define a simple Python function `downloadFile()` that may be built into more automated scripts to fetch and save the file. For our example the file will be a FASTA-format sequence. Firstly a few imports are made. The `fnmatch` module provides a filename-matching function that makes it easy to choose particular file names that have a particular file extension (an ending like `.fna` or `.fasta`), as an alternative to using regular expressions (the `re` module). The `urlopen()` function will handle all of the communication with the remote data server. In Python 3 it is in the `urllib.request` module and in Python2 it is in the `urllib2` module, so we first try to import the former and if that fails try the latter:

```
import os
from fnmatch import fnmatch

try:
  # Python 3
  from urllib.request import urlopen
except ImportError:
  # Python 2
  from urllib2 import urlopen
```

The download function itself takes two arguments, the URL of the remote file we wish to acquire and the file path (file name and directory) of where we save the file locally. In the function we print a message to give the user some information on what is happening, open a connection to the remote location with `urlopen()` and read the contents into the variable `data`.

```
def downloadFile(remoteFile, localFile):

  print('Downloading to %s ...' % localFile)

  response = urlopen(remoteFile)
  data = response.read()
```

Then at the end we open the local output file for writing in binary mode.[9] Note that we are being lazy here and don't make any checks to ensure that we are not overwriting an existing file, but this would be easily handled using `os.path.exists()`.

---

[9] Binary mode is used because `response.read` in Python 3 always gives data bytes and in any case we could be dealing with a compressed file (e.g. GZIP).

```
fileObj = open(localFile, 'b')
fileObj.write(data)
fileObj.close()

print(' ...done')
```

This function is simply tested with the following demonstration code for the *E. coli* genome file mentioned above (saving the file in the current working directory):

```
remoteGenome = FTP_ROOT+'/Bacteria/Escherichia_coli_536_uid58531/NC_008253.fna'
downloadFile(remoteGenome, 'EcoliGenome.fasta')
```

Moving on from fetching single files, the next example downloads several files with related locations. For example, the files may contain the different human chromosome sequences, but are housed in different sub-directories (as is the case for the NCBI's FTP site).

The `downloadGenomeFiles()` function takes a remote directory and a local directory, rather than using a single specific file name, and searches for all the files that end a particular way (by default with 'fna'). Also, in this example we have separated the main (base) location of the remote site into a variable called `url`, given that this is not expected to change much and we can avoid always having to specify the full location of the remote data. After the initial definition line some tidying is done on the input `remoteDir` to ensure that it begins and ends with a '/' character, which is what is expected later in the function.

```
def downloadGenomeFiles(remoteDir, localDir, fileType='*.fna', url=FTP_ROOT):

  if remoteDir[0] != '/':
    remoteDir = '/' + remoteDir

  if remoteDir[-1] != '/':
    remoteDir = remoteDir + '/'
```

The full remote path is then the combination of the base URL and the variable remote path. Note that here we do not use `os.path.join()`, because this would use different slashes under different operating systems ('\' for Windows, '/' for Max OS X and Linux) and is appropriate only for local file systems, not remote Internet resources that expect only '/'.

```
  remotePath = url + remoteDir
  print("Reading %s" % remotePath)
```

Next the `urlopen()` function is used and in this case, because the target is a directory, gets the listing of the file names at the remote location.

```
  response = urllib2.urlopen(remotePath)
  data = response.read()
```

Some empty lists are initialised, which will contain all the file information. The remote data that is read is one long test string containing all the file names, which we split on the newline character to give a list of lines.

```
  fileNames = []
  filePaths = []
  chromosomeDirs = []
  lines = data.split('\n')
```

Then each line from the remote read is considered in a loop, stripping off any whitespace and skipping blank lines:

```
for line in lines:
  line = line.strip()

  if not line:
    continue
```

The actual file name is the last element of an array (`[-1]`) when we split the line into a list according to its whitespace. Then if the entry begins with the letters `'CHR'` it is deemed to be a sub-directory for individual chromosomes that we need to look into to get the sequence files, and this chromosome directory is added to the `chromosomeDirs` list.

```
fileName = line.split()[-1]

if fileName.startswith('CHR'):
  chromosomeDirs.append(fileName + '/')
```

Otherwise, if the file is not a chromosome sub-directory, a check is made to see if it is a sequence file with the prescribed ending. Here we use the imported `fnmatch()` function to see if the file name contains the pattern specified in the input arguments. This pattern defaults to '`*.fna`', which means that the name matches if it ends in `.fna`; the '`*`' matches any number of characters in the file name before the stated ending.

```
elif fnmatch(fileName, fileType):
  fileNames.append(fileName)
  continue
```

We then get the full locations of the individual remote files and download them, to a local file of the same name, using the `downloadFile()` function defined above. Here we do use `os.path.join()` to make a local file name which is consistent with the operating system we are using. The `os.path.abspath()` is used to get the full, long file name, with directories relative to the root of the local file system: this resolves relative directory specifications like '`../`', removes redundancies like '`dirA/./dirB`' and converts any slashes to the right kind for the current operating system.

```
for fileName in fileNames:
  filePath = os.path.join(localDir, fileName)
  filePath = os.path.abspath(filePath)
  filePaths.append(filePath)
  downloadFile(url + remoteDir + fileName, filePath)
```

Given that reading the remote directory may have found some sub-directories for individual chromosomes we then repeat the whole operation on each sub-directory by calling the `downloadGenomeFiles()` function from inside itself, i.e. recursively. This will then go in to each of those other locations and potentially add more matching files to the current list. Note that only the first, main function call will give back files to the user, and all the recursive calls will be absorbed into the `filePaths` list at this point.

```
for chromosomeDir in chromosomeDirs:
  subDir = remoteDir + chromosomeDir
  filePaths += downloadGenomeFiles(subDir, localDir,  fileType, url)

return filePaths
```

At the end the function returns a list of the file paths that were actually saved. We can then test the function on the human genome at the NCBI site used previously. Note that this may take a significant time to download, so only do it for real if you really want all of the human genome data!

```
filePaths = downloadGenomeFiles('H_sapiens','examples','hs_ref*.fa.gz')
```

One further convenience function extracts any compressed GZIP archives (ending in '.gz') so they can be used locally without hindrance. This function simply uses standard Python modules to create open file objects for compressed formats. These can be parsed, line by line, in much the same way as a regular file. Here the example simply writes out the uncompressed lines one by one into a new file, but if there is enough memory the whole file could be read in one go (using .read()). Also, the gzip library could be used to read compressed files directly inside analysis functions, thus avoiding having to store uncompressed files. Note that there are also standard Python modules to handle other compression formats like 'ZIP' and 'BZIP2'; zipfile and bz2.

```
import gzip

def uncompressGzFile(fileName):

  if fileName.endswith('.gz'):
    inFileObj = gzip.open(fileName, 'rb')

    fileName = fileName[:-3]
    outFileObj = open(fileName, 'w')

    for line in inFileObj:
      outFileObj.write(line)

    # Faster alternative, given sufficient memory:
    # outFileObj.write(infileObj.read())

    inFileObj.close()
    outFileObj.close()

  return fileName
```

Next we use an external program that will do the alignment to map the short sequence reads to a genome sequence. Popular open-source software choices for this at the time of writing are BOWTIE[10] and BWA,[11] and we will illustrate the use of the former by wrapping with a convenient Python function, in a similar manner to what was done with BLAST and ClustalW examples in earlier chapters.

Downloading genome sequence data is usually a simple matter, as described above, of accessing an on-line repository. The next step is naturally to consider aligning some short-read data to the genome sequence. For demonstration purposes, there are lots of high-throughput sequencing data sets that are available via on-line services like the Gene

Expression Omnibus,[12] and which can be used with the scripts we describe below. The actual genome alignment in these examples will be done by the program called Bowtie,[13] which is just the open-source example we happen to have chosen for this chapter. Python functions will be illustrated which wrap this external program so that we can use it in larger programs and automated pipeline scripts. Hence the purpose of the examples is to easily use existing (tested and efficient) programs, rather than doing everything from scratch in Python, which would be slower to run and take a long time to describe.

Indexing a genome

Before considering running the actual alignment the genome sequence data first needs to be indexed. This process is critical for the fast short-read alignment program operation. Using a full genome sequence in a textual file format would be exceedingly slow; rather the chromosomes are processed to give a binary index file where the sequence of a short read can be looked up efficiently. The indexing process does take a significant amount of time (and computer memory), but it only has to be done occasionally, when there is a new release of an improved genome sequence.

The first step to get started is to make an import from the standard Python `subprocess` module, which will handle invoking the command to run the external indexing or alignment program. Naturally, we also need to specify the location of the aligner within the local file system; the directory that contains the program executable file(s) is specified in `ALIGNER_PATH`.

```
from subprocess import call
ALIGNER_PATH = '/home/user/programs/bowtie-0.12.7/' # substitute correct path
```

For our example program, Bowtie, the indexing program will be found in the `ALIGNER_PATH` directory specified above. The Python function that controls the indexing, as illustrated below, takes various arguments, which naturally include the location of the genome sequence files and the various options to control the indexing. The options that are used here are specific to the particular program and its current version, so changes would need to be made to use other aligners, albeit using the same kinds of strategy.

The function `indexGenome()` takes the `genomeName`, which will be the identifier for the genome data that will be used later when the actual alignment is made. The `fileNames` are the names of the sequence files and naturally `outputDir` controls where the indexed genome files are placed. The other options control a few of the more important aspects of the indexing process; however, there are many more options that could have been included. The full list of options (normally issued via the command line) is available in the main Bowtie documentation.

```
def indexGenome(genomeName, fileNames, outputDir,
                tableSize=10, quiet=True, pack=True):
```

Firstly, in the function a check is made to decompress any archived genome sequence files, using the function described above, which will not affect sequence files that are not compressed. The names of the uncompressed sequence files are appended to a list, which is then

---

[12] http://www.ncbi.nlm.nih.gov/geo/.
[13] Download via: http://bowtie-bio.sourceforge.net.

joined by commas into one long text string, `fastaFileStr`, and it is in this form that we specify the sequence files when running the indexing program.

```
fastaFiles = []
for fileName in fileNames:
  fileName = uncompressGzFile(fileName)
  fastaFiles.append(fileName)

fastaFileStr= ','.join(fastaFiles)
```

The next stage is to assemble the options (the command line arguments) that will be used when creating the index. These are assembled into the `cmdArgs` list. In essence this list will contain all of the things that would be typed by a person at a command line prompt if the program were to be run directly from the operating system. Accordingly, the first thing in the list is the full file path of the indexing program, which is the `ALIGNER_PATH` joined to the name of the indexing program, which here is `bowtie-build` (and is located in that directory). The '`-f`' option specifies that the input sequence data is hard-wired to be in FASTA format. If this is not always the case in use, then this option could be changed by adding another optional argument in the function definition. The `quiet` and `pack` arguments of the function are used to specify whether the respective '`-q`' and '`-p`' options end up in the list. Here the `quiet` option is used to suppress textual output during the indexing procedure and the `pack` is used to compress data when the index is made; this makes indexing slower, but means it takes less computer memory. Note how the example has substituted a simple command line option and replaced it, as far as running from Python is concerned, with a variable that has a more immediately informative name, e.g. '`-q`' becomes `quiet`.

```
cmdArgs = [ALIGNER_PATH+'bowtie-build', '-f']

if quiet:
  cmdArgs.append('-q')

if pack:
  cmdArgs.append('-p')
```

The final options to add to the list relate to the table size, which is the number of sequence positions from the short query that will be mapped at one time to the genome data, the input sequence files and the name tag for the resultant genome index.

```
cmdArgs += ['-t', str(tableSize), fastaFileStr, genomeName]
```

The competed list of command options is joined, using spaces, into a line that is printed out, to show the user what is being run. This might result in something like '`/home/user/programs/bowtie-build -f -t 10 genomeFasta/NC_008253.fna E_coli`'. Then finally the indexing job is run using `call()` with the program name and option information. Note that the output directory is dictated by setting the current working directory (`cwd`) at the time the call is made, rather than this being a program input option.

```
print(' '.join(cmdArgs))
call(cmdArgs, cwd=outputDir)
```

Under normal circumstance the indexing will only be done occasionally, but beware that the procedure will often take a significant amount of active computer memory. For example, even with the `pack` option being used a mammalian genome will typically require over three gigabytes of free memory to index.

Aligning reads to a genome

Next we create another wrapper to control an external program, to do the alignment with the indexed genome data. There are many more options to control the program this time, as this is highly dependent not only on the source of the data (in terms of which kind of sequence machine produced it) but also on the way in which the sequenced DNA was prepared and how this experimentally informs the biology that we are interested in. The key aspects of the sequences we wish to align are the sequence lengths and within this length what the quality of the called sequence reads is, given this generally diminishes with length. Also when the alignment is actually done we have to consider how mismatches (e.g. polymorphisms or errors in the sequencing) are handled. The full list of options that we pass in, as arguments to the alignment function, is as follows:

`genomeName`	This is the name or tag given to the indexed genome that we wish to align to. This is the same name as was used when building the index with the function described above.
`genomeDir`	The location of the directory that contains the binary genome index files.
`readFiles`	A list containing the names of the files containing the short-read sequences to align.
`pairedReadFiles`	For experiments that use paired read data, a second list of short-read sequence file names that match up (as pairs) with the readFiles.
`outFile`	The name of the output alignment file, which by default will use the SAM format.
`leftTrim`	The number of sequence letters to trim from the start of all the reads; useful to remove known barcode sequences etc., which should not be mapped.
`rightTrim`	How many sequence letters to remove from the end of the sequence reads.
`useSOLiD`	Whether the data comes from an ABI SOLiD sequencing machine.
`qualType`	The type of quality scores to use, which is expected to be 'illumina1.3', 'solexa' or 'sanger'. See the description below of the FASTQ format for more explanation of this.
`maxMismatches`	The maximum number of base-pair mismatches to tolerate in an alignment.
`highQualLen`	How many sequence positions, from the start of the read, are deemed reliable.
`pairSepRange`	The range of separations for paired sequence reads; known from the size selection used to prepare the DNA fragment library.
`showHits`	The number of genome location alignment 'hits' to show for each input read sequence.
`maxHits`	Sets the maximum number of alignments that will be tolerated for a read sequence; otherwise the read is deemed to be unmappable and is discarded.
`enforceBest`	Whether to enforce the display of only the best-matching read, if more than one could be aligned to the genome.
`cpuCores`	If set, this states how many processing jobs to run at one time, making use of multi-core central processing units. The default is to use all available CPU cores.

The short sequence reads that are aligned to the genome are generally stored in a textual format called FASTQ. These files contain sequence entries, one after another, in a form exemplified by the following text:

```
@Annotation
CGGATGATTTTTATCCCATGAGACATCCAGTTCGG
+Annotation
45567778999:9;;<===>?@@@@AAAABCCCDE
```

As illustrated, each sequence read comprises four lines of text. The first and third lines, beginning with '@' and '+' respectively, are textual annotations for the sequence read. In a basic sense these are just identifiers, but the output of most sequencing machines contains a rich variety of information, including details relating to the machine that read the sequence. Importantly, if the FASTQ sequence files consist of read pairs (sequencing from the two ends of DNA fragments) then the annotation for each sequence can be used to unambiguously identify the connected read pairs even if (and as is generally the case) the read pairs are given in two files, one for the sequencing from each end. However, it should be noted that currently in many cases the order of sequences in paired FASTQ files is consistent, so that the pair identity is simply the position in the file. The second line of a FASTQ file is clearly the one-letter code sequence of the DNA base pairs. The fourth line of the entry is an array of quality scores, each character of which relates to the sequence letter at the same read position. It is perhaps unfortunate that there are several slightly different systems of code that are used to state the positional read quality, which is why the quality score scheme is taken as an input to the alignment program. All the quality score systems use a range of textual (ASCII) characters to specify the quality value at each read position and what differs in each system is which range of characters are used. The basic problem is that just using characters from 0 through to 9 does not give enough precision to the scores, so instead more characters are used, which naturally includes letters and other symbols. All the characters are used in the following order:

```
!"#$%&'()*+,-./0123456789:;<=>?@ABCDEFGHIJKLMNOPQRSTUVWXYZ[\]^_
'abcdefghijklmnopqrstuvwxyz{|}~
```

Here the score value in the scale decreases left to right and each sub-sequence character is the next point on the scale. One thing that differs between the different schemes, however, is which characters have the highest and lowest scores. Recent Illumina machines (under the 'sanger' scheme here), for example, use symbols from '!' to 'J' inclusive (older Illumina machines use the 'illumina1.3' scheme). Note that it is possible to also use reads in FASTA file format, rather than FASTQ, although these will not contain any quality information.

Now we come to the Python function that actually does the genome alignment by wrapping the Bowtie aligner program. It should be noted that not all of the possible command line options for Bowtie have been considered (such as the read pair orientation, reverse complementation and backtracking retries), and of course this is specifically designed to fit the requirements of one version of the genome alignment software. The genomeAlign() function takes the arguments described in the above table, noting that we can split the line defining the input alignments to keep things neat and tidy:

```
def genomeAlign(genomeName, genomeDir, readFiles, pairedReadFiles=None,
                outFile=None, leftTrim=0, rightTrim=0, useSOLiD=False,
                qualType='sanger', maxMismatches=2, highQualLen=28,
                pairSepRange=(0,250), showHits=1, maxHits=2,
                enforceBest=True, cpuCores=None):
```

This alignment program requires that the genome directory, containing the binary index files, is specified as an environment variable, i.e. a name needs to be associated with the location at the operating-system level, rather than being specified at run time. To get round this requirement we can simply set the operating system's environment variables within Python, hence the os.environ line which links the variable name to its value using a Python dictionary. The next operation then sets the output file if it was not passed into the function, with the default being the genome name appended with '.sam', given we are making SAM-format files.[14]

```
os.environ['BOWTIE_INDEXES'] = genomeDir

if not outFile:
  outFile = genomeName + '.sam'
```

As with the previous example we collect a list of text strings that are equivalent to the program name with the command line options required to run the program. Here the '-s' option means that the output alignment file format will be SAM, although we could adapt the function to specify something different. The readFilesStr is simply the comma-joined list of input (short-read sequence) file names.

```
cmdArgs = [ALIGNER_PATH+'bowtie', '-S']
readFilesStr = ','.join(readFiles)
```

Next is some code to automatically guess the type of the input file based on the file extension that follows the file name (from the '.' at the end). Here we use the os.path.splitext() function to split the name on the appropriate period/full-stop character to get the trailing characters. Note that this procedure assumes that all the input read files are of the same file format, given we only determine the type from the first in the list, but in reality they really ought to be the same. If the file extension is in the first pre-set list then the input reads are assumed to be in FASTA format, which requires that the '-f' option is used. Otherwise we check whether the format appears to be FASTQ, where we use the '-q' option. And if none of that is true we assume the reads are in 'raw' format, where there is simply a different one-letter sequence on each line.

```
foreName, fileType = os.path.splitext(readFiles[0])

if fileType in ('.fa','.fna','.mfa','.fasta'):
  cmdArgs.append('-f')

elif fileType in ('.fq','.fastq'):
  cmdArgs.append('-q')

else:
  cmdArgs.append('-r')
```

Next is the option for using ABI SOLiD data. Then the quality score type is set, which effectively translates the more informative text string options into '--' style options. Note here that the term 'phred' refers to the original DNA sequence quality score system, from a program of that name, which is adjusted to use the text characters as described above (ASCII with a specified offset).

---

[14] Li, H., Handsaker, B., Wysoker, A., et al.; 1000 Genome Project Data Processing Subgroup (2009). The Sequence Alignment/Map format and SAMtools. *Bioinformatics* 25(16): 2078–2079.

```
if useSOLiD:
  cmdArgs.append('-C')

if qualType == 'illumina1.3': # Phred+64
  cmdArgs.append('--phred64-quals')

elif qualType == 'solexa': # Phred+59
  cmdArgs.append('--solexa-quals')

else: # Sanger, current illumina:  Phred + 33
  cmdArgs.append('--phred33-quals')
```

The options to enforce reporting of only the best-matching genome alignment and setting the number of processing jobs are considered. Note that the number of CPU cores available on the current computer system can be determined using the standard `multiprocessing` library, which is available from Python 2.6 onward.

```
if enforceBest:
  cmdArgs.append('--best')

if not cpuCores:
  import multiprocessing
  cpuCores = multiprocessing.cpu_count()
```

The remaining, numeric, options are converted to text strings and added to the list. Here the `+=` operator is used, which is equivalent to `list.extend()` to add the contents of one list to another.

```
cmdArgs += ['-5', str(leftTrim),
            '-3', str(rightTrim),
            '-n', str(maxMismatches),
            '-l', str(highQualLen),
            '-k', str(showHits),
            '-m', str(maxHits),
            '-p', str(cpuCores),
            '--chunkmbs', '256']
```

If a second list of file names was passed to the function, specifying paired sequence reads from the two ends of the DNA fragments, then we set some further options specifically relating to read pairs. Naturally, we process the input sequence file names, as a comma-joined list like before. Then we split the input separation limits for the pairs from `pairSepRange`; these are input as separate arguments for the aligner. Note that when paired reads are used the '`-1`' and '`-2`' options are employed to set the two file names. Otherwise, i.e. at the `else`, the remaining options just refer to the one `readFilesStr`.

```
if pairedReadFiles:
  pairedReadFiles = ','.join(pairedReadFiles)
  minSep, maxSep = pairSepRange
  cmdArgs += [genomeName,
              '--minins', str(minSep),
              '--maxins', str(maxSep),
              '-1', readFilesStr,
```

```
                    '-2', pairedReadFiles,
                    outFile]

      else:
        cmdArgs += [genomeName, readFilesStr, outFile]
```

And finally at the end we invoke `call()` to run the actual external aligner program, which may take a significant time depending on the number of sequence reads and the available computational power. The output file name is returned from the function.

```
      call(cmdArgs)
      return outFile
```

All of the above can be brought together in the following example of a test, using the demonstration data supplied in the on-line material,[15] which is the single *E. coli* chromosome sequence, although in practice we will only really index the genome if it changes. Note that the test chromosome file is converted to the full (absolute) file-system path using `os.path.abspath`, which is what our indexing program requires.

```
filePath = os.path.abspath('examples/EcoliGenome.fasta')

genomeName = 'E_coli'

indexGenome(genomeName, [filePath,], 'examples')

fastqFile = 'examples/EcoliReads.fastq'

genomeAlign(genomeName, 'examples', [fastqFile], qualType='sanger')
```

The final alignment of this will produce output of the form:

```
# reads processed: 1000
# reads with at least one reported alignment: 685 (68.50%)
# reads that failed to align: 301 (30.10%)
# reads with alignments suppressed due to -m: 14 (1.40%)
Reported 685 alignments to 1 output stream(s)
```

Using the HTSeq library

Now we come to actually using the result of a short-read to genome alignment to illustrate how the results may be analysed. For this section we will be making use of the HTSeq library, which is not a standard Python module, but which should be downloaded and installed separately.[16] All of the following could be done with pure Python and its standard libraries, but by using this extra module we will be making the job much easier, and this will hopefully result in the ability to do more science. Also, we could consider using other packages like Pysam,[17] which can read genomic alignments, though overall HTSeq provides more diverse functionality.

---

[15] http://www.cambridge.org/pythonforbiology.

[16] http://pypi.python.org/pypi/HTSeq. Note that to install this you can use the Python setup tool by issuing the following at the command line: `python setup.py build; python setup.py install`.

[17] A Python wrapper for SAMtools: http://pypi.python.org/pypi/pysam.

To use the HTSeq library, once it is properly installed, we simply import it into our Python programs. The imports give the ability to read several bioinformatics file formats: FASTQ for the sequence reads, SAM for the genome alignment results and GFF files which give the genome annotation information (e.g. where the genes lie in the sequence), to which we can relate the high-throughput sequencing data. The `GenomicArray` and `GenomicInterval` imports are object classes that we will use to hold the sequence information in an efficient and easily accessible manner. Next some imports are made to make graphs with the `matplotlib` library and the usual NumPy module to handle numeric arrays with which we can do the mathematics.

```
from HTSeq import FastqReader, SAM_Reader, GFF_Reader
from HTSeq import GenomicArray, GenomicInterval

from matplotlib import pyplot
from numpy import array
```

Reading sequences from FASTQ files

A FASTQ-format sequence file is read by using the imported reader class `FastqReader` to make an object that represents the open file. It is then a simple matter to extract each of the sequence records that is represented in the file by looping through this object. Each of the `seqRes` objects, as it appears in the loop, is a `SequenceWithQualities` class from the HTSeq library, and as such it comes with lots of inbuilt functionality and some of this is demonstrated below: printing the name, the sequence itself and the reverse complement of the read (here the slice notation `[::-1]` gives the sequence in reverse relative to the main sequence).

```
fileObj = FastqReader(fastqFile)

for seqRead in fileObj:
  print(seqRead.name)
  print(seqRead.seq)
  print(seqRead.get_reverse_complement()[::-1])
```

This gives a result like:

```
r999
AGGATAATGAGGCGAGCCGGGGGAACTGAAANTGG
TCCTATTACTCCGCTCGGCCCCCTTGACTTTNACC
```

Given that these sequence records come from reading a file format that incorporates quality scores we can naturally interrogate those scores. In this example we generate a graph of the mean score along the alignment positions. The `meanQual` initially starts out as `None` and for the first sequence is set to a NumPy array of the scores `seqRead.qual`. For subsequent records the scores are then added (element by element as is the standard NumPy way) to this array, so that at the end the whole array can be divided by `numReads` to give the average value along the sequence, which is plotted with the `pyplot` library.

```
numReads = 0.0
meanQual = None
```

```
for seqRead in fileObj:
  print(seqRead.qual)

  if meanQual is None:
    meanQual = array(seqRead.qual)
  else:
    meanQual += seqRead.qual
  numReads += 1.0

if numReads:
  pyplot.plot(meanQual/numReads)
  pyplot.show()
```

Reading a genome alignment file

The next example illustrates how to read a genome alignment file, which in this case is in SAM format and could have been generated by the `genomeAlign()` defined above. Firstly, the name of the genomic alignment is specified as `alignFile` and a set is defined to collect the chromosome identifier. The SAM file reader is used to create `Alignment` objects, noting here that we choose to loop directly though the reader's output, rather than explicitly defining a variable that represents the open file object. The alignment objects can then be interrogated with the `.read` attribute to access a sequence read object.

```
alignFile = 'examples/EcoliGenomeAlign.sam'
chromosomes = set()

for alignment in SAM_Reader(alignFile):

  if alignment.aligned:
    seqRead = alignment.read
    print(seqRead.name)
    print(seqRead.seq)
```

Also, the alignment's `.iv` attribute (a `GenomicInterval`) provides a means of accessing the location that was aligned in the genome sequence:

```
genomeRegion = alignment.iv
chromo = genomeRegion.chrom
strand = genomeRegion.strand
start = genomeRegion.start
end = genomeRegion.end

chromosomes.add(chromo)
print(chromo, start, end, strand)
```

The set of chromosome identifiers which were collected above are converted into a list so that they can be used to make a `GenomicArray()`, an object that can be used to house any data that we may care to map to an efficient Python representation of the genome, which in this case will store the locations of where short-read sequences aligned. Note that the `typecode` argument dictates what kind of data is housed in the array and can take the values 'd', 'i', 'b' and 'o', respectively representing floating point numbers, integers, Booleans and general Python objects.

```
chromosomes = list(chromosomes)
hitMap = GenomicArray(chromosomes, stranded=True, typecode='i')
```

This genomic array can be filled with the results from a SAM alignment file. A check is made to ensure that the alignment was successful (i.e. that `alignment.aligned` is true) and then the genomic interval `genomeRegion` can be used to set values in the genomic array, noting that we can access the array as if it were a Python dictionary and set the value of `1` for the leading strand and `-1` otherwise.

```
for alignment in SAM_Reader(alignFile):

  if alignment.aligned:
    genomeRegion = alignment.iv

    if genomeRegion.strand == '+':
      hitMap[genomeRegion] = 1
    else:
      hitMap[genomeRegion] = -1
```

To access the results of the `hitMap` genomic array `GenomicInterval` objects are used again, although here they are created from scratch to represent the region of interest. As an example intervals are created for the first chromosome from position zero to 2 Mbp, for different strands.

```
chromo = chromosomes[0]
endPoint = 2000000
plusStrand  = GenomicInterval(chromo, 0, endPoint, '+')
minusStrand = GenomicInterval(chromo, 0, endPoint, '-')
bothStrands = GenomicInterval(chromo, 0, endPoint, '.')
```

The `hitMap` can then simply be queried with the interval object. Converting the result to a list allows the result to be easily plotted in a graph.

```
pyplot.plot(list(hitMap[plusStrand]))
pyplot.plot(list(hitMap[minusStrand]))
pyplot.show()
```

Matching sequence reads to genome data

Lastly we look at how we can relate high-throughput sequence information to annotation information that accompanies a genome. Here the Python example will focus on the data contained within a 'GFF'-formatted text file, although the HTSeq library can be used in an analogous manner for different annotation formats.

We have included an example GFF file with the on-line material that accompanies this book. Also, using the file downloading function defined above, we can obtain GFF files from the same NCBI download site as the genome sequence data.[18]

---

[18] We could also find data that relates to the Ensembl database at an alternative site: ftp://ftp.ensembl.org/pub/current_gtf/; note GTF and GFF formats are generally compatible, the former being a more restricted set of the latter.

```
remoteFileName = '/Bacteria/Escherichia_coli_536_uid58531/NC_008253.gff'
gffFile = 'examples/EcoliGenomeFeatures.gff'
downloadFile(FTP_ROOT+remoteFileName, gffFile)
```

When using genome annotation data it is especially important to make sure that it matches the genome sequence that sequence reads were mapped to. Naturally the annotations should be for the same organism, and any sub-type, as the sequence, but care also needs to be taken to ensure that the same genome assembly (as indicated by the build number) is used for both data sets. For larger eukaryotic genomes, especially if the data is relatively new, there can be some significant improvements between releases.

Assuming the GFF annotation file downloaded successfully, the `GFF_Reader()` import from the HTSeq library makes it easy to access the file data. In this case you can see the file object that is opened and the annotation data can be iterated through in a loop to get `GenomicFeature` class objects, which naturally link a genomic location to a description of the feature.

```
fileObj = GFF_Reader(gffFile)

for genomeFeature in fileObj:
```

Here we fetch the chromosome location information for the feature, by accessing the `genomeFeature.iv` attribute, a genomic interval:

```
genomeRegion = genomeFeature.iv

data = (genomeRegion.chrom,
        genomeRegion.start,
        genomeRegion.end,
        genomeRegion.strand)

print('%s %s - %s (%s)' % data)
```

Similarly there are other attributes that indicate the kind of annotation that goes along with the location. These will include features like genes, introns, exons etc.

```
data = (genomeFeature.name,
        genomeFeature.type,
        genomeFeature.source)

print('%s %s (%s)' % data)
```

One notable attribute is the `attr` dictionary, which contains a variety of information, including cross-links to other databases.

```
print(genomeFeature.attr)
```

This will yield results like the following:

```
NC_008253.1 4933963 - 4935385 (+)
CreC CDS (RefSeq)
{'locus_tag': 'ECP_4785', 'exon_number': '1', 'product': 'sensory histidine
kinase CreC','EC_number': '2.7.3.-', 'note': 'part of a two-component regulatory
system with CreB or PhoB%3B involved in catabolic regulation', 'db_xref':
'GeneID:4190641', 'transl_table': '11', 'protein_id': 'YP_672568.1'}
```

For the next example we will look again at creating `GenomicArray` objects, which will map the annotations on to a Python representation of the chromosomes. Here `geneMap` will link the genomic array directly to the annotation objects and the `genePlot` will hold numbers to plot a graph of gene locations. Note the use of the argument `'auto'`, which means that the data structure will automatically expand to represent all of the chromosome locations that we add.

```
geneMap = GenomicArray('auto', stranded=False, typecode='O')
genePlot = GenomicArray('auto', stranded=False, typecode='i')
```

To fill the `GenomicArray` data we simply use the genome region from the annotation objects as keys in a dictionary-like manner, setting the corresponding values to the `genomeFeature` object and a number. Note that we only do this if the feature is of 'gene' type. Thus, by setting `genePlot` to hold the value `1` at the feature location we will get an intermittent array of values we can turn into a rough graph to show gene positions.

```
for genomeFeature in fileObj:

  if genomeFeature.type == 'gene':

    genomeRegion = genomeFeature.iv
    geneMap[genomeRegion] = genomeFeature
    genePlot[genomeRegion] = 1
```

Once a genomic array is created we can naturally interrogate it and fetch the data we initially mapped. The `steps()` function is useful to provide a mechanism to loop through what is contained, extracting a record of the sequence region covered and whatever data we associated with that location. Here we loop through the positioned `GenomicFeature` objects (although we could have used other types of Python object).

```
for region, feature in geneMap.steps():

  if feature:

    data = (feature.name,
            region.start,
            region.end,
            feature.iv.strand)

    print('%s: %s - %s (%s)' % data)
```

If we wish to make a graph, plotting the locations with the value `1`, which were arbitrarily set for the genes, then we choose a chromosome (from an arbitrary region used above) and construct a 40,000 base `GenomicInterval` to state the region of interest. As before the region is used like a dictionary key to fetch part of the mapped array, and we can convert this to a list for graphing:

```
chromosome = genomeRegion.chrom
region = GenomicInterval(chromosome, 0, 40000, '.')
pyplot.plot(list(genePlot[region]))
pyplot.show()
```

18 Images

Contents

Biological images

Often in biology and medicine the data people use comes in the form of an image. This could be as simple as a photograph of some cells or an image that has been constructed from other data, e.g. from an MRI scan. The images that we will be discussing in this chapter, whatever their source, will be *pixmap* images, also known as *raster* images. They will be constructed as rectangular arrays of colour or grey values, the smallest square element of which we refer to as a *pixel*. We will not be considering the *vector graphics* approach to making pictures, where the data is described in terms of lines and shape outlines. Here we will concentrate on pixel arrays, the kind of image data that comes from our digital cameras and various scientific instruments.

We will deal with pixmap images in a general, slightly mathematical way. It will not matter what the image actually represents for the most part, although we will endeavour to give examples with a biological flavour. Not so long ago images would largely be acquired by using photographic film, but now the digital camera is ubiquitous, and without the need to buy expensive film a scientist can capture as many images as time and storage capacity allow. Thus the examples presented here will often have an emphasis towards automation, and if you need to write programs dealing with biological data this will allow you to construct efficient analytical pipelines.

Pixmaps

A pixmap image can be stored in a variety of different ways on a computer, such as the common file formats like JPEG, PNG or GIF. However, whatever the means of storage, which is often just a cunning way of saving space (or download bandwidth), all pixmaps can be imagined as an array of different colour values. Here the usual convention is that the first pixel

Cell sample   Cell and nucleus   Electrophoresis gel Microarray   Protein crystal

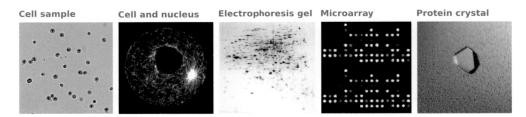

Figure 18.1 (Plate 5). Examples of a variety of different kinds of images used in biology. Shown from left to right are: a microscope image of a mammalian cell culture (courtesy Dr. Anja Winter, University of Leicester); a red-green fluorescence microscope image of an oocyte and its nucleus (courtesy Dr. Melina Schuh, MRC Laboratory of Molecular Biology); a two-dimensional electrophoresis gel of a plant proteome (courtesy Prof. Paul Dupree, University of Cambridge); an image of a DNA microarray (courtesy Karen Howarth, University of Cambridge); a protein crystal that has been grown for structure determination by X-ray crystallography (courtesy Dr. Aleksandra Watson, University of Cambridge). A black and white version of this figure will appear in some formats. For the colour version, please refer to the plate section.

(array position 0, 0) is viewed as the top left of the image, i.e. the other pixels go right and down relative to the first. A pixmap image will have no resolution as such, just one fixed size in terms of points in a matrix; how big it ends up looking is a matter for the display or printer. Each pixel element of such an array will have its own colour specification and placing these all together makes the whole image. There are several common ways of representing colour in computing, some of which are described below. The basic principle is that one or more numbers are allocated to each pixel and these describe the components or properties of the colour. It is then up to the display (or printer) to know how to interpret the pixel's values and to show the colours correctly.

A few of the more common colour models used in computing:

- Greyscale: each pixel is represented by a single value, which determines how bright it is. Zero will represent black and the maximum value will be white, with the grey shades in between. Sometimes greyscale is referred to as *luminance* (although this has a proper meaning in physics).
- RGB: represents each pixel with three numbers which specify the amount of red (R), green (G) and blue (B) component colours that are in the pixel. The mixtures of these components specify other colours. This is similar to the way that most computer screens operate.
- RGBA: this is the same as RGB, but carries an extra number for each pixel called the *alpha* (A) value, which specifies how transparent it is; this is only really useful when making things pretty and overlaying images, to say how much of the background comes through. This is certainly a form to be aware of but not something we usually have to think about too much for science.
- CMYK: represents each pixel with four numbers indicating cyan (C), magenta (M), yellow (Y) and black (K) components. This is a specification useful for printing, where the components match the colours of inks (which are better for mixing on paper than red, green and blue).
- HSV: represents each pixel in terms of hue (H), saturation (S) and value (V). The hue indicates where the pure colour lies in a rainbow (or colour wheel), the saturation specifies how colourful the pixel is compared to grey, and the value says how dark (close to black) the colour is.

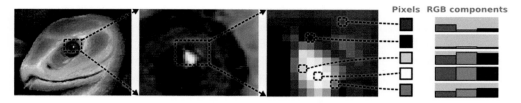

Figure 18.2 (Plate 6). An image, its component pixels and their RGB colour-space values. A section of the book cover picture is shown magnified at different levels to reveal the array of square pixels that the digital image is composed of. For the highest magnification, example pixels with different colours are selected and the component red, green and blue (RBG) values that constitute each colour are shown as histograms. A black and white version of this figure will appear in some formats. For the colour version, please refer to the plate section.

A technical aspect that will impinge on our ability to deal with images is the way that different number ranges are used in different circumstances. Thinking about RGB images, we can imagine the pixels' red, green and blue components as taking values between 0.0 (minimum) and 1.0 (maximum), and this may be convenient for us when doing mathematical manipulations. However, such components are not generally held as floating point values between zero and one, rather they are stored as integers. For example, they commonly range from zero up to 255. For RGB this means using 8 bits for each colour ($2^8 = 256$), which in turn gives rise to the whole image being described as 24-bit (8 red + 8 green + 8 blue). Naturally allowing values to be stored as larger numbers takes up more memory but allows for many more gradations, and so better colour representation. In order to interpret image data correctly we must know what this maximum value is, i.e. whether it is 8-bit, 16-bit etc., otherwise the data will be nonsense.

Once we know how image data is stored we will need to be able to manipulate it. In this chapter we will generally be working with greyscale or the RGB colour model for simplicity, but we will illustrate how to convert to and from the other representations. When performing more mathematical operations with numeric Python we will think of the pixels in an image as being elements of a matrix (a 2D array) and often we can think of the different red, green and blue components as separate layers (also called *channels*) each with a separate matrix, which are then arranged depth-wise to make the full pixmap.

Image manipulation

Much of what is covered in this chapter is about manipulation of pixmaps. Naturally when making changes we have to be mindful of biasing the scientific investigation. It is important to be objective so that we show what is actually there, not just what we expect or what looks pretty. Of course this is not different in principle to any other scientific data, but it is often very easy to manipulate an image and forget to keep the original data. Hence we encourage keeping the original data in important situations. This also allows the development of better analytic methods in the future.

Sometimes when manipulating images we will be working with the pixel data directly in numeric Python. However, such low-level coding is not always required. When performing common operations on an image, such as adjusting contrast or colour balance, we can make good use of existing high-level graphics libraries that work with Python such as the

Python Imaging Library[1] (PIL) and the Python wrapper to *Imagemagick*.[2] Although these libraries are both worth considering we will focus mainly on PIL in this chapter. Generally we will use PIL for high-level functions, especially the ability to display, load and save images, and then use numeric Python for the detailed work. However, for some operations we will illustrate using both PIL and numeric Python and leave it up to the reader to choose which is most convenient or useful.

Basic image operations

The first Python examples in this chapter will illustrate how we can use PIL to get hold of pixmaps from the data stored in various kinds of image file. The data can then be manipulated, if required, and saved back again, potentially to a different type of image file. When working with images we often simply need to convert from one type of image file to another, given computer programs are sometimes particular about the format. Also, by converting we can sometimes improve storage efficiency, e.g. going from an uncompressed TIFF to JPEG.

Python Imaging Library

To work with PIL, the following examples naturally assume that the library is installed.[3] Then to begin we import the `Image` module, which will allow us to construct `Image` class objects.[4]. Inbuilt into this kind of object are lots of useful operations that we can access directly for the image just by calling a function (i.e. a bound method) on the object. The image module can use the `.open()` method to load a file from disk and make an `Image` object. Note that the PIL function automatically guesses at the type of file at load time so that we only have to specify a file name. The example files we are working with here are available in the downloadable data that supports this book.

```
from PIL import Image

img = Image.open('examples/Cells.jpg')
```

With the image object made we can access its properties, and most importantly call the `.show()` method to display it on screen.

```
print(img.size)
print(img.mode)

img.show()
```

A given image object can be saved back to a file using several different file formats. In the example below we use PNG and GIF format. At the time of writing PIL can use any of the

---

[1] As of 2013, PIL has not been ported to Python 3, but an alternative implementation, Pillow, has been: http://python-imaging.github.io.

[2] http://www.imagemagick.org/.

[3] Download PIL via links at http://www.cambridge.org/pythonforbiology.

[4] These `Image` objects sometimes cause problems if they are associated with a local variable, e.g. inside a function, and get *garbage collected* sooner than expected. If this occurs then the general solution is to make sure that the image object has a reference from another object that will persist in memory, for example by putting it in a non-local list or adding it as a `self.` attribute.

following formats: BMP, DCX, EPS, GIF, IM, JPEG, PCD, PDF, PNG, PPM, PSD, TIFF, XBM and XPM. Not all of these will store images in the same way. For the common web formats, JPEG gives the smallest files, but will change the data and may lose quality (it uses *lossy compression*), PNG will preserve all the data but the files will be larger (*lossless compression*). GIF is similar to PNG, but can only handle 256 colours at once (although this *palette* of colours can be chosen from a larger set) so if the image has more colours saving as GIF will lose information.

```
img.save('Cells.png', 'PNG')
```

If we need to have an image which describes its pixel values in a different way we can use the `.convert()` function prior to saving, or some other operation. It is notable that converting to greyscale (code `'L'`) in PIL takes account of the sensitivity of the human eye to colours where, for the same physical intensity, green seems brightest followed by red and then blue. Thus, such a greyscale conversion preserves aesthetic brightness, but this will be a biased average of the pixel values and may not be what we want scientifically.

```
img.convert('CMYK')    #  Cyan, Magenta, Yellow, blacK
img.convert('L')       #  Luminance = greyscale
```

Next we will run though a few of the more basic ways of changing images, which we can admire by using `.show()` or by saving and viewing in another program. The `.crop()` method chops the edges off the pixmap. We need to specify the rectangle to use as left, top, right and bottom edges (in order) and pass these values as arguments in a tuple, not separately. The convention used in the `Image` object is that the pixel with positional indices `(0,0)` is at the top left. Accordingly, in the example the right and bottom edge points are calculated by subtracting from the original width and height. Also, note that we are sending the result back to a variable called `img`, thus we are overwriting the original data, but of course we are free to use a different name if required.

```
w, h = img.size
img = img.crop((10, 10, w-10, h-10))
```

Another easy manipulation is rotation, which is inbuilt. Here we specify the angle of rotation in degrees and then save the rotated pixmap. As before, we are overwriting `img` with new data.

```
img = img.rotate(270)
img.save('CellsAdj.png', 'PNG')
```

To change the size of an image we have one of two options: the first method is `.resize()`, which gives back a new image, preserving the one we passed in. For this operation we need to say how the image will be resized (how to combine the original pixels together to make the new array). Thus we enter what the new width and height will be: here half of the original values. Also, we can optionally supply a resizing method, which specifies which algorithm will be used. The example uses the `Image.ANTIALIAS` option (coming directly from the module `Image`, not the object). Antialiasing is commonly what you would want for making smaller images, though `.BILINEAR`, `.BICUBIC` or `.NEAREST` can all be used for resizing in general.

```
img2 = img.resize((w/2, h/2), Image.ANTIALIAS)
img2.save('CellsHalfSize.png', 'PNG')
```

An alternative way of resizing images is to make smaller preview versions called *thumbnails*. Unlike the previous examples, the `thumbnail()` function actually changes the image in-place. If we do not want the original to be affected we need to make a copy first, which is fortunately easy:

```
img2 = img.copy()
img2.thumbnail((50, 50), Image.ANTIALIAS)
img2.save('CellsThumb.png', 'PNG')
```

Using NumPy for images

Next we move on from the inbuilt PIL methods and place the image data into a numeric Python `array`. Naturally this array will have the same width and height as the image, i.e. we have an array element for each pixel. The array will have different depths depending on the kind of image data that is being interpreted. For example, if the image is greyscale ('black and white'), then we need an array that is only one element deep, to contain the brightness value. If the image is RGB, then we need an array that is three deep; you can imagine this pixmap as consisting of three stacked planes, for red, green and blue layers respectively. Or put another way, a pixel that makes an RGB element is a vector of the form `(red, green, blue)`.

For the following example we will first make the NumPy imports that will be needed. Many of these will be familiar, but `uint8` warrants some explanation. The `uint8` object represents a data type that specifies *unsigned*[5] 8-bit integer numbers. Most integers in Python will be 32-bit and have signs. In essence a `uint8` number will only go from 0 to 255, and this is how many RGB image values are actually stored. To interpret the pixmap data we will need to use this data type, rather than the regular Python number types.

```
from numpy import array, dot, dstack, ones, random, uint8, zeros
```

To make an image from scratch we make an array of the required size. Here we specify a height and width of 200 pixels and use a depth of three, because we want to make RGB pixmaps. Note that when we make the array first we use height `h`, then width `w`, then depth `d`. This may seem counterintuitive to some people, given the custom for using x, y order in coordinates and stating dimensions as width × height. However, putting height first better reflects the way that pixels are stored in image files, and so matches what we will pass on to PIL at the end to save the image.

```
h = 200
w = 200
d = 3
```

The pixmap is constructed as an array using these sizes, and the values that we put in the array will specify the colour of the pixels. All zeros will give rise to black, all ones will become white and `random` could be anything.

```
pixmap = zeros((h, w, d)) # black
pixmap = ones((h, w, d)) # white
pixmap = random.random((h, w, d)) # random colours
pixmap *= 255
```

---

[5] No positive or negative sign.

The arrays constructed above contain values that go from zero to one, but the RGB values will go from 0 to 255. Thus, in the above example $*= 255$ multiplies each value in the array by 255.

Also, the Python arrays usually contain standard integers or floating point numbers, not the required 8-bit variety, so we have to explicitly convert the data type to `uint8`. The converted array can be directly interpreted by PIL to make an `Image` object which we can show on screen, save to file or whatever. Here we encapsulate the conversion operations into a function, `pixmapToImage`, so we can use it later. In the function we check to make sure pixel values do not exceed 255, scaling them back if they do:

```
def pixmapToImage(pixmap, mode='RGB'):

  if pixmap.max() > 255:
    pixmap *= 255.0 / pixmap.max()

  pixmap = array(pixmap, uint8)
  img = Image.fromarray(pixmap, mode)

  return img

img1 = pixmapToImage(pixmap)
img1.show()
```

Next we will consider the construction of images by combining separate matrices for the red, green and blue image components. So, for example, if we wanted to make a pure yellow image, which has maximum red and green components, we can do the following to make component matrices of the same size and use `dstack` to combine them in the right manner:

```
size = (h,w)
redMatrix   = ones(size)
greenMatrix = ones(size)
blueMatrix  = zeros(size)

pixmap = dstack([redMatrix, greenMatrix, blueMatrix])
pixmap *= 255

img1 = pixmapToImage(pixmap)
img1.show()
```

In order to copy existing PIL image data into a numeric Python `array` the handy `.getdata()` method can be used, which is built into all PIL `Image` objects. It is notable that the `unit8` is used again so that the numbers in the image are of the unsigned 8-bit data type. Also, we put in a `.convert()` step to check that the image is of the RGB type. However, it should be noted that our pixmap/image conversion functions should perform many more checks if they were used in real-world scenarios. Initially the `data` will be a plain list of colour data; all pixels will effectively be in a long line. To reconstruct the original pixmap's dimensions `reshape` can be used to rearrange the elements, remembering that the width and height are used in the opposite order to what `.size` gives.

```
def imageToPixmapRGB(img):

  img2 = img.convert('RGB')
  w, h = img2.size
  data = img2.getdata()
```

```
pixmap = array(data, uint8)
pixmap = pixmap.reshape((h,w,3))

return pixmap
```

The loaded image pixmap can now be manipulated as required, although it will be convenient to use PIL to actually visualise what is happening.

As an example of working with image data stored in arrays, below we perform an operation which may be useful to red-green colour-blind people. The notion here is that some types of biological images are bright red and green, because of special fluorescent marker compounds, and must be converted (here to yellow and blue) so that a red-green colour-blind person can view them effectively (see Figure 18.3 for an illustration). The image is loaded and converted to a pixmap in the manner described:

```
img = Image.open('examples/CellNucleusRedGreen.png')
pixmap = imageToPixmapRGB(img)
```

Here it is notable that there is an alternative approach using SciPy (a scientific Python package which is often installed alongside NumPy): the `imread()` function is provided in the `ndimage` module, to create an array directly from the file data. Though it may be convenient to avoid PIL, without an `Image` object we cannot invoke `show()` to easily visualise the loaded data. Also, doing things this way will include the 'alpha' transparency layer, so to get the same kind of array as above we take a slice of the first three (red, green and blue) colour layers with `[:,:,:3]`.

```
from scipy import ndimage
pixmap = ndimage.imread('examples/CellNucleusRedGreen.png')
pixmap = pixmap[:,:,:3]
```

The next task is to define a colour transformation matrix. The way to think of this is that the three rows dictate how to modify red, green and blue respectively and the `[r, g, b]` values within each row specify what the colour will become. Accordingly, in `transform` the first row specifies that the red channel will be transformed into equal red and green (i.e. yellow), the

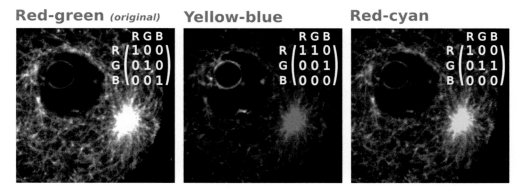

Figure 18.3 (Plate 7). Matrix transformations of pixmap colours. A red-green coloured (i.e. two-channel) fluorescence microscope image of a cell is shown alongside colour-adjusted yellow-blue and red-cyan versions. Inset in each image is the RGB colour transformation matrix relative to the red-green image. A black and white version of this figure will appear in some formats. For the colour version, please refer to the plate section.

second row means that the green channel will become blue and the third row is all zero and so any original blue is removed entirely (not that there is much in the test image). The transformation is applied using a dot product, remembering that the order of the arguments is important.

```
transform = array([[1.0, 1.0, 0.0],
                    [0.0, 0.0, 1.0],
                    [0.0, 0.0, 0.0]])

pixmap2 = dot(pixmap, transform)

img2 = pixmapToImage(pixmap2)
img2.show()
```

Pixmap data can be manipulated as floating point numbers, rather than just integers in the range 0 to 255. In the next example the pixmap is converted to an array of `float` data type, so the numbers fall in the range `0.0` to `255.0`. Dividing by `255.0` the range becomes `0.0` to `1.0`, which is handy for the next step, where the values of the pixmap are squared (an element-by-element operation). Squaring colour values in the range zero to one is a convenient way of changing the brightness of the image; values will move towards zero, but the upper limit is still `1.0`. With the brightness adjusted the array can be converted back into the range `0` to `255` to go back to PIL.

```
pixmap = array(pixmap, float)
pixmap /= 255.0
pixmap = pixmap ** 2
pixmap = array(255*pixmap, uint8)

img2 = pixmapToImage(pixmap)
img2.show()
```

Adjustments and filters

The previous example which adjusts the brightness of a pixmap leads neatly into thinking about more general ways that we can change the global properties of an image. Helpfully many kinds of adjustment, for contrast, brightness, sharpness etc., can be made directly in PIL. This is more convenient than doing things in numeric Python arrays. Nonetheless, we will go on to show some of the equivalent operations with arrays, because this gives more control and teaches something about how the adjustments work at a low level.

The `ImageEnhance` module

To adjust PIL images the `ImageEnhance` module is imported, as well as the `Image` used earlier, and then an example image loaded to experiment with:

```
from PIL import Image, ImageEnhance
img = Image.open('examples/Cells.jpg')
```

Firstly, we will use this image enhancement module to change the contrast in the image. The way that this works with PIL is to create a processing object (here `processObj`) for a given

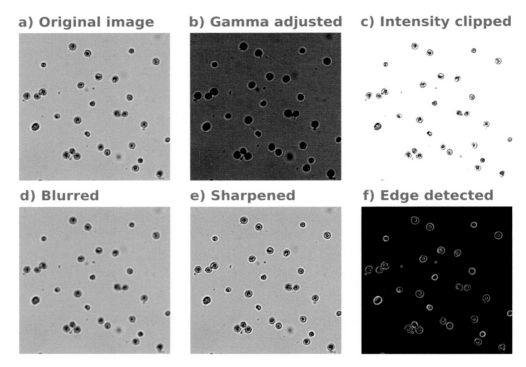

Figure 18.4. Example results from common image-processing operations. An original microscope image of mammalian cells is shown alongside five adjusted versions created with image-processing routines presented here. The gamma adjustment used a factor of 4.0. The blurred, sharpened and edge-detected images were generated by matrix convolution. The intensity clipping set the lightest pixels to mid-grey and then normalised the intensities to the full black-white range.

image, which is then called with the parameters needed to control the image adjustment. Here we make a processing object to adjust the contrast and then call its .enhance() method with a value of 2.0. For all of the kinds of ImageEnhance enhancement calls, a value of 1.0 will preserve the image as it was. Thus here we are increasing the contrast:

```
processObj = ImageEnhance.Contrast(img)
img2 = processObj.enhance(2.0)
img2.show()
```

If we want to perform a different kind of adjustment we need to make a different processing object. The next example is for image sharpness (changing the local contrast at the edges within an image), where we can make things sharper with a value greater than one, or more blurred with a value less than one:

```
processObj = ImageEnhance.Sharpness(img)

img2Sharp = processObj.enhance(4.0)
img2Sharp.show()

imgBlur = processObj.enhance(0.5)
imgBlur.show()
```

Likewise, we can adjust brightness and overall colourfulness, noting that in the next examples we use the method .enhance() straight away, and don't make an explicit processing object; the object is still made after .Brightness(img), but it is not given a named variable.

```
imgBrighter = ImageEnhance.Brightness(img).enhance(0.5)
imgDull = ImageEnhance.Color(img).enhance(0.1)
```

Intensity adjustments using NumPy

Moving to numeric Python, we will show the same kinds of image adjustment for pixmap arrays and also show how things can be taken further. The examples will be constructed as Python functions and testing is demonstrated after the function definitions. As usual, the required NumPy imports are made upfront:

```
from numpy import array, dstack, exp, mgrid, sqrt, uint8
```

The mgrid object imported here may not be familiar. This is used to quickly create arrays that can be used together to form a grid of row and column numbers. For example, mgrid [0:3,0:3] gives the following sub-arrays: [[0,0,0],[1,1,1],[2,2,2]] and [[0,1,2], [0,1,2],[0,1,2]]. This is handy because the first sub-array gives the row number of the elements and the second gives the column number. This can be thought of as being analogous to using the regular range() function, for arrays.

The first example function controls the brightness of an object using what is known as *gamma correction* (see Figure 18.4b). This sort of correction is often used to adjust an image so that it can be presented by different kinds of display, accounting for different innate responses to brightness. The mathematical operation used is very simple: the pixmap values (albeit greyscale, RGB etc.) are converted into the range 0.0 to 1.0 and all the values are raised to the power of gamma. The effect is that if gamma is greater than one the image will look darker, and below one brighter. Taking the mid point, 0.5 as an example, gamma=2.0 changes this to 0.25 and gamma=0.5 gives 0.707. Adjusting the brightness in this way still preserves the extremes (i.e. $0^\gamma = 0$ and $1^\gamma = 1$) but the 'curve' of intermediate values is distorted.

The function takes a pixmap array and the gamma factor as input. Inside the function, the pixmap (which we are assuming takes values from 0 to 255) is scaled, so the maximum possible value is 1.0. The gamma power is applied, and the values are then re-scaled back to their original range. Because we made a new array in the function we pass this back at the return.

```
def gammaAdjust(pixmap, gamma=1.0):

  pixmap = array(pixmap, float)/255.0
  pixmap = pixmap ** gamma
  pixmap *= 255

  return pixmap
```

The next brightness-related function is designed to automatically adjust the values in the pixmap so that they are 'normalised' to take up the full range. So, for example, if we had a dull grey image, with no black or white, the darkest shade would be moved to black (0) and the brightest to white (255). The function works by first subtracting the smallest (.min()) value in the pixmap from all the elements, so that the minimum is set to zero. Next the maximum value is set to be 255, by dividing by the adjusted pixmap's maximum (giving a 0.0 to 1.0 range initially) and then multiplying by `255.0`. Note that we deliberately use the floating point number `255.0` so that the division also gives a floating point result and also that we guard against dividing by a maximum of zero (in an all-black image). The scaled pixmap is then passed back at the end.

```
def normalisePixmap(pixmap):

  pixmap -= pixmap.min()
  maxVal = pixmap.max()

  if maxVal > 0:
    pixmap *= 255.0 / maxVal

  return pixmap
```

The next example is a little more complicated. It sets the minimum and maximum brightness values in an image by clipping, i.e. it only adjusts the edges of the brightness range and doesn't affect the middle. For greyscale images the `clipPixmapValues` function can be used with `normalisePixmap` above to stretch the values to black and white again, thus removing any dark or light image detail, as illustrated in Figure 18.4c.

The function is defined as taking a pixmap and two threshold values. These thresholds have default values so that if they are not set the image does not change, i.e. 0 and 255 are the normal limits and all values will lie between. In the function the pixmap is first copied, so we don't affect the original. Then we define `grey`, which will be a greyscale pixmap (a map of the brightness) by taking an average over all the colour values, i.e. in the depth dimension of the pixmap, hence `axis=2`.[6] It should be noted that, because we take an average of colours, individual red, green and blue components may lie outside the thresholds, so in essence the clipping is according to how close a pixel is to black or white.

With the `grey` pixmap defined, we set any limiting values, first minimum then maximum. In both cases we define `boolArray`, which contains an array of `True` and `False` values depending on whether the test condition was met: if the intensity values of the elements were smaller or larger than the threshold. The arrays of truth values are converted to indices with `.nonzero()`, which pulls out the array coordinates (row and column) of the `True` values. These indices are the ones that are to be changed, and are simply used to set those values in the pixmap to the specified limit.

```
def clipPixmapValues(pixmap, minimum=0, maximum=255):

  pixmap2 = pixmap.copy()
  grey = pixmap2.mean(axis=2)

  boolArray = grey < minimum
  indices = boolArray.nonzero()
  pixmap2[indices] = minimum

  boolArray = grey > maximum
  indices = boolArray.nonzero()
  pixmap2[indices] = maximum

  return pixmap2
```

---

[6] The array axis indices are 0 for the row/height, 1 for column/width and 2 for colour values.

Note that an alternative way of clipping the values of a bitmap would be to use the `.clip()` function of NumPy arrays, e.g:

```
minimum, maximum = 64, 192
pixmap2 = pixmap.clip(minimum, maximum)
```

In contrast to `clipPixmapValues()` this will limit the values in the colour layers separately, rather than the combined, average signals. Naturally which function is more useful will depend on the context.

Ancillary to the above functions that adjust brightness values, it is commonplace to look at a histogram of the values to see what their distribution is. This is a good way of looking at the statistical effect of the operations, and also allows people to make intelligent choices when using thresholds, e.g. to separate signal from noise, or foreground from background. First a grey pixmap of brightness is made by averaging over the depth (colour) axis. This array is then flattened to a one-dimensional array and converted to a regular Python list. This list is passed to the `pyplot.hist()` function from `matplotlib` to make a histogram with 256 bits (or we could use a smaller number for less detail).

```
def showHistogram(pixmap):

  grey = pixmap.mean(axis=2)
  values = grey.flatten().tolist()

  from matplotlib import pyplot

  pyplot.hist(values, 256)
  pyplot.show()
```

The above functions can be tried with a test image, using the PIL `Image` object to take care of loading and display, as discussed previously.

```
from PIL import Image

img = Image.open('examples/Cells.jpg')
pixmap = imageToPixmapRGB(img)

showHistogram(pixmap)

pixmap2 = gammaAdjust(pixmap, 0.7)
pixmap3 = clipPixmapValues(pixmap2, 0, 145)
pixmap4 = normalisePixmap(pixmap3)

pixmapToImage(pixmap4, mode='L').show()
```

Convolving image filters

The next examples move on from pixel brightness to the concept of *filters*. In this context a filter is a way of transforming an image, combining original pixel values together to make new values. A simple example of this is the blurring of an image (see Figure 18.4d). The blurred version of a pixel is constructed by setting its value to be an average of the surrounding pixels. The filters used will be described as matrices. For example, the $3{\times}3$ matrix `[[1,1,1],` `[1,8,1], [1,1,1]]`, can be used to blur an image. The way to think of this is that the centre of the matrix (which has the value `8` here) represents the position of the original pixel, and the other elements are the square of pixels that surrounds it. The values in the filter matrix dictate how much influence each of the pixels has when used to create a new pixel. For the 3×3 example with 1 at the edges and 8 in the centre the new pixel will be an average (in terms of RGB or whatever) of the eight surrounding pixels and the central one, which here has as much influence as all the rest combined. As a consequence the new pixmap will be a blurred version of the original; the pixel values will spread to their neighbours slightly. When applying a filter matrix it is either normalised (elements sum to one) or the image is normalised afterwards so that the final pixel value cannot exceed the image maximum.

Many of the filtering and processing examples that we will illustrate have implementations in the `scipy.ndimage` module.[7] This module is well worth considering, especially in view of its speed and large range of functionality. For example, instead of the Gaussian blurring example that we give below, which uses NumPy alone, the `scipy.ndimage.filters.gaussian_filter()` function can be used instead. However, in this chapter we will mostly use NumPy, to better illustrate what is happening at a low level, and only use SciPy a little for some generic functionality.

The actual application of the filter is mathematically a *convolution*, which we can perform using the handy `ndimage.convolve()` function from the `scipy` module.[8] The convolution operation takes two arrays, which in this instance are the image `pixmap` and the filter `matrix`. One caveat to the `convolve` function is that both input arrays must have the same number of axes (dimensions), so when we are applying a flat matrix (2D) to an RGB or CMYK pixmap (3D) we convolve the 2D matrix separately with each of the colour layers. The check for this is simple given the `.ndim` attribute of the arrays: we insist that the `matrix` is 2D (triggering an exception if not) and that the pixmap is either 2D or 3D. If the pixmap is 2D we can perform the convolution directly. Otherwise, for a 3D pixmap the colour components are convolved separately (extracting each layer with slice notation `pixmap[:,:,i]`) and the transformed colour layers are then stacked in the usual way (depth means colour) and returned from the function as a complete pixmap array.

For the convolution the `mode` can be specified to determine how the limits of the arrays, where the filter would overlap the pixmap edge, are treated. By default this mode is `'reflect'`,[9] which means that the image is effectively extended by using a mirror image at the edge. This kind of edge treatment can introduce processing artefacts, but it at least keeps the size of the output array the same as the input.

```
from scipy import signal

def convolveMatrix2D(pixmap, matrix, mode='reflect'):

  matrix = array(matrix)
  if matrix.ndim != 2:
    raise Exception('Convolution matrix must be 2D')
```

---

[7] SciPy is typically installed with NumPy.
[8] SciPy also has a `signal.convolve()` function, which could be used, though this doesn't deal with the edges of images so well.
[9] The other options include 'constant', which uses a user-defined (`cval=`) constant value outside the image edge; 'nearest', which extends the edge values outwards; and 'wrap', which takes values from the opposite edge.

```
if pixmap.ndim not in (2,3):
  raise Exception('Pixmap must be 2D or 3D')

if pixmap.ndim == 2:
  pixmap2 = ndimage.convolve(pixmap, matrix, mode=mode)

else:
  layers = []
  for i in range(3):
    layer = ndimage.convolve(pixmap[:,:,i], matrix, mode=mode)
    layers.append(layer)

  pixmap2 = dstack(layers)

return pixmap2
```

To process an image with a filter we simply call the above with pixmap and matrix arguments:

```
matrix = [[1, 1, 1],
          [1, 8, 1],
          [1, 1, 1]]

pixmapBlur = convolveMatrix2D(pixmap, matrix)
```

To view the result we need to normalise the image, given that the pixmap was convolved in a way that increased the intensity values by a factor of 16 (8 from the original intensity plus 1 from each of eight neighbouring pixels, according to matrix). Hence, we divide the pixmap so that the intensities of the pixels are put back in the original range.

```
pixmapBlur /= array(matrix).sum()
pixmapToImage(pixmapBlur).show()
```

Sharpen, blur and edge-detection filters

Although we can use any filtering matrix, there are several common operations that are applied to pixmaps, so we will encapsulate some of these in functions. The first example of these sharpens an image, as illustrated in Figure 18.4e. It uses a filter matrix that accentuates the differences between pixels.

```
def sharpenPixmap(pixmap):

  matrix = [[-1,-1,-1],
            [-1, 8, -1],
            [-1,-1,-1]]
```

The procedure is to convert the input pixmap into a grey (brightness) pixmap. The grey pixmap is then convolved with the filter matrix, which increases the contrast at the edges of features (where there are changes in brightness), and normalised to use the full range: 0 to 255.

```
  grey = pixmap.mean(axis=2)

  pixmapEdge = convolveMatrix2D(grey, matrix)
  normalisePixmap(pixmapEdge)
```

The grey pixmap with the enhanced edges has its values centred on the average brightness. So, for example, if the average brightness of `pixmapEdge` is 127, the range of values changes from 0...255 to −127...128. These centred values, either side of zero, represent how much adjustment we will apply to sharpen the original image. Before making the adjustment `pixmapEdge` is stacked so that it is three layers deep, and thus will operate on red, green and blue.

```
pixmapEdge -= pixmapEdge.mean()
pixmapEdge = dstack([pixmapEdge, pixmapEdge, pixmapEdge])
```

The new, sharpened image is created by adding the pixmap edge adjustment to the original pixmap. With the pixels adjusted the `clip` function (inbuilt into NumPy arrays) is used to make sure that adding the pixmaps does not exceed the limits of 0 and 255.

```
pixmapSharp = pixmap + pixmapEdge
pixmapSharp = pixmapSharp.clip(0, 255)

return pixmapSharp
```

The next example is the Gaussian filter, which blurs pixels with a weighting that has a normal ('bell curve') distribution (see Figure 22.4 for an illustration). For this, two values are passed in: `r` is the half-width of the filter excluding the centre and `sigma` is the amount of spread in the distribution. These parameters respectively control the size and strength of the blur. Larger filters with wider distributions (i.e. influence away from the centre) will give more blurring. It is notable that the `mgrid` object is used to give a range of initial grid values for the filter, specifying the separation of each point from the centre in terms of rows and columns; this is similar to using `range()` to generate a list.

```
def gaussFilter(pixmap, r=2, sigma=1.4):

  x, y = mgrid[-r:r+1, -r:r+1]
```

The Gaussian function is applied by taking the row and column values (`x` and `y`), squaring them, scaling by two times `sigma` squared and finally taking the negative exponent of the sum. The exact centre row and column will be zero and so the exponent will be at a maximum here, but the further `x` and `y` row and column values are from the centre the smaller the value is.

```
  s2 = 2.0 * sigma * sigma

  x2 = x * x / s2
  y2 = y * y / s2

  matrix = exp( -(x2 + y2))
  matrix /= matrix.sum()
```

Once the filter matrix is defined it is applied to the pixmap using convolution, to each of the colour components.

```
  pixmap2 = convolveMatrix2D(pixmap, matrix)

  return pixmap2
```

The final filter example is for edge detection and uses what is known as the *Sobel operator.* In essence this is a filter that detects the intensity gradient between nearby pixels (see

Figure 18.4f). It is applied horizontally, vertically or in both directions and gives bright pixels at those edges. As can be seen in the Python code the filter is a 3×3 matrix where there is a line of negative numbers, then zeros, then positive numbers. This matrix is transposed to switch between horizontal and vertical operations. The matrix means that, for a given orientation, a transformed pixel has none of its original value, but rather a value which represents the difference between values on either side.

```
def sobelFilter(pixmap):

  matrix = array([[-1, 0, 1],
                  [-2, 0, 2],
                  [-1, 0, 1]])
```

The Sobel filter matrix is applied to the grey average of the input pixmap. This is done twice for both orientations so we get two edge maps.

```
  grey = pixmap.mean(axis=2)
  edgeX = convolveMatrix2D(grey, matrix)
  edgeY = convolveMatrix2D(grey, matrix.T)
```

The final pixmap of edges is then a combination of horizontal and vertical edge maps. Taking the square root of the sum of the squares of the two edge maps means the values will always be positive; it won't make a difference between an edge going from light to dark or dark to light in an image. The edge-detected pixmap is also normalised so we can see the full range of values and finally it is returned from the function.

```
  pixmap2 = sqrt(edgeX * edgeX + edgeY * edgeY)
  normalisePixmap(pixmap2) # Put min, max at 0, 255

  return pixmap2
```

The filter functions can all be tested with the example image, using `Image.show()` to see the results, after the appropriate array conversions. Note that for the `sobelFilter()` output we pass the `'L'` mode to the PIL conversion function because it is a greyscale image, not RGB.

```
from PIL import Image
img = Image.open('examples/Cells.jpg')
pixmap = imageToPixmapRGB(img)

pixmap = sharpenPixmap(pixmap)
pixmapToImage(pixmap).show()

pixmap = gaussFilter(pixmap)

pixmapGrey = sobelFilter(pixmap)
pixmapToImage(pixmapGrey, mode='L').show()
```

In the above examples we have demonstrated using Python functions, rather than classes (our own kind of Python object), to keep things simple. However, custom object classes can be really convenient when you know what you're doing. So, for the image examples the programmer may consider making a bespoke `Pixmap` class (or whatever name seems best). This could have the ability to work with PIL automatically, doing the right array conversions and perform common operations, i.e. using `Pixmap.sobelFilter()` methods etc.

Feature detection

Lastly in this chapter we show a practical example that aims to automatically extract information about the physical objects that are represented by an image. This touches lightly on the field of *image recognition*, although what we show is simple compared to state-of-the art techniques. Nonetheless it aims to give a basic idea of what kind of thing is possible. The objective will be to count the number of cells in a digital photograph taken using a microscope, which is a fairly routine data-gathering task in biology.

Counting cells

Initially the image is loaded and converted to an array representing the pixmap. Then we apply the Gaussian filter (with default parameters) to blur the image slightly, assigning the result to `pixmap2` to keep the original `pixmap`. The blurring acts to remove the small-scale components of the image; this reduces image noise but does not significantly affect the images of the cells. Then we apply the Sobel edge-detection filter to the image, and normalise. This makes a greyscale image of just the outlines, which we can inspect. An alternative at this point would be to convert the image into black and white (only) using an intensity threshold; however, edge detection will work better where the background colour of the photograph is uneven.

```
from PIL import Image

img = Image.open('examples/Cells.jpg')

pixmap  =   imageToPixmapRGB(img)
pixmap2 = gaussFilter(pixmap)
pixmap2 = sobelFilter(pixmap2)

normalisePixmap(pixmap2)

pixmapToImage(pixmap2).show()
```

Next the `pixmapCluster` function is constructed, which will analyse our pre-processed image by clustering the bright pixels (the edges) so that we can identify blobs that represent cells. The blobs can then be analysed to select those of the required size and shape etc.

Before the clustering a helper function is defined which will find the neighbours of a pixel, investigating those above, below, to the left and to the right, and checking whether each is present in a pre-specified set of points. This `points` set will represent all the bright pixels that come from the edge-detection step. The `check` list specifies the neighbouring locations relative to the input `point` to test, and `neighbours` is the list from among these that are acceptable because they are in `points`. Note that some of the checked positions will be off the edge of the pixmap, but this does not matter because they would never be found in `points` in the first place.

```
def getNeighbours(point, points):

  i, j = point
  check = [(i-1, j), (i+1, j),
           (i, j-1), (i, j+1)]

  neighbours = [p for p in check if p in points]

  return neighbours
```

Next comes the main pixel clustering function. The details of this will not be described here because it is very similar to the simpleCluster() function described fully in Chapter 23. Essentially a threshold value is used to get a list of bright pixel points (using the same indexing strategy as clipPixmap()). These are then grouped into clusters according to whether they are deemed to be neighbours, as determined in getNeighbours. Comparing to simpleCluster() the key differences are that we are working directly with the pixel objects, rather than via indices, and that the neighbour-detecting algorithm does not need to search all pairs of data points; here only a local area of the pixmap needs to be inspected, which is very much quicker. At the end of the clustering, clusters, a list of lists, is passed back, where each sub-list represents all the pixel points (x, y locations) in each detected blob.

```
def brightPixelCluster(pixmap, threshold=60):

  boolArray = pixmap > threshold
  indices = array(boolArray.nonzero()).T
  points = set([tuple(point) for point in indices])

  clusters = []
  pool = set(points)
  clustered = set()

  while pool:
    pointA = pool.pop()
    neighbours = getNeighbours(pointA, points)

    cluster = []
    cluster.append(pointA)
    clustered.add(pointA)

    pool2 = set(neighbours)
    while pool2:
      pointB = pool2.pop()

      if pointB in pool:
        pool.remove(pointB)
        neighbours2 = getNeighbours(pointB, points)
        pool2.update(neighbours2)
        cluster.append(pointB)

    clusters.append(cluster)

  return clusters
```

The bright pixel clustering may then be used on a pre-processed pixmap with highlighted edges. Each cluster of pixels will represent a blob, which we can group into large, medium and small varieties. Here the blob size thresholds were determined by looking at a histogram of the number of points in the clusters:

```
clusters = brightPixelCluster(pixmap2)
sizes = [len(c) for c in clusters]
from matplotlib import pyplot
pyplot.hist(sizes, 40)
pyplot.show()
```

Then the clusters were grouped by size and placed in separate lists for reporting:

```
smallBlobs = []
mediumBlobs = []
bigBlobs = []

for cluster in clusters:
  n = len(cluster)

  if n < 80:
    smallBlobs.append(cluster)
  elif n < 320:
    mediumBlobs.append(cluster)
  else:
    bigBlobs.append(cluster)
print('Found %d small blobs' % len(smallBlobs))
print('Found %d medium blobs' % len(mediumBlobs))
print('Found %d big blobs' % len(bigBlobs))
```

To visualise the clustering results we will add colour codes to a grey version of the original image. Note that the `grey` pixmap has had its edges removed by slicing (`[2:-2, 2:-2]`) because the pre-processed `pixmap2` lost two edge points when it went though the convolution filters. We could improve the filtering process to deal with edges better if required, for example, by extending a pixmap with copied data so that it retains its original size after filtering.

The grey pixmap is stacked three layers deep so we can make an RGB image with the colour codes:

```
grey = pixmap.mean(axis=2)[2:-2, 2:-2]
colorMap = dstack([grey, grey, grey])
```

A list of colours containing (`red`, `green`, `blue`) arrays and the corresponding blob data is constructed:

```
colors = [(255, 0, 0), (255, 255, 0), (0, 0, 255)]
categories = [smallBlobs, mediumBlobs, bigBlobs]
```

Then by going through the clusters in each category we can colour the initially grey pixmap. Each `cluster` contains a list of row and column locations within the pixmap, and by extracting these into two separate lists (x, y) we have a means of selecting a subset of the `colorMap` and setting the colour of blob points to reflect the category. Once it is coloured, we can admire our handiwork with `Image.show()`.

```
for i, blobs in enumerate(categories):
  color = colors[i]

  for cluster in blobs:
    x, y = zip(*cluster)

    colorMap[x,y] = color

Image.fromarray(array(colorMap, uint8), 'RGB').show()
```

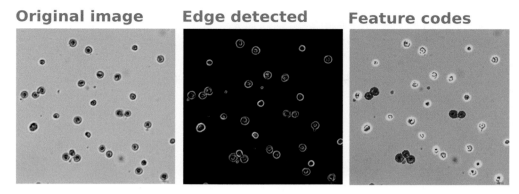

Figure 18.5 (Plate 8). Images from micrograph cell-counting procedure. An original microscope image of mammalian cells is shown alongside the results of the edge detection and a greyscale version where the different cell-edge features are labelled with colours after blob analysis. Isolated cells are yellow, overlapping cells are blue and small fragments are red. A black and white version of this figure will appear in some formats. For the colour version, please refer to the plate section.

The analysis performed on the blobs has only considered their total pixel area, but more sophisticated properties can be used. For example, the shape of the blobs, such as how circular they are, could be measured. Looking at the example blob-detection results (see Figure 18.5) there is an obvious extension to our cell-counting routine, which is to attempt to subdivide the larger blobs to estimate how many cells are overlapped:

```
numCells = len(mediumBlobs)  # initial guess
cellAreas = [len(blob) for blob in mediumBlobs]
meanCellArea = sum(cellAreas) / float(numCells)

for blob in bigBlobs:
  numCells += int( len(blob) // meanCellArea )

print('Estimated number of cells: %d' %  numCells)
```

19 Signal processing

Contents

Signals

In science many different kinds of experiment involve the recording of signals: series of measurements that represent the variation in some kind of underlying physical property. The signal can then be interpreted, based on some theoretical model of the experiment. Commonly the recorded signal is one that varies over time, such as sound or radio waves, but it could also represent a variation in space, or indeed along any other kind of axis. In general a signal is represented by values that are directly recorded by instruments at specific, usually regular, intervals; although in some situations derived data, like a DNA sequence, can also be thought of in terms of signals.

If a signal varies in a regular manner, i.e. oscillates, then it is often the frequencies that occur within the signal that are of interest, rather than the original signal itself. This is because the underlying frequencies are generally characteristic of what made the signal. To take a toy example, if we have a peal of bells, where each bell has a different tone, we can record the variation of the overall sound signal over time. Then, by looking at the component frequencies we can discern the tones of the individual bells that made the sound. As we will illustrate, it is possible to convert the time signal into a spectrum of its component frequencies using what is known as a Fourier transform.

Simulating a signal

We will begin by thinking about how a signal may be simulated in a computer. Naturally, we won't actually be generating a real signal as we won't be taking experimental measurements, but it is nonetheless useful to have code that simulates a signal. This will help give a basis for understanding the signal processing we describe later and allow us to check that the processing code works as expected. A pure frequency signal just oscillates and mathematically this is either a sine or cosine function, or, in terms of complex numbers, an exponential function with an imaginary argument. The signal also has amplitude, which in general is also a complex number. Thus a pure frequency signal (x), in continuous time (t), has the form:

$$x(t) = Ae^{i2\pi\omega t}$$

where ω is the frequency and A is the amplitude.

Note that this is one-dimensional, given there is only one independent variable, t, which can be used to calculate the amplitude at a given time. In many applications higher dimensions may also be considered; imagine calculating the frequency of ripples on the surface of water as an example of two-dimensional waves. An example of a three-dimensional signal would be diffraction patterns in X-ray crystallography, although in this case it is spatial distance rather than time that is the relevant variable. Mathematically there is no extra difficulty with handling signals that have more than one dimension, but visualisation becomes harder.

It is often the case that the signal decays exponentially in time, as illustrated in Figure 19.1. Mathematically we can simulate this by multiplying the signal by an extra decay term, as follows, for decay constant λ:

$$x(t) = Ae^{i2\pi\omega t}e^{-\lambda t}$$

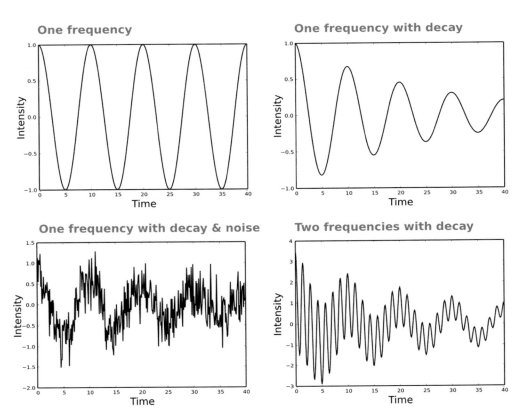

Figure 19.1. Time series for various signal situations. A pure, single frequency signal is shown in various combinations with an exponential decay function, random (normally distributed) noise and a second pure frequency.

In terms of discrete time points (as would generally be recorded by a machine), the signal is described by the same form, but sampled at regular time intervals ($t_k = k\Delta t$, $k = 0, \ldots, N - 1$, for N points). We will allow the superposition of a number of such signals, which can be combined in a simple linear manner, so the total signal will just be the sum of the individual frequency components. We will also assume that the signal has random imperfections, i.e. that it is affected by noise. There are many ways to model noise but we will assume that at each point in time its magnitude is normally distributed (Gaussian) in both real and imaginary components, with the mean at zero amplitude and with a specified standard deviation that is independent of time.

These considerations lead to the following Python code to create a simulated signal:

```
from numpy import arange, zeros, exp, pi, sqrt
from numpy.random import standard_normal as normal

I = 1j    # Square root of -1; complex(0, 1)

def createSignal(parameters, tStep, nPoints, noise):

  sig = zeros(nPoints, dtype=complex)
  t = tStep * arange(nPoints, dtype=float)

  for amplitude, frequency, decay in parameters:
    sig += amplitude * exp(2*pi*I*frequency*t) * exp(-decay*t)

  noise *= sqrt(0.5)
  sig += noise*(normal(nPoints) + I*normal(nPoints))

  return sig
```

The function starts by initialising the signal array `sig` to be filled with zeros for the correct number of time points. The NumPy data type is explicitly set to be `complex` because otherwise it would be regular floating point numbers and the sums would not work (the imaginary parts would be dropped). The function then loops over each component of the signal, each of which is specified by its amplitude, frequency and decay from the `parameters` list passed to the function. Then a noise contribution is added, which is simply sampling a normal distribution (via `normal()`) for both real and imaginary components of the complex number. The factor of `sqrt(0.5)` in the noise calculation is because the real and imaginary components both contribute to the standard deviation. Finally, the function passes back the NumPy array containing the signal.

Displaying a signal

It is easy enough to write a sample test script to run the code, but rather than just look at lists of numbers it is more useful to draw plots of the signals and Fourier transforms. We will use the `pyplot` module from the `matplotlib` library (available with SciPy) to do this. As illustrated in earlier chapters, it can show a graphical plot on the screen as well as save it to a file, although below we will just do the latter. We combine the graph-generating code into one function for convenience:

```
from matplotlib import pyplot

def savePlot(x, y, xlabel, ylabel):

  pyplot.plot(x, y, color='k')    # k means BlacK
  pyplot.xlabel(xlabel)
  pyplot.ylabel(ylabel)
  fileName = ylabel.replace(' ', '')
  pyplot.savefig(fileName)
  pyplot.close()
```

The `pyplot xlabel()` function sets the label for the x axis, and similarly for `ylabel()`. The `savefig()` function saves the figure to a file with the specified name (well, a suffix is appended, e.g. '.png').

In this simple example we assume that there are two pure frequency components of the signal, the first with amplitude `1.0`, frequency `0.1` and decay `0.01`, and the second with amplitude `2.5`, frequency `0.7` and decay `0.05`. A signal is created with 100 points and appropriate time step and noise values.

```
sigParams = ((1.0, 0.1, 0.01), # Amplitude, frequency, decay
             (2.5, 0.7, 0.05))

nPoints = 100
tStep = 1.0
noise = 0.5

sig = createSignal(sigParams, tStep, nPoints, noise)
```

The above plotting function can then be used to draw a graph using the time value for each point on the x axis and the signal intensity on the y axis:

```
times = [i*tStep for i in range(npoints)]
savePlot(times, sig, 'time', 'signal')
```

In the signal plot (see Figure 19.1) we can see the two pure frequency signals ($\omega = 0.1$ and $\omega = 0.7$) fairly clearly. However, if we have lots of pure frequency signals it becomes much harder to see what is going on, which is why in many cases we would go on to investigate the Fourier transform.

Fast Fourier transform

The *Fourier transform* is one mathematical technique for determining the frequencies that exist in a signal.[1,2] In essence, what the Fourier transform does is to work out what combination of pure sinusoidal oscillations the signal is composed of, in terms of both frequency and intensity, so that we can see how much of each frequency is present. The 'transformation' is from the original signal axis, which is typically time, to a frequency axis.

[1] Fourier, J.B.J. (1822). *Théorie Analytique de la Chaleur*. Paris: Chez Firmin Didot, père et fils.
[2] Titchmarsh, E. (1948). *Introduction to the Theory of Fourier Integrals* (2nd edn.). Oxford: Clarendon Press.

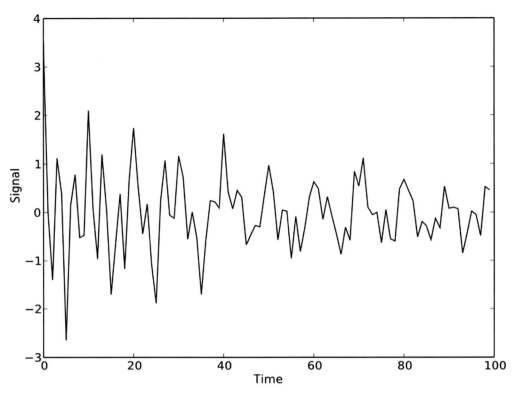

Figure 19.2. An example signal time series generated using Python. A plot of the one-dimensional signal, or time series created using the synthetic signal-generation function. The trace shows two superimposed frequencies with different maximum intensities and decay rates. The jagged nature of the line illustrates how only few points are sampled relative to the period of the signals, which is typical in many real-world situations.

Sometimes the signal oscillates in more than one direction,[3] and so is recorded on two orthogonal axes and is represented by complex numbers, rather than plain (real) numbers. Effectively this means that the signal values are two-dimensional, but this can all be dealt with by the Fourier transform. Although time is a continuous variable, the measurement of the signal is usually made at discrete times, and so the relevant technique to use is the *discrete Fourier transform*.[4]

The Fourier transform can also be used in other circumstances, unrelated to time. For example, a crystal has a periodic three-dimensional atomic structure and as a result the Fourier transform can be used to determine the structure of a molecule in a crystal from X-ray diffraction patterns. Also, the Fourier transform can be used to detect patterns in DNA and protein sequences, e.g. to look at amino acid properties that occur with a periodicity corresponding to alpha-helical structures.

Once a signal has been Fourier transformed, we generally find that only certain frequencies have significant representation, and so various algorithms have been developed to

---

[3] It has polarisation.

[4] http://en.wikipedia.org/wiki/Discrete_Fourier_transform.

determine what those frequencies are. Because of the imprecision of the signal measurement, and perhaps also because of a spread in the phenomenon being observed, normally when there are discrete frequencies the transformed distribution at those points is not perfectly sharp. Instead, for each there is an observed range of frequencies, called a peak, with the maximum value occurring close to the actual frequency. In general we want to determine the centre of each peak and other parameters associated with the peak, such as the intensity (its maximum height) and also the integration of the peak over its frequencies (often called its *volume*). Integration is often used because it allows us to relate the signal strength to the underlying amount of causal phenomenon.

The theory of FFT

Fourier transformation is effectively the weighted summation of sinusoidal waves that reproduces the original, oscillating signal. The transformed frequency plot (often called the *spectrum*) is then the graph of the amount of each frequency found.

If we consider a signal sampled uniformly in time, so that there is time Δt in between every measurement, we can represent the series of values as x_k, where k is the time index ($k = 0, \ldots, N-1$). Assuming that the first measurement is made at time $t = 0$, the signal x_k is measured at time k multiplied by Δt. Here we allow the x_k to be complex numbers, although we could instead restrict them to being real numbers. The discrete Fourier transform of this signal gives the amount of frequency y at frequency index j. This is calculated by measuring the amount of coincidence between the signal values and a pure sinusoid at each frequency. Hence, we sum, over all the time points k, the signal value x_k multiplied by the sinusoid value for frequency index j:

$$y_j = \sum_{k=0}^{N-1} x_k e^{-\frac{i2\pi}{N}jk}$$

Here the minus sign in the exponent is a convention. For each frequency point j there are N terms in the sum and, since j itself has N possible values, there are N^2 operations to determine the entire transformed signal. We use the notation $O(N^2)$ to describe this where the 'O' stands for the 'order', or approximate number, of the operations. In fact we can view this transform as a multiplication of a matrix by a vector, where the vector is x_k and the matrix has components as given by the exponential term in the sum. $O(N^2)$ algorithms are not ideal in computing because if you scale N by say 10 then the number of operations scales by 100.

Note that although we restrict the range of j to be between 0 and $N-1$, one could allow an arbitrary integer j, noting that if we add the total sampled width N (or multiples thereof) to a frequency there is no difference in the sum.[5] For example, sampling 10 points along an oscillation of frequency 1 gives the same result as a frequency of 11, where faster oscillations match the same heights at the sampled points. Thus the transformed signal is periodic with period N.

---

[5] $y_{j+N} = \sum_{k=0}^{N-1} x_k e^{-\frac{i2\pi}{N}(j+N)k} = \sum_{k=0}^{N-1} x_k e^{-\frac{i2\pi}{N}jk} e^{-i2\pi k} = \sum_{k=0}^{N-1} x_k e^{-\frac{i2\pi}{N}jk} = y_j$ since $e^{-i2\pi k} = 1$ for any integer k.

In the 1960s, Cooley and Tukey published a fast algorithm for determining the discrete Fourier transform when N is a power of 2.[6] The number of operations for the algorithm is $O(N \log N)$, which is much better than $O(N^2)$ for large N, and this has become the basis for the widespread use of Fourier transforms in science. Later developments improved the algorithm, for example, to deal with the case when N was not an exact power of 2, and generalisations to other transforming functions than the simple exponential, to give a 'wavelet' transform.

FFT using NumPy

Because this is a numerical problem, the fast Fourier transform (or FFT, for short) is normally implemented in a fast compiled language like Fortran or C.[7] But fortunately NumPy provides a Python wrapper around a C implementation of the FFT:

```
from numpy.fft import fft
freqs = fft(sig)
```

Also, SciPy provides two implementations of the FFT, one actually being the same as the NumPy one, and the alternative being:

```
from scipy.fftpack import fft
freqs = fft(sig)
```

They both accept either a NumPy array or an ordinary Python list or tuple as input, and return a NumPy array as output.

As an example, we can continue the code from the first section, which simulated a signal, `sig`, and calculate its Fourier transform. We can then save plots of the real and imaginary parts of the Fourier transform, and also the magnitude squared of that, which is called the *power spectrum*.

```
freqs = fft(sig)

freqReal = [f.real for f in freqs]
savePlot(times, freqReal, 'freq', 'FT real')

freqImag = [f.imag for f in freqs]
savePlot(times, freqImag, 'freq', 'FT imag')

powerSpec = [abs(f)**2 for f in freqs]
savePlot(times, powerSpec, 'freq', FT power')
```

In the example above we have assumed the intensities we require are represented by the real component of the complex numbers, which is why the frequencies from the imaginary part of the transform fluctuate about zero. In general, however, the real and imaginary parts of the transform are affected by *phasing*. Multiplying the signal by a complex number with magnitude 1 can be thought of as a 'rotation' of the real into the imaginary part and vice versa, hence the term *phase*, which describes the angle of rotation. This multiplication has the

---

[6] Cooley, J.W., and Tukey, J.W. (1965). An algorithm for the machine calculation of complex Fourier series. *Mathematics of Computation* 19: 297–301.

[7] Press, W.H., Teukolsky, S.A., Vetterling, W.T., and Flannery, B.P. (2007). *Numerical Recipes: The Art of Scientific Computing* (3rd edn.). New York: Cambridge University Press.

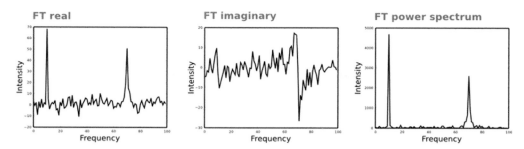

Figure 19.3. The results of a Fast Fourier transform. Illustrated are real and imaginary parts of the Fourier transform (for the time series shown in Figure 19.2) and its power spectrum. The original time signal has been converted into a frequency spectrum where the intensity axis shows the amount of each frequency present in the signal.

same effect on the Fourier transform. But the power spectrum remains unchanged, because it is the magnitude squared, which is one reason it is often used for analysis. However, in some areas of science, for example, NMR (nuclear magnetic resonance), it is important to understand the phasing because it turns out that you can get sharper (narrower) peaks with the real part of a suitably phased transformed signal, versus the power spectrum.

Peaks

Once we have a Fourier transformed signal, which in certain contexts is called a *spectrum*, the next thing to do is to analyse the frequency peaks, or at least the significant ones. These correspond to the underlying frequency components of the signal. We would like to determine the parameters for each peak, namely the amplitude, frequency and decay. If there are many components then the peaks can overlap and this job becomes difficult. To simplify our introduction we will assume here that there is no overlap between the peaks.

     The frequency is the point in the spectrum where the given peak value is at its maximum, but there are a few subtleties to this. The spectrum is specified on a grid, so at equally spaced frequencies. Hence, it's quite likely that the actual underlying frequency does not lie exactly on the point of the sampled grid, but in between two such points. This means that the maximum points need to be interpolated somehow to find the peak frequency positions.

     There is another subtle issue to do with frequency, which comes about because of the discrete time sampling of the signal. If the signal is sampled at time intervals Δt then a pure signal at frequency ω and another one at frequency $\omega + 1/\Delta t$ give the same Fourier transform, since $e^{i2\pi\frac{1}{\Delta t}\Delta t} = e^{i2\pi} = 1$. Indeed, in general we get the same Fourier transform for frequency $\omega + n/\Delta t$ for any integer n. In effect, a signal at one of these frequencies cannot be distinguished from a signal at any of the other frequencies. The frequency with an absolute value less than $1/(2\Delta t)$ is called the *fundamental* frequency and the other frequencies are said to be *aliased* to this one.[8]

---

[8] A related result is the Nyquist-Shannon sampling theorem, which says that if the frequencies are *bandlimited*, that is, $|\omega| < B$ for some B, then the entire signal can be exactly reconstructed from an infinite sequence of samples at interval Δt as long as $B < 1/(2\Delta t)$. Of course in practice an infinite sequence of samples is not taken.

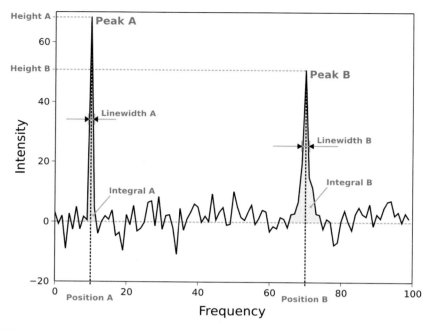

Figure 19.4. Parameters of signal peaks in a frequency spectrum. Two peaks within a frequency spectrum are identified by the frequency positions of their intensity extrema, which here are labelled as their heights. The integrals represent the area under the peak to the zero intensity value, although other integration methods are possible, e.g. if an idealised peak shape is fitted. The linewidths of the peaks are illustrated at half-height values (a common heuristic), and are indicative of the decay of the signal. The data shown is the real part of the Fourier transform also illustrated in Figure 19.3.

The *height* or *intensity* of a peak is the value at its maximum. This does not determine the underlying amplitude by itself, because the observed height is also affected by the decay parameter, and as with the frequency there is also the issue that the peak values are only defined on a grid. The amplitude is proportional to the *volume* (integral) of the peak, which is the summation of the frequency values around the peak, but there is the question of exactly how that is done, such as how far away from the maximum position to include in the sum. The decay parameter is determined by the *linewidth* of the peak, which is roughly speaking how wide the peak is. A common way to measure this is the width at half the peak height. These parameters can be determined by fitting the observed data to a theoretical description of a peak.

A 'Peak' class

One way to store information about peaks would be to use simple data structures, like lists or arrays, which would contain values corresponding to positions, intensities, linewidths etc. However, here we will represent peaks using a custom Python object by constructing a class called Peak. This will link together all of the parameters that relate to the same peak and allow us to associate special functions for the class (i.e. methods) to calculate various peak properties. See Chapter 7 for more detailed discussion about how classes are written, and the role of self. and the constructor function __init__.

In the construction of the `Peak` class we will insist that a peak, at the very least, has a reference to the data it derives from (`self.data`) and a position (`self.position`). The position might be a whole number, where the peak lies at a point on the grid, or it may lie between grid points. The latter indicates that some kind of interpolation has been done, to specify a more accurate position. So that we can store both on-grid and interpolated values `self.point` records the nearest grid point to `self.position`.

Keeping a reference to the underlying (often frequency) data means that we can later interpolate and fit the peak. Hence we include `dataHeight`, the height of the peak, and the `linewidth`. These are optional parameters that may be calculated if not specified in the initial instance (if they are `None`).

```
from numpy import ix_, outer, zeros

class Peak:

  def __init__(self, position, data, dataHeight=None,
               linewidth=None):

    self.position = tuple(position)
    self.data = data
```

The on-grid points for the peaks are initially set up by rounding the input position to the nearest integer, remembering that we have a value for each dimension.

```
    self.point = tuple([int(round(x)) for x in position])
```

If the `dataHeight` is not provided then it is calculated as being the value of the underlying data at `self.point`. A better alternative would be to do some kind of simple quadratic fitting.

```
    if dataHeight is None:
      dataHeight = data[self.point]

    self.dataHeight = dataHeight
```

If the linewidth is not initially specified then it is calculated using the `_calcHalfHeightWidth()` function, which we describe below.

```
    if linewidth is None:
      linewidth = self._calcHalfHeightWidth()

    self.linewidth = linewidth
```

Finally in the constructor we initialise the fitted parameters as null values.

```
    self.fitAmplitude = None
    self.fitPosition = None
    self.fitLinewidth = None
```

The `linewidth` is calculated by the function below as the full width at half the peak height, noting that there is a different width in each dimension. The method to do this goes through each dimension in the data and finds the positions on the two sides of the peak that would be at half the maximum height (see Figure 19.4), using `_findHalfPoints()`. The width is simply the difference between these, which is added to the list that is passed back.

```
def _calcHalfHeightWidth(self):

    dimWidths = []

    for dim in range(self.data.ndim):
        posA, posB = self._findHalfPoints(dim)
        width = posB - posA
        dimWidths.append(width)

    return dimWidths
```

The function _findHalfPoints looks in a given dimension for the places either side of the peak position where the half-height is reached. The first issue to face is that it is possible to hit the edge of the grid, where the data is defined, before the half-height is reached. We have taken a simple approach of just stopping at the edge of the grid, but an alternative would be to wrap round and continue on the other side of the grid. However, if you took the latter approach you would need to make sure the code does not end up in an indefinite loop, going around and around, because the values are possibly all above the half-height.

The initial height is the absolute value of self.dataHeight, and halfHt is half that, i.e. the value we are looking for. Using the absolute value simplifies the code a little because it means we won't need to use separate clauses if the peaks go in the negative direction. We define variables to refer to the underlying data (data) and peak grid position (point) by using self. to access attributes that belong to the Peak class. We are not forced to define new variables here, but as a general practice we minimise use of self. for speed reasons.

```
def _findHalfPoints(self, dim):
    height = abs(self.dataHeight)
    halfHt = 0.5 * height
    data = self.data
    point = self.point
```

The search will be done by investigating the height values associated with a testPoint that varies in position along the tested dimension. Initially this is a copy of the peak's grid point so that the positions are correct in the other (unsearched) dimensions. Note that testPoint is a list because it can be changed internally (unlike a tuple) as we assign different positions in one of the dimensions. The variables posA and posB are the one-dimensional search positions, corresponding to the two sides of the peak, which will be refined and passed back at the end.

```
    testPoint = list(point)
    posA = posB = point[dim]
```

The actual search involves a while loop that tests whether the search position is still within bounds. The first search is backwards, i.e. towards zero, using posA. The value of posA is decreased by one to get the previous point in the data and this is set within the correct dimensional index of testPoint, so that we can access the absolute value of the (possibly multi-dimensional) data. Note NumPy requires that we convert testPoint to a tuple for use as an index of data.

```
prevValue = height
while posA > 0: # Search backwards
  posA -= 1
  testPoint[dim] = posA
  value = abs(data[tuple(testPoint)])
```

At the end of the loop there is a check to see if the half-height has been reached, breaking out of the loop if it has. Linear interpolation is almost always needed, because the half-height is unlikely to occur exactly on a grid point. Hence, if the data value is less than the half-height we know we've over-stepped, and we add back a fraction of a point. This fraction is calculated as the drop in height of `value` below `halfHt` as a proportion of the drop in height from one point to the next.

```
if value <= halfHt:
  posA += (halfHt-value)/(prevValue-value)
  break

prevValue = value
```

And we repeat the procedure for `posB`, searching in the other direction, though this time the last data point in the data is recorded from the `.shape` of the data array (its size) so that we know when to stop. Then at the end we pass back the two positions that correspond to the half-height.

```
lastPoint = data.shape[dim] - 1 # Shape is size
prevValue = height
while posB < lastPoint-1: # Search forwards
  posB += 1
  testPoint[dim] = posB
  value = abs(data[tuple(testPoint)])

  if value <= halfHt:
    posB -= (halfHt-value)/(prevValue-value)
    break

  prevValue = value

return posA, posB
```

Peak picking

Now that we have defined the `Peak` class, we are ready to actually identify or *pick* the peaks from the frequency data. There are many subtle issues here, and we will only provide the simplest peak picker, with some commentary about what can be done to improve it. How best to proceed will depend on the specifics of the type of data being analysed, and it is unlikely that any peak picker will work in all circumstances for all possible data sets.

Here we will just look for local maxima above a specified threshold. This works reasonably well for data that is not especially noisy or crowded. For this kind of simple operation there is existing functionality from SciPy and NumPy that can be used to do most of the work, although the way they are used here is not necessarily speed-efficient. The `maximum_filter()` function will be used to find maxima in the data. Although this is part of SciPy's multi-dimensional image-processing module the data need not actually be an image. It is notable that the `minimum_filter()` function also exists, which could be used to find negative peaks.

```
from scipy.ndimage.filters import maximum_filter
from numpy import argwhere
```

The function `findPeaks` takes an array of data (of arbitrary dimensionality) and a `threshold` value. Only maxima above this threshold will be considered as peaks, and in general the threshold will be set to distinguish the required signals from the noise. The `size` represents the region to consider when searching for maxima. The default value of `3` means to consider a central point (potentially the peak centre) and its nearest neighbours either side, which is reasonable if the data is not too noisy. The `mode` argument specifies how the data is treated at the boundary. When `mode='wrap'` this means that the data is wrapped around at the boundaries, and this is often the most appropriate value for Fourier transformed data, which is naturally periodic. Sometimes, though, the data is truncated, in which case it is more appropriate to use `mode='constant'`, which in effect treats the boundary as the end of the data. (In the previous section we only had code for the equivalent of `mode='constant'`.)

```
def findPeaks(data, threshold, size=3, mode='wrap'):

  peaks = []

  if (data.size == 0) or (data.max() < threshold):
    return peaks
```

For the NumPy array `data` it is possible to filter on a `threshold` just by doing `data > threshold`. This creates an array, with the same size and dimensions as `data`, where each entry is `True` or `False`, depending on whether the corresponding value in `data` is above the threshold.

```
  boolsVal = data > threshold
```

Next we use `maximum_filter()` to find local maxima in the data, considering the specified `size` to inspect. This function returns a copy of the input array where the non-maximal points have been set to zero.[9] Then we create another array of `True` and `False` values, corresponding to whether the filtered points have their original values (`data == maxFilter`), i.e. we will have a `True` at each maximum.

```
  maxFilter = maximum_filter(data, size=size, mode=mode)
  boolsMax = data == maxFilter
```

The peak points that are both above the threshold and local maxima are found as the intersection (logical AND operation) between values from the two arrays of Booleans.

```
  boolsPeak = boolsVal & boolsMax
```

Next we find the indices of the `True` values in the `boolsPeak` array, which correspond to the positions (row, column etc.) of the peaks in `data`. The NumPy function we use to get the

---

[9] We could set non-maximal points to some other value using `origin`, if zero has meaning.

indices is `argwhere()`, which gives an array of the non-zero (true) positions in the form `((row1, col1, ...), (row2, col2, ...) ...)`, i.e. with one group of coordinates for each peak. It is notable that `nonzero()` is often used to find true values in an array, but this would give a result in the form `((row1, row2, ...), (col1, col2, ...) ...)`, which is ideal for indexing array elements but inconvenient here.

```
indices = argwhere(boolsPeak)  #  Position indices for True
```

Finally `Peak` objects are created for these grid positions and appended to the `peaks` list that is passed back.

```
for position in indices:
    position = tuple(position)
    height = data[position]
    peak = Peak(position, data, height)
    peaks.append(peak)

return peaks
```

There are improvements that could be made to the algorithm. Instead of looking in a rectangular region for local maxima, a non-rectangular footprint could be specified. Sometimes it is also worth checking that any putative peak has a shape such that a sufficient drop is made between the maximum height and the minimum value before it turns up again, in any direction. Also, if the data is such that negative height peaks are also of interest, then the algorithm can be modified to look for minima (below zero) by using `minimum_filter()` when the `threshold` is negative.

Peak fitting

In order to fit the peak shape, there has to be a model of what the peak shape should be. For exponentially decaying data, the real part of a continuous Fourier transform of that data has what is called a Lorentzian line shape, in any given dimension:

$$x(\omega) = \frac{A}{1 + 4(\omega - \omega_0)^2/\lambda^2}$$

Here A is the amplitude, ω_0 is the position and λ is the linewidth of the peak, which is the full width of the peak at half-height:

$$x(\omega \pm \tfrac{1}{2}) = \frac{A}{2}$$

In practice, the data is not continuous, it is discrete, and the discrete Fourier transform is carried out instead of the continuous one. This leads to a slightly different, and more complicated, line shape than the above, but in practice the difference is small enough close to the central position that the Lorentzian line shape functional form can be used in the fitting, and that is what will be done here.

Peaks with more than one dimension have a single height (amplitude), but there is a position and a linewidth for each dimension. Thus for a Lorentzian line shape the theoretical value of the data can be modelled as being a product of the amplitude and of the line shape in each dimension (the index i represents one of the dimensions):

$$\frac{1}{1 + 4(\omega_i - \omega_{0i})^2/\lambda_i^2}$$

The `fit()` function is added to the `Peak` class so that it can be called on a `peak` object via `peak.fit(fitWidth)`. As discussed in Chapter 7, because this function belongs inside a class definition it can use the `self` variable to refer to the actual peak object. The first thing the function does is to define a `region` which is `fitWidth` points either side of the integer peak position in each dimension, talking care that the region doesn't fall off the edge of the array, and then it gets `self.fitData`, the data values in the specified `region`. An alternative approach would be for a user to specify the region via a graphical user interface.

```
from scipy import optimize

def fit(self, fitWidth=2):

  region = []
  numPoints = self.data.shape

  for dim, point in enumerate(peak.position):
    start = max(point-fitWidth, 0)
    end = min(point+fitWidth+1, numPoints[dim])
    region.append( (start, end) )
```

The data is normalised by the peak height, to make the fitting work on data that has maximum value at 1, rather than an arbitrary-sized number.

```
  self.fitData = self._getRegionData(region) / self.dataHeight
```

The SciPy function `optimize.fmin` is used to determine the optimum fit. It requires a fitting function and a list of starting values for the parameters. Here the parameters are the `amplitudeScale` (relative to the peak height) and, for each dimension, an `offset` from the peak position and a `linewidth` scale.

```
  amplitudeScale = 1.0
  offset = 0.0
  linewidthScale = 1.0
```

These are inserted into the `params` list. The fitting function is an unnamed `lambda` function (see Chapter 5) that allows the data region to be passed into the fitting function as an argument before its actual execution. An alternative would be to store the region on the peak object. There are several optional arguments for `optimize.fmin`, including `xtol`, which allows a specification of how precise the result needs to be.

```
  ndim = self.data.ndim
  params = [amplitudeScale]
  params.extend(ndim*[offset])
  params.extend(ndim*[linewidthScale])

  fitFunc = lambda params: self._fitFunc(region, params)
  result = optimize.fmin(fitFunc, params, xtol=0.01)
```

The result comes back as a NumPy array, which is then split into the amplitude, offset and linewidth parts using the appropriate slices.

```
amplitudeScale = result[0]
offset = result[1:ndim+1]
linewidthScale = result[ndim+1:]
```

Although they do not have to be, the values are converted to ordinary Python types using the `float()` and `list()` functions and stored as attributes on the `peak` object.

```
peak.fitAmplitude = float(amplitudeScale * peak.dataHeight)
peak.fitPosition  = list(peak.position + offset)
peak.fitLinewidth = list(linewidthScale * peak.linewidth)
```

The function `_getRegionData` does the required extraction of the data in the region from the full data set. The `region` determines the selection by using the standard Python `slice()` function, which allows us to specify an index which goes from `start` to `end`. This is done in each dimension and the result needs to be converted to a tuple for use as a NumPy index.

```
def _getRegionData(self, region):

  slices = tuple([slice(start, end) for start, end in region])

  return self.data[slices]
```

The fit-testing function `_fitFunc` is where the real work is done. The SciPy function `optimize.fmin` calls this repeatedly with a specified choice of `params` to determine how good the fit is, until either convergence to the solution is achieved or the number of iterations reaches its maximum limit. The (fixed) `region` is also passed into the fitting function because of the way we defined it as a `lambda` function. The fitting function first unpacks the `params` list into the corresponding parameters and `sliceData` is initially set with zeros, but will eventually contain the line shapes for each dimension.

```
from numpy import array, zeros

def _fitFunc(self, region, params):

  ndim = self.data.ndim

  amplitudeScale = params[0]
  offset = params[1:1+ndim]
  linewidthScale = params[1+ndim:]
  sliceData = ndim * [0]
```

A `for` loop goes though each data dimension, calculating test values for `linewidth` and `testPos` by using the input test parameters to adjust the values stored on the peak object. The `start` and `end` points for the dimension are extracted.

```
for dim in range(ndim):
  linewidth = linewidthScale[dim] * self.linewidth[dim]
  testPos = offset[dim] + self.position[dim]
  (start, end) = region[dim]
```

A check is made that `linewidth > 0` because there are some situations where the SciPy `optimize.fmin` function will set the `linewidthScale[dim]` parameter to be negative. If it is negative then the line shape will be zero (in all dimensions, given the final product). An alternative would be to use the absolute value of this parameter. The Lorentzian line shape is then calculated (see above equation) for each dimension, in the specified `region`.

```
if linewidth > 0:
  x = array(range(start, end))
  x = (x - testPos) / linewidth
  slice1d = 1.0 / (1.0 + 4.0*x*x)
else:
  slice1d = zeros(end-start)

sliceData[dim] = slice1d
```

The final `heights` for the shape are calculated by multiplying the amplitude by the multi-dimensional outer product (see below) of the one-dimensional shapes contained in `sliceData`. Lastly, the test-fit function must return the difference between the theoretical line shape and the actual data. Here the root-mean-square difference is used, but that is not mandatory (we could just use the square).

```
heights = amplitudeScale * self._outerProduct(sliceData)
diff2 = ((heights-self.fitData)**2).mean()

return sqrt(diff2)
```

The individual line shapes are one-dimensional, and the product of these across the different dimensions is required to get the full theoretical shape. NumPy has a function `outer` which will do the appropriate multiplication in two dimensions. In order to do it in an arbitrary number of dimensions, the `outer` function needs to be called first on the first two dimensions, then next on that result and the shape for the third dimension etc. The result is a two-dimensional array so finally the data needs to be reshaped into the correct number of dimensions. This is all accomplished in the function `_outerProduct()`:

```
from numpy import outer

def _outerProduct(self, data):

  size = [d.shape[0] for d in data]
  product = data[0]

  for dim in range(1, len(size)):
    product = outer(product, data[dim])

  product = product.reshape(size)

  return product
```

There are many improvements that could be made to the code. Different line shapes could be allowed to be fit, for example, a 'Gaussian' one (where the exponential decay is squared). A constant vertical offset (constant baseline), or more complicated descriptions of the baseline, could be fitted to the data. It could be that one or more of the parameters are somehow known, so should not be fitted. If the peaks overlap then more than one peak could be fitted at a time, though the number of parameters is proportional to the number of peaks, and so fitting more than one peak increases the risk of having too many degrees of freedom with too little data.

2D gel peak-picking example

We provide a simple application of using the peak picker to find extrema in an image, which in this case represent spots on a polyacrylamide protein gel. Naturally, this data was not recorded as a signal that required transformation into the frequency domain, but nonetheless it is a biologically relevant test case and is analogous to signals we might observe in a frequency spectrum. The example will use some of the functionality discussed in Chapter 18 to convert between the `Image` objects made by the Python Imaging Library (PIL) and numeric arrays.

```python
from PIL import Image
from Images import imageToPixmapRGB, pixmapToImage

img = Image.open('examples/Gel2D.png')
pixmap = imageToPixmapRGB(img)
```

The RGB (three-component-deep) image is averaged along the last axis, to generate a simple array of values of the same size representing the pixel intensity. This is normalised to be between 0 and 1 by taking away the baseline minimum and dividing by the maximum. Because in the image the spots we want to pick are darker than the background we flip the data array (`1.0 – data`) so that the spots to pick, and not the background, are points of maximum intensity.

```python
data = pixmap.mean(axis=2)
data -= data.min()
data /= data.max()
data = 1.0 - data
```

The peak picking threshold is simply defined as a proportion of the maximum intensity (which here is `1.0`) and then the `findPeaks()` function defined above is used to find the peak locations in the data using a `size` which is appropriate to the scale of the peaks in the image.

```python
threshold = 0.3 * data.max()
peaks = findPeaks(data, threshold, size=7, mode='wrap')
```

Figure 19.5 (Plate 9). Example of 2D peak picking for a protein gel. An image of a two-dimensional polyacrylamide gel, with the peak picked maxima shown as small crosses. The gel has been stained with Coomassie Brilliant Blue dye to show the abundance of different proteins that have been separated according their size and isoelectric value. A black and white version of this figure will appear in some formats. For the colour version, please refer to the plate section.

By looping though the resulting `Peak` objects we can extract the identified positions and use this to mark small crosses in the image pixmap that we initially loaded, here setting the RGB colour for the pixel to be yellow, maximum red and green but no blue. Note the use of the array slice notation `[xIndices, yIndices,:]`, which sets the whole of the last (colour) axis for the required rows and columns (which are specified in the tuples of indices). The pixmap is then converted into a PIL `Image` object so that it can easily be viewed and saved to file.

```
color = (255.0, 255.0, 0.0)
for peak in peaks:
  x,y = peak.position
  xIndices = (x-1, x, x, x, x+1) # X indices for a cross shape
  yIndices = (y, y-1, y, y+1, y) # Y indices for a cross shape
  pixmap[xIndices, yIndices,:] = color

img2 = pixmapToImage(pixmap, mode='RGB')
img2.show()
img2.save('PickedGel.png')
```

If we wish to refine the peak locations, i.e. to interpolate their centres between the integer grid points, it is a simple matter of calling the `.fit()` method of the peak and looking at the the updated attribute values.

```
for peak in peaks:
  peak.fit(fitWidth=3)
  print(peak.fitAmplitude, peak.fitAmplitude, peak.fitLinewidth)
```

20 Databases

Contents

A brief introduction to relational databases

Any collection of data can be considered to be a database, however it is stored or utilised. However, the common use of the word 'database' usually refers to a *relational database*. Relational databases were introduced in the 1970s and model their data in terms of *tables* with rows and columns. There is an associated language, SQL (Structured Query Language), that can be used to send messages to the database to allow the database to be queried and modified: inserting, changing and deleting data elements. SQL also provides the ability to make connections between the data in different tables. In terms of mathematics, relational databases can be thought of as 'first-order predicate logic', and this mathematical underpinning of the principles of relational databases is one reason they are conceptually attractive.

Tables

A table in a relational database has a name and also has some named columns and each row in the table represents one record of data. The type of the data in each column can be specified and the data in any column can be stated to be mandatory, or not. One or more columns in each table define the *key*. Each record in the table must have a unique key; the key identifies the individual record. Sometimes a table has a 'natural' key but sometimes there is no obvious key and so instead a counter (a 'serial' or 'ID' number) is used, which is set (in many database implementations automatically) to 1 for the first record, 2 for the second, etc. At a simplistic level, spreadsheets (e.g. as used in Excel) can be thought of as tables in a weak substitute for a relational database. A table can have one or more columns that refer to one or more other tables, and this is a way that information between tables can be linked. In the database jargon, a query that involves relating information across more than one table is called a *join*.

Many databases (relational or otherwise) provide reliable transaction control, through a mechanism named with the memorable acronym ACID (Atomicity, Consistency, Isolation, Durability), and this, more than anything else, is what makes databases so ubiquitous in the modern world. In effect, multiple agents (people or computers) can reliably access a database at the same time. Relational databases are the most significant and common examples of ACID-compliant databases, although not all relational databases are necessarily ACID-compliant.

Schemas

As with other areas in computing, one of the most important jobs in relational databases is their design, in terms of the definitions of the tables and the relations between the tables. This can be thought of as an exercise in data modelling. In the database context, this is called designing the *database schema*. Sometimes the nature of relational databases forces non-ideal design decisions. In other areas of computing it is quite common to use object-oriented methodology, but the mapping between objects and relational databases is not always easy. In terms of the object-relational mapping, a class of object is mapped onto one or more tables. Each column in a table would represent a property (an attribute or link to another object) of the class. It would be ideal if there could just be one table per class, but that is not always possible because an attribute that is allowed to have multiple possibilities (so a high *cardinality* of more than 1, in data-modelling jargon) requires an additional table in relational databases, and any link between classes that allows multiple possibilities at both ends (so a many-to-many link in data-modelling jargon) also requires an additional table.

Another possibility would be to map more than one class onto one table. This is called *denormalisation* of the database, and it (usually) means that there is redundant information in the table. We could even map all of the classes onto one 'universal' table. Whether this is a good idea depends on how the database is going to be used. If the database is only or is mainly for querying, then a universal table works well. But if the database is frequently updated and the contents modified, then a universal table does not work well. In this chapter we will largely avoid these subtler points but in practice they need to be considered.

Basic SQL

There are many available pieces of software that implement relational databases. Most relational database implementations provide *client* software into which you can type SQL commands. The client usually also allows SQL scripts to be run. Many implementations also have a separate *server*, which receives commands from the client and actually executes them, referring to the data stored in the database, and then gives back the result to the client.

Much of the SQL syntax is well standardised, but most implementations stray slightly from standards and also provide extra functionality that is non-standard. Where such differences matter, we will look at two relational databases in this chapter, MySQL[1] and SQLite.[2] MySQL is the most common open-source relational database. SQLite is a

[1] http://www.mysql.com
[2] http://www.sqlite.org

lightweight implementation that requires no conventional server to be installed or run; it is all handled by the client and it stores the entire database in a single file on disk.

There are Python wrappers around SQL for many of the relational database implementations. With the exception of SQLite, these wrappers are not part of the standard Python distribution, and so require a separate download. For MySQL the Python module is called MySQLdb.[3] Note that for MySQL you will also need the underlying database implementation to be on your computer, in addition to the Python wrapper around it. Many Linux/UNIX distributions will automatically have SQLite installed, but it is less common to have MySQL installed. We will assume in this chapter that the SQLite program (executable) is called `sqlite3` and that the MySQL client program is called `mysql`.

The wrappers to these database implementations allow Python to be used to create or delete tables, and, more commonly, to insert or delete or modify records in the database, or simply to query the database. Unfortunately the interaction methodology is rather weak: the database constructs are not directly modelled, and the way information is passed from the Python world to the database world is just via textual strings. This problem is not unique to Python; other languages, like Java, have the same issue when communicating with an SQL database.

Creating a table

In this section we will work with a very simple database, which we will call `PersonDb`, which will have just one table, called `Person`. The code in the following section is in the SQL language not Python, though later we go on to show how Python can be used to interact via SQL. To create a table, we use the SQL `CREATE` command. The simplest version is where just the column names and types are specified. For example, suppose that in the `Person` table there are three columns, `firstName`, `lastName` (both strings) and `birthYear` (an integer). We could create the table via:

```
CREATE TABLE Person (
   firstName TEXT,
   lastName TEXT,
   birthYear INT
);
```

The SQL keywords are case-insensitive although they are often written in all capital letters. The table and column names might or might not be case-sensitive dependent on the exact SQL implementation being used, but it is probably safest to assume that they are also case-insensitive. The SQL type for text strings is tricky. In most SQL implementations there are several alternatives. `TEXT` is usually one of the alternatives, and it means that the strings are of unlimited length. If we had put `TEXT(30)` (for example) it would mean that the strings are of length no more than 30. And an alternative string type to `TEXT` would be `VARCHAR`, which is somewhat less understandable, but has been around longer in SQL implementations.

If a given column is mandatory, whereby every record has to have a value for that column, then the `NOT NULL` descriptor is used. For example, if the `firstName` and `lastName` were mandatory in the `Person` table then we would have:

```
CREATE TABLE Person (
   firstName TEXT(30) NOT NULL,
   lastName TEXT(30) NOT NULL,
   birthYear INT
);
```

Here we have also changed the strings to have a maximum length of 30.

A *primary key* for a table is a list of one or more columns for which the corresponding values uniquely identify a record, and we can specify which column or columns these are. For example, perhaps we think that the firstName and lastName uniquely identify a person in our database (although that is probably not a very good assumption). Then we would have:

```
CREATE TABLE Person (
   firstName VARCHAR(30) NOT NULL,
   lastName VARCHAR(30) NOT NULL,
   birthYear INT,
   PRIMARY KEY (firstName, lastName)
);
```

It is this final version that we will use here. We have changed the string type from TEXT to VARCHAR because MySQL does not allow the former for any column that is part of the primary key. With SQLite we could have stayed with TEXT(30) or indeed just TEXT. It is these little annoyances that make life difficult if one tries to support more than one SQL implementation.

Running SQL

We can run the SQL program at the operating-system prompt ('>') and then type the command (or, more generally, commands) at the 'SQL prompt'. For example, in SQLite a session might take the form:

```
> sqlite3 PersonDb
...   (information printed out by the program)
sqlite> CREATE TABLE Person (
   ...>      firstName VARCHAR(30) NOT NULL,
   ...>      lastName VARCHAR(30) NOT NULL,
   ...>      birthYear INT,
   ...>      PRIMARY KEY (firstName, lastName)
   ...>   );
sqlite> .quit
```

The last command ('.quit' for SQLite, 'quit' for MySQL) exits us from the SQL prompt back to the operating-system prompt. With MySQL it would be the same except we would probably need to specify a username and enter a password:

```
> mysql -u USERNAME -p PersonDb
Enter password:
...   (information printed out by the program)
mysql> CREATE TABLE Person (
   ->      firstName VARCHAR(30) NOT NULL,
   ->      lastName VARCHAR(30) NOT NULL,
```

```
   ->       birthYear INT,
   ->       PRIMARY KEY (firstName, lastName)
   ->    );
mysql> quit
```

Here USERNAME should be replaced with the appropriate value. And the '-p' flag tells the mysql program to ask for a password before SQL commands can be entered. Alternatively, and this is the normal practice, we could put the command (or more generally, commands) into a file and run this. So suppose we place this CREATE command in a file called createPersonTable. sql, then to create the table in SQLite we would run:

```
> sqlite3 PersonDb < createPersonTable.sql
```

This just runs the script. It does not then leave you at the SQL prompt, but instead back at the operating-system prompt. With MySQL it would be the same except for the issue of the username and password:

```
> mysql -u USERNAME -p PersonDb < createPersonTable.sql
```

Manipulating records

To add records to the database we use the INSERT command. Here we must specify values for the columns that are mandatory (i.e. NOT NULL), but naturally can also specify the values for other columns. For example, the following commands insert three records into the Person table:

```
INSERT INTO Person (firstName, lastName) VALUES ('Mary', 'Jones');
INSERT INTO Person (firstName, lastName) VALUES ('Tom', 'Smith ');
INSERT INTO Person (firstName, lastName, birthYear) VALUES ('Susan', 'Brown',
1723);
```

We can modify the data in existing records, subject to the various constraints, such as data type and uniqueness of the primary key, still being satisfied:

```
UPDATE Person SET birthYear=1942 WHERE lastName= 'Smith';
UPDATE Person SET firstName= 'Ann', birthYear=2001 WHERE lastName= 'Jones';
```

We can query the database and determine which records satisfy specified constraint criteria (here born after 1900):

```
SELECT lastName FROM Person WHERE birthYear > 1900;
```

To remove records from the database we use the DELETE command. We can specify conditions for deletion: for example, the following only deletes records where the firstName is 'Mary':

```
DELETE FROM Person WHERE firstName= 'Mary';
```

And the following deletes records where the lastName is 'Brown' and the birthYear is 1723:

```
DELETE FROM Person WHERE lastName= 'Brown' AND birthYear=1723;
```

We can delete all the records in one go from a table by omitting the WHERE clause:

```
DELETE FROM Person;
```

This does not remove the table, just all the records inside the table. To remove an entire table from the database we use the DROP command:

```
DROP TABLE Person;
```

These last two commands are rather dramatic in their effect and obviously should only be carried out with due care. SQL has much more complexity than we have shown in this section, but this is enough to get started. This is a book about Python after all.

Designing a molecular structure database

In this section we use the model for molecular structures described in Chapter 8 to create a database for structures. Recall that we had four classes (object specifications): Structure, Chain, Residue and Atom. We will create a table for each class, with the name of the table being the same as that of the corresponding class.

For each of the tables (classes) there will be columns (attributes) that could act as natural keys with real meaning, to identify each item of data. However, instead of using these we will use serial number keys (integers) to identify each item, and let the database automatically generate them. Hence they will be unique by construction. Both MySQL and SQLite have the ability to automatically generate serials. Note that these automatically generated serial numbers are unique across the whole table, not just unique relative to a local context (e.g. a parent container item). Although it might seem better to use meaningful natural keys, like the Atom.name that we set up in the structure data model, a serial number key has many advantages, not least of which is that if it is unique across the table (as it is here) then it only requires this one column to specify the key. If we used natural keys, or a serial key that was only unique relative to the parent, then child tables would have to include the full key of the parent table, and so that includes the grandparent key, the great-grandparent key etc., and that gets rather long when you are deep down in the parent-child containment hierarchy. Take, for example, the Atom: again the full key, considering all the parent links, would need to include Structure.pdbId, Structure.conformation, Chain.code, Residue.seqId and Atom.name. We do need another column in the child table, for the serial key of the parent, in order that we know which child goes with which parent, but that is a total of two columns only. That second column is called a *foreign key* because it refers to a key in a different table.

The Structure class has three attributes, name, conformation and pdbId. In the class only name was mandatory, but here we will assume that all three are now mandatory. And then the conformation and pdbId together provide a natural key for the table. When we add a new Structure to the database we want to make sure that (conformation, pdbId) is not already used, and the database will not automatically do that because we are using a serial key, not the natural key. Note that the Structure constructor code cannot check that these are unique because there is no parent class (containing other structures), so there is no way for the constructor to check what other (conformation, pdbId) values have already been used. We could have modelled a parent class for Structure, and called it Database, in which case this check could have been put in the Structure constructor, but we did not do it this way. So instead we have to check the uniqueness of (conformation, pdbId) in our own code that acts as a bridge between the class and the table. In summary, the Structure table will have four columns: id (for the serial key), name, conformation and pdbId, which are all mandatory.

For all the other classes, we do not have this issue with the natural key, because the uniqueness is explicitly checked in the corresponding class constructor. The `Chain` class has two attributes, `code` and `molType`, and a link to the parent `Structure`. As discussed, the key for the `Chain` table will be an auto-generated serial, and the link to the parent structure will be modelled as a foreign key. So the `Chain` table will have four columns: `id` (for the serial key), `structureId` (for the parent serial key), `code` and `molType`. The `Residue` class has two attributes, `seqId` and `code`, and a link to the parent. So the `Residue` table will have four columns: `id` (for the serial key), `chainId` (for the parent serial key) and `seqId` and `code`.

The `Atom` class has three attributes, `name`, `coords` and `element`, and a link to the parent. Here we face an issue, because `coords` has three values (x, y and z), not one, and not all SQL implementations allow many-valued columns (as the ARRAY data type). There are two possible solutions here. We could introduce a new table, just for the coordinates. Having a new table is a possible approach for many-valued attributes. In this approach we could call the new table `AtomCoord`, with columns `id` (for a serial key), `atomId` (for the relevant atom serial key), `dim` (for the dimension being considered, 1, 2 or 3) and `coord` (for the coordinate for that dimension). This works pretty well in the SQL context. But we will take another approach here, relying on the fact that `coords` is always of length 3. So we will stay with one table and split the attribute `coords` into three columns, x, y and z. The code that bridges between the `Atom` class and the `Atom` table will have to deal with translating from `coords` to (x, y, z) and back again. If we had introduced a new table then the bridging code would have had to deal with that, which is more complicated. The `Atom` table will thus have seven columns, `id` (for the serial key), `residueId` (for the parent's serial key), `name` (for the atom name), x, y, z (for the coordinates) and `element` (for the atom element type, e.g. 'C' or 'N').

SQL creation of the database

Although we could use Python to create the database and the tables in the database, this is often done directly in SQL because it is a one-off exercise (except occasional upgrades to the data model) and there are no particular advantages to using Python in this context.

To create a database in SQLite we just need to refer to it when we do anything, so in particular when we create the tables it will automatically create the database if it is not already created. In MySQL you have to create the empty database before you can use it, and often the system administrator has to do this for you, and set up suitable privileges so that you can access it. We will assume here that the database has been created.

Given the model discussed in the previous section, the table creation in SQLite is as follows:

```
CREATE TABLE structure (
    id INTEGER,
    name TEXT NOT NULL,
    pdbId TEXT NOT NULL,
    conformation INTEGER NOT NULL,
    PRIMARY KEY (id)
);

CREATE TABLE chain (
    id INTEGER,
```

Structure

id	name	pdbId	conformation
1	RCC1	1A12	1
2	H1GI	1UST	1
3	H1GI	1UST	2
...			

Chain

id	structureId	molType	code
1	1	protein	A
2	1	protein	B
3	1	protein	C
4	2	protein	A
5	3	protein	A
...			

Figure 20.1. SQL database showing two tables. The table header shows what data is stored in the table, and each row represents one record of data. Here the second table, Chain, has a link, called a foreign key, to the first table, Structure. Thus, for each row in the Chain table there is a unique row in the Structure table (but not vice versa, in general).

```
    structureId INTEGER NOT NULL,
    molType TEXT NOT NULL,
    code TEXT NOT NULL,
    PRIMARY KEY (id),
    FOREIGN KEY (structureId) REFERENCES structure(id)
);
CREATE TABLE residue (
    id INTEGER,
    chainId INTEGER NOT NULL,
    seqId INTEGER NOT NULL,
    code TEXT,
    PRIMARY KEY (id),
    FOREIGN KEY (chainId) REFERENCES chain(id)
);
CREATE TABLE atom (
    id INTEGER,
    residueId INTEGER NOT NULL,
    name TEXT NOT NULL,
    x FLOAT NOT NULL,
```

```
        y FLOAT NOT NULL,
        z FLOAT NOT NULL,
        element TEXT NOT NULL,
        PRIMARY KEY (id),
        FOREIGN KEY (residueId) REFERENCES residue(id)
    );
```

The table creation in MySQL is the same except that we need to explicitly specify the automatic increment feature of the primary keys in each of the four tables:

```
    id INTEGER AUTO_INCREMENT,
```

We use 'CREATE TABLE' followed by the table name, to create that table, and then we list the columns that exist in the table, with their properties. The 'NOT NULL' property means that the corresponding attribute is mandatory. Here, most of our attributes are mandatory. The PRIMARY KEY indicates the column or columns that make up the primary key. And the FOREIGN KEY indicates that the relevant column refers to a column (normally a key) in another table. In particular, on insertion of any record in this table, the SQL implementation should check that there is a record in the other table with a value in the other column equal to that for the column in this table. In essence it's about consistency of information between tables.

If we place these SQL commands in a file called createStructureTables.sql and assume that the database is called StructureDb, we can then create the tables by issuing the following operating-system command for SQLite (at the command line prompt):

```
    > sqlite3 StructureDb < createStructureTables.sql
```

We could also create the equivalent script that deletes (or 'drops') the tables, and that would be accomplished by the much simpler SQL script:

```
DROP TABLE structure;
DROP TABLE chain;
DROP TABLE residue;
DROP TABLE atom;
```

If we place that in the file dropStructureTables.sql, then we can delete the tables via:

```
    > sqlite3 StructureDb < dropStructureTables.sql
```

Obviously one has to be very careful about running such a script. It is good for testing purposes, but in real life it would be unusual to want to delete an entire set of tables like this. The MySQL versions of dropping tables would be the same except for the possible requirement of a username and password.

```
    > mysql -u USERNAME -p StructureDb < createStructureTables.sql
    > mysql -u USERNAME -p StructureDb < dropStructureTables.sql
```

You could also write SQL scripts to insert, modify, query and delete records, but we will do this using Python. If instead of having one table per class we decided to instead have just one table containing all the information, much of it redundant, then we could have the following creation command (in SQLite):

```
CREATE TABLE structure (
    id INTEGER,
    structureName TEXT NOT NULL,
```

```
      structurePdbId TEXT NOT NULL,
      structureConformation INTEGER NOT NULL,
      chainMolType TEXT NOT NULL,
      chainCode TEXT NOT NULL,
      residueSeqId INTEGER NOT NULL,
      residueCode TEXT,
      atomName TEXT NOT NULL,
      atomX FLOAT NOT NULL,
      atomY FLOAT NOT NULL,
      atomZ FLOAT NOT NULL,
      atomElement TEXT NOT NULL,
      PRIMARY KEY (id)
  );
```

In MySQL we would add `AUTO_INCREMENT` to the `id`. Note that the 'ATOM' records in a PDB file could be thought of as representing this kind of table, and indeed in a PDB file there is much redundant information, e.g. the residue code is repeated over and over for each atom in the same residue. As noted previously, this kind of universal table can be desirable if the database is mainly for querying.

Python interaction with the database

In Chapter 8 we read a PDB file and created one or more `Structure` objects as a result. In this section we show how to put a `Structure` object into a database. The Python wrapping around SQLite is handled in a package called `sqlite3`, which is automatically included with the standard Python distribution. The Python wrapping around MySQL is called `MySQLdb`, which requires a separate installation. Mechanically they both work in the same way; the main differences in the Python modules correspond to any differences that exist for the underlying SQL commands.

The first thing that we need to do is to make a connection to the database. For most database implementations this optionally allows a username and password to be supplied, and in a real-world situation this would normally be required. In SQLite the concept of a username and password does not exist, and instead access is totally determined by the user's permissions to the file that contains the database. As an example, with `sqlite3`, to connect to a database, we just have to do:

```
import sqlite3
connection = sqlite3.connect(database)
```

whereas with `MySQLdb` we would instead do, if a username, assumed to be stored in the variable `user`, and password, assumed to be stored in the variable `pwd`, were required:

```
import MySQLdb
connection = MySQLdb.connect(db=database, user=user, passwd=pwd)
```

When we are finished with a connection we can just close it:

```
connection.close()
```

There is a further subtlety with databases, compared to storing data in regular files: at any point you can decide to commit or roll back any changes you have made to the database since the last time you committed changes, or, if you have not previously committed any changes, since you connected to the database. This is an exceedingly powerful ability. So if the changes are acceptable, then you can commit them:

```
connection.commit()
```

and otherwise you can discard them and go back to your previous state:

```
connection.rollback()
```

Note that if you close a connection without doing a commit then all changes will have been lost. For many database implementations, connection is a relatively heavyweight operation to carry out, so normally we would want to make a connection to the database, carry out many commands and then disconnect, rather than connecting and disconnecting for every single command.

A second step is required before any SQL command can be executed, which is that a *cursor* needs to be created for the connection. This is much more lightweight, so can be done on a regular basis. An SQL command is executed using the cursor, and the result is then returned. In some sense, you can think of a cursor as an opaque handle into the database, in much the same way that the Python `open()` command supplies an object to handle a file on disk. To get hold of a cursor from a connection we just have to do:

```
cursor = connection.cursor()
```

When you are done with a cursor you can close it:

```
cursor.close()
```

The cursor allows execution of SQL commands. For example, suppose you want to find the records in the `Structure` table with `pdbId='1AFO'`. Then you can do:

```
stmt = "select * from structure where pdbId='1AFO'"
cursor.execute(stmt)
```

Note that the statement has no semicolon (';') at the end.

The `execute()` function does not return the result from executing the command. There are a couple of further functions you can call to get at the actual result, if there is one (some commands, such as insertions, never return a result). In the way that file handles have two functions, `readline()` and `readlines()` to either read the next line in the file or all the remaining lines, a cursor has the equivalent functionality for the result from a cursor execution. So the function `fetchone()` gives the next record, or `None` if there are no more. And the function `fetchall()` returns all the remaining records, or an empty list if there are no more.

Each record in the result is a tuple, but what is in the tuple depends on the command executed. For example, the above query uses '*' in the query so returns all the columns for the table, which in this case are four in number, so the tuple for every result record will be of length four. Thus we could have the following loop:

```
for (structureId, name, pdbId, conformation) in cursor.fetchall():
  print(pdbId, conformation)  # or whatever
```

The way we have coded it, the above query just uses the constant `'1AFO'` to do the query. In most normal applications the conditions would be provided by variables rather than by constants. This raises a slightly tricky issue. So it would be natural, from a Python point of view, to do the following:

```
stmt = "select * from structure where pdbId='%s'" % pdbId
cursor.execute(stmt)
```

But this is not the recommended methodology, because it makes the application vulnerable to an 'SQL injection attack'. So you, as a developer, might think that the `pdbId` variable is harmless, because it is just a PDB id. But unless you can guarantee the source of the information in this variable, it might contain malicious SQL code. For example, if you ask the user for a PDB id and they enter:

```
"0' or '0'='0"
```

the statement would then become:

```
stmt = "select * from structure where pdbId='0' or '0'='0'"
```

The second condition in the `where` clause is always true so this query returns all the records in the `Structure` table, which is not the intended result. Imagine the trouble that would be caused if instead of a query this were a delete command. The recommendation is instead to use a placeholder for each value, and to set them in the execute function. In `sqlite3` the placeholder is a question mark ('?'):

```
stmt = "select * from structure where pdbId=?"
values = (pdbId,)
cursor.execute(stmt, values)
```

In MySQLdb the placeholder is '`%s`', otherwise the syntax is the same. So in MySQLdb the statement would be

```
stmt = "select * from structure where pdbId=%s"
```

Note that there are no single quotation marks around the placeholder in the statement, which is nice, because it means you do not have to worry about whether a column is a string or not. The second argument to `execute()` has to be a tuple, even if only one condition is being set. As an example with two conditions, consider the query (syntax for `sqlite3`):

```
stmt = "select * from structure where pdbId=? and conformation=?"
values = (pdbId, conformation)
cursor.execute(stmt, values)
```

Adding a structure to the database

The basic ideas of how to deal with the Python wrapper around SQL were discussed in the previous section. Here we provide an implementation of how we can add a structure item, including all of its chains, residues and atoms, into the database. The example below works for both `sqlite3` and `MySQLdb` because we use the simple function `formatStatement()` to insert the correct placeholder text in all the statements, substituting using the normal Python string format code `%s`.

```
def formatStatement(text, placeHolder):

  if placeHolder == '%s':
    return text

  numInserts = text.count('%s')

  return text % numInserts*(placeHolder,)
```

The code below executes one SQL command for each record being inserted into the database. In general SQL commands are rather slow to execute. It would be better to bundle several SQL insert statements into each command. SQLite does not offer this functionality so the method below is the only one used in this case. But MySQL does offer the ability to bundle several insert statements into each command, and after we discuss the generic version then we will modify the code to take advantage of this feature. In real-world applications requiring reasonable performance, the following code would not be acceptable (and so SQLite would not be acceptable). But we illustrate here nonetheless for educational purposes.

The function definition has three arguments: a database `connection`, a `structure` object and a `placeHolder`, which defaults to the MySQLdb value. The function first gets a cursor, with which interactions with the database will take place.

```
def addStructureToDb(connection, structure, placeHolder='%s'):

  cursor = connection.cursor()
```

We first check whether there is already a structure with the same `pdbId` and `conformation` in the database. If so, we raise an error exception.

```
  pdbId = structure.pdbId
  conformation = structure.conformation

  stmt = "select * from structure where pdbId=%s and conformation=%s"
  stmt = formatStatement(stmt, placeHolder)
  cursor.execute(stmt, (pdbId, conformation))

  if cursor.fetchone():
    cursor.close()
    msg = 'structure with (pdbId=%s, conformation=%s) already known'
    raise Exception(msg % (pdbId, conformation))
```

If there is any kind of error when inserting the structure into the database we want to roll back (reverse) the transaction, so we put the entire code into a `try`/`except` block. We then insert the relevant `structure` data into the `structure` table. Note that after a record is inserted into the database, the identifier of that record can be found via `cursor.lastrowid`. This identifier is then used when creating the child records. Here it is the `structureId`.

```
  try:

    stmt = "insert into structure (name, pdbId, conformation) " \
           "values (%s, %s, %s)"
    stmt = formatStatement(stmt, placeHolder)

    cursor.execute(stmt, (structure.name, pdbId, conformation))
    structureId = cursor.lastrowid
```

We then descend down the rest of the hierarchy: `Chain`, `Residue` and `Atom`, in turn. First we add the `chains`.

```
for chain in structure.chains:
  molType = chain.molType
  code = chain.code

  stmt = "insert into chain (structureId, molType, code) " \
         "values (%s, %s, %s)"
  stmt = formatStatement(stmt, placeHolder)

  cursor.execute(stmt, (structureId, molType, code))
  chainId = cursor.lastrowid
```

Then we add the `residues`.

```
for residue in chain.residues:
  seqId = residue.seqId

  # insert residue into database

  stmt = "insert into residue (chainId, seqId, code) " \
         "values (%s, %s, %s)"
  stmt = formatStatement(stmt, placeHolder)

  cursor.execute(stmt, (chainId, seqId, residue.code))
  residueId = cursor.lastrowid
```

Finally we add the `atoms`. This is the longest, and slowest, part of the operation.

```
for atom in residue.atoms:
  # insert atom into database

  (x, y, z) = atom.coords

  stmt = "insert into atom " \
         "(residueId, name, x, y, z, element) " \
         "values (%s, %s, %s, %s, %s, %s)"
  stmt = formatStatement(stmt, placeHolder)

  cursor.execute(stmt,
    (residueId, atom.name, x, y, z, atom.element))
```

If there are no errors then we close the cursor. Technically speaking, Python would eventually close the cursor in any case, but it's better to be explicit. Finally, we commit the transaction.

```
cursor.close()
connection.commit()
```

If there was an error then we again close the cursor. An error exception could occur for a few reasons. For example, the database connection might be lost. Or the `pdbId` might not have been set (this is not required in the `Structure` class constructor). If there has been an exception we do a rollback, to restore the previous state, and afterwards re-raise the originating exception object.

```
except Exception as e:  # syntax from Python 2.6
  cursor.close()
```

```
try:
  connection.rollback()
except:
  pass

raise e # re-raise original exception
```

Now we can test the above on data from a PDB file. We only want to run this test code if the module is run directly, rather than imported from another module, so we use a check on __name__. The user should specify the database name and PDB file, otherwise an error is given and the program exited. We then read the structures from a file, using the function getStructuresFromFile() from Chapter 8.

```
if __name__ == '__main__':

  import sys

  if len(sys.argv) != 3:
    print('need to specify database and PDB file')
    sys.exit(1)

  database = sys.argv[1]
  pdbFile = sys.argv[2]

  from Modelling import getStructuresFromFile
  structures = getStructuresFromFile(pdbFile)
```

We then open a connection and pass that object into the addStructureToDb() function. This is just a matter of taste. We could instead have passed in the database name (and username and password, if needed) and opened, and then at the end, closed the connection inside the function. In sqlite3 we can open the connection just with the database name:

```
import sqlite3
connection = sqlite3.connect(database)
placeHolder = '?'
```

In MySQLdb we also need a username and password. We assume that the user for the database is the same as returned by getpass.getuser(), and similarly for the password. In general this might not be the case, so the test code might need tweaking here. (It is just test code, though.) Accordingly, the MySQLdb alternative to the above block of code would be something like:

```
import MySQLdb
import getpass
user = getpass.getuser()
pwd = getpass.getpass()
connection = MySQLdb.connect(db=database, user=user, passwd=pwd)
placeHolder = '%s'
```

We then add the structures into the database, one after the other. We wrap the code in a try/finally block so that we always close the connection whether or not an error has occurred. Again, Python would do this automatically, but it's better to be explicit. Note that the way the code is written, if we get an error adding some structure, then the previous structures will remain in the database because those transactions will have been committed in `addStructureToDb()`. An alternative would be to do the transaction management here, so that either all the structures are inserted, or none of them is.

```
try:
   for structure in structures:
      addStructureToDb(connection, structure, placeHolder)
finally:
   connection.close()
```

Now we consider how to change the code to allow several insert statements into each SQL command that is executed. This is possible in MySQL but not in SQLite, and the below is how this kind of application should be implemented in practical situations. The simplest change is to insert the atom records for a given residue in one command, since normally there are many atoms per residue, but to leave the other records, for the structure, chains and residues, inserted one at a time. In this case, the only modification required in the code is the innermost loop, which now looks like:

```
values = []
for atom in residue.atoms:
   (x, y, z) = atom.coords
   values.extend([residueId, atom.name, x, y, z, atom.element])

nAtoms = len(values) / 6
atomPlaceHolder = '(%s, %s, %s, %s, %s, %s)'
atomPlaceHolder = nAtoms * [atomPlaceHolder]
atomPlaceHolder = ','.join(atomPlaceHolder)

stmt = "insert into atom" \
       " (residueId, name, x, y, z, element) values " \
       + atomPlaceHolder
cursor.execute(stmt, values)
```

The values that are going to be inserted into the atom table are stored in an array. It would be natural, from a Python point of view, to store this as a list of lists, with each inner list being the six pieces of data relevant for each atom (residue ID, atom name, x, y, z, chemical element). But the way MySQLdb works this is not possible, and instead we have to store everything in one long list. The variable `nAtoms` is the number of atom records being inserted, and this is the length of `values` divided by 6 because there are six pieces of data for each atom. The MySQL syntax for inserting multiple records into a database does expect a list of lists, and this gives rise to the slightly complicated construction of the variable `atomPlaceHolder`. For example, if nAtoms is 2 then `atomPlaceHolder` would be the Python string '(%s, %s, %s, %s, %s, %s),(%s, %s, %s, %s, %s, %s)'.

On a test protein with a total of 20 structures, 40 chains, 1600 residues and 26,440 atoms (around 17 atoms per residue) the time to insert this data into the database was reduced by a factor of around 3.5 in comparison with the original method. So this is definitely worth doing (although as noted, it cannot be done with SQLite). A further optimisation is to insert all the structures in one go, then all the chains in one go etc. This is left as an exercise for the reader. On the test protein this further reduced the time to insert the data by around another 25%, so is a much smaller effect.

Getting a structure from the database

The converse operation from the previous section is to load a structure from the database into a `Structure` object. Standard SQL allows all the data to be fetched in one query of the database, and this is the way it should be done for reasons of efficiency. The `Structure` object is then created from the returned result.

The single query is rather long: for example, in SQLite it would be:

```
stmt = "SELECT structure.name, chain.molType, chain.code,
residue.seqId, residue.code, atom.name, atom.x, atom.y, atom.z FROM
structure, chain, residue, atom WHERE structure.pdbId=? AND
structure.conformation=? AND structure.id=chain.structureId AND
chain.id=residue.chainId AND residue.id=atom.residueId ORDER BY
chain.id, residue.id, atom.id"
```

Note that all the tables are included in the FROM clause and all the parent-child ids are explicitly checked as being equal in the WHERE clause (these are table 'joins'), in addition to the usual check on `pdbId` and `conformation`. Where the column names are not unique we have to include the table name as a qualifier. Here, for the sake of clarity, we have always included the table name, even when the column name is unique.

In the function we first get the `cursor` object.

```
from Structures import Structure, Chain, Residue, Atom

def getStructureFromDb(connection, pdbId, conformation=1,
                       placeHolder='%s'):

  cursor = connection.cursor()
```

We wrap the code in a `try`/`finally` block so that we always close the cursor at the end, whether or not there is an error. We then get the complete record from the database. If that fails then we raise an error exception.

```
  try:
    stmt = "SELECT structure.name, chain.molType, chain.code,
residue.seqId, residue.code, atom.name, atom.x, atom.y, atom.z,
atom.element FROM structure, chain, residue, atom WHERE
structure.pdbId=%s AND structure.conformation=%s AND
structure.id=chain.structureId AND chain.id=residue.chainId AND
residue.id=atom.residueId ORDER BY chain.id, residue.id, atom.id"
    stmt = formatStatement(stmt, placeHolder)

    cursor.execute(stmt, (pdbId, conformation))
    result = cursor.fetchall()
    if not result:
      msg = 'structure with (pdbId=%s, conformation=%s) not known'
      raise Exception( msg % (pdbId, conformation))
```

We then loop over all the records in the result. Each record represents one atom, so the code is similar in style to that in the `getStructuresFromFile()` function in Chapter 8. We create the `structure` object the first time through the loop.

```
    structure = chain = residue = atom = None
    for (structureName, chainMolType, chainCode,
         residueSeqId, residueCode, atomName,
         atomX, atomY, atomZ, atomElement) in result:

      if not structure:
        structure = Structure(structureName, conformation, pdbId)
```

We then create the `residue` and `chain` objects if they are needed.

```
    if not chain or chain.code != chainCode:
      chain = Chain(structure, chainCode, chainMolType)

    if not residue or residue.chain != chain \
        or residue.seqId != residueSeqId:
      residue = Residue(chain, residueSeqId, residueCode)
```

We then (always) create an atom object.

```
    coords = (atomX, atomY, atomZ)
    Atom(residue, atomName, coords, atomElement)
```

At the end we close the `cursor` and return the `structure`.

```
  finally:
    cursor.close()

  return structure
```

We add some test code. For `sqlite3` we have:

```
if __name__ == '__main__':

  import sys

  if len(sys.argv) not in (3, 4):
    print('need to specify database, PDB id, [conformation=1]')
    sys.exit(1)

  database = sys.argv[1]
  pdbId = sys.argv[2]

  if len(sys.argv) == 3:
    conformation = 1
  else:
    conformation = int(sys.argv[3])

  import sqlite3
  connection = sqlite3.connect(database)
  placeHolder = '?'

  try:
    structure = getStructureFromDb(connection, pdbId,
                                   conformation, placeHolder)
  finally:
    connection.close()
```

For MySQLdb we instead have the slightly more complicated:

```
import MySQLdb
import getpass
user = getpass.getuser()
pwd = getpass.getpass()
connection = MySQLdb.connect(db=database, user=user, passwd=pwd)
placeHolder = '%s'
```

The test code does not do anything with the generated `Structure` object but the code could be extended to do so. More generally, once you have a `Structure` object then you can do all the manipulations on a structure that were discussed in Chapter 8. Having all the structures in a database makes doing this on many structures in one go a relatively easy matter.

Querying the database

Instead of loading records from the database into a `Structure` object, and manipulating those, we can just manipulate the records directory. For example, suppose we want to count all the chains in the database of a specified `molType`. Then we can do as follows:

```
def countChainMolType(connection, molType, placeHolder='%s'):

  cursor = connection.cursor()

  try:
    # get matching chain records from database

    stmt = "select * from chain where molType=%s"
    stmt = formatStatement(stmt, placeHolder)

    cursor.execute(stmt, (molType,))
    result = cursor.fetchall()

    count = len(result)

  finally:
    cursor.close()

  return count
```

Alternatively, we could use the SQL statement

```
"select count(*) from chain where molType=%s"
```

which counts the records and returns that result.

The test code, for `sqlite3`, is:

```
import sys

if __name__ == '__main__':

  import sys

  if len(sys.argv) != 3:
    print('need to specify database, molType')
    sys.exit(1)
```

```
database = sys.argv[1]
molType = sys.argv[2]

import sqlite3
connection = sqlite3.connect(database)
placeHolder = '?'

try:
  count = countChainMolType(connection, molType,
                            placeHolder)
  print 'Found %d chain(s)' % count
finally:
  connection.close()
```

Note that we do not even have to deal with the `Structure` table in order to answer this particular query, but just the `Chain` table.

The `Structure` object is useful to have if there is a general application dealing with structures and where it is useful to have the structures all in memory at the same time, and where the database just happens to be the way that the data is stored. But in some applications, e.g. using a browser to display information about structures that are stored in a remote database, it is quite possible that just querying the database and using the data directly without creating a `Structure` is the best way to proceed.

21 Probability

Contents

The basics of probability theory

The theory of probability was based on the observation of random physical events, most notably for games of chance. And naturally, calculating accurate probabilities became especially important for people when money was wagered on the outcome. Probability is a way of ascribing numerical values to the possible outcomes to help us understand a random process more fully. This enables us to ask questions like how much more often one event occurs compared to another, but because of the random nature of what we are studying we can never say what the outcome will definitely be. Rather we tend to think of the process in terms of what the long-term proportions of different outcomes are, if the random experiment were repeated a very large number of times, or perhaps if money is involved what a wager on a particular outcome is worth.

Turning to biological systems, some things in living organisms occur as a result of random processes, like the segregation of a parent's chromosomes among their children or base-pair changes in DNA (such as a result of replication errors or ionising radiation), though, under most circumstances we don't get to see the actual random event. For the most part we just view the outcomes, sometimes billions of years later in the case of DNA sequence changes. Of course a DNA sequence isn't actually random, given that it exists to contain biologically meaningful information representing genes and gene control elements etc. which have been selected for their function during evolution, even if the initial mutations were random. Nonetheless for a sufficiently large and unbiased selection of DNA we can treat the sequence as if it were random in order to ask various questions. For example, how often do I find the sub-sequence AAGCTT in a megabase-long region of DNA?

Probability theory is often also useful in situations where there is no underlying randomness in the biology, but rather an uncertainty in our scientific interpretation. Here a probabilistic treatment of our uncertainty can lead to informative predictions. An example of this would be for the classification of whether two genes have the same function as one another (generally because they have a common ancestor). They either do or do not, and the underlying assignment of this status is not a random process, but our prediction based on the available data does have an uncertain component, and so it can be helpful to treat the situation probabilistically. It is also notable that in biological analyses it may be rare to actually deal with probabilities directly, but probability theory underpins statistical tests which are very commonly used, and we describe some of those in Chapter 22.

Here we will lightly go through some of the fundamentals of probability theory. Being mindful of our expected readership, we will endeavour to avoid going into too much detailed mathematical notation. We won't escape the equations entirely but hopefully these will serve as a primer for further reading.

Sample space

Firstly, we need to define a probabilistic system by knowing what the range of possible outcomes is. In mathematical jargon this means to define the *sample space*. The range of possible outcomes can be fairly straightforward, so for a six-sided die we know that there are simply six outcomes corresponding to the numbers of spots on different faces. If we are thinking about the occurrence of a DNA base at a position in a genome then we know that it must be either G, C, A or T. Often though we are thinking about multiple dice rolls or several positions in a DNA sequence. In these cases we think of the sample space in terms of the combinations of possibilities for each roll or position. Hence for rolling two dice we have six possibilities for the first roll, and then for any given first roll there are a further six possibilities for the second roll. Overall there will be six times six possibilities for the total number of possible outcomes. Naturally if there is a further roll there are six more possibilities for each of the 36 two-roll outcomes. So here the general rule is that the size of the sample space is 6^N, if there are N rolls, i.e. multiplied by six for each roll. The same idea can be applied to sequential positions in DNA. Here there are four nucleotide possibilities at each position and so for a sequence (or sub-sequence) of length N there are 4^N different combinations. Although the nucleotides of a DNA sequence are actually all present in the same molecule, it may be helpful for understanding to fictitiously imagine the sequence being generated by the roll of an imaginary four-sided die.

More generally there can sometimes be the complication that we actually don't have a fixed number of dice rolls or a fixed length of DNA. For example, we may be interested in finding out how many dice rolls we would expect to make, on average, before we roll three sixes. The DNA equivalent of this is to ask what the expected length of DNA (number of positions) is before we find a given small sub-sequence. The latter is quite a relevant question

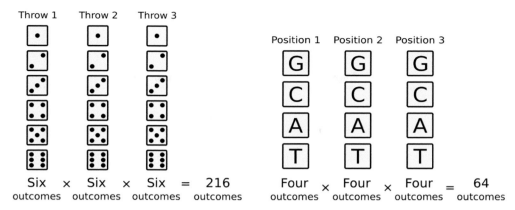

Figure 21.1. Sample space size for sequential events. The calculations to give the size of the sample space are shown for three subsequent rolls of a die and three positions in a DNA sequence. The sample space represents the totality of all possible outcomes. For a sequence of events, where each has a fixed number of possibilities, the size of the sample space is the product of the numbers of possibilities at each point in the sequence.

biologically because the small sub-sequence might be a cut site for a restriction enzyme,[1] where it can be useful to know the average size of DNA fragments the enzyme would generate. In these examples the sample space may be unbounded, or at least very large in the case of a genome. Nonetheless we still have a firm idea of what the range of possibilities is, even if it is technically infinite. For example, even though it may be technically possible to never roll three consecutive sixes if a die were rolled continuously for the history of the universe the odds are so astronomically small (close to zero) that this, and similar extremes, don't have any practical effect.

Probability values

Probabilities are values between zero and one (between impossibility and certainty) that we assign to the outcomes of a random process. The summation of all probabilities, over all possible outcomes, is exactly one. In effect, each outcome occupies a fraction of likelihood from the certainty that something happens in a random process. Hence for a roll of an unbiased die the probability assigned to each of the six possible outcomes is $1/6$, so they all add up to one. As illustrated in Figure 21.2, this can be visualised by each outcome taking a different fraction of a line of unit length, though the order of the resulting regions is unimportant. Note that we introduce the notation Pr(X), which is shorthand to mean the probability of X occurring, whatever occurrence X may be.

For a biological system we may have a good model of how we expect things to behave, i.e. that we know what the probabilities are. For the genetics example illustrated in Figure 21.2 we might assume that the probability is ¾ for having black offspring and ¼ for white offspring, and likewise for a position in DNA we might say that the probability of each base is ¼ and thus for a pair of bases is $1/16$.

---

[1] An enzyme that cuts DNA at specific sub-sequences.

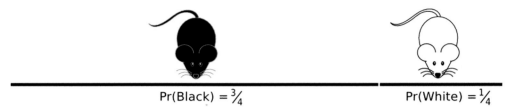

Pr(Black) = ¾   Pr(White) = ¼

Figure 21.2. Probabilities partitioning a unit line. Probability values for the different possible outcomes can be imagined to partition a region of length one. Here we illustrate this for the probabilities for the colour of mouse offspring resulting from a genetic cross (between a pure white strain and a pure black strain) where black is three times more likely than white.

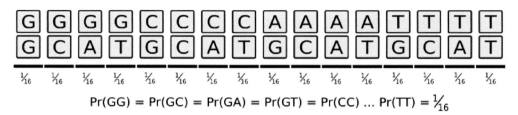

$\frac{1}{16}$ $\frac{1}{16}$ $\frac{1}{16}$ $\frac{1}{16}$ $\frac{1}{16}$ $\frac{1}{16}$ $\frac{1}{16}$ $\frac{1}{16}$ $\frac{1}{16}$ $\frac{1}{16}$ $\frac{1}{16}$ $\frac{1}{16}$ $\frac{1}{16}$ $\frac{1}{16}$ $\frac{1}{16}$ $\frac{1}{16}$

Pr(GG) = Pr(GC) = Pr(GA) = Pr(GT) = Pr(CC) ... Pr(TT) = $\frac{1}{16}$

Figure 21.3. Nucleotide probabilities for two DNA positions. For two positions in a DNA sequence there are 16 possible outcomes, given the four different types of nucleotide. If the probabilities of single nucleotides are equal then the probabilities of all nucleotide pairs are also equal ($^1/_{16}$), and naturally sum to one.

We should not forget, however, that these values stem from idealised models, so that in reality the actual probabilities are not neat whole-number fractions. Accordingly, for a DNA nucleotide position it is only an approximation to say that each base is equally likely. In reality in any given genome there will be more G and C bases than A and T bases, or vice versa. It should be noted that if we consider both DNA strands then because of the base-pairing rules the numbers of G and C will be the same, and thus also the numbers of A and T. Naturally this means that the probability of finding a base at a random position is also equal within these pairs.

As we have alluded to, in order to obtain a realistic probability estimate for different outcomes we generally count the number of occurrences of each in a large data set. Hence for our mouse-breeding example, even if we didn't have a good theoretical model we could cross the two strains, count the different coat colours of the progeny and then express the counts as a proportion of the total. We may do such experiments to validate a given model, which in this case might show something of genetic interest, if the model does not fit. Though, for this kind of hypothesis testing (which is more properly described in Chapter 22) we have to be mindful of how the amount of data affects our confidence. To take an arbitrary example with a mouse cross, just because eight black mice were born in a litter does not mean that the model of a 3:1 black-white ratio is wrong; litters of eight would be all black about 10% of the time (0.75^8). You would need a much larger sample of data to be confident of the probabilities; the more experimental examples we have the closer the experimental ratios will match the long-term probabilities. Likewise for DNA nucleotide probabilities we can count C:G and A:T base pairs we find in a genome,[2] and will get the most accurate results by choosing as large a sample of sequence data as possible. If we want our probabilities to be general for the whole genome we would not want to look at only a small part, which may not be representative.

In Python if we know the number of G:C and the number of A:T pairs for a whole genome then the probability of each, i.e. Pr(G), Pr(C), Pr(A) and Pr(T), at a random position can be calculated as the proportion of the total:

```
counts = {'G':2356491, 'C':2356491, 'A':2283184, 'T':2283184}
total = float(sum(counts.values()))
letterProbs = {}
for letter in counts:
  letterProbs[letter] = counts[letter] / total

print(letterProbs)

# Result: {'A':0.24605, 'C':0.25395, 'T':0.24605, 'G':0.25395}
```

Even though these probabilities are improved from ¼ for all bases it should still potentially be considered as an approximation, depending on the situation at hand. You may have noticed that we have been quite careful to say that this is the probability at a **random** position. If the DNA position we are considering is not random then the above whole-genome average would just be the first approximation.[3] The G:C content of DNA is actually different for different chromosomes and generally varies depending on whether a position is in a gene or non-gene region. We could end up with endless categorisations and qualifications for probabilities. So while it is possible to define the probabilities for C or G being at (to take an arbitrary and complex example) the last position of the first exon of all carbohydrate metabolism genes, we wouldn't want to go into so much detail unless there was a special reason. In general a balance is struck between having accurate general probabilities, supported by large amounts of data, and contextualised probabilities, which may be supported by very little data. In a probabilistic analysis we may wish to account for context, to make more accurate predictions, but naturally we must have data for the different situations and know when to use them. There will be some further discussion of such matters in the Markov chains section below.

Restriction enzyme example

Taking the DNA example a little further, let's consider a restriction enzyme called HindIII that is commonly used in molecular biology and which cuts DNA at the specific sequence AAGCTT. Using a simple probabilistic model about the likelihood of finding a given letter at a given position in an otherwise random DNA sequence, we can estimate various properties, like how often the enzyme cuts or what the size of the fragments will be after cutting. A DNA sequence actually isn't totally random, but the approximation is nonetheless good enough to get useful predictions.

---

[2] If a whole-genome sequence is not available this could also be done the old-fashioned way, by hydrolysing the DNA into its component nucleotides and chemically determining the concentration of the components.

[3] Also, if we are thinking about DNA from different individual organisms which have slightly different sequences, compared to the sequenced genome, then the fractions would vary by a tiny amount, though this would usually be of little concern.

Assuming that the nucleotide at one position does not depend in any way on what the nucleotides are at the other positions (i.e. the nucleotides at different positions are independent), we can calculate the probability of a HindIII site at any six residue sub-sequence to be $Pr(A) \times Pr(A) \times Pr(G) \times Pr(C) \times Pr(T) \times Pr(T)$. This is about one cut in 4096 (4^6) positions, if we assumed equal probabilities for all nucleotides. Hence for a DNA sequence of length N we would expect $N \times \frac{1}{4096}$ restriction enzyme cut sites. Also, on average we could expect the separation to be about 4096 bases. Calculating the probability of the cut site using the non-equal nucleotide probabilities calculated above we get:

```
cutSite = 'AAGCTT'
probSite = 1.0 # Starting value

for letter in cutSite:
  probSite *= letterProbs[letter]

print(probSite)  # 0.00023637 - approx one in 4230
```

Because the occurrence of a site is effectively random we will expect a distribution of different values for the number of cut sites in a given length and also for the lengths of the fragments. In other words because the sites are random, and not regular, the spacing between sites will generally be more or less than 4230. We will consider models for the shape of such probability distributions later in this chapter.

Combining probabilities

The final major rule with probability theory is about how we combine probabilities. If there are mutually exclusive outcomes then the probability of the event that any of these particular outcomes occurs is the sum of their individual probabilities. In mathematical terms we describe an *event* formally as a set of different outcomes. Also, by saying that if an event is defined by one outcome **or** another (or another etc.) then it can be described in set theory terms as the *union* of the outcomes. Taking the roll of two fair dice as an example, as we have already illustrated there are 36 possible outcomes each with probability $\frac{1}{36}$. If we want to know the probability that the total on the two dice is seven, then we first consider which of the outcomes contribute to this event (as illustrated in Figure 21.4) and then add the probabilities for each.

When using dice rolls or DNA bases as examples it is clear that individual outcomes are mutually exclusive, in that a die provides only one number and that only one of the DNA nucleotides is possible in a given position. However, it is also possible to define events that are not mutually exclusive. For example, given two DNA positions we could investigate the event that there is at least one A nucleotide present (7 of 16 possibilities) and the separate event that the two nucleotides are different (12 of 16 possibilities). Clearly these are not exclusive events because some outcomes are present in both sets. The set of common outcomes, which apply to both one event **and** another, is referred to as the *intersection*. Furthermore knowing what the intersection is allows us to calculate things like the probability that one event **but not the other** occurs.

We can use Python to calculate the probabilities for the illustrated events and their intersection in the case where the probabilities of the outcomes are not equal. Here the probabilities of the two-letter outcome are obtained by multiplying the individual probabilities for a nucleotide recorded in `letterProbs`, i.e. the probability of having x at the first position times the probability of having y at the second. The outcomes are then tested and added to the

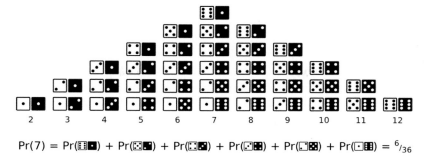

$$Pr(7) = Pr(\boxed{\cdot\,\cdot\cdot}\boxed{\cdot\cdot\cdot}) + Pr(\boxed{\cdot\cdot}\boxed{\cdot\cdot\cdot}) + Pr(\boxed{\cdot\cdot}\boxed{\cdot\cdot}) + Pr(\boxed{\cdot}\boxed{\cdot\cdot}) + Pr(\boxed{\cdot}\boxed{\cdot\cdot}) + Pr(\boxed{\cdot}\boxed{\cdot\cdot}) = {}^6/_{36}$$

Figure 21.4. The range of all possible outcomes for two dice. The outcomes which all contribute to each numeric total are grouped together. The probability of getting a total of seven is then calculated from the sum of the outcomes for that event.

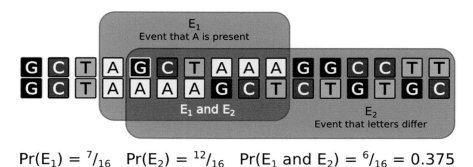

$$Pr(E_1) = {}^7/_{16} \quad Pr(E_2) = {}^{12}/_{16} \quad Pr(E_1 \text{ and } E_2) = {}^6/_{16} = 0.375$$

Figure 21.5. Combining probabilistic events. The first event, that one nucleotide from the 16 pairs contains an **A**, and the second event, that the nucleotides are different, are subsets of the total set of outcomes. The intersection between the two events is the set of outcomes common to both. Probabilities are calculated for the events assuming that all outcomes are equally likely.

appropriate event, which we store as Python sets. We can then use the set operation '&', which generates a new set with the common elements. Finally, the probabilities of the events are simply calculated from the summation of the probabilities within each.

```
probs = {}
letters = ['G','C','A','T']

event1 = set()
event2 = set()

for x in letters:
  for y in letters:
    outcome = (x,y)
    probs[outcome] = letterProbs[x] * letterProbs[y]

    if 'A' in outcome:
      event1.add(outcome)

    if x != y:
      event2.add(outcome)
```

```
intersection = event1 & event2

pEvent1 = sum([probs[xy] for xy in event1]) # 0.43156
pEvent2 = sum([probs[xy] for xy in event2]) # 0.74994
pEvent1and2 = sum([probs[xy] for xy in intersection]) # 0.37102
```

Something that follows from the basic axioms of probability is the notion that we can use the probability of the intersection between events $Pr(E_1$ and $E_2)$ to calculate the probability of the union between events $Pr(E_1$ or $E_2)$:

$$Pr(E_1 \text{ or } E_2) = Pr(E_1) + Pr(E_2) - Pr(E_1 \text{ and } E_2)$$

If there is an intersection between the event E_1 and the event E_2 adding the probabilities for the two will include the overlapping outcomes twice, so subtracting the intersection that both E_1 **and** E_2 happen redresses this. This way each outcome that involves E_1 or E_2 contributes the same. When considering mutually exclusive events the probability $P(E_1$ and $E_2)$ is naturally zero, in which case $Pr(E_1$ or $E_2)$ is just the sum of the independent probabilities.[4] We can show the calculation of $P(E_1$ or $E_2)$ in Python by either creating the appropriate set or by using the above equation:

```
union = event1 | event2 # Set with elements from both
pUnion = sum([probs[xy] for xy in union])

print(pUnion)                      # 0.81049
print(pEvent1 + pEvent2 - pEvent1and2) # 0.81049 - same
```

While we can treat combined dice rolls or DNA positions as discrete outcomes we can also imagine these as arising from a chain of probabilistic selections. In the above examples the trials are independent and the result of the first has no influence on the second, which is reasonable for a fair die. However, for DNA (and many other analogous situations in biology) the probabilities of the occurrence of a nucleotide at each position may not only be different, as discussed before, but the probability for the second position may also vary according to **which** base is present in the first position, or indeed many other positions.

In this case we would say the positions were not independent and the probability of observing the second nucleotide differs, depending on the outcome of the first. To calculate the probability of getting each pair of nucleotides we get the probability of obtaining the first nucleotide and multiply this by the probability of getting the second, **given** the first. This is what is termed a *conditional probability* and in general we would need to know what the probabilities for the four nucleotides were given each particular preceding nucleotide.

Conditional probabilities

Moving to a different kind of example, we will consider probabilities associated with the occurrence of a disease (D) and how this relates to the experimental observation of a particular mutant version (M) of a gene, i.e. with a different DNA sequence. Here the probability that both occur, $Pr(D$ and $M)$, on its own does nothing to suggest whether the two are related. Naturally to investigate the link between the two we would need to know probabilities of the events alone

---

[4] This is actually an axiom: for mutually exclusive or *disjoint* events $Pr(E_1$ or $E_2) = Pr(E_1) + Pr(E_2)$.

(having the disease and having the mutation) and thus whether the intersection of the two is more or less than we would expect if they were unrelated. By doing this we implicitly use the concept of hypothesis testing. As far as medical prediction and diagnosis is concerned it is helpful to consider the *complementary* events. In this case these are the event that there is no disease and the event that there is no mutation. With these we can compare the hypothesis, that the disease and mutations are linked, with an appropriate *alternative* and *null hypothesis* (see Chapter 22).

By counting occurrences of the different situations we can estimate the various combinations of conditional probabilities. For example, we can estimate the probability of having the disease given that the mutation is present, Pr(D given M), and compare it to the probability of having the disease given no mutation Pr(D given noM), i.e. whether the mutation increases or decreases the chance of the disease. Also, if it is established that Pr(D given M) is much greater than Pr(D given noM), i.e. that the mutation is highly correlated with the disease, then knowing the probability of not having the disease given the mutation being present Pr(noD given M) is vital if we hope to use a genetic test to predict the disease outcome; in other words we need to know whether there would be lots of false-positive results.

We can also think of the dependent DNA events in the HindIII restriction enzyme example in terms of conditional probabilities, for example, what the probability of having a cut site (AAGCTT) is in a region of DNA **given** a G:C content greater than 60%. It should be noted that this is a distinct question from asking what the probability of one event **and** another is, though the two are related. For this example the probability that both events occur considers the outcomes from all the possible DNA sequences, while the probability that one occurs given the other does not, it only considers situations where the second event has definitely occurred. The probability that they are both true is the same as the probability of one occurring multiplied by the probability of the second occurring given that we've already got the other. So for two arbitrary events X and Y we have:

$$\text{Pr}(X \text{ and } Y) = \text{Pr}(X) \times \text{Pr}(Y \text{ given } X)$$

And it doesn't matter which way we phrase this, the converse is also true:

$$\text{Pr}(X \text{ and } Y) = \text{Pr}(Y) \times \text{Pr}(X \text{ given } Y)$$

Obviously this only makes sense if Pr(X) and Pr(Y) are not zero. Combining these two formulations we can say that one is equal to the other, i.e. that:

$$\text{Pr}(X) \times \text{Pr}(Y \text{ given } X) = \text{Pr}(Y) \times \text{Pr}(X \text{ given } Y)$$

which is often written in the form:

$$\text{Pr}(Y \text{ given } X) = \text{Pr}(Y) \times \text{Pr}(X \text{ given } Y) / \text{Pr}(X)$$

This is a very important result which is called Bayes' theorem. As we discuss in the next section this formulation is commonly used for hypothesis testing.

Returning to our medical example, for prognosis and appropriate treatment we might want to know the probability of getting the disease given the mutation Pr(D given M). However, it may not be cost-effective to obtain statistics by genetically testing large numbers of people for the mutation, just for the chance that they would get a rare disease. Also, it might be that the disease is difficult to diagnose and doesn't show immediately. Conversely it may be easier to determine Pr(M given D) by testing a limited number of people who definitely do have the disease to discover whether they have the mutation. Using Bayes' theorem we can easily get the probability we want from the other.

Pr(D given M) = Pr(M given D) Pr(D) / Pr(M)

Naturally we must also estimate Pr(D) and Pr(M), the probabilities of disease and mutation in the absence of any other information, from statistical data. However, Pr(D) could simply come from medical records and Pr(M) could come from testing any group of people, whether or not they had the rare disease.

Bayesian analysis

Bayesian analysis is a very powerful and, in the minds of some, the 'proper' way to generally think about scientific matters. Scientific philosophy is largely based upon proving or disproving hypotheses using experimental evidence. Thinking in general and abstract terms with the above example, and introducing the symbolic notation '|' to mean 'given', we have:

$$\text{Pr(Hypothesis}_i \mid \text{Data) Pr(Data)} = \text{Pr(Data} \mid \text{Hypothesis}_i\text{) Pr(Hypothesis}_i\text{)}$$

What this says about the scientific approach may not be immediately clear from a symbolic representation, but a key aspect here is that in science we compare different hypotheses, hence the introduction of the subscript i, to label one hypothesis among others. Essentially what this says is that the interesting *posterior* quantity Pr(Hypothesis$_i$ | Data) is only meaningful in comparison with other, competing hypotheses. The likelihood of a given hypothesis generating the experimental data Pr(Data | Hypothesis$_i$) is a measure of how well the data fits the hypothesis. However, even if one hypothesis seems to fit the experimental data very well, our confidence in this particular hypothesis is naturally diminished if there is a somewhat different hypothesis that also fits the experimental data very well. Conversely if all the hypotheses that fit the data are very similar then the confidence of our answer increases, and we gain an awareness of the width of acceptable solutions: what precision is meaningful in our hypotheses. Accordingly, the Bayesian inferential approach is more objective than a simple deductive approach (where if something fits well it is assumed to be the correct answer), and it has an inbuilt mechanism to quantify the uncertainty associated with a hypothesis.

An aspect of Bayesian analysis which in some situations may not seem particularly scientific is the quantity Pr(Hypothesis$_i$) that represents the *prior* information about the hypothesis, in the absence of any experimental evidence. Indeed the ability of this approach to work well can often depend on a scientist's ability to come up with a good estimate of the prior probability. We can always say that we have no prior information to compare hypotheses in the initial instance, i.e. that the prior is the same for all hypotheses, in which case our analysis effectively becomes a *maximum likelihood* approach. However, it is often possible to do better by thinking about the system under study, which is a general principle when doing mathematical modelling. Thinking of an example about molecular 3D structure, we can use prior probabilities to say that some conformations are more likely than others, considering things like the length of and the angle between chemical bonds. Effectively we are selecting hypotheses that fit what we generally know about molecular structures, disregarding solutions with distorted geometries. You could argue that this is subjective and thus biased, to find solutions that fit our expectations. However, in practice good and useful prior probabilities will generally derive from a well-founded theory or other experimental observations, e.g. about how long different kinds of chemical bonds are on average.

It should be noted that the quantity of Pr(Data) is the same for all hypotheses and so calculating its value doesn't help in determining the best hypothesis. Accordingly it is often ignored and set to a value of 1. However, if an accurate value of $Pr(Hypothesis_i \mid Data)$ is sought then Pr(Data) can be calculated by summing the likelihood over all hypotheses: $\Sigma_i Pr(Data \mid Hypothesis_i)Pr(Hypothesis_i)$.

Random variables

Random variables describe numeric values that relate to the outcomes of a random, probabilistic process. A very simple example of a random variable is the height of individual people in a population. A random variable describes a range of possible values, which we call a *distribution*, and we associate a probability with each value. A random variable can be applied to *discrete* events, like counting the number of G:C nucleotides in a DNA sequence, where the number of outcomes is finite.[5] Alternatively the random variable may be *continuous*, as is the case with our height example. However, in this chapter we will concentrate on discrete random variables. We have implicitly mentioned random variables earlier in this chapter, in relation to what the sequence separation would be for the HindIII restriction enzyme cut site. In this case the random variable represents the variation in the length of the DNA sequence before the cut site, which might be the separation between one cut site and another when cutting a whole genome into small fragments. Here the distribution of lengths, and thus the probability of each, can be modelled with the *geometric distribution*, as we describe below. By matching a random variable to a well-characterised probability distribution we say something about the process that generated it. Going further, if we have a candidate model for a random process that matches the distribution (and thus explains our data) we can then look for deviation from the model. This may suggest a better model or illustrate in what way our data is not random, e.g. if a restriction enzyme cuts at sites that don't match expectations.

Next we will go through some of the more commonly used probability distributions. We aim to give an idea of how they arise and thus what they may be useful for. Practical biological examples are given in Python, often by making use of the SciPy library, which has a module for probability and statistics, `scipy.stats`.

Binomial distribution

Given an event with a fixed probability of occurrence, the *binomial distribution* is the probability distribution of the number of events that occur after a specified number of *independent* trials. A simple example of this would be the event of rolling a six on a die, i.e. with probability $^1/_6$, where after a specified total number of rolls we can count the number of times that a six came up. Repeating the same experiment (with the same total

---

[5] Or more generally countable.

number of rolls) will result in a distribution of different counts for rolling a six. The probability of getting a given count of sixes is described by the binomial distribution. For a given event probability and given number of trials, the probability of a count can be calculated using the formula presented below. This is based on the notion that the probability of a count depends on the number of arrangements in which the count can be obtained. To take the example of rolling a die three times, where there are 216 ($6\times6\times6$) possible outcomes, there is only one way of getting a count of three sixes, but there are 15 ways of getting two sixes (a non-six can occur in three positions, and there are five possibilities for each), 75 ways of getting one six (a six can occur at three positions and there are five times five possibilities for the non-sixes) and 125 ways of getting no six (five possibilities for each position).

The probability $\Pr(k)$ of observing k events from n independent trials given event probability p is:[6]

$$\Pr(k) = \frac{n!}{k!(n-k)!}(1-p)^{n-k}p^k$$

This is often written using $\binom{n}{k}$, which is notation for the combinatorial factor, giving the number of ways of choosing k items from a total of n:

$$\Pr(k) = \binom{n}{k}(1-p)^{n-k}p^k$$

If we seek the probability of getting two sixes from three rolls we multiply the probability of getting two sixes, $p^k = 1/6^2$, by the probability of getting a non-six in the other rolls, $(1-p)^{n-k} = (5/6)^{3-2}$, by the number of ways of choosing two successes from three rolls, $\binom{n}{k} = 3!/2!(3-2)!$, and the result is indeed 15/216.

We can define a function to calculate this in Python, using the handy comb, which we can import from SciPy to calculate the combinatorial factor:

```
from scipy.misc import comb

def binomialProbability(n, k, p):
  return comb(n, k) * p**k * (1-p) ** (n-k)
```

To test this we can again calculate the probability of getting two sixes from three rolls of a die:

```
p = 1/6.0          # Probability of event
n = 3              # Number of trials
k = 2              # Number of events sought

print( binomialProbability(n, k, p) )
# Result is 0.069444444 = 15/216
```

As a biological example we could investigate the distribution in the number of sequencing errors (i.e. calling the wrong nucleotide) we expect when determining a DNA sequence of a given

[6] Here ! means factorial: the product of a number and all the positive integers smaller than it. For example: 6! = $6\times5\times4\times3\times2\times1$.

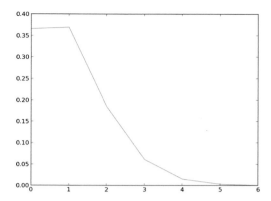

Figure 21.6. An example output of the Binomial distribution. A graph of the output generated from the `binomialProbability()` function, tested for 100 trials with event probability 0.01 and illustrating the probability for discrete numbers of events in the range from 0 to 6. Note that the line is only to guide the eye; because the distribution is discrete, it is only defined for whole numbers.

length. If the sequencing machine has a random error rate of 0.01 and reads the sequence for a total of 100 nucleotides, then the distribution of the number of errors can be plotted as follows:

```
from matplotlib import pyplot

p = 0.01
n = 100
xVals = []
yVals = []

for k in range(7):
  pk = binomialProbability(n, k, p)
  xVals.append(k)
  yVals.append(pk)

pyplot.plot(xVals, yVals)
pyplot.show()
```

This (plotted in Figure 21.6) shows that although the expectation is to have one error every 100 nucleotide positions, around 36% of the time there will be no errors.

It should be noted that the combinatorial factor gets very large as the number of trials n gets large, unless k is near 0 or near n, so the binomial probability can be computationally tricky to calculate. For example, returning to the restriction enzyme HindIII, which may cut a random DNA sequence with a probability of 0.0025, for a total DNA length of 10 megabases we may try to estimate the probability of having 2500 cuts (the mean value) as follows:

```
p = 0.00025
n = 10000000
print( binomialProbability(n, 2500, p) ) # Fails!
```

Although this was a reasonable question to ask the code fails because of the large number of combinations involved. All is not lost, however, because the `scipy.stats` module provides an implementation that is more robust (and quicker). To use this we define a Python object that represents a binomial random variable with given parameters, which here are the number of trials and event probability. This object has various methods (functions bound to it), and one of these is `.pmf()`, which represents the probability mass function, i.e. a function that calculates the probability for a given value of the random variable, just like `binomialProbability()`.

```
from scipy.stats import binom

p = 0.0025
n = 10000000
binomRandomVar = binom(n, p)

print( binomRandomVar.pmf(2500) )# Succeeds: 0.007979577
```

The random variable object also has other handy functions, as described in the SciPy documentation:

```
print( binomRandomVar.mean() ) # Most likely value   : 2500.0
print( binomRandomVar.std() )  # Standard deviation : 49.99
```

Also, as you might expect, the SciPy functions not only operate on single numbers, but also work with NumPy arrays. Hence, we can create a whole array of values for the counts *k* and calculate the array of probabilities for these with `.pmf()`:

```
from scipy.stats import binom
from numpy import array

binomRandomVar = binom(10000000, 0.00025)
counts = array(range(2300, 2700))
probs = binomRandomVar.pmf(counts)

pyplot.plot(counts, probs)
pyplot.show()
```

We can also calculate the cumulative probabilities, the sum of the probabilities from zero to each value of *k* (which is very handy for statistical testing, as discussed in Chapter 22), using `.cdf()`:

```
cumulative = binomRandomVar.cdf(counts)

pyplot.plot(counts, cumulative)
pyplot.show()
```

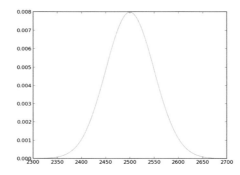

Figure 21.7. Example output of the binomial distribution for a large number of trials. A graph of the output generated using the `binom.pmf()` function from the `scipy.stats` module, tested for 10 million trials with event probability 0.00025. The graph illustrates the probability for discrete numbers of events in the range from 2300 to 2700, covering the mean value at 2500.

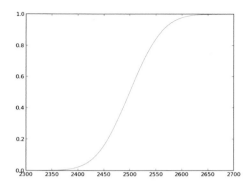

Figure 21.8. Example output of the Binomial cumulative distribution function. A graph of the output generated using the `bionom.cdf()` function from the `scipy.stats` module, tested for 10 million trials with event probability 0.00025 and illustrating the cumulative probability density for discrete numbers of events in the range from 2300 to 2700, covering the mean value at 2500.

Poisson distribution

If we know the **average rate** at which an event occurs, over a large number of independent trials, then the *Poisson distribution* is the probability distribution of the number of events that occur in a time interval. This is closely related to the binomial distribution, but specifying the rate (λ) at which the event occurs means we don't specify the number of trials (n) or the probability of an event (p), though the rate λ is essentially $p \times n$. The Poisson distribution would be used instead of the binomial distribution in situations where the number of trials is not measurable. For example, as we illustrate below, where statistically we observe the average rate of births in a population per day, we can calculate the probability distribution of the number of births per day without knowing the size of the population. The binomial distribution approaches the Poisson distribution as the number of trials (n) becomes large and the event probability (p) becomes small.

For the Poisson distribution the equation for the probability of observing k events given an occurrence rate of λ from independent trials is:

$$\Pr(k) = \frac{\lambda^k}{k!} e^{-\lambda}$$

Here e is the mathematical constant ≈ 2.71828 (Euler's number). It can be shown that the mean of the Poisson distribution is λ and the variance (see Chapter 22) is also λ.

We can implement the Poisson distribution using the `scipy.stats` module, which is quick and robust, compared to calculating the factorials and powers explicitly in basic Python. For an example where in a hospital there is an average of 4.7 births per day the Poisson distribution estimates the probability of observing a given number of births as follows:

```
from scipy.stats import poisson

poissRandomVar = poisson(4.7)

for k in range(10):
  pk = poissRandomVar.pmf(k)
  print('Number of births: %2d probability: %.3f' % (k, pk))
```

We can apply the distribution to the restriction enzyme example we used above, which shows that for large numbers of trials and small event probabilities the Poisson distribution is a very good approximation for the binomial distribution.

```
from scipy.stats import poisson
from numpy import array

rate = 10000000 * 0.00025
poissRandomVar = poisson(rate)
counts = array(range(2300, 2700))
probs = poissRandomVar.pmf(counts)

pyplot.plot(counts, probs)
pyplot.show()
```

Geometric distribution

For an event of specified probability, the *geometric distribution* is the probability distribution for the number of independent trials that **do not** result in the event, until the event is observed once. If we consider the event a 'success', then the distribution is over the number of 'failures'. A good example of this is the distribution of DNA fragment lengths after being cut with an enzyme like HindIII. Here the event is a cut site and the distribution of the number of other DNA positions we observe before finding a cut site is geometric. For the geometric distribution the probability of a number of independent non-event trials (k) required to observe the event with stated probability (p) is given by:

$$\Pr(k) = (1 - p)^{k-1} p$$

Here k takes integer values from one. A different formula arises if k starts at zero, i.e. $\Pr(k) = (1-p)^k \, p$, which represents the number of trials required before the event occurs, i.e. not including the event. As it happens, the scipy.stats.geom function uses the first form, which includes the event. The geometric distribution is related to a more general distribution called

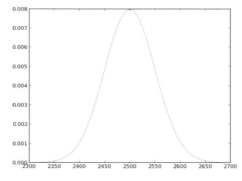

Figure 21.9. An example output of the Poisson distribution. A graph of the output generated using the poisson.pmf() function from the scipy.stats module, tested for an event rate of 2500. The graph illustrates the probability density for discrete numbers of events in the range from 2300 to 2700. This is a very good approximation for the binomial distribution illustrated above (Figure 21.7).

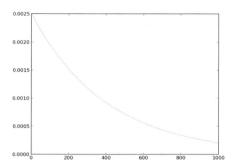

Figure 21.10. An example output of the geometric distribution. A graph of the output generated using the `geom.pmf()` function from the `scipy.stats` module, tested for an event probability of 0.0025. The graph illustrates the probability for the number of independent trials until the event occurs, in the range from 0 to 1000.

the *negative binomial* distribution that represents the probability distribution for the number of trials that occur before an arbitrary number of events of stated probability occur (rather than just one).

Again we can use SciPy to calculate the probabilities for the distribution, here using the probability for a restriction enzyme cut site to get the distribution in the DNA fragment lengths:

```
from scipy.stats import geom
from numpy import array

p = 0.0025
geomRandomVar = geom(p)
lengths = array(range(1, 1000))
probs = geomRandomVar.pmf(lengths)

pyplot.plot(lengths, probs)
pyplot.show()
```

The geometric distribution is the last discrete probability distribution we will describe in detail but there are several other distributions that are easily accessible in Python via their implementation in the `scipy.stats` module. Some of the more notable available probability distributions are as follows:

`dlaplace`: the discrete Laplace distribution; the differences between two independent but identically distributed random variables which themselves have geometric distributions.

`hypergeom`: the hypergeometric distribution, describing the number of successful events occurring after selecting a given number of items from a population without replacement. (With replacement the distribution would be binomial.)

`nbinom`: the negative binomial distribution, a generalisation of the geometric distribution for a variable number of events.

`randint`: the uniform distribution, i.e. where all values are equally likely.

`skellam`: the Skellam distribution, the differences between two independent random variables which themselves have Poisson distributions and different mean values.

Markov chains

When we have referred to the rolling of dice or subsequent DNA positions, we have already been considering a chain of events that occur from subsequent probabilistic trials.[7] Also, until now we have only dealt with situations where the probabilities of the various outcomes are independent of their position in the chain (i.e. that the probabilities don't change for the same kind of outcome at different locations in the chain of trials). As we have mentioned before, this is a simplification for many types of random process. Naturally there could be all sorts of dependencies, where the probabilities of the outcomes vary according to context. In a general sense we can describe this context in terms of conditional probabilities, where a probability is assigned for an event given the case that another event has occurred.

A Markov chain is a model which uses conditional probabilities for a chain of trials with the specific criterion that the probabilities of a trial are conditioned only on the outcome of the previous trial. Phrasing this differently we could say that the chain does not have any memory beyond its current state. This is certainly a simple model, and in reality a probability may actually be dependent on more than just the outcome of the previous trial. However, the fact that we have a whole series of conditional probabilities from the start of the chain means that information is relayed throughout the whole chain. The fact that Markov chains have proven useful in various areas of biology is undeniable.

Markov processes

A Markov chain is defined by a set of possible states that represent the range of possible occurrences for the trials that make up the chain. The probability for a given chain can then be calculated using the conditional probability moving from each state to the next state along the chain. For the simple example of rolling dice, we can use a Markov chain to model the situation where a fair die is occasionally swapped for a loaded (unfair) die, i.e. where the probabilities of the six outcomes are not equal. Hence, according to which die is used the probabilities of the different roll outcomes change and we could model this by having two states: a fair die and a loaded die. Similarly for an otherwise random DNA model we may have different probabilities of observing C:G versus A:T depending on whether the region is a gene or not. Here the states would be gene and non-gene nucleotides, each with associated conditional probabilities. Technically we are describing a *discrete-time homogeneous Markov chain*.[8] It is discrete time because we have fixed positions or trials for the chain, which needn't be true in the general case, and the notion of being homogeneous refers to the fact that the conditional probabilities don't vary along the chain (e.g. vary with 'time').

Expressing this in terms of conditional probability, we model the probability of a trial having a particular state **given** the state of the previous trial. A consequence of this is that we consider all the possible probabilities of going from one state to another. Generally this is described as a matrix of conditional probabilities, which we call the *transition matrix*. Here each element of the matrix (T) is the probability of observing a particular state (State $= j$) at a

---

[7] Or strictly speaking observations in the case of a DNA sequence.
[8] The use of the term 'time' is historic, and the chain needn't represent a temporal process.

position in the chain $(n+1)$ given the occurrence of a potentially different state (State $= i$) at the previous position (n), i.e. we transition from state i to j:

$$T_{i,j} = Pr(State_n + 1 = j | State_n = i)$$

As before we use the typical mathematical notation where the '|' symbol means 'given'. We assume that the transition matrix is the same across the whole Markov chain, i.e. independent of n. Note that since the chain must transition to something, i.e. $State_{n+1}$ must take some value, the summation over all destination states j for starting state i is one:

$$\sum_j T_{i,j} = 1$$

Once we know the probability of going to the next state given the previous one, which would often be derived from statistical observations of real data, we can then use a Markov chain to generate sequences of outcomes or states. In essence this would mean using the transition matrix repeatedly, albeit in a random manner according to its probabilities, to produce a sample from the model.[9] From a relatively simple transitioning model we can then make long-term predictions. For example, we could have a Markov chain that models the reproduction of a population of bacteria, with assigned probabilities for the number of progeny in a generation given the number in the previous generation. Although this process is really continuous we are considering a simplified model with discrete time points by using the notion of a 'generation'. Given different starting populations we can then investigate the different long-term outcomes, e.g. whether the population grows or dies out.

Here we would say the state was the size of the population, and the probability for the size at a next point is predicted only from the current population size. We can repeat this process and take another discrete step along the chain, to predict even further into the future, but naturally if we do this the likelihood of any given outcome becomes even more uncertain. At a glance it may seem futile to extend the chain very far, basing guesses upon guesses, but the real power of the Markov process comes from the ability to compare the relative likelihood of different outcomes and how these may have arisen from a combination of different intermediate states.

For our simplistic model of bacterial population growth, where we assume that generations are distinct, we will use a probability distribution that describes the likelihood of the number of progeny that each individual bacterium can give rise to in a generation. In this case the states that the Markov chain can take are non-negative integers, i.e. numbers zero and above, and we will use a Poisson distribution for the probabilities. A state of zero means that the population dies out. A general transition matrix, which is independent of both the individual bacterium and the generation, then derives from the Poisson distribution being applied to each state. Although the distribution describes the likelihood of progeny for an individual we combine these for all individuals within a given population size.

In this example we do not explicitly calculate the complete transition matrix, given that the population size is unbounded. Rather we will calculate the population of the next

---

[9] Hence the term model, because we are modelling the state generation mechanism using transition probabilities.

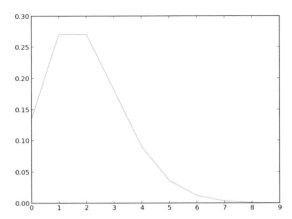

Figure 21.11. Example Markov chain probability distribution. The Poisson probability distribution (in this case rate $= 2.0$) which we use to model the number of progeny for each bacterium in the population.

generation based on the current population by generating random outcomes for each individual bacterium, according to the probability distribution. In subsequent examples where there are more limited states we will describe the whole transition matrix of probabilities upfront. The general form of the transition matrix is as follows, where the subscripts (i and j) respectively represent the current and next population states:

$$\begin{pmatrix} T_{0,0} & T_{0,1} & \cdots & T_{0,j} & \cdots \\ T_{1,0} & T_{1,1} & \cdots & T_{1,j} & \cdots \\ \vdots & \vdots & \ddots & \vdots & \\ T_{i,0} & T_{i,1} & \cdots & T_{i,j} & \cdots \\ \vdots & \vdots & & \vdots & \ddots \end{pmatrix}$$

Naturally $T_{0,j}$ is zero, for all subsequent states j, because once you have no individuals they cannot have any progeny. A bacterial population of one would have the following for the first ten transition probabilities, $T_{1,0}$ to $T_{1,10}$: 0.1353, 0.2707, 0.2707, 0.18046, 0.09028, 0.0361, 0.0120, 0.003, 0.0008, 0.0001, assuming a Poisson distribution with rate 2.0. The calculation of transition probabilities for a population of two ($T_{2,j}$) would take more work, because you have separate outcomes for each individual in the population and these would have to be multiplied for the total number of combinations. However, simulating population growth with this kind of model you fortunately do not need to know the exact value of $T_{j,k}$ for all states j and k but instead consider individuals by themselves and calculate the number of progeny for each.

We can simulate the whole process in Python, given a particular number of individuals in one generation, and with a Poisson distribution randomly determining how many progeny one bacterium will have. We can calculate the number of bacteria in the next generation by considering how many offspring each of the individuals in the current generation has. We define a function to generate the size of the population using a `scipy.stats` random variable object. Note that we use the `.rvs()` method to draw the required number of random samples from this distribution, which in this case would be the numbers of progeny for each bacterium. The population of the next generation is then simply the summation of the numbers of progeny.

```
def getNextGenPop (currentPop, randVar):

  progeny = randVar.rvs(size=currentPop)
  nextPop = progeny.sum()
  return nextPop
```

A random variable object with a Poisson distribution is generated with the required rate (average number of progeny in the generation):

```
from scipy.stats import poisson
rate = 1.02     # A deliberately low rate
poissRandVar = poisson(rate)
```

The Markov process is then simulated by repeatedly recalculating the populations for the subsequent generations. At the low reproductive rate which we use here, you will note that the population sometimes dies out.

```
pop = 25
for i in range(100):
  pop = getNextGenPop(pop, poissRandVar)
  print("Generation:%3d Population:%7d" % (i, pop))
```

Returning to the general situation, the transition matrix determines the properties of the Markov chain. For example, if we define the initial distribution I of the probability of the state j to be:

$$I_j = \Pr(\text{State}_0 = j)$$

then the probability for each state at the next position in the chain is the summation over all the possible starting states multiplied by the conditional probability to get to the next state. In other words we can get to the next state via multiple routes from different initial states, each with a potentially different probability:

$$\Pr(\text{State}_1 = j) = \Pr(\text{State}_1 = j \mid \text{State}_0 = k) \Pr(\text{State}_0 = k) = \sum_k I_k T_{k,j}$$

Thus the distribution at chain position 1 can be determined from that at chain position 0 just by multiplication by the transition matrix (T). It can also be shown, for example, that:

$$\Pr(\text{State}_n = k \mid \text{State}_0 = j) = (T^n)_{j,k}$$

where T^n is the nth power of the transition matrix. Thus in theory we can exactly calculate the distribution of states at any subsequent position (State_n), given the starting state (at position zero) and multiplying by the matrix T^n. Accordingly we have:

$$\Pr(\text{State}_n = j) = \Pr(\text{State}_n = j \mid \text{State}_0 = k) \Pr(\text{State}_0 = k) = \sum_k I_k (T^n)_{k,j}$$

This also allows us to calculate the probability of a whole sequence of specified states (from i_0 to i_n) from the starting state and the product of the individual transition matrix elements which are selected from the knowledge of the states:

$$\Pr(\text{State}_0 = i_0, \text{State}_1 = i_1, \ldots, \text{State}_n = i_n) = I_{i_0} T_{i_0 i_1} T_{i_{n-1} i_n}$$

Lastly, analyses of Markov chains sometimes refer to the *equilibrium distribution*. An equilibrium distribution, π_i, for a Markov chain is a distribution of states which does not vary as the chain evolves for subsequent probabilistic trials, i.e. applying the transition matrix regenerates the previous distribution ($\pi T = \pi$). If an equilibrium distribution exists, then it can be shown, under weak assumptions, that all distributions approach the equilibrium distribution as the number of trials gets large, i.e. $I \times T^n$ approaches π as n gets large, for any starting state. This can be important for biological and predictive systems where the equilibrium distribution represents a long-term description of a system when it has had time to 'settle down'.

Hidden Markov models

A *hidden Markov model* (HMM) is a kind of Markov chain where the states are not directly observable, but a quantity that is directly observable is determined, often probabilistically, by the state at a given point in the sequence. The key idea is that there is not necessarily any direct correspondence between the invisible state and the observable state. For example, if we want to predict the secondary structure of a protein (which we might simply represent as alpha-helix, beta-strand or random coil) then the observable data would be the protein sequence and the hidden states could be modelled as the secondary-structure type at each point in the sequence. Here, by considering the probabilities of changing from one state to another and the probabilities of making an observation, given a particular state, then we can make a prediction about what the hidden states are if we have some observable data, i.e. having a protein sequence allows us to make a prediction of the underlying secondary-structure states. Later in this chapter, once we have covered some key theory and algorithms, we will look at an HMM example in Python that involves protein sequences.

To be more precise, suppose we can obtain some observable data d for each position in the Markov chain. We require a function $e_i(d)$ that gives the probability of generating or 'emitting' the observed data given an underlying state i. It is assumed this emission probability is independent of the position in the chain. Sometimes an HMM model is constructed such that the observable data has a direct correspondence to one of the underlying states, in which case $e_i(d) = 1$ for that specific observation and $e_i(d) = 0$ otherwise; we will demonstrate this idea later. Because for an HMM we distinguish between the hidden state at position n and the data observed (emitted) at position n, we multiply the transition matrices to get the probabilities of the subsequent hidden states (from the previous states) and multiply the emission probabilities to get the likelihood of observing the actual data given this underlying state. Hence, the probability of observing the whole sequence of data, over all the positions of the Markov chain starting with probabilities for the initial state, I_{i_0} can be written as:

$$\Pr(Data_0 = d_0, Data_1 = d_1, \ldots, Data_n = d_n) = I_{i_0} \cdot e_{i_0}(d_0) \cdot T_{i_0 i_1} \cdot e_{i_1}(d_1) \cdot T_{i_1 i_2} \cdots T_{i_{n-1} i_n} \cdot e_{i_n}(d_n)$$

Using Python for hidden Markov models

So far we have covered a fair amount of general theory, so for the remainder of this chapter we turn to some Python versions of key algorithms which will allow you to work with hidden Markov models and go on to show you how these can be practically implemented. Given the definition of an HMM the task naturally turns to extracting some useful information. We usually want to know what the likely underlying hidden states are, given some sequence of observable data. Because we are dealing with a probabilistic model, and bearing in mind a Bayesian approach which considers alternative hypotheses, there will be various competing sequences of hidden states or 'routes' through the Markov chain. Often it is useful to have the most likely sequence of hidden states, hence we cover the Viterbi algorithm. We then go on to the forward-backward algorithm, which will allow us to estimate the probabilities of particular states being present at a particular point in the chain, to compare competing solutions and indicate the points of greatest uncertainty in our predictions.

We will demonstrate the use of these algorithms with a relatively simple example of a hidden Markov model that relates to protein sequences. Specifically we will have states to describe protein residues in sequences as being 'buried' or 'exposed'. These reflect where amino acid residues lie within the 3D structure of a folded protein. The buried residues will be found in the interior of the protein, at its core, and the exposed residues will be on its surface, i.e. exposed to the water solvent. We don't expect state predictions from this HMM to be especially accurate, given that protein folding is so complex, but the example is simple enough for this book. Also, we are basing the HMM on a well-known observation, that the buried cores of residues in proteins have a distinctly different complement of amino acids compared to the exposed surface. The exposed amino acids have a strong tendency to be hydrophilic and have side chains that contain polar and charged atom groups, which can interact with the water. In contrast the buried residues tend to be hydrophobic, and thus have non-polar, uncharged side chains.[10]

A simple way to form the HMM would be to only have two hidden states, for buried and exposed categories. In this case we would calculate the probabilities for changing (or not changing) between states as we move through the protein sequence. For an observed amino acid type there will be probabilities that it came from each of the two states. Hence, the observed data, the protein sequence, would be emitted in a probabilistic manner from these two states.

In practice we will not use this simple HMM, rather we will have many more states. Each state will be a combination of a buried/exposed label **and** an amino acid type, so there will be 40 states in total. The reason behind this is to make the transition probabilities more detailed, so the overall HMM becomes more accurate. By encoding the amino acid type in the hidden state we will be able to incorporate the effect of the amino acid type in the transitions of the chain. For example, where a sequence position has a valine amino acid followed by a serine amino acid we will know more precisely what the probability is to go from a buried or exposed valine to a buried or exposed serine; in effect the transition probability matrix will vary according to the sequence. If we use such combined states the emission probabilities become trivial: the probability of getting the observed amino acid is 1.0 if that amino acid is part of the hidden state and 0.0 otherwise. Naturally the more states we have in the HMM the more statistical data we need to get accurate probabilities; we need data for 40×40 transitions. However, the probabilities will be estimated from a large amount of PDB data (containing protein 3D structures) so there is little concern about getting good statistics here.

---

[10] So typically aliphatic or aromatic hydrocarbon.

Simple two-state model

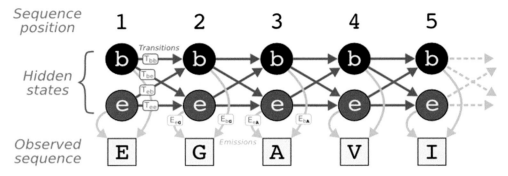

Figure 21.12. A schematic of a simple two-state HMM for buried and exposed protein states. For a model of two hidden states ('b' is buried and 'e' is exposed) we have a transition matrix corresponding to four possible state transitions and an emission matrix specifying the likelihood of each of the 20 amino acids being found in each of the two states.

Forty-state model

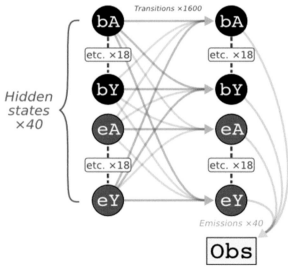

Figure 21.13. A schematic of a 40-state HMM for buried and exposed protein states. In this model there is a separate hidden state for each amino acid and buried/exposed combination (i.e. 2×20 states). This gives rise to a large, 1600-element transition matrix between all the 40 possibilities. The emission matrix in this case, however, is simple, given that an observed amino acid can only come from the two hidden states which involve that specific amino acid, and the non-matching emission probabilities will be zero.

The Viterbi algorithm

The Viterbi algorithm is an efficient means of using the probabilities within an HMM to predict the most likely sequence of hidden states that generates a sequence of known outcomes: our observed data. There are many situations when it can be useful to get a handle on the 'best guess' for the underlying states, and in the example at the end of this chapter we will use the method to estimate the most likely sequence of buried/exposed protein states. However, this is not to suggest that we are always satisfied with a single optimum answer; considering a Bayesian approach it may actually be more relevant to have a distribution of different hidden state sequences, which can tell us something about the reliability of estimates.

The Viterbi algorithm is an example of dynamic programming, similar to the pairwise sequence alignment method shown in Chapter 12, where we compute the path through the hidden state transitions that maximises the overall probability. Rather than multiplying probabilities, which would often result in very small numbers that cannot be represented accurately in floating point representation,[11] we add the logarithms of the probabilities, which will always be in a manageable range.

The procedure starts at the beginning of the HMM chain and calculates the probabilities for all hidden states at subsequent positions until the end, where one 'winning' sequence of hidden states is revealed. Given an array of different states there are several possible transitions (paths) to get to the different states at the subsequent positions in the Markov chain. At a given position where the probability of having each state is recorded, the highest probability for a state in the next position is taken as the maximum of the outcomes when these probabilities are multiplied by the appropriate transition and emission probabilities (i.e. observing the data for the following position). For the protein example this means we multiply by probabilities of going from one buried/exposed state to the next and by the probability of observing a particular amino acid, given the next state.

In the Viterbi algorithm, although there are many sequences of hidden states that can be proposed to yield a particular observed state for a position in the chain, the sub-optimal state sequences are discarded. The procedure can be viewed as using pre-computed solutions for the prior chain positions. The dynamic programming core of the algorithm means that we do not compute all possible routes, only the best ones for each position, thus minimising the amount of computation that has to be done.

The implementation of the Viterbi algorithm is fairly simple using Python with NumPy arrays. We go from one point in the sequence to the next in a `for` loop, given we know the number of observations. Calculating the log-probabilities involves adding arrays for the preceding states (or starting values) to the transition and emission arrays, noting that the emission probabilities are conditioned given the actual observed data for the point. For each step, choosing the best transition to get to each of the states in the subsequent position is a simple maximisation of the summed log-probabilities. Here we use `.argmax()` to give the indices for the highest-scoring transitions. This allows us not only to select the required values, but also to record the list of winning indices at each position. Hence, once the end is

---

[11] Referred to as *underflow*.

reached, the overall highest-scoring path will be determined. Note that we are only recording the best route and log-probability for each state, so this operation is relatively frugal on memory.

For the actual Python example we do the required NumPy imports for arrays and then define a function which takes four arguments: a list of the observed data and log-probabilities for the starting values, transition matrix and emissions. Note that it is assumed that all the probabilities are NumPy arrays and that the logarithms were taken beforehand (e.g. `logArray = numpy.log(myArray)`), so we can add rather than multiply. For simplicity we will often refer to the log-probabilities simply as 'scores'. The observations may be a list or an array and encode the data as indices. So, for example, if the data is a DNA sequence with G, C, A and T letters then `obs` is a list that selects from four indices: `0`, `1`, `2` or `3`. Here the order in which the data states are encoded is arbitrary, but the order must naturally be the same as in the emission matrix `pEmit`.

```
from numpy import empty, array

def viterbi(obs, pStart, pTrans, pEmit):
```

Initially we record the number of hidden states from the input arrays, and then initialise values for the log-probabilities scores. The first value for `scoresPrev` is initialised with the starting scores and the emissions for the first data item. The dictionary `pathsPrev` that records the score-maximising choices, and hence best path through the HMM, uses the index of the state (here `i`) as a key and holds a list of indices, to record the prior states and hence path. The initial value here is naturally `i`, where we start from.

```
    nStates = len(pStart)
    states = range(nStates)

    scores = empty(nStates)
    scoresPrev = pStart + pEmit[:,obs[0]]
    pathsPrev = dict([(i, [i]) for i in states])
```

Next we loop through all of the subsequent observations, i.e. for each position in the Markov chain after the start, noting that we record the observed value as `val`. For each position we redefine the paths based on the previous ones so we create paths, which will extend the previous array of state indices.

```
    for val in obs[1:]:
        paths = {}
```

For the current position we go through each state index `i`, and define the options for the possible outcomes as a sum of log-probabilities. To the previous scores we add the transition to get to state `i` and the probability of emitting the outcome `val`. Note that we take the slice `pTrans[:,i]` so that we get the scores to transition from all states to the current one, and that the emission of the observed value for this state `pEmit[i,val]` is just a single number (and so is added to all array elements). The best score for the state is simply the largest `.max()` and the index for this is extracted using `.argmax()`. The index `bestState` is the winning state that we come from to get to `i`. This index is used to access the path for that state: the winning one up to the previous point. The updated path list is then the extension of the path we came from with the current state.

```
for i in states:

    options = scoresPrev + pTrans[:,i] + pEmit[i,val]
    bestState = options.argmax()
    scores[i] = options.max()
    paths[i] = pathsPrev[bestState] + [i]
```

Once all the states have been considered for this position we store the current paths and scores as the previous ones for the next iteration.

```
pathsPrev = paths
scoresPrev = array(scores)
```

Finally, it is easy to get the best state (or index thereof) and its log-probability score from the arrays we are left with at the end. The winning path and score is returned from the function.

```
endState = scores.argmax()
logProb = scores.max()

return logProb, paths[endState]
```

The forward-backward algorithm

The forward-backward algorithm is commonly used to estimate all the probabilities of the hidden states of the HMM for any sequence of observations. With these hidden state probabilities various things are possible: for example, we can compare the overall probabilities of different hidden state sequences (which may not be the best) for a set of observations; this in turn can be useful to generate a probabilistic distribution of state sequences rather than a single best answer, as we did with the Viterbi algorithm. It is in keeping with the Bayesian approach to consider an ensemble of different solutions, so we have a better idea of the error in the prediction and whether there are competing solutions with similar probabilities.

This method works by expressing the probability of a state at a given position as the product of two components, the forward and backward parts. The forward part is the probability of the states for a given point in the chain given the sequence of observations up to that point. The backward part is the probability of getting the observations for the remainder of the chain after a point given a set of state probabilities. We will not describe the mathematical derivation of the algorithm[12] but in essence the result stems from using Bayes' rule and the fact that, because of the no-memory property of the Markov chain, the observations either side of a position are independent, given the condition that we know the probabilities of the hidden states at that point.

The forward-backward algorithm itself is yet another example of dynamic programming, which makes the process reasonably efficient to compute. Both the forward and backward parts calculate probabilities for subsequent points in the Markov chain, albeit from different directions, by using the previously calculated result from the neighbouring position. This can be thought of as a recursion (although we don't code the Python in that way), where

---

[12] For a better description see Durbin, R.M., Eddy, S.R., Krogh, A., and Mitchison, G. (1998). *Biological Sequence Analysis: Probabilistic Models of Proteins and Nucleic Acids* (1st edn.). Cambridge: Cambridge University Press.

the result for each point extends the previous result and thus includes all the results of all the calculations prior to that one.

The example Python function that implements the algorithm makes use of NumPy to perform matrix algebra, to give a fairly compact result (avoiding loops), using the starting probability matrix `pStart`, transition probability matrix `pTrans` and emission probability matrix `pEmit` discussed before. Naturally these are inputs to the function, together with the sequence of observations `obs` that we wish to estimate the probabilities for. Note that, unlike in the Viterbi algorithm, we don't work with logarithms and thus we will multiply probabilities. Here we re-normalise as we go, to get relative probabilities, and so the values don't get especially small (unlike the probability of a long sequence).

```
from numpy import array, empty, identity, dot, ones

def forwardBackward(obs, pStart, pTrans, pEmit):
```

First in the function a few variables are initialised: the number of observations, the number of hidden states and an identity matrix of an appropriate size.

```
n = len(obs)
nStates = len(pStart)
I = identity(nStates)
```

Next a matrix `fwd` is initialised as an empty NumPy array, which will hold the hidden state probabilities for the forward part of the algorithm. Note that this array is one longer than the number of observations, so that we can include the starting probabilities for the states at position 0.

```
fwd = empty([n+1,nStates])
fwd[0] = pStart
```

Now comes the main loop for the forward part of the algorithm, where we go through each index that references each observation from start to end. The forward probability is calculated by applying the transition matrix to the previous (or starting) hidden state probabilities and then applying the emission probabilities for the observation `val`, all using `dot()` to do the matrix multiplication. Effectively we are saying that a hidden state probability derives from the transitions of the previous states and the likelihood that the state generates the observation. Note the use of the identity matrix `I` to give an array of the right shape to apply the emission probabilities and that we set index `[i+1]`, given that we designed our matrix to begin with the starting probabilities, which we will need later, rather than the result from the first observation. The division by `fProb.sum()` ensures that the probabilities are normalised and so sum to 1.0; this implementation doesn't bother calculating the various probability scaling constants.

```
for i, val in enumerate(obs):
    fProb = dot( pEmit[:,val]*I, dot(pTrans, fwd[i]) )
    fwd[i+1] = fProb / fProb.sum()
```

With the forward part done, we next turn to the backward part. Here we define `bwd`, which will hold the 'backward' hidden state probabilities. Unlike the forward part we don't need to remember the whole array of probabilities (although you could if it was useful) so this is just a simple vector that will be updated for each subsequent step. The starting values in `bwd` are 1.0, from which we can multiply to get subsequent probabilities. The `smooth` array will be filled with the final result, which is the 'smoothed' combination of the probabilities from the forward and backward calculations; the `fwd` and `bwd` probabilities are multiplied and re-normalised. A separate loop is not needed to calculate this as it can be filled in at the same time as `bwd`. Note the last vector of `smooth` probabilities is set upfront as the last of the forward values, given that the `bwd` values here are 1.0 and so multiplication would have no effect.

```
bwd = ones(nStates)
smooth = empty([n+1,nStates])
smooth[-1] = fwd[-1]
```

The backward loop goes through all the positions in reverse order and calculates `bwd`, by applying the transition matrix to the matrix product of the emission probabilities for the current observation `obs[i]` and the previous `bwd` (i.e. for the subsequent, `i+1` point in the sequence). Effectively we are saying that the new `bwd`, the probability of observations (from this point to the end) given the **preceding** (`i-1`) hidden states, can be calculated recursively. We multiply the old `bwd` by the emission probability of the current observation and apply the transition matrix, adding an extra observation and transitioning to the earlier state. The `smooth` probability matrix, which is passed back from the function, is then the normalised product (element-wise) of the forward and backward components.

```
for i in range(n-1, -1, -1):
  bwd = dot(pTrans, dot(pEmit[:,obs[i]]*I, bwd))
  bwd /= bwd.sum()
  prob = fwd[i] * bwd
  smooth[i] = prob / prob.sum()

return smooth
```

Implementing a protein sequence HMM

Finally in this chapter we end with a demonstration of using the above algorithms for handling our example protein sequence HMM, so we can predict buried or exposed status. To do this we must first obtain some probabilities for the transitions between the 40 different possible states. In Python we will label the states in the form (`exposure`, `aminoAcid`). The probabilities are derived from a simple statistical analysis of a subset of the PDB database. The actual subset of 3D structures coordinates used are from the VAST chain set[13] and represent only protein amino acid chains that are dissimilar to each other, within a predefined limit (p-value cut-off 10^{-7}). The idea behind using this non-redundant subset is to try to reduce the amount of bias that comes from having multiple entries for closely related proteins. Overall, we would like the probabilities for our HMM to be representative of proteins in general, and not skewed towards those kinds which have the most structure data. This kind of problem is common when dealing with sequence data in general (protein, DNA or RNA) and is something that a bioinformatician should be wary of.

To determine whether each of the amino acids in our structure database is in a solvent-exposed or buried context an external program was run on all of the PDB-format data files. For reasons of space we will not discuss this in detail but the method is the one described

---

[13] See http://www.ncbi.nlm.nih.gov/Structure/VAST/nrpdb.html.

by Shrake and Rupley.[14] This calculates numerical values representing the exposed surface area for each atom. For our HMM we want to have exposed or buried categories for each residue, so the per-atom values needed to be converted. The algorithm here is to add the values for all the atoms within a residue to get the exposed surface area for the whole residue. This is then divided by the maximum surface area for that kind of residue, which gives a fractional value and eliminates the effect of amino acid size. Buried residues are then defined as those which have an exposed surface area of less than 7% of the maximum.[15] The results from this procedure are represented in a file which accompanies this book: 'PdbSeqExposureCategories.txt' (download via http://www.cambridge.org/pythonforbiology). The lines of this file are in the form illustrated below, with one line of one-letter amino acid codes followed by a second line of the same length representing buried or exposed categories for the sequence. Here a dash '-' represents an exposed position and an asterisk '*' buried, so that the categorisation is easy to see:

```
GSSGSSGHEETECPLRLAVCQHCDLELSILKLKEHEDYCGARTELCGNCGRNVLVKDLKTHPEVCGREGS
-------------------*-----------------**-------------*-------*---*---*
VWSVQIVDNAGLGANLALYPSGNSSTVPRYVTVTGYAPITFSEIGPKTVHQSWYITVHNGDDRAFQLGYEGGGVA
-*-*-----------*-----**-*---*****---*****------------*****-----****---*-*
```

With the exposure category data at hand we move on to calculating the probabilities and generating the HMM. For this example we have plenty of data to derive probabilities from the frequency of observing particular events. However, if data is missing, or otherwise difficult to analyse, we could use a method like the Baum-Welch algorithm, to estimate the emission and transition possibilities, given a sufficient amount of representative training data. The Baum-Welch algorithm can actually be applied when none of the probabilities are initially known, which in effect means that the HMM probabilities can be 'learned' from even proportionately sparse data. In essence this is a variety of machine learning, although we leave further discussion of this topic until Chapter 24. In our example case we simply count the occurrences of the different buried/exposed transitions in the data file.

Initially we get a list of the exposure and amino acid type labels, the combination of which gives the hidden state labels. Although we will be working with NumPy arrays, and thus referring to the states by numeric indices, the order of symbols in these lists relates each state index to a meaningful symbol.

```
expTypes = ['-','*',]
aaTypes = ['A','C','D','E','F','G','H','I','K','L',
           'M','N','P','Q','R','S','T','V','W','Y']
```

The data counts and probabilities will be stored in NumPy arrays, so we define these upfront as empty data structures of the required sizes, and of floating point data type.

```
nExp = len(expTypes)        # Number of exposure categories
nAmino = len(aaTypes)       # Number of amino acid types
```

---

[14] Shrake, A., and Rupley, J.A. (1973). Environment and exposure to solvent of protein atoms. Lysozyme and insulin. *Journal of Molecular Biology* 79(2): 351–371.

[15] Hubbard, T.J., and Blundell, T.L. (1987). Comparison of solvent-inaccessible cores of homologous proteins: definitions useful for protein modelling. *Protein Engineering* 1(3): 159–171.

```
nStates = nExp * nAmino          # Number of HMM states

from numpy import zeros, log

pStart = zeros(nStates, float)              # Starting probabilities
pTrans = zeros((nStates, nStates), float)   # Transition probabilities
pEmit  = zeros((nStates, nAmino), float)    # Emission probabilities
```

Because we will be reading the data from a file containing textual symbols we will need to relate those symbols to the correct indices in the NumPy arrays. Hence we create a dictionary for all the textual state codes so we can quickly look up each `index`. This is a simple matter of looping through all the exposure and amino acid symbols and for each combination making a tuple that acts as a key to the `index`, which is incremented each time in the loop. The additional `stateDict` is made so we can do the reverse look-up later on, converting indices into symbols.

```
indexDict = {}
stateDict = {}
index = 0
for exposure in expTypes:
  for aminoAcid in aaTypes:
    stateKey = (exposure, aminoAcid)
    indexDict[stateKey] = index
    stateDict[index] = stateKey
    index += 1
```

To fill the counts for the state transitions we read through the data file. We open the data file for reading, remembering to use the full path to the file if it is not in the current directory:

```
fileName = 'examples/PdbSeqExposureCategories.txt'
fileObj = open(fileName,'r')
```

The file is read two lines at a time using a `while` loop so that the amino acid sequence comes from `line1` and the exposure code from `line2`.

```
line1 = fileObj.readline()
line2 = fileObj.readline()

while line1 and line2:
  sequence = line1.strip()
  exposure = line2.strip()
```

For each sequence, a second loop goes through all adjacent pairs of amino acids and the corresponding exposure codes. Combining the exposure and amino acid symbols gives a key to look up the correct numeric indices. Note that we use `.get()` to skip situations where we have unusual amino acids (not in our list of 20) and thus have an unknown state key.

```
  n = len(sequence)
  for i in range(n-2):
    aa1, aa2 = sequence[i:i+2]
    exp1, exp2 = exposure[i:i+2]
    stateKey1 = (exp1, aa1)
    stateKey2 = (exp2, aa2)
```

```
index1 = indexDict.get(stateKey1)
index2 = indexDict.get(stateKey2)

if index1 is None or index2 is None:
  continue
```

The indices allow the counts to be incremented in the correct element of the arrays:

```
pStart[index1] += 1.0
pTrans[index1, index2] += 1.0
```

When the inner loop is done, we add a count for the last position (which was otherwise missed because in getting sequence pairs we didn't go to the end) and then get the next lines for the `while` loop.

```
pStart[index2] += 1
line1 = fileObj.readline()
line2 = fileObj.readline()
```

The NumPy arrays contain counts, which are converted to probabilities by dividing by their totals. For `pTrans` we divide the values in each row of the matrix by the total for that row.

```
pStart /= pStart.sum()
for i in range(nStates):
  pTrans[i] /= pTrans[i].sum()
```

Finally we fill the trivial emission probabilities, setting to `1.0` where the amino acid matches the hidden state, and leaving values at `0.0` otherwise. Note that we use the look-up dictionary to get the index for the state (the first dimension in the emission array) but the plain amino acid index simply comes from looping through the list of amino acid types with `enumerate()`.

```
for exposure in expTypes:
  for aminoIndex, aminoAcid in enumerate(aaTypes):
    stateIndex = indexDict[(exposure, aminoAcid)]
    pEmit[stateIndex, aminoIndex] = 1.0
```

With the probability arrays defined we can then test the Viterbi algorithm, noting that we convert the probabilities to logarithms and represent the sequence as numeric indices, rather than code letters. Note the small addition when taking logarithms, to deal with zero probability values.

```
seq = "MYGKIIFVLLLSEIVSISASSTTGVAMHTSTSSSVTKSYISSQTNDTHKRDTYAATPRAH"\
      "EVSEISVRTVYPPEEETGERVQLAHHFSEPEITLIIFGVMAGVIGTILLISYGIRRLIKK"\
      "SPSDVKPLPSPDTDVPLSSVEIENPETSDQ"
obs = [aaTypes.index(aa) for aa in seq]

adj = 1e-99
logStart = log(pStart+adj)
logTrans = log(pTrans+adj)
logEmit  = log(pEmit+adj)

logProbScore, path = viterbi(obs, logStart, logTrans, logEmit)
```

To generate a string of symbols representing the buried or exposed states, the indices from the winning Viterbi path are used as keys to `stateDict`, which gives back the textual symbols for the state (`exposure`, `aminoAcid`) and the first of these (hence `[0]`) is the symbol we want.

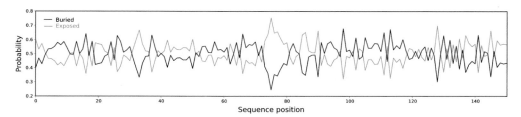

Figure 21.14. Example state probabilities, as calculated by the forward-backward algorithm.
A plot of the probabilities for the buried (black) and exposed (grey) protein amino acid states at each
position of the Markov chain. Where the probabilities are distinctly different we can be more
confident about the category assignments.

```
bestExpCodes = ''.join([stateDict[i][0] for i in path])

print(seq)
print(bestExpCodes)
```

This will give the following output, albeit on one line:

```
MYGKIIFVLLLSEIVSISASSTTGVAMHTSTSSSVTKSYISSQTNDTHKRDTYAATPRAHEVSEISVRT
----*******---**-----------------------*------------------------***---

VYPPEEETGERVQLAHHFSEPEITLIIFGVMAGVIGTILLISYGIRRLIKKSPSDVKPLPSPDTDVPLS
*----------*---------**************************---**------*------------

SVEIENPETSDQ
------------
```

We can also test the implementation of the forward-backward algorithm and plot a graph to
show the underlying probabilities for exposed or buried states at each point in the Markov
chain. For each position we get back an array of smoothed probability values (combining
forward and backward values) and because of the way we created the data arrays we know
that the first 20 values correspond to the exposed category, and the remainder the buried. The
sum of each half of values will give the total probability for the category, although for our
example HMM only one amino acid (for the observed type) will have a non-zero value.

```
smooth = forwardBackward(obs, pStart, pTrans, pEmit)

buriedList = []
exposeList = []

for values in smooth:
  exposeList.append(sum(values[:20]))
  buriedList.append(sum(values[20:]))

xAxisValues = list(range(len(exposeList))) # Sequence positions

from matplotlib import pyplot

pyplot.plot(xAxisValues, exposeList, c='#A0A0A0')
pyplot.plot(xAxisValues, buriedList, c='#000000')
pyplot.show()
```

22 Statistics

Statistical analyses

In this chapter we look at the analysis and interpretation of collections of data in a mathematical way. In order to understand the basics of statistics we will assume some familiarity with the basics of probability, as discussed in Chapter 21.

Generally when we gather numerical measurements we don't get identical results, rather we get a spread of values. The underlying reason for this variation could be a natural variation in what we are measuring, an error in the way we make the measurements or, as is almost always the case, a combination of both of these. Statistics helps us to make sense of variations in numerical data and commonly we are asking the question whether what we measure is statistically significant, according to some prior hypothesis. Depending on the result this naturally then drives further investigations, based on a belief of a hypothesis being true or untrue. Statistics is a vast subject, so in this chapter we can only cover a few of the more important aspects that we either refer to elsewhere in this book or that are otherwise commonly used in biology.

Samples and significance

One of the key principles, which underpins most statistical analyses, is the idea that the data we collect contains a limited number of samples from some kind of underlying probability distribution. This probability distribution can be thought of as the mechanism by which the data values are generated, but naturally the data arises due to some physical process and by ascribing a probability distribution we are merely forming a mathematical model, which is often significantly simplified, to approximate the data-generation process.

For a given situation, if we have an idea of what type of underlying probability distribution would be appropriate, then by looking at the observed data we can begin to estimate what the parameters of the distribution are, such as where its centre is and how much it spreads. Given parameter estimates we can then begin to answer questions which relate to the probabilistic model, such as how likely it is that a given value is generated by the model. In virtually all cases the answer provided is not certain, rather the answer is given as being true with a certain probability, which for parameter estimation is often called a *confidence level*. It is often the case that a 95% probability is considered a suitable confidence level for inferring significance, but of course even at this seemingly strict level, 5% (1 in 20) of the sampled values would lie outside the quoted range.

In Python several of the commonly used probability distributions are represented, including in the `scipy.stats` module, which we will routinely refer to in this chapter, and also in the `numpy.random` module, which allows us to draw random samples from a distribution. Here we illustrate creating random samplings with different numbers of points, selecting from a normal distribution using `random.normal`, which we then show as a histogram:

```
from matplotlib import pyplot
from numpy import random

mean = 0.0
stdDev = 1.0

for nPoints in (10, 100, 1000, 10000,100000):
  sample = random.normal(mean, stdDev, nPoints)
  pyplot.hist(sample, bins=20, range=(-4,4), normed=True)
  pyplot.show()
```

Predictions from a probability distribution are often coupled to the idea of a competing hypothesis. Here the probability distribution is often a model of what we expect at **random** and the competing hypothesis would mean that something significantly non-random was happening. Hence, rather than drawing significance if this model appears to fit the data, we assert that there is significance if the random model is unlikely to explain the data samples; that our data does not fit the probability distribution of the random situation. So by applying a probabilistic model we are generally not assuming that we actually have a good physical model for our data, but rather that there is a mathematical approximation to the data-generation process, which is nonetheless useful for making predictions and for understanding key aspects of what we are studying.

Lastly, it is important to note that even in situations where the underlying probability distribution is not known we can nonetheless estimate some statistical parameters. In the simplest situation, we might simply try and estimate the mean (average) or standard deviation

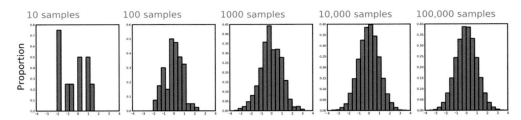

Figure 22.1. An example of results from drawing different numbers of random samples from an underlying probability distribution. The samples illustrated are taken randomly from a normal distribution with a mean of 0.0 and a standard deviation of 1.0, where the sampling was done using the Python example.

(spread) of the distribution, given the data, and not worry too much about what the distribution is.

Null and alternative hypotheses

Instead of trying to show whether something is likely to be true, in statistics we often aim to show whether something is likely to be false. The basic reason for this is that it is often easier to know what the underlying probability distribution would be if the observed measurements were generated by a random process, rather than know what the situation would be in an as yet unproven, non-random situation. In essence, a random process is often used as a scientific control. In this case we hypothesise a statement called a *null hypothesis* and we try and demonstrate, to a certain confidence level, whether it can be rejected. For example, when analysing whether a new drug is beneficial for a medical condition the null hypothesis might be that there is no difference in the treatment outcome compared to a placebo. If we can demonstrate that this hypothesis is likely to be wrong, then we have evidence that the new drug has a worthwhile effect. The *alternative hypothesis* is the negation (the opposite) of the null hypothesis, and thus if the null hypothesis is rejected then the alternative hypothesis is accepted.

Type I and II errors

Because statistics deals with predictions and probability, some mistakes will be made. Thus, it is important to consider how we can quantify the errors of predictions made on the basis of statistical analysis. With a mathematical formulation based on testing a null hypothesis (a random control), the analysis can lead to two types of errors:

- A Type I error is the situation where the null hypothesis has wrongly been rejected when it is really correct.
- A Type II error is the situation when the null hypothesis has not been rejected when it should have been, because it is really incorrect.

Alternatively, we might say that these errors are respectively false positive and false negative as far as the alternative hypothesis is concerned. As an example of a Type I error, a statistical analysis of a specific trial might lead us to believe that a drug has a beneficial effect, thus favouring the alternative hypothesis and rejecting the null hypothesis, when actually there was no real benefit and the null hypothesis was true. The end result is that further development of the drug would be needlessly undertaken. Conversely for a Type II error, a new drug might have a worthwhile effect but statistics of a specific trial might lead us to believe that it does not, incorrectly accepting the null hypothesis, meaning that further development of a promising drug might be stopped.

Naturally, both kinds of errors are potentially serious. In a given situation, with a given sample size, you could specify in advance what Type I or Type II error you are willing to tolerate.[1] However, by decreasing the probability of a Type I error you generally increase the probability of a Type II error, and vice versa. For example, if we incorporate a new medical test that makes it less likely to detect a spurious effect from a drug we will get fewer false positives, but we will tend to miss more marginal cases and have more false negatives. Hence, it is commonplace to have a balance between Type I and Type II errors, depending on things like financial cost and danger etc. Also, for a specified Type I error level, you can decrease the probability of a Type II error by increasing the sample size, and vice versa. This is because having more data values means we get more precise estimates of the null hypothesis parameters, so for a fixed Type I error probability the Type II error is less affected by the variation due to the random data sampling.

Statistics in Python

Although you can use knowledge of the underlying mathematics to design and code your own statistics software, for anything beyond the simplest analysis it is common to use existing statistical routines. For the examples in this chapter we will primarily be using data held in NumPy arrays in combination with the SciPy[2] module `scipy.stats`. At the time of writing the statistical package R[3] is very commonly used in biology. R has its own language, but there is a Python wrapper around it called RPy,[4] so you can use R tools in Python programs, including with NumPy arrays. In the Appendix we will illustrate the use of RPy for some of the most common statistical tests.

Simple statistical parameters

Our first practical analyses of data involve measuring some of the fundamental statistical parameters. These can be calculated without any knowledge of what the underlying probability distribution might be. However, if we do consider the underlying distribution these sample-based parameters will be estimates for their probabilistic equivalents. The theory regarding whether we are truly estimating the underlying probabilistic parameters in an

---

[1] The probability of a Type I error is usually denoted by α, and the probability of a Type II error by β. We sometimes speak of the power of a statistical analysis, which is the probability that the null hypothesis has been rejected when it really is false, which is equivalent to $1 - \beta$.
[2] The examples in this chapter were tested with SciPy 0.7.1.
[3] http://www.r-project.org/
[4] http://rpy.sourceforge.net/

unbiased way is left to further reading. Here we will simply introduce the simple statistical parameters and how we usually calculate them.

Mode, median and mean

For a collection of values, one of the most useful measures is to estimate where the centre of the distribution is. The general idea here is that we get a single value that is most representative of the data set as a whole. There are a few different ways that are generally used to get such a measure, which are called *mode*, *median* and *mean*. Naturally these have different properties and are useful in different situations, although the mean is the most common parameter used.

The *mode* is the most commonly occurring value in a set of data. For example, the mode of the list of values [1,2,2,3,2,1,4,2,3,1,0] is 2, because the number 2 appears most often. Naturally, if each value only occurs once then the mode tells you nothing. Hence, for this to be a useful measure the amount of data and whether the values are represented with a specific precision are important. This is especially true when using floating point numbers, where repeated values can be unlikely, in which case it is commonplace to represent the data as a histogram. If the values are assigned to suitable ranges the shape of the distribution can become more apparent and the mode will be the histogram bin with the most values.

Using standard Python we can calculate the mode of the values in a list using the list's `.count()` method. We use a list comprehension to build a `counts` list containing `(count, val)` pairs, noting that we use `set()` to remove any repeats in the values. Using `max()` on these pairs will find the one with the largest count, although the mode will be the second item of the pair; the value that went with the count.

```
values = [1,2,2,3,2,1,4,2,3,1,0]
counts = [(values.count(val), val) for val in set(values)]
count, mode = max(counts)

print( mode )
```

Calculating the mode is easier to do with SciPy, as there is a pre-constructed `stats.mode()` function that works with NumPy `array` objects, though this also gives back an array, hence we take the `[0]` item from the result:

```
from scipy import stats
from numpy import array

valArray = array(values, float)
mode, count = stats.mode(valArray)

print('Mode:', mode[0] ) # Result is 2
```

The *median* represents the middle-ranked value when the data is placed in its sorted order. Or put differently, the median is the 50th percentile point that separates the top and bottom halves of the values. Taking the example [1,2,2,3,2,1,4,2,3,1,0] again, sorting this gives [0,1,1,1,2,**2**,2,2,3,3,4] and the middle value is 2. If there is an even number of points the median is generally represented as the average of the two middle points. The median is a fairly robust statistic to use, including where the underlying probability distribution is not known, because the middle ranking will be insensitive to outlier points (with extreme values).

We can calculate the median in standard Python by sorting the values and selecting the middle index, though if there is an even number of values (`nValues % 2 == 0`) we take the average of the central two:

```
def getMedian(values):

  vSorted = sorted(values)
  nValues = len(values)

  if nValues % 2 == 0: # even number
    index = nValues//2
    median = sum(vSorted[index-1:index+1])/2.0

  else:
    index = (nValues-1)//2
    median = vSorted[index]

  return median

med = getMedian(values)
```

Calculating the median is easy using NumPy, given its `median()` function:

```
from numpy import median

med = median(valArray)
print('Median:', med)    # Result is 2
```

The *mean* is the numerical average of a set of values. It is analogous to the centre of 'mass' of the distribution. In simple terms the sample mean is calculated by adding up all the values and dividing by the number of values. The mean of $[1,2,2,3,2,1,4,2,3,1,0]$ is $^{21}/_{11} = 1.909$. In terms of an underlying probability distribution, the mean of a random variable, X, is referred to as the expectation of the random variable, written $E(X)$, because it represents the value that represents the long-term average, considering an unlimited amount of data, and thus also the most representative centre value for the distribution. It should be noted that in this chapter we will be considering two types of mean value. The first is the true mean value of the underlying probability distribution,[5] and for a random variable X we will give it the label μ_x. The other kind of mean is the sample mean, labelled $\bar{x}$, which often acts as an estimate for the true mean, and which is calculated as an average value of a series of measurements, x_i, as one might expect:

$$\bar{x} = \frac{1}{n}\sum_{i=1}^{n} x_i$$

We can readily calculate the sample mean in standard Python:

```
values = [1,2,2,3,2,1,4,2,3,1,0]
mean = sum(values)/float(len(values))
```

or using NumPy arrays, noting that `mean()` is both a stand-alone function and a method bound to `array` objects:

---

[5] If the distribution is a large data set which we know in its entirety the true mean may be referred to as the *population mean*.

```
from numpy import array, mean

valArray = array(values, float)

m = valArray.mean()
# or
m = mean(valArray)

print('Mean', m)    # Result is 1.909
```

It is handy that these NumPy functions also take an `axis` argument, so that in a multi-dimensional array you can calculate the mean across rows or columns of values etc:

```
valArray2 = array([[7,9,5],
                   [1,4,3]])

print(valArray2.mean())
# All elements - result is 4.8333

print(valArray2.mean(axis=0))
# Column means - result is [4.0, 6.5, 4.0]

print(valArray2.mean(axis=1))
# Row means - result is [.0, 2.6667]
```

For most named probability distributions the mean is either a fundamental parameter that is used in the description of the distribution (e.g. for Gaussian) or is readily derived from the fundamental parameters (e.g. binomial, geometric). However, there are some curious cases where the mean is undefined, e.g. for the Cauchy distribution.[6]

Variance, standard deviation and skew

The standard deviation and variance (the square of the standard deviation) are measures of the spread in the range of values. These can be calculated for any given sample of values. However, and in a similar manner to the mean, the parameters calculated for a given set of samples are only an estimate for the underlying probability distribution. Here we label the true (or population) standard deviation as σ and true variance as σ^2, whereas the sample standard deviation is s and sample variance is s^2.

The variance is a measure of how far the values are spread from the mean. Mathematically it is the expectation of the squared differences from the mean. For a sample it is calculated as the sum of the square differences from the mean divided by the number of values (n) **minus one**:

$$s^2 = \frac{1}{n-1} \sum_{n=1}^{n} (x_i - \bar{x})^2$$

This is an *unbiased estimate*[7] of the underlying variance, but it is also commonplace to simply divide by the number of values, which for a large sample size makes little difference, though strictly speaking it is biased:

---

[6] Cauchy probability density function: $\frac{1}{\pi(1+x^2)}$, see http://en.wikipedia.org/wiki/Cauchy_distribution.
[7] http://en.wikipedia.org/wiki/Bias_of_an_estimator.

$$s^2 = \frac{1}{n} \sum_{n=1}^{n} (x_i - \bar{x})^2$$

We can calculate the variance in standard Python if we need to:

```
values = [1,2,2,3,2,1,4,2,3,1,0]

n = float(len(values))
mean = sum(values)/n
diffs = [v-mean for v in values]
variance = sum([d*d for d in diffs])/(n-1) # Unbiased estimate
```

Although, as you might expect, there is a handy `var()` function in NumPy, which is also built into `array` objects. Similar to the mean function, we can also specify an `axis` to get variances across rows and columns of multi-dimensional arrays. It should be noted that `var()` takes a `ddof` argument,[8] which should be set at `1` for the unbiased estimate; the default value is zero for the biased estimate.

```
from numpy import array
valArray = array(values)

variance = valArray.var()  # Biased estimate
print('Var 1', variance)   # Result is 1.1736

variance = valArray.var(ddof=1) # Unbiased estimate
print('Var 2', variance)        # Result is 1.2909
```

The biased variance equation can be rearranged as follows, as the mean (the expectation) of the squared values minus the square of the mean:

$$s^2 = \left(\frac{1}{n} \sum_{n=1}^{n} x_i^2 \right) - \bar{x}^2$$

Formulating the variance in this way can be handy in various situations because it involves fewer computational steps; we don't need to do a subtraction for every data point (which additionally may incur floating point errors).

Given that the standard deviation is the square root of the variance it is useful because it gives a measure of spread in the same units of measurement as the data. This is handy when describing statistical samples, so, for example, the height of a population may be described as the mean plus or minus the standard deviation: e.g. 1.777 ± 0.075 metres. The standard deviation is trivial to obtain using a square root operation and the above equations for variance, though there is also a handy `std()` function that is also inbuilt into NumPy arrays, noting again that we set `ddof=1` to use the unbiased estimate of the variance (although even in this case, `std(ddof=1)` does not give a truly unbiased estimate of the standard deviation):

```
from numpy import std, sqrt

stdDev = sqrt(variance)
stdDev = std(valArray)          # Biased estimate   - 1.0833
stdDev = valArray.std(ddof=1)  # "Unbiased" estimate - 1.1362

print('Std:', stdDev)
```

---

[8] 'Delta degrees of freedom'.

Related to the standard deviation is a value called the standard error of the mean (SEM). Given that the mean of a sample is only an estimation of the underlying mean there will naturally be some variation in its calculation. The SEM represents the standard deviation in the sample mean that results from different samplings of the underlying probability distribution. Scientifically it can be important to acknowledge that the sample mean is an estimate, and when supporting theories with a mean value it is often helpful to show the SEM, for example, on a graph, to indicate the confidence in the argument. The standard error of the mean is the standard deviation in x (s_x) divided by the square root of the number of values:

$$SE_{\bar{x}} = \frac{s_x}{\sqrt{n}}$$

This may be calculated in Python from the standard deviation and also by using a function from `scipy.stats`:

```
stdErrMean = valArray.std(ddof=1)/sqrt(len(valArray))

from scipy.stats import sem

stdErrMean = sem(valArray, ddof=1) # Result is 0.3426
```

The skewness of a distribution is a measure of asymmetry or lopsidedness. Though perhaps not as commonly used as the other parameters, estimating the skewness can be a useful test if you believe the underlying probability distribution ought to be symmetric. The skewness is commonly estimated for a sample as the mean cubed difference of the data from the mean divided by the standard deviation cubed:

$$g = \sum_{i=1}^{n} \left(\frac{x_i - \bar{x}}{s}\right)^3$$

We can illustrate this in Python using a random sample drawn from the asymmetric gamma function.

```
from scipy.stats import skew
from numpy import random

samples = random.gamma(3.0, 2.0, 100) # Example data

skewdness = skew(samples)
print( 'Skew', skewness )  # Result depends on random sample
```

Alternatively, as a very rough measure of skew, the non-parametric skew is easy to calculate as the difference between the mean and the median divided by the standard deviation, with the general idea being that the mean and the median will be the same when the distribution is symmetric.

Statistical tests

Significance and hypotheses

Many statistical tests are tests of significance, but we will require different levels of certainty in different contexts before we say that a group of observations is 'significant'. This is often dependent on the cost of the outcome if we're wrong. More generally though, statistical tests

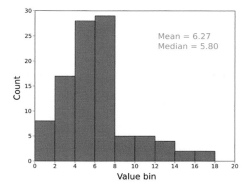

Figure 22.2. A histogram of samples taken from a skewed distribution. The above data were generated by randomly selecting 100 values using a gamma distribution, which is asymmetric, as illustrated in the Python sample. The difference between the mean and median of a distribution compared to the standard deviation can be used as a rough measure of asymmetry, though the proper skewness is easy to calculate using SciPy.

are about probabilities, and indeed it can be useful to estimate the probability of an event occurring, given a hypothesis, without formulating any notion of what is, or is not, significant.

When a statistical model has an approximating probability distribution then we can calculate probabilities with reference to that distribution. And, as we mentioned above, the probability distribution could be a null hypothesis, which represents a random expectation. In this case, our notion of significance would be based on the likelihood of the null hypothesis generating the observed data, though this is often also compared to the likelihood of the same data being generated by a competing hypothesis, such as the alternative hypothesis.

Tailed tests and p-values

A simple test to perform with a probability distribution is to evaluate the likelihood that a value was generated by the distribution, which is a matter of knowing the mathematical form of the distribution. Although this may be helpful for some discrete probability distributions (as illustrated in the previous chapter), as the number of possible outcomes becomes larger the probability of specific outcomes gets ever smaller. If we are trying to test whether an observation was generated by a probability distribution, it often doesn't help to know that a particular observed outcome is itself highly unlikely. This notion is taken to the extreme for continuous probability distributions where the probability that any single, precise real number value was generated by a probability distribution is zero, because technically there are an infinite number of values that could have arisen. Hence, rather than thinking about the likelihood of a single value we instead tend to test the probability of a value falling within a given range. Mathematically we can formulate this as an integral of the probability mass function (introduced in Chapter 21), i.e. the area of a region under the curve defining the probabilities for each value (see Figure 22.3).

The most common kind of region, and hence integral, to choose for statistical tests is the tails of the probability distribution. Here, by 'tail' we mean the area at the extreme edge of the distribution that is bounded by a specified threshold. Many probability distributions will

Probability mass function
Binomial distribution; p=0.5, n=1000

Cumulative distribution

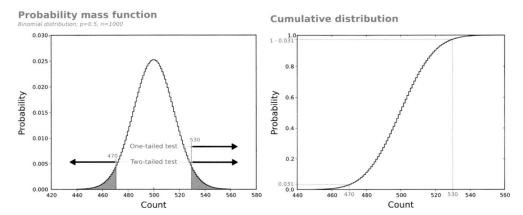

Figure 22.3. Calculating tail probabilities in the binomial test. The probability mass function and the regions considered in probabilistic tail tests are illustrated for an example binomial distribution, where probability of an event (p) is 0.5 and the number of trials (n) is 1000. For a one-tailed test the probability considered corresponds to the area under the curve from the test value, which here is 530, and above. For the two-tailed test the probability considered corresponds to the area under the curve in two regions above and below the mean (530 and 470 here) that start at the same separation from the mean and go outwards. The cumulative distribution is the running total of the probabilities for all the counts up to a particular value (i.e. starting at zero) and can be used to look up the areas of the tails from the probability mass function. For the right-hand threshold (530) the cumulative probability is the sum **up to** that value, but for the tail area we require the cumulative probability **from** that value, hence we subtract the cumulative probability before the right-hand threshold from 1.0 to get the required value.

have two tails, for extremely small and extremely large values respectively, and in a statistical test we may choose to consider either or both of these tails, depending on the situation. Accordingly, you will see both one-tailed and two-tailed tests used in scientific contexts. In biological research **p-values** are frequently quoted, which is simply the application of one-tailed tests to a null hypothesis. Put more formally, a p-value is the probability of values at least as extreme as the observation being generated by the null hypothesis.[9] (It is not the probability that the null hypothesis is correct or the probability that a value is random 'noise'.) While it is possible to define other regions of the distribution for testing purposes, using the tails of the distributions has the advantage of having no extra parameters to consider (such as a region width) and their integral is easy to calculate using a cumulative density function, which represents the running total of probabilities up to a specific value. We will illustrate one-tailed and two-tailed tests using the discrete (i.e. whole number) binomial and Poisson probability distributions that we introduced in Chapter 21, as well as for the continuous normal distribution.

First we consider the binomial test, which is concerned with the number of occurrences of an event that has a fixed probability of occurring, given a certain number of trials. If we apply this to the example of a DNA sequence of specified length (a specified number of trials) where we believe that the probability of G:C base pairs occurring is equal to the probability of A:T pairs we can ask whether an observed G:C versus A:T count fits with our model. If we look at a DNA sequence of length 1000, then following the random hypothesis we would expect that on average around 500 of the nucleotides would be G or C, and

---

[9] The more extreme values can either be smaller or larger than the observation, depending on the context.

500 would be A or T. If we get significantly more or fewer than 500 (of either pair) then we can conclude that the G:C content is likely to be not random. The binomial test formalises these considerations: if the probability of the G:C event is 0.5 then the expected number of events in 1000 trials is $0.5 \times 1000 = 500$. Furthermore, the binomial distribution allows us to associate a probability to every number of G:C events that may be observed in 1000 trials and hence also calculate the area under the tails of the distribution.

The exact form of the one-tailed test depends on whether the observed number of events is higher or lower than the expected number of events. If it is higher, then we calculate the probability that the observed number is that value or larger. If it is lower, we instead consider the probability that it is that value or smaller. In both cases the two-tailed test is the same. If in our example the actual number of observed G:C pairs is 530, then the one-tailed test considers the probability that we would observe **at least** 530 G:C nucleotide pairs in the sequence of length 1000. The two-tailed test not only considers the probability of observing at least 530 G:C (30 more than the mean) but also considers the probability of observing **at most** 470 G:C (30 fewer than the mean). Conversely, if the actual number of observed G:C pairs is 470, then the one-tailed test considers the probability that we would observe **at most** 470 G:C nucleotide pairs in the sequence of length 1000.

The two-tailed binomial test is available as a function in SciPy:

```
from scipy.stats import binom_test

count, nTrials, pEvent = 530, 1000, 0.5
result = binom_test(count, nTrials, pEvent)
print('Binomial two tail', result)
```

However, we can also create our own function using the explicit probability distribution. We will allow the input `counts` argument to be a list, tuple or a NumPy array. With a bit of extra effort we could also allow the argument to be a simple integer. We first convert `counts` to be a NumPy array, because that allows for calculations to be done on the entire array in one go, rather than having to loop over the elements. (For a discussion about NumPy, see Chapter 9.)

We also include an option to specify one-tailed tests. The one-tailed test is complicated because, as noted above, the calculation is different depending on whether the counts are greater or less than or equal to the mean. Here we test whether elements are greater than the mean (`counts > mean`), which yields an array of `True` and `False` values. Then the NumPy `nonzero()` function is used to obtain the indices of the true elements: those that were greater than the mean. Conveniently, we can access the corresponding elements of a NumPy array using these indices. This is faster than looping over the individual elements of the array, at the cost of requiring more memory for the indices. The values that are less than the mean will be the opposite selection: we also get these from the same test, but it is noteworthy that a NumPy array cannot be negated with the ordinary Python `not` Boolean operator. Rather the '~' symbol negates each element of the array (swaps `True` with `False`).

```
from scipy.stats import binom
from numpy import array, zeros

def binomialTailTest(counts, nTrials, pEvent, oneSided=True):

  counts = array(counts)
  mean = nTrials * pEvent
```

```
if oneSided:
    result = zeros(counts.shape)
    isAboveMean = counts > mean
    aboveIdx = isAboveMean.nonzero()
    belowIdx = (~isAboveMean).nonzero()
    result[aboveIdx] = binom.sf(counts[aboveIdx]-1, nTrials, pEvent)
    result[belowIdx] = binom.cdf(counts[belowIdx], nTrials, pEvent)

else:
    diffs = abs(counts-mean)
    result = binom.cdf(mean-diffs, nTrials, pEvent)
    result += binom.sf(mean+diffs-1, nTrials, pEvent)

return result
```

For counts less than or equal to the mean, we use the cumulative distribution function, `.cdf()`, which exactly provides the probability that the value is less than or equal to the count. For counts higher than the mean we instead use `.sf()`, the 'survival' function, as an alternative to calculating 1.0 minus the cumulative distribution function, which would be required to calculate the integral of the probabilities greater than or equal to the count. The subtraction of `1` from the counts in the `.sf()` call is the main oddity here, and this is because the `binom.sf()` function calculates the probability that the number of successes is greater than that argument, rather than greater than or equal.

For the two-tailed test we subtract the test values (`counts`) from the distribution's mean value to get separations from the mean. These differences are then used to get the upper and lower test thresholds. The probability sums for the lower thresholds are obtained via `.cdf()`, and the upper thresholds with `.sf()`, like the one-tailed test.

Continuing with the G:C content example, where we observe 530 G:C events from 1000 positions, to do the one-tailed test to get the probability that we observe at least 530 we simply call the function with the appropriate arguments:

```
counts = [530]
result = binomialTailTest(counts, 1000, 0.5, oneSided=True)
print( 'Binomial one tail', result))
```

and this gives a probability of 0.0310. Given that the binomial distribution was the null hypothesis this one-tailed probability is the p-value. Similarly, we can do the two-tailed test to calculate the probability that we observe at least 530 or at most 470 events, remembering that we set `oneSided` to be `False`:

```
result = binomialTailTest(counts, 1000, 0.5, oneSided=False)
print( 'Binomial two tail', result)
```

which gives a result of 0.0620. These results illustrate the subtlety of statistical analysis. Often it is deemed that something is significant if it has less than a 5% probability of originating from the null distribution. Given the one-tailed probability here is 3.1% one might want to conclude from this that it is likely that the DNA sequence is non-random (doesn't fit the null hypothesis). However, the two-tailed probability is 6.2%, so one might equally say that one cannot conclude this. Fundamentally it is the probability itself that matters, including how it is defined (one-tailed or two-tailed) and no matter what threshold we use, the probability itself tells us how likely we are to be right or wrong about any specific statement. Significance thresholds like 5% are merely conveniences for standardisation and not always applicable in every situation.

Although we have illustrated the tailed test for the binomial distribution, we can use the same strategy for other distributions that are provided by SciPy. In the next example, we illustrate for the Poisson distribution, noting that the function (like the binomial one) can accept an array of values to calculate probabilities for:

```
from scipy.stats import poisson
from numpy import abs, array

def poissonTailTest(counts, eventRate, oneSided=True):

  counts = array(counts)

  if oneSided:
    result = poisson.sf(counts-1, eventRate)

  else:
    diffs = abs(counts-eventRate)
    result = 2*poisson.cdf(eventRate-diffs, eventRate)

  return result

counts = [2300, 2400, 2550]
result = poissonTailTest(counts, 2500, oneSided=False)

result = poissonTailTest(counts, 2500, oneSided=False)
print( 'Poisson two tail', result)
# Result is [0.00002655, 0.022746, 0.1611]
```

The only differences to the binomial example are that the mean value is simply the event rate (the distribution's only parameter) and because the distribution is symmetric either side of the mean the one-tail calculation is simpler, and we can double the integral for one tail to get the probability corresponding to both tails.

Next we move on to an example that uses the *normal distribution*, otherwise known as the Gaussian distribution. In this case we are dealing with a continuous probability distribution (rather than a discrete one), i.e. where the outcome is a value from the continuum of real numbers. The concepts largely carry over from discrete to continuous distributions, with the important difference that discrete sums become integrals.[10] And as part of that difference, for a continuous random variable, X, we now have a *probability density function*, $f(x)$.[11] The probability density function for the normal distribution is as follows:

$$f(x) = \frac{1}{\sigma\sqrt{2\pi}}e^{-\frac{(x-\mu)^2}{2\sigma^2}}$$

---

[10] So the total probability, $1 = \int f(x)dx$ where the integral is over the region where X is defined. The mean, $\mu = E(X) = \int x\, f(x)dx$ and the variance is still $E(X^2) - E(X^2)$, where $E(X^2) = \int x^2\, f(x)dx$.

[11] Here $f(x)$ can be thought of as the probability that the random variable X takes the value x, although that is not quite correct mathematically, since in general the probability that X takes any precise value is 0, given there are an infinity of real numbers in a given range. More correctly, $f(x)$ is the probability that X takes values in the range $(x-\delta, x)$, then divided by δ, and in the limit that δ approaches 0.

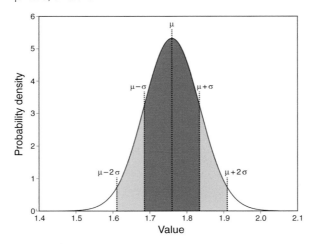

Figure 22.4. A normal distribution with mean and one and two standard deviations marked.
Corresponding to the example of human heights, the graph is the probability density function for a
normal (or Gaussian) distribution with a mean (μ) of 1.76 and a standard deviation (σ) of 0.075.
The values corresponding to one and two standard deviations above and below the mean value are
marked.

We use the normal distribution to model the distribution of heights in a population
of male humans, where we might say the mean (μ) is 1.76 and the standard deviation (σ) is
0.075, and we plot $f(x)$. Here we illustrate with Python using the `stats.norm()` function
directly, with the required mean and standard deviation, to obtain an object representing the
random variable (`normRandVar`). We can then use this object to calculate the probability
density function for an array of values, which here represent a range of the horizontal x axis
that we will plot (see Figure 22.4 for a labelled version of the output). It is notable that the
function we call is `.pdf()`, which is distinct from the `.pmf()` used for the discrete
probability distributions.

```
from matplotlib import pyplot
from scipy.stats import norm
from numpy import arange

mean = 1.76
stdDev = 0.075
stepSize = 0.0001
normRandVar = norm(mean, stdDev)   # Random variable object

xVals = arange(1.42, 2.1, stepSize)   # Graph range
yVals = normRandVar.pdf(xVals)        # Note PDF not PMF

pyplot.plot(xVals, yVals, color='black')
```

Then we can also plot the regions corresponding to one and two standard deviation widths
using the handy `pyplot.fill_between()` function, which in this case fills the area from the tail
region of the curve to the x axis (y = 0).

```
xVals = arange(mean-2*stdDev, mean+2*stdDev, stepSize)
yVals = normRandVar.pdf(xVals)
pyplot.fill_between(xVals, yVals, 0, color='lightgrey')

xVals = arange(mean-stdDev, mean+stdDev, stepSize)
yVals = normRandVar.pdf(xVals)
pyplot.fill_between(xVals, yVals, 0, color='grey')

pyplot.show()
```

The normal distribution is important because of the *central limit theorem*, which says that, under fairly weak assumptions, the distribution of the average of a number of independent and identically distributed random variables approaches a normal distribution, as the number of random variables increases. Considering two random variables, if for each point in the distribution for the first we superimpose the spread that arises from the distribution of the second, then the summation is a 'smoothed' probability distribution. The more independent random variables we add the closer the overall density gets to the normal distribution.

This commonly applies in science because observed values are often complicated combinations of multiple random variables, i.e. different factors, that all contribute to an observed distribution of values, and data samples are often assumed to be independent and identically distributed. Hence, the central limit theorem is often invoked to justify considering the measurement of some property to be distributed normally. For the example of the heights of (male) humans the independent factors that contribute to the final value may be things like multiple genetic factors (each with probability density functions for outcomes), nutrition, mother's weight etc., and it is the combination of all these random factors that gives rise to the single statistic of height.

In the same manner as for the discrete probability distributions, we can create a simple function to do one-tailed and two-tailed probability tests for the normal distribution using functions from the `scipy.stats` module.

```
def normalTailTest(values, meanVal, stdDev, oneSided=True):

  normRandVar = norm(meanVal, stdDev)

  diffs = abs(values-meanVal)
  result = normRandVar.cdf(meanVal-diffs) # Distrib is symmetric
  if not oneSided:
    result *= 2

  return result
```

We can test this for an array of test values (i.e. human heights):

```
mean = 1.76
stdDev = 0.075
values = array([1.8, 1.9, 2.0])

result = normalTailTest(values, mean, stdDev, oneSided=True)
print( 'Normal one tail', result)
# Result is: [0.297, 0.03097, 0.000687]
```

Assuming the normal distribution and its parameters are a good model for male human height, the results estimate that 29.7% are 1.8 metres or taller, 3.1% are 1.9 metres or taller and 0.069 % are over 2.0 metres.

Z-scores and Z-test

A Z-score (also called the standard score) is the number of standard deviations an observed value is different from the mean. So for our human height example, values of 1.685 metres and 1.910 metres have Z-scores of -1.0 and 2.0 because they are respectively 1σ below and 2σ above the mean. This is formalised in the following equation, i.e. subtract the mean and divide by the standard deviation:

$$z = \frac{x - \mu}{\sigma}$$

If we apply this to a whole distribution of values then we will centre it (the mean) at zero and give it a standard deviation of 1.0, e.g. to create the *standard normal* distribution, whose random variable is often labelled Z. We can easily calculate a Z-score in Python, here taking parameters from the human height example:

```
from numpy import abs

mean = 1.76
stdDev = 0.075
values = array([1.8, 1.9, 2.0])
zScores = abs(values - mean)/stdDev

print('Z scores', zScores)
```

Thus we estimate that 1.8, 1.9 and 2.0 metres respectively correspond to about 0.5, 1.9 and 3.2 standard deviations from the mean. Note that SciPy provides the `stats.zscore()` function, but it operates differently because it estimates its own sample mean and sample standard deviation from the input values:

```
from scipy.stats import zscore, norm

samples = norm.rvs(mean, stdDev, size=25)   # Values for testing
zScores = zscore(samples, ddof=1)           # Unbiased estimators

print('Est. Z scores ', zScores)
```

A related concept to this is the Z-test, which can be used when we have samples that are taken from a normal distribution where the true mean and standard deviation are known. The Z-test is effectively the calculation of a Z-score for a sample mean. A common situation for use of the Z-test is where a large population is known to have a mean, μ_0, and standard deviation, σ, and where some other population of size n is measured to have a sample mean, $\bar{x}$, and the same standard deviation. We want to know whether this is significantly different and the null hypothesis would be that the two populations have the same mean. For the Z-test the Z-score is defined as:

$$z = \frac{\bar{x} - \mu_0}{\sigma / \sqrt{n}}$$

As discussed above, in the context of the standard error of the mean, the standard deviation of the sample mean is a factor of $1/\sqrt{n}$ smaller than the standard deviation of the distribution. The analysis also works if the distribution is not normal but the number of samples, n, is large, by the central limit theorem (assuming the conditions for the theorem are satisfied). If the standard deviation is not known, then the T-test described in the next section should be used instead.

Given a standard normal distribution ($\mu = 0, \sigma = 1$), the probability of observing a Z-score or worse is a two-tailed test. If this probability is low then the two populations are deemed to have a significantly different mean, and the null hypothesis is rejected. If z were positive we could also consider a one-tailed test, which is the probability of observing a result at least this positive. For the Z-test there is no direct SciPy function to perform the whole calculation of tail probabilities. Hence we need to take specific steps to find the integral of the probability distribution from the Z-score. Fortunately this is partly solved by having a cumulative distribution available: the summation up to a threshold of the probability density function. The cumulative distribution of the standard normal (Φ) is required for the tailed test. This is easily calculated in Python using the *error function*[12] available in SciPy, which is related to cumulative distribution of the standard normal :$\Phi(z) = 1/2 + 1/2\,erf(z/\sqrt 2)$, and thus solves the integral we require without too much hassle.

The code to calculate the Z-test probability in SciPy involves calculating the Z-scores for the standard error of the means and then using the error function `erf()` to derive the cumulative probability:

```
from numpy import sqrt
from scipy.special import erf

def zTestMean(sMean, nSamples, normMean, stdDev, oneSided=True):

  zScore = abs(sMean - normMean) / (stdDev / sqrt(nSamples))
  prob = 1-erf(zScore/sqrt(2))

  if oneSided:
    prob *= 0.5

  return prob
```

The calculation of the probability involves a trivial bit of arithmetic, remembering that we want $1-\Phi$, the tail of the cumulative distribution of the standard normal, and noting that the initial cumulative probability calculation is the two-tailed result (i.e. twice Φ), which we halve for the one-tailed result. This can be tested with some example data values which are roughly normal:

```
samples = array([1.752, 1.818, 1.597, 1.697, 1.644,  1.593,
                 1.878, 1.648, 1.819, 1.794, 1.745,  1.827])
mean = 1.76
stDev = 0.075
result = zTestMean(samples.mean(), len(samples),
                   mean, stdDev, oneSided=True)

print( 'Z-test', result) # Result is 0.1179
```

[12] $erf(x) = \frac{2}{\sqrt\pi}\int_0^x e^{-t^2}dt.$

The resulting probability of the sample mean coming from the normal distribution is 11.8%, so we generally wouldn't want to reject the notion that the samples were generated from it.

As another example, suppose we have a large database of DNA sequences and the G:C content of sequences in the database has mean 0.59 and standard deviation 0.1. The G:C content would not usually be modelled using a normal distribution, but if we have 100 sequences not in the database, and measure the G:C content of each, then we could still reasonably apply the Z-test, thus informing us whether they are likely to be from the same population of sequences. Suppose that the average G:C content in these 100 sequences is 0.61. The one-tailed test is given by

```
result = zTestMean(0.59, 100, 0.61, 0.1)
```

with result 0.023. The two-tailed test gives twice this, so 0.046. In both cases, if 5% is the significance level used, then the null hypothesis is rejected, and it is concluded that the 100 sequences have a significantly different G:C content than the sequences in the database.

T-tests

The Z-test we described relied on knowledge of a distribution's standard deviation (or having a good estimate from a large population). However, in many situations we do not know the underlying mean and standard deviations of the probability distributions. This is often the natural outcome of having small statistical samples. Nonetheless, we may still want to evaluate whether statistical samples are significantly different from one another. This is where the idea of T-tests comes in.

T-tests are based on the notion of the T-statistic, which is similar to the Z-score discussed before. Accordingly, the T-statistic is the measure of the number of standard errors a measured parameter value is from its true value. In many cases the parameter we're interested in is the mean of a normal distribution, in which case the T-statistic could be the number of standard errors that the sample mean ($\bar{x}$) lies from the true mean (μ_x):

$$t = \frac{\bar{x} - \mu_x}{s \Big/ \sqrt{n}}$$

The important difference compared to the Z-score is that the sample standard deviation (s) is used. Because the sample standard deviation is only an estimate for the true standard deviation (σ), given it is calculated from a limited number of samples, then when we wish to perform a statistical test we will get an extra uncertainty that accounts for the sampling process. It should be noted that there are many different formulations of T-statistics that can be used, depending on the question being asked. However, they all have the form of a difference from a parameter estimate, divided by the corresponding standard error in the sample. For example, a T-test can be applied to determine whether two sets of samples, each from a normal distribution, have the same underlying mean, assuming that they have same variance (which might not be known). In this case the T-statistic is formulated for the two independent sample means $\bar{x}$ and $\bar{y}$, with sample sizes n_x and n_y respectively:

$$t = \frac{\bar{x} - \bar{y}}{s_{x,y}\sqrt{\frac{1}{n_x} + \frac{1}{n_y}}}$$

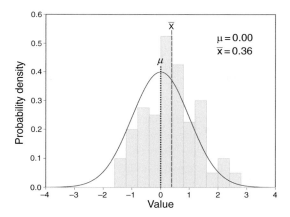

Figure 22.5. Comparing the underlying mean of a probability distribution and the sample mean.
A standard normal distribution, with a mean, μ, of 0.0 is superimposed on a set of data with a sample
mean, $\bar{x}$, of 0.36. Given that the sample mean has an associated error depending on the number of
samples taken, we can use a T-test to assess whether the separation between the two means is significant,
and thus whether the probability distribution is a good model for the data.

Here $s_{x,y}$ is the sample estimate for the standard deviation that is shared between the two
samples, irrespective of whether they have the same mean. This may be calculated from the
individual sample variances for the two distributions (s_x^2 and s_y^2) as follows:

$$s_{x,y} = \sqrt{\frac{(n_x - 1)s_x^2 + (n_y - 1)s_y^2}{n_x + n_y - 2}}$$

This is the standard deviation we would obtain if we combined both samples together, with an
adjustment because we have two sample means, one for each group. Effectively we replace
the $1 / (n - 1)$ fraction in the unbiased sample variance with $1 / (n_x + n_y - 2)$ to consider both
samples.

In mathematical parlance, for the two-sample T-test the T-statistic turns out to follow
a T-distribution with $n_x + n_y - 2$ degrees of freedom. Similarly the one-sample T-test has a
T-statistic with $n - 1$ degrees of freedom. In statistics the notion of 'degrees of freedom' can
be a somewhat tricky concept, but the principle is to know the number of independent data
points that can truly vary. To take an arbitrary but simple example, where there are three
sample values that have a mean of zero, once two values are known then there is no choice
about the third, because we know it must give the known mean. Hence, in general, for a
statistical analysis the number of degrees of freedom is the number of independent sample
values, minus the number of restraining parameters.

Once we have an appropriate T-statistic, with an appropriate number of degrees of
freedom, in order to perform a statistical test we use the distribution of how the T-statistic
itself varies when taking different samplings. This probability distribution is the *Student
T-distribution*,[13] which assumes the samples are independent and have the same normal

---

[13] 'Student' was a pseudonym of William Gosset, who discovered the distribution while working for the Guinness
brewery in the early twentieth century.

distribution. We won't go into details of this distribution, only to say that it is a bell shape, like the normal distribution but with thicker tails. For Python the `scipy.stats` module has some pre-packaged T-test functions as well as facilities to access the Student T-distribution, which allow us to easily make probability estimates to assess the variation due to sample variance.

The complete T-test functions available in `scipy.stats` are `ttest_1samp`, `ttest_ind` and `ttest_rel` and these accept samples, represented as arrays of values, and perform the appropriate two-tailed test to estimate a probability. Also, because the T-distribution is symmetric, for a one-tailed test we can simply halve the probability. Illustrating each of these functions in turn, `ttest_1samp` finds the probability of a sample mean being the same as the true mean from a distribution (e.g. a null hypothesis), and thus uses the T-statistic for $\bar{x} - \mu_x$ described above. Note that the T-statistic as well as the two-tailed probability are passed back by the function:

```
from scipy.stats import ttest_1samp

trueMean = 1.76
samples = array([1.752, 1.818, 1.597, 1.697, 1.644, 1.593,
                 1.878, 1.648, 1.819, 1.794, 1.745, 1.827])

tStat, twoTailProb = ttest_1samp(samples, trueMean)
# Result is: -0.918, 0.378
```

The function `ttest_ind` performs the two-sample T-test, testing whether two independent samples have the same underlying, true mean, based on their respective sample means, described as $\bar{x}$ and $\bar{y}$ above:

```
from scipy.stats import ttest_ind

samples1 = array([1.752, 1.818, 1.597, 1.697, 1.644, 1.593])
samples2 = array([1.878, 1.648, 1.819, 1.794, 1.745, 1.827])

tStat, twoTailProb = ttest_ind(samples1, samples2)
# Result is: -2.072, 0.0650
```

There is an extra option to this function, to relax the requirement that both samples have the same variance, in which case the test is called Welch's T-test,[14] though the difference for our test case is slight:

```
tStat, twoTailProb = ttest_ind(samples1, samples2, equal_var=False)
# Result is: # -2.072 0.0654
```

Lastly, the `ttest_rel` function again works with two samples in the same way as above, but assumes that the samples are dependent, i.e. that the values in the pair of samples are related to one another (they must have the same variance, hence there is no `equal_var` option). An example of this would be to take some measure from a group of people as the first samples and then to take repeated measurements for the same people at a different time (perhaps after some treatment) or using a different method.

---

[14] The T-statistic for Welch's T-test is: $t = (\bar{x} - \bar{y}) / \sqrt{\frac{s_x^2}{n_x} + \frac{s_y^2}{n_y}}$.

Probability intervals

So far in this chapter we have been using tailed tests to calculate the probability of obtaining a given value from a statistical sample, and then on the basis of this probability we can decide whether we deem the sample to be significantly different from a null hypothesis using a threshold probability, say 5%. However, we can also take the reverse approach and use a probability threshold upfront to calculate what the equivalent test statistic would be for this limiting value. In turn this then leads to a corresponding interval in the actual measurements.

Returning to the one-sample T-test example of comparing a sample mean, $\bar{x}$, with the true mean, μ_x, we can determine a confidence interval for the true mean, related to a specified probability, given the sample mean and the unbiased sample standard deviation. Mathematically we want to determine the interval size I such that there is a specified probability that μ_x is within I of $\bar{x}$.

This is a two-tailed test, and the one-sided equivalent would be the probability that $\bar{x} - \mu_x$ is larger or smaller than some value. To calculate the interval we say that the probability of the absolute difference between means is the same as the probability that the magnitude of the T-distribution is less than the interval divided by the standard error, which simply comes from rearranging the formula for the T-statistic:

$$Prob(|\bar{x} - \mu_x| < I) = Prob\left(|T| < \frac{\sqrt{n}}{s}I\right)$$

We need to invert this function to determine I given a probability. To do this practically we use a function called the *quantile function* or *percent point function*. This does the inverse job to the cumulative distribution function, so we pass in a probability and get out a threshold value that the random variable will be bounded by (at or below). Fortunately for Python the percent point function is available for all the common probability distributions described in the scipy.stats module, so we generally don't have to worry about its precise formulation. When we have calculated the inverse for a given probability we then simply multiply by an appropriate factor, representing the standard error, to obtain the measurement interval.

We now provide a Python function to calculate the value of the interval, given the probability, or confidence that the samples were drawn from the distribution. The input can be a list or a NumPy array of samples, and a confidence level (e.g. 0.95 for 95% confidence). The result is the sampleMean and the interval. For the two-sided test this means that the actual mean is between sampleMean-interval and sampleMean+interval with the probability given by the confidence level.

```
from numpy import mean, std, sqrt
from scipy.stats import t

def tConfInterval(samples, confidence, isOneSided=True):

  n = len(samples)
  sampleMean = mean(samples)
  sampleStdDev = std(samples, ddof=1) # Unbiased estimate

  if not isOneSided:
    confidence = 0.5 * (1+confidence)

  interval = t(n-1).ppf(confidence) * sampleStdDev / sqrt(n)

  return sampleMean, interval
```

Inside the function, if the test is two-tailed we adjust the confidence value so that the tail probability used is half that for a single tail. For example, for an input 95% confidence (5% tail probability) we will find the interval corresponding to a one-tailed confidence of 97.5% (2.5% tail probability) because there will be two tail integrals that both contribute. Next, using scipy.stats.t we pass the appropriate number of degrees of freedom (n-1) in to the T-distribution and use the percent point function for this with ppf(). The value obtained is actually $I \cdot \sqrt{n}/s$, so we scale this by $s/\sqrt{n}$ to get the required interval. The function can be tested with our previous example, using a sample of human heights:

```
from numpy import array
samples = array([1.752, 1.818, 1.597, 1.697, 1.644,  1.593,
                 1.878, 1.648, 1.819, 1.794, 1.745,  1.827])

sMean, intvl = tConfInterval(samples, 0.95, isOneSided=False)

print('Sample mean: %.3f, 95%% interval:%.4f' % (sMean, intvl))
```

Note that the double '%%' in the print() statement is because Python treats a single '%' as the first character in a format string.

Hence, the difference to the mean that we would accept for a 95% confidence limit, when accepting an underlying probability distribution, is an interval of 0.0615 metres. If the mean of our null hypothesis distribution is actually 1.76 metres, then we would accept the sample mean of 1.734 metres because it is 0.0257 metres away from the mean, and thus lies within the interval.

Chi-squared and G-tests

The statistical tests discussed so far in this chapter have considered whether individual sample parameters or data values fit with probability distributions. For example, we have illustrated tailed tests, to compare values with a null hypothesis distribution and thus obtain a p-value. However, there are various ways that we can compare multiple parameters and even whole distributions with one another, and naturally this can involve a distribution for hypothesis testing in what is termed a *goodness-of-fit* test. If two distributions have the same mean but otherwise have different shapes, then such an analysis will convey a clear advantage. Though, as before we must naturally account for the error associated in taking random (and potentially small) samples from an underlying probability distribution.

The first method to compare multiple variables we will cover is *Pearson's chi-squared test*. This test is based on the chi-squared statistic (χ^2), which is defined as follows for observed frequencies of events (o_i) and the expected frequencies of events (e_i), which is generally based on the null hypothesis:

$$\chi^2 = \sum_{i=1}^{n} \frac{(o_i - e_i)^2}{e_i}$$

The test can be applied to categorical selection, and by extension to histograms of counts which may be used to approximate any arbitrary distribution. The assumption for the test is that each pair of expected and observed counts is derived from a normal random variable. The chi-square distribution, which the statistical test is based upon, is the distribution of such a sum of squares resulting from different random samplings in each of the variable categories. For the chi-square distribution (as with the T-distribution) we will need to know the number of degrees of freedom, which in general will be the number of observed values (e.g. categories) minus the number of restraining parameters (e.g. totals).

Returning to the example G:C versus A:T content of a 1000 DNA base pairs, we can apply the chi-square statistic to these two base-pair categories, though naturally they are restrained because they must sum to a given total. However, we treat them as independent observations for the calculation of the statistic and then consider the appropriate number of degrees of freedom. Hence if we have 530 G:C pairs and 470 A:T pairs and the expected count for each is 500, then the statistic is:

$$\chi^2 = \frac{(530 - 500)^2}{500} + \frac{(470 - 500)^2}{500} = 3.6$$

After calculating the chi-square statistic, comparing observed and expected counts, the next stage is to evaluate the statistical significance of the resulting value (3.6 in the above example). For this we use the chi-square distribution. The number of degrees of freedom here is the number of random variables (which in the above case is the number of categories) minus one; we lose a degree of freedom because the total is fixed, so the A:T count is not random given the G:C count. We can use the cumulative density function chi-square distribution for one degree of freedom to generate a p-value (i.e. do a one-tailed test compared to the null hypothesis) for the observed chi-squared statistic. Fortunately in SciPy this is all handled in one neat `chisquare` function. This will assume the number of degrees of freedom is $(n-1)$, though in other situations we could pass in `ddof`, representing the difference in the number of degrees of from the default:

```
from scipy.stats import chisquare

obs = array([530, 470])
exp = array([500, 500])
chSqStat, pValue = chisquare(obs, exp)

print('DNA Chi-square:', chSqStat, pValue) # 3.6, 0.05778
```

The result for this is the anticipated chi-square statistic of 3.6 and a test probability of 0.058. It should be noted that the chi-square test is almost always a one-tailed test because we are normally interested in whether the fit is worse than the expected fit, and not concerned if the fit is better than expected (i.e. too good).

Moving on from the simple DNA example, we can think of samples from a probability distribution where the resulting values have been binned into a histogram (see Figure 22.6 for an example histogram). This will give a discrete set of categories, one for each range, and we can treat each category as a separate, independent sampling and compare it to the expectation from the null hypothesis. In this case the expected count will be the area of the probability for reach range (a region of the probability density function) multiplied by the total number of observations.

We can illustrate this for the heights of people, using a null hypothesis with a normal distribution, and then test the goodness of fit for different height categories. Here `bins` represents the centres of the histogram ranges and `obsd` contains the corresponding number

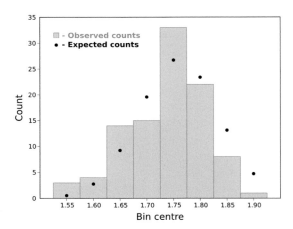

Figure 22.6. A histogram of observed and expected counts for a G-test. The illustrated histogram represents the observed counts for a sample of 100 values that have been binned into regions of width 0.05. The expected counts, represented by black spots, are calculated from the probability density funtion of the normal distribution, to which the histogram approximates. For the G-test the different bins are treated as independent random variables, and overall we estimate the goodness of fit of the sample histogram to the normal expectation. It should be noted that because some counts are small (<5) it would be inappropriate to do the chi-square test.

of observed counts for each bin. The expected counts `expd` are calculated for each of the `bins` values by applying the `.pdf()` for the normal distribution (with appropriate mean and standard deviation) and then scaling so that the total is the same as the number of observations.

```
from scipy.stats import norm
from numpy import array

bins = array([1.65, 1.7, 1.75, 1.8, 1.85,])
obsd = array([ 14, 15, 33, 22, 8,])

mean = 1.76
std = 0.075

nObs = obsd.sum()
expd = norm.pdf(bins, mean, std)
expd *= nObs / expd.sum()

# Expected counts: 9.196, 19.576, 26.720, 23.385, 13.122

chSqStat, pValue = chisquare(obsd, expd)

print('Chi square A', chSqStat, pValue)
```

The result is a chi-square statistic of 7.14 and corresponding p-value of 0.129 (12.9%). Hence, we would not reject the null hypothesis. As an alternative method we could use the SciPy distribution `chi2` and do the statistical test for a one-tailed value as we illustrated previously, noting that here we have to explicitly specify the number of degrees of freedom for the chi-square distribution (which is one fewer than the number of categories):

```
from scipy.stats import chi2

degFree = len(obsd)-1
pValue = chi2.sf(chSqStat, degFree) # Result is 0.129
```

Not only can the chi-square test be applied to discrete categorical counts as we illustrate above, but the process can also be applied to the fundamental parameters of a probability distribution. For example, the null hypothesis may be a specific probability distribution which can be described by a few fundamental parameters (μ, σ etc.) and we can compare that with the alternative hypothesis where the parameters are unrestricted. In the case of mean and standard deviation there would be three degrees of freedom in the test (two of which come from the fundamental parameters).

Pearson's chi-square test is commonly used, and thus implemented in SciPy, but there are limitations to its use, especially concerning low counts. A popular heuristic is that at least five observed counts are required for each random variable (each category). Although corrections can be applied to account for this, in general where counts are small it may be easier to apply a different test. Indeed the chi-square statistic is really only an approximation, developed in the days before computers for ease of calculation, to the *likelihood ratio* statistic. In the modern computer age we have little problem implementing likelihood ratio tests, such as the G-test that we describe next, which is less sensitive to small counts. The G-test uses the following statistic:

$$G = 2 \sum_i o_i \ln\left(\frac{o_i}{e_i}\right)$$

The statistic is a more accurate version of chi-square so we still do statistical tests using the chi-square distribution with the same degrees of freedom as the equivalent chi-square test. The G-statistic is not only useful for deriving probabilities, but also has a useful meaning in itself. It is the same thing as twice the Kullback-Leibler divergence[15] or relative entropy: a measure of the extra information required to get the observed distribution from the expected distribution. Unfortunately the G-test is not implemented directly in a single SciPy function, but we can easily do the required steps, making use of the chi-square distribution's survival function (`chi2.sf()`) to obtain a p-value. Firstly, we gather some test data, which is similar to before, although we have included a few more histogram bins with small counts. Then we calculate the expectation from the normal distribution:

```
bins = array([1.55, 1.6, 1.65, 1.7, 1.75, 1.8, 1.85, 1.90])
obsd = array([   3,   4,   14,    15,   33,   22,    8,    1])

mean = 1.76
std = 0.075

nObs = obsd.sum()
expd = norm.pdf(bins, mean, std)
expd *= nObs/expd.sum()

#Expected: 0.535, 2.769, 9.194, 19.571, 26.713, 23.379, 13.119, 4.720
```

---

[15] Mentioned in Chapter 11 to measure the repetitiveness of a sequence.

The G-test is easily calculated for the arrays by taking the natural logarithm (base e) of the required ratio, multiplying by the observations and then finding twice the sum of the resulting array. The resulting G-value is then applied to the chi-square survival function, with the appropriate degrees of freedom:

```
from numpy import log

g = 2.0 * sum(obsd * log(obsd/expd))
degFree = len(obsd)-1
pValue = chi2.sf(g, degFree)

print('G test', g, pValue)   # Result is: 17.34, 0.015
```

As illustrated, the result is a one-tailed probability of 1.5%. If we compare this with the chi-square test, even though it is not appropriate to do so with such small counts, we get:

```
chSqStat, pValue = chisquare(obsd, expd)
print('Chi-square', chSqStat, pValue)   # Result: 21.98, 0.00256
```

With the chi-square test the probability is now estimated to be 0.26%, which is substantively different from the G-test value.

Correlation and covariance

For the last part of this chapter we move from studying distributions of one type of measurement to the comparison of two different types, each with a different random variable. We can imagine the random variables to correspond to different dimensions or axes. Hence, a data point will be composed of two values, one for each axis. An approach here might be to apply statistical tests to a two-dimensional, joint probability distribution, employing the methods already discussed. However, we are often interested in the relatively simple question of whether the values for the two axes vary together in some way. In other words if the value of one measurement increases we would like to know whether the other measurement also increases, decreases or stays the same overall. This is what we call correlation. Naturally, this is also subject to significance testing because the variation associated with sampling of the probability distributions impinges on our measures of correlation. In particular, because of the variation arising from a small number of samples we may observe an apparent correlation and need to know the likelihood that it was generated by a random process.

Covariance

Covariance is a measure of whether two random variables vary simultaneously as their values increase or decrease. The covariance is calculated by subtracting the means of the random variables, so they are effectively centred on zero, and then finding the average product of the two coordinates. Hence for two probability distributions, described by random variables X and Y with sample points x_i and y_i respectively, the covariance may be written as:

$$\sigma_{xy} = \sum_{i=1}^{N} \frac{(x_i - \mu_x)(y_i - \mu_y)}{N} = E((X - \mu_x)(Y - \mu_y))$$

The idea is that if there is a correlation then the positions from both axes will be on the same side of their means, giving consistently positive products. If there is no correlation the products will be both positive and negative, averaging towards zero. In Python there is the handy `numpy.cov()` function to do the work for us. Here we illustrate with two test combinations for random `xVals`: `yVals1` is completely random and `yVals2` is derived from `xVals` by adding an offset, gradient and small random deviations:

```
from numpy import random, cov

xVals = random.normal(0.0, 1.0, 100)
yVals1 = random.normal(0.0, 1.0, 100) # Random, independent of xVals

deltas = random.normal(0.0, 0.75, 100)
yVals2 = 0.5 + 2.0 * (xVals - deltas)    # Derived from xVals

cov1 = cov(xVals, yVals1)

# The exact values below depend on the random numbers

# Cov 1: [[0.848, 0.022]
#          [0.022, 1.048]]

cov2 = cov(xVals, yVals2)
# Cov 2: [[0.848, 1.809]
#          [1.809, 5.819]]
```

The result here is the covalence matrix, rather than just a single value. This is just a generalisation of the process, where if you pass in several arrays it will give back a matrix of the covariance for all possible pairs. Hence for our two input arrays we will get a matrix with four values, i.e. $\begin{pmatrix} \sigma^2_x & \sigma_{xy} \\ \sigma_{yx} & \sigma^2_y \end{pmatrix}$, so the diagonal is simply the variances for X and Y and the other values are equal to the covariance we generally want. Here the interesting covariances are 0.022 and 1.809 for `yVals1` and `yVals2` respectively.

Correlation coefficient

As the covariance calculation derives from products of the two variables, the magnitude of the measure is dependent on the scaling of both X and Y. Correspondingly, the units of covariance will be the product of the X and Y units. We can remove the effect of scaling by dividing the covariance by the standard deviations of the two random variables and thus obtain a dimensionless value that lies between -1 and 1. A value of -1 will mean perfectly anti-correlated, that the quantities change together but in opposite directions; a value of zero will mean completely uncorrelated and 1 means fully correlated. This is what is called *Pearson's correlation coefficient* and it gives us a handy measure of how well two quantities are linearly correlated, irrespective of any different scales or units they may have. We can write the equation for this as:[16]

$$r = \frac{\sigma_{xy}}{\sigma_x \sigma_y}$$

---

[16] It also follows that r^2 is the coefficient of determination: the fraction of the variance in Y that is explained by X.

r=-0.949   r=0.023   r=0.814   r=-0.057   r=-0.082

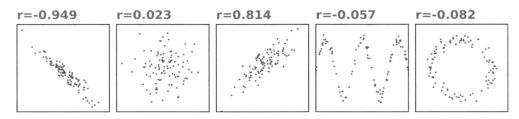

Figure 22.7. Pearson's correlation coefficient values (*r*) for a variety of different data samples.
The coefficient represents the degree of linear covariance in the two quantities and is scaled so that the
value lies between −1 (for negative correlation) and +1 (positive correlation). Values near zero indicate
the quantities are non-linearly correlated, although there may be other patterns or forms of non-linear
correlation, which would not be exposed by this test.

Although we have written this for the true standard deviations of the random
variables the calculation is the same for the sample correlation coefficient, except that the
unbiased sample standard deviation is used. If we apply this to various data sets, plotted as
graphs in Figure 22.7, we can see how the value of the correlation coefficient corresponds to
the degree of linear correlation.

The correlation coefficient is readily calculated in Python using the `numpy.corrcoef()`
function, and as with the covariance function we get back a matrix of values, for all pairs of
inputs. Testing on the previously used values we get:

```
from numpy import corrcoef

r1 = corrcoef(xVals, yVals1)[0, 1] # Result is: 0.0231
r2 = corrcoef(xVals, yVals2)[0, 1] # Result is: 0.8145
```

Hence we can see that `xVals` has almost no correlation with `yVals1`, but has a large positive
correlation (0.8145) with `yVals2`, as we might expect. If we wished we could naturally also
derive the correlation coefficient from the previously calculated covariance, remembering that
we use the unbiased sample standard deviation (`ddof=1`):

```
from numpy import std

cov2 = cov2[0,1] # X-Y element
stdDevX = std(xVals, ddof=1)
stdDevY = std(yVals2, ddof=1)

r2 = cov2 / (stdDevX*stdDevY)
```

Although the correlation coefficient is insensitive to different sample means and
variances for the quantities, it should not be forgotten that it is only a test of a linear
relationship. There may be a distinct non-random, non-linear relationship between the quantities which will not be picked up by the test, although in some instances it is possible to
transform a quantity (e.g. by taking a logarithm) so that the relationship becomes linear.

We can subject the correlation coefficient to significance tests if we consider an
uncorrelated null hypothesis, i.e. where the underlying correlation coefficient is 0. The basic
idea here is that even if distributions are really uncorrelated they can appear to be correlated
(points are coincidentally linear), especially if the size of a sample is small. If the underlying

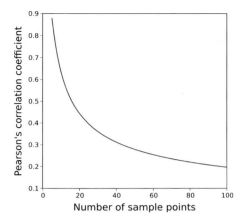

Figure 22.8. The Pearson correlation coefficient that rejects an uncorrelated hypothesis at the 5% level for different sample sizes. For two uncorrelated random variables the likelihood of coincidentally obtaining a Pearson correlation coefficient (r) substantially different from zero diminishes with increasing sample size. The graph shows, for different sample sizes, the threshold value of r that will be exceeded 5% of the time by truly uncorrelated normal distributions.

distributions are normal then it can be shown that the null hypothesis can be rejected at the 0.95 confidence level if the test statistic

$$t = r\left(\frac{n-2}{1-r^2}\right)^{1/2}$$

is larger than the corresponding T-distribution percent point function with confidence level 0.975 (because our test is two-tailed) and $n-2$ degrees of freedom. Here n is the number of sample points in each of X and Y. We can invert the above function, and solve for the correlation coefficient r as a function of n.

$$r = \left(\frac{t^2}{n-2+t^2}\right)^{1/2}$$

Accordingly we can plot the correlation coefficient as a function of the sample size, as illustrated in Figure 22.8. If r is larger than the value then the null hypothesis is rejected. This is readily done in Python using the `.ppf()` function of the `scipy.stats.t` distribution object and applying the above equation:

```
from numpy import sqrt
from scipy.stats import t

nVals = range(5, 101)
rVals = []

for n in nVals:
    tVal = t(n-2).ppf(0.975)
    tVal2 = tVal * tVal
    rVal = sqrt(tVal2/(n-2+tVal2))
    rVals.append(rVal)

pyplot.plot(nVals, rVals, color='black')
pyplot.show()
```

Simple linear regression

Related to the linear correlation coefficient is the notion of the line of best fit. Although there are various ways to calculate such a line, Pearson's correlation coefficient provides a handy way of doing a line fit called *simple linear regression*. For reasons we won't go into, it can be shown that the covariance of X and Y divided by the variance of X is the gradient of the line that minimises the sum of the squared differences along the Y axis, from the fit line to the data points. Hence the gradient β can be written as this ratio, which in turn may be derived from the correlation coefficient:

$$\beta = \frac{\sigma_{xy}}{\sigma_x^2} = r\frac{\sigma_y}{\sigma_x}$$

Following from this, the intersection of the Y axis, α can be calculated as:

$$\alpha = \mu_y - \beta\mu_x$$

This is all very easy to do in Python using NumPy functions, noting that as before we test with some artificial values derived from a random sample:

```
from numpy import cov, var, mean, random

xVals = random.normal(0.0, 1.0, 100)
yVals = 2.0 + -0.7 * xVals + random.normal(0.0, 0.2, 100)

grad = cov(xVals, yVals)/var(xVals, ddof=1)
yInt = mean(yVals) - grad * mean(xVals)
print('LR 1:', grad, yInt) # Result for one run was: -0.711 2.04
```

Accordingly, the fitted line has a gradient and Y-intercept close to the artificial test values of -0.7 and 2.0. There is also a handy function in the `scipy.stats` function `linregress` which calculates lots of useful things in one fell swoop, including the correlation coefficient

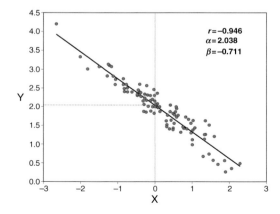

Figure 22.9. **An example of simple linear regression applied to a linearly correlated data set.** The two random variables X and Y have been fitted by the minimisation of the difference, at a specified X-value, between the Y-value of each point and the Y-value at the line. The Y-intercept, α (where $x = 0$) and gradient, β were calculated as illustrated in the Python example shown in the text.

and the two-tailed probability of the hypothesis that the gradient is zero. Here we do the linear regression and then plot the best-fit line on the same graph as the input data (like Figure 22.9):

```
from scipy.stats import linregress
from matplotlib import pyplot

grad, yInt, corrCoeff, pValue, stdErr = linregress(xVals, yVals)

print('LR 2:', grad, yInt, corrCoeff, pValue, stdErr)
# Result for one run was: -0.711, 2.04, -0.949, 9.639e-51, 0.0240

xValsFit = [xVals.min(),xVals.max()]
yValsFit = [yInt + x*grad for x in xValsFit]

pyplot.plot(xVals, yVals, 'o')
pyplot.plot(xValsFit, yValsFit)
pyplot.show()
```

Tailed tests and confidence intervals may be calculated more generally for the gradient using a T-test (with $n-2$ degrees of freedom). Here the T-statistic is calculated as follows, where $\hat{\beta}$ is the estimated gradient and $\hat{y}_i$ is a y value estimated by the fit line:

$$t = \frac{\hat{\beta} - \beta}{\sqrt{(n-2)\dfrac{\sum_{i=1}^{n}(y_i-\hat{y}_i)^2}{\sum_{i=1}^{n}(x_i-\bar{x})^2}}}$$

23 Clustering and discrimination

Separating and grouping data

When dealing with biological information, the question at hand often relates to the ability to separate a pool of data into different groups. This may be a simple two-way split, for example between people who do or do not have a disease, or it may involve many more data categories. Sometimes, however, the number of groups may not be known and it may not even be appropriate to think in terms of rigidly defined groups. Rather, it might be better to first determine the most discriminating features that separate the data and then investigate afterwards whether groups are present, and if so how many. Any kind of discrimination exercise naturally requires some form of information on which a judgement may be based, such as the results from an experiment, which can even include things like DNA sequences. Implicit in this sort of analysis is the notion that units of data are being separated, but each unit may relate to several pieces of information. For example, if a unit of data corresponds to a person they may be diagnosed by several different parameters and test measurements, or if a unit is a biological molecule it may be categorised by many different properties and experimental results.

Whatever the situation and type of data, sometimes the question being asked tries to place each unit of data in one group or another, where there is no possibility of something being in more than one group. Naturally, whether this is a valid assumption will depend on context and the formulation of the problem. In reality, a hard boundary between groups might not actually be as useful as a more fuzzy membership. Referring again to the problem of diagnosing a condition in people using experimental test results, it may be that two people with identical test results have different outcomes; there may not be a simple dividing line between groups. We may have official values to distinguish between 'underweight', 'normal' and 'overweight' people to help guide healthcare, but of course it is a continuous scale, so it may be sufficient to merely separate people (e.g. using height, weight and gender information) and be able to make more flexible decisions, not based on rigid categories.

Where there are discrete groups, identification and classification will sometimes be based on rich, well-studied data, e.g. people who definitely do or do not have a condition, but the groupings may then be used to make predictions with more limited information, where there is no certainty. In such situations, it may be appropriate to approach the classification of a unit of data, within one group or another, in a probabilistic manner. While Chapter 21 deals with the concept of probability, here we focus on the process of separating data, in terms of both making groups and determining the most discriminating information. We refer to the formation of discrete groups by bringing together units of data as *clustering*, and use *discrimination* to mean how we find the best combination of the different kinds of data feature to perform separation.

Vector data units

This chapter will consider all units of data (the things that are being separated) in an abstract way. All data in a given unit will be represented numerically and all different kinds of information will be placed together in the same data structure. Doing things in this manner enables us to think in a more general, mathematical way and the computational methods that we consider will work on any input data, whatever its origin. To this end we will refer to each separable unit of data as a *feature vector*. A *feature* refers to a different kind of measurement, whether weight, length, height, x-coordinate, y-coordinate or whatever. A *vector* refers to the placement of all of the features that go together into particular slots of an array. For example, a colour may be described as an array consisting of red, green and blue component values, i.e. `color = (red, green, blue)`. Vectors are often used to describe positions in three-dimensional space, and in the same way a more general feature vector can be thought of as a position in a *feature space*. The only difference is that the axes of a feature space don't necessarily represent spatial position; the axes represent whatever is being measured and can have any number of 'dimensions', one for each feature. Just as distances can be measured between points in space, distances can also be measured in a feature space. We will often be measuring such distances for the purposes of separating data, as a means of measuring the degree of similarity between units of data. This is not to suggest that the usual Cartesian distance (square root of the sum of square axis differences) is always the best measure; the distance criterion should be chosen to be appropriate to the problem.

Discrimination

Some of the computational methods presented in this chapter are used to determine which aspects of which features best separate the data. For example, if there are many different kinds of tests that can be performed on a person to diagnose a disease, each of the different kinds of measurement will have a different degree of importance to the outcome. Also, certain combinations of measurements may be important for classification, in either a positive or negative sense. In essence, with such information we wish to determine the best view of the feature vectors to observe the correlations and distinctions. To take a very simple three-dimensional example, imagine the problem was to distinguish blurred points of light by taking photographs. Here you would not expect to be able to separate the different lights if the camera view meant that one light lay directly behind the other; the best separation for two lights would be a view perpendicular to the line between them. Generalising the problem for any feature space we would seek to find a projection (view) of the data where differences or groups are most obvious. Implicit in this reasoning is the tactic of mapping several different kinds of features into a simpler, flatter representation, otherwise known as *dimensional reduction*.

Taking a photograph of real objects involves going from three dimensions to a two-dimensional projection, so this is an example of dimensional reduction, although for the purposes of data discrimination we would not take just any view, but rather the one that gives optimal separation. If there are only two data categories that are to be separated, we could draw a line through the 'centre' of one category to the other. Although we know where this line is in the feature space of the data, the line itself is only a one-dimensional object that charts the transition of going from one group to the other. By transforming multi-dimensional data (lots of features) to points on an optimally positioned one-dimensional line we automatically create an axis for separation; a decision boundary would be a point on the line between the groups. It is noteworthy that although dimensional reduction can often simplify a problem involving large numbers of features, including giving human beings the kinds of graphs and 2D pictures they can visually appreciate, this simplification is not a prerequisite for separating data items. Many methods allow data to be grouped and separated in its original high-dimensional, feature vector form. Where it is possible, separating the unmapped data should be considered first, given that dimensional reduction loses information, which may obscure separation.

In this chapter we will look at two forms of data discrimination with different approaches. These are *principal component analysis* (PCA) and *linear discriminant analysis* (LDA), and either may be used to work out a best-separating projection (view) of data represented as feature vectors. Accordingly they may also be used as a means of dimensional reduction.

Figure 23.1. How many seashells in how many groups? Some views of data are better at distinguishing items and clusters than others.

Clustering

Clustering relates to the process of partitioning data units into discrete groups. Such an operation requires that the similarity (or difference) between units is measured and then the members of each group are allocated to give the arrangement that maximises the association of similar items and the separation of dissimilar ones. In practice most of the clustering methods presented here will not be able to give an immediate analytical solution to this optimisation problem, rather the process will be an iterative one, with several cycles of improvement until a stable solution is found. As mentioned above, clustering may operate on data items which have a high dimensionality, represented as feature vectors. However, if the analysis is too slow or too complicated the original data may be transformed (projected) into a set of lower-dimensionality data by methods like PCA prior to the clustering operation.

Depending on the situation, the process of clustering may work with prior knowledge about the number of clusters, e.g. what the underlying data categories are. Alternatively, the number of clusters may be completely unknown. If the numbers of clusters is not known then this number must be deduced or optimised. Generally, several different trials are run, each of which involves a different number of clusters. Within each trial there is a separate optimisation for how the data items are allocated within that number of clusters. The best number of clusters is then determined from the best overall arrangement from all the trials. It would be possible to place each data item in a separate cluster, thus giving maximum separation, but the objective is to give the best balance between the number of clusters and the degree of separation, rather than only maximising separation.

Once clusters are defined the result may then be used as a means of predicting classification, i.e. estimating in which cluster a previously unseen piece of data lies. Making a prediction may be as simple as finding which cluster is closest. Alternatively, more advanced

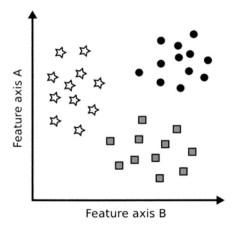

Figure 23.2. An illustration of three clusters of points in two dimensions. Each cluster is shown by a different kind of shape. Points are plotted at different positions relative to the two axes, which represent two different quantities relating to the data. The data categories may be known in advance, in which case the plot can be used to illustrate the degree of separation in the data. Alternatively, the clusters in the data may be inferred by analysis of the spatial relationship of the data points, by grouping points with similar axis locations (i.e. features).

approaches, such as the supervised machine learning methods described in Chapter 24, can be used where classification is not so easy. These can learn patterns from training data with known, fixed classifications before predictions are made. One of the machine learning methods presented later, the *self-organising map*, is notable because it is *unsupervised* (needs no prior classifications) and thus can be viewed as an alternative to the linear clustering methods presented in this chapter.

Clustering methods

The following Python examples demonstrate a range of clustering methods. Overall we have tried to give an overview of the different approaches and have chosen examples that are reasonably easy to explain (and write in Python) but which should also prove useful in many different situations.

Simple threshold clustering

The first Python example in this chapter involves grouping items of data into clusters in a fairly simple manner, based upon how close (similar) the items of data are. This method is somewhat naïve and suffers from a few problems, but is a good place to start in the subject and we will subsequently show improvements. The algorithm works by initially considering an item of data as a potentially separate cluster and then expands the cluster by adding any other items that are sufficiently close, i.e. the distance between data points is less than a threshold. The general problem with this kind of approach is that it is sensitive to the arrangement of data points; two genuine clusters can be merged into one by the presence of outlying points that just happen to span the gap between the clusters. The method is sometimes described as being 'too greedy'.

Firstly, in the Python example we define a simple function that calculates the regular, Euclidian distance between two items of data (feature vectors). Although this is a reasonable general choice for measuring the similarity between data items, there are some situations where something else may be more appropriate. A good example of using a different kind of measure would be for biological sequences, where it is possible to estimate how similar the sequences are, assuming an evolutionary process.

The euclideanDist function will assume the data items are NumPy array objects, so that we can do some quick manipulations without having to go through loops. The distance is calculated by first finding the difference between two vectors: another vector, diff. The square root of the sum of the squares of the values in the diff vector is then the distance we want. We use the dot() function to do the summing and squaring for us. (See Chapter 9 for an explanation of dot().)

```
from numpy import dot, sqrt

def euclideanDist(vectorA, vectorB):

  diff = vectorA-vectorB

  return sqrt(dot(diff,diff))
```

Next is another small helper function that we will use in the clustering. In this case the findNeighbours function is built to look through all pairs of data items, calculate the distance between the items (for example, using the euclideanDist function) and record those items that are sufficiently close to be considered as near neighbours. The three input arguments are naturally the data itself, the function to get the distance value and a threshold for defining neighbours. The results will be stored in a Python dictionary neighbourDict, so this is defined and initially filled with empty lists, one for each item of data.

```
def findNeighbours(data, distFunc, threshold):

  neighbourDict = {}

  n = len(data)
  for i in range(n):
    neighbourDict[i] = []

  for i in range(0,n-1):
    for j in range(i+1,n):
      dist = distFunc(data[i], data[j])

      if dist < threshold:
        neighbourDict[i].append(j)
        neighbourDict[j].append(i)

  return neighbourDict
```

The remainder of the function involves going through all pairs of data items. Hence we have two loops, one with the index i and one with the index j, to get two items. Note how we have deliberately chosen the ranges for these indices to avoid repetition, i.e. j are the indices larger than i. The distance is calculated by using the function that was passed in and then if the distance is less than the threshold we record the data for i and j indices to be neighbours; the indices go in each other's list of neighbours, held in neighbourDict.

Finally we get to the actual simpleCluster clustering function. This takes a list of data, each item of which is assumed to be a NumPy array, a distance threshold and the function to make the distance measurement. Initially, the dictionary of neighbours is filled using findNeighbours, then the remainder of the function works to find clusters by combining neighbours together into groups.

```
def simpleCluster(data, threshold, distFunc=euclideanDist):

  neighbourDict = findNeighbours(data, distFunc, threshold)
```

For the clusters we first define an empty list, which will represent items of data in discrete groups. The pool variable is a set of numbers, the indices of data items that we have not yet processed. The aim will be to go through all the pool and place the indices in the appropriate cluster. We are using Python sets here because detecting something in a set is quicker than looking though a list.

```
clusters = []
pool = set(range(len(data)))
```

Now the task is to process the data, which involves taking an item from the 'to-do' pool (remembering pop() removes something from a set so the pool shrinks) and looking through its neighbours, which we fetch from the `neighbourDict`. Because the item was in the pool it will not yet be in a cluster, thus we define a new one as an empty set and add the item index `i`. The item is now considered processed. It may seem dumb to already consider the item processed, but as will become apparent most items will be added in a different way; this is just a start for the cluster.

```
while pool:
    i = pool.pop()
    neighbours = neighbourDict[i]
    cluster = set()
    cluster.add(i)
```

Once the new cluster is defined we then add all of the neighbours which are known to be close to the same cluster. To do this a new pool `pool2` is defined from the indices `j` of the neighbours.

```
    pool2 = set(neighbours)
    while pool2:
        j = pool2.pop()
```

If the neighbour has not yet been processed (`j` still in `pool`) then we process it by removing it from the main pool, and add it to the current cluster. Its own neighbours then go into `pool2`, the set that we are currently working on. In this way neighbours of neighbours are placed in the same cluster. Note the use of the `update`, which will expand a Python set to include any new values from another group.

```
        if j in pool:
            pool.remove(j)
            cluster.add(j)
            neighbours2 = neighbourDict[j]
            pool2.update(neighbours2)
```

This inner loop ends when no more neighbours are placed in the pool for the current cluster and `pool2` is fully processed (empty), whereupon the cluster is added to the list of clusters. Another cluster will be created by the outer `while` loop as long as the pool still contains something to process: those that are not a member of any clusters seen so far.

```
    clusters.append(cluster)
```

The final output of the clusters will be given as a list of lists; each sub-list will contain all of the data items that go together in the same cluster. So far the clusters are only defined in terms of index numbers, so it is a simple matter to go through each set of clustered indices and add the actual data items they correspond to (`data[i]`) to the `clusterData` list which will be passed back at the end of the function.

```
clusterData = []
for cluster in clusters:
    clusterData.append( [data[i] for i in cluster] )

return clusterData
```

For testing we will use some random data clusters with a normal (Gaussian) distribution, readily created using `random.normal`. Here each data cluster is centred on a different point but has the same spread and size: 100 points in two dimensions. These are then combined into a single array of vectors (300 points on two dimensions) using `vstack`, which combines the arrays of vectors along their first axis (rows), and shuffled to mix up the vectors from the different clusters.

```
from numpy import random, vstack

spread = 0.12
sizeDims = (100,2)
data = [random.normal(( 0.0, 0.0), spread, sizeDims),
        random.normal(( 1.0, 1.0), spread, sizeDims),
        random.normal(( 1.0, 0.0), spread, sizeDims)]

data = vstack(data)
random.shuffle(data) # Randomise order

clusters = simpleCluster(data, 0.10)
```

The following can then plot the results with different symbols and colours. Note that we plot all the small 'noise' clusters that have fewer than three items as black spots. Also, the `zip` function is handy here because the `scatter` plotting function requires two separate lists of x values and y values, but our data has (x,y) pairs. Hence `*cluster` expands the one list of (x,y) pairs into multiple small lists, and `zip` groups the separate x and y values together (see Chapter 10).

```
from matplotlib import pyplot

colors = ['#F0F0F0','#A0A0A0','#505050',
          '#D0D0D0','#808080','#202020']

markers = ['d','o','s','>','^']

i = 0
for cluster in clusters:
   allX, allY = zip(*cluster)

   if len(cluster) > 3:
     color = colors[i % len(colors)]
     marker = markers[i % len(markers)]
     pyplot.scatter(allX, allY, s=30, c=color, marker=marker)
     i += 1

   else:
     pyplot.scatter(allX, allY, s=5, c='black', marker='o')

pyplot.show()
```

Density-based clustering

The threshold clustering method works fairly quickly to group our data, but we will now consider a way of improving the method so that it takes account of the density of data points, and thus the method is less likely to have intermediate points that greedily combine otherwise

Optimal threshold   **Large threshold**

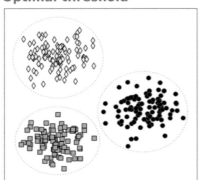 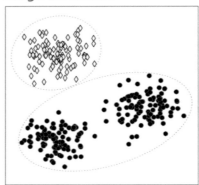

Figure 23.3. Example output of simple associative clustering at two thresholds. A simple associative clustering method works by grouping points together into a cluster if the separation between them (from one point to any other, considering all axes) is below a specified threshold. Assuming the data is well separated and an appropriate threshold is chosen, this method can be useful and is very fast. However, simple associative clustering will naturally not separate overlapping clusters or situations where there are outliers, within the clustering threshold, that can bridge between what otherwise appear to be distinct groups.

separate clusters. The method we have chosen here is referred to as DBSCAN[1] and is commonly used in scientific analyses. We might have also considered the OPTICS[2] method, but we omit this only for reasons of brevity. The basic principle here is that distance alone is not sufficient to define the clusters; the number of neighbours should be considered too. Hence cluster membership requires support, i.e. a certain density of points. The method can be criticised as being subjective, given that the threshold distance and minimum number of supporting neighbours have to be specified. Nonetheless the method is free from any assumptions about how the data is distributed and copes well when the clusters are irregular shapes. Also, there is the inherent concept of 'noise' points: data items that are disconnected from the clusters.

The dbScanCluster function resembles simpleCluster but we add checks for the number of neighbours and accordingly add any disconnected points to a list of noise, rather than a cluster. The inputs now include minNeighbour, which as the name suggests says how many neighbours an item should have at minimum.

```
def dbScanCluster(data, threshold, minNeighbour, distFunc=euclideanDist):
```

As before, the dictionary of neighbours is calculated and we initialise some variables, including a set to hold the indices of the noise items.

[1] Ester, M., Kriegel, H.-P., Sander, J., and Xu, X. (1996). A density-based algorithm for discovering clusters in large spatial databases with noise. In E. Simoudis, J. Han, and U.M. Fayyad (eds.), *Proceedings of the Second International Conference on Knowledge Discovery and Data Mining (KDD-96)*. AAAI Press. pp. 226–231.
[2] Ankerst, M., Breunig, M.M., Kriegel, H.-P., and Sander, J. (1999). OPTICS: ordering points to identify the clustering structure. *ACM SIGMOD International Conference on Management of Data*. ACM Press. pp. 49–60.

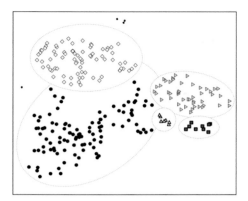

Figure 23.4. Example results of density-based clustering, illustrating clusters and noise points.
Density-based clustering can be imagined as an extension of associative clustering. Points are joined
into the same cluster, but with the extra requirement that there needs to be a minimum number of
close points. This requirement reduces the occurrence of sparse points bridging clusters. Also,
because of the requirement of a minimum number of close neighbours for a point to be included
in a cluster, some points will be excluded from clusters below threshold size and thus marked
as 'noise'.

```
neighbourDict = findNeighbours(data, distFunc, threshold)

clusters = []
noise = set()
pool = set(range(len(data)))
```

Then we go through the work pool, popping out indices and adding to clusters. However,
we now check that the item (index `i`) has enough neighbours. If it does not (`len(neigh-
bours) < minNeighbour`), then the index is added to the noise set and clustering is skipped
for that item.

```
while pool:
  i = pool.pop()
  neighbours = neighbourDict[i]

  if len(neighbours) < minNeighbour:
    noise.add(i)
```

Otherwise the clustering proceeds. A new cluster is defined and the index is added to
the set.

```
  else:
    cluster = set()
    cluster.add(i)
```

Next a new pool is defined from the neighbours of the current item. However, we do not
automatically accept all neighbours into the current cluster. Those with too few neighbours
themselves are added to the noise set instead. All points, be they noise or not, are removed
from the work pool, so they are only ever considered once.

```
pool2 = set(neighbours)
while pool2:
  j = pool2.pop()

  if j in pool:
    pool.remove(j)
    neighbours2 = neighbourDict.get(j, [])

    if len(neighbours2) < minNeighbour:
      noise.add(j)

    else:
      pool2.update(neighbours2)
      cluster.add(j)

clusters.append(cluster)
```

Finally, the function collates the data items based upon whether their indices ended up in the noise set (isolated items) or in one of the main clusters.

```
noiseData = [data[i] for i in noise]

clusterData = []
for cluster in clusters:
  clusterData.append( [data[i] for i in cluster] )

return clusterData, noiseData
```

This function can be tested in the same manner as the simple clustering:

```
clusters, noise = dbScanCluster(data, 0.10, 2)
```

K-means clustering

The next Python example for clustering is an algorithm known as *k*-means clustering. Here the '*k*' refers to the number of clusters that we wish to group our list data items into. The 'means' refers to the concept of describing each cluster as the central average position: a geometric average of its members. There are implementations of the *k*-means algorithm in the Scientific Python library (`scipy.cluster.vq.kmeans` and `scipy.cluster.vq.kmeans2`), but we will create one from scratch here to illustrate the methods and to give a basis for modification. Also, there are variants of this clustering method that can be used, like '*k*-medoids' or '*k*-medians', which can also be considered, but which we will not go into here for reasons of brevity. Compared to the threshold-based methods already discussed, the *k*-means method has the advantage that you do not have to choose a threshold. Although you do have to say how many clusters (*k*) to make, it is possible to try several values for *k* and take the best one. Compared to DBSCAN, *k*-means will not work as well with data clusters that have irregular or intermingled shapes, but will allow separation of regular ('blobby') clusters even when they are overlapped.

Initially the cluster centres are defined randomly, basing them on the locations of random data items. Each of the input data items will then become a member of a cluster by simply determining which centre is closest. The effect is that the discrimination boundaries between clusters are straight lines.[3] Most likely the initial guess for the cluster centres will be bad, so the algorithm goes through a number of iterative cycles to improve the locations of the centres. Each cycle consists of re-appraising which data items are members of which cluster and then moving the centres of the clusters to the average of their memberships. When a cluster centre moves the membership of items may change for the next cycle, which in turn leads to a new centre, and so the cycle repeats. The k-means algorithm is simple in that there is no recorded statistic that is being optimised. Instead the iteration continues until a stable situation is reached. For data sets where there are (k) obvious clumps of items the algorithm will be fairly reproducible. However, where the data items are more evenly spread the final clustering will depend on the initial guess of the centres, i.e. there can be more than one stable solution. If there isn't a stable solution you should consider whether it is meaningful to cluster the data in this way; imagine dividing a regular grid of points into two, where any straight line that makes two groups of equal size is a possible solution.

The Python function that performs k-means clustering takes an array of input data vectors and the number k, to specify how many clusters there will be. We also have the option of passing in some initial guesses for the cluster centres, but by default this is None, and we choose random centres. If unspecified, to define the initial guess for centers we use the sample function from Python's standard random number module, which will choose k different items from the data, and then convert the resulting list into a NumPy array.

```
from numpy import array, random
from random import sample

def kMeans(data, k, centers=None):

  if centers is None:
    centers = array( sample(list(data), k) ) # list() not needed in Python 2
```

With the initial cluster centres defined we now come to the iterative part. Here we define the amount of change, which is initially large (proportionately speaking). The while loop will continue as long as the change between subsequent cycles is big enough, otherwise we deem the clustering to have reached a stable solution.

```
  change = 1.0

  while change > 1e-8:
```

The membership of the clusters is defined by putting each data item into a list, one for each cluster, according to which cluster centre is closest. Hence clusters is defined as a list of lists which collects the data items. Note that the membership of clusters may change each cycle, so we have to repeatedly re-appraise the memberships. The for loop goes though each data item, here called vector, and subtracts the centres; this will give an array of difference vectors from the data item to all of the centres. Note that this subtraction assumes we are working with NumPy arrays, hence subtracting two arrays gives another array. The array of differences is then squared and then summed up for each vector, so that we have a list of (square[4]) distances, with one value for each centre; i.e. an array of (x, y, z) items

[3] It will be a Voronoi diagram.
[4] There is no point taking the square root given that we only want to find the closest.

becomes (x^2, y^2, z^2) and then $(x^2+y^2+z^2)$. The closest centre is then the index of the smallest value in the array `dists.argmin()`. The data vector is added to the appropriate list in `clusters` with the `closest` index.

```
clusters = [[] for x in range(k)]
for vector in data:
    diffs = centers - vector
    dists = (diffs * diffs).sum(axis=1)
    closest = dists.argmin()
    clusters[closest].append(vector)
```

With the membership of the clusters defined, the next loop recalculates the centres of the clusters based on the data vectors that belong to each. We simply go through the list of clusters and, after converting to a NumPy array, we calculate the average data vector, taking a summation for each data dimension and dividing by the length of the cluster. The difference between the old and new cluster centre is calculated by subtraction, then added as the sum of squares (i.e. length squared) to the total amount of change for this cycle. Finally in the loop the new centre replaces the old one: `centers[i] = center`.

```
change = 0
for i, cluster in enumerate(clusters):
    cluster = array(cluster)
    center = cluster.sum(axis=0)/len(cluster)
    diff = center - centers[i]
    change += (diff * diff).sum()
    centers[i] = center

return centers, clusters
```

At the end of the function, once the change is small enough, we pass back the last values of the cluster centres and the list which contains the data items grouped according to cluster membership.

To test the data we can use a random spread of 1000 data points in two dimensions, grouping into three clusters:

```
testDataA = random.random((1000,2)) # No clumps

centers, clusters = kMeans(testDataA, 3)
```

Or we can add two distinct clumps of two-dimensional points:

```
from numpy import vstack

testDataB1 = random.normal(0.0, 2.0, (100,2))
testDataB2 = random.normal(7.0, 2.0, (100,2))
testDataB = vstack([testDataB1, testDataB2]) # Two clumps

centers, clusters = kMeans(testDataB, 2)
```

For each test we can display the result in a scatter graph using the `matplotlib` library. Note that to make the plot we extract the x and y coordinates for the data items separately using `zip`, because `matplotlib.pyplot` expects two separate lists (rather than (x,y) points). After plotting the data with separate colours we also make a scatter plot of the centres of the clusters with black circles.

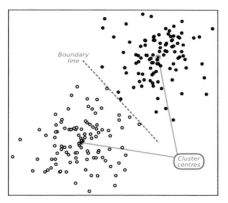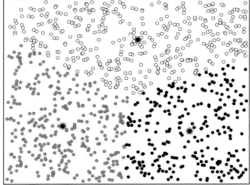

Figure 23.5. Example output from the *k*-means clustering method, for both sparse and dense data. By using an iterative algorithm, the *k*-means clustering method partitions data points into a specified number of clusters (*k*). The cluster centres are initially chosen at random, and the membership of each data point is determined by its closest cluster centre. Subsequent cluster centres are defined as the geometric mean of the cluster members. The cluster assignment and centre recalculation steps are repeated until convergence. Because membership is based on distance to the cluster centres, the boundaries between clusters will be straight lines representing the points of equal distance between two centres.

```
from matplotlib import pyplot
colors = ['#FF0000','#00FF00','#0000FF',
          '#FFFF00','#00FFFF','#FF00FF']

for i, cluster in enumerate(clusters):
  x, y = zip(*cluster)
  color = colors[i % len(colors)]
  pyplot.scatter(x, y, c=color, marker='o')

x, y = zip(*centers)
pyplot.scatter(x, y, s=40, c='black', marker='o')
pyplot.show()
```

Improving *k*-means

The *k*-means algorithm can be improved by having a better guess at the starting cluster centres. The idea is that by choosing cluster centres that are spaced further apart the optimisation converges more quickly and is less likely to get stuck in sub-optimal positions, which are often characterised by two centres being close and dividing a real cluster at the expense of a globally better solution. One method that uses better starting positions is called *k*-means++, which works by choosing centres on a probabilistic basis, where the chance of a point becoming an initial centre increases with the square of the distance to the other centres specified so far. However, here we will consider another, slightly more directed approach with a similar aim.

For the `kMeansSpread` function we guess one cluster centre by taking a random data point and then choose the centres of subsequent clusters by selecting points that are furthest away from those defined so far. Each cluster centre is chosen by creating an `index`, which is placed in the set `indices`, which is used at the end to select corresponding data items and thus create an array of centres. The index for the first cluster is a random integer between zero (the first data item) and `n-1` (the last data item), selected using the standard random number Python module: `random.randint(0, n-1)`.[5] The remaining indices are added in a `while` loop, until a total of `k` is achieved, by choosing subsequent points which have minimum radial influence from the centres already chosen.

```python
from numpy import zeros, ones, vstack
from random import randint

def kMeansSpread(data, k):

  n = len(data)
  index = randint(0, n-1)
  indices = set([index])

  influence = zeros(n)
  while len(indices) < k:
    diff = data - data[index]
    sumSq = (diff * diff).sum(axis=1) + 1.0
    influence += 1.0 / sumSq
    index = influence.argmin()

    while index in indices:
      index = randint(0, n-1)

    indices.add(index)

  centers = vstack([data[i] for i in indices])

  return kMeans(data, k, centers)
```

Key to this approach is the `influence` array. Initially starting with zeros, each time we add an `index` for a new centre we calculate the difference from that cluster centre (`data[index]`) to all the data, thus creating `diff`. Each value in the `influence` array is then increased by one over `sumSq`: the sum of this difference squared. Note that `sumSq` has `1.0` added to each element to avoid dividing by zero, i.e. where a data point coincides exactly with the previous centre. Also, we do an element-wise division when we divide the simple number `1.0` by the whole array `sumSq`, so we get an array of reciprocals. In effect each centre has a diminishing radial influence and we choose the index where the sum of all these influences `influence` is minimised. In some circumstances an index can actually be picked twice, but we guard against this with a second `while` loop which checks for repeats and chooses a random index until an unused one is found. Once `k` indices are collected, a list of data items, representing the cluster centres, is created using a list comprehension and stacked into a NumPy array (`vstack` operates on arrays much like `append` does on lists). Overall this method is analogous to the concept of repelling charges. In the end the 'repelled' centres are a fair starting point for *k*-means, which is called at the end so its results are passed back directly at the `return`.

---

[5] Note that unlike the `range(a, b)` function the `randint(a, b)` nomenclature will give numbers including b.

The function is used, and can be tested, in the same way as `kMeans()`. If speed of execution becomes important then, aside from faster implementations (e.g. using a C-language extension, as described in Chapter 27), further improvements can sometimes be made to the algorithmic efficiency. For example, if the dimensionality of the data items is not too large (e.g. three-dimensional) techniques like kD-trees[6] can be used to more efficiently calculate the closest centre for each data item. Also, if the data set to be clustered is large, significant time savings can be made by initially running the algorithm on a smaller, random subset of the data. The cluster centres from the smaller subset can then be used as good starting values in a second round of clustering using the whole data set. A point of note with k-means and similar methods is that they tend to only work well where the clusters are of about the same size and density and are convex (blobs).

The 'jump' method

Next we will consider the situation where you do not know in advance how many clusters the data should be grouped into, but do not wish to use a threshold-based method. We will consider an augmentation on top of the k-means method to try various numbers of clusters and determine the most meaningful. The example function uses what we will term the 'jump' method.[7] It will perform k-means clustering with different values of k and then assesses which of the steps between increasing values of k represents the best compromise between the number of clusters and complexity of the solution.

For a given trial value of k, the result of a trial clustering attempt involves the calculation of what we term the 'distortion', which in essence is a measure of the average spread, from the cluster centres, of data points over all the clusters. This spread is adjusted for the covariance[8] in the data, which indicates how much variation there is in the values for the various axes, such that the relative scale of the different features does not matter.

The `jumpMethodCluster` function is constructed to take a set of data, as an array of arrays, an optional range of k values (numbers of clusters) to test and `cycles`, which determines how many times to repeat the clustering. This cyclic repetition is important because different clustering attempts, involving random numbers, may give variable results.

```
from numpy import cov, linalg

def jumpMethodCluster(data, kRange=None, cycles=10):
```

Firstly, we extract the length of the data and the number of dimensions (features). This is simply the size (`.shape` in NumPy) of the array. Then we define the range of k values if none were passed in; by default the range is between two and the number of data items. Note that the range is specified by an upper limit, which will not be included.

---

[6] A completely different k to the one used in these clustering algorithms.

[7] Sugar, C.A., and James, G.M. (2003). Finding the number of clusters in a data set: an information theoretic approach. *Journal of the American Statistical Association* 98: 750–763.

[8] Covariance is discussed more fully in Chapter 22. Here the covariance matrix has elements that show the covariance, and hence scaled correlation, between the different dimensions or features of a data set.

```
n, dims = data.shape

if kRange is None:
  start, limit = (2, n+1)
else:
  start, limit = kRange
```

Next we define the `power` variable, which will be used to adjust the calculated distortion values according to the number of dimensions in the data (see the primary reference for an explanation of this). The distortions are collected in a dictionary, keyed by *k* values, so this is initially defined as blank. The `invCovMat` is a matrix representing the covariance in the dimensions of the input data. Here we can make use of NumPy's `cov` and `linalg.pinv` functions to estimate the covariance and then invert the matrix.[9]

```
power = dims/2.0
distortions = {}
invCovMat = linalg.pinv(cov(data.T))
```

Then a loop goes through all of the values of `k` which are to be tested. For each we define an array, initially at zero, of average distortions (scaled distances). We will have one value for each cycle, with the aim being to find the cycle with the best clustering and hence minimum distortion.

```
for k in range(start, limit):
  meanDists = zeros(cycles)
```

The next `for` loop exists to provide several clustering attempts, to guard against clustering where *k*-means gets stuck in a stable but sub-optimal situation. For each attempt we define a sum of distortions `sumDist`, and use the `kMeansSpread` function to actually generate the clusters for this cycle and value of *k*.

```
for c in range(cycles):
  sumDist = 0.0
  centers, clusters = kMeansSpread(data, k)
```

With the clustering attempt made the results are analysed. We go through each cluster in turn and calculate `diffs`, the difference between the data items in the cluster and the cluster centres that the *k*-means algorithm gave.

```
for i, cluster in enumerate(clusters):
  size = len(cluster)
  diffs = array(cluster) - centers[i]
```

The differences are used to calculate the average distortion. At a basic level this requires that the `diff` arrays are squared using a dot product (`dot(diff.T, diff)`), but we also apply the inverse covariance matrix to scale the values according to the spread of data in the dimensions (i.e. treat each data feature equivocally). The distortion, which can be thought of as a measure of distance, is then divided by the size of the cluster, to get an average for the items in the cluster, and added to the total.

```
for j, diff in enumerate(diffs):
  dist = dot(diff.T, dot(diff, invCovMat))
  sumDist += dist / size
```

---

[9] See Chapter 9 for details about such matrix manipulations.

The final summation of average distortions in the cycle is then averaged again, over the number of clusters and dimensions, so that we get the average for the whole clustering solution.

```
meanDists[c] = sumDist / (dims * k)
```

Finally, after all trial cycles are done the minimum value of the averaged distortions is selected: the trial that gave the best clustering. This value is adjusted by being raised to the negative power -power, which scales the result according to how the distortion is expected to vary according to the number of data dimensions (see reference above).

```
distortions[k] = min(meanDists) ** (-power)
```

With the distortions calculated, it only remains to go through the values to find the value of k (cluster size) that corresponds to the biggest improvement. The largest jump in the distortion value will occur when we reach the optimum value of k, the one that give the most notable improvement in the relative separation between the data items and the cluster centres.

```
maxJump = None
bestK = None

for k in range(start+1, limit):
  jump = distortions[k] - distortions[k-1]

  if (maxJump is None) or (jump > maxJump):
    maxJump = jump
    bestK = k

return bestK
```

To visualise how the jump method works, imagine three real clusters, as illustrated in Figure 23.6. There will be some large spreads from the cluster centres if we try to use only two clusters. Using four clusters will tend to split a real cluster, leading to only a modest

Two clusters

Three clusters

Four clusters

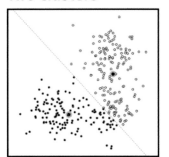
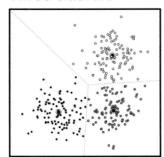
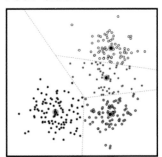

Figure 23.6. Finding the optimum number of clusters for the *k*-means method. For a given data set, the *k*-means clustering method can be applied to partition the data into different numbers of clusters. If the number of clusters (*k*) is not known different numbers can be tried so that an optimum can be found. A simple way to evaluate the optimum number of clusters is to measure the relative improvement of the separations of the data points from their nearest cluster centre. Having more clusters will always give smaller separations but there will be the most significant change near the optimum number. For the illustrated example, going from two to three clusters has a bigger jump in minimising separations than going from three to four clusters. Using four clusters only improves separations to the centres mildly.

reduction in the spread. Scaled properly, according to the effect of dimensionality, the spread measure (distortion) can be used to suggest the value of k.

We can test the jump method on a simple normal data set as before. The underlying clusters are stacked and shuffled randomly and passed in to the function.

```
data = [random.normal(( 0.0, 0.0), spread, sizeDims),
        random.normal(( 1.0, 1.0), spread, sizeDims),
        random.normal(( 1.0, 0.0), spread, sizeDims)]

data = vstack(data)
random.shuffle(data)

k = jumpMethodCluster(data, (2, 10), 20)

print('Number of clusters:', k)
```

Data discrimination

Now we will move on from clustering data to consider data discrimination, i.e. finding the view (projection) of the data that gives the best separation. We will consider two examples. The first is *principal component analysis* (PCA), which can be used to show trends in a data set without first having to group it into categories, although clustering will often be performed after PCA. The second example is *linear discriminant analysis* (LDA), which operates on two categories of data and illustrates how rules can be set up to discriminate between the groups, e.g. to perform classification on unseen data.

Principal component analysis

Principal component analysis (PCA) is a relatively simple but widely used technique to extract the innate trends within data. The principal components of data are the vectors (in the same feature space of the data) that give the best separation of the data items in terms of their covariances. Mathematically, PCA gives the eigenvectors of the covariance matrix. Taking the eigenvalues of these in size order we can find the most significant, principal components that account for most of the variance in the data. Taking fewer principal components than the number of dimensions present in the input data set allows for a lower-dimensionality representation of the data (by projecting it onto these directions) that still contains the important correlations.

When we calculate the principal components of a data set we obtain vectors, directions in the data, which are orthogonal (perpendicular) to one another. Thus each component vector represents an independent axis and there can be as many components as there are data dimensions. As some axes are more significant than others, for separating the data, those are the ones we are usually interested in. We can also consider the principal component vectors as a transformation, which we can apply to our data: to orient it and stretch it along these orthogonal axes.

Principal component analysis does have limitations, which the programmer should be aware of, but it is quick and easy, so often worth a try. A classic example where PCA fails is for 'checkerboard' data, i.e. alternating squares of two categories, where there are no simple axes in data that separate the categories. In such instances more sophisticated, non-linear techniques, such as *support vector machines* (see Chapter 24), may be used.

In the Python example for PCA we first make the NumPy imports and define the function which takes data (as an array of arrays) and the number of principal components we wish to extract.

```
from numpy import cov, linalg, sqrt, zeros, ones, diag

def principalComponentAnalysis(data, n=2):
```

First we get the size of the data, in terms of the number of data items (samples) and the number of dimensions (features).

```
    samples, features = data.shape
```

We calculate the average data item (effectively the centre of the data) by finding the mean along the primary data axis. We then centre the input data on zero (on the feature axes) by taking away this average. Note that we then transpose the data with `.T`, turning it sideways, so we can calculate the covariance in its dimensions.

```
    meanVec = data.mean(axis=0)
    dataC = (data - meanVec).T
```

The covariance is estimated using the `cov()` function and the eigenvalues and eigenvectors are extracted with the `linalg.eig()` function. Here the inbuilt NumPy functions for dealing with arrays and linear algebra really show their value.

```
    covar = cov(dataC)
    evals, evecs = linalg.eig(covar)
```

The resulting eigenvalues, which represent the scaling factors along the eigenvectors, are sorted by size. Here we use the `.argsort()` function, which gives the indices of the array in the order in which the values increase. We then use the cunning trick `[::-1]` for reversing NumPy arrays.[10] The eigenvectors are then reordered by using these indices, to put them in order of decreasing eigenvalue, i.e. most significant component first.

```
    indices = evals.argsort()[::-1]

    evecs = evecs[:,indices]
```

We can then take the top eigenvectors as the first `n` principal components, which we call `basis`, because these are the directions that we can use to map our data to. The `energy` is simply a measure of how much covariance our top eigenvalues explain, which is useful for detecting whether more principal components should be considered.

```
    basis = evecs[:,:n]
    energy = evals[:n].sum()

    # norm wrt to variance
    #sd = sqrt(diag(covar))
    #zscores = dataC.T / sd

    return basis, energy
```

---

[10] See Chapter 10.

Extracting a principal component

As the calculation of all of the principal components of a data set is often not required, e.g. we may be only interested in the first two principal components, the next example function can more efficiently extract the principal components one at a time (most significant first) without calculating a full covariance matrix. Each time a component is extracted the data may be transformed, eliminating that component, so the next most important component can be extracted. This is an iterative method that converges on the principal component, hence we pass in a precision value to state how long we iteratively cycle to improve the convergence.

```
from numpy import random, dot, array, outer

def extractPrincipalComponent(data, precision=1e-9):
```

As before, we extract the number of samples (data items) and features from the input data. We then calculate the mean (centre) of the data so that it can be moved to the zero point.

```
    samples, features = data.shape
    meanVec = data.mean(axis=0)
    dataC = data - meanVec
```

The initial guess at the principal component `pc1` is constructed as a random array, with the same length as the number of features. We also initialise `pc0`, which will be the value from the previous cycle, so that we can check for convergence. Initially `pc0` is arbitrarily set to be different to `pc1`.

```
    pc1 = random.random(features)
    pc0 = pc1 - 1.0
```

The main optimisation involves a `while` loop that checks the two subsequent estimates for the principal components. If their difference, summed up over all dimensions, is sufficiently close (within the `precision` value) the loop will stop.

```
    while abs((pc0-pc1).sum()) > precision:
```

Inside the loop the principal component vector is improved upon for the next cycle. This involves going through all the items in the (centred) data, scaling each according to the projection onto the guess for the principal component (with dot product) and adding them to a total vector, `t`. When normalised (scaled to length one) this total vector will be the new estimate for the principal component. Here the dot product gives the coincidence of the data vectors with the current principal component estimate. The summation will give an average vector which is weighted according to the data items that have most coincidence. The convergence occurs because the largest correlation in the data will have a biased influence on the weights, increasing the size of `t` each cycle until there is maximum overlap with `pc1`. At this point `pc1` represents a fundamental axis in the data, along which most convergence occurs.

```
        t = zeros(features)
        for datum in dataC:
            t += dot(datum, pc1) * datum
```

At the end of each cycle the previous principal component estimate is stored as `pc0` and the new estimate is found by scaling the `t` vector to length one.

```
        pc0 = pc1
        pc1 = t / sqrt(dot(t,t))
```

Once there is little improvement in the principal component the cycles stop and the estimate is returned.

```
return pc1
```

We will test both the full and quick PCA functions with the same data set. This will be a random data set of 100 points in two dimensions, which is transformed by a shearing matrix, in order to produce a visible correlation in the data. Naturally we are not limited to only two-dimensional data, but this is easier to illustrate here.

```
testData = random.normal(0.0, 2.0, (100,2))

shear = array([[2,1],[1,0]])

testData = dot(testData, shear)
```

The PCA is performed by the two functions and the results are compared.

```
pc1 = extractPrincipalComponent(testData)
print('Quick PC1:', pc1)

basis, energy = principalComponentAnalysis(testData, n=2)
print('Full PCA:', basis, energy)
```

The test data can be plotted in the usual way and we can visualise the first principal component (pc1) by drawing a line along its direction. We make a line from pc1, which we can plot, by scaling the vector (by a factor of ten to make it visible) either side of the centre.

```
from matplotlib import pyplot

x,y = zip(*testData)
pyplot.scatter(x, y, s=20, c='#F0F0F0', marker='o')

x,y = zip(-10*pc1, 10*pc1)
pyplot.plot(x, y)
```

To illustrate that the basis matrix transforms the data in the right way we can apply it to the test data, so that the principal component axes are aligned with the axes of our graph.

```
transformed = dot(testData, basis)

x,y = zip(*transformed)
pyplot.scatter(x, y, s=10, c='#000000', marker='^')

pyplot.show()
```

Linear discriminant analysis

While PCA finds the directions of maximum variance within a data set, LDA attempts to determine the line of separation between different data sets. What we show here is sometimes referred to as *Fisher's linear discriminant*.[11] This particular approach assumes a normal distribution within the data.

The basic idea is that we wish to determine a discrimination matrix that represents how to transform the data sets into an orientation that best separates them. This can be

[11] Fisher, R.A. (1936). The use of multiple measurements in taxonomic problems. *Annals of Eugenics* 7(2): 179–188.

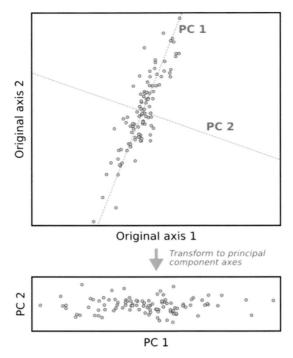

Figure 23.7. Principal component analysis of 2D data and its projection onto a principal axis. As illustrated for a two-dimensional example, the first principal component (PC 1) is the single linear combination of features (i.e. a direction relative to the axes) that explains most of the variance in the data. Subsequent principal components represent orthogonal directions (i.e. at right angles to the other components) that maximise the remaining variance not covered by earlier principal components; though, for this two-dimensional example PC 2 is determined by PC 1, since it is the only possibility. Projecting a data set onto its most important principal component axes allows for the dimensionality of the data set to be reduced, while still preserving as much linear correlation as possible. This can be useful for visualisation and to reduce the complexity of high-dimensionality data sets.

imagined as rotating the data sets so that they maximally align with an axis of separation, equivalent to a line between the two sets. If the data sets had the same scatter (variance) then this line is simply the line between the averages (means) of the two data sets.

The LDA optimisation is achieved by finding the matrix (and hence orientation) that maximises the separation (scatter) between the data sets, relative to the separation within each data set. The scatter within is simply the weighted covariances of the data sets separately, and the scatter between is the difference between their means. In essence we scale the line between the two means of the data sets by the combined size of the data scatter for each dimension. The resultant matrix can be used to transform the data into a new orientation for easy discrimination.

Firstly a Python function is defined which takes two data sets. These could be the results of a clustering operation or a previously known classification.

```
def twoClassLda(dataA, dataB):
```

Firstly, we find the averages (centres) of the data sets, by summing along the major axes, i.e. adding the vectors together and dividing by the total number.

```
meanA = dataA.mean(axis=0)
meanB = dataB.mean(axis=0)
```

Then the `cov()` function is used to calculated the covariance matrix for each data set (the size of the correlation between the data dimensions).

```
covA = cov(dataA.T)
covB = cov(dataB.T)
```

Then the number of points in each data set, less one, is calculated for the data sets. The subtraction is present effectively because having only one data point does not give a scatter, so we use the number of points above this first one.

```
nA = len(dataA)-1.0
nB = len(dataB)-1.0
```

The scatter within each category is simply defined as the sum of the covariance matrices, weighted according to the size of the data sets. The scatter between categories is simply the separation between the data sets, i.e. the difference from one data centre to another.

```
scatterWithin = nA * covA + nB * covB
scatterBetween = meanA - meanB
```

The discrimination matrix between data sets is the line between centres (`scatterBetween`) divided by the scatter within the data (multiplied by inverse matrix) for each dimension.

```
discrim = dot(linalg.inv(scatterWithin),scatterBetween)
```

The data sets are transformed using the discrimination matrix, reorienting them along the line of best separation. These are passed back at the end for inspection.

```
transfA = dot(dataA, discrim.T)
transfB = dot(dataB, discrim.T)
```

The best guess for the dividing point that separates the data sets is the average of the two data centres reshaped (transformed) to lie along the discriminating direction.

```
divide = dot(discrim,(meanA+meanB))/2.0

return transfA, transfB, divide
```

Here we test the LDA function with two normally distributed, random data sets. One has a small spread and is moved to the side (by adding `array([-10.0,5.0])`) and the other has a wider spread. The two sets should overlap (intermingle).

```
testData1 = random.normal(0.0, 2.0, (100,2)) + array([-10.0,5.0])
testData2 = random.normal(0.0, 6.0, (100,2))
```

The test sets can be visualised with `matplotlib` in the usual way:

```
from matplotlib import pyplot

x, y = zip(*testData1)
pyplot.scatter(x, y, s=25, c='#404040', marker='o')

x, y = zip(*testData2)
pyplot.scatter(x, y, s=25, c='#FFFFFF', marker='^')

pyplot.show()
```

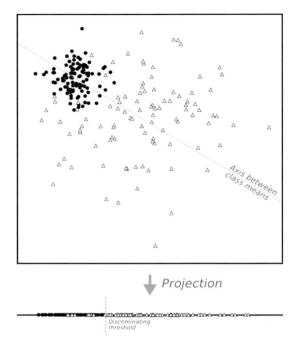

Projection

Discriminating
threshold

Figure 23.8. Results from linear discriminant analysis of 2D data to separate two data classes.
Linear discriminant analysis is a method to find the linear combination of features, and hence projection,
that best separates different data classes. Here the separation is measured as the ratio of the separation
between class means to the combined covariance. This separation is maximised by a line that goes
through the sample means of the two data classes, and a class discriminating point on this line
(technically a hyperplane in the original dimensions) can be estimated from the average of the two
class means.

Running the function on these data sets we get two arrays proj1, proj2 containing transformed data and a dividing value, div. These arrays are one-dimensional and represent the
projection of our two data sets onto the discrimination line. We can plot them as points along a
line, which here we separate for clarity with y-values at 0.5 and -0.5. The dividing value div
can be used to draw a line to show where the LDA has estimated the best boundary between
categories (assuming a symmetric, normal distribution).

```
proj1, proj2, div = twoClassLda(testData1, testData2)
print(div)

x = proj1
y = [0.5] * len(x)
pyplot.scatter(x, y, s=35, c='#404040', marker='o')

x = proj2
y = [-0.5] * len(x)
pyplot.scatter(x, y, s=35, c='#FFFFFF', marker='^')

pyplot.plot((div, div), (1.0, -1.0))

pyplot.show()
```

24 Machine learning

Contents

A guide to machine learning

When using computers to solve scientific problems there can be situations where you have some measured data and a related property of the data, but there is no known or fixed formula to link the two. Sometimes the link between the two sets of data may be easy for a human to see, but otherwise difficult to encode in a computer algorithm. A simple example of this would be in the reading of handwriting; humans do not write in a fixed typeface, every letter of a given kind will be written slightly differently, and yet we can read most other people's handwriting without much effort. When we look at writing we attempt to recognise the letters and words, and where there is ambiguity we can use our intelligence to infer what was intended by using the context of what the writing means, or any other clues that we can glean. Writing a computer program to read handwriting is difficult, and not nearly as reliable as a person would be. Nevertheless it can be done, and is put to good use in the mechanised sorting of mail by postal (zip) code. The common trick to getting a computer to perform tasks like this is not to program it with a designed and elaborate rule, but rather to bestow a computer program with a degree of *artificial intelligence* so that it can come up with its own rules and learn. The exercise whereby a program comes up with its own rules to solve a problem is often referred to as *machine learning*. It should be noted, however, that we usually don't expect a computer to learn a task perfectly; if perfection were possible we generally wouldn't have to resort to such means. Instead it is best to think of machine learning algorithms as making predictions, and as such the predictive power should be tested before we make reliance upon it. There are two kinds of machine learning which are commonly discussed, *supervised learning* and *unsupervised learning*, and we will give examples of both in this chapter.

Supervised machine learning

The supervised kind of machine learning involves having a computer algorithm that we can train on some known data. Here you would have some input data and knowledge of what each piece of data corresponds to. Using the postal code handwriting example, the input data might be examples of handwritten marks and the output would be knowledge of which of the 10 numerals or 26 letters of the alphabet were written. To take a biological example you may have a set of DNA or protein sequences and associate each with an experimentally determined category. Initially, during the training stage, our special computer algorithm learns to associate each input with the correct, known, output by adapting: essentially changing some internal parameters so that if you present an input to the algorithm it gives an output that is as close to the known result as possible. The objective is usually to have a single set of parameters that performs the job in a general way, adapting to all of the data to learn the overall properties of the problem, rather than optimising performance for some data at the expense of others. In this way, after the training stage the parameters of the algorithm are fixed and the algorithm may be applied to input data it has not seen before and come up with an output prediction: a prediction made following the rules learned during the training phase.

Broadly speaking you will encounter supervised machine learning being used in two major ways: to generate discrete alternative outputs, for example to perform classification as with the reading of handwritten letters and numbers; and also to predict the value of a continuous variable, which is often referred to as *functional approximation*. Predicting a

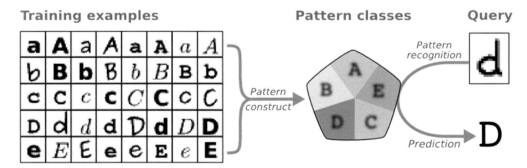

Figure 24.1. A simple schematic illustrating the application of machine learning to the classification problem of text character recognition. A training data set consists of many known examples for each type of letter, represented as a grid of image pixels with different intensities. Using the training data, generalised pattern classes representing the different characters are constructed. This is often achieved by the calculation of decision boundaries between the different possibilities. The identity of query characters is predicted by finding the best-matching category pattern. This prediction is often quick because the match is performed on the generalised class features, rather than comparing to all the training data.

continuous variable would occur when you are predicting something on a sliding scale, like temperature, energy etc. It is perhaps fortunate that we can often use the same basic kinds of computational algorithm, albeit with a degree of modification, whether we are predicting a discrete classification or a continuous variable. In the case of artificial neural network methods, which we introduce below, the distinction between the two types of situation can be as simple as restricting values to 0.0 or 1.0 if the problem is of the discrete kind.

Unsupervised machine learning

Unsupervised learning is where we have an algorithm that isn't shown a known data set for training. Instead the algorithm proceeds and optimises of its own accord without influence from a human-imposed set of standards. Usually the objective is for the algorithm to organise or arrange some input data according to relatively simple rules, but where there is nevertheless no straightforward answer. Where the data has been arranged so that similar inputs have been placed near to one another the end result can be a classification. Unsupervised methods are often used as a means of *dimensional reduction*, where the input data has many independent qualities (dimensions) but you wish to represent the data in only a few dimensions, say as a two-dimensional map. A biological example might be if you have some medical data where you have an abundance of different test measurements (lots of dimensions) but wish to categorise patients into a few discrete groups, each of which will have a different treatment regime; patients within each group will have similar test results, even though the results are multi-factorial and it is not easy for a human to derive the groupings.

Machine learning algorithms

There are some properties of machine learning algorithms that make them attractive to use, even when there are alternative approaches that could be used to achieve the same task. Firstly, you don't have to think about the precise details of what is going on. As long as you can train a program to do a good job, then that may be enough; although in some cases people will be distressed at the lack of a proper 'reason why'. Secondly, most machine learning methods are indifferent to the kind of data that is being input or output, as long as it can be encoded numerically; all sorts of disparate kinds of input data may be combined if they improve the prediction being made. Thirdly, machine learning methods are able to make non-linear and contextual decisions, which is to say that they can make predictions when the relationships between data items are not straightforward, including, for example, when two sets of input are generally very similar but some subtle correlation causes a completely different result.

In this chapter we will cover four different machine learning examples, and you can try these for your own computational problems where appropriate. For each, we describe a simple Python implementation (or as simple as we can make it while still being useful) and aim to point out the advantages and disadvantages of each method. We start with the *k-nearest neighbour* algorithm, which is perhaps the simplest of all machine learning algorithms and relatively easy to understand. Despite its simplicity, however, in some situations it can make good classifications with relatively little effort. Also, it introduces some of the principles, like vector input, that will be discussed in the other methods. Next we will describe a *self-organising map* as an example of an unsupervised method, and then go on to two supervised methods: a feed-forward *artificial neural network* and a *support vector machine*. Both of these methods can be used in a large number of different situations where there is training data available. The support vector machine is the more recent invention and holds certain advantages over neural networks: it is generally deterministic, giving the same result on the same training data, and it is much less susceptible to *over-training*, where a method 'learns' the training data too well and is not general enough to make the best predictions on data not seen before. Nonetheless, we include the feed-forward neural network because it is easier to implement, especially for multi-option classification, and often a good place to start in order to judge whether machine learning is an effective strategy for any given situation.

Feature space

The *k*-nearest neighbour algorithm described below is a very simple yet surprisingly powerful method to perform classification. The idea is that you describe the known, classified data as points, with coordinate locations, in a *feature space*. Just like real three-dimensional space, a feature space is defined by separate axes and a location in that space is defined by coordinates on each of the axes. A simple example of a feature space is colour, which may be described[1] by red, green and blue axes. Any colour within and including the extremes of black (zero red, green and blue) and white (maximum red, green and blue) is described by a certain amount of red, green and blue, in other words positions on these axes, and thus a colour may be described as a location vector (red, green, blue). We may also measure the distance between colour vectors, which is effectively to say how similar the colours are. Note that with this colour example we don't have the same kind of unbounded continuum that we have in real three-dimensional space: it is meaningless to have a negative colour value and we don't let the values exceed the maximum intensity; nothing can be whiter than white.

When dealing with feature spaces in general you can have a mixture of both bounded and unbounded axes, and as many numbers of 'dimensions' as is required to describe your data, although it is often a good idea to normalise the numeric data so that it fits in a limited range of values (typically -1 to $+1$). A good example of high-dimensionality data occurs in the use of biological sequences. For example, if you have input data that consists of DNA sequences that are five base pairs long, then because we can have one of four independent nucleotide residue types at each position along the length we need a vector of length 20 (20 'dimensions') to describe a five-letter sequence; 5 positions times 4 residue types. In this case with axes for G, C, A and T, the sequence 'TATGA' would have the vector (0, 0, 0, 1, 0, 0, 1, 0, 0, 0, 0, 1, 1, 0, 0, 0, 0, 0, 1, 0), where each sequential group of four numbers corresponds to one position in the DNA sequence. Note that 1 indicates the presence of a letter on its axis, and 0 the absence. If we moved from describing pure sequences to alignment profiles (i.e. proportions of the residue type at each position) we could have a sequence with fractional values instead of the plain 0 or 1. You might be tempted to encode DNA sequences with a numbered system where, for example, A = 0, T = 1, C = 2, G = 3, so the above example would be (1,0,1,3,0). However, this would not work because the different nucleotides are entirely alternative states and not points on a continuum where one type is meaningfully 'closer' to another.

[1] As far as human visual appreciation of light is concerned.

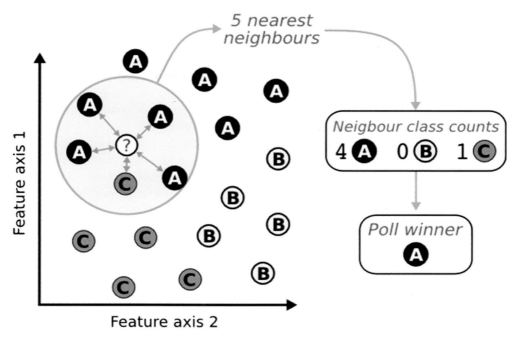

Figure 24.2. An overview of the k-nearest neighbour method. Data items are represented as vector locations within a space where the axes correspond to different features of the data. The training data vectors have known classifications and the classification of a query point is predicted from its *k* (in this case five) nearest neighbour points. A poll is taken of the categories present among the neighbour points and the most common category is the prediction for the query point.

k-nearest neighbours

If you can describe your data as position vectors in a feature space, and have a good way of calculating the distance or similarity between points, then the *k*-nearest neighbour method can be used to classify an unknown data vector by comparing it to data vectors for which there is a known classification. In essence this is like taking a voting poll. For an unclassified query point you look for a given number,[2] *k*, of nearest neighbours in the feature space that you have a classification for. You then assume that the classification of the query is the same as the majority of its *k* neighbours. Implementing this method successfully is dependent on having reasonable training data; you must have sufficient numbers of well-dispersed vectors of known classification so that any query is not too far from a known data point, and one class of data should not be significantly more abundant than any other, otherwise the more populous classification will have a positive bias for being a neighbour to the query. It should also be noted that this method can be fairly slow, given that you have to calculate distances to many classified points; however, if the method works well you can optimise later, for example, by using constructions like *k-dimensional trees*[3] to find nearest neighbours without having to check every data point.

---

[2] A certain number of points best determined by experimentation.
[3] Just an unfortunate coincidence in terminology, this is a different *k* to before.

Distance between feature vectors

Firstly, before introducing the main kNearestNeighbour function we will consider a small function that will be used to measure the distance between two points of data. Effectively this is to say how similar two pieces of input data are; the smaller the distance between the feature measurements that make up two data points the more similar they are. The distance measurement demonstrated here is simply the summation of the squared differences between each corresponding pair of values. For example, if we were measuring the 'distance' between two colours this calculation means finding the squares of the difference in the red, green and blue values. Note that we need not take the square root of the summation, as you might do in the calculation of the conventional distance, because the square root operation takes more calculation time and is not really needed; we only need to find the closest points of data and the smallest distance will also have the smallest squared value. If the following function is not such a good measure of similarity/difference[4] in data points then you can easily replace it with one that is, without affecting the main function.

The 'distance' function simply takes two feature vectors as inputs, which represent two data points. We use the inbuilt zip() function to extract equivalent feature values from both inputs at the same time, e.g. a and b would be the two red, green or blue values in turn, if the inputs were colours. Then we calculate the difference for each pair of values, square the difference and add it to the total. The total is then given back at the end at the return statement so that it can be picked up by the calling function.

```
def getFeatureDistance(vector1, vector2):

  distance = 0.0
  for a, b in zip(vector1, vector2):
    delta = a-b
    distance += delta * delta

  return distance
```

k-nearest neighbours in Python

Next is the main *k*-nearest neighbour function that takes an input of known data, feature vectors with a known classification, and a query vector that we wish to determine the classification of. It also accepts a value for k, which dictates how many of the nearest neighbours to the query point are considered when making the classification estimate. Depending upon the problem the best value of k to use will vary, and the user should optimise this in a fair way, although in many problems the value can be surprisingly small. This next function has not been optimised in terms of speed of execution, given we are primarily interested in clarity here. However, in the on-line material we also include a speed-optimised version which uses NumPy arrays to avoid loops and repeated calls to getFeatureDistance(). A notable simplification in the code below is that we only make a simple decision if there is a tie in the scores, where competing classifications have equal numbers among the nearest

---

[4] Alternative metrics might be things like Hamming distance, the minimum number of substitutions to convert one sequence to another, or non-Euclidian distances like the surface distance between points on a sphere.

neighbours. Here a tie is broken by taking the category with the closest single point to the query. Also, when there are only two categories using an odd value for k would avoid this issue given that ties would then be impossible.

The function code involves the definition with the input arguments. We then perform a check to make sure k is small enough for the data set. After this, the next step is to fill the starting values of the dists list, which records the distances and categories for all the known (already classified) data points. The list is appended with small tuples of distance and category (dist, cat), with the distance being first so that when we sort the list we sort according to distance, but the categories remain paired with their corresponding distances. The small tuple could also contain the feature vector from the known data input, if we need the function to report what the closest known data points actually are, rather than just the best classification. After the dists list is filled it is sorted so that it is in order of increasing distance. The k closest of the known categories is then simply taken from the start of the list using the appropriate slice notation dists[:k].

```
def kNearestNeighbour(knowns, query, k=7):

  if k >= len(knowns):
    raise Exception('Length of training data must be larger than k')

  dists = []
  for vector, cat in knowns[:k]:
    dist = getFeatureDistance(vector, query)
    dists.append( (dist, cat) ) # Vector could be included

  dists.sort()
  closest = dists[:k]

  counts = {}
  for dist, cat in closest:
    counts[cat] = counts.get(cat, 0) + 1

  bestCount = max(counts.values())
  bestCats = [cat for cat in counts if counts[cat] == bestCount]

  for dist, cat in closest:
    if cat in bestCats:
      return cat
```

The remainder of the function involves looking at the k closest data points to the query and determining what the most popular category is. This is achieved by making a dictionary for category counts, and then as we loop through the closest of the known points the count is made by adding one to the count for each category encountered. Note that the .get() function is used so that the starting value for a category's count is zero; there is no previous entry in the dictionary to add one to. With categories of the closest points tallied the best count is determined as the maximum of the list values. The best categories are then determined by finding all those that have this maximum count, using Python's compact list comprehension notation. More than one category may have a maximal count, indicating a tie. Potential ties are resolved by the final loop where the closest matches are gone through in order, remembering that they are sorted to give the closest first. The first category in the list from those with maximum count is then returned as the best category prediction (whereupon the loop ceases). The first point in the closest list is not necessarily the winner if other categories within the closest k are more populous.

We can test the functions with some crude fictitious data. Here we have colour vectors that are placed into only two named categories. We have tried to have about equal numbers of well-spaced points for each category, so that the choice of inputs doesn't introduce much bias. We then test by running the function on a query colour. This example shows how the input of known data with classifications is expected to be a list of smaller lists (or tuples), each of which contains first a feature vector, and second a textual category classification. Other Python data structures could be used, but they should naturally match the programming of the prediction function. Also, in many situations you would need to check the form and validity of the input before running the calculation.

```
knownClasses = [((1.0, 0.0, 0.0), 'warm'), # red
                ((0.0, 1.0, 0.0), 'cool'), # green
                ((0.0, 0.0, 1.0), 'cool'), # blue
                ((0.0, 1.0, 1.0), 'cool'), # cyan
                ((1.0, 1.0, 0.0), 'warm'), # yellow
                ((1.0, 0.0, 1.0), 'warm'), # magenta
                ((0.0, 0.0, 0.0), 'cool'), # black
                ((0.5, 0.5, 0.5), 'cool'), # grey
                ((1.0, 1.0, 1.0), 'cool'), # white
                ((1.0, 1.0, 0.5), 'warm'), # light yellow
                ((0.5, 0.0, 0.0), 'warm'), # maroon
                ((1.0, 0.5, 0.5), 'warm'), # pink
                ]

result = kNearestNeighbour(knownClasses, (0.7,0.7,0.2), k=3)

print('Colour class:', result)
```

Self-organising maps

The next machine-learning example is an illustration of *unsupervised learning*, where we don't pass input data with any associated known classification or value. Instead, the algorithm will simply organise the data, in this case putting similar points of data near one another and separating dissimilar points. The example we give is called a *self-organising map*, and this form is also known as a *Kohonen map*.[5] The reason for organising data in this way naturally depends upon the kind of problem being addressed. Once the data has been rearranged into an organised form, the 'map' of the data, it can be divided into different regions to define categories. Also, because the organising mechanism may consider the similarity of large numbers of features from the input data (large vectors) but creates a low-dimensionality (often two-dimensional) map, the organisation process performs a kind of *dimensional reduction*; we can more easily visualise the major differences and similarities in the data without having to think in *n*-dimensional space.

The self-organising map presented here is the first example of an *artificial neural network*. Although this name originally stems from analogies to how brain cells interact with one another, in the computational sense a neural network may be imagined as a network of

---

[5] Kohonen, T. (1982). Self-organized formation of topologically correct feature maps. *Biological Cybernetics* 43(1): 59–69.

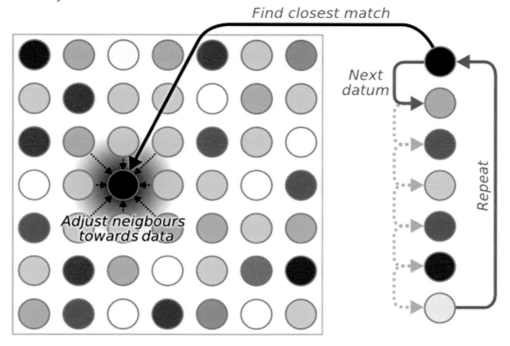

Map of feature vectors
Initially random values

Data vectors

Find closest match

Next datum

Adjust neigbours towards data

Repeat

Figure 24.3. A schematic of the learning process of a self-organising map. A regular array of initially random feature vectors is constructed. The feature vector for each real data item is compared to the array and the most similar point in the map is found. This closest matching vector in the map and its neighbours are adjusted to better match the data item. The matching and adjustment is then repeated for all the other data items before a new cycle considers all the data points again. The process continues for a large number of cycles or to convergence.

interconnected data points or *nodes*. Each node may be connected to several others and the strength of the connections is determined by a weighting. What a data node means depends somewhat on the kind of neural network being used. For the self-organising map example each node will represent a position on a rectangular[6] grid that makes up the map dimensions. Each grid node will possess a feature vector that is of the same kind as the input data (in the test example this will be a colour). The features of different nodes will be moved towards and away from the input data vectors so that different kinds of input are mapped to different spots on the grid. Here it should be noted that the strength of the connections between the grid nodes does not change; effectively a node will always stay in the same grid position, but how each node maps to the input data varies. For the *feed-forward neural network* described later the strength of connections between nodes does change.

The organisation process of the self-organising map occurs by repeatedly exposing the input data to the map of nodes, which initially have random similarities to the data. For

---

[6] In general the grid need not be rectangular. For example, it could be hexagonal.

each input data point the single node on the grid that best matches is determined: how well the input features (e.g. colour vector) match the features stored for the node. This best node is then pulled closer to the input point along with a few of the surrounding grid nodes, so that the feature vectors of that region of the grid more closely resemble that input. Different input points will pull different parts of the grid, in terms of feature vectors, towards themselves and away from dissimilar points. After many rounds of adjustment to the nodes, similar feature vectors will cluster together on the grid; similar input points will map to nodes that are close on the grid. After sufficient iterations for the features of the grid nodes to stabilise the 'learning' process stops. To help with this stabilisation the strength of the pull from the inputs on the grid features is gradually diminished, so that towards the end of the process only minor adjustments are made. Any data point, even if not seen before, may then be mapped onto the grid by finding the closest node. Naturally for any given input there need not be an exact match on the grid, even for the values used in the training data; the grid is of a finite size and nodes may represent compromises between competing data. Nevertheless, by finding a matching node, data may be mapped, reducing dimensionality and allowing categorisation if the map is divided into regions.

A Kohonen map in Python

The Python function below encodes a simple self-organising map. It makes extensive use of array functionality imported from NumPy, which in this instance reduces calculation time for large maps and, perhaps more importantly for this book, improves brevity by avoiding loops. Although some readers may find the use of vector/array operations disconcerting to start with, the approach does focus attention on the higher-level operations and reasoning that underpin the method.

Firstly, we import a couple of modules to give access to array operations and the exponential function. Then we define the function that takes a list of `inputs` (feature vectors defined as `numpy.array` objects) and something we have called `spread`, which represents the strength of connectivity between close nodes on the grid; this is a square array (a matrix) of weights where the centre represents the amount of pull from the input data on the best-matching grid node during learning and the off-centre values determine how much this influence spreads to adjacent grid positions. The `size` input is the number of rows and columns in the grid map and `steps` is simply the number of learning cycles that will be performed.

```
import numpy

from math import exp

def selfOrganisingMap(inputs, spread, size, steps=1000):
```

We extract the number of rows and columns for the grid from the input size and then determine `vecLen`, which represents the size of the feature vectors (e.g. three for a colour containing red, green and blue values). These numbers are then used to make the initial version of the map grid, which is named `somap`. This grid naturally has the required number of rows and columns, but also has a depth axis so that each grid location contains a number of feature values (a feature vector), hence the input axis sizes for the map are `nRows`, `nCols`, `vecLen`.

```
nRows, nCols = size
vecLen = len(inputs[0])
somap = numpy.random.rand(nRows, nCols, vecLen)
```

The next thing to initialise is an array that determines the influence (pull) of an input data vector on a region of the map grid. The input `spread` array gives the basic form of this; however, this needs to be applied to all of the feature values in the map, however deep the feature vectors; all the values along the vector are moved closer to the corresponding values from the best-match input. The `spread` array is only one value deep, but the map is `vecLen` deep, hence we define the `influence` array, which uses the values from `spread`, repeated `vecLen` times. Building a deep array of this kind upfront means we do not have to loop through the features to apply the influence weighting. The `influence` array is simple to create; by using the '*' operation on a list the spread array is replicated,[7] and the `dstack()`[8] function stacks the arrays on top of one another along the depth axis. The `infWidth` is calculated to find the radius of the influence, as needed later in the calculation. Then to improve clarity and speed we define the function `makeMesh`, which is simply a renaming of the cryptic `numpy.ix_` function; this takes lists of row and column indices and creates a 'mesh' where the indices intersect, and hence allows the extraction of a sub-matrix from a larger matrix.

```
influence = numpy.dstack([spread]*vecLen) # One for each feature
infWidth = (len(spread)-1) // 2
makeMesh = numpy.ix_    # Ugly
```

With the initialisation done the code now goes through the main learning steps. In Python 2, the `xrange()` function is used instead of `range()` because the number of steps may be fairly large and the latter would make a large list in memory, not just yield a step number, which is all we need. In Python 3, the `range()` function behaves like `xrange()` in Python 2. For each step the `decay` variable is calculated as the exponent of the proportion we have gone through all steps; this is then used to diminish the pull of the input data on the map vectors so that initially they can change a large amount but later stabilise. Then for each step we begin a loop through the array of input data vectors.

```
for s in range(steps):  # xrange in Python 2

  decay = exp(-s/float(steps))

  for vector in inputs:
```

Inside the loop, for each input, we calculate the difference of the current map to the vector. This is an array operation in compact NumPy notation where same vector is taken away from each of the feature vectors across all of the rows and columns of the grid; the result is an array the same size as the grid, with a difference vector for each grid position. This difference array is then squared in an element-wise manner (not matrix multiplication), then the square differences are added up along the depth (feature) axis. The result `dist2` is effectively a distance (squared) for each point on the grid to the current input, to give a measure of similarity.

---

[7] The square brackets are vital so the whole array is 'multiplied' by repetition, rather than the elements inside being numerically multiplied.

[8] There are also `vstack()` and `hstack()` functions for the other axes.

```
diff  =  somap-vector
diff2 = diff*diff
dist2 = diff2.sum(axis=2)
```

The `argmin()` function quickly finds the index in the array representing the minimum vector distance (most similar grid position); unfortunately, however, the index given back is the position in the array as if it were flattened into one long list. Thus, we then have to perform more calculations to determine what the column and the row of this index are; the row is found using integer division (the number of complete rows covered by the index), and the column is a remainder of the index from the start of the row.

```
index = dist2.argmin()
row = index // nRows
col = index % nRows
```

Given the row and column of the best-matching feature vector in the grid, we determine lists of rows and columns which cover the area of the map that will be 'pulled' towards the input. These lists are a range of values that goes `infWidth` positions either side of the best-matching row or column. Note the second argument for `range()` has one added to it, because the second argument is a limit that is not included on the output. The row and column numbers (x and y) are subject to a modulo '`%`' operation[9] so that if values from the range fall off the edge of the grid (less than zero or greater than last position) the calculated remainder continues the row/column on the opposite side. Effectively this joins the edges of the grid so that it wraps round in a continuous way; it needn't be done like this, but edge effects are helpfully removed.

```
rows = [x % nRows for x in range(row-infWidth, row+1+infWidth)]
cols = [y % nCols for y in range(col-infWidth, col+1+infWidth)]
```

Finally in the loop, the rows and columns are used to make a mesh, where they intersect, that is then used to extract sub-matrices for the grid map and difference array. The sub-matrix part of the map is the region that will be influenced by the input vector. The sub-matrix of the map is moved towards the input feature vector by subtracting the corresponding sub-matrix of the difference array (reducing the difference between map and input). Note that the sub-matrix of difference values (`diff[mesh]`) is scaled by two things: the influence of the input vector on the grid points near to the best-match position and the scaling factor that decays during the course of the learning. The `influence` array is the same size and shape as the sub-matrix, but `decay` is a simple number. The scaling means the amount of adjustment of the map grid towards the input vector diminishes both with the in-map distance from the best-match point and over the course of repeating the procedure. Finally, after the loop the grid array `somap` (of mapped feature vectors) is passed back.

```
mesh = makeMesh(rows,cols)
somap[mesh] -= diff[mesh] * influence * decay

  return somap
```

To test the function we first define a `numpy.array` which will determine how far across the self-organising map (grid) the input influence spreads. The centre (scoring 1.0 here) represents the best-match position in the map.

---

[9] The remainder after integer division, e.g. 13 % 10 = 3 and 3 % 10 = 3.

Start
(random pixels)   **1 iteration**   **10 iterations**   **100 iterations**

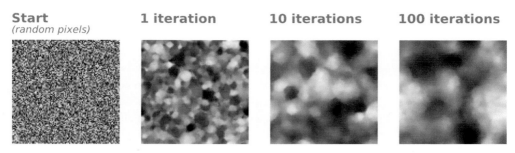

Figure 24.4 (Plate 10). Example self-organising map output. Results from of an initially random 100×100 colour pixel map (left) and the effect of the self-organising map on clustering the colours after 1, 10 and 100 iterations. A black and white version of this figure will appear in some formats. For the colour version, please refer to the plate section.

```
spread = numpy.array([[0.0, 0.10, 0.2, 0.10, 0.0],
                      [0.1, 0.35, 0.5, 0.35, 0.1],
                      [0.2, 0.50, 1.0, 0.50, 0.2],
                      [0.1, 0.35, 0.5, 0.35, 0.1],
                      [0.0, 0.10, 0.2, 0.10, 0.0]])
```

As a test we will make some random input; an array 400 (20×20) vectors of length 3. Each feature vector will represent a different colour in terms of red, green and blue values between zero and one.

```
rows, cols = 20, 20
testInput = numpy.random.rand(rows * cols, 3)
```

The self-organising map is run for 100 iterations, and will arrange the values on a 20×20 grid. Note that the input is of length rows × cols so that we have exactly one data point (input colour vector) for each map position. This needn't be the case, but is good to use for this example because it allows a 1-to-1 mapping of inputs to grid positions; thus the operation effectively performs shuffling of the inputs to make a 2D map of colour similarity.

```
som = selfOrganisingMap(testInput, spread, (rows, cols), 100)
```

We can then view the results of the organised map (look at the feature vectors for each node in the grid) by converting the `numpy.array` into a real colour image that we can view on screen. The conversion of an array of numbers to a displayable, graphical image is discussed in Chapter 18.

```
from PIL import Image

colors = som*255
colors = colors.astype(numpy.uint8)
img1 = Image.fromarray(colors, 'RGB')
img1.save('som.png', 'PNG')
img1.show()
```

Feed-forward artificial neural networks

The next machine learning example that we will cover is another kind of artificial neural network, but this time it is one that will undergo *supervised learning*. This means that when the network 'learns', it takes input data (more feature vectors) and changes its internal weights so that it can reproduce a known answer. The supervisory process whereby the programmer adjusts the network so that it gives the right answer, or as close to the right answer as possible, for some known data is usually referred to as *training*. Naturally, when training a neural network it is important to have as large and as representative a set of training data as possible. The predictive power comes from the fact that the neural network can accept input from data that it has not seen before, that was not used in the training. Predictions can be made for unseen data because inputs that resemble those that were used during the initial training will give similar outputs. In this regard it doesn't actually matter very much what the input or output data represents, the patterns and connections between them can be learnt nonetheless.

The neural network that we describe below is composed of a series of nodes arranged into three layers. The prediction of this network will proceed by a *feed-forward* mechanism, whereby input (often referred to as 'signal') is entered into the first layer of nodes. This input data is then moved to the middle or *hidden* layer to which it is connected, before finally reaching the last output layer of nodes. It is possible to construct feed-forward networks with more than three layers (i.e. more hidden layers). However, these can be more difficult to train and it has been shown that for many situations three layers are sufficient to do everything that more layers can do[10] (although the number of nodes will differ). The number of nodes in the three-layer network depends on the problem being addressed. The number of input nodes represents the size of the input vector; the value of each feature goes to a different input node. For example, if the input was a colour with red, green and blue features, there would be three input nodes. If the input was a DNA sequence composed of four base letters, there would be four input nodes for each position of the sequence analysed, thus a sequence of length ten would need 40 inputs. The number of output nodes depends on the problem, but there is some flexibility to represent the data in different ways. For example, if the network is used to predict an angle then the output could be a single number or it could be the sine and the cosine of the angle separately. When being used for categorisation, then there would be as many output nodes as there are categories. If the neural network was instead being used to approximate a continuous function, then the output will have a variable number of nodes, depending on how many axes are required. The number of hidden nodes used will depend on the type and complexity but will normally be optimised to give the best predictions. Numbers between three and ten are common. The smaller the number of nodes the quicker it is to optimise the network during training, but the fewer the number of patterns that can be detected in the data. The optimum number of hidden nodes can often be smaller than the number of inputs but is usually larger than the number of outputs. A convenient way to think of things is that the number of hidden nodes represents the complexity (dimensionality) of the problem, which is not necessarily related to the size of the input or output.

The three layers of nodes in our feed-forward network will be connected together. Each node will be connected to **all** of the others in a neighbouring layer. Thus, each input node is connected to all hidden nodes; each hidden node is connected to all of the input and output nodes; and each output to each hidden node. The properties of a neural network emerge because the strength of the connection between nodes can vary during the learning process; so some nodes become more or less well connected. If a connection ends up having a zero weight then its linked nodes are effectively disconnected; thus the network can represent a large number of possible internal organisations. A node will be connected to many others to varying degrees, but

---

[10] This is described by the universal approximation theorem; see http://http://en.wikipedia.org/wiki/Universal_approximation_theorem.

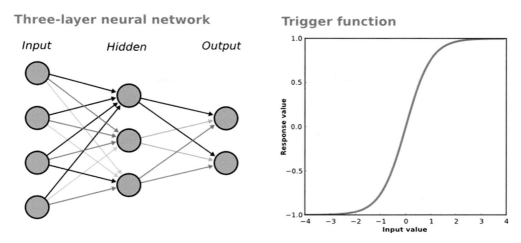

Figure 24.5. Components of a feed-forward artificial neural network. The organisation of nodes in a three-layer artificial neural network (left) where the node in each layer is connected to all the nodes in the next layer, albeit with different connection strengths. The hyperbolic tangent is often used as a trigger function to modulate each node's output (right). For the trigger function the x axis corresponds to the total input, which is the weighted sum of the inputs from all the connections. The y axis represents the node output that may be sent to the next layer.

the actual feed-forward action of the network that is used to make predictions (generate output) uses what is known as a *trigger function* to adjust the response. In essence, a node collects input signals on one side and has to combine these in some manner to generate output on the other side, which could be an intermediate or final output signal. The input signals are added together, but the strength of the resulting output which is sent to any nodes in the next layer is altered. Firstly, the summation of the combined inputs is scaled to be within certain minimum and maximum bounds for practical purposes. And secondly, the input is applied to the trigger function to increase or decrease the effect that certain amounts of input have. Sometimes the trigger function is a two-state switch where smaller input values produce very little response, but above a particular threshold the response is very strong; this is perhaps analogous to the firing of a neuron inside a brain. However, many types of trigger functions are possible, and the one we employ here is the popular hyperbolic tangent function (`tanh`; see Figure 24.5). Using the sigmoid-shaped hyperbolic tangent curve means that in mid ranges the strength of a node's output is roughly proportional to its input, but at the high and low input extremes the output is attenuated towards limits. This function also benefits from having an easily calculated gradient (required for training) and has successfully been used in many diverse situations.

Training a neural network by back propagation

The artificial neural network presented here is trained via a mechanism known as *back propagation*. This is a fairly efficient general solution for training, but other ways of finding network connection weights are possible, like the slower but more rigorous Markov chain Monte Carlo (see Chapter 25). The back-propagation mechanism takes the known output values for the input training data and adjusts the connection weighting between the nodes, working backward layer by layer from output, via hidden to input. The objective at each stage is to minimise the error between the fixed, known result and the actual network output (the prediction). The weights are adjusted a little to minimise the error of each bit of training data in turn, although it is often important to randomise the order of the data. Because different examples of training data may compete with one another (pull weights in different directions) and because a given node is influenced by many others we can really only guess at how to adjust the weights to make output match. Hence training can be a slow and cautious process, repeatedly going through all the training data many times, while the connection weights settle into a hopefully stable pattern. The actual amount that weights are adjusted for each bit of data in each cycle will naturally depend on the kind of trigger function used by the nodes, but in general the idea is that the gradient of the function indicates in which direction the inputs to a node should be adjusted to better match the output.

The programmer should always be cautious when training an artificial neural network, and it can only legitimately be used to make predictions if the performance is properly tested on data that it has never seen before; it is commonplace to hold back some of the training data set for testing. Also, these networks can suffer from over-training, where the network learns to associate the training input and output too well; it becomes too specialised and performs poorly on data it has not seen before. Over-training can be minimised by selecting a widely spread set of training examples, optimising performance by testing on some data that has never been seen before, and not worrying too much about small improvements in the connection weight optimisation. Even considering these things though, the user also has to be mindful, as with any machine learning, that the problem being addressed is well formulated. There is the anecdotal example of the military neural network that was designed to automatically distinguish between pictures of friendly and enemy tanks. In training, this neural network seemed to work very well, but in the real world it performed poorly. It turned out that pictures of friendly and enemy tanks generally had different kinds of backgrounds and the network has learned the classification based upon the (easier to distinguish) terrain type, not on the tanks. Putting an enemy tank in front of some trees made it look friendly, at least as far as the neural network was concerned. The moral here is to only use input that is unbiased and relevant.

A Python neural network

The feed-forward neural network example in Python has been split into two functions: one that makes predictions and one that does the training. It would also be possible to construct this neural network using classes (custom kinds of Python objects), and this may hold certain advantages, like the ability to make adapted subclasses. However, using functions makes it simpler to describe the principles of what is happening.

The first function is called `neuralNetPredict`, which takes some input data for the first layer of network nodes, applies the first weighted connections and trigger functions to pass the signal to the hidden layer of nodes and then applies the second weights and triggers to generate some output. This is used both during the training of the network, to set up the connection weights, and to make predictions on unseen data. Initially some mathematical functions are imported from the NumPy library, so that we can express the operations concisely as arrays and matrices.[11]

```
from numpy import array, tanh, zeros, ones, random, sum, append
```

---

[11] Although using NumPy can actually be slower for small neural networks.

Then we define the function name and its input arguments: an array of input features (inputVec) and two matrices that represent the connection weights. The matrix weightsIn represents the strength of connection between the input nodes (which include the bias node we describe below) and the hidden nodes. Likewise, weightsOut represents the strengths between the hidden and the output nodes. The weights are represented as matrices so that the rows correspond to a set of nodes in one layer and the columns represent the set of nodes in the other layer, to connect everything in one layer to everything in the other. For example, if the network has four input, five hidden and two output nodes, then weightsIn will be a 4×5 matrix, and weightsOut will be a 5×2 matrix. Inside the function the first step is to define the signalIn vector for the network. This is simply a copy of the input features array with an extra value of 1.0 appended to the end. This extra, fixed input is what is known as a *bias node*, and is present so the baseline (the level without meaningful signal) of an input can be adjusted. This gives more flexibility at the trigger function used for the hidden layer of nodes, which improves learning. The weight matrices must be of the right size to account for the bias node, and although weights from the bias node are still adjusted by training they are naturally not affected by the input data. A bias connection going to each hidden node enables the input to that node to be offset, effectively shifting the centre of the trigger function about so that it can better distinguish the input values; the upshot of this is that the programmer doesn't have to worry about centring input feature values (e.g. making their mean values zero).

```
def neuralNetPredict(inputVec, weightsIn, weightsOut):

  signalIn = append(inputVec, 1.0) # input layer

  prod = signalIn * weightsIn.T
  sums = sum(prod, axis=1)
  signalHid = tanh(sums)  # hidden  layer

  prod = signalHid * weightsOut.T
  sums = sum(prod, axis=1)
  signalOut = tanh(sums)  # output  layer

  return signalIn, signalHid, signalOut
```

The main operation of the function involves multiplying the input vector, element by element, with the columns of the first matrix of weights. As a result of the training process we describe later, the weight matrix is arranged so that there is a column for each of the hidden nodes. Given we want to apply the input signal to each hidden node, we use the transpose (.T) of the weight matrix so that columns are switched with rows for the multiplication. This is a requirement because element multiplication of a one-dimensional NumPy array with a two-dimensional array is done on a per-row basis. Next we calculate the summation of the weighted input down each column (axis=1), so we get one value for each hidden node. Then to get the signal that comes from the hidden layer we calculate the hyperbolic tangent of the sums, applying the sigmoid-shaped trigger function to each. This whole operation is then repeated in the same manner for going from the hidden layer to the output layer; we apply weights to the signal vector, sum over columns and apply the trigger function. The final output vector is the prediction from the network. At the end of the function we return all the signal vectors, and although only the output values are useful in making predictions the other vectors are used in training the network.

The second Python function for the feed-forward neural network is a function to train it by the back-propagation method, to find an optimal pair of weight matrices. The objective is to minimise error between the output vectors predicted by the network and the target values (known because this is training data). Here the error is calculated as the sum of the squared differences, but other methods may be more appropriate in certain situations. The function is defined and takes the training data as an argument, which is expected to be an array containing pairs of items: an input feature vector and the known output vector. The next argument is the number of nodes in the hidden layer; the size of input and output layers need not be specified because they can be deduced from the length of the input and output vectors used in training. The remaining arguments relate to the number of training steps (cycles over the data) that will be made, a value for the learning rate that governs how strongly weights are adjusted and a momentum factor that allows each training cycle to use a fraction of the adjustments that were used in the previous cycle, which makes for smoother training. In practice the learning rate and momentum factor can be optimised, but the default values are generally a fair start.

```
def neuralNetTrain(trainData, numHid, steps=100, rate=0.5, momentum=0.2):
```

Within the function a few values are initialised. The numbers of nodes in the input and output layers are extracted from the size of the first item (index zero) of training data, noting that the number of inputs is then increased by one to accommodate the bias node. The error value which we aim to minimise starts as None, but will be filled with numeric values later.

```
numInp = len(trainData[0][0])
numOut = len(trainData[0][1])
numInp += 1
minError = None
```

Next we make the initial signal vectors as arrays of the required sizes (a value comes from each node) with all elements starting out as 1 courtesy of numpy.ones(). The input will be the feature vector we pass in and the output will be the prediction.

```
sigInp = ones(numInp)
sigHid = ones(numHid)
sigOut = ones(numOut)
```

The initial weight matrices are constructed with random values between −0.5 and 0.5, with the required number of rows and columns in each. The random.random function makes matrices of random numbers in the range 0.0 to 1.0, but by taking 0.5 away (from every element) we shift this range. This particular range is not a strict requirement, but is a fairly good general strategy; too small and the network can get stuck, but too large and the learning is stifled. The best weight matrices, which is what we are going to pass back from the function at the end of training, start as these initial weights but then improve.

```
wInp = random.random((numInp, numHid))-0.5
wOut = random.random((numHid, numOut))-0.5
bestWeightMatrices = (wInp, wOut)
```

The next initialisation is for the change matrices, which will indicate how much the weight matrices differ from one training cycle to the next. These are important so that there is a degree of memory or momentum in the training; strong corrections to the weights will tend to keep going and help convergence.

```
cInp = zeros((numInp, numHid))
cOut = zeros((numHid, numOut))
```

The final initialisation is for the training data: pairs of input and output vectors. This is done to convert all of the vectors into `numpy.array` data type, thus allowing the training data to be input as lists and/or tuples. We simply loop through the data, extract each pair, convert to arrays and then put the pair back in the list at the appropriate index (x).

```
for x, (inputs, knownOut) in enumerate(trainData):
    trainData[x] = (array(inputs), array(knownOut))
```

With everything initialised, we can then begin the actual network training, so we go through the required number of loops and in Python 2 use `xrange()` so that a large list doesn't have to be created. Note we don't use a `while` loop to check for convergence on the error because a neural network is not always guaranteed to converge and sometimes it can stall before convergence. For each step we shuffle the training data, which is often very important for training; without this there is a bias in the way the weights get optimised. After the shuffle, the error starts at zero for the cycle.

```
for step in range(steps):  # xrange() in Python 2
    random.shuffle(trainData)  # Important
    error = 0.0
```

Next we loop through all of the training data, getting the input feature vector and known output for each example. We then use the current values of the weight matrices, with the prediction function described above, to calculate the signal vectors. Initially the output signal vector (the prediction) will be quite different from the known output vector, but this will hopefully improve over time.

```
for inputs, knownOut in trainData:
    sigIn, sigHid, sigOut = neuralNetPredict(inputs, wInp, wOut)
```

Given the neural network signals that come from the current estimates for weight matrices we now apply the back-propagation method to try to reduce the error in the prediction. Thus we calculate the difference between the known output vector and the signal output from the neural network. This difference is squared and summed up over all the features (`diff` is an array) before being added to the total error for this cycle.

```
diff = knownOut - sigOut
error += sum(diff * diff)
```

Next we work out an adjustment that will be made to the output weights, to hopefully reduce the error. The adjustment is calculated from the gradient of the trigger function. Because this example uses a hyperbolic tangent function, the gradient at the signal value is one minus the signal value squared (differentiate $y = \tanh(x)$ and you get $1 - \tanh^2(x)$ which equals $1 - y^2$). The signal gradient multiplied by the signal difference then represents the change in the signal before the trigger function, which can be used to adjust the weight matrices. Note that all these mathematical operations are performed on all the elements of whole arrays at once, courtesy of NumPy.

```
gradient = ones(numOut) - (sigOut*sigOut)
outAdjust = gradient * diff
```

The same kind of operation is repeated for the hidden layer, to find the adjustment that will be made for the input weight matrix. Again, we calculate a signal difference and a trigger function gradient and multiply them to get an adjustment for what goes into the trigger function. However, this time we can't compare output vectors, so instead we take the array of signal adjustments just calculated and propagate them back through the network. Thus the signal difference for the hidden layer is calculated by taking the signal adjustment for the output later and passing it through the output weight matrix, i.e. backwards through the last layer.

```
diff = sum(outAdjust * wOut, axis=1)
gradient = ones(numHid) - (sigHid*sigHid)
hidAdjust = gradient * diff
```

With the adjustments calculated it then remains to make the changes to the weight matrices, and hopefully get an improvement in the error. The weight change going from hidden to output layers requires that we calculate a change matrix (the same size as the weights), hence we take the vector of adjustments and the vector of hidden signals and combine them; each row of adjustments (one per output) is multiplied by a column of signals (one per hidden node) to get the new weights. Note how we use the `reshape()` function to convert the array of signals, a single row, into a column vector; it is tipped on its side so that the multiplication can be made to generate a matrix with rows and columns.

```
# update output
change = outAdjust * sigHid.reshape(numHid, 1)
wOut += (rate * change) + (momentum * cOut)
cOut = change
```

In the same manner the changes are made to the input weight matrix.

```
# update input
change = hidAdjust * sigIn.reshape(numInp, 1)
wInp += (rate * change) + (momentum * cInp)
cInp = change
```

Then finally in the training cycle, we see if the minimum error has been improved on. During the first cycle the minimum error is None, so we always fill it with the first real calculated error value in that case. Each time we find a new minimum error we record the best weight matrices (so far) by taking copies of the current versions, using the handy `.copy()` function of NumPy arrays. Then finally at the end of all of the training cycles, the best weight matrices are returned.

```
if (minError is None) or (error < minError):
  minError = error
  bestWeightMatrices = (wInp.copy(), wOut.copy())
  print("Step: %d Error: %f" % (step, error))

return bestWeightMatrices
```

We can test the feed-forward neural network by using some test training data. As a very simple example, the first test takes input vectors with a pair of numbers which are either one or zero. The output corresponds to the 'exclusive or' (XOR) logic function: the output is 1 if either of the inputs is 1, but not both. This test data is a list of [input, output] pairs. Note that even though the output is just a single number it is nonetheless represented as a list with a single item.

```
data = [[[0,0], [0]],
        [[0,1], [1]],
        [[1,0], [1]],
        [[1,1], [0]]]
```

The number of hidden nodes used here is simply stated as 2, but in practical situations several values will need to be tried, and their performance evaluated. Then we run the training function in the data to estimate the best weight matrices for the neural network.

```
wMatrixIn, wMatrixOut = neuralNetTrain(data, 2, 1000)
```

The output weight matrices can then be run on test data for evaluation. At the very least they ought to do a reasonable job at predicting the output signals for the training set, although in practice these really ought to be for data that has not been used in the training.

```
for inputs, knownOut in data:
  sIn, sHid, sOut =  neuralNetPredict(array(inputs), wMatrixIn, wMatrixOut)
  print(knownOut, sOut[0])
```

A neural network for biological sequences

Next we move on from the trivial neural network test example to illustrate how feature vectors may be generated from biological sequences (and category data generally), so that they can be used in such machine learning programs. The example will predict the secondary structure of a residue in the middle of a five-amino-acid sequence. Both the amino acid sequence and the output secondary-structure categories will be represented initially as code letters, but they will be converted into numbers (zeros and ones) before being passed into the feed-forward neural network. Although this example uses protein sequences an analogous procedure can be used for DNA and RNA.

The test data that will be illustrated for the example is very small, simply to give a taste of the data and still have it fit on the page. As a result, a neural network trained on this data would be totally useless at making secondary-structure predictions in practice. However, in the on-line material a file (SecStrucTrainingData.tsv, available via http://www.cambridge.org/pythonforbiology) with a large data set containing many thousands of sequences is available. Using this as input would give a vastly superior result. The test data is presented as a list of 2-tuples; each tuple has a five-letter residue protein sequence and a secondary-structure code. Three secondary-structure codes are used to represent three general conformations of protein backbone geometry. These codes are 'E' for extended conformations (mostly beta-strand), 'H' for helices (mostly alpha-helix) and 'C' for random coil or unstructured stretches (everything else).

```
seqSecStrucData = [('ADTLL','E'),
                   ('DTLLI','E'),
                   ('TLLIL','E'),
                   ('LLILG','E'),
                   ('LILGD','E'),
                   ('ILGDS','E'),
                   ('LGDSL','C'),
```

```
('GDSLS','H'),
('DSLSA','H'),
('SLSAG','H'),
('LSAGY','H'),
('SAGYR','C'),
('AGYRM','C'),
('GYRMS','C'),
('YRMSA','C'),
('RMSAS','C')]
```

Before the above data can be used it will need to be converted from text strings into numeric feature vectors (arrays of numbers). All of the feature vectors that represent either sequence input or prediction output will contain numbers to represent the presence or absence of a particular category of item. In this example the feature vectors will contain ones to indicate the presence and zeros to represent the absence of an amino acid (for input) or of a secondary-structure code letter (for output). Other numbers could have been used instead with equal success (e.g. ± 1 or ± 0.5), although the two values chosen for presence or absence should naturally be distinct and lie in the range of the trigger function where there is a steep gradient; for the hyperbolic tangent example used here the range from -1 to $+1$ is usually best. Note that the size of the vector generated is the length of the input sequence multiplied by the number of possible letters; each element of the vector represents a different residue at a different sequence position. For a five-letter protein sequence the vector will have 20 elements (one for each amino acid) for each of the five sequence positions, and thus the total length will be 100.

To make the feature vectors we first define dictionaries so that a letter can be used as a key to look up the position (index) in the vector that should be set to 1.0. The actual order that we go through the letter codes is unimportant, but it should be consistent for a given program. Here we could have used the `list.index(letter)` form to get a position index, but this is slower, especially if the number of possible letters and amount of training data are large. To make the index-look-up dictionary for the amino acids we loop through a list of possibilities and associate the residue letter code with the index i, which in this case was generated with `enumerate()`:[12]

```
aminoAcids = 'ACDEFGHIKLMNPQRSTVWY'
aaIndexDict = {}
for i, aa in enumerate(aminoAcids):
  aaIndexDict[aa] = i
```

The same sort of thing is repeated for the secondary-structure codes, albeit with a smaller number of possible letters:

```
ssIndexDict = {}
ssCodes = 'HCE'
for i, code in enumerate(ssCodes):
  ssIndexDict[code] = i
```

Now we actually do the conversion of the training data from the text strings to numeric vectors that can be used in the neural network routine. To help with this the `convertSeqTo-Vector` function defined below is constructed to take a sequence of letters `seq`, and

---

[12] Though it is less obvious what is happening we could also do the following and avoid the loop: `aaIndexDict = dict(enumerate(aminoAcids))`.

indexDict, which can convert each letter to the correct position in the code alphabet. The vector initially starts filled with zeros, but then selective positions are converted to ones, depending on which sequence letters are observed. The actual index in the vector that needs to be set is determined by the index-look-up dictionary for that letter, indexDict[letter], and the start point for that sequence position, pos * numLetters. For example, if the third letter (index 2, counting from 0) in seq is 'F', this adds a one at position 45; forty (2 × 20) to get to the start of the block that represents the third sequence position and five more because 'F' is at index 5 in the protein sequence alphabet.

```
def convertSeqToVector(seq, indexDict):

  numLetters = len(indexDict)
  vector = [0.0] * len(seq) * numLetters

  for pos, letter in enumerate(seq):
    index = pos * numLetters + indexDict[letter]
    vector[index] = 1.0

  return vector
```

The actual training data for the neural network is made by looping through all of the pairs of protein sequence and secondary-structure code, and for each of these using the above function to make the feature vector from the text. The trainingData list is constructed as linked pairs of input and output feature vectors. Note that because the secondary-structure code ss is only a single letter the output vector will be of length three. Specifically, for the three codes the output vectors will be as follows: 'E': [0.0, 0.0, 1.0]; 'C': [0.0, 1.0, 0.0]; 'H': [1.0, 0.0, 0.0].

```
trainingData = []
for seq, ss in seqSecStrucData:

  inputVec = convertSeqToVector(seq, aaIndexDict)
  outputVec = convertSeqToVector(ss, ssIndexDict)

  trainingData.append( (inputVec, outputVec) )
```

The number of hidden nodes is set to three, by way of example. The training data and network size are then passed into the main training function, in order to generate the predictive weight matrices.

```
wMatrixIn, wMatrixOut = neuralNetTrain(trainingData, 3, 1000)
```

After training, the neural network can make secondary-structure predictions for any five-letter protein sequence, but, as before, this must be converted into the numeric vector form. Because the prediction operates on a whole numpy.array the test vector is converted into an array (here with one element).

```
testSeq = 'DLLSA'
testVec = convertSeqToVector(testSeq, aaIndexDict)
testArray = array( [testVec,] )
```

The weight matrices from the training are used to make predictions on the test array and the output signal, sOut, is interrogated to find the position of the largest value (i.e. the position predicted to be one not zero). This index represents the best secondary-structure code, and the code itself can be obtained by using the index with the original set of codes to get a letter back.

```
sIn, sHid, sOut =  neuralNetPredict(testArray, wMatrixIn, wMatrixOut)
index = sOut.argmax()
print("Test prediction: %s" % ssCodes[index])
```

Support vector machines

The final example of a machine learning method in this chapter moves away from neural networks to a different and more recent approach, described generally as *kernel methods*, and the specific example given here is known as a *support vector machine* (SVM). The SVM can be thought of as being related to the *k*-nearest neighbour idea, where we visualise known points of data in a vector space and aim to make predictions on previously unseen input by looking where a query lies in this space in relation to the known, training data. Whereas with the *k*-nearest neighbour method a prediction is made by looking at a number of known values in the vicinity of a query, for an SVM the training data vectors are used to define the location of a boundary that separates the vector space into two regions. A prediction is made for an unseen query vector by seeing on which side of the decision boundary it occurs. For two dimensions such a decision boundary would be a line, and in three dimensions it is a plane, but for most SVMs we are working in more, often many more, than three dimensions (data features), in which case the boundary is a *hyperplane*. The name support vector machine derives from the *support vectors*, which are the known, training data points that are found on the edge of this boundary, defining the boundary position.

One of the useful properties of support vector machines compared to neural network techniques comes from the fact that they maximise the margin between data categories. If you imagine the training data to be two distinct, fixed groups the operation finds the widest region that separates them. The boundary between these two groups is placed in the centre of the separating region, and the prediction of previously unseen data is made by finding out on which side of the boundary the query point lies. With a neural network the effective boundary between data categories will not always be fairly placed in the middle. This is a particular issue when the training data is not evenly spread (e.g. one prediction category is significantly more populous) and if it has been over-trained. The decision line may be much closer to one category or the other, with the result that unseen query data that lies between the two will tend to be classified inappropriately, in only one way. With a support vector machine it is only the support vectors at the edge of the maximised margin between categories that define the boundary, thus over-training is not possible and the amount of data on either side does not affect the decision line so much.[13]

A support vector machine effectively defines the widest boundary between two categories of data vector by doing linear optimisation, similar to finding a line of best fit. However, the input data may not naturally lie in two discrete regions and thus be separable with a simple plane boundary. Imagine trying to decide which applicants should be allowed to join the army based on their weight measurements. You wouldn't expect to be able to sensibly use a single decision line to separate those who pass and those who fail, given that you have to distinguish between both those people who are overweight and those who are underweight. The solution to this kind of problem can nonetheless be solved using a single

---

[13] It can, strictly speaking, affect the likelihood of finding a support vector.

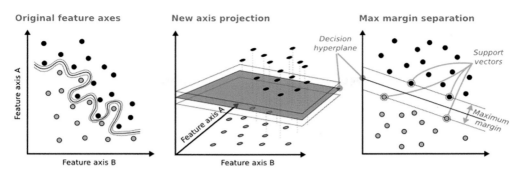

Figure 24.6. A schematic overview of support vector machines. Feature vectors representing the training data are separated into different regions by a boundary line, which in the original feature space can follow a complex path. The support vector machine finds the boundary line by finding the linear hyperplane that best separates the data in a higher number of dimensions. The decision hyperplane is in the middle of the widest margin between the data classes, and this margin is itself determined by the support vectors: data items which border the decision zone.

line, and thus by a support vector machine, because the decision boundary can be placed in a space of higher dimensionality than the original problem. Referring again to the army recruits, if instead of having a single weight axis you put the weight measurements in a semi-circle then you can now separate the pass and fail categories above and below a single line. It is simply that the data is now in two dimensions, rather than the original one. The extra dimension has nothing to do with the original data, it has simply spread it out in a predictable way and you can still retrieve the original values if required: in this case from the distance along the curve. Using a known function to spread data points into more dimensions is often referred to as the *kernel trick*[14] and gives power to the SVM to operate on nonlinear problems (e.g. with context dependency) while using mathematically robust, linear means to make an optimum decision. Although the simple example mentioned uses two dimensions to achieve separation, a general support vector machine can use any number of extra dimensions. Although at first this might seem to arbitrarily complicate the problem, the really clever part of the kernel trick is that within the algorithm to find the separating boundary the high-dimensionality locations of the data points do not have to be stored or even calculated. The optimisation can be performed by merely considering what the projection of the high-dimensionality points onto the original feature axes (i.e. using the dot product) would be. For a more mathematical explanation of support vector machines and the kernel trick see the reference given below.[15]

The particular algorithm that our example uses in its learning procedure is called successive over-relaxation. This is a means of efficiently solving the linear equations that govern the location of the decision hyperplane between two categories of data. The objective of this algorithm is to define which of the feature vectors in the training data are support

---

[14] Aizerman, M., Braverman, E., and Rozonoer, L. (1964). Theoretical foundations of the potential function method in pattern recognition learning. *Automation and Remote Control* 25: 821–837.

[15] Boser, B.E., Guyon, I.M., and Vapnik, V.N. (1992). A training algorithm for optimal margin classifiers. In D. Haussler (ed.), *5th Annual ACM Workshop on COLT.* Pittsburgh, PA: ACM Press. pp. 144–152.

vectors, and thus define the hyperplane direction. We will not discuss the mathematical detail of this method or of SVMs in general here, we will merely give a flavour of what is happening. However, the keen and more mathematically inclined readers can investigate the specified literary references.

The support vector machine example given here will learn and predict classifications between two categories of vector data, here encoded internally as $+1$ and -1 respectively. Obviously there are often situations where there are more than just two categories that we wish to predict. In these cases multiple support vector machines can be used to make separate two-way decisions. Imagine that you have three categories of data A, B and C: the first support vector machine might distinguish A from everything else, i.e. the other category is B **and** C, and the second SVM will distinguish between the remaining B and C. It should be noted, however, that where there is overlap between the different categories, the order of two-way decisions may be important; in general you would try different combinations and tend to make the most secure predictions first. Although we will only be discussing an SVM that can be used for classification into two discrete categories, there is a closely related method, *support vector regression*, which may be used to predict continuous values. Here the vectors of training data have a range of different numeric values and the support vectors are used to give a line of best fit to these in high-dimensionality space. This line will yield predictions by interpolation; calculating the position of a query along the line of known slope gives an estimate of the associated numeric value.

A Python support vector machine

The Python example of a support vector machine starts with the import of various array and mathematical NumPy functions that will be needed. As with the neural network example the use of array operations aims to focus attention on the high-level operation being performed, and thus the reasoning involved in the method. Helpfully, the NumPy routines are also quicker to execute than looping through large arrays. There are several NumPy imports and it should be noted that many of these have the same name as the equivalent in the standard `math` module, although as you might expect these versions work on whole arrays, not just single numbers:

```
from numpy import exp, power, array, zeros, sqrt
from numpy import nonzero, random, abs, sum, dot
```

Next, because we will be using some random numbers we initialise the random number generator in NumPy. This initialisation is done using a *seed* number, which in this instance is the system time in seconds[16] because it is easy to access and constantly changing. Initialising random numbers in this way is always good if you do not want to have exactly the same result each time. Although, if you do want to have the same result, say when doing debugging of the code, you can use the same seed number; so that although a stream of numbers come out apparently randomly it will be the same numbers each time the program is run.

```
from time import time
random.seed(int(time()))
```

---

[16] Since midnight 1 January 1970: the beginning of 'UNIX time'.

Next we define the kernel functions, which are used to replace the explicit calculation of dot products between feature vectors in high-dimensionality space, as a measure of the coincidence between two data points. We give only two simple examples for kernel functions here, although more could be considered. Both functions take in two feature vectors (data points) and give a single value measure of their similarity or coincidence. The Gaussian kernel (normal curve) is perhaps the most general and widely used form. The calculation for the Gaussian kernel is fairly simple and involves only a single parameter, sigma, which dictates the width of the function. First the difference between two feature vectors is calculated, then the sum of the differences squared (a single scalar value) is conveniently calculated using the dot product. The number is finally scaled and the exponent of its negative returned. This kernel will give a measure of coincidence that is always positive and how quickly this diminishes will depend on sigma.

```
def kernelGauss(vectorI, vectorJ, sigma=1.0):

  sigma2 = sigma * sigma
  diff = vectorI - vectorJ
  dotProd = dot(diff,diff)

  return exp( -0.5 * dotProd / sigma2 )
```

The linear kernel is also very simple; the function calculates the difference of each vector to the vector mean of the input data (which we calculated beforehand) and returns the dot product of these differences. Effectively the value generated is a measure of the coincidence of the input vectors relative to the centre of the data. It should be noted that the output can be negative if the input vectors go in opposite directions relative to the mean.

```
def kernelLinear(vectorI, vectorJ, mean):

  diffI = vectorI - mean
  diffJ = vectorJ - mean

  return dot(diffI, diffJ)
```

Next we get to the main support vector machine training algorithm, the details of which follow the method of 'successive over-relaxation' as described in the reference given below.[17] In essence this will find the optimal boundary between two sets of input data. The inputs to the function comprise an array of known classifications containing +1 or −1 for the two classes, an array of training feature vectors (data) which are typically normalised to be between 0.0 and 1.0, a kernel function, like those described above, a set of parameters that are used with the kernel function and three numeric values. It should be noted that all of the input arrays are expected to be of the numpy.array type. The limit parameter is the upper limit to the amount of 'support' that a data vector can provide towards the optimisation of the classifying boundary plane. The maxSteps number is simply the maximum number of cycles that we will allow during the training of the SVM. The last argument (relax) takes values between 0 and 2.0 and is known as the 'relaxation parameter'. This governs the rate at which the successive over-relaxation technique performs the optimisation. Under many circumstances this does not need to be changed from its default value.

[17] Mangasarian, O.L., and Musicant, D.R. (1999). Successive overrelaxation for support vector machines. *IEEE Transactions on Neural Networks* 10(5): 1032–1037.

```
def svmTrain(knowns, data, kernelFunc, kernelParams,
             limit=1.0, maxSteps=500, relax=1.3):
```

Next we do some initialisation to set up various numbers and arrays in the calculation. Firstly, we define the size of the problem from the dimensions of the input data array. The dimensionality of the feature vectors that comprise the training data is n and the number of training data points is m. Next we initialise the supports array, which is the thing we are trying to optimise by the successive over-relaxation technique that the training is based on. The supports array will define which of the training data are support vectors, and hence where the classification hyperplane lies. Initially the support values are set as zeros using the handy zeros() function from NumPy, noting that we pass in float (an inbuilt Python data type) as an argument to state that the array is composed of floating point numbers, rather than integers. The change value will be used to record how much change is made to the support array during the optimisation, testing for convergence. Its starting value is 1.0, which is arbitrary but large enough to cause the first optimisation cycle to proceed.

```
m, n = data.shape
supports = zeros(m, float)
change = 1.0 # arbitrary but big start
```

The next initialisation is to pre-calculate the kernelArray. This square array represents the coincidence (similarity) between all of the m pairs of feature vectors that make up the input training data. Then the kernel function is used for all pairs of data vector to calculate the coincidence values, which are stored in the array using indices. Once the kernel array is filled we can quickly look up values using appropriate indices, rather than having to calculate a value each time. It should be noted that in order to keep this example relatively simple, and improve calculation speed, we calculate all of the kernel values at the start, but this requires memory to store the information. If the training data set were very large, compared to the amount of available memory, then more on-the-fly calculations and a smaller memory cache can be used. Because of the successive over-relaxation method used to perform the optimisation, the kernel array has all of its elements incremented by 1.0 (the diagonal will carry 1.0 values as that was not set from the kernel function).

```
kernelArray = zeros((m,m), float)
for i in range(m):        # xrange() in Python 2
  for j in range(i+1):    # xrange() in Python 2
     coincidence = kernelFunc(data[i], data[j], *kernelParams)
     kernelArray[i,j] = kernelArray[j,i] = coincidence

kernelArray += 1
```

The main part of the function performs the actual optimising loops that employ the method of successive over-relaxation. We use a while loop to continue the optimisation of the supports array, until it converges (the change is small) or we hit the maxSteps limit. The first line inside the loop makes a copy of the supports array so that we can work out how much change has occurred during the cycle.

```
steps = 0
while (change > 1e-4) and (steps < maxSteps):
  prevSupports = supports.copy()
```

Next we build `sortSup`, an array of values and index numbers from the `supports`. Naturally, for the first optimisation cycle all the values will be zero, but that will quickly change for subsequent iterations. We get these indices by using the inbuilt `enumerate()` function inside a list comprehension, where we collect the index `i`. The `sortSup` array is then sorted in order of decreasing value, which will be according to the first element in each `(val,i)` tuple; this follows the published successive over-relaxation procedure, but other implementations[18] favour using a random order, the code for which is shown in the comment in the code example below.

```
sortSup = [(val,i) for i, val in enumerate(supports)]
sortSup.sort(reverse=True)

#random.shuffle(sortSup) - also possible
```

Eventually we get to the actual successive over-relaxation optimisation. A precise explanation of what is being done will not be given here; this is only a programming book with limited space after all. However, the keen more mathematical reader can investigate the SVM section of the book reference given in footnote 18. What we will aim to do is merely give a hint at why things are done. The next loop is the first of two optimisation stages. At this point we consider all of the data vectors, or more precisely all of the indices in `range(m)`. For each index, representing one of the training data points, a summation named `pull` is calculated by multiplying the `supports` values by the required row of the `kernelArray` matrix and the values for the known classification. Here the kernel array will specify how similar the data vector at index `i` is to all of the other vectors. The `knowns` array contains values of $+1.0$ or -1.0, thus each individual multiplication that goes into the final summation will pull the value to the positive or negative side, depending on the classification of the training data vector. Note that we are making use of NumPy arrays here and that when we do multiplication it is being performed in an element-by-element manner within the arrays.

```
for support, i in sortSup:
    pull = sum( supports * kernelArray[i,:] * knowns )
```

The `pull` value, towards the negative or positive, is used to calculate an adjustment for the support at this position. The product `knowns[i] * pull` will be positive if both the pull and the known classification have the same sign, i.e. go in the same direction. Taking 1.0 from this means that when the values have the same sign the adjustment will be closer to zero compared to when they have a different sign. Dissimilar signs mean that the current `supports` values do not separate the training data vectors in the right way. Note that negative values of `adjust` will actually increase the `supports[i]` value, so that mismatching training vectors get more influence on the placement of the classification boundary. The actual adjustment to the `supports` values is scaled relative to the maximum amount of support for that data point (i.e. the kernel array value at the diagonal; `kernelArray[i,i]`) and multiplied by the `relax` parameter, which governs the rate of change and thus convergence. Note that this adjustment can overshoot the optimum, so a bigger `relax` is not always better. Finally, the adjusted value for the support at position `i` is bounded by the upper and lower limits; it must be at least zero and at most `limit`.

---

[18] Press, W.H., Teukolsky, S.A., Vetterling, W.T., and Flannery, B.P. (2007). *Numerical Recipes. The Art of Scientific Computing* (3rd edn.). Cambridge: Cambridge University Press.

```
adjust = knowns[i] * pull - 1.0
supports[i] -= adjust * relax / kernelArray[i,i]
supports[i] = max(0.0, min(limit, supports[i]))
```

At this point we could go through another round optimising all the support values. However, we will get quicker convergence if we do some more optimising in this cycle for only a subset of the data points: those with a support value that is above zero. Hence, we calculate the nonZeroSup subset of points which have at least some support (val > 0) in a similar manner to what was done for sortSup, by using enumerate to get the value and the index, so that later we can sort according to the value. Note that if there is none of this type of support we skip to the next cycle with continue. It is reasonable to suggest that these points on the edges of the separation margin need further attention and more optimisation, and in practice this really does help.

```
nonZeroSup = [(val,i) for i, val in enumerate(supports) if val > 0]

if not nonZeroSup:
  continue

nonZeroSup.sort()
```

We collect the indices inds for the non-zero support positions, because this is handy for the next stage, and then estimate how many extra optimisation cycles (niter) to do on the supporting values based on the number that remain. This is not a precise calculation but the general principle is that the more marginal support vectors there are the more cycles are required. If there are no non-zero supports we use a minimum value, but otherwise make a guess based on the square root of the number of non-zero supports.

```
inds = [x[1] for x in nonZeroSup]
niter = 1 + int(sqrt(len(inds)))
```

Next we repeat exactly the same kind of support value optimisation as before, but this time only using the indices of the non-zero, supporting points. These extra sub-cycles will often be significantly quicker than the previous optimisation that goes through all the points, i.e. the number of supports is relatively few. Note how we use slice indexing of the form 'knowns[inds]' to perform operations on only the required subset of the NumPy arrays.

```
for i in range(niter):  # xrange in Python 2
  for j in inds:
    pull = sum(kernelArray[j,inds] * knowns[inds] * supports[inds])
    adjust = knowns[j] * pull - 1.0
    supports[j] -= adjust * relax / kernelArray[j,j]
    supports[j] = max(0.0, min(limit, supports[j]))
```

Finally, after all the optimisation is done for this cycle we see how much change we have conferred to the support value and increase the step count. The change is simply the square root of the summation of the difference between old and new values squared.

```
diff = supports - prevSupports
change = sqrt( sum(diff * diff) )
steps += 1
```

At the end of the function we pass back the optimised array of support values, the number of steps that were taken during the optimisation and the array containing the kernel values (the measures of coincidence between the data vectors), to save calculating it again when investigating the decision hyperplane.

```
return supports, steps, kernelArray
```

Support vector machine predictions

To use the SVM to make a prediction involves working out which side of the decision hyperplane, determined during training, a query feature vector lies. Naturally the prediction function takes a query vector as input, together with the training data and its known categories. We also pass in the function and parameters that allow the calculation of the coincidence of feature vectors using a kernel.

```
def svmPredict(query, data, knowns, supports, kernelFunc, kernelParams):

  prediction = 0.0
  for j, vector in enumerate(data):
    support = supports[j]

    if support > 0:
      coincidence = kernelFunc(vector, query, *kernelParams) + 1.0
      prediction += coincidence * support * knowns[j]

  return prediction
```

The SVM prediction is made by going through all of the training data points and finding those that are support vectors (support > 0). When a support vector is found its coincidence with (similarity to) the query is found using the kernel function. The degree of coincidence is multiplied by the amount of support for that training vector and the known classification. Given that the known classification of the data vector is +1.0 or −1.0 this will either add or subtract from the prediction total; effectively each support vector pulls the summation to the positive or the negative size. In the end whether the predictSum value is finally positive or negative determines the category of the query.

This next function, svmSeparation(), is used to test whether the training data was well separated into two categories, i.e. reproducing the known classification. We don't use the above prediction function because we can reuse the pre-calculated kernelArray for speed. As before, the known classification is in the form of an array containing values of +1.0 or −1.0.

```
def svmSeparation(knowns, supports, kernelArray):

  score = 0.0
  nz = [i for i, val in enumerate(supports) if val > 0]

  for i, known in enumerate(knowns):
    prediction = sum(supports[nz] * knowns[nz] * kernelArray[nz, i] )

    if known * prediction > 0.0: # same sign
      score += 1.0

  return 100.0 * score / len(knowns)
```

Making the prediction is done using the same logic as described for svmPredict(), although here we do it in one line using NumPy array operations, given that we don't have to call the kernel function and can use the pre-calculated array instead. It is also notable that we calculate nz, a list of the indices for the non-zero support values, upfront to help reduce the number of calculations. With each prediction value, to actually test whether the classification is correct we see if the prediction is the same sign as the known classification. At the end the function gives back a percentage of correct classifications for the training data.

To test out the support vector machine code we will make a fairly simple example that contains a discontinuous patchwork of points in a two-dimensional plane that have been placed into one of two categories, each in distinct regions. The following code goes through a grid of x and y positions, which are normalised to be between 0.0 and 1.0, to make an alternating chequerboard pattern for the categorisation (−1 or +1), except for the middle square, which is flipped the other way, resulting in a central cross. This will give a recognisable shape in the data that we can look for afterwards.

At each grid location the random.normal function from NumPy is used to make a cluster of points by specifying a set of values for the x and y axes. The category and the x and y value for each point are placed in the main catData list. This list is then shuffled to introduce a random order. The list of known categorisations is extracted as the last index (-1) for all catData items and the training feature vectors as everything up to the last index ([:,:-1]).

```
numPoints = 20
catData = []

for x in range(1,6):
  for y in range(1,6):
    xNorm = x/6.0        # Normalise range [0,1]
    yNorm = y/6.0

    if (x == 3) and (y == 3):
      category = -1.0

    elif (x%2) == (y%2):
      category = 1.0

    else:
      category = -1.0

    xvals = random.normal(xNorm, 0.2, numPoints)
    yvals = random.normal(yNorm, 0.2, numPoints)

    for i in range(numPoints):  # xrange in Python 2
      catData.append( (xvals[i], yvals[i], category) )

catData = array(catData)
random.shuffle(catData)

knowns = catData [:,-1]
data = catData [:,:-1]
```

Running the SVM on this data involves passing in the known classifications, training data, a Gaussian kernel function and the parameters for the kernel. After training the svmSeparation() function can be used to assess how well the SVM separates the known categories.

Training data and support vectors   Category map

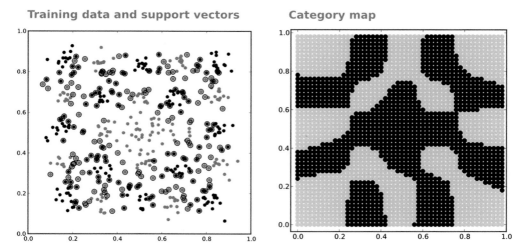

Figure 24.7. Example output for an SVM classifier with two data dimensions. The input SVM training data is shown with black and grey circles representing the two data categories (left). The support vectors are indicated as the circled points. The prediction result is displayed over the whole data range, to illustrate the classification boundaries (right). The category for each point is illustrated by being a black or grey shade.

```
params = (0.1,)
supports, steps, kernelArray = svmTrain(knowns, data, kernelGauss, params)

score = svmSeparation(knowns, supports, kernelArray)
print('Known data: %5.2f%% correct' % ( score ))
```

The following Python code tests the trained SVM on a whole range of different points in the plane between 0 and 1, thus showing the shape of the categorisation boundaries. The distinction between the categories is made according to whether the prediction is greater than zero (positive) or not. Note that the x and y values for a given query are appended to lists so that the data can be displayed as graphical scatter plots, using the helpful pyplot library (from MatplotLib).

```
from matplotlib import pyplot

ds1x = []
ds1y = []
ds2x = []
ds2y = []

x = 0.0
while x < 1.0:

  y = 0.0
  while y < 1.0:
    query = array( (x,y) )

    prediction = svmPredict(query, data, knowns, supports,
                            kernelGauss, params)
```

```
    if prediction > 0:
      ds1x.append(x)
      ds1y.append(y)
    else:
      ds2x.append(x)
      ds2y.append(y)

    y += 0.02
  x += 0.02

pyplot.scatter( ds1x, ds1y, color='grey' )
pyplot.scatter( ds2x, ds2y, color='black' )
pyplot.show()
```

25 Hard problems

Contents

Solving hard problems

This chapter deals with problems that cannot be readily solved with a straightforward, deterministic algorithm. This includes problems that computer scientists would describe as NP and not P (non-deterministic in polynomial time, but not solvable in polynomial time), which is a way of saying that a problem is not efficiently solvable. Whether a problem is straightforward to solve will depend on the complexity of the system. To take a classic example, solving the gravitational equations for two orbiting masses, like the Sun and Earth, is fairly easy, but adding more masses, e.g. the Moon, Mars etc., makes the problem much harder. The basic equations of the system do not have to be complicated though. Another famous (NP-hard) problem is the *travelling salesman problem*. Here the objective is to find the shortest route on a tour that goes through all the places on the salesman's list. The problem is easy to describe, and it is easy to calculate the length of a solution (a route), but the number of combinations grows very quickly with the number of places to visit and so finding the best solution can be difficult. This is somewhat different to a classic optimisation problem, e.g. finding the minimum of a function, where you can typically follow gradients to home in on the answer.

When it comes to biological information there are many situations of this kind, because biology frequently deals with large and interacting systems. For example, determining the structure of a protein generally involves several thousands of atoms and in general we can only 'solve' the structure with good experimental data (e.g. from high-resolution X-ray crystallography); it is not sufficient to start with unstructured atoms and a physical model. However, for a complex problem like this, and in a similar vein to measuring a travelling salesman's route, testing a given solution to see if it is better or worse can be proportionately straightforward. Referring again to protein structures, there are many methods that can quickly calculate the likelihood (or energy) of a structural model. An obvious approach to working out how a protein folds into a structure would be to approach the situation in reverse: generate an array of possible solutions and then use the easier testing methods to see if any of these look reasonable. Unfortunately, the number of possible combinations is enormous, and there

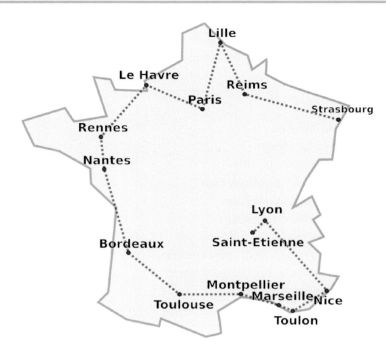

Figure 25.1. The travelling salesman problem as an example of a hard problem. Using the map coordinates of cities in France as an example, the route that minimises the total journey distance between all cities is an example of a hard problem that cannot be solved by classic minimisation techniques. Here the number of possible routes, with no fixed starting or end points, is 15! = 1,307,674,368,000, or half that number if we consider reverse routes to be the same. Because it is often impractical to test all possible routes we can use methods like Monte Carlo, or Monte Carlo in combination with simulated annealing, to find a good solution in a reasonable time. The illustrated route is the optimum solution found by the Python functions described later in this chapter.

simply wouldn't be time to exhaustively search through all possibilities. We can be smarter than this though, for example, taking a smaller number of random solutions and only testing those, and then basing a subsequent round of guesses on the best solutions from the previous round. This leads us into the realm of Monte Carlo and Markov chains, which we describe below. Also, it is often helpful to use *heuristics*, which for big problems may provide smart guesses about which kinds of solution should never be tested, thus saving time. A heuristic for solving a protein structure might be that the dihedral angles along its backbone have unstrained conformations.

Problems of this kind are a vast topic that we can only touch upon lightly. However, using Python examples we will illustrate a few core techniques, which are commonly used to solve complex problems in general. Accordingly we will describe the Monte Carlo method, simulated annealing, particle dynamics and Markov chains. These will then be applied to situations that a biologist will be passingly familiar with. Unfortunately there will not be room to include descriptions of other methods, such as genetic algorithms or bounded tree traversal. It is worth noting, however, that the methods described are complementary to the machine learning approaches described in Chapter 24, which are good at recognising patterns and approximating functions in complex systems: useful for getting good heuristics.

The Monte Carlo method

The *Monte Carlo method*[1] means to randomly test data points to approximate or guess a solution, thus getting a good idea of what is going on by sampling only a few points, rather than many. This approach is often coupled with targeting the selection of data points towards the most promising or important results. This leads directly into the Markov chain idea, where we can base a new selection on the last one, and a famous implementation of this is the Metropolis-Hastings algorithm,[2,3] for making the next guess, for an effectively random walk with probabilistic selection that echoes thermodynamic energy.

Key to the implementation of Monte Carlo and related methods is the ability to generate random, or almost random, numbers. Fortunately in Python we can exploit the `random` module and its equivalent in NumPy. Usually we do not require absolutely random numbers so the usual *pseudorandom* generators will be sufficient. The important thing is that the distribution of the emitted numbers is approximately correct (typically uniform; level throughout a range) and that there is no significant selection bias. Having truly random numbers that are entirely unpredictable doesn't matter. Indeed, being able to restart a Monte Carlo run with the same set of pseudorandom numbers can be useful for testing.

Simple Monte Carlo integration

The first and simplest Python example that uses a Monte Carlo approach is one that aims to perform integration, i.e. to determine the area bounded by some condition. Here we will use a circle, and the result will not depend upon knowing what the value of the constant π is. Indeed, this example actually provides a means of estimating a value for π, based upon the measurement for the area of a circle. As illustrated in Figure 25.2, the method works by taking a square area where x and y axis values range between -1 and 1. Random data points (i.e. `(x,y)` values) are added to this square and then the number of points that fall inside a circle are compared to the total number of points. Here it is easy to tell which points are inside the circle because they will be within a fixed distance from the centre. For this example this distance (the radius of the circle) is 1.0 and the centre is at the origin of the coordinates. After a large number of points have been added, the number of points that have randomly fallen inside the circle will be proportionate to the area of the circle. This then leads to an estimation of the area compared to the total square, and thence to an estimated value for π.

Firstly, we import the `random` module from NumPy and make a few definitions. The `uniform` function is defined upfront so that we are not repeatedly using the dot notation inside loops (which is slower). This will generate pseudorandom numbers with an even distribution in our desired range. The number of random points sampled for the example is defined as

---

[1] The name Monte Carlo is from the Monte Carlo Casino in Monaco, and the name was given by the physicist John von Neuman for the idea that Stanisław Ulam had for a way to estimate neutron collisions. Both were working at Los Alamos at the time, developing the atomic bomb. Monte Carlo was a secret code phrase and had notions of probability and chance, albeit related to gambling.

[2] Metropolis, N., Rosenbluth, A.W., Rosenbluth, M.N., Teller, A.H., and Teller, E. (1953). Equation of state calculations by fast computing machines. *Journal of Chemical Physics* 21(6): 1087.

[3] Hastings, W.K. (1970). Monte Carlo sampling methods using Markov chains and their applications. *Biometrika* 57(1): 97–109.

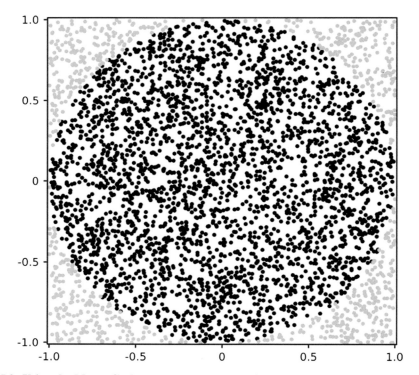

Figure 25.2. Using the Monte Carlo method to estimate π from the area of a circle. By uniformly selecting random points in a square interval we can estimate the area of a circle from the proportion of points that fall within a given distance of the centre. This in turn can be used to give an estimate for the mathematical constant π: for a circle of radius 1, and hence area π, the area of the bounding square is 4. Thus the ratio of circle area points to total points approaches $\pi/4$ as the sample size increases. This is a simple example of the Monte Carlo approach, which uses random samples of test points to solve a problem.

100,000, but it is interesting to play with this to see the effect on the accuracy of estimating π. The number of points inside the circle `numInside` is naturally defined as zero at the start.

```
from numpy import random

uniform = random.uniform
numSamples = 100000
numInside = 0
```

Then comes a loop with the required number of cycles where we define the random points. Here we use the `range()` function, which works in both Python 2 and Python 3, although for the latter, rather than generating a whole list of values, it creates an iterable object that gives the sequence of numbers on demand (you can use `xrange()` in Python 2 for this behaviour). Note that the first two arguments to the function `uniform()` define the limit to the random values (i.e. between plus and minus one) and the last argument indicates how many values to yield; here we need two, for x and y.

```
for i in range(numSamples):      # could use xrange in Python 2
  x, y = uniform(-1.0, 1.0, 2)
```

With the coordinates of the random point defined we now test whether the point is inside the circle; if the sum of coordinates squared is less than one (squared) the point is sufficiently close to the centre. If the coordinates represent a point for which this holds then the count for internal points is increased by one.

```
if (x * x) + (y * y) < 1.0:
    numInside += 1
```

At the end the estimate for π is simply four times the proportion of internal points, relative to the total number of points. This follows from the area of the region being 4.0, i.e. a circle of radius 1.0 is bounded by a square with sides of length 2.0.

```
pi = 4.0 * numInside / float(numSamples)
print(pi)
```

This is not an especially efficient means of estimating π[4], but the same idea can be applied to any shape or function that can be defined, however many dimensions it might have.

Function minimisation by Monte Carlo

The next example will be to minimise a fairly simple two-dimensional function: $f(x,y) = (1-x)^2 + 100(y-x^2)^2$. This is known as the Rosenbrock test[5] and is sometimes used to test optimisation performance. The function has a crescent-shaped valley. The minimum at $(1,1)$ is in a flat region of the valley and is fairly hard to pin down exactly.

In Python the test function is defined below. This will be used to estimate the value of the function (which can be imagined as the 'height') for each of the test points. Simply the coordinates are accepted as arguments and the value at that point is passed back.

```
def testFunc(point):
    x, y = point
    a = 1.0 - x
    b = y - (x * x)
    return (a * a) + (100 * b * b)
```

We will start with a simplistic approach to minimising this function, which will then be built upon to give something resembling a more normal Monte Carlo search. For the most basic approach we will define a range of values for the coordinates (here between -5 and 5) and choose random points from within that. Naturally when trying to find the answer to an optimisation problem it helps to know something about the range of sensible answers. It is important that the range of tested points spans the optimum answer but smaller ranges result in quicker searches. For combinatorial problems (such as travelling salesman) the question is generally well bounded but for an arbitrary function there may be no known limits, so it is up to the programmer to choose sensible bounds for variables and perhaps perform preliminary tests to establish these.

---

[4] It can be shown that the standard deviation of this estimate for π is $\sqrt{\frac{\pi}{n}(1-\pi)}$, where n is the number of samples, so with 100000 samples this gives 0.013.

[5] http://en.wikipedia.org/wiki/Rosenbrock_function. Rosenbrock, H.H. (1960). An automatic method for finding the greatest or least value of a function. *Computer Journal* 3: 175–184.

The search begins by initialising the best point (the one with the smallest value) as a random coordinate. Here we do this with `uniform` and record the initial `bestValue` as the value of the function at this point.

```
bestPoint = uniform(-5, 5, 2)
bestValue = testFunc(bestPoint)
```

Next there's a loop, for a given number of steps, which defines a random point between the pre-set limits. The value of the function at that point is tested. If the value of the latest point is smaller than the current best value then the best point and corresponding best value are redefined. In this example we print a line to the screen when a better point is found so that the progress can be visualised.

```
numSteps = 100000
for i in range(numSteps):
  point = uniform(-5, 5, 2)
  value = testFunc(point)

  if value < bestValue:
    bestPoint = point
    bestValue = value

    x, y = point
    print('%5d x:%.3f y:%.3f value:%.3f' % (i, x, y, value))
```

The next example illustrates an improvement on the above loop. Rather than choosing a completely random point to test next the subsequent points are based on the best point so far; this is done using the `normal()` function, which selects the next point in a random Gaussian spread. The variable `bestPoint` is passed to this function as the centre of the distribution, i.e. the test point is based on the previous best. The values `0.5` and `2` represent the spread of the distribution and the number of values (dimensions) respectively. The rest of the loop remains the same, testing whether the value is an improvement. Because the test points now follow the good solutions, this example tends to reach the optimum point more efficiently than the previous example.

```
normal = random.normal
mumSteps = 100000
for i in range(numSteps): # could use xrange in Python 2
  point = normal(bestPoint, 0.5, 2)
  value = testFunc(point)

  if value < bestValue:
    bestPoint = point
    bestValue = value

    x, y = point
    print('%5d x:%.3f y:%.3f value:%e' % (i, x, y, value))
```

Metropolis-Hastings Monte Carlo

After the gentle introduction, the next example illustrates a further improvement to the Monte Carlo optimisation approach, giving what is termed the Metropolis-Hastings approach. The Metropolis-Hastings algorithm is formally described in terms of a probability distribution, so it fits well with optimisation problems that have a statistical nature; it takes samples from a probability distribution. The example problem used here is not actually probabilistic, but we can pretend that it is and the method will work well nonetheless.

The algorithm works by estimating what is effectively a probability ratio, comparing the value at a previous good point to a new test point. If the test point is better it is always accepted, but if the test point is worse it may still be accepted, in a random way, depending on how much worse it is. This is the key difference between the Metropolis-Hastings algorithm and the previous one, which only accepted new points (solutions to the problem) if they are better. By sometimes accepting worse points the algorithm has an opportunity to jump out of what may be only a local optimum, to search for a better global optimum. Nonetheless, the algorithm will still effectively home in on good solutions. This approach is a way of taking samples from a probability distribution (or other function) and is sometimes solely used for that purpose; finding the absolute optimum is not always the objective. However, here we define a Monte Carlo function that does record the optimum point found so far. Hence, in the example below we record `bestPoint`, the globally best point so far, as well as `prevPoint`, which is the previous 'good' point that we are currently searching from, but which may not be the absolute best.

The next example is more general than the previous one, and is defined as a Python function rather than a bare loop. Thus we could use it for multiple different problems by defining the test function and dimensionality appropriately. Compared to the full Metropolis-Hastings algorithm the function has a simplification, given that we will assume that the probability of a jump from the previous point to the test point is the same as the reverse jump; the 'proposal density' is symmetric. For some problems this need not be the case, but for this example it keeps things simple and means that the acceptance probability (`prob`) is based only on the values of the function, and not on the way in which search points are generated with a normal distribution.

The appropriate import of `exp` to calculate exponents is made and the function `monteCarlo` is defined, taking arguments representing the number of search steps, the function to test, the spread of the normal distribution, which generates the next test point, and the number of dimensions of the problem (defaults to 2, i.e. for x and y axes).

```
from math import exp

def monteCarlo(numSteps, testFunc, spread=0.1, nDims=2):
```

The initial coordinates for the best point so far and for the previous test point (which may not be the best) are defined, here as the same random vector, with coordinates in the range -1.0 to 1.0 (although other values could be used). The corresponding value of the function being optimised is calculated for these points. Initially the best and previous points and values are the same, but this will change after the main search starts.

```
    bestPoint = uniform(-1.0, 1.0, nDims)
    prevPoint = bestPoint

    bestValue = testFunc(bestPoint)
    prevValue = bestValue
```

The main `for` loop is constructed for the specified number of steps. As before, the test point is based on a previous one, using the `normal()` function and the input amount of spread. The value is then simply the value of the input function at this test point.

```
for i in range(numSteps):
  testPoint = normal(prevPoint, spread, nDims)
  value = testFunc(testPoint)
```

The acceptance probability is the exponent of the difference between the test value and the previous value; this is equivalent to the ratio between two probabilities ($e^{(a-b)} = e^a/e^b$), so we are effectively saying that the exponent of the value given by the test function is proportional to the probability. Naturally, if proper probabilities are available for a particular problem these should be used instead.

```
prob = exp(prevValue-value)
```

Next we test whether the 'probability' score is greater than a random number (in the range 0.0 to 1.0). If the test point gives a value that is better than the previous point the value of `prob` will be greater than 1.0, so this test will definitely be passed. If the `prob` is less than 1.0 (the test point is not as good as the previous one) the value may be accepted, although with less likelihood the further `prob` is from 1.0. Effectively this can be viewed as `uniform()` generating a threshold, and if `prob` exceeds this the test point is accepted and the previous point for the next cycle is redefined.

```
if prob > uniform():
  prevPoint = testPoint
  prevValue = value
```

A second check, subject to passing the first, determines whether the tested value is the best overall so far inspected, and, if it is, this value and the corresponding point are recorded. As before, we print out the best points to give an indication of optimisation progress. Although, because the number of dimensions of the points in any given situation can vary, a Python list comprehension converts the numbers to strings and is then joined with `', '.join()` to make the coordinates into a Python string. At the end of the function the best point and its corresponding value are passed back. Alternatively, all of the accepted points (`prevPoint`) could be recorded to give an indication of the 'trajectory' that the Monte Carlo search took; this may be plotted using `matplotlib`.

```
    if value < bestValue:
      bestPoint = testPoint
      bestValue = value

      coordinates = ', '.join(['%.3f' % v for v in testPoint])
      print('%5d [%s] value:%e' % (i, coordinates, value))

  return bestValue, bestPoint
```

The function is simply tested on the Rosenbrock function using a large number of points. It is worth experimenting to see how repeat runs locate the minimum (1.0,1.0); some search paths locate the global minimum more quickly than others:

```
monteCarlo(100000, testFunc, 2)
```

The travelling salesman problem

The Python examples now move on from Monte Carlo optimisation of a mathematical function to combinatorics: the travelling salesman problem. Specifically, we will show an example that can determine the best route between 15 of the largest cities in France (see

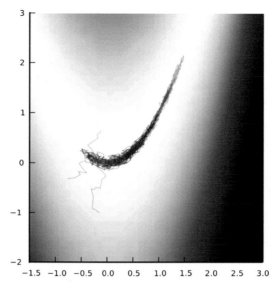

Figure 25.3. Trajectories of subsequent trials from Metropolis-Hastings Monte Carlo searches to minimise a shallow function. In this plot the lighter the background shade the smaller is the value of a two-dimensional test function. The trajectories of test points that result from applying the Metropolis-Hastings Monte Carlo method are visualised as darker lines. This method works by selecting subsequent test points (after a random start) relative to the position of the previous point, but only accepting the new solution if it is better and occasionally if it is worse, with a probability specified by the Metropolis-Hastings criterion. In this way the routes of accepted points will follow the function to its minimum but have occasional jumps, which may be important in overcoming local minima.

Figure 25.1). There are a total of 1,307,674,368,000 possible ways of ordering the cities into a route, although our solution will search only a fraction of these to come up with the optimum answer. We don't mean to suggest that Monte Carlo is the best way to solve this particular problem, but it gives a clear example to work with. Likewise, we could also choose to solve a different problem, but hopefully by keeping things simple you will see how analogous situations that involve the ordering of data items can be solved.

The travelling salesman problem solely involves considering the distances between cities, but in other, more biological, situations there may be more information to reduce the complexity of the problem. For example, in nuclear magnetic resonance (NMR) it is a common operation to assign signals that occur in a spectrum to the amino acid residues of a protein chain. The problem is that you don't know which signal corresponds to which amino acid in the sequence. Specific NMR experiments allow you to determine which signals are likely to be neighbours in a sequence (although not unambiguously so) and also what kinds of amino acid a signal could represent. Thus, although you can solve the problem is the same basic way as the travelling salesman, to determine the best data sequence you have considerably more information and can exclude many possibilities; many data points are known to not be neighbours, and certain data points can be excluded from particular locations in the sequence. Generalising for other biological problems, the more information you have, to give better probabilities, the more robust the solution.

To solve the travelling salesman problem with a simple Python function we will first make a few handy imports and then define a helper function, which has the job of determining the total distance for a particular route between all the cities.

```
from math import sqrt, exp
from random import shuffle, randint
from numpy import array
```

The function getRouteLength() takes a dictionary containing the distances between points (cities in our example) and the route to calculate the distance for. It should be noted that if the distances between all of the relevant points can be determined in advance then this will generally lead to a quicker program, given that we won't have to repeat the distance calculation. However, caching the distances in this way requires memory, which may be insufficient for large problems. The innards of the function are fairly simple. The initial distance is initialised as zero, then we go through each pair of subsequent points visited in the route; the order of cities in the list defines the route. The inbuilt enumerate() is used to extract the index and each location of the route at the same time. Note that this loop only goes up to the penultimate point in the route (route[:-1]) so that we always have room to fetch the next point in a pair: route[i+1]. The two neighbouring points are then used as a key to look up the distance between them in the distanceData dictionary passed in, which is then added to the total. If it is not practical to have a look-up table of distances then this function could calculate the distances from the data points themselves. Construction of the dictionary of city distances is illustrated in the function calcCityDistances() discussed later.

```
def getRouteLength(distanceData, route):

  distance = 0.0

  for i, pointA in enumerate(route[:-1]):
    pointB = route[i+1]
    key = frozenset((pointA, pointB))
    distance += distanceData[key]

  return distance
```

Next we come to the function that does the actual solving of the travelling salesman problem using the Monte Carlo method. The general idea is that we will start with a random route (order of cities) and then swap random pairs of cities on the route to generate the next test route. A test route will be accepted or rejected in the same manner that we did for the previous function optimisation. Swapping pairs of cities is not the only way that we could generate new routes (states) to test: for example, random neighbours could be swapped. However, swapping completely random points works well for the test problem and generally allows a wide range of solutions to be sampled. Swapping only neighbours gives less drastic changes and may be better at refining good solutions, but will not be as quick to search a wide variety of routes. Subsequently this method that uses random swaps will be refined, using *simulated annealing*, so that we get an initially wide and unbiased search that subsequently narrows to a refined solution.

The function travellingSalesman() is defined as taking the distanceData dictionary that provides the distance look-up, a list of the cities involved and the number of search steps to use. Naturally, the number of steps to use will depend on the number of cities that are considered and acceptable values may be determined by experimentation.

```
def travellingSalesman(distanceData, cities, numSteps=10000):
```

The number of cities in the problem is recorded and the initial bestRoute list is a randomly shuffled copy of the input list of cities.

```
n = len(cities)
bestRoute = cities[:]
shuffle(bestRoute)
```

The acceptance scores we use below will depend on the differences in route lengths, but the magnitude of these distances will vary according to the situation at hand (in our example we happen to be dealing with thousands of kilometres). Hence, so that it doesn't matter what the scale of the distances are, we calculate a normalising factor, which we call scale. The heuristic we use here is to take half of the standard deviation of the inter-city distances as the scale factor. This is easily calculated with the function std() inbuilt into NumPy arrays:

```
dists = list(distanceData.values()) # list() not needed in Python 2
scale = 0.5 * array(dists).std()
```

The corresponding initial bestDistance uses the helper function getRouteLength() and the dictionary of distance information to calculate the length of the initial route. Copies of the initial route and distance are then made, which will be the starting point for the Monte Carlo search.

```
bestDistance = getRouteLength(distanceData, bestRoute)
prevRoute = bestRoute
prevDistance = bestDistance
```

A for loop is defined to go through each of the search steps. Two indices a and b are defined as random numbers, between zero and the index of the last city. These will be the indices of the two cities that will be swapped in the route. The test route for this cycle is initially defined as a copy of the previous 'good' route. Then the entries at the two random indices are swapped; position a in the old route is b in the new one and vice versa.

```
for i in range(numSteps):
    a = randint(0, n-1)
    b = randint(0, n-1)

    route = prevRoute[:]
    route[a] = prevRoute[b]
    route[b] = prevRoute[a]
```

With the test route defined, the length going through the cities is calculated. The score (equivalent to a Metropolis-Hastings acceptance probability) is calculated as the exponent of the difference in distances between the test route and the prior one, divided by the normalising distance scale factor. The result of this is that if a test route length is longer by the value of scale there is an e^{-1} (37.8%) chance of acceptance.

```
distance = getRouteLength(distanceData, route)
score = exp((prevDistance-distance)/scale)
```

The score is compared against a random number in the range 0.0 to 1.0. If the new route is shorter the score will be greater than 1.0 and it will definitely be accepted, otherwise it is a matter of chance depending on how much longer the distance is. If the test route is accepted it is recorded as `prevRoute` so that it will be the state to go from in the next cycle.

```
if score > uniform():
    prevRoute = route
    prevDistance = distance
```

A second test determines whether the distance is the overall shortest so far measured. If so the route is remembered as the best one. The best distance and step number are printed to indicate progress.

```
if distance < bestDistance:
    bestRoute = route[:]
    bestDistance = distance
    print('%5d %.5f' % (i, distance))

return bestDistance, bestRoute
```

To finally test the function some distance information is required. Here this is done using a dictionary that contains the coordinates (latitude and longitude in degrees) of the cities concerned. Of course, this will only allow us to calculate approximate distance across the Earth's globe, rather than practical road or rail distances. Nonetheless, this is good enough to illustrate the procedure.

```
cityCoords = {'Paris':(48.856667, 2.350833),
              'Marseille':(43.296386, 5.369954),
              'Lyon':(45.759723, 4.842223),
              'Toulouse':(43.604503, 1.444026),
              'Nice':(43.703393, 7.266274),
              'Strasbourg':(48.584445, 7.748612),
              'Nantes':(47.21806, -1.55278),
              'Bordeaux':(44.838611, -0.578333),
              'Montpellier':(43.61194, 3.87722),
              'Rennes':(48.114722, -1.679444),
              'Lille':(50.637222, 3.063333),
              'Le Havre':(49.498889, 0.121111),
              'Reims':(49.26278, 4.03472),
              'Saint-Etienne':(45.434722, 4.390278),
              'Toulon':(43.125, 5.930556)}
```

The next function uses the city coordinates to calculate a dictionary containing the distances between each pair, hence we use two `for` loops to get all city pairs. The distance metric derives from the length of the great arc between the cities, given that we are dealing with positions on the surface of the Earth rather than a flat plane. Hence we first calculate the central angle between the cities (in radians) and then multiply this by the radius of the Earth. Note that the key to the `distances` dictionary is a Python `frozenset` object, thus the key to get a distance value does not depend on the order of the two cities in the key. If we used a Python tuple as a key then we would have to store both `(cityA, cityB)` and `(cityB, cityA)` or always sort the key items. Here a simple Python set cannot be used, it must be the frozen kind as dictionary keys cannot have modifiable values.

```
from math import acos, cos, sin, radians

def calcCityDistances(coordDict):

  cities = list(coordDict.keys())
  n = len(cities)
  distances = {}

  for i in range(n-1):
    cityA = cities[i]
    latA, longA = coordDict[cityA]
    latA = radians(latA)
    longA = radians(longA)

    for j in range(i+1, n):
      cityB = cities[j]
      latB, longB = coordDict[cityB]
      latB = radians(latB)
      longB = radians(longB)

      dLong = abs(longA - longB)
      angle = acos(sin(latA)*sin(latB) + cos(latA)*cos(latB)*cos(dLong))
      dist = angle * 6371.1 # Mean Earth radius (km)

      key = frozenset((cityA, cityB))
      distances[key] = dist

  return distances
```

Using the distance information, the best route between the cities can hopefully be found using the Monte Carlo method.

```
distances = calcCityDistances(cityCoords)
cities = list(cityCoords.keys())                # Use all the cities

dist, route = travellingSalesman(distances, cities, 1000000)
print('%.3f %s' % (dist, ', '.join(route)))
```

If all goes well, the test will give the following route, or its equally good reverse, with the distance of 2465.56 km:

```
Strasbourg, Reims, Lille, Paris, Le Havre, Rennes, Nantes, Bordeaux, Toulouse,
Montpellier, Marseille, Toulon, Nice, Lyon, Saint-Etienne
```

Simulated annealing

The next section of this chapter relates to an adjustment to the Monte Carlo method termed *simulated annealing*,[6,7] and this may be applied to both the combinatorial and function optimisations already described. Here the main principle is to have the same kind of random selection and acceptance rules as before, but the degree of acceptance of non-improving states

---

[6] Kirkpatrick, S., Gelatt Jr., C.D., and Vecchi, M.P. (1983). Optimization by simulated annealing. *Science* 220(4598): 671–680.
[7] Černý, V. (1985). Thermodynamical approach to the traveling salesman problem: an efficient simulation algorithm. *Journal of Optimization Theory and Applications* 45: 41–51.

diminishes. As the data sampling proceeds the acceptance criterion becomes stricter, and the range of accepted steps effectively becomes narrower. This enables an initially wide search that will hopefully sample enough to explore close to the globally best solution, but which will later settle on a refined optimum. Using simulated annealing will counter some of the later moves which would otherwise cause the state to jump out of a globally optimum solution. For some problems the simulated annealing approach need not be used, as it does not sample as widely as plain Monte Carlo, in the same number of steps, and has a greater tendency to get stuck in sub-optimal minima. However, when used with care it makes a better refinement protocol, and at the end of the search the last state will tend to be close to an optimum (local or otherwise), which can remove the need to record good solutions during the search. Often simulated annealing will be used in situations where it is not required to have the absolute best solution, just a reasonably good one in a short time.

The term simulated annealing reflects a similar physical process that occurs in materials science, when a solid is heated and slowly cooled to create a more ordered state, e.g. with bigger crystals. The term that we introduce into a Monte Carlo search to diminish the likelihood of accepting less optimal states is analogous to temperature in physical annealing. Here we introduce the variable `cool`, which represents the sampling 'temperature', or more generally the *annealing schedule*. The Python examples will only use a simple exponential decay to provide the value of `cool`, but in other situations it is commonplace to have more complex annealing schedules. For example, the solutions to some problems that have been caught in sub-optimal states may be rescued by restarting the annealing from a recorded earlier state and temperature, or by re-raising the temperature ('shake and bake'). Another approach that is worth mentioning, but which we do not have room to describe in detail, is *ensemble*-based methods, where multiple, separate copies of a Monte Carlo search are done in parallel. Each of the *replicas* in an ensemble of concurrent Monte Carlo searches will find different solutions and may operate under different conditions. A method called *parallel tempering* or *replica exchange Monte*

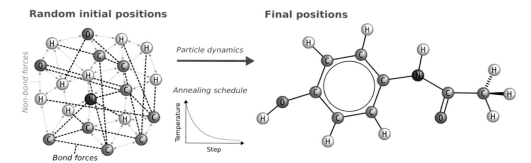

Figure 25.4 (Plate 11). An overview of the simulated annealing procedure, as applied to a molecular structure. The method of simulated annealing can be applied to the problem of computing a molecular structure given information relating to the bonding and non-bonding forces between atoms. Starting from random positions the forces that result from a particular arrangement of atoms can be estimated using knowledge of the forces' energy potentials. In turn this leads to a dynamics simulation where the forces, momenta and thus changes of atomic position can be modelled at discrete time steps. By applying an annealing schedule with a decreasing effective temperature, the simulated kinetic energy of the atoms will initially be large, so that large differences in conformation can be explored, but end up making only small adjustments, to better home in on a precise solution. A black and white version of this figure will appear in some formats. For the colour version, please refer to the plate section.

Carlo will perform the separate searches at different temperatures and occasionally swap solutions between the different temperatures, searching both widely and precisely while maintaining *detailed balance*. Another common ensemble approach is to use a *genetic algorithm*, whereby different replicas exchange part of their states: the genetics analogy is with the mutation and crossover of DNA strands when parents produce offspring.

Simulated annealing of the travelling salesman

The first simulated annealing example in Python is the same as the `travellingSalesman()` described above but with the introduction of the `cool` variable, which can act as a kind of temperature. We will use an exponential cooling decay based on the proportion of the current step through the whole run: the current step relative to the total (i/m). The only three differences in the Python function are the introduction of `m`, the total number of steps as a floating point number; `cool`, the current effective temperature in the annealing schedule; and a different `score`: this is now divided by `cool` before the exponent is taken. The effect of this is that as the search proceeds `cool` diminishes exponentially from 1.0 to almost 0.368 (e^{-1}), and the difference in distances, as they affect the acceptance criterion, is magnified. Thus the `score` will be closer to zero for the same distance difference in later cycles. Better distances will still always be accepted at any stage, but it will be increasingly less likely for other scores to be accepted. Hence, the state will tend not to jump out of a good solution towards the end, but initial exploration is unaffected.

```python
def travellingSalesmanSimAnneal(distanceData, cities, numIter=10000):

  n = len(cities)
  bestRoute = cities[:]
  shuffle(bestRoute)

  dists = list(distanceData.values())
  scale = 0.5 * array(dists).std()

  bestDistance = getRouteLength(distanceData, bestRoute)
  prevRoute = bestRoute
  prevDistance = bestDistance

  m = float(numIter)              # Use to calculate cool
  for i in range(numIter):        # could use xrange in Python 2
    cool = exp(-i/m)              # annealing schedule 'temperature'

    a = randint(0, n-1)
    b = randint(0, n-1)

    route = prevRoute[:]
    route[a] = prevRoute[b]
    route[b] = prevRoute[a]

    distance = getRouteLength(distanceData, route)
    score = exp( (prevDistance - distance) / (scale*cool) ) # Adjusted score

    if score > uniform():
      prevRoute = route
      prevDistance = distance
```

```
        if distance < bestDistance:
          bestRoute = route[:]
          bestDistance = distance

          print('%5d Dist:%.5f Temp:%.5f' % (i, distance, cool))

  return bestDistance, bestRoute
```

The function may be tested in exactly the same manner as the non-annealing version.

```
distances = calcCityDistances(cityCoords)
cities = list(cityCoords.keys())

dist, route = travellingSalesmanSimAnneal(distances, cities, 1000000)
print('%.3f %s' % (dist, '-'.join(route)))
```

With the appropriate number of search steps this will generally give the same result as the previous function `travellingSalesman()`, although the last tested route will usually be much closer to the best route than for the non-annealing version. Testing different numbers of cycles will show that if the number of search steps is too small the annealing will 'cool' before a reasonable number of states have been sampled and the final result will not be the best. Accordingly, it is critical to give the algorithm enough steps if the absolute optimum is required. However, an advantage of simulated annealing compared to plain Monte Carlo is its speed, so if all you want is a reasonably good solution in a reasonable time extra steps may not be needed.

Function minimisation by simulated annealing

In the next example we return to the function minimisation previously discussed. Here by using simulated annealing the final solutions will focus much more precisely on the optimum value compared to plain Monte Carlo, given that there will be more chance of refinement and less chance of sub-optimal jumps. The Python is very similar to the previous `monteCarlo` function. Again, there is only the introduction of a few extra values to calculate and the application of the cooling schedule: n is the number of iterations, cool is the effective temperature and prob is now different because the difference in values is scaled by cool prior to exponentiation.

```
def simAnneal(numIter, testFunc, spread=0.1, nDims=2):

  n = float(numIter)
  bestPoint = uniform(-1.0, 1.0, nDims)
  bestValue = testFunc(bestPoint)
  prevPoint = bestPoint
  prevValue = bestValue

  for i in range(numIter):
    cool = exp(-i/n)

    testPoint = normal(prevPoint, spread, nDims)

    value = testFunc(testPoint)

    prob = exp( (prevValue-value) / cool )   # Adjusted acceptance score.

    if prob > uniform():
```

```
        prevPoint = testPoint
        prevValue = value

      if value < bestValue:
        bestPoint = testPoint
        bestValue = value

        pointStr = ' '.join(['%.3f' % p for p in testPoint])
        print('%5d T:%.3f %s value:%e' % (i, cool, pointStr, value))

    return bestValue, bestPoint
```

We can test with both the plain Monte Carlo and the simulated annealing optimisers to compare and contrast their performance:

```
numSteps = 100000

simAnneal(numSteps, testFunc)
monteCarlo(numSteps, testFunc)
```

Particle dynamics simulation

We now move on to dynamics simulations that are somewhat analogous to simulated annealing. The method shares the concept of going to a new solution to the problem based on a prior one and also a notion of cooling. However, dynamics simulations move away from using purely random numbers to determine the search trajectory. Rather, we create what is called a *force field* that will push and pull the items into more optimal locations. Naturally, this sort of approach only works if you have some idea of the kind of influences on the items of data, e.g. a physical model. The advantages of having a simulated computer model of a physical system are numerous, but as far as optimisation problems are concerned they can be efficient search methods that are more directed than Monte Carlo. Also, we may benefit from knowing that the real physical system actually finds a solution (e.g. a protein folds) so in theory a model stands at least some chance of also finding a solution. Here we will use a very simple model system, of repulsive atoms bonded in a chemical structure, to determine an extended conformation for a molecule. The molecule will not be large and the force field will be very simplistic (definitely not the proper physics equations), but it will give you the general idea and will be good enough to give a reasonable picture of the molecule.

The force field used in the Python example will consist of a simple bond model and a simple repulsive force between atoms. The bonds will pull atoms to a single set distance (in reality bond lengths vary and we have double bonds and bond orbitals etc.), and will attract or repel accordingly when atoms are too close or too distant. The repulsive force will be calculated between non-bonded atoms and will diminish with an inverse power relation to the distance between atoms; close atoms will repel strongly but distant atoms will have little effect. The test molecule itself is the common pharmaceutical paracetamol (also known as acetaminophen). This is complex enough to present a challenge, but not too unwieldy for this book. A Python dictionary is used to contain the chemical bond information, i.e. which atoms are bound to which. The bond dictionary for paracetamol is given below. Here we have used a system where each atom is identified by a unique name. There is a key in the dictionary for each atom and the values for that atom are the names of the other atoms with which it shares chemical bonds. An alternative here would be to use just index numbers, one for each atom, so that we didn't have to decide upon the names of atoms (which follow no special rule). However, with names it is perhaps easier to visualise the setup.

```
chemBonds = {'H1': ['O1',], 'O1': ['H1', 'C1'],  'C1': ['O1', 'C2', 'C6'],
             'C2': ['C1', 'H2', 'C3'], 'H2': ['C2'],  'C3': ['C2', 'H3', 'C4'],
             'H3': ['C3'], 'C4': ['C3', 'N7', 'C5'], 'C5': ['C4', 'H5', 'C6'],
             'H5': ['C5'], 'C6': ['C5', 'H6', 'C1'], 'H6': ['C6'],
             'N7': ['C4', 'H7', 'C8'], 'H7': ['N7'], 'C8': ['N7', 'O8', 'C9'],
             'O8': ['C8'], 'C9': ['C8', 'H9a', 'H9b', 'H9c'], 'H9a': ['C9'],
             'H9b': ['C9'], 'H9c': ['C9']}
```

The main dynamics function is written to take a chemical bond dictionary, the number of dynamics steps and the length of chemical bonds to aim for. It should be noted that the scale of any measurements is not explicitly stated, but for a good physical model the bond lengths would be of the order of 10^{-10} metres. Firstly, we make some extra NumPy imports that are required and then define the function name.

```
from numpy import zeros, sqrt

def chemParticleDynamics(bondDict, numSteps=5000, bondLen=1.0, timeStep=0.01):
```

The list of atoms is determined as the keys of the dictionary that contains the bond information. The number of atoms is the length of this list and the atomic coordinates are initialised with random values, using a uniform distribution over a reasonable range. Note that the last argument to `uniform`, when defining the initial coordinates, defines the number of rows and columns to yield values for. Here we want a vector of length three (x, y, z) for each atom. If developing this method further it may be useful to use custom Python classes to define `Atom` objects, as illustrated in Chapter 8, rather than using lists.

```
atoms = list(bondDict.keys())
numAtoms = len(atoms)
atomCoords = uniform(-10.0, 10.0, (numAtoms, 3))
```

The list `indices` is simply the numbers of all the atoms. This is defined upfront so that we don't have to recreate it repeatedly in the search steps. Likewise n is the floating point version of the number of steps, and is used to generate the temperature factor in the annealing schedule.

```
indices = range(numAtoms)
n = float(numSteps)
```

The main loop goes through the specified number of search steps and for each of these the effective temperature `temp` is defined as in other examples.

```
for step in range(numSteps):
    temp = exp(-step/n)
```

For the given step, with its temperature factor, the dynamics simulation requires that we go through all atoms in the molecule and calculate the effect of the other atoms according to the force field. The `for` loop goes through all the indices for the atoms, so that we can get hold of the atom's name (`atom`) and the coordinates. Note that we can ignore the first atom in this loop (hence `[1:]`), which doesn't need to move; the others will move around it. The variable `velocity` is defined initially as a zero vector and will represent the change that is made to the atom's position along the x, y and z axes, once we have considered the forces.

```
for i in indices[1:]:
  atom = atoms[i]
  coords = atomCoords[i]
  velocity = zeros(3, float)
```

For the primary atom (at index `i`) we then need to consider all of the other atoms that it interacts with, according to the force field. Hence we define a second loop and corresponding atom index `j`, skipping the loop if we come across the primary atom.

```
for j in indices:
  if i == j:
    continue
```

Then comparing each of the secondary atoms with the primary (index `i` with index `j`) `delta` is calculated as the difference vector between their coordinate positions. Then the sum of this squared is `dist2`, the distance between atoms squared.

```
delta = coords - atomCoords[j]
delta2 = delta * delta
dist2 = delta2.sum()
```

Next we apply the actual force field. Because we are keeping the demonstration simple we are using only two terms and these are mutually exclusive. Proper molecular force fields will have many more, complex terms for things like electrostatic charge, van der Waals force, bond angle etc. Anyhow, here we test whether the primary and secondary atoms are directly bound. If they are bound the name of one atom will be in the bonded list, which is extracted from the `bondDict` using the other atom as a look-up key. If they are bound a simple 'force' value is calculated as the difference between the input bond length and the distance between atoms.[8] Note that we only calculated the square distance up to this point and only now calculate the square root when we need to. The `force` value could also be scaled separately to other forces, which in general is a way of balancing the different, competing terms. Also, any scaling dictates how much movement can be gained in each step, although here we use a general scale factor `timeStep` to control the step size. The scale factors for the two force terms in this example just happen to work well when they are the same (1.0).

```
bound = bondDict[atoms[j]]
if atom in bound:
  force = bondLen - sqrt(dist2)
```

If the atoms do not share a direct chemical bond we apply the repulsive term, which is inversely proportional to the distance between atoms raised to the fourth power (the square distance squared). Using this power law is quite arbitrary, but works nicely here.

```
else:
  force = 1.0 / (dist2*dist2)
```

---

[8] When the force is proportional to distance the term is described as 'harmonic'.

With the `force` variable defined for a bond or repulsion, we do a check to make sure it is not too great. This is important because a force field can potentially produce some very large numbers (e.g. repelling atoms that are very close) which would otherwise throw an atom too far away for practical purposes, and may also lead to numerical Python errors. The bounded force value is then multiplied by `timeStep`, `temp` and `delta`, and added to the total for the `velocity`. The temperature factor is part of the annealing process discussed previously and the time step determines how much movement can occur for each iteration. Longer time steps will move things more quickly but can lead to atoms overshooting optimal positions. The `delta` represents the vector between the two atoms, and thus applies the force in the correct direction, i.e. the force is between atoms.

```
force = min(max(-200.0, force), 200.0)
velocity += delta * force * temp * timeStep
```

After all the interacting atoms have been considered the `velocity` variable will represent the residual push and pull on the primary atom. At an optimised location all the competing forces will balance out and this will be a zero vector, but otherwise the residual will move the atom. Accordingly, `velocity` is added to the coordinates for the atom.

```
atomCoords[i] += velocity
```

After all of the dynamics steps the atom coordinates are centred (for ease of inspection), by subtracting the average position, and then returned from the function.

```
center = atomCoords.mean(axis=0)
atomCoords = atomCoords-center

return atomCoords
```

Testing is simply a matter of using the chemical bonding data defined earlier (the connection topology) with the particle dynamics function. The resulting atom coordinates may be printed out, or even displayed in a graphical interface. Naturally, it is only the relative values of the

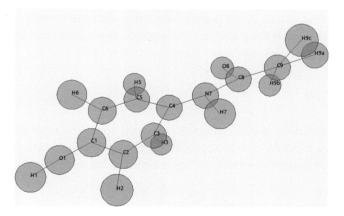

Figure 25.5. A simple graphical representation of the chemical structure of paracetamol, with atom coordinates determined by the particle dynamics algorithm. Python code for the `ChemView` class used to generate such graphics is available in the on-line material, http://www.cambridge.org/pythonforbiology.

atom coordinates that are important, rather than their exact values. If all the positions are rotated or translated by the same amounts it is still the same structure.

```
coords = chemParticleDynamics(chemBonds)
print(coords)
```

Graphical inspection of the atomic coordinates reveals a bent aromatic ring and thus reaffirms that the force field used is very simple and naïve. However, the result shows that the approach is working and we have a reasonable extended conformation, given the specified rules.

26 Graphical interfaces

Contents

An introduction to graphical user interfaces

At some stage when writing your own programs there may come a time when you want others to be able to use what you have created without them necessarily having to know anything about programming or Python. Should this happen the next step is to consider writing a more friendly interface to the program. Once upon a time in computing everything was text-based and the user had to type commands to get things to work. Fortunately things have moved on and we are now usually presented with graphics and a pointing device, either a mouse or a touch screen, and the user can interact with graphical objects like menus and buttons.

When building a *graphical user interface* (GUI) the programmer must be mindful of various factors, which are sometimes antagonistic, forcing us to make compromises. For example, the designer has to strike a balance between on the one hand giving lots of functionality and on the other hand keeping things simple for novices and intuitive to use. In this chapter we will aim to give some general advice about the programming, but we leave you to make the tough choices. We wish to be clear that this chapter deals with making graphical interfaces that run on the users' local computer. We will not venture into the world of Internet-based applications, although these are becoming increasingly important, and the Pyjamas library,[1] which is available for Python programmers, works in a remarkably similar way to the graphical libraries discussed here.

Widgets and graphics libraries

Most graphical interfaces are composed of discrete graphical objects like buttons, menus and text boxes etc. Collectively these are referred to as *widgets*.[2] In general each on-screen widget will have a corresponding computational object. Such widget classes allow the programmer to build graphical interfaces from common building blocks, without having to worry about how the object is drawn on screen. For example, if we add a clickable button to a program we usually only have to think about its general placement and what operation it triggers. The graphical library that lies behind the widget will take care of exactly where and how to draw

---

[1] http://pyjs.org/.
[2] Borrowing from the term used to describe small manufactured items of indeterminate function.

its pixels. Occasionally, however, the standard graphical objects are not sufficient to meet our needs, and we may have to create customised or even completely novel widgets, although the common graphics libraries have tools to do this without too much pain.

Before attempting to construct a GUI the programmer must first make the choice about which graphics library to use, i.e. which system of widgets the interface will be constructed with. Graphical systems are generally not directly available within the core Python libraries, but there are various capable choices and some of the more popular include PyGtk, Tkinter, PyQt and WxPython. Because of the limited space in this book we will focus on just a couple of these: Tkinter and PyQt. Both of these libraries are cross-platform and will run on Linux, Mac and Windows operating systems. The Tkinter library, which is Python wrapping around a system known as Tcl/Tk, is not the most sophisticated or most modern graphical system, but it is the de facto standard for Python and is usually bundled with the standard Python installation for Windows. Hence, if you want to do something quick and simple then Tkinter may be the best choice. PyQt and PySide, an alternative implementation, are Python wrappers to the Qt libraries, written in C++. This is a more sophisticated, state-of-the-art choice and includes support for OpenGL (for fast 3D graphics), embedded web browsing and basic multimedia support, to name only a little. Thus we would recommend Qt for larger projects. Especially helpful is the Qt Designer system, which allows you to lay out widgets in a graphical (non-programmatic) way, although in this chapter we demonstrate only the direct programming route.

GUI construction

When constructing a GUI there are a few main areas that the programmer must think about above and beyond the appearance of the graphical items. At the simplest level the widgets that constitute a GUI will collect information from the user, trigger various actions (i.e. function calls) and display information. Thus the graphical objects need a means of interacting with the other parts of the program, which is commonly kept somewhat separate from the GUI code. When a user interacts with a widget, e.g. to enter some text into a box, the program needs to access that data and perhaps change the state of the graphics accordingly. With these interactions in mind, we use the term *callback* to describe the process of graphical widget calling a function in our main program and the term *update* to describe the setting of the widgets to display new information. The term *signal* is general and will refer to both callback and update operations. To give a more specific example, if a graphical interface is constructed with a tick box (check button) we can specify a function to be called when the state changes, i.e. when the box is ticked. This callback from the widget then invokes some change in our program. Conversely, we may wish to send a signal in the opposite direction so that the tick box is kept updated to reflect the state of the main program after some event. Inherent in this sort of thinking is the notion that the GUI forms a separate layer to the rest of the program.

A GUI is programmed by selecting which kinds of widgets are to be used, how they signal to interact with the rest of the program and, of course, how the objects are displayed. When it comes to display, a widget is either placed directly in the context of the screen or inside the borders of another widget. A *top-level widget* is the term that we will use for a widget that is not graphically contained by another. The usual form of top-level widgets in today's operating systems is a window. The geometry of a top-level widget, in terms of position and size, is

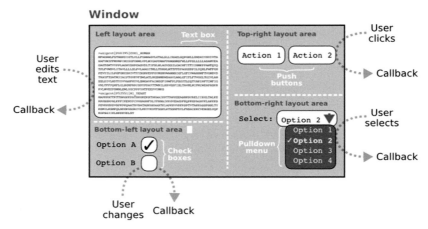

Figure 26.1. User interaction with a program via its graphical interface. Various concepts involved in the design and operation of a graphical interface are illustrated for a mock example of a graphical interface window. In general all of the graphical items, or widgets, are placed within the context of a larger area using some form of geometry management: here we illustrate with left and right layout areas, which are then subdivided top to bottom. Some graphical objects, like text labels, are generally static and cannot be changed by the user. However, others like text boxes (for typing into), buttons (to press), check boxes (to tick) and pull-down menus (to select within) can be manipulated by the user via a pointing device and keyboard etc. When a graphical object is manipulated the underlying program is usually informed of any changes, via a 'callback' function, so that it can respond to the new state, or in other words perform the functionality that the user expects.

something that is often adjusted by the user. However, the placement of sub-widgets that go inside the top level is not usually subject to user control. The positioning of such widgets is a large part of the GUI's design, and the items are generally placed to make operations convenient and intuitive. Placement of internal widgets is usually done in a fairly flexible way so that the interface can cope when the top-level window is resized. Accordingly, it is more common to specify that a widget sticks to the edge of a window, or expands to fit the current size, rather than specifying its location absolutely. Graphical libraries will have a *layout* or *geometry management* system as a means of directing where widgets are placed. For example, we often just specify that widgets are stacked vertically, lie horizontally or are arranged in a grid and the layout manager handles the fine placement automatically (e.g. making things line up).

Python GUI examples

In this chapter we will give a simple introduction to the Python versions of the Tkinter and Qt graphics systems. Such a task could fill a whole book, so for this chapter we will keep things very simple and aim to give an impression of the basic principles and what kinds of things are possible: to help a programmer get started, rather than give an in-depth description of every possible widget. The examples construct a simple graphical interface for some of the biological sequence analysis that was discussed in Chapter 11. Indeed, we will import functions defined there.[3] We will only show a few of the more basic widget types, but the

---

[3] Available as a Python file in the downloadable material, http://www.cambridge.org/pythonforbiology.

interface will be functional. The same basic interface will be constructed using both Tkinter and Qt, so that you can compare and contrast the differences. There might be a more elegant way of handling things in either example, but we have tried to keep the equivalence reasonably close.

Compared to most of the other chapters, the reader will notice that the examples are constructed by using classes, which generate a customised version of a graphical Python object. Whereas in most of the examples in this book we can illustrate the principles by sticking predominantly to simpler Python functions (and leave it to the programmer to decide whether to use classes), here for the GUI code we actively encourage using classes from the outset. The reason for this is the high degree of interconnectivity and data sharing that goes on with graphical items and the underlying data they represent or manipulate. Making a class for a graphical interface, which usually at least requires a subclass of a top-level window, allows all of the component widgets to be automatically grouped. Also, it is less sensitive to the order in which the code is written and minimises the arguments that have to be passed; internal functions can access class attributes (variables that belong to the GUI object). A GUI can be written with functions alone, but it is harder work.

The interfaces that form the examples are pictured in Figure 26.2. Each example consists of a single top-level window and inside this we place the sub-widgets: small text labels to provide headings, large text boxes to display DNA sequences and output, buttons to allow the user to trigger actions and a small text entry box for a user to type into. It is deliberate that we have accepted the default styling for the widgets and not tried to customise the look; that comes later if you can find the time. The examples only demonstrate a few of the available widget types, but a more complete listing is available in the on-line documentation for Tkinter and Qt.

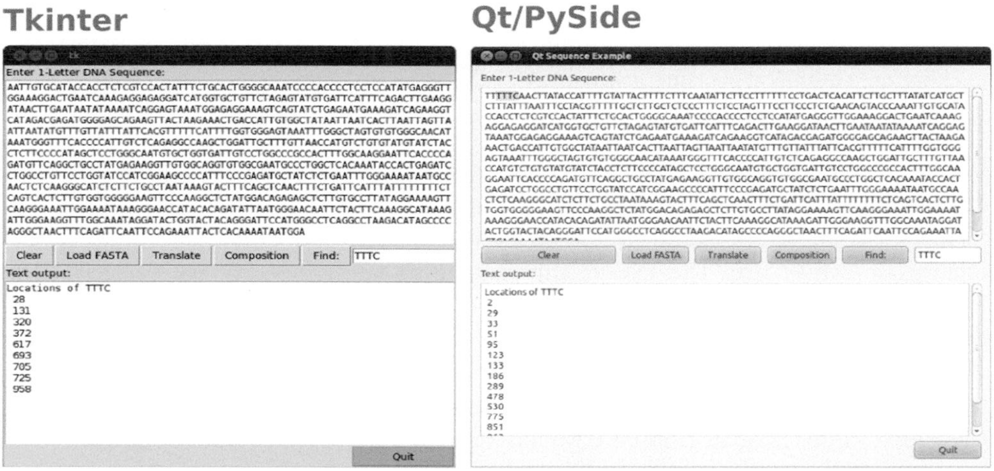

Figure 26.2. An example of graphical interfaces to operate simple DNA sequence analysis functionality. The two windows result from the Python examples constructed using the Tkinter (left) and Qt/PySide (right) graphical object libraries. The widget styles are the defaults for each situation and reflect what happens under the same Linux operating system.

Using Tkinter

Before getting into the main example we will initially demonstrate 'hello world' code for Tkinter. Naturally the examples assume that the Tkinter library is installed and available to Python. See http:www.cambridge.org/pythonforbiology for Tkinter download and install instructions, as well as links to full documentation. The module is called `Tkinter` in Python 2 and `tkinter` in Python 3, so we try the latter and if that does not work then try the former. We then create the top-level `rootWindow`, which is a `Tk()` class of object.

```
try:
  import tkinter
except:
  import Tkinter as tkinter

rootWindow = tkinter.Tk()
```

Then we make a widget to put inside the window, which in this case is a simple text label, of the `tkinter.Label` class. Note that we construct the label object using the `rootWindow` as the first argument, which is the means of specifying that the label belongs to the window; in GUI speak `rootWindow` is the parent and `label` is the child.

```
label = tkinter.Label(rootWindow, text='Hello World')
```

Once the label is created we must specify where in the window it will go. Here the `pack` geometry manager is used (because it is simple), which by default adds widgets to their parent from top to bottom. Unless we use a geometry manager the label will not appear, because Tkinter will not know where to draw it.

```
label.pack()
```

Then to actually see the result the `mainloop()` function call is issued from the top-most parent widget (often called the 'root'). If we did not issue this function call the Python interpreter would make all the graphical objects but the program would then immediately end, without displaying anything. By invoking a main graphics loop the system is informed that it should not end the program. Instead Tk remains active and waits to detect graphical events, like clicking on a button or resizing a window.

```
rootWindow.mainloop()
```

With these basic principles in mind we move on to the definition of the graphical interface class for simple DNA sequence analysis. As usual, we first make the appropriate imports. The `re` module is imported because we will do a precautionary check of the DNA sequences, to remove any whitespace. The Tkinter imports now include `filedialog` and `messagebox` in Python 3, or equivalently `tkFileDialog` and `tkMessageBox` in Python 2, which are pre-constructed Tk elements, for finding files and displaying pop-up messages. These larger compound widgets exist to easily perform some of the most common operations. The remaining imports are from the `Sequence` module, which refers to the examples from Chapter 11 in this book that can be downloaded from the supporting on-line material.

```
import re

try:
  import tkinter
  from tkinter import filedialog, messagebox
```

```
except:
  import Tkinter as tkinter
  import tkFileDialog as filedialog
  import tkMessageBox as messagebox
```

```
from Sequences import proteinTranslation, STANDARD_GENETIC_CODE
```

To construct the GUI a new class is defined as a subclass of `tkinter.Tk`. Thus it inherits all of the properties of this `Tk` main window. Little of the `Tk` class will be changed, but rather we will augment the new object definition to embed sub-widgets (text boxes, button etc.) and add a number of bound methods, which include function calls to actually do the specialist science operations. Immediately after the class statement the `__init__` function is redefined (which is called when a new object of this type is made). All Python objects will have an `__init__()`, so here we are overwriting the one from the Tk superclass. However, we invoke the `__init__` for `tkinter.Tk` directly on `self` (which represents the current instance of an object) as the first task, so the superclass initialisation is still done. The reason to overwrite the function in this way is to keep the original functionality, i.e. actually making a GUI window in this case, but at the same time create a place where customisation can occur. Hence, in the remainder of the `__init__` function we add extra code that creates this specialist window, including adding any internal sub-widgets.

```
class SequenceTkGui(tkinter.Tk):

  def __init__(self):

    tkinter.Tk.__init__(self)
```

Unlike the simple 'hello world' example above we will use a different geometry manager called `grid` to create the layout of the widgets inside the top-level window. Using a grid is an easier way to manage things overall (in the authors' own experience) because it is easier to predict the results. As the name suggests, using `grid` means that we will be placing widgets in the main window by specifying the row and column they lie within. Also, when required, a widget can be made to span multiple rows or columns,[4] which adds lots of flexibility.

The next command configures the behaviour of the rows and the columns of the grid system within which the graphical widgets will be placed. By default all rows and columns have `weight=0`, which means that they do not expand to fill any extra space beyond the immediate size of the item they contain. Setting the `weight=1` below specifically for column 5, row 1 and row 4 (counting from zeros) means that these will be the expanding rows and column in our window. Hence, when the main window is resized these will resize too. If required, weights greater than 1 could be used if one part needs to expand more than another.

```
    self.grid_columnconfigure(5, weight=1)
    self.grid_rowconfigure(1, weight=1)
    self.grid_rowconfigure(4, weight=1)
```

---

[4] Similar to the way tables are set up in HTML.

The first graphical widget that is added will be a text label, as was demonstrated above. In this example widgets are added to the window class in display order, from top to bottom and left to right. This is just good practice to make visual inspection easier and not an absolute requirement. The `tkinter.Label` is created as belonging to `self` and having the required text. The object is assigned to the `self.label1` variable so that we can access it anywhere inside the class (without `self` it would only be accessible in the immediate function). In keeping with the intention to use a grid layout we invoke the `.grid()` call, which is available to all Tk widgets. As the arguments indicate, the label is placed at grid position `row=0`, `column=0` and spans six columns, i.e. the whole of the top row. The last `sticky` argument states how the widget inside the grid will adhere to the edges of its cell. The system Tkinter uses is based on the cardinal compass coordinates, i.e. North, South, East and West. This can seem a bit odd, given that compass directions depend on which way you are facing, but can be imagined if you are facing a map with North at the top. Accordingly the specification `tkinter.EW` here means to stick to both the left- and right-hand edges.

```
self.label1 = tkinter.Label(self, text='Enter 1-Letter DNA Sequence:')
self.label1.grid(row=0, column=0, columnspan=6, sticky=tkinter.EW)
```

The next widget is a `tkinter.Text`, which will allow us to display multiple lines of text. It is placed into the grid on the next row with NSEW stickiness, i.e. to stick to all four edges of the grid cell.

```
self.seqTextBox = tkinter.Text(self)
self.seqTextBox.grid(row=1, column=0, columnspan=6,
                     sticky=tkinter.NSEW)
```

Below the text box comes a row of buttons that the user can 'push' by clicking with the mouse cursor.

The button objects are defined using the `tkinter.Button` class and assigned to respective `self.` variables. The arguments for constructing the buttons are `self` (the parent), some `text` to display on the button and a `command`. The `command` is the name of a Python callback function which will be triggered when the user pushes the button. Here the callback functions are custom ones that will be defined later in the class structure. Naturally the text of the buttons reflects the functions they call. The functions take no arguments, but arguments may be added via a `lambda` function.[5] All of the buttons are placed in separate columns within row 2, sticking to the left of the grid cell (`tkinter.W`).

```
self.clearButton = tkinter.Button(self, text='Clear',
                                  command=self.clearSeq)
self.clearButton.grid(row=2, column=0, sticky=tkinter.W)

self.loadButton = tkinter.Button(self, text='Load FASTA',
                                 command=self.loadFasta)
self.loadButton.grid(row=2, column=1, sticky=tkinter.W)

self.transButton = tkinter.Button(self, text='Translate',
                                  command=self.seqTranslate)
```

---

[5] Constructing buttons with `command=func(args)` will call the function too early, but `command=lambda: func(args)` will not. See Chapter 5.

```
self.transButton.grid(row=2, column=2, sticky=tkinter.W)

self.compButton = tkinter.Button(self, text='Composition',
                                 command=self.seqComposition)
self.compButton.grid(row=2, column=3, sticky=tkinter.W)

self.findButton = tkinter.Button(self, text='Find:',
                                 command=self.seqFind)
self.findButton.grid(row=2, column=4, sticky=tkinter.EW)
```

The last widget in row 2 is a `tkinter.Entry` rather than a button. An `Entry` object allows the user to type in a small piece of text. This will be used to enter a query DNA sequence, which will be searched for within the main sequence.

```
self.findEntry = tkinter.Entry(self)
self.findEntry.grid(row=2, column=5, sticky=tkinter.EW)
```

The next two rows contain another `Label`, giving the title for the section, and second large `Text` box to display the textual output for the user. Both of these widgets span all six columns, remembering that columns 0 to 5 inclusive were filled above.

```
self.label2 = tkinter.Label(self, text='Text output:')
self.label2.grid(row=3, column=0, columnspan=6, sticky=tkinter.W)

self.outTextBox = tkinter.Text(self)
self.outTextBox.grid(row=4, column=0, columnspan=6,
                     sticky=tkinter.NSEW)
```

The final widget is placed in a row on its own. This is another `Button` and it calls the `self.destroy` function. This function is inbuilt into all `Tk()` objects and provides a means of removing the main window, which stops the Tkinter `mainloop()` and so causes the program to end.

```
self.closeButton = tkinter.Button(self, text='Quit',
                                  command=self.destroy)
self.closeButton.grid(row=5, column=5, sticky=tkinter.EW)
self.closeButton.config(bg='yellow') # Yellow background
```

With the widget construction done the remainder of the class involves defining the functions that underpin the graphics to make things work. All of the functions at least take `self` as an argument so they can access the `self.` names from within the object. Although, as discussed previously, the `self` is not passed in brackets when calling the function; rather it is implicit because of the dot notation.

First are functions to clear and set the DNA sequence text within `self.seqTextBox` (the upper text area). Note that Tk uses a string (`row.column`) based system to identify parts of the text within the Text widget; `'0.0'` is the beginning, and `tkinter.END` represents the end, wherever that is. Thus clearing the sequence means applying `delete()` to all the text. When setting the sequence the text box is cleared before the `text` that was passed in as an argument is added.

```
def clearSeq(self):

    self.seqTextBox.delete('0.0', tkinter.END)
```

```
def setSequence(self, text):

  self.clearSeq()
  self.seqTextBox.insert(tkinter.END, text)
```

The function to get the DNA sequence from the upper box extracts all the widget text, between start and end points. The `re` (regular expression module; see Appendix 5) is used to tidy the sequence by removing any whitespace, including tabs and line returns. The sequence is also forced to be upper case. Mostly these checks are present as examples to remind us that whenever the user provides input that is supposed to have some meaning (so here it should be a DNA sequence not a shopping list) our program should aim to detect or remove nonsense. At the end of the function the curated sequence string `seq` is returned.

```
def getSequence(self):

  seq = self.seqTextBox.get('0.0', tkinter.END)
  seq = re.sub('\s+','',seq)
  seq = seq.upper()

  return seq
```

Two functions control the contents of the lower text area, `self.outTextBox`. The `showText()` function is for adding text to the box. This is similar to `seqSequence()` but we do not clear the text area first. Also, an explicit check is made to ensure that all added text ends with a `'\n'` (new line) character; this function adds a new line each time. The `clearOutput()` function removes all output text using the `Text.delete()` call with ranges, as mentioned for `clearSeq()`.

```
def showText(self, text):

  if text[-1] != '\n':
    text += '\n'
  self.outTextBox.insert(tkinter.END, text)

def clearOutput(self):

  self.outTextBox.delete('0.0', tkinter.END)
```

With the functions to control the text areas defined, attention now turns to the functions that are called by pressing the buttons, i.e. callbacks connected via `command`. The first of these is a function to load a sequence from a FASTA-format file. It uses functionality from BioPython, as discussed in Chapter 11, to read the entries. Here we only take the first sequence from the file, i.e. there is a `break` in the loop, but we could take more sequences if the GUI was adjusted accordingly. The notable part of this function is that it uses the `filedialog`, which comes with Tkinter and allows us to easily create a widget that lets the user select a file. The `.askopenfile()` call actually displays the file-requesting widget and gives back an open file object (same as if using the `open()` keyword), although we have to check for `None` if no file was selected.

```
def loadFasta(self):

  fileObj = filedialog.askopenfile(parent=self, mode='rU',
                                     title='Choose a FASTA file')

  if fileObj:
    from Bio import SeqIO
    for entry in SeqIO.parse(fileObj, 'fasta'):
```

```
     self.setSequence(entry.seq)
     break

   fileObj.close()
```

Next comes the first scientific function of the class. As the name hints `seqTranslate()` will translate the DNA sequence (in the upper panel) into three-letter protein sequences that are displayed in the lower text panel. The `self.getSequence()` function is called to extract the currently displayed DNA sequence. The output area is cleared and we use `showText()` to display a title. Then comes a `for` loop inside which the sequence translation occurs. A loop is used so that we can define `indent` (as 0, 1, 2) to specify where in the DNA sequence we start translating, remembering that a protein's amino acids are coded by three DNA bases. Thus by using the loop we will get translations of all three forward reading frames.[6]

```
def seqTranslate(self):

  seq = self.getSequence()

  self.clearOutput()
  self.showText('DNA sequence')
  self.showText(seq)
  self.showText('Protein sequence')

  for indent in range(3):
```

In the loop the protein sequence is obtained by calling the `proteinTranslation()` defined earlier in the book. We translate with the standard genetic code, so that is passed in. The GUI could be expanded so that the user may select from among several genetic codes. The translated protein sequence is initially a list of Python strings, but is then joined into one long line of text. The variable `spaces` is defined, which will act as padding in the output, to move the indentation of each subsequent translated reading frame one space to the right, i.e. so the amino acid codes are staggered and lie exactly under their DNA codon triplet. At the end of the loop the elements are combined to give the output `text`, which is displayed in the GUI using `showText()`.

```
    proteinSeq = proteinTranslation(seq[indent:], STANDARD_GENETIC_CODE)
    proteinSeq = ''.join(proteinSeq)
    spaces = ' ' * indent

    text = 'Reading frame %d\n%s%s' % (indent, spaces, proteinSeq)

    self.showText(text)
```

A second scientific function is one that gets the DNA sequence and counts the different letters. Each letter is used as a key to the `counts` dictionary. The letters are then sorted and for each kind the average composition, as a percentage, is calculated. The data is then used to make a line of text and passed to `self.showText()` for display.

```
def seqComposition(self):

  self.clearOutput()
  seq = self.getSequence()
```

---

[6] See Chapter 11 for explanations of *codons* and *reading frames*.

```
n = 0.0
counts = {}
for letter in seq:
  counts[letter] = counts.get(letter, 0) + 1
  n += 1.0

letters = counts.keys()
letters.sort()

text = "Composition:"
for letter in letters:
  text += ' %s;%.2f%%' % (letter, counts[letter] * 100 / n)

self.showText(text)
```

The last function in the `SequenceTkGui` class is used to locate a query sub-sequence within the main DNA sequence. The `query` sequence is obtained using the `.get()` call that goes with the `self.findEntry` widget; this gives back the contents of the box. Any whitespace at the edges of the query is removed with `.strip()`. Then a check is made to ensure that we are not searching with something blank. Thus if `query` is empty we use `messageBox` to create a pre-constructed Tk widget and inform the user that the search could not be done. After a warning the `return` statement immediately quits the function. Otherwise, if the search query was defined, the main sequence, `seq`, is fetched. Then it is a relatively simple matter to see if the query sequence is present. If it is we loop through the main sequence to find all occurrences, i.e. `query` is compared with `seq[i:i+win]`, where `i` is the position and `win` is the query width. Whether the search made a match or not is indicated by the text that is passed into `self.showText()`.

```
def seqFind(self):

  self.clearOutput()

  query = self.findEntry.get()
  query = query.strip()

  if not query:
    messagebox.showwarning("Warning", "Search sequence was blank")
    return

  seq  = self.getSequence()

  if query in seq:
    text = "Locations of %s" % (query)
    self.showText(text)
    win = len(query)

    for i in range(len(seq)-win):
      if seq[i:i+win] == query:
        self.showText(' %d' % i)
  else:
    text = "Sub-sequence %s not found" % (query)
    self.showText(text)
```

Finally, at the end of the class and function definitions we can write testing the code. Note that this is subject to the __name__ == '__main__' clause, which only runs the test if the Python file is used directly. This allows for the `SequenceTkGui` to be imported by other Python modules without the test code being run. The testing is done by creating `window` as a `SequenceTkGui` class object, and then calling the `Tk()` main loop, which the class inherits from, to view the graphics.

```
if __name__ == '__main__':

    window = SequenceTkGui()
    window.mainloop()
```

Using Python Qt

The same DNA sequence GUI example will be shown again, but using Qt-based libraries to compare and contrast with the Tkinter system. There are currently several choices for a Python-based Qt graphics library, PyQt4, PyQt5 and PySide, though PyQt5 only arrived after most of this book was written. For the most part these are almost identical and Python code that works in one will work with the other with only minor adjustments. For the example we will use PySide because we feel it has a more Pythonic[7] way of dealing with GUI signals (widget callbacks etc.), and also less restrictive licence conditions. Unfortunately PyQt5 arrived too late to be considered in this book, though this is the system the authors would use in the future, given that it has the most active development and the future of PySide is somewhat uncertain. See http://www.cambridge.org/pythonforbiology for PySide and PyQt download and install instructions, as well as links to full documentation.

Compared to Tk the Qt libraries for Python have a larger variety of widgets, although we will only use a few in the example. Also, Qt naturally exposes a greater variety of signals, although Tk can be expanded somewhat by the use of `.bind()` calls, to connect events to widgets. For example, a `Tkinter.Button` is generally only used with a single callback for when the button is pressed. The nearest Qt equivalent, `QtGui.QPushButton`, automatically comes with `clicked`, `pressed`, `released` and `toggled` signals.

Compared to other libraries used in this book the Python bindings to Qt feel a little different, because of the way the underlying C++ code is wrapped. For example, most Qt objects are created with few initial arguments, and the object is configured with various calls after it is created. Also nomenclature like '`QtCore.Qt.AlignLeft | QtCore.Qt.AlignRight`' can seem a bit unfriendly at first. Here the vertical line is the logical OR operator and is simply a means of combining the two options for left and right alignment, which are represented as separate bits in a binary number (a typical C++ style of doing things). This example, albeit using binary, is actually just equivalent to `1 + 2` in Python.

For the GUI example we make the initial imports. If PyQt4 is used instead the `connect()` calls in the class must be adjusted, as mentioned below. Qt is separated into discrete modules which relate to different aspects of the system and here we import the required `QtCore` and `QtGui`.

```
import re

from PySide import QtCore, QtGui # or from PyQt4

from Sequences import proteinTranslation, STANDARD_GENETIC_CODE
```

---

[7] Is more in keeping with the Python philosophy.

The `SequenceQtGui` definition is a subclass of the basic `QtGui.QWidget` and when we initialise this class we also initialise the superclass for `self`. Although `QWidget` is more directly comparable to Tk used earlier, the programmer can also consider using `QMainWindow` because this comes prepared with slots for a main menu, tool bar and status bar etc.

```
class SequenceQtGui(QtGui.QWidget):

  def __init__(self):

    QtGui.QWidget.__init__(self, parent=None)
```

In Qt it is fairly common to use simple horizontal and vertical layouts to arrange sub-widgets in a window, but here we will use a grid, as in the previous example. Rather than the grid being inbuilt into the widget system we need to make a layout object, `QGridLayout`, and say that this belongs to `self`, i.e. it is the layout for the main window.

```
    grid = QtGui.QGridLayout(self)
```

The layout object `grid` can then be used to configure the size behaviour of the rows and columns. As before column 5, row 1 and row 4 are set to expand with weight 1.

```
    grid.setColumnStretch(5, 1)
    grid.setRowStretch(1, 1)
    grid.setRowStretch(4, 1)
```

As an equivalent to `Tkinter.EW`, to refer to the left and right edges of a grid cell, we define `leftRight` for later use by combining the binary `Qt.AlignLeft` and `Qt.AlignRight` options.

```
    leftRight = QtCore.Qt.AlignLeft | QtCore.Qt.AlignRight
```

Next we make the actual sub-widgets. A `QtGui.QLabel` is constructed, which is the equivalent of `Tkinter.Label`. Instead of a simple label the `QtGui.QGroupBox` could be used, which provides a title and a border to group sub-widgets. Once the widget is made it is then added to the `grid` layout at position `0,0`; the first row and column.

```
    self.label1 = QtGui.QLabel(text='Enter 1-Letter DNA Sequence:',
                               parent=self)
    grid.addWidget(self.label1, 0, 0)
```

Next to be constructed is the upper text box. This will be from the `QtGui.QPlainTextEdit` class (there is also `QTextEdit` which supports rich text mark-up). This goes in the next row of the grid (`1,0`) and we include the row and column span arguments (`1,6`) to cover one row and six columns. The `align` argument dictates how the widget sticks to the grid sides. Note that compared to the Tk equivalent the Qt text box has more functionality, e.g. there is a context menu (right mouse click) with editing options, which we do not use here.

```
    self.seqTextBox = QtGui.QPlainTextEdit(parent=self)
    grid.addWidget(self.seqTextBox, 1, 0, 1, 6, align=leftRight)
```

The `QtGui.QPushButton` class is used for simple buttons. Unlike Tk, where we just specify the callback for the button with `command`, for Qt we have to select which kind of user action we want it to respond to. Here we want the callback to be triggered from the `clicked` action (technically a *slot* in Qt). The connecting of an action slot to a callback function is usually done differently for PyQt4 and PySide libraries. For this PySide example the `self.clear-Button.clicked` object represents the button clicking, and this has a `connect()` function to link it to whichever callback function we desire. Thus the button click calls `self.clearSeq`. A comment is included below to illustrate the equivalent for PyQt4. This involves actually creating a `SIGNAL` object using a text string (not very Pythonic) and then the `connect()` call to link the target of the signal to the function comes from a Qt widget (`self` in this case).[8]

```
self.clearButton = QtGui.QPushButton(text='Clear', parent=self)
self.clearButton.clicked.connect(self.clearSeq)
grid.addWidget(self.clearButton, 2, 0)

#PyQt4 uses:
#self.connect(self, QtCore.SIGNAL('clicked()'), self.clearSeq)
```

The other buttons are made in a similar way, and placed in separate grid columns:

```
self.loadButton = QtGui.QPushButton(text='Load FASTA', parent=self)
self.loadButton.clicked.connect(self.loadFasta)
grid.addWidget(self.loadButton, 2, 1)

self.transButton = QtGui.QPushButton(text='Translate', parent=self)
self.transButton.clicked.connect(self.seqTranslate)
grid.addWidget(self.transButton, 2, 2)

self.compButton = QtGui.QPushButton(text='Composition', parent=self)
self.compButton.clicked.connect(self.seqComposition)
grid.addWidget(self.compButton, 2, 3)

self.findButton = QtGui.QPushButton(text='Find:', parent=self)
self.findButton.clicked.connect(self.seqFind)
grid.addWidget(self.findButton, 2, 4)
```

The last widget in row 2 is the small text box for the user to enter a query DNA sequence. The class used is `QtGui.QLineEdit` and we will not connect any callback to the widget, although we could use `returnPressed` or `editingFinished` action slots to do something when the user changes the text.

```
self.findEntry = QtGui.QLineEdit(parent=self)
grid.addWidget(self.findEntry, 2, 5)
```

Lastly we add the second label, the lower text area for output and the quit button (which is connected to the inbuilt `QWidget.destroy()`). Note that in order to configure the text widget, in this case to say how the text wraps around at the end of a line, the function `.setLineWrapMode()` is used after the object is created.

```
self.label2 = QtGui.QLabel(text='Text output:', parent=self)
grid.addWidget(self.label2, 3, 0, 1, 6)

self.outTextBox = QtGui.QPlainTextEdit(parent=self)
self.outTextBox.setLineWrapMode(QtGui.QPlainTextEdit.NoWrap)
grid.addWidget(self.outTextBox, 4, 0, 1, 6, align=leftRight)

self.closeButton = QtGui.QPushButton(self, text='Quit')
self.closeButton.clicked.connect(self.destroy)
grid.addWidget(self.closeButton, 5, 5, align=leftRight)
```

---

[8] PySide can do it this way too, if you have to.

The next functions after the initialisation involve clearing and updating the text areas. This is a little simpler than with Tk, given that `.clear()` is inbuilt in to `QtGui.QPlainTextEdit`. As you might expect, the names and the actions of the functions differ between the two systems. For example, `setPlainText()` replaces all the text, so we don't have to clear first. Likewise, using `appendPlainText()` does not need newline characters to be added to `text`.

```
def clearSeq(self):

  self.seqTextBox.clear()

def setSequence(self, text):

  self.seqTextBox.setPlainText(text)

def showText(self, text):

  self.outTextBox.appendPlainText(text)

def clearOutput(self):

  self.outTextBox.clear()
```

Fetching the DNA sequence is simple and employs `.toPlainText()` rather than Tk's `.get()`. The tidying of the DNA sequence is done as described earlier.

```
def getSequence(self):
  seq = self.seqTextBox.toPlainText()
  seq = re.sub('\s+','',seq)
  seq = seq.upper()

  return seq
```

The `loadFasta` function uses the compound widget `QtGui.QFileDialog` to get the file name. This looks somewhat prettier than the Tk equivalent, and will have a style that represents the native operating system. Note that `getOpenFileName` does not give back a file object. Rather it gives back the location of the file, so we use `open()` to get the actual file object that BioPython works with.

```
def loadFasta(self):

  msg = 'Choose a FASTA file'
  filePath, filtr = QtGui.QFileDialog.getOpenFileName(self, msg)

  if filePath: # Something was selected
    fileObj = open(filePath, 'rU')

    from Bio import SeqIO
    for entry in SeqIO.parse(fileObj, 'fasta'):
      self.setSequence(str(entry.seq))
      break

    fileObj.close()
```

The two functions `seqTranslate()` and `seqComposition()` are unchanged compared to the Tk equivalent, so we will not repeat them. However, it is worth pointing out that the lack of any difference shows that we have separated the more graphical functions from the scientific functions, which is generally a good plan.

For finding a query DNA sub-sequence the `seqFind` function is similar to before. The differences are that the query entry box uses `.text()` not `.get()` and that we have naturally swapped to `QtGui.QMessageBox` to display warnings to the user. One extra thing that has been included here is `self.seqTextBox.find(query)`. This highlights successive instances of the query string in the upper text area. Although `Tkinter.Text` has a `search()` function this merely finds text, but does not highlight it.

```python
def seqFind(self):

  self.clearOutput()

  query = self.findEntry.text()
  query = query.strip()

  if not query:
    QtGui.QMessageBox.warning(self, "Warning",
                              "Search sequence was blank")
    return

  seq = self.getSequence()

  self.seqTextBox.find(query)

  if query in seq:
    text = "Locations of %s" % (query)
    self.showText(text)
    win = len(query)

    for i in range(len(seq)-win):
      if seq[i:i+win] == query:
        self.showText(' %d' % i)

  else:
    text = "Sub-sequence %s not found" % (query)
    self.showText(text)
```

To test the `SequenceQtGui` class code we have a little work to do. A Qt graphical interface only works if there is a `QApplication` instance to control the flow of the GUI program and carry its main settings (this does some of the job that the Tk root does). The application object is accordingly made and assigned to the `app` variable. The `window` object is made using the `SequenceQtGui` class described above and will become the top-level graphical object. The `.show()` call is required to actually see something; although this might seem a little tedious it is really handy to be able to control the visibility of widgets (all `QWidgets` have `.show()`). The final line looks a little odd, but is merely running the `QApplication`, and is equivalent to `Tk.mainloop()`. The function is called 'exec_' not 'exec' because the latter is an inbuilt keyword of Python. This call is wrapped by `sys.exit()`, which is required for the program to exit cleanly when done.

```python
if __name__ == '__main__':

  import sys

  app = QtGui.QApplication(['Qt Sequence Example'])

  window = SequenceQtGui()
  window.show()

  sys.exit(app.exec_())
```

27 Improving speed

Running things faster

This chapter is all about how to make Python programs run faster. We will discuss optimising existing routines so that they take a shorter amount of time to run, above and beyond the simple Python tips and tricks discussed earlier. Initially parallel computing, where a job is split into parts and run concurrently on separate processors (or processing cores), is discussed in a basic way. For this we use modules that are available from Python 2.6 and above, which allow programs to take account of multiple processing cores present in a single computer. For the remainder of the chapter we will deal with improving the performance of a single processing job.

    At the end some timing results will be given so that the reader can see how much was gained for the effort. For mathematical routines involving lots of loops it is not uncommon to get speed improvements of better than tenfold. The fine details about the logic and underlying algorithms of the examples used here will not be described; an example will be taken from earlier in the book where such things are described fully. Also, which particular example we have chosen is not especially important, other than the fact that it is a computationally intensive one that takes a noticeable time to run. It should be noted that this chapter comes with a 'health warning' for novice programmers, because the mainstay of the optimisation will be to move away from Python. Some of the focus will be on the low-level compiled language C, although it will be used in a way to provide a module that can still be used directly inside Python programs. The details of the C language, and how to compile it, will not be discussed and to actually learn to program in C we recommend further reading.[1] Nonetheless, if you have no experience with C we hope that we can provide a basic appreciation of how it can help. We also consider Cython,[2] a C-like extension to Python, which has made it possible to

---

[1] There are many books for learning C: for example, Kelley, A., and Pohl, I. (1997). *A Book on C: Programming in C.* Addison Wesley.

[2] http://cython.org.

benefit from the speed of C without having to necessarily deal with all the complexities of C. This is particularly powerful in combination with using NumPy arrays.

Naturally, for an analysis of program speed we will be starting from something that is already working correctly. Although with experience it is certainly possible to write speedy code in the first instance, it is common to write a simple, slower pure Python version of a program first and then optimise it afterwards: effectively making a working prototype. Writing in regular Python is relatively efficient in terms of human effort, and there is little point in optimising something that doesn't do what is intended. Also, we should be mindful that it may not be necessary to optimise at all; if a program takes only 0.1 seconds to run, the effort to improve speed may have no noticeable effect. Where a program does take a while to run we generally do not need to optimise the speed of a whole program. Usually there will be particular bottlenecks that can be optimised to good effect, and optimising the remainder would make no significant difference. Commonly we will find that regular Python is left unaltered to do the main program control, but that the computationally intensive, more mathematical parts are optimised. Often this means writing a separate fast module that is then embedded.

Parallelisation

Most computers these days have more than one central processing core. If we have a job that can be split up into parts that can be run at the same time, we can exploit the multiple processing capabilities and run things in parallel and so complete the overall task more quickly.[3] Naturally this adds complexity to our code, but from Python version 2.6 there is a convenient standard means of doing this using the `multiprocessing` module. When considering running tasks with this module, we should consider how much communication is needed between the parallel parts. In some cases, where a job frequently needs to exchange information with other jobs, or is dependent on jobs being synchronised, then we do not expect to get the full advantage of multiple processes, i.e. using two cores does not always make things twice as fast. Thus the best speed gains come (and are worth the extra coding effort) where a procedure can be easily split into fairly discrete units.

Using the '`multiprocessing`' module

Below we give a couple of simple examples that illustrate how `multiprocessing` can be used. The examples are simple in the sense that the calculation jobs are completely separate and each parallel process doesn't need to communicate much, other than to accept input arguments and pass back the result. However, this is not an uncommon situation in biological computing; for example, things like molecular structure calculations generate a range (an *ensemble*) of separate solutions to a problem, and in many circumstances jobs can be run separately on independent input data.

For our examples we make three imports, one for the `Process` class, which is the simplest way of generating a parallel job, another for `Queue`, which allows data to be shared

---

[3] Although we don't go further than this here, that natural next step would be to distribute jobs over many separate computers.

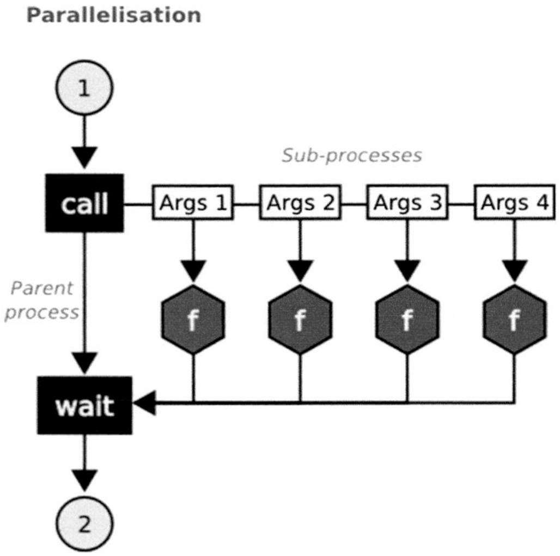

Parallelisation

Figure 27.1. A schematic of how a program may be executed in parallel by spawning sub-processes.
Parallelism in programming is often achieved by invoking a function call from an originating 'parent'
process, which then waits for the outcomes from the 'child' sub-processes that it started. In general the
same program code will be executed by the child processes (here labelled 'f'), although they will
commonly operate on different input arguments. For the examples illustrated in this chapter the sub-
processes will be run on the same computer, although it is also possible to interact with other computers
via a network. The ability to run several jobs in parallel on one computer will depend on the number
of processing cores that a computer has, although on most modern operating systems it is possible to
initiate more jobs than there are cores. In this case the different processing tasks (the threads) will be
shared at a fine scale by the processor(s), and will not all run at their maximum speed.

between jobs, and `Pool`, which provides a convenient way of matching a fixed number of
processing cores with a variable number of sub-tasks; the pooling mechanism collects jobs to
be run and then allocates them to the processor when there is a free slot.

```
from multiprocessing import Process, Queue, Pool
```

Firstly, we define a function that will do the actual work, and this will be called separately for
the different jobs. This is a regular Python function, and for the example we perform an
arbitrary mathematical calculation that takes a few seconds to run and prints out the job status,
if only to illustrate the principles. The function `calcFunc` takes two numbers as input
arguments, but we could send any input data to the function as long as that data is *pickleable*,
i.e. can be converted from an in-memory to a serial representation; this includes all the regular
Python data structures.

```
def calcFunc(n, m):

  print("Running %d %d" % (n,m))
  result = sum([x*x for x in range(n) if x % m == 0])
  print("Result %d %d : %d " % (n, m, result))

  return result
```

To set up the parallel jobs we first use the `Process` class to create objects that represent each of the parallel jobs. The `target` argument is the name of the function that will be called upon to do the work and `args` is a tuple of the values that will be input to the function when it is actually run. We create two job specifications, running the same operation, but with different inputs thus:

```
job1 = Process(target=calcFunc, args=(8745678, 2))
job2 = Process(target=calcFunc, args=(2359141, 3))
```

Each of the jobs can be started as required, so this is where the main flow of the Python script separates from the sub-jobs:

```
job1.start()
job2.start()
```

Then it is a simple matter of having the main, parent process wait until the sub-jobs are complete before continuing any further operations. The waiting is easily done using the `.join()` method, which optionally takes a time in seconds to wait before proceeding regardless (this is called a *timeout*).

```
job1.join()
job2.join()
```

Using data queues

Note that with the above example we fire off jobs but don't actually take any measures to collect the results back into the main program. We could, however, collect results using a `Queue` object, which can be passed as an argument, to be filled by the sub-jobs. To illustrate we define a modified calculation function that takes an extra argument `queue` to collect results using `queue.put`. Note that in this we record `(n, m, result)` so that we know which results go with which inputs, given that jobs will not necessarily finish in any particular order.

```
def calcFuncWithQ(queue, n, m):
  result = sum([x*x for x in range(n) if x % m == 0])
  queue.put( (n, m, result) )
```

We can run things in a similar manner to before, but this time we create the `Queue` object upfront and can use its `.get()` method at the end to fetch the results:

```
queue = Queue()

job1 = Process(target=calcFuncWithQ, args=(queue, 8745676, 2) )
job2 = Process(target=calcFuncWithQ, args=(queue, 2359461, 3) )

job1.start()
job2.start()

job1.join()
job2.join()

print("Result", queue.get())
print("Result", queue.get())

queue.close()
```

Perhaps a simpler way of managing parallel jobs, without having to worry about queues, is to use a `Pool` object to manage the processing. As mentioned above, this also has the advantage of allocating an arbitrary number of jobs to a fixed number of processors/cores.

As an example, we consider a list of input values and `pool` is created to organise the allocation of work. By default a `Pool` will use the total number of central processor cores that are available on the computer as the maximum number of jobs to run at one time, but the number of parallel jobs may be passed in instead (the default comes from `multiprocessing.cpu_count()`).

```
inputList = [37645, 8374634, 3487584, 191981, 754967, 12345]
pool = Pool()
```

With the worker pool created, we next set up the individual jobs via `pool.apply_async()`, which as the name suggests will start the calculations in an *asynchronous* manner. As with the earlier examples the basic point is to associate a worker function with a set of arguments for that sub-job. The `job` objects themselves are collected in a list so that we can get the `return` result of the `calcFunc` call, which in this case is collected for us without having to take any special measures:

```
jobs = []
for value in inputList:
  inputArgs = (value, 2)
  job = pool.apply_async(calcFunc, inputArgs)
  jobs.append(job)
```

We can then collect the results directly from the job objects using `.get()`:

```
results = []
for job in jobs:
  result = job.get()
  results.append(result)
```

And finally we print the result and do some clean up, to close the worker pool and to make sure that the main program does not proceed until the sub-processes have fully terminated.

```
pool.close()
pool.join()

print(results)
```

If speed improvement is still really critical after optimising (cf. Chapter 10) and perhaps parallelising Python, then it might be worth not using Python at all, or at least not entirely. Python code can only be sped up to a point, after which you can try one of the compiled languages like C, C++ or Fortran for the slow bits, which may then be interfaced with the main Python program. With regard to parallel execution, the libraries like OpenMP[4] will allow C and C++ modules to control efficient, fine-grained parallel processing on a computer system with a single, shared memory.

---

[4] http://openmp.org.

Writing faster modules

At one time doing anything especially intensive in Python was not practical. Fortunately though, the main implementation of Python is itself written in C and has a specified, although not extensively documented, way of interfacing C code with the Python world. Thus the idea is to write numerically intensive code in C (or C++, or even Fortran suitably wrapped in C) and then call that routine from inside Python. As we discuss in the last section of this chapter, one way to do this is by using the `ctypes` module. This lets Python create C-compatible data types and call C functions, which is especially useful to interact with established C libraries. However, in the initial instance we will make the language interface in C. In theory you can also access Python data types directly in C, given that's what they are written in, but the most useful approach is to keep most of the C code general (i.e. not Python-specific) and then to wrap a thin layer of specialised interface code in C that connects the Python world to your C code. This interface layer is often described as *boilerplate code*, referring to the fact that you have to write the same kind of interface for each module. Accordingly, and as illustrated in Figure 27.2, when a specific C function gets called from the Python code, the Python variables are converted into variables that your C code understands, the compiled C code is called, the returned variable is converted to a Python data type and the function passes that back to the Python world. You can even use a tool to write this *boilerplate code*, the best known being SWIG (Simplified Wrapper and Interface Generator).

   More recently there has been less reason to go down the C route. For much of the numerical work you can use the NumPy and SciPy modules. Of course, the Python modules they provide are actually also wrappers around C, to make it quick, so the library authors have simply done the hard work for you. However, it's possible that you require the use of some algorithm that is not easily expressed in functionality that NumPy and SciPy provide. After all

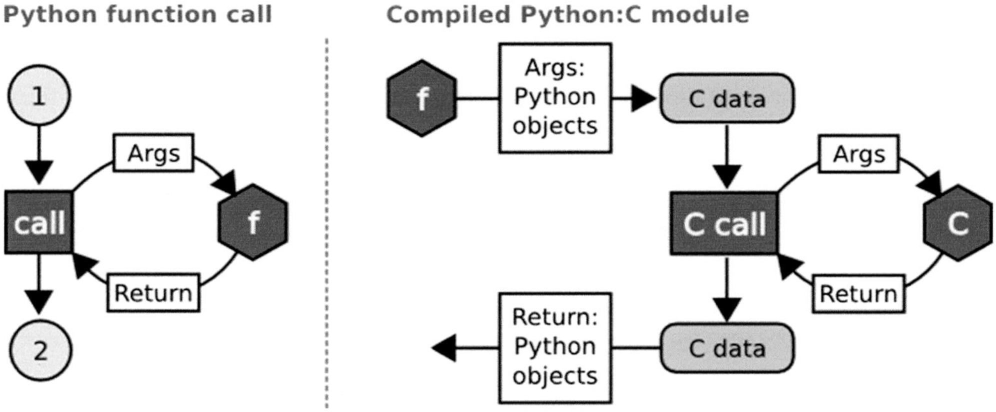

Figure 27.2. An overview of how a module written in C may be wrapped so it can be called from Python. To increase the speed at which calculation-intensive Python programs run we can write fast modules in the compiled language C and encapsulate them so that they can be called like normal Python functions. To do this the C module must be written in such a way as to accept input as Python objects and also to send back any return values as Python objects. This interface can be written by directly accessing the Python data structures using Python's own C library or by using a system like Cython, which can automatically convert between the two systems. Otherwise, once the data is routed to C data structures a fast C routine can be constructed in the normal way.

these libraries naturally provide mathematical and array operations that are general. Fortunately, there are various ways to write C-like code in Python itself, and this provides another way of optimising code for speed. In this chapter we discuss one of these, called Cython. Cython is actually a separate language, although it is very similar to Python and unmodified Python code will normally work directly without any alteration. What Cython offers is a way to mix Python code with some elements of the C language and then to automatically convert this friendly language into pure C code, which is then compiled in the normal C manner, usually to make a Python-compatible module.

There is still at least one good reason why you might want to interface directly to C code, and that is if you have an existing extensive library that you do not want to have to rewrite in NumPy or Cython. It is also possible that a C version of the code is just that much faster that it is worth writing. Unfortunately it is difficult to know for sure which way is actually best until the C code is written. Note that one of the main problems with using C code is that it has to be compiled for each type of machine architecture[5] on which you want it to run. The same is actually true of Python and NumPy, but generally someone else has already done the compiling for you. This is an advantage of Python code, and should not be underestimated, especially if you are intending to distribute your code to other people.

In this chapter we use the self-organising map from Chapter 24 as an example, which should be looked at before reading this chapter in detail. In that chapter a solution to the problem was implemented by using NumPy arrays. Here we will re-implement it in plain (non-NumPy) Python, C and Cython, to give a comparison between the various approaches.

Pure Python implementation

The self-organising map deals with matrices that are potentially large, and thus it would be expected that a pure Python implementation will be pretty slow in comparison with the NumPy implementation. A notable reason for keeping a pure Python implementation is to avoid the requirement of having NumPy installed, given that it is not part of the official Python release. For mainly numerical code it also turns out that a pure Python implementation is fairly close in style and even in syntax to a C implementation. So once we have a pure Python implementation it is going to be fairly easy to write the C version. The main additional burdens in the C implementation will be memory allocation, and wrapping Python around the C code.

The arguments to the main function, `selfOrganisingMap`, will be the same as for the NumPy code, except that we will use lists of lists instead of NumPy arrays. The `inputs` argument is of size the number of inputs times the `depth`, which in our example will be 3 because the data represents an RGB colour image. The `spread` argument is a list of lists of size `width` times `height` describing how far across the self-organising map (grid) the input influence spreads. The `steps` argument is the number of iterations for the algorithm. We need the Python `random` module for generating random numbers, and the `math` module for taking the exponential of a number.

---

[5] Dependent on what kind of CPU is present.

```
import math
import random

def selfOrganisingMap(inputs, spread, size, steps):

  nrows, ncols = size
  depth = len(inputs[0])
```

We then initialise the map to be random numbers between 0 and 1. We use the Python `random.random` function for this. We just loop over all rows and columns and the depth.

```
somap = nrows * [0]
for i in range(nrows):
  somap[i] = ncols * [0]
  for j in range(ncols):
    somap[i][j] = depth * [0]
    for k in range(depth):
      somap[i][j][k] = random.random()
```

We then iterate the number of times specified, updating the map for each cell (row, column). The decay parameter determines the relative influences of the map and the input value in updating the map.

```
for step in range(steps):
  decay = math.exp(-step / float(steps))
  for vector in inputs:
    updateSomap(somap, vector, spread, decay)

return somap
```

The code that updates the map of course implements the same functionality as with the NumPy version but is much more tedious because it all has to be done with explicit looping. First we determine the relevant sizes.

```
def updateSomap(somap, vector, size, spread, decay):

  nrows, ncols = size
  depth = len(vector)
  width = len(spread)
  height = len(spread[0])
```

Then we determine the cell with the smallest difference between the map and the input.

```
imin = jmin = -1
diff2min = 0
for i in range(nrows):
  for j in range(ncols):
    diff2 = 0.0
    for k in range(depth):
      diff = somap[i][j][k] - inputs[k]
      diff2 += diff * diff

    if (imin == -1) or (diff2 < diff2min):
      imin = i
```

```
jmin = j
diff2min = diff2
```

Then this cell and surrounding cells, determined by the `spread` matrix, are updated, using a weighting determined by the `decay` parameter.

```
halfWidth  = (width-1) // 2
halfHeight = (height-1) // 2

for k in range(width):
  i = (imin + k - halfWidth + nrows) % nrows
  for l in range(height):
    j = (jmin + l - halfHeight + ncols) % ncols
    alpha = decay * spread[k][l]
    for m in range(depth):
      somap[i][j][m] = (1.0-alpha) * somap[i][j][m] + alpha * vector[m]
```

NumPy implementation

With NumPy a lot of the pain is removed from matrix manipulations, and in particular there are fewer explicit loops. In particular, the `updateSomap` function in the pure Python implementation becomes just one line in the NumPy version. See Chapter 24 for further discussion of this implementation, since the code is just a repeat of what is there.

```
def selfOrganisingMap(inputs, spread, size, steps=1000):

  nRows, nCols = size
  vecLen = inputs.shape
  somap = numpy.random.rand(nRows, nCols, vecLen)

  influence = numpy.dstack([spread]*vecLen) # One for each feature
  infWidth = (len(influence)-1) // 2
  makeMesh = numpy.ix_    # Ugly

  for s in range(steps):

    decay = numpy.exp(-s/float(steps))

    for vector in inputs:

      diff =  somap-vector
      diff2 = diff*diff
      dist2 = diff2.sum(axis=2)

      index = dist2.argmin()
      row = index // nRows
      col = index % nRows

      rows = [x % nRows for x in range(row-infWidth, row+1+infWidth)]
      cols = [y % nCols for y in range(col-infWidth, col+1+infWidth)]

      mesh = makeMesh(rows,cols)
      somap[mesh] -= diff[mesh] * influence * decay

  return somap
```

C implementation

The C implementation is very similar to the pure Python implementation except that there are memory management issues to address. These do not occur in Python because there is automatic *memory allocation* and freeing of unused memory (called *garbage collection*). Further, if you want to be able to call the C functionality from Python then you need to write additional code to accomplish this, to convert from Python's own data structures to more normal C types on the way in, and convert the result C types to Python's own data types on the way out. This approach assumes that you have an existing C library that you want to use without having to rewrite it. If instead you were writing a C implementation from scratch then an alternative strategy would be to use Python's or NumPy's own types throughout your C code. Whether or not this is a good idea depends on the details of the code, and what else you might want to do with it, though the C types generally give quicker execution. Here we will write the core part of the C implementation to use standard C types rather than Python's or NumPy's own types. We will not explain all the vagaries of the C syntax but just the important differences with Python.

      In C it is standard practice to have two files go together: the first (suffix '.h') gives a specification of the interface to the functionality, in particular stating what functions you want to make publicly accessible, and the other file (suffix '.c') is the actual implementation of the functionality. Here we will call the former file `somap.h` and the latter `somap.c`. In general there might be lots of functions that you want to specify as being publicly accessible. Here we will only have one function, so `somap.h` has:

```
extern double ***selfOrganisingMap(double **inputs,
                int ninputs, int depth, int nrows, int ncols,
                double **spread, int width, int height, int nsteps);
```

In C we have to specify types explicitly. Here we have '`int`' for integers and '`double`' for *double-precision* floating point numbers.[6] The two-asterisk `**` and three-asterisk `***` syntaxes are effectively the C way of specifying that the corresponding variable is a list of lists or a list of lists of lists, respectively. The function `selfOrganisingMap` is expecting nine arguments. The extra arguments compared with the Python implementation are to specify the sizes of the lists, because in C the list variables themselves do not contain this information (one of the common sources of bugs in C programs). The result returned has to be specified and here it is a list of lists of lists of double-precision floats, which represents the constructed map. The '`extern`' just says that the function is implemented elsewhere (i.e. in this case in the file `somap.c`).

      The implementation of `selfOrganisingMap` is very similar to the Python version. In `somap.c` we declare the function, just repeating what is given in `somap.h` but without the '`extern`':

```
double ***selfOrganisingMap(double **inputs,
                int ninputs, int depth, int nrows, int ncols,
                double **spread, int width, int height, int nsteps)
```

In C, blocks of code are delineated not by whitespace but by curly brackets, '{' and '}'. In C we also need to explicitly declare the data type of any variable that is being used. And in C statements have to end with a semicolon ';'.

---

[6] Python float numbers have this double precision as standard.

```
{
  int step, i;
  double decay;
  double ***somap;
```

We then set up the initial value for `somap`. This is similar to what happens in Python except that here we also require explicit memory allocation, and to be tidy we put it in a separate function, `randomMap`. In C the equivalent (as such) of `None` is called `NULL`. For the memory allocation a result of `NULL` means the allocation failed. This can be checked with the short expression `!somap`. If that happens there is not much we can do except to give up.

```
somap = randomMap(nrows, ncols, depth);
if (!somap)
  return NULL;
```

We then repeat the same loops as in the Python version:

```
for (step = 0; step < nsteps; step++)
{
  decay = exp(-step / (double) nsteps);
  for (i = 0; i < ninputs; i++)
    updateSomap(somap, inputs[i], depth , nrows, ncols,
                spread, width, height, decay);
}

return somap;
}
```

The `randomMap` function does the memory allocation and setting the initial values to random numbers. The 'static' means that this function is not visible outside this file, so is private. The function which allocates memory is called `malloc` and you have to tell it how many bytes of memory you want allocated, and `sizeof` provides the relevant multiplier to go from the number of objects to the number of bytes.

```
static double ***randomMap(int nrows, int ncols, int depth)
{
  int i, j, k;
  double ***x;

  x = (double ***) malloc(nrows * sizeof(double **));
  if (!x)
    return NULL;

  for (i = 0; i < nrows; i++)
  {
    x[i] = (double **) malloc(ncols * sizeof(double *));
    if (!x[i])
      return NULL;

    for (j = 0; j < ncols; j++)
    {
      x[i][j] = (double *) malloc(depth * sizeof(double));
      if (!x[i][j])
        return NULL;
```

```
        for (k = 0; k < depth; k++)
           x[i][j][k] = randomNumber();
    }
  }

  return x;
}
```

The `randomNumber()` function should return a uniformly sampled random number between 0 and 1. On most computers there is a function provided for this; for example, on many Unix systems there is one called `drand48()`.

The `updateSomap()` function is very similar in C to the pure Python implementation, except here we have to declare the type of variables.

```
static void updateSomap(double ***somap, double *input,
                int depth, int nrows, int ncols,
                double **spread, int width, int height, double decay)
{
  int halfWidth, halfHeight, i, j, k, l, m, imin, jmin;
  double diff, diff2, diff2min, lambda;

  imin = jmin = -1;
  diff2min = 0; // will change in first pass
  for (i = 0; i < nrows; i++)
  {
    for (j = 0; j < ncols; j++)
    {
      diff2 = 0.0;
      for (k = 0; k < depth; k++)
      {
        diff = somap[i][j][k] - input[k];
        diff2 += diff * diff;
      }

      if ((imin == -1) || (diff2 < diff2min))
      {
        imin = i;
        jmin = j;
        diff2min = diff2;
      }
    }
  }

  halfWidth = (width-1) / 2;
  halfHeight = (height-1) / 2;

  for (k = 0; k < width; k++)
  {
    i = (imin + k - halfWidth + nrows) % nrows;
    for (l = 0; l < height; l++)
    {
      j = (jmin + l - halfHeight + ncols) % ncols;
```

```
      lambda = decay * spread[k][l];
      for (m = 0; m < depth; m++)
         somap[i][j][m] = (1.0-lambda) * somap[i][j][m] + lambda * input[m];
   }
 }
}
```

The wrapper around this C code that converts to and from Python data types is the boilerplate code. We will name the file `py_somap.c`, but it could be called anything. The way that Python interfaces to C code changed between Python 2 and Python 3, and if you want to write this wrapper code to be compliant with both then you can check whether the defined constant `PY_MAJOR_VERSION` is $>= 3$, and if it is then use the Python 3 form of the code, and otherwise the Python 2 form. C has convenient macros `#ifdef` / `#else` / `#endif` which can be used to selectively include code in compilation.

In order to access the module in Python there must be a publicly available function. In Python 2 it is named as '`init`' concatenated with the module name. Here we will name the module `somap` and so the function must be named `initsomap`. This function returns nothing, which in C means `void`. We first need to call the internal Python function (albeit via C) `Py_InitModule` in order to specify the module name and functions that will be made available to Python.

In Python 3 the publicly available function instead has '`PyInit_`' concatenated with the module name, so here `PyInit_somap`. It returns the `module` object, which is of type `PyObject *`. And to initialise the module we call the function `PyModule_Create()`, which expects an argument of type `PyModuleDef *`, which in turn contains the name of the module and the functions that will be made available to Python.

Then we need to call `import_array` in order to initialise NumPy and also create an exception object, `ErrorObject`, so that we can report errors that occur when using this module. In Python 2 we return nothing, but in Python 3 we return the `module` object.

```
#if PY_MAJOR_VERSION >= 3
static struct PyModuleDef module_def = {
    PyModuleDef_HEAD_INIT,
    "somap",                /* m_name */
    NULL,                   /* m_doc */
    -1,                     /* m_size */
    Somap_type_methods,     /* m_methods */
    NULL,                   /* m_reload */
    NULL,                   /* m_traverse */
    NULL,                   /* m_clear */
    NULL,                   /* m_free */
};
#endif

#if PY_MAJOR_VERSION >= 3
PyObject *PyInit_somap(void)
#else
void initsomap(void)
#endif
```

```
{
  PyObject *module;

  /* create the module and add the functions */
#if PY_MAJOR_VERSION >= 3
  module = PyModule_Create(&module_def);
#else
  module = Py_InitModule("somap", Somap_type_methods);
#endif

  import_array();  /* needed for numpy, otherwise it crashes */

  /* create exception object and add to module */
  ErrorObject = PyErr_NewException("somap.error", NULL, NULL);
  Py_INCREF(ErrorObject);
  PyModule_AddObject(module, "error", ErrorObject);

  /* check for errors */
  if (PyErr_Occurred())
    Py_FatalError("can't initialize module somap");

#if PY_MAJOR_VERSION >= 3
  return module;
#endif
}
```

Everything else in the module will be 'static', i.e. private in the C world (although the specified functions will be available in the Python world), and should be placed above this function in the file.

The exception object is just a simple variable. All Python objects in C are of type either `PyObject*` or extensions thereof.

```
static PyObject *ErrorObject;
```

The functions available to the module are passed as the second argument to `Py_InitModule` and here this is called `Somap_type_methods`. This is a list with each entry containing four items: the name of the function (method) as it will be called in Python, the function itself, the calling convention and a documentation string, here called `somap_doc`. The list should be terminated with an entry that contains a `NULL` value for the function name. Here we only have one function, which we also call `somap`.

```
static char somap_doc[] = "Creates a self-organising map";

static struct PyMethodDef Somap_type_methods[] =
{
  { "somap",   (PyCFunction) somap,   METH_VARARGS,   somap_doc },
  { NULL,      NULL,                  0,              NULL }
};
```

The most common convention is `METH_VARARGS`, where the functions expect two arguments, the first being the object itself (by convention called `self`, as in Python code) and the second being a tuple containing the values passed to the function in Python (by convention called `args`). For other calling conventions see the Python documentation (see links at http://www.cambridge.org/pythonforbiology).

The function `somap` is what does the work converting from the Python world to the C world, calling the C function and then converting back from the C world to the Python world. It first unpacks the arguments, and checks that they are of the correct data type. The function `PyArg_ParseTuple` unpacks the `args` tuple. Here the tuple is expected to be of length five, because there are five arguments that are passed in from the Python code. The first element of the tuple is the `inputs` object, the second the `spread` object, the third the number of rows, the fourth the number of columns and the fifth the number of steps for the algorithm. The 'O!' in the call to `PyArg_ParseTuple` says that the corresponding element in the tuple must be Python objects of the specified type, here `PyArray_Type`, which is the NumPy array type. The 'i' in the call to `PyArg_ParseTuple` says that the last three elements must be integers. Next it is checked that `inputs_obj` and `spread_obj` are both of the required type and shape as NumPy arrays. The `RETURN_ERROR` C *macro* is a shorthand for creating an error report (exception) that will be passed back to Python.

```
#define RETURN_ERROR(message) \
        { PyErr_SetString(ErrorObject, message); return NULL; }
static PyObject *somap(PyObject *self, PyObject *args)
{
  int ninputs, nsteps, nrows, ncols, depth, width, height, i, j, k;
  PyArrayObject *inputs_obj, *spread_obj;
  PyObject *somap_obj;
  double **inputs, **spread, ***somap;
  npy_intp dims[3];

  if (!PyArg_ParseTuple(args, "O!O!iii", &PyArray_Type, &inputs_obj,
               &PyArray_Type, &spread_obj, &nrows, &cols, &nsteps))
    RETURN_ERROR("need 5 args: inputs, spread, nrows, ncols, nsteps");

  if (!PyArray_Check(inputs_obj))
    RETURN_ERROR("inputs needs to be NumPy array");

  if (PyArray_NDIM(inputs_obj) != 2)
    RETURN_ERROR("inputs needs to be NumPy array with ndim 2");

  if (!PyArray_Check(spread_obj))
    RETURN_ERROR("spread needs to be NumPy array");

  if (PyArray_NDIM(spread_obj) != 2)
    RETURN_ERROR("spread needs to be NumPy array with ndim 2");

  if (PyArray_TYPE(inputs_obj) != NPY_DOUBLE)
    RETURN_ERROR("inputs needs to be array of doubles");

  if (PyArray_TYPE(spread_obj) != NPY_DOUBLE)
    RETURN_ERROR("spread needs to be array of doubles");
```

Next the function determines the size of `inputs_obj` and `spread_obj` and then copies these two NumPy arrays into standard C arrays.

```
  ninputs = PyArray_DIM(inputs_obj, 0);
  depth = PyArray_DIM(inputs_obj, 1);
```

```
width = PyArray_DIM(spread_obj, 0);
height = PyArray_DIM(spread_obj, 1);

if (!(inputs = copyArray2(inputs_obj)))
  RETURN_ERROR("getting inputs as C array");

if (!(spread = copyArray2(spread_obj)))
{
  freeArray2(inputs, nrows, ncols);
  RETURN_ERROR("getting spread as C array");
}
```

If there are any problems then the function returns an error message. The `copyArray2` function includes memory allocation, and there are of course corresponding memory-freeing functions. We also need a `freeArray3` function for later use.

```
static void freeArray2(double **array, int m)
{
  int i;

  for (i = 0; i < m; i++)
    free(array[i]);

  free(array);
}

static void freeArray3(double ***array, int m, int n)
{
  int i;

  for (i = 0; i < m; i++)
    freeArray2(array[i], n);

  free(array);
}

static double **allocArray2(int m, int n)
{
  int i;
  double **array;

  array = (double **) malloc(m*sizeof(double *));
  if (array)
  {
    for (i = 0; i < m; i++)
      array[i] = (double *) malloc(n*sizeof(double));
  }

  return array;
}

  static double **copyArray2(PyArrayObject *array_obj)
  {
    int i, j, m, n;
    double **array;
```

```
  m = PyArray_DIM(array_obj, 0);
  n = PyArray_DIM(array_obj, 1);
  array = allocArray2(m, n);

  if (array)
  {
    for (i = 0; i < m; i++)
    {
      for (j = 0; j < n; j++)
        array[i][j] = *((double *)
                        PyArray_GETPTR2(array_obj, i, j));
    }
  }

  return array;
}
```

Getting back to the somap function, up to this point we have just been converting the Python data types into C data types. Next the function calls the straight C function that we wrote above, which is the whole point of the exercise and is the one line of code that actually does anything directly useful.

```
somap = selfOrganisingMap(inputs, ninputs, depth, nrows, ncols,
                          spread, width, height, nsteps);
```

Finally the function copies the result back into a Python data type and frees memory and returns the result to Python.

```
dims[0] = nrows;
dims[1] = ncols;
dims[2] = 3;
somap_obj = PyArray_SimpleNew(3, dims, NPY_DOUBLE);
for (i = 0; i < nrows; i++)
  for (j = 0; j < ncols; j++)
    for (k = 0; k < depth; k++)
      *((double *) PyArray_GETPTR3(somap_obj,i,j,k)) = somap[i][j][k];

freeArray3(somap, nrows, ncols);
freeArray2(inputs, ninputs);
freeArray2(spread, width);

return somap_obj;
}
```

Once written, the C code needs to be *compiled* before it can be used, to convert the textual source code into binary code that can be executed on a computer platform. Further information on how to compile and run the C code examples is given at http://www.cambridge.org/pythonforbiology.

Cython implementation

The Cython language allows C-like constructs, including data type information, to be introduced into what otherwise looks like Python code. This allows for much faster execution at the expense of more complicated coding, but compared to regular C using

Cython greatly simplifies the conversion between the C and Python worlds. Cython code is conventionally placed in a file ending with `.pyx`, and the Cython compiler then converts that into a compiled C library which can then be imported in Python code, the same way as with directly compiled C code. The compilation is normally done via a Python `setup.py` file that controls a regular C compiler (like `cc` or `gcc`). For details see the Cython documentation (see links at http://www.cambridge.org/pythonforbiology).

Pure Python code is valid Cython, and you can usually get a modest speed-up just running ordinary Python code through the Cython compiler, although this will naturally restrict portability because the code will only run on the processor architecture for which it was compiled. The Cython implementation of the self-organising map could be made very similar to the C implementation. As in the Python+C implementation, you could convert from Python objects to their C equivalent, then do the calculation, then convert the result back from C objects to Python objects. In particular, a list in Python would get converted to a C pointer and require memory allocation and freeing. However, we will avoid the complication of data conversion and memory allocation and freeing by instead using NumPy data throughout the code. This makes the code simple to write and avoids a tedious aspect of writing code in C.

We will not explain all the Cython constructs, but the main difference in syntax with ordinary Python is that C data types are specified using `cdef`. Though Python objects can also have a type specified, and in general even that makes for faster execution. In addition, imports of C libraries are done using `cimport` instead of `import`. Here we need the NumPy C `ndarray` data type. We also need `random` and `exp` from the standard NumPy. Finally, we need the import of `cython` itself.

```
from numpy cimport ndarray
from numpy import random, exp
import cython
```

The main function in the Cython implementation is different from other functions we have seen in that it has some additional Cython-specific typing for some of the arguments. In particular, we want the `inputs` and `spread` arguments to be two-dimensional NumPy arrays. They both have type `double`, which is the default NumPy type for floating point (real number) data.

```
def selfOrganisingMap(ndarray[double, ndim=2] inputs,
                      ndarray[double, ndim=2] spread,
                      size, nsteps):
```

We then define the incidental variables we use in the code to all be C types rather than Python types. This includes the map itself, which will be a three-dimensional NumPy array. It is important to use C data types to make Cython code fast, although some variables are less important than others for speed.

```
cdef int nrows, ncols, ninputs, depth, width, height
cdef int step, i, j
cdef double decay
cdef ndarray[double, ndim=3] somap
```

We then initialise the variables. The `somap` variable can be initialised using the standard NumPy random number generation functionality.

```
nrows, ncols = size
ninputs = len(inputs)
depth = len(inputs[0])
width = len(spread)
height = len(spread[0])

somap = random.random((nrows, ncols, depth))
```

We then iterate over `nsteps` in the same way as previously.

```
for step in range(nsteps):
  decay = exp(-step / float(nsteps))
  for i in range(ninputs):
    updateSomap(somap, inputs[i], spread, nrows, ncols, depth,
                width, height, decay)
```

Finally, we return `somap`.

```
return somap
```

The important work is done in the `updateSomap()` function. In order to make Cython as quick as possible, we turn off some checking that would otherwise be done on arrays, using function decorators from the `cython` module (with the '@' syntax; see Chapter 5). This is dangerous to do and should only be done if you are very confident that your code has no array-indexing bugs in it. Python handles such errors graciously, but C does not.

```
@cython.boundscheck(False)
@cython.nonecheck(False)
@cython.wraparound(False)
cdef void updateSomap(ndarray[double, ndim=3] somap,
                      ndarray[double, ndim=1] inputs,
                      ndarray[double, ndim=2] spread,
                      int nrows, int ncols, int depth,
                      int width, int height, double decay):
```

We then define the variables we will use.

```
cdef int halfWidth, halfHeight, i, j, k, l, m, imin, jmin
cdef double alpha, diff, diff2, diff2min
```

The rest of the code is very similar to the C code. It is important to use the bracket notation `[i, j, k]` for array access rather than `[i][j][k]` because otherwise sub-arrays are constructed, which significantly reduces the speed. It is also important that these indices be C integers rather than Python integers.

```
halfWidth = (width-1) // 2
halfHeight = (height-1) // 2

imin = jmin = -1
diff2min = 0.0
for i in range(nrows):
  for j in range(ncols):
    diff2 = 0.0
```

```
    for k in range(depth):
      diff = somap[i,j,k] - inputs[k]
      diff2 += diff * diff

    if ((imin == -1) or (diff2 < diff2min)):
      imin = i
      jmin = j
      diff2min = diff2

for k in range(width):
  i = (imin + k - halfWidth + nrows) % nrows
  for l in range(height):
    j = (jmin + l - halfHeight + ncols) % ncols
    alpha = decay * spread[k,l]
    for m in range(depth):
      somap[i,j,m] = (1.0-alpha) * somap[i,j,m] + alpha * inputs[m]
```

We could use the following NumPy code, which is similar to code we used in the NumPy implementation, to replace the calculation of `imin` and `jmin`:

```
cdef ndarray[double, ndim=3] diff
cdef ndarray[double, ndim=2] diff2
diff = somap - inputs[i]
diff2 = (diff*diff).sum(axis=2)
cdef int index = dist2.argmin()
imin = index // nrows
jmin = index % nrows
```

However, it turns out that this is much slower.

Before the above functions can be used, the code must be transformed into binary code that can actually be run. This is a two-step process in Cython. Initially the Python-like code is converted into the equivalent C and then the C code is compiled into binary code that can be executed on a particular type of computer system.[7] These conversions are fairly easy to do using the Cython system. In essence, we run a setup script with Python, which from an operating-system prompt would look something like:

```
>python setup.py build_ext --inplace
```

The 'setup.py' script instructs Cython which Python-like `.pyx` files to convert into fast, compiled Python modules, and what the names of those modules should be. We give full details of the setup script and how compilation is practically achieved on different computer systems at http://www.cambridge.org/pythonforbiology. Our example uses 'somapCython.pyx' to create the 'somapCython' module (the actual compiled file is 'somapCython.so'), which can then be imported into standard Python code:

```
from somapCython import selfOrganisingMap
```

And, if we set appropriate values for the input arguments, the function can be called from Python in the normal manner:

```
mapOut = selfOrganisingMap(inData, spreadArray, mapSize, numSteps)
```

---

[7] It will be compiled so that it runs on a particular processor architecture, such as x86, x86-64, ARM etc.

Speed comparison of implementations

It is often difficult to compare the exact speed of implementations because the details of how something is compiled can make a large performance difference. For example, with the C implementation, you can get a difference in speed of a factor of four just by choosing different optimisation options for the compiler. For the data given here we use the GCC compiler version 4.6.3 with the -o2 flag,

It is also important to understand how the running time scales with the size of the problem: in this case how big the input data arrays are. All the implementations scale in the same basic way, with the slight exception of the NumPy implementation, where some of the array calculations are proportionally more efficient for larger arrays; although once large enough to be efficient the NumPy scaling will be about the same. Here we have to consider the size of the inputs array (parameters ninputs and depth) and the size of the spread array (parameters width and height) and the number of steps (parameter steps), in addition to the size of the output self-organising map (parameters rows, cols and depth). For testing we have taken ninputs to be rows × cols. In many problems it is possible that the spread array is of a fixed size (e.g. 5×5) and the depth is of a fixed size (e.g. 3) and that we really might consider varying just the three parameters rows, cols and steps. Looking at the algorithm we see that the number of operations scales proportionately to steps, but with the squares of both rows and cols. Typically the rows and cols are the same, such that doubling each multiplies the number of operations by (approximately) 16. This is not a fast algorithm.

There are three C-like implementations that can be tested and compared to the Python versions. Firstly, we can test the C implementation with code itself written entirely in C (and quite possibly this might be how such an implementation would have been used originally). Secondly, we can test the C implementation using the Python wrapping around it that we have provided above. Thirdly, we can test the Cython implementation. Not surprisingly, these all come out at around the same speed. Note that all of these tests depend on how the C code and the Python code are compiled. Naturally, the relative values are specific to the actual program being tested, so the speed-up factors are only a rough guide to the general situation.

The pure Python implementation is unsurprisingly the slowest, by more than two orders of magnitude compared with the fastest.

Table 27.1. *Comparing Python, Cython and C module execution speeds.* *Speed test results for various implementations of the self-organising map; testing 100 iterations of a 100 by 100 input matrix. The Python version used during testing was 2.7.3 and the C compiler used was GGC version 4.6.3 with the −O2 option.*

Implementation	Average run time (seconds)	Speed-up factor
C	82	227
C+Python	85	219
Cython	81	230
Python+NumPy	658	28
Python	18616	1

The '`ctypes`' module

Python has a module, called `ctypes`, which lets Python code interact directly with C data types and call C functions. This is not necessarily much easier than writing Python/C wrapper code, as was done above, but it does save having to compile the code, although you still have to be able to understand the C specification of whatever functionality you are using. It is also important to understand who (the C world or the Python world) 'owns' dynamically allocated memory, otherwise it could potentially cause problems with memory leaks, or crashes from using memory that has been freed. For complete documentation and further discussion of the many issues that need to be considered, see the documentation page for the `ctypes` module on the Python website, http://www.python.org. We will illustrate a few examples here using the C runtime library, because that is available on most computer systems. The functionality we use as an example is already available in Python, but it nonetheless shows the beginnings of how the `ctypes` module works.

The `ctypes` module has an object, `cdll`, for loading dynamically linked libraries, and so the first thing to do is to import that:

```
from ctypes import cdll
```

On Windows machines the C runtime library is available directly as an attribute of `cdll`. On other platforms, like Linux and OSX, there is instead a function, `LoadLibrary()`. This has one argument, which is the file name of the library (so including the suffix). The file name is platform specific. But there is a utility function, `find_library()`, which allows the user to find the file name of the standard libraries such as the C runtime library. To determine if we are under a Windows operating system, we can use `sys.platform`, and if it starts with `"win"` we assume we are using Windows.

```
import sys
if sys.platform[:3] == "win":
  libc = cdll.msvcrt
else:
  from ctypes.util import import find_library
  fileName = find_library("c") # "c" for C runtime library
  libc = cdll.LoadLibrary(fileName)
```

Once we have the handle to the C runtime library we can call its available functionality. For example, to call the C `time()` function we just do:

```
print("time = %d" % libc.time(None))
```

This prints the number of seconds since 1 January 1970. Here the argument `None` to `time()` represents the C null pointer.

The standard C print function, `printf()`, is also available, and this illustrates how to deal with C types. Only the following restricted set of Python data types can be passed directly to C functions: `None`, integers (and longs, in Python 2) and bytes objects (and strings in Python 2). Other types need converting. For example, for Python floating point numbers there are three corresponding C data types: 'float', 'double' and 'long double'. Respectively, these have their own conversion functions: `c_float()`, `c_double()` and `c_longdouble()`. Accordingly, to print a Python `float` as a C 'double', to three decimal places, we can do:

```
from ctypes import c_double
x = 3.14159
libc.printf(b"x = %.3f\n", c_double(x))
```

In Python 3 the 'b' converts the string to a bytes object, in Python 2 it is not needed but it works (for Python 2.6 and 2.7). In C the `printf()` function returns the number of characters written, and so the above `printf()` gives two lines of output when called from the Python prompt:[8]

```
x = 3.142
10
```

In addition to the standard types, C also allows user-defined data types. These are just a list of attributes, and for each attribute a type. The `ctypes` module has a Python class called `Structure`, and by subclassing this you effectively get the Python version of a C data type. In this class you specify the attribute `_fields_` (only one underscore before and after, not two). This is a list and each element of the list contains a 2-tuple, where the first element of the tuple is the name of the associated datum, and the second element of the tuple is its type. The names are your choice, but it is good practice to use the C names, which can be found from reading the C documentation.

We will illustrate use of `Structure` with a calendar example. For working with calendar time there is a C data type, 'struct tm', which stores the second, month, hour, day and so on. You have to read the C documentation to know exactly how it is stored in order to be able to use it via `ctypes`. For 'struct tm' the data type of all the attributes is the C 'int' so here we use `c_int`. This leads to:

```
from ctypes import Structure, c_int

class TimeStruct(Structure):
  _fields_ = [ \
    ('tm_sec', c_int),   # seconds
    ('tm_min', c_int),   # minutes
    ('tm_hour', c_int), # hours
    ('tm_mday', c_int), # day of the month
    ('tm_mon', c_int),   # month
    ('tm_year', c_int), # year
    ('tm_wday', c_int), # day of the week
    ('tm_yday', c_int), # day in the year
    ('tm_isdst', c_int) # daylight saving time
  ]
```

In the following example we will fetch the current time in seconds, using the `libc.time()` function mentioned above, and then use the function `libc.localtime()`, which will take an input time in seconds and convert it into a `TimeStruct`. Here we need to set the return data type of `localtime()`, otherwise Python will interpret it as an integer (the default). This is done by setting the attribute `restype` of the function.

```
from ctypes import POINTER , c_long, byref
libc.localtime.restype = POINTER(TimeStruct)
```

---

[8] The count of 10 includes the newline character, '\n'.

When we fetch the time, one technical detail is that `time()` returns a Python integer but we need it to be a `c_long`, so we convert:

```
t = libc.time(None)
t = c_long(t)
```

Then the time in seconds has to be passed into the `localtime()` function using `byref()`, which requires that the calling argument should be passed by reference rather than by value.

```
resultPtr = libc.localtime(byref(t))
```

From reading the C documentation we know what the return data type is: it is what C calls a 'pointer', which in this case is to the `TimeStruct` object. Here, this can be thought of as a Python list of length one, where the one and only element is the actual `TimeStruct` object. Hence we take index `0`:

```
result = resultPtr[0]
```

Finally, we print out the result using `TimeStruct` attributes. The year starts at 1900, so we add that to turn it into the usual convention. And the month starts at 0 rather than 1, so we also add that.

```
print("day = %04d %02d %02d, time = %02d:%02d:%02d" %
        (result.tm_year+1900, result.tm_mon+1, result.tm_mday,
         result.tm_hour, result.tm_min, result.tm_sec))
```

Appendices

These appendices contain simple explanations and definitions for a subset of the Python language, its standard libraries and a few of the key modules used in this book. The objective is not to give a complete description of every possible option, which is already documented (and will be more up-to-date) on the Internet. Rather, the aim is to cover all of the components used in this book as well as a few extra useful details in relatively plain English, to help with learning the language. Hence, it is deliberate that we have simplified or omitted certain details to avoid obfuscating the main points for novice programmers. While we describe most of the core components of standard Python, for some of the libraries we will only highlight some parts we have found particularly useful. In some cases, where we don't describe individual components, we will describe what a library or a module is generally used for, in order to guide further investigation.

In addition to the material presented here, the website http://www.cambridge.org/pythonforbiology provides links to full, in-depth documentation for Python and the associated libraries that are used throughout this book.

APPENDIX 1
Simplified language reference

Keywords

The following table gives basic descriptions of the Python keywords, which are a reserved part of the language and so cannot be used for variable names.

Keyword	Description	Example
and	Performs the logical AND operation (conjunction) on two values, yielding a true value if both are true, and otherwise a false value.	`if x and y:` `    print("Both are true")`
as	Used in conjunction with the `with` keyword to define a variable that holds a context manager object. Alternatively used with `import` statements to create synonyms for external components.	[See examples for 'with' and 'import']
assert	Checks whether a statement is true, and if not triggers an exception. Often used to check the validity of input values. The command is ignored if Python is run in optimised mode.	`assert(isinstance(x, int))` **Note:** triggers an `AssertionError` exception if the test fails.

break	Causes the program execution within current, innermost, iterative loop (invoked with `for` or `while`) to stop immediately.	`for value in valueList:`
		`    if value > 0:`
		`        print('Loop stopped')`
		`        break`

class	Defines a class: the construction of a named type of Python object, which may be based on one or more other classes.	`class Biologist(Scientist):` `    field = 'Biology'`

continue	Causes program execution within current, innermost, iterative loop (invoked with `for` or `while`) to skip to the next iteration.	`for value in valueList:`
		`    if value > 0:`
		`        print('Value skipped')`
		`        continue`

def	Defines a named Python function (subroutine), the code for which is in an indented block.	`def dotProduct(vec1, vec2):`
		`    pairs = zip(vec1, vec2)`
		`    p = sum([a*b for a,b in pairs])`
		`    return p`

del	Deletes a variable or item, directly from memory or from an enclosing data structure.	`p = ['Neptune','Uranus','Pluto']`
		`del p[2]`
		Note: deleted list item with index 2 (Pluto).

Keyword	Description	Example

elif	Used after an `if` statement to define a further block of code that is conditionally executed if a different condition holds.	```if x > 0:```
		``` print("x is positive")```
		```elif x < 0:```
		``` print("x is negative")```

else	Used after an `if` statement to define a final block of code that is executed if none of the conditions after the `if` or `elif` statements hold. Also used after `try` to define a block that is executed if no error occurs, and after `for` to define a block that is executed if the `for` loop has not been exited because of a `break`.	```if x % 2:```
		``` print("x is odd")```
		```else:```
		``` print("x is even")```

except	Used after the `try` keyword to provide a code block that catches and deals with particular kinds of error exception.	[See example for `try`]
exec	Executes Python statements that are contained within text strings or a file; allows the creation of dynamic code.	```text = 'sum(range(1,100,3))'``` ```exec(text)```

finally	Used after the `try` keyword in order to specify a block of code that is always executed after exception handling, typically to perform clean-up operations.	```fileObj = open(name)```
		```try:```
		``` print(fileObj.readlines())```
		```finally:```
		``` fileObj.close()```

for	Creates a repeating, iterative loop where a variable is assigned values in turn from a collection or iterable object.	```total = 0.0```
		```for value in [1,3,5,7,9,11]:```
		``` total += value ** 0.5```

from	Used in a statement to specify which module to import a component from.	```from math import sqrt``` ```y = sqrt(2.0)```

global	Makes named variables accessible in a global context, i.e. throughout the whole file, even if they only appear within inner, local code blocks.	```def func(text):```
		``` global value```
		``` value = len(text)```
		```func('hello')```
		```print(value) # 5```

if	Used to execute a block of code only on condition that a statement holds. May be used in conjunction with `elif` and `else`.	```if x > 0:``` ``` print("Here if x is positive")```

import	Imports a named external module so that it is available within the current program. Can be used with `from` to import a particular sub-component and with `as` to create a synonym for the module.	```import math```
		```y = math.sqrt(2.0)```
		
		```import numpy as np```
		```a = np.array([2,0,4,1])```

in	Used in various situations. On its own used in tests whether an item is a member of a collection, giving `True` if it is and `False`	```letters = ['G','C','A','T']```
		```if x in letters:```
		``` print('x is in the list')```

Keyword	Description	Example
	otherwise. Also used after `for` and `exec` keywords.	[See also `for` and `exec` keywords]
`is`	Compares two Python objects to test whether they are the same object, yielding `True` if they are and `False` otherwise.	`if x is None:` `print("x not defined")`
`is not`	Compares two Python objects to test whether they are not the same object, yielding `True` if they are different and `False` if they are the same object.	`if x is not None:` `print("x is defined")`

`lambda`	Creates an unnamed (anonymous) function. Useful for defining small functions in place in larger expressions.	`data = [('A',4), ('B',9), ('C',2)]`
		`data.sort(key=lambda x: x[1])`
		Sorts on the second item.

`not`	Performs a logical NOT operation (negation) on a value, giving `True` if the value is false and `False` if the value is true.	`if not x:` `print("x is false")`
`or`	Performs the logical OR operation (disjunction) on two values, yielding a true value if one or other is true, and otherwise a false value.	`if x or y:` `print("At least one is true")`

`pass`	A keyword that doesn't actually do anything except form part of a valid Python syntax. Used as a placeholder where code may be filled in later.	`if x is None:`
		  `pass # Add code later`
		`else:`
		  `x += 2.0`

`print`	Displays a line of text to screen (standard out), with an option to redirect the data, e.g. to a file. In Python 2 `print` is a statement or a function. In Python 3 it is a function.	`print 'Hello'`
		`print >> fileObj, "Data line"`
		Note: above lines are for Python 2.
		`print('Hello')`
		Note: above line is for Python 2 or 3.
		`print('Data line', file=fileObj)`
		Note: above line is for Python 3.

`raise`	Manually triggers an error exception. May be used on its own for custom exceptions or to re-throw error objects caught with `try:` and `except:`.	`if x == 0.0:` `raise Exception("Detected zero")`

`return`	Exits from a named function and optionally passes back one or more items to the point where the function was called.	`def countWords(text):`
		  `n = len(text.split())`
		  `return n`
		`print(countWords('Hello world'))`

`try`	Encapsulates a block of code so that if an illegal state triggers an exception the error can be caught and dealt with in a special way. Used in conjunction with `except` and or `finally` keywords.	`try:`
		  `x += a / b`
		`except ZeroDivisionError:`
		  `print("Ignored zero division")`

`while`	Continues the repeated, iterative execution of a block of code while a certain condition holds.	`x = 2`
		`while x < 1000:`
		  `print(x)`
		  `x *= 2`

Keyword	Description	Example

`with`	Encapsulates a block of code using a context manager object, which has dedicated methods to deal with starting and cleanly ending the context. Generally used so that the context can have clean-up before it exits or an exception occurs. Optionally uses the `as` keyword.	```
		with open(fileName) as fileObj:
		    for line in fileObj:
		        print(line)
		``
		The `with` statement encapsulates the file reading block so the file is closed at the end or if an error occurs.

`yield`	Used inside a function to pass back a value like `return`, but allows the function to be re-entered, keeping the previous state. Used to make generator functions.	```
		def generateSquare():
		    for x in range(1,10):
		        yield x*x
		print(generateSquare()) # 1
		print(generateSquare()) # 4
		``

Escape sequences

The following are some of the most commonly used escape sequences in Python, which allow you to specify textual characters or control codes when they are not available as regular symbols.

Code	Description	Example
`\\`	A backslash character, which needs to be forced when the following character would otherwise form an escape code.	`text = '\\title'` Text is `'\title'` and does not have a tab (`\t`) code inside.
`\'`	A single quote, which may need to be escaped in situations where it should not be considered as the start or end of a string.	`text = 'Don\'t do that!'` **Note:** not required when a string is defined with double quotes.
`\"`	A double quote, which may need to be escaped in situations where it should not be considered as the start or end of a string.	`text = "Shout \"Help!\" loudly."` **Note:** not required when a string is defined with single quotes.
`\n`	A newline (linefeed) control character. Used to separate lines of text in Unix- and Linux-based computers	`text = 'Line A\nLine B\n'` Text value is split into two lines on Linux and Unix machines.
`\r`	A carriage return control character. Used in combination with `\n` on Windows-based systems to separate lines of text.	`text = "Line A\r\nLine B\r\n"` Text is split into two lines on Windows machines.
`\t`	A tab character, providing indentation with whitespace to pre-set stop points.	`text = 'Col 1\tCol 2\tCol 3\n'` Tabs indent to form three columns.

Code	Description	Example
\u····	Specifies a Unicode character using a 16-bit hexadecimal value.	`text = u'\u03b1-helix'` Creates 'α-helix', e.g. for graphical displays.
\x··	Specifies a character using a hexadecimal value.	`text = '\x48\x65\x6C\x6C\x6f'` Text is hexadecimal code for 'Hello'.

Constants

The following are the commonly used named constants. From Python 3 these become keywords, and thus are a reserved part of the language and may not be redefined.

Name	Description	Example

True	A Python object that represents truth in Boolean logic.	`foundPositive = False`
		`for value in numbers:`
		`    if value > 0:`
		`        foundPositive = True`
		`        break`

False	A Python object that represents falsehood in Boolean logic.	[As above]

None	A Python object that represents an absence, or that something is undefined.	`value = myDict.get(key)`
		`if value is None:`
		`    print("key was not in dictionary")`
		`    print("(or its value was None)")`
		Note: the `.get()` method of Python dictionaries gives `None` if a key is absent.

Mathematical operations

The following table lists the standard mathematical operations that are inbuilt into Python. Some of these are specified with operator symbols, while others use standard named functions.

Operation	Description	Example
x + y	Addition: x plus y.	`revenue = profit + expenses`
x - y	Subtraction: x minus y:	`income = profit - taxes`
x * y	Multiplication: x times y.	`area = volume * height`
x / y	Division: x divided by y, within the precision allowed by the system. Note that in Python 2 dividing two integers gives an integer (it gives a floating point number in Python 3).	`mean = (x + y + z) / 3.0`

Operation	Description	Example
x // y	Integer floor division: x divided by y but giving the nearest whole number rounded downwards.	a = 13.0 // 5.0 The value of a is 2.0.
x % y	Modulus: find the remainder when dividing x by y.	a = 13 % 5 The value of a is 3.
-x	Negate the value of x, equivalent to multiplying by −1.	a = 5 b = -a * 3 The value of b is −15.
a + bj	Specifies a complex number a + bi: (a is the real part; b is imaginary). Here b must be an explicit number, not a variable.	x = 1.0 + -1.0j print(x * x) Gives 0-2i (-2j in Python)
abs(x)	Gives the absolute value of x, irrespective of sign, i.e. makes values positive.	a = abs(-7) The value of a is 7.
int(x)	Convert x into the integer (whole number) equivalent; gives an integer data type, rounding towards 0 as required. A second argument is the base in which to interpret x.	int(' 21 ') int(3.0 * 7.0) int('10101', 2) All these values are 21.
long(x)	Convert x into a long integer (i.e. of arbitrary length): this is mostly redundant. Not available in Python 3.	x = long(12345123451234512345) Note: int() will automatically create long integers as needed.
float(x)	Convert x into the equivalent floating point number: a fixed precision number with significant digits and an exponent.	c = float(' 6.0322e23 ') d = float(12)
round(x[,n])	Round x to the nearest whole number or optionally accepts n, stating how many decimal places to round to.	a = 1.5555 b = round(a) # 2.0 c = round(a, 2) # 1.56
complex(a,b)	Creates an imaginary number with real part a and imaginary part b, i.e. a + bi.	c = complex(1, -2) d = c.conjugate()
divmod(x, y)	Divide x by y giving the integer floor and remainder as separate values.	a, b = divmod(13.0, 5.0) The value of a is 2.0 and b is 3.0.
pow(x, y) or x ** y	Raise x to the power of y, i.e. x^y.	a = 2 ** 3 The value of a is 8.

Comparison operators

The following table lists the standard item comparison operations that are built into Python, to compare the value or identity of objects. Note that in Python 3 inequality comparisons can only be made on objects of a comparable type. In Python 2 you can make the comparison 2 < 'a' (even though it does not really make sense) but this is illegal in Python 3.

Operator	Description	Example

==	Tests whether two operands (either side of the operator) have the same value, giving `True` if they do and `False` otherwise.	`x = 3 * 5`
		`if x == 15:`
		`    print("Equal")`

!=	Tests whether two operands have different values, giving `True` if they do and `False` otherwise.	`seq = 'GCGC'`
		`if seq != 'TATA':`
		`    print("Not equal")`

>	Tests whether the value of the first operand is **greater** in value than the second, giving `True` if it is and `False` otherwise.	`x = 2**10`
		`x > 1024`
		Evaluates to False; 2^{10} is exactly 1024, not greater.

<	Tests whether the value of the first operand is **smaller** in value than the second, giving `True` if it is and `False` otherwise.	`x = 2**10`
		`x < 1025`
		Evaluates to True.

>=	Tests whether the value of the first operand is **greater or equal** in value than the second, giving `True` if it is and `False` otherwise.	`x = 2**10`
		`x >= 1024`
		Evaluates to True.

<=	Tests whether the value of the first operand is **smaller or equal** in value than the second, giving `True` if it is and `False` otherwise.	`x = 2**10`
		`x <= 512`
		Evaluates to False.

is	Tests whether two operands represent the same Python object, giving `True` if they do and `False` otherwise.	`x = True`
		`if x is True:`
		`    print("Success")`

is not	Tests whether two operands represent different Python objects, giving `True` if they are different and `False` if they are the same.	`value = myDict.get(key)`
		`if value is not None:`
		`    print('Found key')`

Binary operators

The following are standard bitwise binary operations that are built into Python.

Operator	Description	Example

x & y	The bitwise binary AND operation, giving 1 if both bits at a position are 1, and 0 otherwise.	`x = 0b11111000`
		`y = 0b10011111`
		`print(bin(x & y))`
		Result is `0b10011000`.

x | y	The bitwise binary OR operation, giving 1 if any of the bits at a position are 1, and 0 otherwise.	`x = 0b11010000`
		`y = 0b00001101`
		`print(bin(x | y))`
		Result is `0b11011101`.

Operator	Description	Example

x ^ y	The bitwise binary XOR operation, giving 1 only if one of the bits, but not both, at a position is 1, and 0 otherwise.	`x = 0b11111000`
		`y = 0b00011111`
		`print(bin(x ^ y))`
		Result is `0b11100111`.

x << n	The left-shift bit operation. Moves the bits of x by n places to the left. The same as multiplying x by 2**n.	`x = 0b00010101 # 21 in decimal`
		`print(bin(x << 2))`
		Result is `0b1010100`; 84 in decimal.

x >> n	The right-shift bit operation. Moves the bits of x by n places to the right. The same as the integer part of dividing x by 2**n.	`x = 0b01010101 # 85 in decimal`
		`print(bin(x >> 1))`
		Result is `0b101010`; 42 in decimal.

~x	The bitwise binary NOT operation; forms the complementary binary number by flipping 1 for 0 and 0 for 1. Note this flips all bits in the full binary representation whether they are directly represented or not (i.e. have implicit leading zeros).	`from numpy import int8`
		`x = int8(0b11000011)`
		`print(bin(~x))`
		Result is `0b111100`. Note use of 8-bit integers from NumPy to avoid implicit zeros in normal Python integers.

Inbuilt functions

The following lightly describes selected, standard Python functions which require no imports to use. See the standard Python documentation for more technical descriptions and details of all arguments. Optional arguments are specified in italics.

Function	Description	Example

abs(val)	Generates the absolute value of a number, i.e. its magnitude, irrespective of sign.	`numbers = [-3,-2,-1,0,1,2,3]`
		`print([abs(x) for x in numbers])`
		Result is `[3, 2, 1, 0, 1, 2, 3]`.

all(vals)	Determines whether all the items in a collection (or other iterable) have true values. Gives `True` if they do and `False` otherwise. This does not work on multi-dimensional NumPy arrays.	`all([True, False, True])`
		`# False`
		`all([1,2,3,4]) # True`

any(vals)	Determines whether any of the items in a collection (or other iterable object) have true values. Gives `True` if so and `False` otherwise. This does not work on multi-dimensional NumPy arrays.	`any([True, False, True]) #`
		`True`
		`any([False, None, 0.0]) #`
		`False`

bin(val)	Creates a string representing the binary version of a number; 0 and 1 characters prefixed with `'0b'`. Available from Python 2.6 onwards.	`x = 341`
		`print(bin(x))`
		Result is `'0b101010101'`.

Function	Description	Example

`bool(val)`	Determines whether a value is logically true or false, i.e. converts a value into a Boolean object.	`bool(0.0) # False`
		`bool(3.0) # True`
		`bool('abc') # True`

`callable(obj)`	Determines whether an object is callable, i.e. can be invoked like a function. Giving `True` if so and `False` otherwise.	`y = 1.47`
		`x = abs # The inbuilt function`
		`print(callable(y), callable(x)) # False, True`

`chr(code)`	Gives the ASCII character string for an integer code number. Performs the inverse operation to `ord()`.	`i = ord('A')`
		`print(chr(i+1))`
		Result is `'B'`.

`cmp(val1, val2)`	Compares two objects and gives back 1 if the first is larger, −1 if the second is larger or 0 if they are equal. Not available in Python 3.	`cmp(1,2)  # -1`
		`cmp(2,2)  #  0`
		`cmp(2,1)  #  1`

`complex(real, imag)`	Creates a complex number using separate real and imaginary components.	`x = complex(1,-1.4142)` Result is $(1-1.4142j)$; $1-1.4142i$.

`dict(vals)`	Creates a dictionary from a specified object, e.g. to copy another dictionary or to convert a list of keyword:value pairs.	`x = [(1,1),(2,4),(3,9),(4,16)]`
		`y = dict(x)`
		Result is `{1:1, 2:4, 3:9, 4:16}`.

`dir(obj)`	Gives a list of variable names that are present within the current program scope, or if object is specified the names associated with that object as attributes.	`print(dir())`
		`print(dir(''))`
		First line shows imports and declarations. Second line shows attributes of string types.

`divmod(val1, val2)`	Divide x by y giving the integer floor and remainder as separate values.	[See Mathematics section]

`enumerate(vals, start)`	Generates sequential pairs of (`number`, `item`) from a collection or other iterable object, e.g. extracting objects and their indices from a list. Creates an iterator object. Takes a start number as a second argument.	`letters = 'GCAT'`
		`enum = list(enumerate(letters))`
		Result is `[(0,'G'), (1,'C'), (2,'A'), (3,'T')]`.

`eval(exprn, locals, globals)`	Evaluates a string as an expression in Python syntax, i.e. to allow the dynamic generation of expressions. Allows optional dictionary arguments to specify global and local scope variable names.	`text = 'x*7+5*y'`
		`x = 4.7`
		`y = 8.1`
		`eval(text)`
		Result is `73.4`.

`execfile(fileName, locals, globals)`	Like `exec()` (see above) but operates on a named file. Dynamically reads and executes the Python code in the file. Allows optional dictionary arguments to specify global and local scope variable names. Not available in Python 3.	`execfile(fileName)`

Function	Description	Example
`file(name, `*`mode`*`)`	Creates a file object. Little used because `open()`, described below, is generally used instead, but useful for checking data types. Not available in Python 3.	`if isinstance(obj, file):` `  print("Object is a file")`
`float(`*`val`*`)`	Converts a number or string into its best floating point equivalent.	`print(float('7'))` `print(float('+18.57e12') / 2.0)`

`format(val,`
`specification``)`
	Formats a string value according to a specified representation. See Appendix 4 below for new-style formatting codes.	`format(3.14159,'E')`
		`format('DNA', '>4s')`
		Results are `'3.141590E+00'` and `'DNA'`.

`frozenset(`*`vals`*`)`	Creates a frozen set from a given object; an immutable unordered collection of non-repeating items. Sometimes used to convert sets into the immutable equivalent so they may be used as keys in dictionaries.	`l = [0,1,2,3,1,2,3,2,1,0,1,3]`
		`x = frozenset(l)`
		`print(x)`
		Result is `frozenset([0, 1, 2, 3]).`

`getattr(obj,`
`name,`
`defaultVal``)`
	Retrieves a particular named attribute from a specified Python object. Sometimes used as an alternative to the dot notation when the attribute name is a variable.	`import math`
		
		`print(getattr(math, 'pi'))`
		`print(math.pi)`
		Results are both
		`3.141592653589793.`

`globals()`	Gives a dictionary of variable name and object pairs that are available in the outermost, global context of the program's execution.	`varDict = globals()` `print(varDict)`

`hasattr(`*`obj`*`,`
`name)`
	Determines whether a Python object has a particular named attribute. Gives `True` if it does and `False` otherwise.	`import math`
		`hasattr(math,'sin') # True`
		`hasattr(math,'apple')`
		` # False`

`help(`*`obj`*`)`	Gets help documentation for a given Python object.	`print(help(float))`
		`fileObj = open('a.txt')`
		`print(help(fileObj))`

`hex(val)`	Creates a text string representing the hexadecimal (i.e. base 16) version of an integer number.	`print(hex(2**24-1))` Result is `'0xffffff'.`

`id(obj)`	Gives the unique number that identifies a particular Python object. Such numbers do not change within a Python session, but they will (usually) be different for a new session.	`print(id(7))`
		`print(id(float))`
		`print(id(None))`

`input(`*`prompt`*`)`	Prompts the user at the command line for keyboard input, i.e. so the entered values can be used in the program.	`x = input('Enter value:')` `print("Value is: ", x)`

Function	Description	Example
`int(val, base)`	Converts a number or string into an integer representation. Optionally takes the radix number for which base to use.	`print(int(34.96))` # Base 10 `print(int('1000100',2))` # Base 2 Results are 34 and 68.
`isinstance(obj, class)`	Determines whether one object is derived from a specified class (or subclass thereof), returning `True` if it is and `False` otherwise.	`isinstance(7, int)` # True `isinstance(7, float)` # False
`issubclass (class1, class2)`	Determines whether one object class is a subclass of another, returning `True` if it is and `False` otherwise.	`issubclass(Protein, Molecule)`
`len(vals)`	Gives the size of an object, e.g. the number of items in a collection.	`letters = ['G','C','A','T']` `print(len(letters))` # 4

`list(vals)`	Creates a Python list collection from a specified object, which must be iterable. Can be used on a list to make a copy.	`t = ('V','I','L','A','M','P')`
		`x = list(t)`
		`y = list(enumerate(range(7,11)))`
		First example converts tuple to list. Second converts iterator, giving `[(0, 7), (1, 8), (2, 9), (3, 10)]`.

`locals()`	Gives a dictionary of variable name and object pairs that are available in the innermost, local context at a point in the program's execution.	`for x in range(5):`
		  `varDict =  locals()`
		  `print(varDict)`
		Results show x changes locally, in the loop.

`long(val, base)`	Converts `val` into a long integer (i.e. of arbitrary length): this is mostly redundant. Not available in Python 3.	[See Mathematics section]

`map(func, vals, ...)`	Takes an iterable object, like a list, and applies a function to each item, generating a new list (in Python 2) or a map iterator (in Python 3). This function is largely redundant and it is more commonplace to use a list comprehension instead.	`vals = [30.0, 31.0, 28.25]`
		`ints1 = map(int, vals)`
		`ints2 = [int(x) for x in vals]`
		Note: `ints1` and `ints2` are the same in Python 2, but in Python 3 `ints1` is a map iterator.

`max(vals)` or `max(val1, val2, ...)`	Finds the minimum value of a collection (or other iterable object) or list of arguments. This does not work on multi-dimensional NumPy arrays.	`print(max(3,11, 9, 5))` # 11 `l = [30.0, 31.0, 28.25]` `print(max(l))` # 31.0
`min(vals)` or `min(val1, val2, ...)`	Finds the minimum value of a collection (or other iterable object) or list of arguments. This does not work on multi-dimensional NumPy arrays.	`print(min(8, 12, 3, 34))` # 3 `l = [30.0, 31.0, 28.25]` `print(min(l))` # 28.25

Function	Description	Example
`object()`	Creates a basic, blank, featureless object. The class of the object is the superclass of all Python objects.	`x = object()` `print(dir(x))`
`oct(val)`	Creates a text string representing the octal (i.e. base 8) version of an integer number.	`print(oct(2**24-1))` Result is `'077777777'` in Python 2 and `'0o77777777'` in Python 3.

`open(fileName, mode)`	Opens a file on disk for reading or writing by creating a file type object. The second argument is a mode string that specifies whether the file is for reading (`'r'`), writing (`'w'`), appending (`'a'`) etc. The default mode is reading.	`fObj1 = open(inName, 'rU')`
		`line = fObj1.readline()`
		
		`fObj2 = open(outName, 'w')`
		`fObj2.write('Hello world\n')`

`ord(char)`	Gives the ASCII code number for a character string. Performs the inverse operation to `chr()`.	`i = ord('Z')-1`
		`print(i, chr(i))`
		Result is `89`, `Y`.

`pow(val1, val2, modulo)`	Raises a number to a given power, equivalent to using the '`**`' operator.	`x = pow(2, 8)` `x = 2**8  # Same`

`print(text, sep, end, file)`	This function is present in Python 2 and from Python 2.6 is the same as the function in Python 3 by using `from __future__ import print_function`. It replaces the print statement which is in Python 2 but not in Python 3. Prints a textual representation of one or more Python objects to the screen or file.	`x = 'Some data'`
		`print(x)`
		`print(x, file=fileObj) #Python 3`
		`print(1,2,3)`
		
		Result is `1 2 3`.
		`print(1, 2, 3, sep=';',`
		`          end='*\n')`
		Result is `1;2;3*`.

`property(getFunc, setFunc, delFunc, docStr)`	Allows attribute-style dot notation access (`classObj.attrName`) for getter and setter functions in 'new-style' class definitions (which inherit from `object`). This means that the code looks cleaner and allows extra functionality, for example, validation in the setter function. Optionally there is a third argument which is a function for attribute deletion, and a fourth argument, which is a document string.	`class DemoObj(object):`
		`    def __init__(self):`
		`        self._x = None`
		`    def getX(self):`
		`        return self._x`
		`    def setX(self, value):`
		`        self._x = value`
		`    x = property(getX, setX)`
		`d = DemoObj()`
		`d.x = 55 # same as d.setX(55)`
		`print(d.x) # same as d.getX()`
		
		Result is `55`.

`range(end)`
or
`range(start, end, step)`
	Gives a range of integers from a start up to (but not including) an end value with a regular increment. By default the range starts at 0 and increments	`range(7) # [0,1,2,3,4,5,6]`
		`range(3,8) # [3, 4, 5, 6, 7]`
		`range(3,10,2) # [3, 5, 7, 9]`
		`range(5,0,-1) # [5, 4, 3, 2, 1]`

Function	Description	Example
	by 1. In Python 2 `range()` creates a list, but in Python 3 becomes an iterable object, more like `xrange()`, and `xrange()` itself disappears.	In Python 3 the result is not a list but instead a range iterable.
`raw_input` `(prompt)`	Prompts the user at the command line for keyboard input, i.e. so the entered values can be used in the program. Not available in Python 3, use `input()` instead.	`x = raw_input('Enter value:')` `print("Value is: ", x)`
`reload(module)`	Reloads a Python module, assuming it was previously imported. Not available in Python 3.	`import math` `math.pi = 3.0` `print(math.pi) # 3.0` `reload(math)` `print(math.pi) #` `3.14159265359`
`repr(obj)`	Creates a formal, textual representation of a Python object. Similar to `str()`, but gives unambiguous (in terms of identifying the original object), albeit sometimes less readable, text.	`from numpy import array` `a = array([1,2,3])` `str(a)  # '[1 2 3]'` `repr(a) # 'array([1, 2, 3])'`
`reversed(seq)`	Creates an iterator object that provides items in the reverse order, based on an ordered collection.	`for x in reversed(range(7)):` `print(x)`
`round(val,` `places)`	Rounds `val` to the nearest whole number or optionally accepts `places`, stating how many decimal places to round to.	[See Mathematics section above]
`set(vals)`	Creates a set from a given object; an unordered collection of non-repeating items.	`l = [0,1,2,3,1,2,3,2,1,0,1,3]` `x = set(l)` `print(x)` Result is `set([0, 1, 2, 3])`.
`setattr(obj,` `name, val)`	Sets a named attribute of a specified object with a given value. This is often an alternative to the `object.attr = value` notation where the name of the attribute can vary.	`p1 = Molecule()` `setattr(p1, 'name', 'c-Myc')` `p1.name = 'c-Myc' # Same`
`sorted(vals,` `comparator, key,` `reverse)`	Creates a new list with items in sorted order from the items of a collection or other iterable object.	`list1 = ['G','C','A','T']` `list2 = sorted(list1)` `print(list2)` Result is `['A','C','G','T']`.
`str(obj)`	Converts a Python object into an informal textual string representation. Numeric values may be rounded for display.	`str(7.500000000001) # '7.5'` `str([5,7,11]) # '[5, 7, 11]'` `str(type(1)) #  "<type 'int'>"`
`sum(vals, start)`	Adds all the items of a collection or other iterable object. For NumPy arrays it sums along the first axis.	`values =` `[7,69,31,99,53,16,72]` `print(sum(values))  # 347`

Function	Description	Example
| `super(objType, obj)` | When a function is overridden in a subclass this allows calling of a superclass function without mentioning the superclass by name. An optional object (or class) may be passed in as a second argument. | ```
class C(B):
  def f(self, arg):
    super(C, self).f(arg)
    # same as doing:
    # B.f(self, arg)
``` |
| `tuple(vals)` | Creates a Python tuple collection from a specified object, which must be iterable. Can be used on a tuple to make a copy. Sometimes used to convert a list, so that it may be used as a dictionary key. | ```
l = ['D','E','R','K','H']
t = tuple(enumerate(l))
```<br>Result is `((0,'D'), (1,'E'), (2,'R'), (3,'K'), (4,'H'))`. |
| `type(obj)` | Gives an object representing the type of a specified object. | `type(1) # <type 'int'>`
`type('a') # <type 'str'>`
`type(type(1)) # <type 'type'>`
In Python 3 the result is `<class 'int'>` etc. |
| `unichr(code)` | Gives the Unicode character string for an integer code number. Performs the inverse operation to `ord()` for Unicode strings. Not available in Python 3 since all strings are Unicode. | ```
alpha = u'\u03b1'
i = ord(alpha)
print(unichr(i+1))
```<br>Result is `u'\u03b2'`; 'β'. |
| `unicode(obj, encoding)` | Converts a Python object into a Unicode string representation in a manner similar to `str()`. The optional second argument allows the encoding type to be specified, e.g. when converting plain text. Not available in Python 3, since all strings are Unicode. | ```
text1 = unicode(3.141)
x = '\xce\xb1-helix'
text2 = unicode(x, 'utf-8')
```<br>Last result is `u'\u03b1-helix'`; 'α-helix'. |
| `xrange(end)` or `xrange(start, end, step)` | Used for looping through large ranges of numbers. Compared to `range()`, `xrange()` doesn't create a whole list and thus saves memory; instead it creates an iterable object. Not available in Python 3 since `range()` behaves like `xrange()`. | ```
n = 1000000
y = [x**0.5 for x in xrange(n)]
``` |
| `zip(vals1, vals, ...)` | Takes items in sequence from a number of collections (or other iterable objects) to make a list of tuples, where each tuple contains one item from each collection. Often used to group items from lists which have the same index. Note the inverse operation is achieved with `zip(*c)`. In Python 3 the result is a zip iterator rather than a list. | ```
a = [1,2,3]
b = ['x','y','z']
c = list(zip(a,b))
d = list(zip(*c))
```<br>Note: `list()` conversion not necessary in Python 2. Result:<br>c is `[(1,'x'), (2,'y'), (3,'z')]`<br>d is `[(1,2,3), ('x','y','z')]` |

APPENDIX 2
Selected standard type methods and operations

Operations common to strings, lists and tuples

The table below lists most of the functions that may be applied to Python sequence (ordered) collection types. For Python 2 the sequence types are `buffer`, `bytearray`, `list`, `str`, `tuple`, `unicode` and `xrange`. For Python 3 these types are `bytearray`, `bytes`, `memoryview`, `list`, `range`, `str` and `tuple`.[1]

Operation	Description	Example
x in seq x not in seq	Determines whether an item is or is not in a collection, giving True or False accordingly.	myList = [3,1,4,1,5,9] 2 in myList # False 8 not in myList # True
seqA + seqB	Generates a new collection which is the combination of two collections of the same type, in order.	tupleA = ('G','C') tupleB = ('A', 'T') tupleA + tupleB ('G', 'C', 'A', 'T')
seq * n n * seq	Generates a new collection based by repeating the items of a collection a number of times.	'AB' * 4 # 'ABABABAB' 5 * [0] # [0,0,0,0,0]
seq[i]	Accesses an item in an ordered collection by using a positional index, which starts from zero.	text = 'Banana' text[0] # 'B' nums = [1,4,9,16,25] nums[2] # 9
seq[i:j]	Generates another, usually smaller collection, by accessing a range of items from an ordered collection, from a starting positional index, up to **but not including** a second index. If unspecified, the starting index defaults to zero and the end defaults to the end of the collection. Negative index numbers count from the end.	text = 'Banana' text[1:5] # 'anan' x = [1,4,9,16,25] x[2:4] # [9,16] x[:4] # [1,4,9,16] x[1:] # [4,9,16,25] x[1:-1] # [4,9,16]
seq[i:j:k]	Generates another, usually smaller collection, by accessing a range of items from an ordered collection, using positional indices (see above) and a step size to skip certain indices. Step may be negative for reverse direction.	x = [1,4,9,16,25] x[0:5:2] # [1,9,25] x[::2] # [1,9,25] x[::-1] # [25,16,9,4,1]

[1] Compared to Python 2, in Python 3 all strings are Unicode and the type `bytes` is used for immutable binary data, rather than strings.

Operation	Description	Example

`len(seq)`	Gives the total number of items in the collection; its size.	`len('Banana') # 6`
		`len((1,4,9,16)) # 4`
		`len([]) # 0`

`min(seq)`	Retrieves the item from a collection that has the smallest value. This does not work on multi-dimensional NumPy arrays.	`x = [10, 5, 1, 10, 3]`
		`min(x) # 1`
		`min('ABCDE') # 'A'`

`max(seq)`	Retrieves the item from a collection that has the largest value. This does not work on multi-dimensional NumPy arrays.	`x = (10, 5, 1, 10, 3)`
		`max(x) # 10`
		`max('abcdefgh') # 'h'`

`seq.index(item)`	Retrieves the (first) positional index at which an item is found in a collection. Assumes the item is present and gives a `ValueError` if not.	`text = 'Banana'`
		`text.index('a') # 1`
		`x = [10,5,1,10,3]`
		`x.index(3) # 4`

`seq.count(item)`	Counts the number of occurrences of a given item within a collection.	`text = 'Banana'` `text.count('a') # 3`

String methods

The table below describes selected methods that relate to standard Python textual string objects, which you would call using the dot notation `string.method(arg)`. Separate descriptions for string formatting (both old and new style) and the regular expression module `re` are given in Appendix 4 and Appendix 5 respectively. Note that most objects can be converted to a string representation[2] with `str(x)`.

It should be noted that there are big changes in Python between version 2 and version 3 with regard to handling Unicode. In Python 3 **all** strings are Unicode, so there is no need for `unicode(x)` or `u''` syntax;[3] any '\u' escaped codes in a string will naturally be recognised as Unicode characters. Also, the binary data type is introduced, which may be created with `bytes ()`, which is immutable, or `bytearray()`, which is mutable. The `bytes` type may be used for general binary data, which may include binary **encoded** Unicode. Before Python 3 the normal `string` type was used for binary data.

Method	Description	Example

`s.capitalize()`	Generates a copy of a string where the first character is capitalised, if it is a lower-case letter.	`'abc'.capitalize()  # 'Abc'`
		`'Abc'.capitalize()  # 'Abc'`
		`'1abc'.capitalize() #'1abc'`

---

[2] Or to Unicode before Python 3 with `unicode(x)`.
[3] Note that the `u''` syntax was removed in Python 3.0 to 3.2, but reintroduced in 3.3.

Method	Description	Example
s.count(substr, *start, end*)	Counts the number of occurrences (without overlap) of a substring within a larger string. Optional index range arguments.	text = 'Bananarama' text.count('a') # 5 text.count('q') # 0 text.count('an') # 2 text.count('ana') # 1
s.endswith (substr, *start, end*)	Determines whether a string ends with a given substring. Optional position range arguments.	'xyz'.endswith('z') # True 'xyz'.endswith('y') # False 'xyz'.endswith('yz')# True 'xyz'.endswith('y',0,2) # True
s.find(substr, *start, end*)	Gives the starting index of the **first** occurrence of a substring within a larger string, or −1 if none is found. Optional index range arguments.	text = 'Bananarama' text.find('r') # 6 text.find('a') # 1 text.find('q') # -1 text.find('am') # 7
s.format(*args, **kwd)	Generates a specially formatted version of a string.	[See formatting in Appendix 4 below]
s.index(substr, *start, end*)	Gives the starting index of the **first** occurrence of a substring within a larger string. Creates a ValueError if none is found. Optional index range arguments.	text = 'Bananarama' text.index('a') # 1 text.index('am') # 7 text.index('q') # Fails!
s.isalnum()	Determines whether a string contains only alphanumeric characters. Gives True if so, and False otherwise.	'ab12'.isalnum() # True 'ab/12'.isalnum() # False
s.isalpha()	Determines whether a string contains only letters of the alphabet.	'abc'.isalpha() # True 'abc?'.isalpha() # False
s.isdigit()	Determines whether a string contains only numeric digit characters. Gives True if so, and False otherwise.	'10'.isdigit() # True '1.0'.isdigit() # False
s.islower()	Determines whether a string contains letters that are all lower case. Gives True if so, and False otherwise.	'abc'.islower() # True 'Abc'.islower() # False 'ab@#12'.islower() # True '@#12'.islower() # False
s.isspace()	Determines whether a string contains only whitespace characters. Gives True if so, and False otherwise.	' '.isspace() # True 'a'.isspace() # False ' a'.isspace() # False '\t\n '.isspace() # True
s.isupper()	Determines whether a string contains letters that are all upper case. Gives True if so, and False otherwise.	'ABC'.isupper () # True 'Abc'.isupper () # False 'AB@#12'.isupper() # True '@#12'.isupper() # False

Method	Description	Example

`s.join(iterable)`	Uses one string to join the items (which must also be strings) that come from a sequence collection, or other iterable object, to form a new string.	`x = ['G','C','A','T']`
		`sep = ','`
		`sep.join(x) # 'G,C,A,T'`
		`'; '.join(x) # 'G; C; A; T'`
		`'/'.join(('AC','DC'))`
		`# 'AC/DC'`

`s.lower()`	Generates a copy of a string where any upper-case characters are converted to lower case.	`'Ala258'.lower() # 'ala258'`
		
		`'ALA258'.lower() # 'ala258'`

`s.lstrip(chars)`	Generates a copy of a string where specified characters at the start (left) are removed. Optional string argument to specify which characters to consider, otherwise it is leading whitespace which is removed.	`'  X Y  '.lstrip() # 'X Y  '`

`s.replace(old, new, maxNum)`	Generates one string from another by replacing all occurrences of one substring with another substring. Optional maximum number of replacements.	`rna = 'AUGCAUAGCA'`
		`dna = rna.replace('U','T')`
		`Result is 'ATGCATAGCA'.`

`s.rfind(substr, start, end)`	Like `find()`, but gives the starting index of the **last** occurrence of a substring within a larger string, or -1 if none is found.	`text = 'Bananarama'`
		`text.rfind('r') # 6`
		`text.rfind('a') # 9`
		`text.rfind('q') # -1`

`s.rindex(substr, start, end)`	Like `index()`, but gives the starting index of the **last** occurrence of a substring within a larger string. Creates a `ValueError` if none is found.	`text = 'Bananarama'`
		`text.rindex('a') # 9`
		`text.rindex('am') # 7`
		`text.rindex('q') # Fails!`

`s.rstrip(chars)`	Generates a copy of a string where specified characters at the end (right) are removed. Optional string argument to specify which characters to consider, otherwise it is trailing whitespace which is removed.	`'  X Y  '.rstrip() # '  X Y'`

`s.split(sep, maxNum)`	Generates a list of strings by splitting a string into parts where a substring is found. If substring is not specified then split on whitespace.	`t = 'G, C, A, T'`
		`t.split(', ')`
		`# ['G','C','A','T']`
		`t.split(':')`
		`# ['G, C, A, T']`

`s.splitlines(keepends)`	Generates a list of strings by splitting a string into parts where newline (`'\n'`, `'\r'`, `'\r\n'`) characters are found.	`text = 'AC\nDC'` `text.splitlines() # ['AC', 'DC']`

Method	Description	Example

s.startswith(prefix, *start, end*)	Determines whether a string starts with a given substring. Optional position range arguments.	`'xyz'.startswith('x') # True`
		`'xyz'.startswith('y') # False`
		`'xyz'.startswith('xy') # True`
		`'xyz'.startswith('y',1) # True`

s.strip(*chars*)	Generates a copy of a string where specified characters at either end are removed. Optional string argument to specify which characters to consider, otherwise it is whitespace which is removed.	`'  X Y '.strip() # 'X Y'`
s.title()	Generates a copy of a string where the first character of each group of letters is capitalised.	`' hi joe '.title() # ' Hi Joe '`

s.translate (table, *delChars*)	Generates one string from another by using a 256-character translation table to map characters, and optionally also delete characters.	`from string import maketrans`
		`table = maketrans('U','T')`
		`# v2`
		`rna = 'AUGCAUAGCA'`
		`dna = rna.translate(table)`
		Result is `'ATGCATAGCA'`.
		In Python 3 second line instead is:
		`table = str.maketrans`
		`('U','T')`

s.upper()	Generates a copy of a string where any lower-case characters are converted to upper case.	`'ala258'.upper() # 'ALA258'` `'Ala258'.upper() # 'ALA258'`
s.zfill(*width*)	Fill a string with zero characters '0', up to a given total width.	`'7'.zfill(3) # '007'` `'7'.zfill(1) # '7'`

List operations

The following table describes some of the operations and inbuilt methods for Python lists and related sequence containers, which are generally represented as `seq`. It should be noted that a collection object (set, tuple, other list …) can be converted to a list using `list(collection)` and that `list()` generates an empty list, just like `[]`.

Operation	Description	Example

len(seq)	Determines the number of items in the list; its length.	`letters =`
		`['G','C','A','T']`
		`len(letters) # 4`

Operation	Description	Example

`seq[i] = obj`	Sets the item at a given index position in a list.	`letters =`
		`['G','C','A','T']`
		`letters[3] = 'U'`
		`# ['G','C','A','U']`

`seq[i:j] = vals`	Sets a number of items covering a range of positional indices; from a starting positional index and up to **but not including** a second index.	`nums = [1,2,3,4,5,6]`
		`nums[2:] = [3,2,1]`
		`# [1, 2, 3, 2, 1]`

`vals = seq[i:j]`	Extracts number of items covering a range of positional indices as a new list; from a starting positional index and up to **but not including** a second index.	`nums = [1,2,3,4,5,6]`
		`new1 = nums[3:]`
		`# [4,5,6]`
		`new2 = nums[:] # Copy all`

`del seq[i:j]`	Deletes an item or range of items from a list of specified indices.	`nums = [1,2,3,4,5,6]`
		`del nums[1:5]`
		
		`# [1, 6]`

`seq[i:j:k] = vals`	Sets a number of items covering a range of positional indices with a given step.	`myList = [0,0,0,0,0,0]`
		`myList[1::2] = [1,1,1]`
		`# [0, 1, 0, 1, 0, 1]`

`del seq[i:j:k]`	Deletes a number of items covering a range of positional indices with a given step.	`letters =`
		`['A','B','C','D','E']`
		`del letters[::2]`
		`# ['B', 'D']`

Method	Description	Example

`seq.append(val)`	Adds a single item to the end of a list.	`x = [True, False]`
		`x.append(None)`
		
		`# [True, False, None]`

`seq.extend(vals)`	Adds one or more items from a collection onto the end of a list.	`x = [1,2]`
		`y = [3,4]`
		`x.extend(y)`
		`# x is [1, 2, 3, 4]`

`seq.count(val)`	Counts the number of occurrences of a given item.	`x = [7,9,2,7,1]` `x.count(7) # 2`

`seq.index(val, start, end)`	Determines the index of the **first** occurrence of an item, within an optional index range.	`x = [7,9,2,7,1]`
		`x.index(7) # 0`
		`x.index(7,1) # 3`

`seq.insert(index, val)`	Inserts an item at a given positional index within a list, increasing the length of the list by one.	`x = [1,2,2,1]`
		`x.insert(2,4)`
		`# [1, 2, 4, 2, 1]`

Operation	Description	Example

`seq.pop(`*index*`)`	Removes and passes back an item from a list, reducing the length of the list by one. Defaults to remove the last item, otherwise an index may be specified.	`x = ['a','b','c','d']`
		`x.pop() # 'd'; x now`
		`['a', 'b', 'c']`
		`x.pop(1) # 'b'; x now`
		`['a', 'c']`

`seq.remove(val)`	Removes the **first** occurrence of an item from a list.	`x = [1,2,3,2,1]`
		`x.remove(2)`
		`# [ 1, 3, 2, 1]`

`seq.reverse()`	Reverses the order of items of the list. Does not generate a new list.	`x = ['a','b','c','d']`
		`x.reverse()`
		`# ['d', 'c', 'b', 'a']`

`seq.sort`
`(`*comparator*, *key*,
reverse`)`
Above for Python 2
`seq.sort(`*key,*
reverse`)`
Above for Python 3
	Sorts the items of a list, using optional comparison function (in Python 2) and key attribute. Option to sort in reverse order.	`x = [7,9,2,7,1]`
		`x.sort()`
		`# [1, 2, 7, 7, 9]`
		`x.sort(reverse=True)`
		`# [9, 7, 7, 2, 1]`

Set operations

The `set` and the immutable equivalent `frozenset` are Python's standard unordered non-repeating collection data types. The following table lists common operations and methods for sets (and where appropriate frozen sets). Sets may be specified with the inbuilt `set()` function, operating on another collection, e.g. `set([a,b,c])` and also (from Python 2.7 onwards) by using the curly brace notation `{a,b,c}`, which differs from dictionaries due to the absence of colons. Using `set()` with no arguments or on an empty collection creates an empty set, but `{}` does not; it is an empty dictionary. For the descriptions below we list methods of sets in the form `s.method(args)` and also any equivalents that can be performed with a symbolic operator syntax. For example, `set3 = set1.union(set2)` is equivalent to `set3 = set1 | set2`.

Operation/Method	Description	Example

`len(s)`	Determines the number of items within a set; its size.	`s = {1,2,3,2,1}`
		`len(s) # 3 - no repeats`
		`len(set()) # 0 - Empty`

`x in s`
`x not in s`
	Determines whether an item is within a set, or not.	`s = {'G','C','A','T'}`
		`'G' in s # True`
		`'G' not in s # False`

`s.isdisjoint(other)`	Determines whether one set has no items in common with another.	`s = {1,2,3}`
		`t = {4,5,6}`
		`s.isdisjoint(t) # True`

Operation/Method	Description	Example

`s.issubset(other)`
or
`set <= other`
	Determines whether all of the set's items are contained within another set.	`s = {1,2,3,4}`
		`t = {2,4}`
		`s.issubset(t) # False`
		`t <= s # True`
		`s <= s # True`

`set < other`	Determines whether all of the set's items are contained within another set and the sets are not the same.	`s = {1,2,3,4}`
		`t = {2,4}`
		`t < s # True`
		`s < s # False`

`s.issuperset(other)`
or
`set >= other`
	Determines whether all the items of another set are contained in this set.	`s = {1,2,3,4}`
		`t = {2,4}`
		`s.issuperset(t) # True`
		`t >= s # False`
		`s >= s # True`

`set > other`	Determines whether all the items of another set are contained in this set and the sets are not the same.	`s = {1,2,3,4}`
		`t = {2,4}`
		`s > t # True`
		`t > s # False`
		`s > s # False`

`s.union(other, ...)` or `set	other	...`	Generates a new set that contains all the items that are present in any of a group of sets.	`a = {1,2}` `b = a.union({2,3},` `{3,4})` `c = {1,2}	{2,3}	{3,4}` `# b,c both {1,2,3,4}`

`s.intersection`
`(other, ...)`
or
`set & other & ...`
	Generates a new set that contains only items that are present in all the sets in a group.	`{1,2,3}.intersection`
		`({2,3,4})`
		`# {2,3}`
		`{1,2} & {2,3} & {3,4}`
		`# set() - Empty`

`s.difference`
`(other, ...)`
or
`set - other - ...`
	Creates a copy of a set with any items that are common to other sets removed.	`{0,1,2,3}.difference`
		`({2,3,4})`
		`# {0,1}`
		`{1,2,3,4,5} - {2,5} -`
		`{2,3}`
		`# {1, 4}`

`s.symmetric_`
`difference(other)`
or
`set ^ other`
	Generates a new set from two sets by selecting only items that appear in one but not both of the sets.	`{0,1,2,3} ^ {2,3,4} ^`
		`{2,5}`
		`# {0, 1, 2, 4, 5}`

`s.copy()`	Creates a copy of a set; a separate Python collection with the same items. Same as `s2 = set(s1)`.	`s = {1,2,3}`
		`t = s.copy()`
		`t.add(4)`
		`# t is {1, 2, 3, 4}`

Operation/Method	Description	Example

`s.update(other, ...)`
 `set |= other
	...`	Adds any items which are not already present from one or more other sets.	`s = {0}`
			 `s.update({1,2})`
			 `s |= {2,4,9}`
			 `# {0, 1, 2, 4, 9}`

`s.intersection_update (other, ...)`
 or
 `set &= other & ...`
	Modifies a set by adding any items that are common to all the sets in a group.	`s = {1,2,4,9}`
		 `s &= {1,2,3,4}`
		 `# s is {1, 2, 4}`

`s.difference_update (other, ...)`
 or
 `set -= other
	...`	Modifies a set by removing any items that are present in a group of other sets.	`s = {1,2,4,9}`
			 `s -= {1,2,3,4}`
			 `# s is {9}`

`s.symmetric_difference_ update(other)`
 or
 `set ^= other`
	Modifies a set so it contains items that are only present in one, but not both, of the two sets.	`s = {1,2,4,9}`
		 `s ^= {1,2,3,4}`
		 `# s is {9,3}`

`s.add(obj)`	Adds a single item to a set, if it is not already present in the set.	`s = {1,2,3,4}`
		 `s.add(5)`
		 `s.add(3)`
		 `# s is {1, 2, 3, 4, 5}`

`s.remove(obj)`	Removes a single item from a set, assuming it is present in the set; gives a `KeyError` if not.	`s = {1,2,3,4}`
		 `s.remove(2)`
		 `# s is {1, 3, 4}`

`s.discard(obj)`	Removes a single item from a set, if it is present in the set. If the item is not in the set there is no error.	`s = {1,2,4,9,16}`
		 `s.discard(5) # No effect`
		 `s.discard(1)`
		 `# s is {16, 9, 2, 4}`

`s.pop()`	Removes and passes back a single, arbitrary item from a set, making the set smaller. If the set is empty it gives a `KeyError`.	`s = {1,2,4,9,16}`
		 `s.pop()`
		 `# 16(for example)`
		 `# s becomes {9, 2, 4, 1}`

`s.clear()`	Removes all the items from a set, generating an empty set.	`s = {1,2,4,9,16}`
		 `s.clear()`
		 `# s is set() - empty`

Dictionary operations

The following table lists the common operations that may be used with dictionaries. Dictionaries may be created from other collections (containing `key`, `value` pairs) using the inbuilt `dict(collection)` or explicitly using the curly brace notation `{k1: v1, k2:v2}`.

Using `dict()` with no arguments creates an empty dictionary, as does `{}`. It should be noted that only *hashable* objects, which do not allow modification of their innate value, can be used as dictionary keys; this excludes lists, sets and other dictionaries but includes tuples, frozen sets, strings, integers, floating point numbers and most other Python objects.

Operation	Description	Example

`len(d)`	Determines the number of (key: value) pairs in a dictionary	`d1 = {'G':3,'C':3,'A':2,'T':2}`
		`len(d1) # 4`
		`d2 = {'pi':3.141, 'e':2.718}`
		`len(d2) # 2`

`d[key]`	Retrieves the value from a dictionary that is associated with a given key, giving a `KeyError` if the key is not present.	`d = {'FR':33,'DE':49,'GB':44}`
		`d['GB'] # 44`
		`d['DE'] # 49`

`d[key] = value`	Sets the value for a specified key.	`d = {'FR':33,'DE':49,'GB':44}`
		`d['ES'] = 34`
		`# d is {'FR': 33, 'DE': 49,`
		`# 'GB': 44, 'ES': 34}`

`del d[key]`	Removes a specified (key:value) pair from the dictionary.	`d = {1:'G', 2:'C',`
		`     3:'A', 4:'T'}`
		`del d[4]`
		`# d is {1:'G', 2:'C', 3:'A'}`

`key in d`
`key not in d`
	Determines whether a key is used by a dictionary, or not.	`d = {1:'G', 2:'C',`
		`     3:'A', 4:'T'}`
		`1 in d # True`
		`1 not in d # False`
		`'A' in d # False - not a key`

`iter(d)`	Creates an iterator object from a dictionary, which provides an alternative way of looping through all the keys.	`d = {'G':3,'C':3,`
		`     'A':2,'T':2}`
		`iterObj = iter(d)`
		`for key in iterObj:`
		`    print(key)`

Method	Description	Example

`d.clear()`	Removes all key:value pairs from a dictionary, leaving an empty dictionary.	`d = {'pi':3.141, 'e':2.718}`
		`d.clear()`
		`# {} - Empty`

`d.copy()`	Makes a copy of a dictionary; a new, separate Python object with the same key:value pairs. Same as `d2 = dict(d1)`.	`d1 = {'pi':3.141, 'e':2.718}`
		`d2 = d1.copy()`
		`d2['r2'] = 1.414`
		`# d2 is {'pi': 3.141,`
		`#     'e':2.718, 'r2': 1.414}`

Operation	Description	Example

`d = dict.`
`fromkeys(seq,`
`value)`
	Generates a new dictionary using keys from a specified collection. All values will be set to `None` or an optional default.	`d = dict.fromkeys('ABC', 0)`
		`# {'A': 0, 'C': 0, 'B': 0}`

`d.get(key,`
`default)`
	Retrieves the value from a dictionary that is associated with a given key, and if the key is not in the dictionary then it gives `None` or the specified optional default.	`d = {'pi':3.141, 'e':2.718}`
		`d.get('pi') # 3.141`
		`d.get('mu') # None`
		`d.get('mu', 0.0) # 0.0`

`d.has_key(key)`	Determines whether a specified key is used in a dictionary. **Deprecated**, use `key in dict` instead. Not available in Python 3.	`[Deprecated, use "key in dict"` `instead.]`

`d.items()`	In Python 2, generates a list of tuples containing (`key`, `value`) pairs from a dictionary. In Python 3, gives an iterable view object instead of a list.	`d = {'G':3,'C':3,'A':2,'T':2}`
		`d.items()`
		`# dict_item([('A', 2), ('C',`
		`3), ('T', 2), ('G', 3)])`

`d.iteritems()`	Generates an iterator object that can loop through (`key`, `value`) pairs from a dictionary. Not available in Python 3, use `d.items()` instead.	`d = {'G':3,'C':3,`
		`    'A':2,'T':2}`
		`iterObj = d.iteritems()`
		`for key, val in iterObj:`
		`  print(key, val)`

`d.iterkeys()`	Generates an iterator object that can loop through all keys from a dictionary. The method `dict1.iterkeys()` does the same as `iter(dict1)`. Not available in Python 3, use `d.keys()` instead.	`d = {'G':3,'C':3,`
		`    'A':2,'T':2}`
		`iterObj = d.iterkeys()`
		`for key in iterObj:`
		`  print(key)`

`d.itervalues()`	Generates an iterator object that can loop through all values from a dictionary. Not available in Python 3, use `d.values()` instead.	`d = {'G':3,'C':3,`
		`    'A':2,'T':2}`
		`iterObj = d.itervalues()`
		`for value in iterObj:`
		`  print(value)`

`d.keys()`	In Python 2, generates a list containing the keys from a dictionary. The items in the list are in no particular order. In Python 3, gives an iterable view object instead of a list.	`d = {'pi':3.141, 'e':2.718}`
		`k = d.keys()`
		`print(k)`
		`# dict_keys(['pi', 'e'])`

`d.pop(key,`
`default)`
	Passes back a value associated with a specified key and removes the (key: value) pair from the dictionary, substituting an optional default value if a key is not present.	`d = {'pi':3.141, 'e':2.718}`
		`d.pop('e') # 2.718`
		`d.pop('mu', 0.0) # 0.0`
		`# d is {'pi': 3.141}`

Operation	Description	Example
`d.popitem()`	Removes and passes back an arbitrary key:value pair, as a tuple, from the dictionary.	`d = {'G':3,'C':3,` `    'A':2,'T':2}` `item = d.popitem()` `# d might be` `{'C':3,'T':2,'G':3}` `# item is then ('A', 2)`
`d.setdefault` `(key, default)`	Retrieves the value from a dictionary that is associated with a given key and if a key is not present adds it to the dictionary with a value of `None` or optional default.	`d = {'G':3,'C':3}` `d.setdefault('A') # None` `d.setdefault('G') # 3` `# {'A':None,'C':3,'G':3}`
`d.update` `(otherDict)`	Adds all the (key:value) pairs from one dictionary to another, replacing any values for keys that are already present.	`d = {'G':3,'C':3}` `d.update({'A':2})` `# {'A':2, 'C':3, 'G':3}`
`d.values()`	In Python 2, generates a list containing the values from a dictionary. The items in the list are in no particular order. In Python 3, gives an iterable view object instead of a list.	`d = {'pi':3.141, 'e':2.718}` `v = d.values()` `print(v)` `# dict_values([3.141, 2.718])`

File objects

The following table lists commonly used methods and attributes of `file` objects, which are generally created with `open(fileSystemPath, readWriteMode)` in order to read data from and/ or write data to a file system, e.g. hard disk, DVD etc.

Method	Description
`f.close()`	Closes a file object so that it is no longer used for reading or writing. This often happens implicitly when the handle on the object falls out of scope (e.g. returning from a function).
`f.flush()`	Flushes out and writes any cached file data (held in the file buffer). Note use of `os.fsync(fileObj.fileno())` may additionally be required to force the operating system to write immediately.
`f.fileno()`	Gives a number that identifies the particular file object to low-level operating-system procedures.
`f.read(size)`	Reads a specified number of bytes from a file object as a string or (from Python 3) as bytes if in binary mode, although this may be limited by the end of the file. With no size argument reads all data from the file.
`f.readline(size)`	Reads a single line from file, optionally specifying the maximum number of bytes to read (thus potentially truncating the line).

Method	Description
`f.readlines(targetSize)`	Reads all the lines from a file object. With optional argument, no more lines are read once total size of lines read so far exceeds the given number of bytes.
`f.seek(offset, code)`	Moves the file object's current read/write position to a given byte position. Generally more useful for 'binary' files than text files. Optional second argument is `0`, `1` or `2` to seek relative to the start (default), current position or file end respectively.
`f.tell()`	Retrieves the file's current read/write position.
`f.truncate(size)`	Truncates the file to the current read/write position, or to an optionally specified size.
`f.write(text)`	Writes a specified text string or bytes (if Python 3 binary mode) to file. Note that the data may not actually be written to disk until the buffering cache is full.
`f.writelines(texts)`	Writes an ordered collection of text strings or bytes (if Python 3 binary mode) to file.

Attribute	Description
`f.closed`	Determines whether the file object is closed for reading/writing.
`f.mode`	Fetches the read/write/append mode that the file object was created with.
`f.name`	Gives the name of the file, as represented on disk.

APPENDIX 3
Standard module highlights

The 'string' module

The table below lists attributes which can be imported from the string module. It should be noted that many things that were previously handled using the string module in very old versions of Python are now generally built into the string class, e.g. a = b.upper(), so we only list a few of the more useful string constants below.

Constant (string.)	Description	Value
ascii_letters	All ASCII letters both upper and lower case, in order.	'abcdefghijklmnopqrstuvwxyz ABCDEFGHIJKLMNOPQRSTUVWXYZ'
ascii_lowercase	All ASCII lower-case letters, in order.	'abcdefghijklmnopqrstuvwxyz'
ascii_uppercase	All ASCII upper-case letters, in order.	'ABCDEFGHIJKLMNOPQRSTUVWXYZ'
digits	The characters that are used to represent whole numbers.	'0123456789'
hexdigits	The characters that are used in hexadecimal representations of numbers.	'0123456789abcdefABCDEF'
letters	All letters both upper and lower case, in order. Not available in Python 3, use string.ascii_letters instead.	'abcdefghijklmnopqrstuvwxyz ABCDEFGHIJKLMNOPQRSTUVWXYZ'
lowercase	All lower-case letters, in order. Not available in Python 3, use string.ascii_lowercase instead.	'abcdefghijklmnopqrstuvwxyz'
octdigits	The characters that are used in octal represntations of numbers.	'01234567'
punctuation	All punctuation characters, i.e. not letters, digits or whitespace.	'!"#$%&\'()*+, -./:;<=>?@[\\]^_`{|}~'
printable	All printable characters that have an on-screen effect.	'0123456789abcdefghijklmno pqrstuvwxyzABCDEFGHIJKLMNO PQRSTUVWXYZ!"#$%&\'()*+, -./:;<=>?@[\\]^_`{|}~\t\n \r\x0b\x0c'
uppercase	All upper-case letters, in order. Not available in Python 3, use string.ascii_uppercase instead.	'ABCDEFGHIJKLMNOPQRSTUVWXYZ'
whitespace	All whitespace characters.	'\t\n\x0b\x0c\r '

The '`math`' module

The following table lists some of the more commonly used mathematical operations that may be used by importing from the `math` module. Most descriptions will also apply to equivalents in the `cmath` module, which is used for complex numbers. It should be noted that although the methods are listed in the form `math.method()`, if many mathematical operations are being performed (inside loops), and execution speed is important, then it is generally better to import the function directly, i.e. `from math import method`, and use `method(args)` rather than repeatedly use the dot notation `math.method(args)`.

Method/Attribute (`math.`)	Description
`ceil(x)`	Gives a floating point value that is rounded to the next integer value towards infinity. Equivalent to `math.floor(x)+1.0`.
`factorial(x)`	For an integer, the product of all natural numbers up to and including that value: *x!*. Noting that *0!* is defined to be 1.
`floor(x)`	Gives a floating point value that is rounded to the next integer value towards negative infinity.
`isinf(x)`	Determines whether a value is an infinity object.
`isnan(x)`	Determines whether a value is a 'not a number' object.
`exp(x)`	Finds the exponent; *e* to the power of a value.
`log(x, base)`	Finds the logarithm of a value. Defaults to a natural logarithm, but an optional base may be specified.
`pow(x, y)`	Raises one value to the power of another. Equivalent to `x ** y`.
`sqrt(x)`	Finds the square root of a non-negative value.
`acos(x)`	Finds the inverse cosine (arccosine).
`asin(x)`	Finds the inverse sine (arcsine).
`atan(x)`	Finds the inverse tangent (arctangent).
`atan2(y, x)`	Finds the inverse tangent of *y/x* (arctangent).
`cos(x)`	Finds the cosine of an angle (in radians).
`hypot(x, y)`	Finds the length of the hypotenuse of a right-angled triangle (2D), given the length of the other two sides. Equivalent to `math.sqrt(x*x + y*y)`.
`sin(x)`	Finds the sine of an angle (in radians).
`tan(x)`	Finds the tangent of an angle (in radians).
`degrees(x)`	Converts an angle in radians to degrees.
`radians(x)`	Converts an angle in degrees to radians.
`acosh(x)`	Finds the inverse hyperbolic cosine.
`asinh(x)`	Finds the inverse hyperbolic sine.
`atanh(x)`	Finds the inverse hyperbolic tangent.

Method/Attribute (`math.`)	Description
`cosh(x)`	Finds the hyperbolic cosine.
`sinh(x)`	Finds the hyperbolic sine.
`tanh(x)`	Finds the hyperbolic tangent.
`pi`	The constant Pi (π); 3.141592653589793...
`e`	The constant Euler's number (e); 2.718281828459...

The '`random`' module

The following table lists some of the more general methods of the `random` module, which is used to generate pseudorandom numbers. It should be noted that for testing purposes it may be useful to use `random.seed()` with a fixed value to get a reproducible set of apparently random numbers.

Method (`random.`)	Description
`seed(x)`	Sets an initialising seed number for the (pseudo)random number generator. Defaults to use the current time as the seed. Using a fixed seed allows for the same pseudorandom sequence to be generated.
`randint(start, end)`	Generates a random integer value from a specified range. Note that the start and end of the range is included.
`choice(vals)`	Retrieves a random item from a list or other ordered collection.
`shuffle(vals, func)`	Randomly changes the order of items in a list (or other mutable sequence). Optional argument to specify a specific random number [0–1] generating function.
`sample(vals, num)`	Randomly selects a number of items from a list or generator, without replacement. Leaves the original list unaltered.
`random()`	Generates a random floating number between zero and one (interval [0,1]), using a uniform probability distribution.
`uniform(start, end)`	Generates a random floating number from between specified bounds, using a uniform probability distribution.
`normalvariate(mu, sigma)`	Generates a random numbers with a normal/Gaussian probability distribution of specified mean and standard deviation.

The '`os`' module

The `os` module is used for things that depend on the operating system (whether Windows, Mac OS X, Linux etc.) and presents them in a standardised way. For example, `os.path` is used to interact with the file system. We've skipped the description of lots of functionality involving processes, devices etc., concentrating only on a few of the more commonly used, general functions. This module should not be confused with the sys module, which deals with Python interpreter information and not the operating system.

Method (os.)	Description
chdir(path)	Changes the current working directory; for specifying relative file-system paths.
getenv(varname, *default*)	Gets an environment variable of a given name, with optional default value if it doesn't exist.
putenv(varname, value)	Sets an environment variable of a given name, with a given value.
uname()	In Python 2 gives a tuple containing five items detailing the current operating-system type. In Python 3 gives a uname_result object, representing the same kind of information.
unsetenv(*varname*)	Deletes a named environment variable.
tmpfile()	Creates an unnamed, temporary file object with mode 'w+b' (write binary) which only exists for the current session.
access(path, mode)	Gets status information for a file-system path, using a specified mode to check if the path exists, is readable, writeable or executable.
chdir(path)	Changes the current working directory (for relative paths) to the specified path.
getcwd()	Retrieves a path string representing the current working directory; for relative paths.
chmod(path, mode)	Changes read, write and execution permissions for a given path. Equivalent to UNIX 'chmod'.
chown(path, uid, gid)	Changes the user and group ownership or a given path. Equivalent to UNIX 'chown'. Note this can be used on Windows systems but only sets the read-only status.
link(target, linkName)	Creates a (hard) file link connecting a stated source/target path with a specified name.
listdir(path)	Creates a list of file and directory names for a specified file-system path.
mkdir(path, *mode*)	Creates a new, empty directory with a specified file-system path. Optional argument to specify the file access mode.
makedirs(path, *mode*)	Recursively creates a new, empty directory with a stated path, including any intervening directories if they do not exist. Optional argument to specify the file access mode.
remove(path)	Deletes a file of specified path. Gives an OSError if the path is a directory (use rmdir() for deleting directories).
removedirs(path)	Recursively removes a directory, if empty, and then parent directories that are empty. Gives an OSError if the directory is not empty.
rename(old, new)	Change the name of a file or directory.
rmdir(path)	Removes a specified directory path from a file system, if it is empty, otherwise it gives an OSError.

Method (`os.`)	Description
`stat(path)`	Generates a status object that has attributes detailing information about a given path, including access permissions, device, owner, group, file size, access time, modification time etc.
`symlink(target, linkName)`	Creates a symbolic (soft) link to a given path.
`unlink(path)`	Same as `os.remove()`.
`walk(top, topDown, onError, followLinks)`	Creates a generator that yields (`parentPath, dirNames, fileNames`) tuples that recursively walk through a directory and any sub-directories.
`system(command)`	Issues a command to the operating system as if it were from a command line prompt. It is recommended to use `subprocess.call()` instead.

The '`os.path`' sub-module

The sub-module `os.path` contains further functionality that applies to file-system paths. The term path refers to a particular location of a file, directory or link within the hierarchical organisation of a file system, e.g. on a hard disk or DVD. All the functionality presented here works on strings to represent the file-system paths, e.g. '`/home/user/file.py`'. Given that Python strings are immutable, the relevant methods will generate new path strings etc. rather than modify existing ones. All of the examples below assume the appropriate import has been made in the form: `from os.path import method`, although we do not mean to preclude other import styles. From Python 3.4 there is a module, `pathlib`, which provides an object-oriented approach to file paths.

Method (`os.path.`)	Description
`abspath(path)`	Generates a full absolute path from a relative path, i.e. by joining it to the current working directory.
`basename(path)`	Gives the file or directory name at the end of a path, removing any leading directories.
`commonprefix(list)`	Generates the longest string possible which has characters that matches the start of all the paths in a list.
`dirname(path)`	Gives the directory name or names for a path, removing any file or directory name at the end.
`exists(path)`	Determines whether a file or directory path exists in the file system, giving `True` if it does and `False` otherwise. Broken symbolic links give False.
`lexists(path)`	As above for `exists()`, but broken symbolic links give `True`.
`expanduser(path)`	Generates a path where any '~' or '~user' stand-ins for a home directory are expanded to the current user's full home directory.
`expandvars(path)`	Generates a path where any substrings representing environment variables of the form '$env' or '{$env}' are filled in with the textual value of the environment variable (if it is known).
`getatime(path)`	Gets the last access time for a path in seconds, assuming it exists.*
`getmtime(path)`	Get the last modification time for a path in seconds, assuming it exists.*

Method (os.path.)	Description
getctime(path)	Gets creation time for a path in seconds, assuming it exists.*
getsize(path)	Gets the number of bytes used to store a path, assuming it exists, on a file system. Generally use to get the size of a file.
isabs(path)	Determines whether a path is an absolute path (or otherwise is a relative path). For UNIX-based systems an absolute path will start from the root '/' and on Windows from '\' or a drive specification like 'C:\'.
isfile(path)	Determines whether a path represents the location of a normal file, or a symbolically linked file, i.e. exists and is not a directory.
isdir(path)	Determines whether a path represents the location of a directory, or a symbolically linked directory, i.e. exists and is not a file.
islink(path)	Determines whether a path represents a symbolic link, whether to a file or directory.
ismount(path)	Determines whether a path represents a location that points to a different device (hard disk, CD ROM, USB drive) compared to the parent directory, i.e. whether it is a mount point.
join(path1, path2,...)	Generates a longer path string by combining multiple other paths (e.g. sub-directories and a file name) in order using the appropriate separator for the system ('\' or '/').
normcase(path)	Generates a path with letters of consistent case; only has an effect on systems without case sensitivity (in which case it converts to lower case). Has no effect on UNIX-like systems.
normpath(path)	Generates a normalised version of a path which has any redundant directory specifications removed (e.g. involving '../').
realpath(path)	Provides the real, underlying file-system path for the destination of a symbolic link. Gives back the input path unaltered if it is not a symbolic link.
relpath(path, start)	Given a path creates a relative path to that file-system location from the current working directory, or optionally from another specified directory.
samefile(path1, path2)	Determines whether two path specifications represent the same file or directory, giving True if they do and False otherwise.
sameopenfile(fileObj1, fileObj2)	Determines whether two file objects represent the same file or directory, giving True if they do and False otherwise.
split(path)	Creates a split version of the path, separating any leading directory from a final file or directory name. Gives a 2-tuple in the form (leading, final). Same as (dirname(path),basename(path)).
splitdrive(path)	Creates a split version of the path, separating any leading drive specification from the rest of a path, yielding a 2-tuple. Generally only useful on Windows systems.
splitext(path)	Creates a split version of the path, separating any trailing file extension (e.g. '.jpg', '.py' etc.) from the remainder of the path, which generally includes the directories and the rest of the file name. Gives a 2-tuple in the form (leading, extension).

* Since the system zero time, which is somewhat arbitrarily 00:00:00 UTC on 1 January 1970, the start of the computing 'epoch'.

The 'pickle' module

The module `pickle` can be used to serialise Python objects, to convert them to/from a textual representation, which may be saved to disk etc. All the normal Python data types as well as module-level classes and functions can be serialised in this way. Custom classes may also be serialised if the class is imported into the current name-space of a program. It should be noted that in Python 2 there is also a `cPickle` module, which is a faster C-language implementation of `pickle`, and normally `cPickle` is the one to use unless you are defining `pickle` subclasses. In Python 3 there is no `cPickle` module, and `pickle` automatically uses the C-language implementation when it is available.

Method (`pickle.`)	Description
`dump(obj, file, protocol)`	Saves a copy of a Python object to disk, as a serialised Pickle file, using a specified file object.
`load(file)`	Loads a Python object saved as a serialised Pickle file into memory.
`dumps(obj, protocol)`	Creates a serialised Pickle string that represents a given object.
`loads(string)`	Creates a Python object from a serialised Pickle string version.

The 'sys' module

The `sys` module contains information that relates to the Python interpreter and its runtime environment. As with the `os` module (which is for operating-system interaction) we only focus on some of the more commonly used functionality.

Attribute/Method (`sys.`)	Description
`argv`	Retrieves a list that contains the command line arguments that were passed in when Python was run, starting with the name of the Python script.
`exit(code)`	Quits the current Python session/program, optionally passing back a specific exit status code to the system.
`modules`	Retrieves a dictionary that contains all of the currently imported modules as `(module name: file path)`, key, value pairs.
`path`	Retrieves a list of strings representing the Python module look-up paths. Starts with the value from the PYTHONPATH environment variable. This list of paths can be modified programmatically.
`platform`	Retrieves a string indicating what basic kind of computer system Python is running on, i.e. indicating Windows, Mac OS X, Linux etc.
`stdin`	Retrieves a file object that represents standard input, where textual input is taken from, which by default is the keyboard.
`stdout`	Retrieves a file object that represents standard output to the screen, where textual output, in the absence of a specific file, is sent. This is what `print` writes to by default.

Attribute/Method (`sys.`)	Description
`stderr`	Retrieves a file object that represents standard error output; usually to screen, unless otherwise redirected.
`version`	Retrieves a string indicating which version of Python is being used.

The '`time`' module

The following are a few simple methods from the `time` model. There is much more functionality in this module, and in the related `datetime` module, e.g. for time string representation, formatting and daylight savings, which is described in the standard on-line Python documentation.

Method (`time.`)	Description
`ctime(secs)`	Generates a human-readable string from a time represented in seconds (since the arbitrary zero time: 00:00:00 UTC on 1 January 1970) . Defaults to the current time if no argument is specified.
`sleep(secs)`	Suspends the execution of the program for a given number of seconds.
`time()`	Retrieves the time in seconds since the system zero time (00:00:00 UTC on 1 January 1970).

The '`shutil`' module

The `shutil` module is used to manipulate whole files, providing functionality that is not present in `os`.

Method (`shutil.`)	Description
`copyfileobj(srcOb, dstOb [, chunkSize])`	Copies the contents of one open file object to another, optionally specifying the number of byte 'chunks' to buffer while copying.
`copyfile(source, dest)`	Copies the data contained in a file from one path to another **full file path** (not just a directory name). Does not copy permissions, access times etc.
`copymode(source, dest)`	Copies read, write and execute permissions from a source path (e.g. file or a directory) to a destination path.
`copystat(source, dest)`	Copies the permissions and access/modification/creation times from a source path to a destination path.
`copy(source, dest)`	Copies a file from one path to a destination file path or directory. If the destination is a directory then the source file name will be used. File permissions will be copied, but not access time etc.
`copy2(source, dest)`	Like `copy()` above, but will copy access/modification/creation times.

Method (`shutil.`)	Description
`copytree(source, dest,` `symlinks, ignore)`	Copies the contents of an entire directory tree recursively to a destination directory. Options to also copy symbolic links and callable argument that can be used to exclude items.
`rmtree(path, errorIgnore,` `func)`	Deletes an entire directory tree recursively. Options to ignore any errors and a callable argument to use in the event of an error.
`move(source, dest)`	Moves a file or directory (recursively) from one path to another.

Miscellaneous modules

Below we list some of the standard Python modules in terms of their general function and give a few simple examples. Full documentation for these modules can be found at the official Python website at http://docs.python.org/2/library/ or http://docs.python.org/3/library/. These are modules that will need to be imported into a Python script, but which should be present as standard in the Python installation. It should be noted that the `re` module is described in extensive detail later in Appendix 5.

Module	Description	Example
`argparse`	A module that helps interpret command line options/arguments, i.e. information typed after the name of a program, as available in `sys.argv`.	`from argparse import` `ArgumentParser` `parser = ArgumentParser` `(prog='MyProgram')` `parser.add_argument('-x',` `type=float, help='Help for X')` `parser.add_argument('-y',` `type=int, nargs='?',` `default=1)` `parser.print_help()`
`array`	A packed numeric array object, i.e. an ordered collection containing all numbers or characters, with a given data type (specified with a one-letter code). Not as capable as `numpy.array` but part of standard Python and efficient for interpreting 'binary' data.	`from array import array` `data = [9,7,5,4]` `x = array('i', data) # Int type` `len(x) # 4`
`copy`	Creates a new Python object by copying an existing object. Can create shallow or deep copies, where any object contained by an object is itself also copied.	`from copy import copy, deepcopy` `x = [[1,2], [3,4]]` `y = copy(x) # Same _contents_` `z = deepcopy(x) # All new` `x[1].append(5)` `print(y) # [[1, 2], [3, 4, 5]]` `print(z) # [[1, 2], [3, 4]]`

Module	Description	Example

`cStringIO` (Python 2) or `io` (Python 3)	Used to create an object that can be used to read string data as if it were a file.	Python 2:
		`from cStringIO import StringIO`
		`obj = StringIO('Start\nMid\nEnd\n')`
		`obj.readline() # 'Start\n'`
		Python 3:
		`from io import StringIO`

`datetime`	A module that contains `date`, `time`, `timedelta` and `datetime` objects to represent temporal information. Deals with daylight savings, date formatting, time string interpretation etc.	`from datetime import datetime`
		`text = '07/May/1945 02:41'`
		`format = '%d/%b/%Y %H:%M'`
		`dt = datetime.strptime(text, format)`
		`dt.month # 5`
		`dt.ctime()`
		`# 'Mon May 7 02:41:00 1945'`

`fnmatch`	Provides file name matching using UNIX-like wild cards, i.e. patterns that include '*' and '?' rather than regular expressions.	`from os import listdir`
		`from fnmatch import fnmatch`
		`for file in listdir('.'):`
		   `if fnmatch(file,'*.txt'):`
		      `print(file)`

`ftplib`	Used to send and receive files using the File Transfer Protocol.	`from ftplib import FTP`
		`ftpSession = FTP('ftp.ncbi.gov', username,password)`
		`ftpSession.cwd('genomes')`
		`ftpSession.dir()`
		`ftpSession.retrbinary('RETR: remoteFile')`
		`ftpSession.storLines('STOR localfile')`
		`ftpSession.quit()`

`gzip, bz2, zipfile, tarfile`	Libraries that deal with creating and extracting compressed and/or archived files.	`import gzip`
		`fileObj = gzip.open('data.gz')`
		`for line in fileObj:`
		   `print(line)`

`httplib` (Python 2) or `http.client` (Python 3)	Used to send and receive information across the Internet using the Hypertext Transport Protocol. A lower-level library than `urllib/urllib2`.	Python 2:
		`from httplib import HTTPConnection`
		`conObj = HTTPConnection("www.python.org")`
		`conObj.request("GET","/index.html")`
		`resp = conObj.getresponse()`
		`print(resp.status)`
		`print(resp.read())`

Module	Description	Example

		Python 3:
		```from http.client```
		```import HTTPConnection```

multiprocessing	Runs Python code as separate, parallel, processes. Generally used on multiple core/processor systems.	```from multiprocessing import```
		```Process```
		```job1 = Process(target=calcFunc,```
		``` args=work1)```
		```job2 = Process(target=calcFunc,```
		``` args=work2)```
		```job1.start()```
		```job2.start()```
		```job1.join()```
		```job2.join()```

platform	Used to get information about the current computer and its architecture.	```import platform```
		```platform.processor()```
		```# e.g. 'x86_64'```
		```platform.python_version()```
		```# e.g. '2.7.3'```
		```platform.architecture()```
		```# e.g. ('64bit', 'ELF')```
		```platform.node()```
		```# e.g. 'MyPC'```

re	Regular expressions; used to find and replace substrings using pattern matching. See Appendix 5.	See Appendix 5.

sqlite3	Allows interaction with a lightweight SQL database called SQLite.	```import sqlite3```
		```conn = sqlite3.connect('myDb')```
		```cursor = conn.cursor()```
		```stmt = "select * from structure```
		```where pdbId='1AFO'"```
		```cursor.execute(stmt)```
		```result = cursor.fetchall()```
		```cursor.close()```
		```conn.close()```

subprocess	Runs an external program as a separate job/process and connects any input/output data streams.	```from subprocess import call```
		```command = 'clustalw seq.fasta'```
		```call(command, stdIn=filObj)```

threading	Runs Python code in separate threads. These will not run concurrently on multiprocessor systems (use multiprocessing instead for that), but can be useful to process intermittent data streams.	```from threading import Thread```
		```job1 = Thread(target=calcFunc,```
		``` args=work1)```
		```job2 = Thread(target=calcFunc,```
		``` args=work2)```
		```job1.start()```
		```job2.start()```
		```job1.join()```
		```job2.join()```

Module	Description	Example

`urllib, urllib2` (Python 2) or `urllib.request,` `urllib.parse,` `urllib.error` (Python 3)	Used to send and receive information across the Internet: a higher-level, and so often more convenient, library than `httplib`. Handles web proxies, redirection, passwords, cookies etc. Often used to interact with web services and databases.	Python 2:
		`import urllib, urllib2`
		`optionDict = {'format':'PDB',`
		`'compression':'None'}`
		`optionStr = urllib.urlencode`
		`(optionDict)`
		`url = 'http://www.rcsb.org/pdb/`
		`cgi/export.cgi/1OUN.pdb'`
		`req = urllib2.Request(url,`
		`optionStr)`
		`resp = urllib2.urlopen(req)`
		`print(resp.read())`
		
		Python 3:
		`import urllib, urllib.request`
		`optionDict = {'format':'PDB',`
		`'compression':'None'}`
		`optionStr = urllib.parse.`
		`urlencode(optionDict).encode`
		`('utf-8')`
		`url = 'http://www.rcsb.org/pdb/`
		`cgi/export.cgi/1OUN.pdb'`
		`req = urllib.request.Request`
		`(url, optionStr)`
		`resp = urllib.request.urlopen`
		`(req)`
		`print(resp.read().decode('utf-8'))`

`zlib`	Used to compress data into more compact representation, using the zlib algorithm. Can be useful for caches and undo functions.	`import zlib`
		
		`x = zlib.compress('Bananarama')`
		`print(x, zlib.decompress(x))`

Numerical Python: 'numpy'

The table below details a subset of the functionality available from the `numpy` module. For brevity we have only included some of the more commonly used aspects as well as those mentioned in this book. Often we have not included all the possible arguments for a function, instead only focussing on what we deem most important. Fuller documentation may be found at the NumPy and SciPy website: http://docs.scipy.org/doc/. It should be noted that the `numpy` module is not part of the standard Python library, and as such is installed separately.

Although not listed in the table, the `numpy` module includes many mathematical operations, e.g. `abs`, `sqrt`, `exp`, `log`, `power`, `cos`, `sin`, `tan`, `arccos`, `arcsin`, `arctan cosh`, `sinh` and `tanh`, which sometimes share the same names as standard `math` methods. These methods accept single values (so may be used instead of `math` methods) and also operate on arrays, where they act in an element-wise manner.

NumPy has more numeric data types than standard Python, which may be used to represent numbers with various numbers of bits (and thus also levels of precision). Examples include: `int`, `int8`, `int16`, `uint8`, `uint16`, `float`, `float32` etc. The types `int` and `float`, without a bit specification, will correspond to the same data type as regular Python (although this is specific to the system, i.e. 32 or 64 bit).

In NumPy the most important class is the container object `array` and many of the functions which may be independently imported from `numpy` are also bound methods of `array`. Thus, for example, to calculate a dot product we could do direct imports:

```
from numpy import array, dot
a1 = array([1, 2, 3])
a2 = array([4, 5, 6])
dp = dot(a1, a2)
```

Or use a bound method:

```
from numpy import array
a1 = array([1, 2, 3])
a2 = array([4, 5, 6])
dp = a1.dot(a2)
```

We list both approaches in the table, which naturally assumes the appropriate modules are imported for each case. It should also be noted that the input arguments generally do not themselves need to be NumPy `array` (or `matrix`) objects. In many cases data can be input in the form of an 'array-like' sequence of values, which would typically include tuples and lists of numbers.

Constructor (`numpy.`)	Description	Example

`array(data, dtype)`	Creates a numeric array object from another array or array-like data. All values will be converted to the same data type. Data type will be inferred from the input values if not specifically stated. An array may have more than one dimension/axis, e.g. constructed using a list of lists.	`a = array([[1, 2, 3],`
		`  [4, 5, 6]])`
		`a.shape # (2,3) - rows, cols`
		`a.ndim # 2 - axes`
		`a.tolist() # Make Python list`
		
		`b = array([3.0, 0, 4.7, 1], float)`
		`# Floats: 3.0, 0.0, 4.7, 1.0`

`matrix(data, dtype)`	Creates a numeric matrix object from an array or array-like data. Provides an alternative to `array()` that has more matrix-oriented operations, e.g. multiplication is not element-wise. Data type will be inferred from the input values if not specifically stated.	`m1 = matrix([[1,2], [1,0]])`
		`m2 = matrix([[3,0], [0,3]])`
		
		`print(m1 * m2)`
		
		`# matrix([[3, 6],`
		`#         [3, 0]])`

Method	Description	Example
`append(data1, data2, dtype)`	Add one or more values on the end of an array to make a longer array.	`a = array([[1, 2, 3],` `        [4, 5, 6]])`

Constructor (numpy.)	Description	Example

	Unless an axis is specified the result will be a flattened, 1D array. If an axis is specified the array added must have the same number of dimensions, and naturally must match the shape along that axis. Note it will generally be quicker to use `empty()` and then fill array values by index rather than call `append()` multiple times.	`b = append(a, [7,8,9])`
		# Makes flat 1 x 9 array
		`c = append(a, [[7,8,9]], axis=0)`
		# Makes 3 x 3 array
		`d = append(a, [[4],[7]], axis=1)`
		# Makes 2 x 4 array

arange(*start*, end, *step*, *dtype*)	Generates an array of equally spaced numbers, from a start point, up to (but not including) an end point using a given step size. The start defaults to zero and the step to 1. The data type will be inferred from the arguments unless specifically stated. Similar to range/xrange in standard Python, but for more data types.	`arange(3.0)`
		# Floats: array([0.0, 1.0, 2.0])
		`arange(7, 3, -1)`
		# Ints: array([7, 6, 5, 4])
		`arange(0, 6 ,2, float)`
		# Floats: array([0.0, 2.0, 4.0])
		`arange(3.2, 3.5, 0.1)`
		# Floats: array([3.2, 3.3, 3.4])

argmax(arry, *axis*)
or
a.argmax(*axis*)
	Gets an array of indices that specify the position of the maximum value in an array, either along a specified axis or in the flattened 1D version of the array if no axis is specified.	`a = array([[1,2,8], [9,1,1]])`
		`argmax(a) # 3`
		# Position of max in 1D
		`argmax(a, axis=0)`
		# array([1, 0, 0])
		# Which row has max for each col
		`a.argmax(axis=1)`
		# array([2, 0])
		# Which col has max for each row

argmin(arry, *axis*)
or
a.argmax(*axis*)
	As above, but finds the indices for the minimum value. Also an inbuilt method of numpy.array.	`a = array([[1,2,8], [9,1,1]])`
		`a.argmin(0) # axis=0`
		# array([0, 1, 1])
		# Which row has min for each col

argsort(arry, *axis*)
or
a.argsort (*axis*)
	Compares the values in an array to generate an array of their indices in value-sorted order. If the axis is not specified the result refers to a flattened 1D version of the array.	`a = array([5,9,3,4,1,8])`
		`sortIndices = a.argsort()`
		`print(sortIndices)`
		# [4,2,3,0,5,1] – min val at a[4]
		`print(a[sortIndices])`
		# [1,3,4,5,8,9]

argwhere(arry)	Finds the indices of the non-zero (or true) elements of an array. Unlike nonzero(), the returned indices are grouped by element, rather than by axis.	`a1 = array([[8, 0, 0],`
		`            [0, 0, 9]])`
		`indices = argwhere(a1)`
		# array([[0, 0], [1, 2]])
		# Positions of non-zero elements

Constructor (numpy.)	Description	Example

cov(data1, data2)	Makes an array representing the sample covariance matrix for given array or array-like data. Each data row is taken to be a different variable, and each data column represents a separate data point/vector. The covariance matrix represents the correlations between different pairs of variables (data dimensions).	```x = [0.0, 1.0, 2.0, 3.0]```
		```y = [0.4, 2.2, 3.9, 5.8]```
		```cov(x) # 1.666667```
		```cov(y) # 5.342500```
		```cov([x, y])```
		
		```# array([[1.666667, 2.983333],```
		``` [2.983333, 5.342500]])```

cross(arry1, arry2)	The cross-product between two vector arrays. The arrays must be 2 or 3 dimensional. For 3D arrays the result is a vector array that is perpendicular to both input vectors (where possible). For 2D arrays the result is the determinant of the associated 2×2 matrix.	```cross([0,1,0], [0,0,1])```
		```# array([1, 0, 0])```
		
		```cross((3,1),(1,2))```
		```# 5```

dot(arry1, arry2) or a.dot(arry)	The dot product or scalar product between two arrays. For 2D matrices the result is matrix multiplication. For 1D vectors the result is a scalar, equivalent to the magnitude of the projection of one vector on to the other.	```m1 = array([[1, 2],```
		``` [-2, 0]])```
		```m2 = array([[-1, 1],```
		``` [2, 3]])```
		```m3 = dot(m1, m2)```
		```# array([[3, 7],```
		```# [2, -2]])```
		```v1 = array([1,2,3,4])```
		```v1.dot(array([4,3,2,1])) # 20```

dstack(arrays)	Combines a sequence (e.g. list) of arrays into a larger array by stacking them depth-wise, along their third axis. For example, combines separate 2D matrices into a 3D tensor.	```m1 = array([[1,2,3],```
		``` [4,5,6]])```
		```m2 = array([[7,8,9],```
		``` [0,1,2]])```
		```m3 = dstack([m1, m2])```
		```# array([[[1, 7],[2, 8],[3, 9]],```
		``` [[4, 0],[5, 1],[6, 2]]])```
		```print(m3.shape)```
		```# (2, 3, 2)```

empty(sizes, dtype)	Generates a new empty array, of specified size. The array will actually be filled with arbitrary (but small) numbers, rather than zero. The data type may be specified, but otherwise defaults to floating point. The size can be an integer or a tuple of integers, one for each axis, specifying number of rows, columns etc.	```a = empty((2,1))```
		```# 2 x 1 array, arbitrarily:```
		```# array([[1.34634549e-316],```
		```# [1.81895505e-317]])```

eye(nRows, nCols)	Creates an identity matrix of a specified size; a 2D array with ones on the diagonal and zeros elsewhere. If	```eye(3)``` ```# array([[1.0, 0.0, 0.0],```

Constructor (numpy.)	Description	Example
	the number of columns is not specified the matrix is square.	`#          [ 0.0, 1.0, 0.0],` `#          [ 0.0, 0.0, 1.0]])`

`histogram (arry, bins, valRange, density)`	Given an input array of values generates a histogram (counts and edges arrays) by counting values within range bins. The `bins` can be specified as a number of bins or as a list of boundary values. If unspecified the value range is from the minimum to maximum values of the input. Option to normalise the histogram so its summation is one: `density=True`.	`data = [0.1, 0.5, 1.5, 1.3, 1.0]`
		`hist = histogram(data, 3, (0,3))`
		`vals, edges = hist`
		
		`print(vals)`
		`# [2, 3, 0]`
		`print(edges)`
		`# [ 0., 1., 2., 3.]`

`hstack(arrays)`	Combines a sequence (e.g. list) of arrays into a larger array by stacking them column-wise, along their first axis. For example, combines 1D vectors into a single long 1D vector.	`v1 = array([0,1,2])`
		`v2 = array([7,8,9])`
		`v3 = hstack([v1, v2])`
		`# array([0, 1, 2, 7, 8, 9])`

`inner(arry1, arry2)`	Calculates the inner product of two arrays, scalars or one of each. Equivalent to `.dot()` for 1D vectors, but for 2D and above the result is the sum of the products over the last axes (rather than last and penultimate in matrix multiplication).	`m1 = array([[ 1, 2],`
		`             [-2, 0]])`
		`m2 = array([[-1, 1],`
		`             [ 2, 3]])`
		`m3 = inner(m1, m2)`
		`# array([[ 1, 8],`
		`#        [ 2, -4]])`

`ix_(indices1, indices2, ...)`	Given a list of indices for each axis/ dimension generates a mesh of indices: a tuple of arrays that can be used to index the selected rows, columns etc. of another array. Can be used to extract sub-matrices.	`rows = [0,1]`
		`cols = [0,2]`
		`mesh = ix_(rows, cols)`
		`a1 = array([[0, 1, 2],`
		`             [3, 4, 5],`
		`             [6, 7, 8]])`
		
		`a2 = a1[mesh]`
		`# array([[0, 2],`
		`#        [3, 5]])`

`max(array, axis)` or `a.max()`	Gives the maximum values of an array along a given axis, or if no axis is specified the maximum value in the whole array.	`m1 = array([[-1, 1],`
		`             [ 2, 3]])`
		`m1.max()`
		`# 3`
		`m1.max(axis=0)`
		`# array([2, 3])`
		`m1.max(axis=1)`
		`# array([1, 3])`

Constructor (numpy.)	Description	Example

mean(array, *axis*) or a.mean()	Gives the mean (average) values of an array along a given axis, or if no axis is specified the mean value of the whole array.	`m1 = array([[-1, 1],`
		`            [2, 3]])`
		`m1.mean()`
		`# 1.25`
		`m1.mean(axis=0)`
		`# array([0.5, 2.0])`

mgrid[slice1, *slice2, ...*]	An *N*-dimensional grid object which can be indexed to generate multi-dimensional ranges of indices. Provides similar functionality to range() over multiple dimensions.	`m = mgrid[0:3,0:3]`
		`#array([[[0,0,0],`
		`#        [1,1,1],`
		`#        [2,2,2]],`
		`#       [[0,1,2],`
		`#        [0,1,2],`
		`#        [0,1,2]]])`

min(array, *axis*) or a.min()	Gives the minimum values of an array along a given axis, or if no axis is specified the minimum value in the whole array.	`m1 = array([[-1, 1],`
		`            [2, 3]])`
		`m1.min()`
		`# -1`
		`m1.min(axis=0)`
		`# array([-1, 1])`
		`m1.min(axis=1)`
		`# array([-1, 2])`

nonzero(arry) or a.nonzero()	Gives the positional indices for elements in an array that are non-zero, or true in a Boolean sense.	`a1 = array([[8, 0, 0],`
		`            [0, 0, 9]])`
		`rows, cols = a1.nonzero()`
		`# array([0, 1]), array([0, 2])`
		`# Nonzero at (0,0) and (1,2)`

ones(sizes, *dtype*)	Generates a new array, of specified size, filled with ones. The data type may be specified, but otherwise defaults to floating point. The size can be an integer or a tuple of integers, one for each axis, specifying number of rows, columns etc.	`ones(4, int)`
		`# Ints: array([1, 1, 1, 1])`
		
		`ones((2,3))`
		`# 2 x 3 array`
		`# array([[1.0, 1.0, 1.0],`
		`#        [1.0, 1.0, 1.0]])`

outer(arry1, arry2)	Calculates the outer product or tensor product of two array vectors. The outer product of two vectors will create a matrix with elements that are the product of the elements from each vector, where the first vector is applied across rows and the second across columns.	`v1 = array([0,1,2])`
		`v2 = array([7,8,9])`
		
		`m = outer(v1,v2)`
		`# array([[ 0,  0,  0],`
		`#        [ 7,  8,  9],`
		`#        [14, 16, 18]])`

radians (degrees)	Converts an angle value in degrees to radians: multiply by $\pi/180$.	`radians(120.0)` `# 2.0943951023931953`

Constructor (numpy.)	Description	Example

`ravel(arry)` or `a.ravel()`	Creates a flattened, one-dimensional version of an array.	`a = array([[0, 1, 2],`
		`             [3, 4, 5],`
		`             [6, 7, 8]])`
		`a.ravel()`
		`# array([0,1,2,3,4,5,6,7,8])`

`reshape(arry, sizes)` or `a.reshape (sizes)`	Rearranges the elements of an array to make a new array with specified number of positions along each axis (rows, columns etc.).	`a = array([1,2,3,4,5,6])`
		`reshape(a, (2,3)) # Rows, cols`
		`# array([[1, 2, 3],`
		`#        [4, 5, 6]])`
		`a.reshape(3,2)`
		`# array([[1, 2],`
		`#        [3, 4],`
		`#        [5, 6]])`

`std(arry, axis)` or `a.std(axis)`	Calculates the standard deviation of the values in an array along a specified axis. If the axis is not specified the standard deviation is over all values, i.e. the array is flattened.	`data = array([[0.1, 0.2, 0.4],`
		`              [1.5, 1.3, 1.0]])`
		`data.std()`
		`# 0.543905629069`
		`data.std(axis=1)`
		`# array([0.12472191,`
		`0.20548047])`

`sum(arry, axes)` or `a.sum(axes)`	Calculates the summation of the values in an array, along a specified axis (or tuple of axes). If no axis is specified the summation is over all values in the array.	`data = array([[0.1, 0.2, 0.4],`
		`              [1.5, 1.3, 1.0]])`
		`data.sum()`
		`# 4.5`
		`data.sum(axis=0)`
		`# array([1.6, 1.5, 1.4])`

`var(arry, axis)` or `a.var(axis)`	Calculates the variance of the values in an array along a specified axis. If the axis is not specified the variance is over all values, i.e. the array is flattened.	`data = array([[0.1, 0.2, 0.4],`
		`              [1.5, 1.3, 1.0]])`
		`data.var()`
		`# 0.295833333333`
		`data.var(axis=1)`
		`# array([0.01555556,`
		`0.04222222])`

`vstack(arrays)`	Combines a sequence (e.g. list) of arrays into a larger array by stacking them row-wise, along their second axis. For example, combines 1D vectors into a 2D matrix.	`v1 = array([0,1,2])`
		`v2 = array([7,8,9])`
		`m = vstack([v1, v2])`
		`# array([[0, 1, 2],`
		`#        [7, 8, 9]])`

`zeros(sizes, dtype)`	Generates a new array, of specified size, filled with zeros. The data type may be specified, but otherwise defaults to floating point. The size can	`zeros(5)`
		`# Floats: 0.0, 0.0, 0.0, 0.0, 0.0`
		`zeros(4, int)`

Constructor (numpy.)	Description	Example

	be an integer or a tuple of integers, one for each axis, specifying number of rows, columns etc.	`# Ints: array([0, 0, 0, 0])`
		
		`zeros((2,3))`
		`# 2 x 3 array`
		`# array([[0.0, 0.0, 0.0],`
		`#         [0.0, 0.0, 0.0]])`

Module	Description	Example

`fft`	A module with methods to perform Fourier analysis, such as the discrete Fourier transform and its inverse.	`from numpy import exp, pi, arange`
		`I = complex(0, 1)`
		`t = arange(0.0, 5.0, 0.1)`
		`decay = 0.3`
		`x = exp(2*pi*I*t) * exp(-decay*t)`
		`w = fft.fft(x)`
		`z = fft.ifft(w)`
		`# z = x up to numerical error`

`linalg`	A module with methods to perform array operations common in linear (matrix) algebra, for example, to calculate determinant, eigenvalues, eigenvectors, inverse, single value decomposition etc.	`m1 = array([[-1, 1], [ 2, 3]])`
		
		`linalg.eig(m1)`
		`linalg.det(m1)`
		`linalg.inv(m1)`
		`linalg.svd(m1)`

`random`	A module with methods for generating and sampling (pseudo-) random numbers.	`vals = random.rand(6)`
		`# Six floats from 0.0 to 1.0`
		`vals = random.randint(2,9,5)`
		`# Five ints in range 2 to 5`
		`x = [1,2,3,4]`
		`random.shuffle(x)`
		`# New order`
		`random.seed()`
		`# Init pseudorandom generator`

APPENDIX 4

String formatting

Old-style formatting

The following table describes old-style string formatting codes, which are used in Python strings which are followed by the '%' format operator, such as '%8.3f' % value. These follow the general system used by sprintf/printf in the standard C library. The general form of the formatting code syntax is:

'%<prefix><minWidth><.numDigits><format>' % data

where data is either a single value or a tuple of values. This formatting system is described at the end of Chapter 3.

Format code	Description	Example
%	A literal percentage sign.	'%.2f%%' % 32.1 # '32.10%'
c	Character, specified directly or using an ASCII code.	'%c-%c' % (0x61, 'a') # 'a-a'
d or i	Integer (decimal).	'%d %d' % (4, -7) # '4 -7'
e or E	Exponential floating point. Lower case and upper case respectively.	'%.2e' % (7,) # '7.00e+00' '%.2E' % (-1.3e-7) # '-1.30E-07'
f or F	Floating point.	'%.3f' % (7) # '7.000' '%6.2f' % (7.3999) # ' 7.40'
g or G	General floating point, using an exponent above or at the precision number (after the point) and also below 10^{-4}. Lower case and upper case respectively.	'%.2g' % (100,) # '1e+02' '%.3G' % (100,) # '100'
o	Octal (base 8).	'%o' % (100,) # '144'
r	String, converting objects with repr().	'%r %r' % (None, 'abc') # 'None "abc"'
s	String, converting objects with str().	'%s %s' % (True, 'abc') # 'True abc'
x or X	Hexadecimal (base 16). Using lower-case a–f or upper-case A–F respectively.	'%x' % (1023,) # '3ff'

Control options	Description	Example
0	Pads numeric values with leading zeros.	`'%05i' % (7,)` `# '00007'`
-	Left justifies the string, so spaces occur to the right.	`'%-5s' % ('abc')` `# 'abc '`
<Space>	Places a space before a positive number.	`'% .2f' % (123)` `# ' 123.00'`
+	Puts a plus sign before a positive number.	`'%+.2f' % (123)` `# '+123.00'`
#	For certain formats, uses an alternative form.	`'%#x' % (1023,)` `# '0x3ff'`

New-style formatting

From Python 2.6 a new style of formatting was introduced with the `.format()` method for strings. The general idea is that curly braces are placed inside the string and the method inserts the stated values into the appropriate places, formatting appropriately. The braces accept an identifier stating which of the values to place in which slot. For example, here the values 6 and 7 are placed in slots '{0}' and '{1}' respectively, using the index of each argument as an identifier:

```
'X:{0} Y:{1}'.format(6, 7) # Gives 'X:6 Y:7'
```

From Python version 2.7 the identifiers are optional and the order of inserted values is the same as the order of arguments, so we could also do the following to get the same result as before:

```
'X:{} Y:{}'.format(6, 7) # From Python 2.7
```

However, by using identifiers, the order in which the values are placed into the string can be controlled, so, for example, we could reverse the order by swapping the identifier numbers:

```
'X:{1} Y:{0}'.format(6, 7) # 'X:7 Y:6'
```

The system also accepts named identifiers, where the argument name is the same as the name inside the curly brace where it is to be placed:

```
'X:{a} Y:{b}'.format(a=5, b=2) # 'X:5 Y:2'
```

Identifiers can also be a mixture of argument indices and names:

```
'X:{0} Y:{b}'.format(5, b=2)   # 'X:5 Y:2'
```

If a sequence of values is passed in (e.g. a list or tuple), then the sequence can be indexed inside the curly brace specification, so, for example, here the initial '0' identifiers refer to dList and the square brackets refer to the index of elements within that list.

```
dList = [5,7]
'X:{0[0]} Y:{0[1]}'.format(dList) # 'X:5 Y:7'
```

The sequence identifier can also be a named argument, which perhaps makes for more readable code:

```
'X:{data[0]} Y:{data[1]}'.format(data=dList)
```

If the argument is a Python object then named attributes of that object can be used in the braces:

```
import math
'Constant Pi:{0.pi}'.format(math) # 'Constant Pi:3.14159265359'
```

For the examples given so far we have let the formatting system display the values in a standard, default manner. However, there are many additions that we can make to the format specification inside the braces to control exactly how the output string should be formatted. Hence, for example, we could repeat the above but display the floating point number to only four decimal places using the formatting code after the colon:

```
'Constant Pi:{0.pi:.4f}'.format(math) # 'Constant Pi:3.1416'
```

The general form of the format specification is {<identifier><!converter><:formatting>}. So, in the above example <identifier> is '0.pi', <!converter> is not specified and <:formatting> is ':.4f'. The <!converter> part is generally not used, in which case it is the equivalent of using the inbuilt str() function. If instead we use the '!r' code then the conversion is equivalent to repr(), which is especially handy if we have an object that has a special string representation method (__repr__()) which is different from a plain string conversion.

```
'Text: {0}'.format('abc')  # Using str() - 'Text: abc'
'Text: {0!r}'.format('abc') # Using repr() - "Text: 'abc'"
```

The <:formatting> part of the specification itself may have several sub-components, and the general form of this is <:(fill)align><sign><0><#><width><,><precision><type>. We will describe some of these options, although a more complete list of the possibilities is listed in the table below.

An alignment code is used when the substring may be smaller than the insert width. In such cases we can choose to align the value with the left, right or middle of the insert. Here we use the '<' code to push the value to the left of the insert and the value '5' to specify the width of the insert:

```
'Text:{0:<5}'.format('a') # 'Text:a    '
```

Also, we could explicitly say that the insertion is of string type (which is the default anyhow) using the 's' data type code.

```
'Text:{0:<5s}'.format('a') # 'Text:a    '
```

Similarly we could format to the right of the insert with a decimal integer (type code 'd') value, although right alignment is the default so need not be specified:

```
'Value:{0:>5d}'.format(-7)  # 'Value:   -7'
```

Or the alignment code '^' could be used to place the value in the centre:

```
'Value:{0:^5d}'.format(7)  # 'Value: 7 '
```

There is one further alignment option '=', where any numeric sign is maximally separated from a number:

```
'Value:{0:=5d}'.format(-7) # 'Value:-   7'
```

As standard, when we are aligning format values within a larger width, any extra room is padded out with space characters. However, we may like to use something else as padding, in which case a different padding character is placed before the alignment code. So, for example, if we wanted to pad with asterisks:

```
'Value:{0:*>5d}'.format(7)  # 'Value:****7'
```

Numeric signs will normally only be shown for negative numbers, but we can force the appearance of a plus symbol for the positives with '+' following the alignment code (where present):

```
'Value:{0:>+5d}'.format(7)  # 'Value:  +7'
```

Likewise we can use a single space as a sign code, so that positive numbers have an extra space before, which can be handy to keep alignment with negative numbers:

```
'Value:{0:< 6d}'.format(1023)  #'Value: 1023 '
'Value:{0:< 6d}'.format(-1023) #'Value:-1023 '
```

Also for similar purposes, numeric values are easily padded with leading zeros to fill the specified width:

```
'Value:{0:05d}'.format(123)  # 'Value:00123'
```

A more recent addition to the scheme (from Python 2.7 onwards) is to use commas in the format output to separate each 'thousands' block of three digits. As you might expect this uses a ',' code:

```
'Value:{0:,d}'.format(123456789) # 'Value:123,456,789' – Python 2.7+
```

A numeric value need not be specified as a decimal integer or floating point, as we have shown so far. There are several other codes for different base systems, so, for example, we can format as hexadecimal:

```
'Value:{0:4x}'.format(1023)  # 'Value: 3ff'
```

For floating point numbers, as mentioned above, it is generally useful to specify the precision (number of decimal places), which we achieve with the dot notation:

```
'Value:{0:.3f}'.format(123)  # 'Value:123.000'
```

And as you may expect this can be combined with the required total width specification. Here, for example, we make sure that at least ten characters are inserted for a value specified to three decimal places:

```
'Value:{0:10.3f}'.format(123)  # 'Value:   123.000'
```

As with the old-style formatting, floating point numbers can be represented in the exponent format 'e':

```
'Value:{0:.2e}'.format(7)      # 'Value:7.00e+00'
'Value:{0:.2e}'.format(0.0012) # 'Value:1.20e-03'
```

And also using the general number format 'g' which only uses the exponent form if the exponent is small (< -4) or greater than the precision number:

```
'Value:{0:.4g}'.format(7.0)    # 'Value:7'
'Value:{0:.4g}'.format(10234)  # 'Value:1.023e+05'
```

It should be noted that the number of decimal places can be set dynamically from a variable, should it need to change in a program. This can be achieved by having one formatting code inside another. Hence here the number of decimal places is inserted into the '{1}' (from the second argument) to complete the outer formatting code, which in the example below would be equivalent to '{0:.7f}':

```
nPlaces = 7
'Value: {0:.{1}f}'.format(123, nPlaces) # 'Value: 123.0000000'
```

Lastly, we can use this tactic of having braces inside braces to control other aspects of a format specification: here, for example, setting the alignment code with a named attribute:

```
'Value: {0:{align}5d}'.format(5, align='<') # 'Value: 5    '
```

The table below summarises the various alignment, sign and data type codes used with Python's new-style string formatting system.

Alignment code	Description
<	Left alignment.
>	Right alignment.
^	Centre alignment.
=	Separate sign and number.

Sign code	Description
+	Shows positive and negative signs.
-	Negative signs only (default).
<Space>	Uses a space before positive.

Type code	Description
b	Binary (base 2).
c	Character, using a numeric ASCII code.
d	Decimal integer (base 10).
e	Exponential, with lower-case 'e'.
E	Exponential with upper-case 'E'.
f	Floating point.
g	General number, formatted with exponential if the exponent is less than -4 or greater than the precision number.
G	General number, as above, but using upper-case 'E'.
n	Number with local settings, e.g. for thousand separator, decimal point.
o	Octal (base 8).
x	Hexadecimal (base 16), with lower-case letters a–f.
X	Hexadecimal with upper-case letters A–F.
%	Converts to a percentage (multiply by 100.0) and adds a '%' sign.

APPENDIX 5

Regular expressions

The term 'regular expression' refers to the specification of a particular pattern of characters which may occur within a text string. Given a pattern a program can then look for certain kinds of character being present in particular locations. As an example we may wish to search for email addresses within a larger piece of text. In reality this is fairly complex,[1] but the basic pattern for this would consist of a leading group of (mostly) alphanumeric characters, without spaces, before the '@' symbol, then more characters before ending with a dot '.' and a final group of letters. Such a pattern would expect to identify strings like 'mickey@disney.com' or 'h.j.simpson@springfield.ac.uk'.

In practice, regular expressions in Python are specified using a system where special codes are used to represent general or ambiguous kinds of character (it is actually a very similar system to what is used in other languages like Perl and C). For example, '\d' is used to match any digit character in the range '0' through to '9'. Given a regular expression the essential task is to determine whether a pattern does or does not occur within a string. If it does, often we will also want to extract the substrings that matched and perhaps replace them with some other text. It should be noted that some simple text string queries and manipulations, which certainly could be done with regular expressions, are often more easily performed using the standard string methods. Thus, when dealing with only a few exact substrings we recommend considering string methods like text.find(subString), text.replace(old, new) in preference to regular expressions.

A guide to regular expressions

Moving on to practical matters, the following is an introductory guide to the basic use of regular expressions in Python. To use regular expressions in Python we must first import the re module:

```
import re
```

or the required sub-modules:

```
from re import search, compile, sub
```

We use the former approach here. To begin, we will consider looking for a particular substring within a larger string using re.search. This takes the general form:

```
matchObj = re.search(regExpPattern, textString)
```

And in practice we would do something like the following, where for illustrative purposes we use an exact string as a pattern to search for:

```
text = 'Antidisestablishmentarianism'
matchObjA = re.search('establish', text)  # Present - gives MatchObject
matchObjB = re.search('banana', text)     # Absent - gives None
```

---

[1] And in practice it is often best to use the Python module email to deal with this sort of thing.

As you can see, the search gives back a special `MatchObject` if successful or `None` if the search failed. A `MatchObject` may then be interrogated to determine where the match occurred, and what the substring was etc.

If the search pattern doesn't change often compared to the number of searches then it may be quicker to compile the pattern from the regular expression string once at the start, before then applying the pattern repeatedly. In this instance to 'compile' a regular expression means to interpret the pattern specification (which is initially just text) and create a `Regex-Object`, which has methods (bound functions) to perform searches etc. on that particular pattern. So adapting an example from above we could do the following, noting the `.search()` now comes from `regexObj`, the compiled regular expression object:

```
regexObj = re.compile('establish')
matchObj = regexObj.search(text)
print(matchObj)
```

Various useful functionalities are associated with the match object:

```
print(matchObj.group())   # The substring that was found/matched
print(matchObj.start())   # index in the string of the start of match
print(matchObj.end())     # index in the string for just after end of match
print(matchObj.span())    # (start, end+1)
```

In general we need to check that a search was successful, i.e. that it did not give `None`, before proceeding to interrogate any match object:

```
regexObj = re.compile('Green')
texts = ['Green tomatoes', 'Red brick house']

for text in texts:
  matchObj = regexObj.search(text)

  if matchObj is None:
    print('Pattern does not match')
  else:
    print(matchObj.span())
```

Considering the above search pattern 'Green', we may wish to be less specific and also accept lower-case 'green'. The regular expression string for this could be '[Gg]reen', so that we accept either upper- or lower-case letters at the start of the word.[2]

```
regexObj = re.compile('[Gg]reen')
print(regexObj.search('Green door'))
print(regexObj.search('Fried green tomatoes'))
```

As you may expect the above regular expression was made more general by creating a group of accepted characters using square brackets `[]`. Hence, square brackets have a special meaning when they are in a regular expression. There are other characters with special meanings, and the complete list is:

```
. ^ $ * + ? { } [ ] \ | ( )
```

---

[2] This can also be done with the `re.I` flag for case-insensitive searching, but that would also match "gReen" etc.

See the table below for what these are used for. If we want to have these characters used in a literal way then we have to put a '\' in front so that it escapes any special interpretation (in the jargon the character is said to have been 'escaped').

```
regexObj = re.compile('\[abc\]') # Match the bracket characters
regexObj.search('Text with exactly [abc] inside')
```

While we can explicitly define groups of characters by stating all possibilities, for a range of consecutive characters we can use a shorter notation. Thus instead of doing the following to match any grade from A through to E:

```
text = 'passed the exam with grade D'
regexObj = re.compile('grade [ABCDE]')
matchObj = regexObj.search(text)
```

we could use a range specified with the group '[A-E]'.

```
regexObj = re.compile('grade [A-E]')
matchObj = regexObj.search(text)
```

In a similar way you can define range groups like '[A-Z]', '[a-z]' or '[a-z0-9]', the last of which would match lower-case letters or digits. For some of the more general, regularly used character groups there are some even simpler codes. Hence instead of the complete digit character group '[0-9]' the code '\d' can be used instead:

```
regexObj = re.compile('grade \d')
regexObj.search('Wizard grade 1')
regexObj.search('Wizard grade 2')
```

As described in the table below the commonly used group codes are:

```
\s \d \w \S \D \W
```

Respectively these represent whitespace, digit, alphanumeric/underscore and their opposites; non-whitespace, non-digit, non-alphanumeric/underscore. Accordingly, in one of the above examples we could use '\w' instead, and this will match both letters and numbers:

```
regexObj = re.compile('grade \w')  # 'Wordy' character [0-9a-zA-Z_]
regexObj.search('grade D')
regexObj.search('grade 1')
```

Sometimes a regular expression cannot be expressed in terms of simple character groups and codes. For example, we may wish to accept alternative words. In such cases we can simply specify complete substring alternatives using '|', which acts as an OR operator.

```
regexObj = re.compile('\s(trousers|pants)\.')
matchObj = regexObj.search('I got mud on my trousers.')
```

So far we have only considered groups and codes for matching single characters. Naturally we often want to find more than one character from a given group. This is achieved using the '+' symbol, which means to match one or more (as many as are available). For example, to match multiple digits:

```
regexObj = re.compile('\d+')
matchObj = regexObj.search('In the year of 1949.')
print(matchObj.group()) # Print the part that matches
```

In this case we would match the multiple digits of the substring `'1949'`. Note that this matching is 'greedy', so gets all of the sequence of digits. Multiple character groups and codes can be used in the same regular expression. So adapting the above example we could also match one or more non-digit characters, specified with '`\D+`' before any number of digits '`\d+`'.

```
regexObj = re.compile('\D+\d+') # One or more non-digit, one or more digit
matchObj = regexObj.search('I arrived in 1949 from Cuba.')
print(matchObj.group())
```

The result here is `'I arrived in 1949'`, so you can see that the '`\D+`' matched all the initial characters and the '`\d+`' matched the year in the combined pattern. As an alternative we could use the code '`.+`', which means one or more of any character, but it should be noted that this will also match to digits. Hence, if we do the following:

```
regexObj = re.compile('.+\d+') # One or more of anything, one or more digit
matchObj = regexObj.search('I arrived in 1949 from Cuba.')
print(matchObj.group())
```

The final result is the same as before but the '`.+`' code actually matches `'I arrived in 194'`; it is 'greedy' and matches as many characters as possible before the next code '`\d+`', which here only matches the single character `'9'`. As another example, you may wish to match digits only if they are preceded by a space (or other whitespace character like '`\t`', '`\r`' or '`\n`'). In this case you could do:

```
regexObj = re.compile('\s\d+')
regexObj.search('Year 2013') # Success
regexObj.search('Year2178')  # Gives None: no whitespace before digit
```

Note that here we could also use '`\s+`', but searching for multiple spaces doesn't make any difference if all we require is one. If the pattern really must have a space, and nothing else, then `re.compile(' \d+')` can be used.

As far as the output is concerned, the examples so far don't distinguish the different parts of the regular expression. However, we may wish to segregate the different parts of the matched string so that we can extract it separately using the `.group()` or `.groups()` method of the match object. Hence, considering the extraction of numbers from the above example we can use round brackets `()` to define a group so that although the match must include whitespace we can access the digits separately:

```
regexObj = re.compile('\s(\d+)')
matchObj = regexObj.search('In the year of 1949.')
print(matchObj.group(1)) # '1949' - digits only
print(matchObj.groups()) # ('1949',) - all match groups as a tuple
```

Note that using `.group(0)` gives the complete match substring. Naturally, if there are more groups these take subsequent numbers, as in the following example, where '`\D`' means non-digit:

```
regexObj = re.compile('(\d+)\D+(\d+)')
matchObj = regexObj.search('The 14th day of January 1865.')
print(matchObj.group(0)) # '14th day of January 1865'
print(matchObj.group(1)) # '14'
print(matchObj.group(2)) # '1865'
print(matchObj.groups()) # ('14', '1865')
```

There are some tricky details when using special character codes. Consider the following, for example:

```
text = 'C:\data'
regexObj = re.compile('\data')
matchObj = regexObj.search(text)
print(matchObj.group()) # Fails
```

The problem arises because '\d' is a special code for digit characters and is not interpreted as a literal backslash character followed by a 'd'. Now, given what we mentioned above by escaping characters by prepending them with '\' you might expect the following to work to treat the slash literally.

```
regexObj = re.compile('\\data')
print(regexObj.search(text).group()) # Still fails!
```

To get this to work as we intended we need the following completely horrid pattern:

```
regexObj = re.compile('\\\\data')   # Works - Yuck!
print(regexObj.search(text).group())
```

The problem we have encountered here occurs because in reality there are actually two rounds of string interpretation and in both rounds '\\' is an escape code. The first interpretation is the normal Python string handling, and here '\\' also means a literal backslash (remembering that a backslash is used for whitespace codes like '\n', '\t' etc.). The second interpretation is the interpretation as a regular expression, which has its own set of escape codes. Thus the first round of string interpretation in compile('\\data') replaces the double backslash code with a single literal backslash, so that by the time the string gets to the regular expression interpretation the double backslash is removed and we're back at '\d'. At this point it should be noted that we didn't have this problem with the '\[' or '\]' in previous examples because these only act as escape codes in regular expressions, not in regular Python strings.

As you can see, we can add yet more backslashes so that the removal of double slashes in the string interpretation leaves enough for the regular expression. However, there is a much more palatable way of doing things by disabling the first round of escape character interpretation using what are called *raw strings* so all characters are treated literally, which uses the r'' or r"" syntax. Hence we can do:

```
regexObj = re.compile(r'\\data')        # Raw string
print(regexObj.search(text).group()) # Success!
```

Another aspect of regular expressions which should be noted is that the matches are done as close to the start of the queried string as possible. Hence, if there are two possibilities which both match in theory it is the first that is matched in practice. For example:

```
regexObj = re.compile('\d+')
matchObj = regexObj.search('In the years 1949 and 1954.')
print(matchObj.group()) # '1949' - First match
```

Where there are multiple matches for the pattern we can use .findall() to get multiple match occurrences, noting that this gives back the matches' substrings rather than a MatchObject (we could get such objects using .finditer(), as we show later):

```
regexObj = re.compile('\d+')
matchStrs = regexObj.findall('In the years 1949, 1954 and 1963.')
for matchStr in matchStrs:
  print(matchStr)
```

So far we have only considered regular expressions where a particular character code or group must occur. However, there are situations when a group of characters may sometimes be absent. For example, consider the following strings where we wish to extract the numeric data after equal signs but where there may or may not be multiple spaces before the digits.

```
s1 = 'first=123457'
s2 = 'second=   6'
s3 = 'third= 8768'
```

All of these numeric substrings can be extracted with a single regular expression. Here '*' means zero or more (as applied to the preceding character) so we have flexibility with regard to the presence of spaces before getting one or more digits:

```
regexObj = re.compile('= *\d+')
print(regexObj.search(s3).group())
```

Taking this kind of example further, the extraction of numbers may be further complicated with the presence or absence of minus signs and decimal points. However, we only accept the presence of a single minus sign and/or a single decimal point, so we use '?' to mean zero or one (and not more). Considering the following string:

```
line = 'p1=123.457, p2=  1.80, delta1= -7.869, delta2=-10'
```

A regular expression to match all the numbers must account for zero or more spaces '*', an optional minus sign '-?', one or more digits '\d+', an optional decimal point '\.?' (remembering the backslash because a plain dot is a code for any character) and then any optional remaining digits '\d*'. The resulting regular expression may seem somewhat unreadable at first glance, but it is readily broken down into its component parts:

```
regexObj = re.compile('= *(-?\d+.?\d*)')
for match in regexObj.finditer(line): # iterates through all match objects
  print(match.group(1))
```

Note that by bracketing the part of the character specification that includes the numbers and any minus sign we can get just the numeric part with .group(1). So far we have considered codes for zero or one '?', zero or more '*' and one or more '+', but naturally there are other possibilities, such as allowing between two and four, but no more or less. In this case we use the curly brace specification in the form '{minAllowed, maxAllowed}'. Hence to allow two, three or four whitespace characters '\s' before digits we could do:

```
regexObj = re.compile('=\s{2,4}\d+') # From two to four, inclusive
```

If only one number is given in braces then there must be exactly that number of characters for a match:

```
regexObj = re.compile('=\s{2}\d+') # Exactly two
```

If the first number is omitted, with a comma still present, then the minimum number defaults to zero. So the following accepts up to two whitespace characters, but no more:

```
regexObj = re.compile('=\s{,2}\d+') # Zero to two
```

If the second number after the comma is omitted the maximum number of occurrences is unlimited. The following accepts two or more whitespace characters:

```
regexObj = re.compile('=\s{2,}\d+') # Two or more
```

Moving on from simply matching and extracting substrings, the `re` module and `RegexpObject` have a substitution method `.sub()`. Here if the pattern matches the matching substring is replaced with another substring, yielding a new string. In the following example any negative integer numbers are replaced with 'neg!':[3]

```
text = 'N: -9 4 -2 7 8 -8'
regexObj = re.compile('-\d+')
newText = regexObj.sub('neg!', text) # Gives 'N: neg! 4 neg! 7 8 neg!'
```

Alternatively the replacement substring can simply be empty, so that the matches are removed. Here we remove any negative numbers and preceding whitespace:

```
text = 'N: -9 4 -2 7 8 -8'
regexObj = re.compile('\s+-\d+')
newText = regexObj.sub('', text) # Gives 'N: 4 7 8'
```

If we wish to keep the digits we found after the minus sign we can capture them in a group and then recall them in the replacement text using the '\1' etc. (a second group would be '\2'). Hence the following matches both the minus sign and digits, but puts the digits back into the new string after a space:

```
text = 'N: -9 4 -2 7 8 -8'
regexObj = re.compile('\s+-(\d+)')
newText = regexObj.sub(r' \1', text) # Gives 'N: 9 4 2 7 8 8'
```

Note that here we use the raw string notation `r''` so that Python uses the characters literally and does not attempt to interpret '\1' as an escaped control character.

Another useful operation involving regular expressions is to split a string according to a pattern. The method for this (of both the `re` module and `RegexpObject`) is `.split()`. This is like the string method with the same name, in that it gives a list of strings by breaking a long string at the points where a separator matches, but the matching is done with a regular expression, not just an exact substring of characters. The following is an example of splitting with a regular expression, but take special note that the pattern uses '.+?'. If you try the example without the question mark you will see that the '.+', meaning one or more of any character, is too greedy and will match all of the rest of the string, up to the last angle bracket '>'. Rather what we want is for the pattern to match conservatively and only go up to the next '>', hence we use '+?' which means a minimalistic search for one or more:

```
text = '<p>Paris<br>London</p>Berlin<br>New York'
regexObj = re.compile('<.+?>')
print(regexObj.split(text)) # ['', 'Paris', 'London', 'Berlin', 'New York']
```

There are many more subtleties and options with regular expressions in Python. Many of these are detailed in the following tables, but we recommend reading the main on-line Python documentation for the complete picture.

---

[3] Note that the first argument to `sub()` is the substring replacement and the second argument is the string being modified.

Regular expression codes

The following table summarises a subset of the code syntaxes which have special meanings within regular expression specifications. Note that regular expressions also accept the escaped control characters which are present in general Python strings, such as \n \r \t \\ etc.

Regexp code	Meaning
.	Matches any character, except newline (unless in flag=re.DOTALL mode).
^	Matches the beginning of a string, or the beginning of a line if flag=re.MULTILINE.
$	Matches the end of a string.
*	Requires zero or more matches.
+	Requires one or more matches.
?	Requires zero or one match.
*?, +?, ??	As above, but matching minimally (non-greedy) before the next term.
{n}	Requires exactly n number or matches.
{min,max}	Requires between min and max matches, inclusive.
{min,max}?	As above, but matching minimally (non-greedy) before the next term.
[abc]	Matches a set of explicit characters. Control codes like '.' and '+' are treated literally.
[a-e]	Matches a set made with a consecutive range of characters, from a to e, inclusive. May be combined with explicit characters.
[^abc]	Matches the inverse character set, i.e. if the character is not a, b or c.
a|b	Matches either specification a or b.
(expression)	Captures part of a match as a group which may be recalled.
(?...)	An extended system allowing more complex rules. See http://www.python.org for full documentation.
\1, \2, ... \99	Recalls the match from a numbered group. Can be used to repeat a match or include group contents in a substitution.
\A	Matches the beginning of a string.
\b	Matches a word boundary; between a word (\w) and non-word (\W) character. Used to match the edges of a word without specifying what's outside.
\B	Matches a non-word boundary.
\d	Matches digit characters. Equivalent to the set [0-9].
\D	Matches non-digit characters.
\s	Matches whitespace characters. Equivalent to the set [\n\r\t\v\f].
\S	Matches non-whitespace characters.

Regexp code	Meaning
\w	Matches alphanumeric characters and underscore ('word' characters). Equivalent to the set [0-9a-zA-Z_].
\W	Matches non-word characters.
\Z	Matches the end of string.

The regular expression module '`re`'

The table below describes selected functions that are in the `re` module and bound methods of the `re.RegexpObject` class, which we represent here as `reObj`. Note in many cases the `RegexpObject` methods accept optional start and end points to specify which region of the `text` string to search within. Also, all of the `re` methods below that accept a pattern can include optional flags to specify special behaviour: e.g. `re.compile(patt, flags=re.I)` is used for case-insensitive searching. These are listed at http://www.python.org.

Method	Description
`re.compile(patt, `*flags*`)`	Creates a `RegexpObject` from a pattern string.
`re.escape(text)`	Creates a string where non-alphanumeric characters are escaped, by putting a backslash in front.
`re.findall(patt, text)` or `reObj.findall(text, `*start, end*`)`	Creates a list of substrings that match a pattern.
`re.finditer(patt, text)` or `reObj.finditer(text, `*start, end*`)`	Like .findall() but generates an iterator that gives back objects of type `MatchObject`.
`re.match(patt, text)` or `reObj.match(text, `*start, end*`)`	Creates a `MatchObject` if a pattern matches the start of a string, gives `None` otherwise.
`re.search(patt, text)` or `reObj.search(text, `*start, end*`)`	Create a `MatchObject` if a pattern matches within a string, gives `None` otherwise.
`re.split(patt, text, `*maxNum*`)` or `reObj.split(text)`	Splits a string into a list of strings according to where a pattern matches. Optional maximum number of splits to perform.
`re.sub(patt, repl, text, `*maxNum*`)` or `reObj.sub(repl, text, `*maxNum*`)`	Creates a new string by replacing all matches to a pattern with a replacement substring. Optional maximum number of replacements can be given.
`re.subn(patt, repl, text, `*maxNum*`)` or `reObj.sub(repl, text, `*maxNum*`)`	As above but also reports the number of substitutions made. Output is a tuple (`newString`, `numSubs`).

The following tables lists selected methods of the `re.MatchObject`, as given by `re.search()` etc.

Method	Description
`group(id)` or `group(id1, id2)`	Gives the substring matched by a group given one or more group identifiers (typically numbers from 1). No identifier or an identifier of 0 gives the whole matched substring.
`groups (nullMatch)`	Gives all the substring matches back as a tuple. The optional argument allows a value to be specified for pattern groups that legitimately match nothing, otherwise these give `None`.
`start(id)`	The start position of a given group within the queried string.
`end(id)`	The end position of a given group within the queried string.
`span(id)`	The start and end positions of a given group within the queried string.

APPENDIX 6

Further statistics

RPy and the R statistical package

The R statistical package[1] is one of the most commonly used ones for analysing statistical data. It has its own language. There is a Python wrapper around it called RPy.[2] The main reason to use RPy would be if you have lots of existing R code that you wish to interface to in Python.

There are a few things to keep in mind when using RPy. Standard Python collection types or NumPy arrays have to be converted into special RPy data types, and results that are returned from R have to be suitably interpreted. Reading the R documentation is crucial to using RPy.

We will illustrate the use of R via RPy for a few standard examples.

Binomial test

First we consider the binomial test, which is concerned with the number of occurrences of an event that has a fixed probability of occurring, given a certain number of trials. R has a method, 'binom.test', to do the binomial test. We create a function, binomialTailTest(), which calls this method via RPy, and which has the same arguments as in our previous version of the function in Chapter 22, which used SciPy.

First we need to import the RPy module, rpy2.robjects, which we call R below. This has an object inside it, R.r, which is what we use to get hold of R methods, using dictionary syntax keyed on the name of the R method. Here we want to use the R method binom.test, and so R.r['binom.test'] is the Python version of this R method.

The R documentation tells us that this function has four arguments, x, n, p and alternative, which correspond to our arguments count, nTrials, pEvent and oneSided, although alternative is a string in R rather than a Boolean. In fact alternative can take three values in R, 'greater', which is for our one-tailed calculation, 'two.sided', which is for our two-tailed calculation and 'less', which would give 1.0 minus the one-tailed calculation, so we do not need that here.

In R, there is an optional fifth argument, conf.level, which defaults to 0.95 and which is used if you want to calculate a confidence interval, for example. Here we are just calculating a probability. We are using the four arguments in the expected R order, so in fact here we do not need to use the 'key=value' syntax, we can just list the values.

The one oddity is how to extract what we want from the returned result. The R output contains a lot of information. It turns out that the probability is item 2 of the result

---

[1] http://www.r-project.org/.

[2] http://rpy.sourceforge.net/; code was tested on versions 2.0.8 and 2.2.7.

considered as a collection. That is not obvious, and can only be determined by looking at the output. And further we need to access item 0 of that, because it is an RPy collection type of length one. This leads to the strange-looking `result[2][0]`.

```
import rpy2.robjects as R

def binomialTailTest(count, nTrials, pEvent, oneSided=True):

  alt = 'greater' if oneSided else 'two.sided'

  func = R.r['binom.test']
  result = func(x=count, n=nTrials, p=pEvent, alternative=alt)

  return result[2][0]
```

We can now test the function.

```
count = 530
nTrials = 1000
pEvent = 0.5

result = binomialTailTest(count, nTrials, pEvent, oneSided=True)
print('Binomial one tail', result)

result = binomialTailTest(count, nTrials, pEvent, oneSided=False)
print('Binomial two tail', result)
```

Two-sample T-test

The two-sample T-test tests whether two independent samples have the same underlying, true mean. In R there is a method, 't.test', to carry this out. There are a few extra subtleties in comparison with the RPy code for the binomial test.

The T-test involves comparing two collections of samples. These need to be converted to RPy data types using the converter `R.FloatVector()`. The R method, 't.test', has an optional argument, `var.equal`, which specifies whether the two samples should be treated as having the same variance or not. In Python, `var.equal` cannot be used as a key in a function argument because it is not valid syntax. In RPy we can often use these arguments by changing the dot (.) to an underscore (_). But replacing `var.equal` with `var_equal` does not work here and instead we use a bit of Python trickery. In Python you can specify 'key=value' arguments by using a dictionary. But the trickery is that the dictionary itself is not passed in (the function is not expecting a dictionary as an argument) but instead we use a magic ** in front of the dictionary, which means that the dictionary contents are unwrapped into `key=value` pairs. This is the safest way to pass arguments with dots in them in RPy.

We pass back both the T statistic, which happens to be item 0, and the probability, which happens to be item 2, of the `result`.

```
def tTest(x, y, sameVariance=False):

  func = R.r['t.test']
  argDict = {'var.equal': sameVariance}
  result = func(x=R.FloatVector(x), y=R.FloatVector(y), **argDict)

  return result[0][0], result[2][0]
```

We can now test the function.

```
from numpy import array

samples1 = array([1.752, 1.818, 1.597, 1.697, 1.644, 1.593])
samples2 = array([1.878, 1.648, 1.819, 1.794, 1.745, 1.827])

print('Same variance result', tTest(samples1, samples2, sameVariance=True))
# Result is: -2.072, 0.0650

print('Not same variance result', tTest(samples1, samples2, sameVariance=False))
# Result is: # -2.072 0.0654
```

Glossary

This is a simple English glossary to aid understanding. These are not formal definitions, but rather where you can turn if a particular term makes no sense.

Absolute path *(computing)*: The full specification of a file or directory location within a computer's hierarchical file system, starting from the highest level that contains all files. On Linux and OS X systems this starts with the root file system '/' and in Windows systems with a drive letter, e.g. 'C:\'.

Absolute value *(mathematics)*: The magnitude of a number, irrespective of sign. Calculating an absolute value effectively means to give the positive version of the number.

Acid dissociation constant *(chemistry)*: A physical constant which describes the affinity of a chemical group for hydrogen ions in aqueous solutions, and thus the strength of an acid. By definition this number is the concentration of free component times the concentration of hydrogen ions divided by the concentration of hydrogen-bound component: $[A^-][H^+]/[HA]$.

Acidic group *(chemistry)*: A chemical group which gives up hydrogen ions, H^+, in water (aqueous solution).

Active site *(molecular biology)*: The part of a biological molecule's structure that catalyses a chemical reaction by binding the reacting substances in a specific conformation.

Affine transformation *(mathematics)*: A mathematical operation that transforms vectors (e.g. coordinates) by using a linear transformation (such as a rotation, scale or shear) and then a translation (movement). For example, this may be used to change the orientation and position of a shape.

Aliased *(signal processing)*: Where the measured frequency of a signal is different from its actual frequency because the range of measured frequencies is limited and frequencies outside that range are mapped, in a periodic fashion, into that range.

Alignment *(bioinformatics)*: The placement of macromolecule sequences (codes for different kinds of component chemicals in a chain) next to one another so that equivalent residues within the sequences line up. Usually an alignment is constructed to maximise the similarity between residues by placing gaps in the sequences.

Alpha-globin *(molecular biology)*: One of the protein components of the haemoglobin particle which carries oxygen in blood. Each haemoglobin particle contains two copies of alpha-globin and two copies of the similar beta-globin. Each goblin protein binds a red haem compound which in turn binds oxygen.

Alpha channel *(graphics)*: The component of an image or pixel that describes transparency. Usually a maximum alpha value means fully opaque, where no background shows through, and zero means fully transparent.

Alternative hypothesis *(mathematics)*: In statistics, the opposite of the null hypothesis, i.e. where there is a non-random relationship between data.

Annealing schedule *(computing)*: The specification of a series of temperature values that control the range of search space explored in a *simulated annealing* protocol (see below). Typically a schedule starts with a high temperature, corresponding to a wide initial search, which then diminishes to a narrow but precise search at the end.

Argument *(computing)*: A value that is passed into a function or subroutine. An argument is represented by a variable inside a general function definition, but will represent a specific value when the function is used.

Array *(computing)*: A collection of data items, of the same data type, arranged in a linear manner, i.e. in order with a first and last position.

Artificial intelligence *(computing)*: The concept of using computers to mimic or approximate the actions of a person, albeit usually for a specific task. For example, an artificial intelligence (AI) software system can be used to represent non-human players in a computer game. AI methods are often involved in pattern recognition.

Artificial neural network *(computing)*: A computer program that consists of a network of interconnected data nodes that are capable of machine learning, commonly used to perform data classification and as substitutes for analytical functions. Learning occurs when a network is trained, by changing the strength of connections between nodes.

Asynchronous *(computing)*: Where events occur independently of one another and the main thread of a program. In parallel processing this means new jobs may start before previous jobs have finished.

Attribute *(computing)*: A named variable that belongs to a computer object of a given type.

Back propagation *(machine learning)*: A means of training an artificial neural network in a supervised way. A known, correct result is used to adjust the layers of the network in a retrograde manner, from output to input, so that the output of the network better matches the known result.

Bandlimited *(signal processing)*: The recording of a signal where only a finite range of frequencies is measured. In practice all signal recordings are bandlimited because recording devices cannot detect and represent all possible frequencies.

Base *(chemistry)*: A chemical group which accepts hydrogen ions, H^+, in water (aqueous solution); the opposite of an acid.

Base *(molecular biology)*: In reference to nucleic acids like DNA or RNA, the part of a nucleotide that contains nitrogenous aromatic rings.

Basic *(chemistry)*: The characteristic of being a chemical base (as above), the opposite of acidic.

Beta-globin *(biochemistry)*: One of the protein components of the haemoglobin particle which carries oxygen in blood. Each haemoglobin particle contains two copies of beta-globin. Each goblin protein binds a red haeme compound which in turn binds oxygen.

Bias node *(computing)*: With reference to machine learning, an extra node in an artificial neural network which is in the same layer as regular nodes but which receives no variable input; it usually has a fixed signal strength of 1.0. The strengths of a bias node's connections are trained in the usual way and serve to introduce signal offsets to the next layer of nodes. In practical terms the bias connections set a baseline signal that the other, input-responsive signals adjust.

Bilayer *(molecular biology)*: A double layer of fatty lipid molecules that form cell membranes. Bilayers are arranged in continuous sheets, with oily, hydrophobic groups facing the interior and hydrophilic groups facing the exterior. Lipid bilayers form a barrier that surrounds and defines cells and sub-cellular compartments.

Binary number *(mathematics)*: Representing numbers using base 2, i.e. using only zeros and ones. In computing such ones and zeros are represented by the presence or absence of an electrical signal.

Binomial distribution *(mathematics)*: A discrete probability distribution that models the number of successes in a specified number of trials, where the probability of success for each trial is independent of previous successes and independent of the trial number.

Binomial test *(mathematics)*: A statistical test that determines whether the number of observations of some property follows a binomial distribution with a specified parameter.

Bit *(computing)*: One part of a number represented in binary form, either taking the value zero or one.

Bitwise *(computing)*: To perform an operation on the binary representation of a number where each 0 or 1 digit position is considered separately. For example, a bitwise operation 101000 OR 100110 gives 101110, where the answer gives a 1 if either of the input numbers have a 1 at the same position.

Block *(computing)*: A discrete part of a computer program where commands form an ordered group that are/is executed together. Blocks occur naturally when using program flow control statements, such as clauses, loops and functions: for example, an `if` statement conditionally executes a block of code.

Boilerplate code *(computing)*: A generic form of program code that needs to be written time and again, generally performing a mundane task.

Boolean value *(computing)*: A value representing whether something is logically true or logically false. In Python there are special `True` and `False` computer objects to represent these two truth values.

Call *(computing)*: The occasion of executing a general function (running a subroutine) with specific data (including none).

Callback *(computing)*: A function that is executed in response to the actions of the user, typically interacting via a graphical interface.

Camel case *(computing)*: The use of mixed capitalisation within a conjoined word. Used in programming to join separate words into a single variable name, e.g. `rootMeanSquare`: an alternative to concatenating with underscores, e.g. `root_mean_square`.

Cardinality *(computing)*: In data modelling, the number of items that is represented by an object attribute. An attribute may represent a collection, and thus the cardinality is the size of that collection. For example, an `Atom` object may have three values in its `coordinates` attribute, to specify its 3D (`x`, `y`, `z`) position.

Central limit theorem *(mathematics)*: A theorem that says that the mean value of a large number of measurements of independent and identically distributed random variables is approximately normal (follows a Gaussian distribution).

Centroid *(mathematics)*: The geometric centre of a shape (defined by a set of locations). If weightings are included, such as atomic masses in a molecule, the centroid may be the centre of mass.

C extension *(computing)*: The means of extending the functionality of a high-level programming language, like Python, by creating a module in the low-level C programming language (generally quicker to run) which can be incorporated directly in the high-level language using normal syntax.

Changeable *(computing)*: The ability to manipulate a data item so that its value or content changes internally, but leaving the handle (reference) to the data item unaffected.

Channels *(graphics)*: The separate components of an image, e.g. red, green and blue, which when combined form a complete image.

Channels *(signal processing)*: The separate signal streams of a composite data source.

Character *(computing)*: The smallest, indivisible part of a piece of text that represents letters, digits and punctuation etc. Computationally each character is represented by a number within a specific scheme.

Chemical character *(biochemistry)*: When referring to biological molecules, its physical and chemical properties, e.g. whether it is hydrophobic (oily), polar, charged etc.

Chromatin *(molecular biology)*: A section of DNA which is bound by proteins such as histones that package it into a more compact form and regulate its expression, i.e. how it is read.

Chromosome *(biology)*: A large, double-stranded DNA chain that is the particle of inheritance in organisms. A chromosome is bound by proteins and contains genes (sub-regions of DNA) that encode the biological functions required for life. Each species has a distinct number of chromosomes within its cells, and the total of all chromosomes constitutes the genome.

Class *(computing)*: The abstract template that specifies how a specific kind of computer object is constructed. Classes are used to create custom data structures in programs, which frees the programmer from having to only use the inbuilt data types.

Class function *(computing)*: In object-oriented programming, a function in a class that can be defined independently of the properties of any object constructed with the class.

Client *(computing)*: The part of a computer application that interacts with the user, typically as a result of using a service on a remote server.

Clustering *(mathematics)*: For a collection of data items, the process of grouping similar items together into a number of clusters. The number of groups may or may not be known ahead of time but a data item is allocated to only one cluster. Clustering works by changing the membership of data items in cluster groups to maximise similarity of items within clusters and minimise the similarity of items between clusters.

Coding strand *(molecular biology)*: In a double-stranded DNA chain that contains a gene, the coding strand is the one that is reproduced as RNA, in terms of sequence, when the DNA is read.

Codon *(molecular biology)*: A group of three adjacent nucleotide residues that specify the incorporation of a particular kind of amino acid into a given position of a protein chain.

Collection *(computing)*: The grouping of otherwise separate data items into a container data structure, which may be referred to as a distinct entity. In Python the usual collection types are lists, sets, tuples and dictionaries.

Colour space *(graphics)*: The notion of describing colours as vector locations, relative to colour axes, which describe the component properties (e.g. red, green and blue) of colours.

Comparative modelling *(bioinformatics)*: The process by which the three-dimensional structures of proteins may be predicted by comparison to similar proteins with experimentally determined structures. If two proteins are inferred to be homologous (evolutionarily related) by virtue of sequence similarity, then because structure is more conserved than sequence the known structure of one protein can be used as a template to determine the fold of the other.

Compilation *(computing)*: The process of converting textual source code, which is generally independent of any particular kind of computer, into executable binary code that may be run on computers with a specific type of processor architecture.

Complement *(mathematics)*: In set theory, the elements that are not in a specified set.

Complementarity *(molecular biology)*: The relationship between two molecules where the chemical properties of one complement those of the other, and so allow a tight interaction. This is often used to describe aromatic base pairs between nucleotides, which form hydrogen bonds across a DNA duplex. Accordingly A and T residues are complementary and likewise C and G.

Complex *(molecular biology)*: Two or more biological molecules that bind to one another in a specific manner to form a larger particle.

Composition *(mathematics)*: The application of two transformations, e.g. by matrix multiplications, one after the other.

Conditional probability *(mathematics)*: The probability that one event occurs given the condition that some other event has occurred.

Conditional statement *(computing)*: A means of controlling the flow of a program's execution, doing different things depending on whether some test turns out to give true or false. In Python this is done with `if`, `elif` and `else` statements.

Confidence level *(mathematics)*: In statistics, the probability that a measurement will be within some specified range of values.

Conformation *(chemistry)*: For a molecule, one specific three-dimensional shape from amongst a range of many possible shapes. Different conformations may be similar to one another, e.g. if a protein has a compact fold, and a set of conformations can represent the dynamics or uncertainty of a structure.

Consensus sequence *(bioinformatics)*: A consensus sequence represents the average of all the sequences in a sequence alignment, and thus is not usually a real biological sequence. Each position in the consensus may reflect the most common residue observed at that position or the average residue properties of that position.

Conservation *(molecular biology)*: The tendency of a biological feature, such as a sequence or a 3D structure, to resist changes during evolution, resulting in similar forms. Some evolutionary changes are more conservative than others, e.g. DNA changes that don't affect protein sequence or amino acid changes that conserve chemical characteristics.

Constructor *(computing)*: A function that is part of the definition of an object class, which is executed whenever a new object of that type is made. Thus a constructor may be used to set up the initial state of a computer object. In Python the constructor function is named __init__.

Containment hierarchy *(computing)*: A means of describing the links and relationships between computational data items, especially in a data model, where one kind of object belongs to, or is contained by, another, producing a tree-like system of categorisation.

Continuous *(mathematics)*: Where a quantity is modelled in terms of the real numbers instead of a discrete series, i.e. taking any value, not just whole numbers.

Convolution *(mathematics)*: The modification of one function given another, whereby for different offsets (sliding one relative to the other) a new function is created as the summation of the product of the two. Commonly used in image adjustment where original pixel values are modified by a convolution matrix that specifies how pixels of the adjusted image are formed by the weighted combinations of surrounding pixels.

Covalent bond *(chemistry)*: A strong chemical bond between two atoms that are held together in a discrete molecule.

Cross-product *(mathematics)*: In three dimensions, a vector which is orthogonal to both of two specified vectors, and with length given by the area of the parallelogram determined by the two vectors.

Ctrl *(computing)*: The control key on the keyboard, usually used in combination with other keys so the user can issue specific commands. The equivalent for German keyboards is 'Strg'.

Cursor *(computing)*: The point on a computer screen where typed keyboard characters appear. Often indicated by a graphical marker so the user can see where text will be extended or inserted.

Cursor *(databases)*: A way of accessing the data records in a relational database via programming, to process it in terms of its rows.

Database schema *(computing)*: A way of specifying the layout of the information that will be stored in a relational database.

Data model *(computing)*: Using data object definitions, as formalised in classes, to create a computational representation of a network of interconnected concepts. A data model will often, but not always, contain classes that correspond closely to real-world entities, such as atoms, molecules or even people. The purpose of the model is to provide a convenient way of representing data and its connections. The model itself is an abstract concept, and so does not contain any specific data. Rather the model is a specification of how to represent data, which in turn is used to create real instances of computer objects.

Decorator *(computing)*: A Python function that is used to wrap and modify another function (similar to annotations in Java). A modifying decorator function is declared before the target function it operates on (using the @ symbol), but is equivalent to calling the modifying function on the result of the target.

Degrees of freedom *(mathematics)*: The number of parameters that are free to vary in a mathematical model; the minimum number of parameters that are needed to fully specify the state of a system.

Denormalisation *(computing)*: The redundant storage of data in a relational database so that fewer tables are used, i.e. by repeating values.

Destructor *(computing)*: A function that is part of the definition of an object class, which is executed whenever an object is deleted, e.g. to perform any required clean-up.

Detailed balance: When considering the smallest component process of a dynamic system, detailed balance is reached when the system is at equilibrium and the reverse of each micro-process restores equilibrium. In Markov chains, when using the Metropolis-Hastings criterion, detailed balance is used to test whether the chain has reached the required equilibrium state.

Determinant *(mathematics)*: A number that can be calculated from a matrix that gives the volume of the region obtained when the matrix is applied as a transformation to a region with unit volume.

Dictionary *(computing)*: A collection data type in Python (equivalent to a Perl hash and a Java hash map) where data items are held in pairs such that a unique key is used to look up a corresponding value from the dictionary.

Diffraction pattern *(chemistry)*: The primary experimental data collected in X-ray crystallography, which may be used to determine the three-dimensional structures of molecules. The diffraction pattern is created when a beam of X-ray radiation irradiates a regular crystal lattice, effectively reflecting from the different atomic planes in the crystal.

Dihedral angle *(mathematics)*: Also called torsion angle, is a measure of twist between two planes. A dihedral angle can be measured for an ordered list of four points; the first three points define one plane and the last three the second plane, such that the angle of twist is around the central two points which the planes share.

Dimension *(mathematics)*: The number of independent axes (or features) that describe a location within a given parameter space. A dimension may also refer to one of those independent axes, e.g. the 'depth' dimension.

Dimensional reduction *(mathematics)*: The practice of representing a set of data using fewer dimensions than the original data, for example, by using projection. Often the aim of dimensional reduction is to render the data more amenable to analysis while still preserving the essential features of the data.

Discrete *(mathematics)*: Where a quantity is represented in terms of the whole numbers, rather than as a continuum of real values.

Discrete Fourier transform *(computing)*: The Fourier transform (converting a signal from time to frequency) applied to a data series with discrete, rather than continuous, time points.

Discrete-time homogeneous Markov chain *(computing)*: A Markov chain with discrete, rather than continuous, positions in the chains (time), and where the transition probabilities between states do not depend on chain position (or time).

Discrimination *(computing)*: The process of allocating data points into distinct categories or clusters. In computing this may be a predictive method rather than an unambiguous classification.

Disjoint *(mathematics)*: In probability, two events that do not intersect, and so share no common outcome.

Distance matrix *(mathematics)*: A rectangular array of values representing the distances between items, where each item corresponds to a different row and/or column in the matrix. The distance may be a conventional geometric distance or some other metric that measures the similarity or separation between data items.

Distribution *(mathematics)*: An assignment of a probability to each possible, measureable outcome.

Disulphide *(chemistry)*: A chemical group consisting of two covalently linked sulphur atoms. Disulphide groups are often present in proteins found outside cells (an oxidising environment), linking the side chains of two cysteine residues.

DNA *(biology)*: 2′-Deoxyribonucleic acid, the long-chain molecule that is the main store of genetic information; the template for making RNA and proteins and the means by which genes are inherited.

DNA sequence *(molecular biology)*: The sequential order of the four different types of nucleotide compound that make up a linear DNA chain. Although DNA is a double strand, containing two tightly bound chains, because the residues of the chains are complementary the sequence of one strand is automatically known from the other. When referring to genes it is conventional to refer to the DNA sequence of the coding strand.

Domains *(molecular biology)*: A domain is an autonomous part of a protein that has a discrete structure, function and evolutionary history, relative to other parts of a protein. Domains are often separate globules that can fold independently. A family of related domains, with similar structure and function, may occur in proteins that are otherwise very different: shuffling and recombining domains is a major evolutionary mechanism.

Dot product *(mathematics)*: Also called the scalar product. A scalar value, calculated as the sum of the component-wise product of two vectors (or arrays) which represents the size of the projection of one vector on to the other. Equivalent to the product of the length of two vectors and the cosine of the angle between them.

Double precision *(computing)*: A numerical data type that typically uses 64 binary bits to store the value; basically two of the traditional 32-bit memory slots.

Drosphila melanogaster *(biology)*: A fruit fly that is commonly used for biological experimentation, particularly for genetic studies.

Duplex *(molecular biology)*: The double-stranded form of DNA, which is structurally a double helix. The duplex is formed of two anti-parallel DNA chains (going in opposite directions) with complementary nucleotide sequences that cause the chains to bind tightly along their length.

Dynamic programming *(computing)*: An algorithm which is commonly used to align pairs of biological sequences. The algorithm is much more efficient than an exhaustive search and works by disregarding many possible solutions at an early stage, where there is a known, better solution to an alignment sub-problem.

Electron density *(chemistry)*: Often with reference to its detection by X-ray crystallography, a spatial map of the density of electrons in a molecule. An electron density map may be fitted with a chemical structure to give the three-dimensional structure of a molecule, including for proteins and nucleic acids.

Ensemble *(bioinformatics)*: With regard to structures, a short form of *structure ensemble* (see below). In general an ensemble is a collection of similar but distinct arrangements, conformations or examples for a particular system: e.g. an ensemble of related solutions to a mathematical problem.

Equilibrium distribution *(mathematics)*: In (well-behaved) Markov chains, the long-run proportion of time that the Markov chain spends in each state.

Eukaryotic *(biology)*: Pertaining to the kinds of organisms that have cells with nuclei, i.e. plants, animals, fungi, amoebae etc. In essence this is all cellular organisms except bacteria (true bacteria and archea).

Event *(mathematics)*: In probability, a possible set of outcomes.

Exception *(computing)*: An error condition indicating that something, often unexpected, has gone wrong in a computer program while it is running.

Excision *(molecular biology)*: The act of cutting to remove part of something, used to describe the removal of sections of DNA or RNA by cutting with enzymes. This occurs naturally during the repair of DNA damage and when viruses replicate.

Exon *(molecular biology)*: A part of a gene that is transcribed from DNA into RNA and which remains after *introns* are removed.

Expected value *(mathematics)*: In probability, the average value of a function of a random variable.

Expression *(computing)*: A specific combination of elements in a program (variables, values, operators etc.) that is used to compute and give back values.

Expression *(molecular biology)*: The realisation of genetic code to produce biologically functional protein or RNA molecules.

Family *(molecular biology)*: Or more specifically, a homologous family. When referring to genes or proteins a family is a group of related genes or proteins that share a common evolutionary ancestor. Members of a family will share many, but not all, characteristics (e.g. in terms of sequence, structure or function) by virtue of inheritance, with more closely related members generally having more in common.

Feature *(mathematics)*: An independent property or axis that is used to describe an item of data, e.g. hue saturation and value can describe colours.

Feature space *(mathematics)*: The range of possible values that the independent features of a data item may take. A given item of data will effectively have a position within this space.

Feature vector *(mathematics)*: The location vector that describes the position of a data item within a feature space, in terms of its coordinates on independent feature axes.

Feed-forward neural network *(computing)*: A type of artificial neural network (a machine learning method) where nodes are arranged in distinct layers, from input to output, and the network operates for recall or prediction by passing signals from one layer to the next.

File handle *(computing)*: A computational object that represents a stream of external data so that it can be accessed within a program. This data stream often corresponds to the contents of a file as it is stored on disk and typically involves accessing the data line by line.

File pointer *(computing)*: A synonym for file handle.

Filter *(graphics)*: A pixel transformation that is applied to an image to change particular features. For example, filters can be used to change colours, blur images or detect edges.

Filter *(signal processing)*: A method for modifying a signal: for example, to suppress undesired frequencies.

Fisher's linear discriminant *(mathematics)*: A method to separate two or more classes of data by calculating a linear combination of their features (an axis of projection) that best separates the data.

Float *(computing)*: A colloquial term for floating point number.

Floating point error *(computing)*: A kind of numeric error that occurs with digital computers, because they cannot represent all fractional and decimal values precisely. This results in a small error in the least significant digit. This error may become significant under some circumstances, e.g. when looking at very small differences between values.

Floating point number *(computing)*: A data type that is used to represent real (non-integer) numbers such as decimals. The value is represented in terms of a fixed number of digits and an exponent that scales the value, e.g. to a given power of ten. Such numbers are of limited precision but cover a very large range of values.

Floor division *(computing)*: When dividing one number by another, rounding the result down to the previous integer; the largest integer less than or equal to the actual real value of the division. Denoted in Python by `//`.

Fold *(biochemistry)*: The general three-dimensional arrangement of a biological molecule, especially a protein. In proteins folds are usually described in terms of the composition and relative orientation of secondary-structure elements, such as coiled alpha-helices and extended beta-strands.

Force field *(computing)*: A computational model of a physical system constructed by specifying the equations that describe the relative energy or force between items (the spatial gradient of energy). An example is a molecular force field, describing the motions and interactions between atoms, which simulates molecular folding and dynamics.

Foreign key *(computing)*: In databases, a column or group of columns in one table that uniquely identifies a row in another table.

Format *(computing)*: The way that computational information is arranged when stored, i.e. using a regular or organised scheme so that the meaning of each part of the data is understood.

Fourier transform *(mathematics)*: A method for converting data in a time series to the equivalent in frequencies.

Frozen *(computing)*: Describes a data structure that is fixed and cannot be modified after creation, for example, a tuple or frozen set in Python.

Function *(computing)*: Sometimes called a subroutine, a function is part of a computer program that represents an encapsulated set of commands in a block that is separate from the main flow of the program, but which may be invoked from different points within the program. A function is usually written to perform a general operation on an abstract set of inputs.

Functional approximation *(computing)*: Representing a continuous mathematical function (which is typically highly complex and/or unknown) for which some data points are known, in an approximate way using a combination of simpler functions that have been fitted to the data.

Fundamental frequency *(signal processing)*: A frequency in the range of the recorded frequencies, which might not be the true frequency of a signal because of aliasing (see above).

Gamma correction *(graphics)*: The process of changing the brightness of an image using a power law. This does not affect the black and white extremes of brightness, but intermediate values are scaled to lighten or darken an image, depending on the gamma value: above 1.0 will darken and below 1.0 will lighten.

Garbage collection *(computing)*: Freeing allocated computer memory that is no longer being used.

Gaussian distribution *(mathematics)*: Also called the normal distribution, a special continuous probability distribution with a 'bell' shape that is determined by its mean and standard deviation and that occurs in many applications of statistics.

Gel electrophoresis *(molecular biology)*: The process of separating dissolved molecules by passing them through a gel substance using an electric current. Typically this is done to separate the molecules according to their size or electric charge.

Gene *(molecular biology)*: The unit of inheritance in biological systems, which is formed by a particular DNA sequence within the genome of an organism. Genes are a persistent store of information and are read to make functional RNA and protein molecules that control an organism's biochemistry.

Gene duplication *(molecular biology)*: The process of copying a gene inside a given genome. Duplication arises from errors in DNA replication and from mobile genetic elements (like viruses). The copying of a gene allows for evolutionary diversification, as the two genes can adopt different roles as they diverge.

Generator *(computing)*: A memory-saving method for functions that give back a list, by only returning each element in the list as it is needed, rather than a full list.

Genetic algorithm *(computing)*: An iterative optimisation method that seeks to find better combinations of parameters by mimicking the way that a population of genes change and are selected in evolution. The algorithm involves a population of solutions to the problem which are improved upon by mutation: small changes in parameters, crossover; swapping large blocks of parameters and then propagation of the 'fittest' solutions to the next round. This is an alternative to *Monte Carlo* and *simulated annealing* approaches.

Genetic code *(molecular biology)*: The rules for translating DNA and RNA sequences into protein sequences; for going from three adjacent nucleotide residues in a codon (64 possibilities) to the 20 standard amino acids or a sequence stop. A genetic code is specified by tRNA molecules and the synthetase enzymes that associate a particular amino acid to a particular codon. Animals have a slightly different genetic code in the nucleus compared to their mitochondria.

Genome *(biology)*: The entirety of an organism's (or cell's) genetic information, stored in its chromosomes as a sequence of nucleic acids (generally DNA but RNA in some viruses).

Genome build *(bioinformatics)*: The particular version of an assembled genome sequence, i.e. the current state of sequence knowledge.

Geometric distribution *(mathematics)*: A discrete probability distribution that models how long it takes to have a successful trial, where the probability of success for each trial is independent of previous successes and independent of the trial number.

Geometry management *(graphics)*: When considering graphical user interfaces, the process of arranging graphical items by controlling certain aspects of their placement and size, the idea being to specify something general about how items are arranged, such as whether they go to the left or right and whether they expand to fit empty space, and so the programmer avoids having to give fine details of exactly how something is drawn.

Getter *(computing)*: A function that gets (fetches) a particular attribute of an object. The attribute might be stored directly in the object, or it might be calculated from other attributes.

Goodness of fit *(mathematics)*: In statistics, a measure of how two distributions compare, usually one measured and the other provided by a model.

Graphical user interface *(computing)*: A means by which the user can interact with a computer program via graphical items displayed on screen or other dynamic visual medium.

Hard problem *(computing)*: A problem that is difficult to solve using straightforward, deterministic means. Although it may be difficult to find the best solution to such a problem, testing the merit of a given solution may be significantly easier.

Hashable *(computing)*: Whether an unchanging hash value (see below) can be assigned to an object.

Hash symbol *(computing)*: The '#' key of a keyboard, often called the pound sign in America (distinct from the unit of weight or currency) and elsewhere the number sign.

Hash value *(computing)*: An integer that can be used as a short identifier for an object in a look-up data structure. Strictly, this does not have to uniquely identify the object but works best in practice if not many objects share the same hash value.

Height *(signals)*: The extreme of magnitude for a signal peak, in contrast to the volume, which is an integral or summation.

Heuristic: A rule, determined by experience, which is designed to simplify or speed up a process. A heuristic may be an approximation and cause errors, but for a good heuristic the benefits will outweigh the drawbacks.

Hidden layer *(computing)*: With reference to machine learning, one of the middle layers of an artificial neural network, i.e. not the input or output.

Hidden Markov model *(mathematics)*: A Markov chain where the states are not directly observable.

High cardinality *(computing)*: In data modelling, for an object attribute that is a collection, the maximum number of items that have to be contained in the value.

High-throughput sequence analysis *(biology)*: An experiment that studies large numbers of nucleotide sequences in a biological sample, not limited to merely discovering the sequence of a genome. For example, an analysis may be used to determine the identities and amounts of expressed RNA molecules.

Homogeneous: Describes something as having relatively even or uniform properties.

Homology *(biology)*: The notion that two biological entities have similarities because they share a common ancestor and are thus related by evolution. Often used when describing gene and protein families. When dealing with biological sequences homology is often inferred by virtue of sequence similarity.

Homology modelling *(bioinformatics)*: A synonym for comparative modelling.

Hydrogen bond *(chemistry)*: A relatively weak chemical bond that occurs between the hydrogen of a donating group and an electronegative acceptor atom, such as oxygen or nitrogen. Hydrogen bonds are common in biological molecules, forming the base-pair interactions between DNA strands and the backbone secondary structures of proteins.

Hydrophilic *(chemistry)*: A water-loving chemical or chemical group. Such groups are polar or charged and readily dissolve in water by forming stabilising interactions.

Hydrophobic *(chemistry)*: A water-hating chemical or chemical group. Such groups are non-polar, typically consisting of hydrocarbons. A lack of favourable interactions with water means hydrophobic groups tend to cluster together, for example, in the cores of globular proteins and in lipid bilayers.

Hyperplane *(mathematics)*: A flat surface in a space with a high number of dimensions (or independent features).

Hypervariable site *(molecular biology)*: When talking of variation in protein or DNA sequences, a position where the rate of substitution, from one residue to another, during evolution is significantly higher than average. Hypervariable sequences are useful in distinguishing individuals for DNA fingerprinting.

Identity matrix *(mathematics)*: A matrix which when applied to a vector or other matrix (by matrix multiplication) leaves the subject unaltered. The elements of the matrix are 1 along the diagonal but otherwise 0.

Image recognition *(graphics)*: The analysis of an image to determine properties about the physical items that the image represents, for example, to count items or determine their shape.

Implementation *(computing)*: The actual code in a computer program that provides some desired functionality.

Import *(computing)*: With reference to a computer program code, to incorporate the functionality from a separate program module that exists elsewhere, e.g. saved in a different file.

Independent *(mathematics)*: With two or more random variables, whether the values of any subset of these depend in any way on the values of the remaining ones.

Index *(computing)*: The position of a data item within an ordered collection; in a containment data structure. In Python indices are integers that start at 0 and negative numbers count backward from the end.

Influenza *(biology)*: The RNA virus that causes the infectious disease of the same name (or simply 'flu').

Information entropy *(mathematics)*: Otherwise known as Shannon entropy, is a measure of the amount of randomness in a variable quantity, which in turn is a measure of information content. The further a set of data is from a random distribution (or any null hypothesis) the more information it contains compared to that distribution. High entropy corresponds to low information content.

Inheritance *(computing)*: When considering computational class definitions, inheritance is the mechanism by which a subclass will automatically gain the same definitions (of attributes and methods etc.) as a parent superclass. The inherited properties of the subclass may differ from the superclass, but only if they are explicitly overwritten in the subclass.

Inner product *(mathematics)*: The sum of the component-wise product of two vectors, or arrays. Generally synonymous with dot product (see above).

Instantiation *(computing)*: To make a real, in-memory representation of something that was initially defined in abstract terms. For example, a class is an abstract definition of an object which is instantiated to make an actual in-memory object.

Integer *(computing)*: A whole-number data type, without a decimal point. Integers can have a positive or negative sign. In Python the number 3 is of integer type whereas 3.0 is not; although these numbers have the same mathematical value their internal representation is different.

Intensity *(signal processing)*: A measure of the strength of a measured signal. This is a somewhat ambiguous term; for signal data it can refer to signal height or volume. For radiant energy, such as light or sound, the intensity is a measure of energy output for a given spatial region.

Interface *(computing)*: A specification for the functionality provided by a class or module, without worrying about how that functionality is actually implemented.

Intersection *(mathematics)*: The common elements between two collections, usually sets.

Intron *(molecular biology)*: The non-coding parts in the middle of genes. Intron sequences are initially transcribed into RNA, but are removed to make mature RNA molecules, and so do not contribute to the final product. Introns separate a gene's exons, which do form the mature RNA and thus may encode protein sequence.

Isoelectric point *(chemistry)*: The pH at which a molecule has no overall charge; naturally only relevant for molecules that contain acidic and/or basic chemical groups.

Iterable *(computing)*: A property whereby an object can provide the individual members from a group of items, e.g. from a list. In Python an iterable object can be used to provide a sequence of items, such as in a 'for' loop. This is a concept distinct from the **iterator** object (see below), though all iterable objects can be converted into iterators using the iter() function.

Iterate *(computing)*: The process of repeating the same basic operation a number of times, such as extracting all the items from a list in sequence. In the context of algorithms this might be to improve the solution to a problem, where the solution to one cycle serves as a foundation for the next.

Iterator *(computing)*: An object that emits a succession of items, by continually asking for the 'next' item until all items have been considered. In Python an iterator is a special kind of object that provides the next item in a data sequence via its __next()__ method (in Python 3; next() in Python 2). In this context an iterator is distinct from an **iterable**, which does not support the __next()__ method but which can nonetheless be used in loops etc. All iterators are iterable but not vice versa.

Java *(computing)*: A compiled, object-oriented programming language commonly used for database and Internet applications.

Join *(computing)*: With regard to databases the way information can be linked across different tables.

K_a *(chemistry)*: A symbolic abbreviation for the acid dissociation constant.

K-dimensional trees *(computing)*: A way of grouping data with K dimensions (independent data axes) into a tree-like structure, i.e. a hierarchy of branching nodes. Such an arrangement can make particular operations, such as finding data points within a given radius, more efficient. The basic notion is that the tree arrangement means that only some of the data needs to be checked.

Kernel methods *(computing)*: A type of algorithm used for machine learning (pattern recognition) where the original data are placed in a *vector space* so that the data may be analysed in terms of the distance between points. Such methods often use a kernel function to efficiently calculate the coincidence or similarity (inner product) between data points.

Kernel trick *(mathematics)*: The means by which a kernel method uses a kernel function to effectively map a pattern-recognition problem to one of higher dimensionality, e.g. to help separate data into categories. This is achieved by calculating inner products, as a measure of coincidence in the data, and eliminates the need to explicitly represent the data in the higher dimensions.

Key, value *(computing)*: The pairs of items that constitute a dictionary data structure. The key is a unique, unmodifiable item that is used to access the corresponding value.

k-nearest neighbour *(computing)*: A simple machine learning algorithm that classifies data by finding the *k* nearest neighbours in the feature space, representing the closest data points with known classification. The classification is assigned by taking a poll of the classes represented in the neighbouring points.

Kohonen map *(computing)*: A synonym for a self-organising map (see below), named after Teuvo Kohonen.

Layout *(graphics)*: With reference to graphical user interfaces, the relative arrangement of graphical items on screen.

Likelihood ratio test *(mathematics)*: A statistical test to compare two distributions, one of which is a special case of the other.

Linear algebra *(mathematics)*: The mathematics that describes linear transformations, typically using matrices. A transformation is linear if applying it to the sum of any two vectors is the same as the sum of the result of applying the transformation to the two vectors individually.

Linear discriminant analysis (LDA) *(mathematics)*: A method to classify points within a set of data using linear combinations of the features to separate the data.

Linewidth *(signal processing)*: The width of a specific well-defined part of a signal, often measured as the difference, or 'full width', at the positions which are half the maximum value in both directions from the maximum.

Lipid *(chemistry)*: A variety of fatty, hydrocarbon-rich molecules that are the main component of biological membranes, i.e. forming a lipid bilayer. Typical lipid molecules include cholesterol, long-chain fatty acids and phospholipids.

List *(computing)*: A data structure containing an ordered collection of items. In Python a list may be modified, e.g. to add and remove items, and can contain items of any data type.

List comprehension *(computing)*: A kind of program syntax in which a list is constructed in its entirety in one expression using another data list. In Python this is in the basic form `newList = [item for item in oldList]`.

Literal *(computing)*: A stated quantity in a computing program that is a specific constant number or string rather than a variable.

Long integer *(computing)*: A data type to represent whole (integer) numbers that are larger in value than standard integers (which use a limited amount of memory).

Loop *(computing)*: A program control structure where the same block of code is used repeatedly a number of times, usually considering different values each time, for example, to process the contents of a list. In Python this is done with `for` and `while` statements.

Lossless compression *(computing)*: A means by which data is transformed into a different, smaller representation, without affecting the fidelity of the underlying data, i.e. uncompressing the data recreates the original data exactly.

Lossy compression *(computing)*: A means by which data is transformed into a different, smaller representation, where the compressed form is only an approximation of the original. Uncompressing the data does not normally recreate the exact details of the original data, rather it gives something similar.

Low cardinality *(computing)*: In data modelling, for an object attribute that is a collection, the minimum number of items that have to be contained in the value.

Luminance *(imaging)*: The brightness of an area of an image. When applied to pixmaps this usually means the brightness of a pixel.

Machine learning *(computing)*: A method whereby a computer program learns to perform a task (usually classification or functional approximation) based on a set of input training data. Such methods may be described as supervised if the training data has known values/categories.

Macro *(computing)*: In C, a fragment of code which has a specific name and optionally arguments, and which can then be used by invoking the name.

Macromolecule *(molecular biology)*: A large biological molecule that is a polymer of smaller components, typically taken to mean DNA, RNA or protein chains with more than a few residues.

Markov chain *(mathematics)*: A random process that moves in time (or position) between some specified states and where the next state only depends on the current state and not on the previous states.

Matrix *(mathematics)*: A two-dimensional, rectangular array of numbers, i.e. arranged into rows and columns.

Matrix *(molecular biology)*: Inert, solid material that is used to support or attach a component of interest.

Maximum likelihood *(mathematics)*: The estimate of parameters of a probability distribution which maximises the probability that the data is observed given the hypothesised distribution.

Mean *(mathematics)*: The average value of a collection of data.

Memoization *(computing)*: The process by which a function remembers the results from previous executions, so that a cached result may be returned if an operation is repeated for the same input data, thus eliminating the need to perform the entire operation again. This often improves program speed, at the expense of using more memory.

Memory allocation *(computing)*: The allocation of a specific chunk of memory for use by a program, which some (typically older) programming languages usually have to do explicitly.

Meoisis *(biology)*: The separation of chromosomes within the nuclei of eukaryotic cells (i.e. not bacteria) into two sets, each containing half of the chromosomes. After cell division this results in daughter cells which have only one copy of each kind of homologous chromosome. For example, the original cell will have two copies of chromosome 1, but after divisions the cells have one copy each.

Method *(computing)*: A *function* that belongs to a computational object, and which is generally specified in the object's class definition.

Microarray *(biology)*: A kind of multiplexed experiment, where many miniature experiments, which are of the same type but use different reagents, are performed at the same time within a rectangular array on a solid support.

Mitosis *(biology)*: The reproduction and separation of chromosomes within the nuclei of eukaryotic cells (i.e. not bacteria) that results in two identical sets of chromosomes. Mitosis is usually followed by cell division, i.e. to create two copies of the original cell.

Model *(bioinformatics)*: In relation to three-dimensional structures, a *conformation* (see above) that is consistent with the known data.

Model *(computing)*: An abbreviation of data model.

Module *(computing)*: A discrete unit of computer code that is stored as a separate file, but which may have its functionality imported into the program code represented in other files.

Monte Carlo method *(computing)*: A means of solving a problem by using random numbers to create test points or solutions. The Monte Carlo method is used as an unbiased way of getting a representative picture of a problem without having to have any innate understanding of the underlying mechanics and while only considering a small proportion of possible values.

Motif *(molecular biology)*: A small, recurring feature within biological molecules, such as a group of closely related sequences or sub-structures, that can be related to a particular biological function. For example, a protein may bind to a given DNA sequence motif to activate a number of different genes.

Multiple inheritance *(computing)*: When considering object orientation and the construction of class definitions, the process by which one class is based on two or more parent superclasses and thus inherits properties from all of them.

Multiple-sequence alignment *(bioinformatics)*: An alignment (see above) where there are more than two macromolecule sequences. Sequence elements are arranged, by placing gaps, to maximise the similarity between residues in each column.

Mutable *(computing)*: The ability to manipulate a data item so that its value or contents changes internally, but leaving the handle (reference) to the data item unaffected.

Native conformation *(molecular biology)*: The properly folded and functional arrangement of a biological molecule's structure, which is generally believed to be close to the lowest-energy form.

Negative binomial distribution *(mathematics)*: A discrete probability distribution that models the number of successful trials before a specified number of failed trials occurs, where the probability of success for each trial is independent of previous successes and independent of the trial number. A geometric distribution is a special case of the negative binomial distribution where number of failures sought is one.

Nested *(computing)*: An arrangement of computer code where one control structure is placed inside the block of another. For example, a loop may be defined, to repeat the execution of a bit of code, and that bit of code itself may contain another loop, controlling an inner block. In this case one loop is said to be nested in the other.

Newline character *(computing)*: A special character, not resulting in a printed symbol, which causes the flow of text to begin at a new line. On Linux and Mac OS X systems the newline character was originally a linefeed command, from the days of typewriters; represented in Python by the escape code '\n'. On Windows systems a new line is given by a carriage return and a linefeed control character; in Python '\r\n'.

Nodes *(computing)*: A point of connectivity within a network, which performs a process and which is associated with input and output data streams.

Non-deterministic polynomial (NP) *(computing)*: A problem whose solution, if given, can be verified in polynomial time (polynomial in the size of the problem).

Normal *(mathematics)*: A vector that is perpendicular (at right angles) to a plane or another vector.

Normal distribution *(mathematics)*: Also called the Gaussian distribution, a special continuous probability distribution with a 'bell' shape that is determined by its mean and standard deviation and that occurs in many applications of statistics.

Null hypothesis *(mathematics)*: When assessing a hypothesis, a competing hypothesis that represents what would be expected to happen given random chance or in the absence of discriminating information. The null hypothesis is useful to give an objective baseline to predictive theories.

Object *(computing)*: A computational data structure, built according to a class definition, that contains other items of data (as attributes) and which may have bound functions (methods).

Object attribute *(computing)*: A bound variable that represents a property of an object.

Object orientation *(computing)*: A general method of writing computer programs, representing the data items as objects, often with interconnectivity.

One-tailed test *(mathematics)*: A statistical test that determines the probability of an observed value having at least a given separation from the mean value, on one side only. For example, the probability that an observation is at or above one standard deviation from the mean.

Optimisation *(mathematics)*: A method to select the best values (or other elements) from a range of possibilities. Often this means to find the parameters which give the minimum or maximum value of a function.

Order (*computing, mathematics*): How the number of operations required to perform a computational task scales with the size of the problem.

Orthologues *(molecular biology)*: Entities which are homologous, sharing a common ancestor, because of the separation that occurs as separate species evolve. Effectively different versions of the same protein or gene from different organisms.

Over-training *(computing)*: Pertaining to the pattern recognition of supervised machine learning methods, the problem of applying the training data too much, so that the method is less general and performs sub-optimally on unseen data.

Paired-end reads *(molecular biology)*: A pair of experimentally determined DNA sequences (reads) which are known to come from either end of the same fragment of DNA.

Palette *(graphics)*: A limited array of colours used as a look-up table to construct an image. Using a colour palette gives smaller images because colours are referred to by index, rather than by value, but the number of colours available for the image is limited.

Parallel tempering *(computing)*: Also called replica exchange, is a variant of the Markov chain Monte Carlo method that uses parallel implementations at different 'temperatures', i.e. where higher temperatures allow greater variation between subsequent steps. The temperatures of parallel computations are swapped so that each can explore a large number of states in an unbiased manner.

Paralogues *(molecular biology)*: Entities which are homologous, sharing a common ancestor, because of a gene duplication event. Effectively finding closely related genes or proteins inside one organism.

Parent-child relationship *(computing)*: When referring to data models, the relationship between objects stating that one is formally contained by the other; the parent contains the child. For example, a `molecule` object may contain `atom` objects. This containment hierarchy usually involves objects of completely different types and is distinct from the super- to subclass relationship.

Parsimony *(bioinformatics)*: The principle of the most likely evolutionary route, which, for example, has generated divergent sequences, being the one that involves the fewest changes, i.e. the most frugal.

Path *(computing)*: The location of a file stored in a file system, e.g. on disk, described as a path through a hierarchical directory structure.

Pearson's chi-squared test *(mathematics)*: In statistics, a goodness-of-fit test, first discussed by Karl Pearson. The test considers a null hypothesis that one distribution, determined from a measured sample, is consistent with another distribution provided by a model.

Pearson's correlation coefficient *(mathematics)*: In statistics, a measure of the linear correlation between two distributions, usually provided by paired measurements of a sample.

Percent point function *(mathematics)*: A function that determines the confidence interval around the sample mean, given a specified confidence level.

Perl *(computing)*: A programming language, which like Python is high level and interpreted. Perl is an older language than Python, and is popular in the bioinformatics community.

pH *(chemistry)*: The negative logarithm of the hydrogen ion concentration of a solution; a measure of the acidity where a value of 7.0 is neutral; lower than this is acidic and greater than this is alkaline.

Phase *(signal processing)*: In certain applications, this describes the amount by which the real and imaginary parts of a complex signal need to be rotated in order to put them into a form that provides a purer description of the signal, which can help in interpretation of the signal.

Phasing *(signal processing)*: The process whereby a complex signal has its phase corrected in some situations.

Phylogenetic tree *(bioinformatics)*: A hierarchical data structure used to describe the evolutionary relationship between biological entities, such as whole organisms or single macromolecules.

pI *(chemistry)*: The isoelectric point; the pH at which a molecule has no overall charge, where positive and negative charged chemical groups balance. The pI is a property of a molecule that depends on the number and relative strengths of its acidic and basic groups.

Pickle *(computing)*: The Python method of serialising data, automatically converting in-memory data structures into a stream of data that may be stored on disk, typically a plain text file.

Pixel *(graphics)*: The smallest square area of colour with which a computer image is constructed.

Pixmap *(computing)*: A rectangular array of colour intensity values that describe the pixels of an image.

pK_a *(chemistry)*: A numerical constant describing the strength of an acidic group; equivalent to the pH at which the acidic group is half ionised.

Plain text *(computing)*: A means of representing data using a restricted set of standard, textual characters.

Poisson distribution *(mathematics)*: A discrete probability distribution that is used to describe the number of events that occur in a specific interval, assuming the events occur independently and with a common rate.

Polypeptide *(chemistry)*: A long polymer molecule consisting of a string of amino acid residues, linked via peptide bonds. Polypeptides are the primary component of proteins.

Position-specific scoring matrix *(bioinformatics)*: A table describing the abundance of different kinds of residue at each position within the alignment of a family of homologous genes or proteins.

Positive selection *(biology)*: The tendency for the evolutionary process, in some situations, to select gene sequences that show differences, i.e. that are actively changing in sequence.

Posterior *(mathematics)*: The probability that a hypothesis is true given some experimental data.

Post translational *(molecular biology)*: Relating to processes occurring after a protein has been made, e.g. modification of the standard amino acids by joining sugars or lipids.

Pound symbol *(computing)*: The '#' key of a keyboard, often called the hash symbol in the UK and elsewhere the number sign.

Power spectrum *(signal processing)*: The magnitude squared of a complex valued signal, as a function of frequency.

Preservation *(molecular biology)*: The tendency for a biological feature to be preserved during the process of evolution, e.g. protein or DNA sequences that do not change.

Proteomics (biology): The study of the total protein complement of cells or organisms.

Primary key *(computing)*: In a relational database, a list of one or more columns of a table for which the corresponding values uniquely identify a record in the table.

Principal component analysis (PCA) *(mathematics)*: A method to find the innate directions within data, represented as vectors, which explain most of the variance within the data. Taking the first few principal components of a high-dimensionality data set provides a method for simplifying the data (dimensional reduction) while preserving an optimal amount of variation.

Prior *(mathematics)*: The probability that a hypothesis is true before experimental data is obtained.

Probability density function *(mathematics)*: For a continuous probability distribution this can be thought of as the probability that the distribution takes a specified value. Technically, it is the probability that the distribution takes a value in a small region around the specified value, divided by the size of the region, in the limit that the latter goes to zero.

Product *(mathematics)*: The result of a multiplication operation.

Profile *(bioinformatics)*: When used in the context of biological sequences, a profile describes the abundances of different residues at the various positions within a multiple alignment.

Progressive pairing *(bioinformatics)*: A fast, heuristic method of generating a hierarchical, tree-like data structure by progressively joining the groups of data items, starting with the most similar pair.

Prompt *(computing)*: The symbols printed at the start of a textual, command-line interface, indicating where to type commands to control a computer.

Protein *(molecular biology)*: A biological macromolecule primarily composed of one or more polypeptide chains. Proteins are the diverse, functional molecules of living organisms that catalyse and control biological processes.

Protein sequence *(molecular biology)*: The order in which the different kinds of amino acid residues are joined together, to create the polypeptide chain of a protein.

Pseudorandom number *(computing)*: A value generated by an algorithm to mimic the effect of having random numbers. Pseudorandom numbers are not truly random (they may eventually repeat, for example) but often appear to be sufficiently random to perform various computational tasks, such as the *Monte Carlo method*.

Purifying selection *(biology)*: Also referred to as negative selection, the tendency for the evolutionary process, in some situations, to select gene sequences that do not change, i.e. to suppress sequence variation.

Quantile function *(mathematics)*: A function that determines the confidence interval around the sample mean, given a specified confidence level.

Quantile normalisation *(statistics)*: A technique for mapping the statistical properties of one sample distribution to another distribution. This is achieved by replacing values for data points in one distribution with equivalent values from the other distribution that share the same rank (preserving order of relative magnitude).

Ramachandran plot *(bioinformatics)*: A two-dimensional plot of the backbone (phi and psi) dihedral angles in a polypeptide, showing how the chain twists and indicating common conformations like alpha-helices and beta-strands.

Random variable *(mathematics)*: A mathematical construct to describe random behaviour, usually for some specified model of the randomness. A way of assigning a label to a property that has its value selected at random, according to a given probability distribution.

Rank *(mathematics)*: With reference to matrices, the maximum number of linearly independent rows or columns.

Raster *(graphics)*: To compose an image from a regular array of pixels, i.e. as a pixmap, in contrast to using vector graphics where images are described in terms of geometric shapes.

Raw string *(computing)*: A Python string where the standard Python escape characters (i.e. prefixed with '\') are not active and instead the string is considered literally as typed.

Read (molecular biolog*y*): In the context of sequencing, the experimental determination of one, often small, section of DNA sequence; which may then be combined with others.

Readable *(computing)*: Whether the data stored in a file, e.g. on disk, can be accessed.

Reading frame *(biochemistry)*: The relative offset or phase of a nucleotide sequence interpreted as codons. There are three possibilities for a strand and each offset selects a different grouping of nucleotide triples and hence a different amino acid translation.

Real *(mathematics)*: A number from the unrestricted continuum of negative and positive values, i.e. not a complex number.

Reflections *(crystallography)*: A pattern of regularly spaced spots created when X-ray radiation diffracts through a crystal (an atomic lattice). The spots are points of constructive interference between the X-rays.

Regular expression *(computing)*: The specification of a particular pattern of characters that may be sought within a text string. The pattern itself is a string, which can contain both literal characters and metacharacters, with the latter describing ambiguous character groups, and thus the type of pattern to be searched for.

Relational database *(computing)*: A way of storing certain types of structured data that allows fast querying of logical relationships in the data.

Relative entropy *(mathematics)*: A measure of whether two variable quantities have the same distribution, and equivalently a measure of the information content one distribution holds above another. This is related to Shannon information entropy, where the comparison is instead with respect to a random distribution. The relative entropy is useful because it indicates whether data is significant and different.

Replica exchange Monte Carlo *(computing)*: See parallel tempering.

Replicas *(computing)*: Multiple copies of a simulation or search process that all relate to the same data, for example, so that an optimisation is done several times in parallel from different starting points.

Residue *(biochemistry)*: One of the component monomer chemicals at a particular position in a biological polymer sequence, i.e. an amino acid at a location in a polypeptide chain or a nucleotide in a nucleic acid chain (DNA or RNA).

Resolution *(bioinformatics/graphics)*: A measure of the limit in the precision of a measurement, i.e. what size features can be seen or resolved. With reference to biomolecular structures the resolution reflects the uncertainty in the positions of atoms, and depends on the experimental technique used and the quality of data. When referring to images the resolution is usually stated simply as the number of component pixels, often giving width and height separately.

Retroviruses *(biology)*: A virus that has an RNA genome, which uses a reverse transcriptase enzyme to convert the RNA into DNA, the opposite of the normal biological information flow. The conversion is often for the purpose of inserting the viral DNA into the DNA of the host's genome, a means of lying dormant and evading the host's defence mechanisms. HIV, which causes AIDS, is an example of a retrovirus.

Return *(computing)*: To finish the execution of a function (a subroutine) and pass a value back to the previous point in the program flow, where the function was called from.

Return key *(computing)*: The key on a keyboard which is used to start a new line or complete the entry of data. Often synonymous with the Enter key and frequently symbolised by a bent arrow, going down and left.

Reverse complement *(molecular biology)*: A DNA or RNA sequence that has the opposite kinds of base and reversed order compared to another sequence, such that the two nucleic acids could form complementary base pairs, but are both viewed in the same orientation (usually $5'$ to $3'$). Given A:T and G:C base pairs, the reverse complement of the sequence ATCCGACTCAG would be

CTGAGTCGGAT, so that the first position for the first sequence pairs with the last position of the second sequence.

Ribosome *(molecular biology)*: The large enzyme particles inside cells, containing both protein and RNA, which catalyse the synthesis of new proteins (i.e. polypeptide chains) by joining individual amino acids (delivered on a tRNA) and a messenger RNA that specifies the sequence.

Ribozymes *(molecular biology)*: Biological catalysts that use RNA to increase the rate of a reaction, i.e. enzymes that use RNA rather than protein at the active site.

RNA *(molecular biology)*: Ribonucleic acid, biological polymer modules that are copies of parts of DNA sequences and which have various roles in cells, including enabling the creation of proteins (transferring DNA sequence information and delivering amino acids), acting directly as functional molecules alongside proteins and controlling the stability of other RNA molecules.

RNA sequence *(molecular biology)*: The sequential order of the four different types of nucleotide compound that make up a linear RNA chain, which is generally a molecule with a single strand, unlike DNA.

Root *(computing)*: When referring to locations within a file system, the root is the very top of the containment hierarchy which contains all of the other sub-directories. On Unix, Linux and Mac OS X-based systems this is represented with a single forward slash symbol '/'. On Windows-based systems there is no single root, rather there are separate drive letters like 'C:\'.

Root-mean-square deviation *(mathematics)*: A measure of the overall difference between two sets of paired values. It is calculated as the square root of the average squared difference between the paired values. Commonly used as a measure of coordinate difference between three-dimensional structures.

Rooted tree *(bioinformatics)*: A hierarchical branching data classification where there is a base or root. For a phylogenetic tree this means that there is an ancestor node.

Rotation matrix *(mathematics)*: A matrix which, when applied to a vector or another matrix (of suitable size) by means of matrix multiplication, causes a transformation of coordinates that is a pure rotation.

Runtime *(computing)*: The time when a computer program is running.

Sample mean *(mathematics)*: The observed mean, or average value, of a property from a sample of data.

Sample space *(mathematics)*: In probability, the specification of the set of possible outcomes which can occur.

Sample variance *(mathematics)*: The observed variance of a property from a sample of data.

Scalar product *(mathematics)*: Also called the dot product. A scalar value, calculated as the sum of the component-wise product of two vectors (or arrays) which represents the size of the projection of one vector on to the other. Equivalent to the product of the length of two vectors and the cosine of the angle between them

Scope *(computing)*: Short for lexical scope, the context within the commands of a computer program for which a particular variable is defined. Often discussed in terms of a named variable being local or global, with respect to a block of code. For example, a variable defined in a Python function is local to the function, i.e. its name only binds to a value inside the function and using the same name outside the function actually refers to a different thing.

Scripting *(computing)*: The use of a high-level interpreted (i.e. not explicitly compiled) computer language, like Python, Perl or Unix shell commands, to create a simple computer program to perform a simple task. The distinction between what is a script and what is a program is somewhat blurred for a fully functional language like Python or Perl, but a script is usually a program to provide an automated and convenient alternative to something that could be done by other, more tedious means.

Search path *(computing)*: The specification of which locations should be investigated to find a given item, usually within a hierarchical data structure, such as a file system. With particular reference to Python modules the search path is effectively a list of directories that contain the files describing the modules which can be loaded into a program to provide extra functionality.

Secondary structure *(molecular biology)*: The classification of the local conformation of a biological polymer chain, which generally corresponds to a particular backbone hydrogen-bonding pattern with characteristic backbone dihedral (twist) angles. The term is generally used for proteins, where the common secondary-structure types are (right-handed) alpha-helix and extended beta-strand (which form beta-sheets).

Seed *(computing)*: With reference to random numbers, the seed is an arbitrary number which is used to start a pseudorandom number-generating algorithm. The seeds themselves are often quite random, but the use of the same seed, in what otherwise appears to be a random process, is important for determinism, i.e. the same seed gives reproducible pseudorandom output, which is useful for testing software.

Self-organising map *(machine learning)*: Also called a Kohonen map. A kind of unsupervised neural network that operates by arranging data, represented as vectors, in a grid (i.e. a map of low dimensionality) to arrange the data into a simpler form that preserves the distinction and similarity between items. It can be used to automatically cluster data and map it to fewer dimensions; this is similar to what principal component analysis does, but does not assume any underlying distribution in the data.

Sequence annotation *(bioinformatics)*: The mechanism of ascribing a biological sequence with further information, typically describing what its function and relation to other sequences are.

Sequence complexity *(mathematics)*: A measure of how random or non-random a sequence is. This is related to the concept of information entropy (see above). A repetitive sequence, where only a few kinds of residue from the total available are present, is deemed to have low complexity; less information is needed to encode it and so it is easier to generate at random.

Sequence identity *(bioinformatics)*: A measure of the number or proportion of residues in two aligned biological sequences that are exactly the same, i.e. have the same kinds of residue at equivalent positions.

Sequence similarity *(bioinformatics)*: A measure of how alike two aligned protein sequences are, accounting for the similarities and differences in the component amino acids. Often the similarity is derived from a measure of how substitutable different residues are in evolution.

Serialise *(computing)*: Automatically convert in-memory data structures into a stream of data that may be stored on disk, typically a plain text file. In Python the inbuilt method of serialising data is called pickling.

Server *(computing)*: The part of a computer application that does the work, e.g. calculations, or putting together the contents of a web page, which is distinct, and often physically remote, from the client with which the user interacts.

Set *(mathematics)*: A collection of items which is unordered and where no item is repeated.

Setter *(computing)*: A function that sets a specified attribute of an object to a specified value. This can include verification that the value is allowed.

Side chains *(molecular biology)*: When referring to the amino acid compounds that make up proteins, the side chain is the part of an amino acid that varies between the different types, and which sticks out from the backbone (the part that is linked together to form chains).

Signal *(computing)*: With regard to graphical interfaces, particularly those based on the Qt libraries, a signal is a notification sent out from a graphical item to indicate that something has happened, e.g. the item was clicked or resized. In Qt signals are connected to slots, which are functions that are triggered in response to the signal.

Silent substitutions *(molecular biology)*: A change within the protein-coding region of a gene's DNA sequence which does not result in a change in the protein sequence. This occurs because an amino acid may be represented by multiple different three-base codons in the genetic code.

Similarity matrix *(bioinfromatics)*: See *substitution matrix*.

Simple linear regression *(mathematics)*: In statistics, a linear fit of one distribution, Y, against a single other distribution, X, with both usually provided by measurements. Accordingly, we find a (y-axis intercept) and b (gradient) such that $Y = a + b\,X$ best fits the data, using a least-squares calculation.

Simulated annealing *(computing)*: A global optimisation procedure, i.e. for finding the best overall solution to a problem, that involves having a search strategy which bases the next trial solution on the previous solution using a parameter analogous to temperature. At a high 'temperature' subsequent solutions can be quite different from one another (analogous to large thermal motion) and sample the parameter space widely, but as the temperature is cooled the differences become smaller, allowing the solution to refine and approach a precise optimum. Simulated annealing can often find good solutions to complex problems in a reasonable time.

Single-nucleotide polymorphism (SNP) *(molecular biology)*: A point of variation in a DNA sequence, often when comparing individuals of the same species, consisting of a single base-pair change.

Singular value decomposition (SVD) *(mathematics)*: A useful technique for decomposing a matrix into its fundamental parts: a rotation, a scaling and another rotation. Commonly used to find the optimal rotation which minimises differences between spatial coordinates, i.e. to superpose/ align structures.

Slice *(computing)*: A means of accessing a range of elements in an ordered data collection (like a Python list, text string or tuple) by specifying the extent of the selection, rather than each item individually.

Slot *(computing)*: Referring to graphical user interfaces, especially for the Qt libraries, a slot is a function belonging to one graphical object that is invoked in response to a notification signal emitted from another. This mechanism allows the coordination of graphical items in response to a common set of data.

Sobel operator *(images)*: Also called a Sobel filter. A matrix which is convolved with a pixmap image for the purpose of detecting contrast changes (i.e. edges) along a horizontal or vertical direction. Effectively this creates a secondary image where the signal gives an approximation of the gradient of intensity changes in the primary image.

Spatial restraints *(bioinformatics)*: Bits of geometric information that relate to the spatial conformation of an object, such as a protein structure, which when considered in combination help determine the overall three-dimensional shape.

Spectrum *(signal processing)*: An experimental signal viewed in terms of frequency, rather than time or space.

Spin active *(chemistry)*: With reference to atoms, a nucleus which exhibits nuclear magnetic resonance, i.e. absorbs characteristic radio frequencies when placed in a magnetic field.

Splice *(molecular biology)*: To join together two biopolymer chains. Splicing notably occurs in the processing of messenger rRNA modules, where the unwanted parts, introns, are removed and the remainder (which generally contains the protein-coding region) are joined together.

Stack *(computing)*: A collection of items that are processed in the order 'last in, first out'.

Standard deviation *(mathematics)*: A measure of the spread of values from their mean. Calculated as the square root of the average square difference between the values and the mean.

Standard normal distribution *(mathematics)*: A normal distribution with mean 0 and standard deviation 1.

Statement *(computing)*: The smallest logical elements of a computer program.

Stem-loops *(molecular biology)*: Small regions of double-stranded nucleic acid, usually RNA, where a single chain folds back and base-pairs with itself.

Strand *(molecular biology)*: One nucleic acid chain molecule. DNA usually has two nucleic acid strands joined along their length by hydrogen-bonded base pairs; the strands go in the opposite directions, in terms of the chemical bonds.

String *(computing)*: A computational data type representing an ordered collection of characters that form a piece of text.

Structure ensemble *(bioinformatics)*: A representation of a molecule's three-dimensional structure, consisting of a collection of alternative models (i.e. sets of coordinates) which together represent the variability or uncertainty in the data. Often the models of an ensemble represent alternative conformations which all fit with the experimental data.

Structural environment *(bioinformatics)*: The categories of local three-dimensional conformation within a biological molecule. Generally with reference to the different features found in protein structure that affect the nature of residue substitution during evolution. Environment categories typically correspond to secondary structure, hydrogen bonding and solvent accessibility states.

Student's T distribution *(mathematics)*: Used in various statistical tests including determining confidence intervals on the mean of a distribution given a sampling, i.e. whether two means are significantly different.

Subclass *(computing)*: With reference to object-oriented programming, a subclass is a type of object definition that is based on, i.e. inherits from, another type of object, called the superclass.

Substitutable *(molecular biology)*: The ability of two different types of residue to swap with one another (in a DNA or protein chain) during evolution.

Substitution *(molecular biology)*: The swapping during evolution of one residue type for another in a given position of a biological polymer molecule.

Substitution matrix *(bioinformatics)*: Also called a similarity matrix. A means of representing the equivalence between different types of residue in a sequence alignment. Described in the form of a table, so that a score may be obtained for each possible pair of residue type. This provides the basis to calculate a score for an alignment and thus the method of generating optimised alignments. The equivalence scores usually indicate how readily different residue types substitute for one another.

Superclass *(computing)*: With reference to object-oriented programming, a superclass is a type of object definition that lends its specification as the basis for another kind of object (a subclass).

Supervised learning *(computing)*: A type of machine learning method where the pattern-matching algorithm is trained to associating known inputs with known outputs so that afterwards it may make predictions in a general way on unseen data. See also *training* below.

Support vector *(computing)*: With regard to machine learning, one of the data points considered by a support vector machine that lies on the edge of, and thus defines, the boundary used in the separation of data classes.

Support vector machines *(computing)*: A kind of machine learning algorithm that works by performing a linear separation of data mapped into a space of higher dimensionality, using the *kernel trick*. The widest margin region between different data classes is found in a regression-like process, which defines a decision boundary that can be used to separate unseen data points in a general way. This separation margin is defined by the *support vectors* (see above) and is less susceptible to over-training compared to methods like artificial neural networks.

Support vector regression *(computing)*: Using support vector machines to approximate continuous functions, e.g. to predict where points lie on a graph. This is in contrast to using support vector machines to classify data points into discrete categories.

Systems biology: The study of the emergent properties of a biological system considered as a whole, i.e. things that would not be evident by looking at individual components in isolation.

Tab *(computing)*: Derived from a 'tab stop' on a typewriter. A tab is a key on the keyboard and the resulting special (whitespace) character that causes indentation of text. A tab is often longer than several normal spaces.

Tables *(computing)*: Relational databases store data in tables, with each column representing an attribute of the data and each row representing one record of data. This is a similar notion to the usage of tables in spreadsheets.

Tautomerism *(chemistry)*: The ability of certain chemical compounds to be able to swap between two different forms that contain the same atoms. Often this involves an equilibrium where both chemical structures are present to some degree and continuously inter-convert.

Taxa *(biology)*: The different hierarchical classifications of organisms within the evolutionary tree of life, i.e. the named groups of related organisms.

Template *(bioinformatics)*: With specific reference to protein structures, a template is a protein of known (experimentally determined) structure which is used as the basis for predicting the structure of another similar protein: generally a homologue with a similar sequence.

Template *(molecular biology)*: When referring to nucleic acid strands the template is the strand that is the basis for the creation of a new strand using base-pairing rules, for example, in DNA and RNA synthesis.

Thread *(computing)*: A sequence of the smallest segments of program instructions that can be scheduled to run independently on a computer. The notion behind this is that a processor's stream of execution can rapidly switch between different computing tasks. A result is that multiple programs can effectively be run at the same time on a single processor, though on a small scale the execution may merely be interlaced. Threading is a concept distinct from true parallel processing, which uses separate processor cores, though most modern multi-core processors implement both concepts.

Timeout *(computing)*: A specified time that is allowed to elapse before a specific event is triggered, often used as a failsafe in the case that various asynchronous tasks do not complete in a reasonable time.

Token *(computing)*: A string of one or more characters that when taken together represent the smallest meaningful parts of a computer language.

Top-level widget *(graphics)*: A graphical object in a user interface that can contain other graphical items, but which is not itself contained by any other items. Such objects are typically free-floating windows and have an iconified representation when fully minimised.

Top object *(computing)*: The top of an object hierarchy in a data model, i.e. the class of object that contains all others.

Torsion angle *(mathematics)*: See *dihedral angle* above.

Training *(computing)*: With reference to supervised machine learning methods, the stage where data with known outcome (e.g. there is a known classification for a given set of input values) is used to adjust internal parameters so that the method may be used to recognise patterns and make predictions in a general way on unseen data. For a neural network training involves setting the connectivity weights between nodes so that the output signals are close to the known values for the given input.

Transcribe *(molecular biology)*: The process of converting a DNA sequence into an RNA sequence, effectively making a copy of the coding DNA strand by forming a chain that is complementary, following base-pairing rules, to the template DNA strand.

Transformation *(mathematics)*: A function that maps elements from one set to another, generally preserving any structure that may be present in the data. When applied to spatial coordinates, transformations commonly involve scaling, rotation, translation and shearing.

Transition matrix *(mathematics)*: The matrix that specifies how a Markov chain evolves from one moment in time to the next; the probability of transitioning from one state to another.

Translation *(molecular biology)*: The process of synthesising a protein's polypeptide chain where the sequence of amino acids is determined by subsequent groups of three bases (codon) in a messenger RNA molecule. The directed polymerisation reaction is catalysed by a large multi-component particle called the ribosome.

Transmembrane *(molecular biology)*: Something that resides in and spans the lipid bilayers that form cellular membranes. Usually used to refer to hydrophobic protein domains that are embedded in lipids so that the protein is exposed on both sizes of the membrane, in contrast to a membrane-associated protein which is only attached on one side.

Transpose *(mathematics)*: A matrix obtained by swapping rows with columns, i.e. flipping about the diagonal.

Travelling salesman problem *(computing)*: A classic example of a computationally hard problem, involving the determination of the most efficient route for a person to visit a given set of points on a map, i.e. which order the visits should be made in.

Trigger function *(computing)*: In machine learning, when referring to the nodes of a neural network, the trigger function is the operation associated with a node that converts the summation of all the (weighted) input signals into an output signal. The trigger functions are typically step-like or S-shaped (e.g. the sigmoidal hyperbolic tangent) so that only inputs above a threshold cause a significant output and so that the maximum output is limited.

Trypsin *(molecular biology)*: A protein-cleaving enzyme (a protease), originally isolated from the pancreas of many animals where it works to digest food. Artificially synthesised trypsin is a common tool in molecular biology for cutting up protein chains at predictable locations.

Tuple *(computing)*: A data structure containing an ordered collection of items. In Python a tuple cannot be modified after creation and may contain any kind of item, including mixtures of different types.

Two-sample T-test *(mathematics)*: A statistics test which indicates whether two samplings which are assumed to be normally distributed with the same variance have the same mean.

Two-tailed test *(mathematics)*: A statistical test that determines the probability of an observed value having at least a given separation from a mean value (in either direction); for example, the probability that an observation is separated from the mean by at least one standard deviation.

Type *(computing)*: Short for data type, indicating what kind of computational data structure is being referred to, whether a simple one like a number, sequential ones like text strings or lists, or complex arbitrary objects.

Type checking *(computing)*: The process of ensuring that the data types of any values are correct for the kind of operation being performed, e.g. to ensure that mathematical operations are only performed on numbers.

Unbiased estimate *(mathematics)*: For a probability distribution which is described by a certain parameter and where an estimate of that parameter is made by observing the random variable. The estimate is unbiased if there is no difference between its expected value (long-term average) and the true parameter value.

Underflow *(computing)*: The result of an arithmetic calculation that is too small in magnitude to store in computer memory.

Uniform distribution *(mathematics)*: A totally flat, even distribution of values. For example, in the context of generating random numbers all values are equally likely.

Union *(mathematics)*: With reference to sets of items, the total obtained by combining all of the items into a larger set.

Unit vector *(mathematics)*: A vector of length one. Unit vectors are often constructed to represent the direction, but not magnitude, of another vector.

Unrooted tree *(bioinformatics)*: A hierarchical branching data classification where there is no root or base. For a phylogenetic tree this means that the ancestor node is not explicitly represented.

Unsigned *(computing)*: With reference to numeric data types, a way of representing a value without any sign (positive or negative) information.

Unstructured *(molecular biology)*: A biological molecule is unstructured if it doesn't adopt any particular conformation, e.g. a protein which is highly dynamic and not folded into a compact shape.

Unsupervised learning *(computing)*: The process by which some machine learning methods can recognise patterns in data without the requirement for training data of known classification or value.

Update *(graphics)*: With reference to graphical interfaces, to refresh the display to reflect the state of the underlying data, i.e. to have a graphical display that responds to changes.

Van der Waals *(chemistry)*: The non-bond interaction that occurs between atoms by virtue of their radii. At close distances the van der Waals force is repulsive, as the atoms' electrons resist being superimposed, but at larger distances the interaction is attractive (albeit diminishing), due to transient, uneven charge distributions.

Variable *(computing)*: A data slot that can contain many different values. Generally a variable is a named item, i.e. a means of allocating a value with a label so that it can be referred to in an abstract way within a program.

Variable *(molecular biology)*: With reference to a biological molecule, the property of changing rapidly during evolution, i.e. giving variations in a protein or DNA sequence.

Variance *(mathematics)*: A measure of the amount of spread or variation in a set of values, which is the square of the standard deviation. Calculated as the average square difference between the values and the mean.

Vector *(mathematics)*: An array of numbers used to specify the location of a point in a space.

Vector graphics *(computing)*: A way of creating a graphical display by describing the underlying geometry of items (e.g. using points, lines and areas) so that they can be rendered into an image of any size by considering a particular view (lighting and orientation etc.).

Vector space *(mathematics)*: An abstract space where any position may be reached using a combination of vectors that span the different axes. For many computational methods, independent qualities of data items may be expressed in vector form, within a vector space.

View *(computing)*: A dynamic representation of a collection of items, which avoids the need of having to make detached copies of the collection while still supporting membership tests and iteration. In Python 3 a view is the result of a dictionary method, such as `keys()`, which is updated when the dictionary is updated.

Volume *(signal processing)*: The integral or summation of a signal peak, representing the strength of the signal.

Whitespace *(computing)*: Textual characters which don't have a symbol and control the space or separation between other characters. Examples include simple spaces, tab stops and line returns.

Widgets *(computing)*: On-screen graphical objects that are used to construct a graphical user interface. Examples include buttons, check boxes and menus.

Window *(computing)*: A discrete area within a graphical interface containing smaller graphical items, usually a box that can be moved and resized within the display.

XML *(computing)*: Extensible Markup Language, a textual data format used to store information in a hierarchical, object-oriented manner. The format comprises data tags that have attributes and may contain other tags.

Index

@ syntax, 76

__class__ attribute, 114
__init__() function, 108, 121
__init__.py file, 42
__name__ attribute, 62, 179

3D structure. *See* structures

active substitutions, 275
alignments
  amino acid properties, 257
  BLAST, 225
  ClustalW, 241
  conservation, 255
  gap penalties, 219
  guide tree, 269
  mapped genomic, 357
  multiple, 232, 241, 269
  pairwise, 221
  profile, 235–236
alternative hypothesis, 456
amino acids, 182
AND operation, 49
arithmetic operations, 25
arrays
  hierarchical clustering, 338
  normalisation, 328
  NumPy. *See* NumPy arrays
  quantile normalisation, 333
  reading image files, 321
  text files, 319
  value clipping, 373
  writing text files, 325
  Z-score normalisation, 330
atan2() function, 139
attributes
  checking, fetching, 113
  frozen, 136
  getters, setters, 132
  listing, 113
  privacy, 134

bar charts, 142
Bayes' theorem, 429
best-fit line, 485
binary operations, 613
binomial distribution, 431
binomial test, 464
BioPython, 205
  multiple alignments, 242
  reading sequence files, 205
  structures, 312

  writing sequence files, 206
BLAST, 225
BLOSUM matrix, 216
bool() function, 49
Boolean data type, 24, 47
Bowtie program, 349
break statement, 53

C language
  calling functions, 603
  Cython, 598
  interface, 587
  modules, 591
  Python library, 594
  Python values, 596
calling external programs, 225, 242
cardinality, 122
catching exceptions, 58
cell counting, 378
central limit theorem, 469
character ranges. *See* slicing: strings
chi-square. *See* Pearson's chi-squared test
chi-square distribution, 477
classes, 100
  __class__ attribute, 114
  __dict__, attribute dictionary, 112
  __init__() function, 108
  __name__ attribute, 114
  attribute getters, setters, 132
  attributes, 105, 113
  constructor, initialisation, 108
  defining, 102
  destructors, 125
  functions, 103
  hierarchy, 119
  importing, 103
  inheritance, subclassing, 102, 105
  instantiation, 109
    hierarchical, 129
  methods, 103
  parent-child links, 124
  private attributes, 134
  property attributes, 135
  self attributes, 106
  self-referencing, 103
  superclass function, 110
  type checking, 115
classification
  linear discriminant analysis, 507
  neural network, 526
  support vector machine, 536
ClustalW, 241
clustering, 489

Printed in the United States
By Bookmasters